Microwave-Enhanced Chemistry

Microwave-Enhanced Chemistry

Fundamentals, Sample Preparation, and Applications

H. M. (Skip) Kingston, Editor
Duquesne University

Stephen J. Haswell, Editor
University of Hull

American Chemical Society, Washington, DC

Library of Congress Cataloging-in-Publication Data

Microwave-enhanced chemistry / [edited by] H. M.
  "Skip" Kingston, Stephen J. Haswell.
      p. cm.—(ACS professional reference book)
    Includes bibliographical references and index.
    ISBN 0-8412-3375-6 (alk. paper)
    1. Microwave heating. 2. Chemical reactions.
I. Kingston, Howard M.   II. Haswell, S. J. (Stephen
John), 1954– . III. Series.

QD655.M53   1997
541.3'9—dc21   97-5983
                                                         CIP

The paper used in this publication meets the minimum requirements of American
National Standard for Information Sciences—Permanence of Paper for Printed
Library Materials, ANSI Z39.48-1984.

PRINTED IN THE UNITED STATES OF AMERICA

Advisory Board

About the Editors

H. M. (Skip) Kingston is Professor of Analytical Chemistry at Duquesne University in Pittsburgh, Pennsylvania. From 1976 to 1991 he was a Supervisory Research Chemist in the Analytical Research Division of the National Institute of Standards and Technology (NIST). While at NIST he specialized in sample preparation in the certification of standard reference materials and had more than 135 certifications. He also conceived and managed the Consortium on Automated Analytical Laboratory Systems (CAALS) at NIST, which was dedicated to developing U.S. laboratory standards. For the past 15 years, Dr. Kingston has been actively involved in advancing the area of microwave sample preparation through fundamental research and method development, including standardization of procedures. As a pioneer in the development of new analytical methods, he is the recipient of three R&D 100 awards for the development of the "Microwave Dissolution System" (1987), "Chelation Chromatography System" (1989), and "Speciated Isotope Dilution Mass Spectrometry" (1996). He conceived, co-edited, and co-authored the first American Chemical Society (ACS) Professional Reference Book on microwave sample preparation. He has taught the ACS Laboratory Microwave courses for the past 10 years. In sample preparation research he has received the "Pioneer in Laboratory Robotics" award (1988), the Department of Commerce Bronze Medal (1990), the NIST Applied Research

Award (1990), the Award of Merit from the Federal Laboratory Consortium for Technology Transfer (1991), and the Presidents Award for Excellence in Scholarship from Duquesne University (1995). Dr. Kingston has participated in drafting or updating more than 10 national and international standard methods for agencies such as the Environmental Protection Agency, based on microwave-enhanced chemistry. Dr. Kingston also serves on a number of academic, industrial, and governmental advisory boards.

Stephen J. Haswell is a Senior Lecturer in Analytical Chemistry at the University of Hull, United Kingdom. After working for five years in the food and plastics industries, he undertook a Ph.D. in the area of atomic spectroscopy, graduating in 1983 from the University of Plymouth. A dynamic figure in the field of analytical chemistry, his current research activities are in the areas of microwave-enhanced reaction chemistry, microreactor systems design, trace elemental speciation, and chemometrics. He is the author of more than 100 research papers, books, and patents, and he has presented numerous national and international lectures on his research activities. In 1993 he established the first U.K. Centre for Chemometrics and Process Analytical Science and is an analytical consultant to various multinational corporations.

He sits on the editorial boards of three international analytical journals and is an active member of numerous Royal Society of Chemistry Committees, recently chairing the Biannual National Atomic Spectroscopy Symposium (BNASS), the United Kingdom's premier international meeting in atomic spectroscopy. In addition to his successful academic and research career, he is known for his passion for traditional music and is often found entertaining conference delegates with a few tunes on his squeeze box.

Table of Contents

ALTERNATIVE LABORATORY
MICROWAVE INSTRUMENTS

CHEMISTRY APPLICATIONS

BIOCHEMISTRY APPLICATIONS

ACCESSING SAMPLE PREPARATION INFORMATION ON THE WWW

LABORATORY MICROWAVE SAFETY

INDEX

Contributors

David R. Baghurst *pages 3, 523*
Department of Chemistry, Imperial College of Science, Technology and Medicine, South Kensington, London SW7 2AY, United Kingdom

Gary Bond *page 551*
Department of Chemistry, University of Central Lancashire, Preston PR1 2HE, United Kingdom

Mathilde E. Boon *page 641*
Leiden Cytology and Pathology Laboratory, P. O. Box 16084, 2301 GB Leiden, Netherlands

Jason Brown *page 667*
Center for Communications and Information Technology, Duquesne University, Pittsburgh, PA 15282

Stuart Chalk *pages 55, 223, 667*
Department of Chemistry and Biochemistry, Duquesne University, Pittsburgh, PA 15282–1503

W. Gary Engelhart *pages 613, 697*
8305 S.E. Ketch Court, Hobe Sound, FL 33455

E. Hasty *page 569*
CEM Corporation, 3100 Smith Farm Road, P. O. Box 200, Matthews, NC 28105

Stephen J. Haswell   *page 401*
School of Chemistry, University of Hull, Hull HU6 7RX, United
Kingdom

Rodgers Hicks   *page 507*
Department of Chemistry, The University of Georgia, Athens, GA 30602

L. Jassie   *page 569*
CEM Corporation, 3100 Smith Farm Road, P. O. Box 200, Matthews, NC
28105

P. Kettisch   *page 423*
Anton Paar Company, Kärntner Straße 322, A–8054 Graz, Austria

T. Kierstead   *page 569*
CEM Corporation, 3100 Smith Farm Road, P. O. Box 200, Matthews, NC
28105

H. M. (Skip) Kingston   *pages 55, 223, 667, 697*
Department of Chemistry and Biochemistry, Duquesne University,
Pittsburgh, PA 15282–1503

Günter Knapp   *page 423*
Department for Analytical Chemistry, Micro- and Radiochemistry,
Graz University of Technology, Technikerstraße 4, A–8010 Graz, Austria

L. P. Kok   *page 641*
Institute for Theoretical Physics, University of Groningen, Nijenborgh
4, 9747 AG Groningen, Netherlands

Dirk Link   *page 223*
Department of Chemistry, Duquesne University, Pittsburgh, PA 15282

Elke Lorentzen   *page 223*
Department of Chemistry, Duquesne University, Pittsburgh, PA 15282

George Majetich   *pages 455, 507*
Department of Chemistry, The University of Georgia, Athens, GA 30602

Henryk Matusiewicz   *page 353*
Department of Analytical Chemistry, Politechnika Poznanska, 60–965
Poznan, Poland

S. Matz *page 569*
Millenium Petrochemicals Inc., 11530 North Lake Drive, P. O. Box
429566, Cincinnati, OH 45249

Kristen McQuillin *page 667*
Center for Communications and Information Technology, Duquesne
University, Pittsburgh, PA 15282

J. M. Mermet *page 371*
Laboratory of Analytical Sciences, University Claude-Bernard-Lyon I,
F-69622 Villeurbanne Cedex, France

D. Michael P. Mingos *pages 3, 523*
Department of Chemistry, Imperial College of Science, Technology and
Medicine, South Kensington, London SW7 2AY, United Kingdom

Richard B. Moyes *page 551*
School of Chemistry, University of Hull, Hull HU6 7RX, United
Kingdom

Edwin Neas *page 507*
Summit Biotechnology, P. O. Box 270793, Fort Collins, CO 80527–0793

F. Panholzer *page 423*
Department for Analytical Chemistry, Micro- and Radiochemistry,
Graz University of Technology, Technikerstraße 4, A–8010 Graz, Austria

Patrick J. Parsons *page 697*
Wadsworth Center, New York State Department of Health, P. O. Box
509, Albany, NY 12201–0509

R. Revesz *page 569*
CEM Corporation, 3100 Smith Farm Road, P. O. Box 200, Matthews, NC
28105

A. Schalk *page 423*
Anton Paar Company, Kärntner Straße 322, A–8054 Graz, Austria

Peter J. Walter *pages 55, 223, 667, 697*
Department of Chemistry and Biochemistry, Duquesne University,
Pittsburgh, PA 15282–1503

Karen Wheless *page 455*
Department of Chemistry, The University of Georgia, Athens, GA
30602

K. E. Williams   *page 401*
School of Chemistry, University of Hull, Hull HU6 7RX, United
Kingdom

Foreword

Much of chemistry involves energy. In the right place, right time, and right amount, energy drives the reactions, separations, and other functions of the chemical universe. In the wrong place, wrong time, or wrong amount, energy can ruin a batch, create unwanted by-products, or otherwise interfere with the desired outcome. Thus, chemical research is intimately involved with the control of the type, amount, and timing of energy introduction into a system.

The supply and control of energy has not generally received appropriate attention, nor has energy enjoyed glamour within many sectors of the chemistry community. Glamour has been reserved for the exotic synthesis, complex spectrometers, structure identification, high-tech analytical instruments, and the like. There have been exceptions, to be sure (Bunsen comes to mind), but, in general, we take "heating", "refluxing", and "digesting" for granted. Publications often ignore important details. Appreciation and skills are often not taught. Energy is often one of the most significant uncontrolled (or at best improperly controlled) variables in an experiment. Microwaves have the potential to change that, partly by the fundamental way the energy is delivered to the molecules of interest and partly by the electronic controls in the apparatus, facilitating the setting and documentation of the time, energy level, final temperature, and so forth.

Microwaves have revolutionalized chemistry in the kitchen during the past two decades. Consumers love the speed and control as they heat leftovers, thaw food, or cook entire meals. It is precisely this factor of speed and control that attracts the chemical community. As in the kitchen, where many recipes are still best prepared in a traditional oven, microwave energy is not a panacea for all energy needs in laboratory and industrial chemistry. Understanding the pros and cons of microwave energy is important in deciding when and where it is the choice for an energy source.

Microwave applications are penetrating many chemical and allied disciplines. Chemists involved in synthesis can use microwaves to drive reactions, and physical chemists use them to study molecular properties. Analytical chemists have discovered microwave energy for its ease and versatility, particularly in sample digestion and separation. For example, microwave dissolution is applied to biological, environmental, geological, and metallurgical matrices for elemental analysis. The control and repeatability of microwave digestions and separations reduce the technique-dependent art and improve analytical precision. The control of conditions reduces the number of artifacts generated by methods that use more brute force, reducing analytical interferences and improving synthetic yields.

Teachers are finding microwaves to be a safe and effective method for heating samples in the teaching laboratory. A chapter is devoted to describing microwave applications in the undergraduate organic chemistry laboratory.

The breadth of topics covered within the general field is an asset to the book. We all bring our disciplinary biases to reading this book—mine is analytical chemistry—and we probably know something about microwave applications within our immediate areas of interest. The breadth of coverage within this book provides easy access to applications in fields far removed from our daily research interests. I welcome the opportunity to at least peruse and become aware of research, applications, and success stories in my cousin disciplines and trust that many other readers feel likewise.

Understanding how microwave energy can enhance chemical reactions or separations requires a good working knowledge of not only the fundamentals of microwaves but also the practical advantages and limitations of the available apparatus and techniques—the tools. This book addresses both areas: An initial chapter reviews the theory of microwaves, and the remaining 15 chapters discuss applications.

Microwave-enhanced chemistry is an emerging field. Chemists are becoming increasingly aware of the versatility of this energy source. More and more studies in journal publications address or employ microwave energy, either as a research topic or in a routine laboratory apparatus. Manufacturers are improving the instrumentation. Vessels, reagents, flow systems, and other apparatus are being adapted to and improved for microwave energy. Analytical methods are being written about microwaves. A comprehensive treatise on the topic is in order, and this book fills that need.

In the late 1990s, publishing is undergoing a revolution with the increased availability of electronic media. We now have not only paper journals and books but also electronic journals, electronic appendices to print articles, and postings of World Wide Web sites. True to this publishing dynamic, this book supplies links between the paper and electronic worlds at several points. Chapter 15 is a "how-to" chapter on accessing electronic media related to microwave chemistry. Other chapters (e.g., Chapters 2 and 16) contain references to electronic access to emerging information, color renditions of the figures, and other electronic publishing features. This electronic linkage opens a door for the reader to maintain currency in this rapidly evolving field as new concepts and techniques emerge. The window into the electronic world will help readers keep up-to-date in the future as this book becomes increasingly dated.

The popularity of microwaves as a chemical laboratory technique is growing rapidly, as judged by instrument sales, publications, and presentations at meetings. This book provides a needed broad perspective on the area and has the added advantage of linking to electronic information sources, so the book will continue to be timely into the future.

Acknowledgment

This work was supported by the U.S. Department of Energy, Assistant Secretary for Environmental Management, under contract W–31–109–Eng–38.

MITCHELL D. ERICKSON
Environmental Chemistry Group
Environmental Research Division
Argonne National Laboratory
9700 South Cass Avenue
Argonne, IL 60439
MDErickson@anl.gov

Preface

The first publication on the use of microwave ovens in chemical analysis appeared in 1975; it was the beginning of the application of microwave radiation for sample treatment. The majority of the publications of the next decade were devoted to the decomposition (in open systems) of environmental samples, foodstuffs, oils, ores, minerals, metals, alloys, and so on. The results of these investigations and fundamental sample-preparation chemistry were systematized in the excellent book entitled *Introduction to Microwave Sample Preparation: Theory and Practice*, edited by H. M. (Skip) Kingston and Lois B. Jassie (American Chemical Society: Washington, DC, 1988). This book has played an important role in the development and dissemination of microwave sample preparation.

However, significant changes have taken place during the years since the first book. It is clear now that microwave radiation can be used in many other fields of chemistry such as synthesis of organic, organometallic, and inorganic compounds or catalysts. The application sphere of sample preparation continues to expand. Numerous microwave applications are published on decomposition, fusion, dry and wet mineralization, ashing, and extraction. In our laboratories attention was paid to acceleration of chemical reactions for substitutionally inert coordination compounds and to effect increases in the sorption rate in preconcentration of traces of platinum metal, particularly in flowing systems. At present we are investigating the effect of microwave radiation on complex formation, formation of colored and luminescent analytical forms in solutions and in "solid matrix solutions", and on chromatographic processes. Similar diverse investigations are being performed in other laboratories. Our new emphasis is on nonthermal effects and a thorough understanding of microwave interaction with substances.

Introduction to Microwave Sample Preparation: Theory and Practice was a significant landmark in the development of the field. *Microwave-Enhanced Chemistry: Fundamentals, Sample Preparation, and Applications* is an indication of the formation of microwave chemistry as a new branch of chemical science. This work was conceived and prepared by H. M. (Skip) Kingston and Stephen J. Haswell, who have also drawn from leading specialists in microwave applications; this book is not only an impressive result but also a stimulus to the further development of microwave chemistry.

Undoubtedly, this book will be used by many chemists, including analysts in Russia and Europe. The first book was translated into Russian and Chinese in 1991 and 1992, respectively. In Russia the importance of the effect of microwave analysis on chemical systems is being investigated at the Vernadsky Institute of Geochemistry and Analytical Chemistry and Moscow University.

We thank the authors and the editors for this excellent book that will be used by many investigators for applications in chemical laboratories.

Yu. A. Zolotov
Kurnakov Institute of Geochemistry and Inorganic Chemistry
Russian Academy of Sciences
31 Leninskii Prospect
117907 GSP–1, Moscow, Russia
7–095–952–0224 or 7–095–256–5327 (telephone)
7–095–954–1279 (fax)

Nikolai M. Kuzmin
V. I. Vernadsky Institute of Geochemistry and Analytical Chemistry
Russian Academy of Sciences
Kosygin str., 19
SU–117975 Moscow, Russia 117975
7–137–86–15 (telephone)
700095–938–2054 (fax)
411633 TERRA SU (telex)

Fundamentals of Microwave
Application

Chapter 1

Applications of Microwave Dielectric Heating Effects to Synthetic Problems in Chemistry

D. Michael P. Mingos and David R. Baghurst

Dr. Johnson's Dictionary of 1775 defined chemistry as "an art whereby sensible bodies contained in vessels...are so changed by means of certain instruments, and principally fire, that their several powers and virtues are thereby discovered, with a view to philosophy or medicine". Although fire is rarely used these days the heating of chemicals in containment vessels still remains the primary means of stimulating chemical reactions which proceed slowly under ambient conditions. Since Johnson's time, however, photochemical, catalytic, sonic, and high pressure techniques (1) have been added to the chemist's repertoire for accelerating chemical reactions. In this review we describe an alternative to conventional conductive heating for introducing energy into reactions. The microwave dielectric heating effect uses the ability of some liquids and solids to transform electromagnetic energy into heat and thereby drive chemical reactions. This in situ mode of energy conversion has many attractions to the chemist, because its magnitude depends on the properties of the molecules. This allows some control of the material's properties and may lead to reaction selectivity. This application of microwaves should not be confused with gas phase reactions where a microwave discharge can create a plasma with a very high temperature and can cause dramatic fragmentation and recombination reactions.

This chapter has been reproduced in full from "Applications of Microwave Dielectric Heating Effects to Synthetic Problems in Chemistry" (Tilden Lecture), D. M. P. Mingos and D. R. Baghurst, in *Chemical Society Reviews, 1991,* **20,** 1, by kind permission of the Royal Society of Chemistry, Cambridge, UK.

A reliable device for generating fixed frequency microwaves was designed by Randall and Booth at the University of Birmingham as part of the development of RADAR during the Second World War. This device, the magnetron (2), was produced in large numbers during the war with the aid of United States industrial expertise. Even in those early days it was recognised that microwaves could heat water in a dramatic fashion, and domestic and commercial appliances for heating and cooking foodstuffs began to appear in the United States in the 1950s. The widespread domestic use of microwave ovens occurred during the 1970s and 1980s as a result of effective Japanese technology transfer and global marketing. The purpose of this review is to explain, in a chemically intelligible fashion, the origins of microwave dielectric heating effects and to describe some recent applications to problems in synthetic chemistry. Although microwave spectroscopy forms a common component of the great majority of undergraduate chemical courses, microwave dielectric heating effects are generally neglected. In our opinion this lack of definition between the two subjects has resulted in the relative neglect of the chemical applications of microwave heating effects. While the mass production of microwave ovens has made them readily available at a cheap price their utilisation in chemical laboratories has been sparse.

Microwaves and Their Interactions with Matter

The microwave region of the electromagnetic spectrum (*see* Figure 1) lies between infrared radiation and radio frequencies and corresponds to wavelengths of 1 cm to 1 m (frequencies of 30 GHz to 300 MHz respectively). The wavelengths between 1 cm and 25 cm are extensively used for RADAR transmissions and the remaining wavelength range is used for telecommunications. In order not to interfere with these uses, domestic and industrial microwave heaters are required to operate at either 12.2 cm (2.45 GHz) or 33.3 cm (900 MHz) unless the apparatus is shielded in such a way that no radiation losses occur. Domestic microwave ovens generally operate at 2.45 GHz.

Since some confusion may persist in the mind of chemists concerning the distinction between microwave spectroscopy and microwave dielectric heating effects the differences will be discussed in some detail below. In microwave spectroscopy molecules are studied in the gas phase and the microwave spectrum of a molecule shows many sharp bands (3) in the frequency range 3–60 GHz. The sharp bands in the spectrum arise from transitions between quantised rotational states of the molecule defined by the following relationship:

$$\text{Rotational energy, } E_j = J(J + 1)h^2/8\pi I^2 \tag{1}$$

Where J is the rotational quantum number, I is the moment of inertia and h is Planck's constant. In order to observe a pure rotational spectrum an

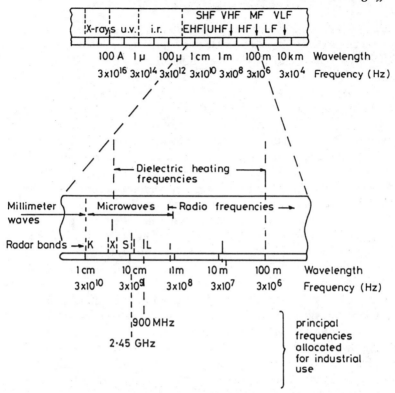

Figure 1. The electromagnetic spectrum indicating the important microwave frequencies for dielectric heating.

oscillating dipole has to be associated with the molecular rotation. To a first approximation this corresponds to the molecule possessing a permanent dipole moment (4).

The microwave spectra of small molecules are sharp and lines less than 1 MHz apart can often be distinguished. Therefore, microwave spectroscopy provides an excellent fingerprinting technique for identifying molecules in the gas phase and has, for example, been used to confirm the presence of a wide range of molecules in outer space. Additionally, the accurate B values [$B = J/(8p^2 Ic)$] available can, by the judicious application of isotopic substitution, lead to the precise determination of bond lengths and angles. Isotopic substitution has also led to the determination of some of the most accurate estimates of relative atomic masses. The departure of the molecule from the ideal rigid rotator can be studied by the evaluation of the distortion constants D_J and $D_{J,K}$, leading to calculations of approximate force constants, and the application of the Stark effect allows the determination of precise values of molecular dipole moments.

In the gas phase, at low pressure, the lifetime of the excited state produced by exciting a particular rotational state is long. However, as a consequence of Heisenberg's uncertainty principle at pressures of around 10^{-1} mmHg the reduction in the lifetime due to collisions broadens the spectral peaks. In liquids and solids, where the molecules are generally not free to rotate independently, the spectra are too broad to be observed. It is to these phases that microwave dielectric loss heating effects are relevant and need to be distinguished from the spectroscopic effects. A material can be heated by applying energy to it in the form of high frequency electromagnetic waves. The origin of the heating effect produced by the high frequency electromagnetic waves arises from the ability of an electric field to exert a force on charged particles. If the particles present in the substance can move freely through it, then a current has been induced. However, if the charge carriers are bound to certain regions they will move until a counter force balances them and the net result is a dielectric polarisation. Both conduction and dielectric polarisation are sources of microwave heating and we will deal with them in turn. The microwave heating effect depends on the frequency as well as the power applied, but unlike microwave spectroscopy the effect does not result from well spaced discrete quantised energy states, but with a broad band bulk phenomenon which can be treated classically.

The theory of microwave heating has been developed by many workers, among them Debye (5), Frohlich (6), Daniel (7), Cole and Cole (8), Hill (9), and Hasted (10). A detailed analysis is beyond the scope of this review. We shall nevertheless discuss the important principles and indicate their chemical significance.

Dielectric Polarisation

One source of microwave dielectric heating lies in the ability of an electric field to polarise charges in a material and the inability of this polarisation to follow rapid reversals of an electric field.

The total polarisation is the sum of a number of individual components:

$$\alpha_t = \alpha_e + \alpha_a + \alpha_d + \alpha_i \tag{2}$$

where, α_e, the electronic polarisation, arises from the realignment of electrons around specific nuclei; α_a, the atomic polarisation, results from the relative displacement of nuclei due to the unequal distribution of charge within the molecule; α_d, is the dipolar polarisation, resulting from the orientation of permanent dipoles by the electric field; and α_i is the interfacial polarisation, or Maxwell–Wagner effect, which occurs when there is a build up of charges at interfaces.

With an oscillating electric field, such as that associated with electromagnetic radiation, the response of a material depends on the time scales of the orientation and disorientation phenomena relative to the frequency of the radiation. The time scales for the polarisation and depolarisation of α_e, and α_a, are much faster than the microwave frequencies and therefore these effects do not contribute to the dielectric heating effect. In contrast the time scales for polarisation associated with the permanent dipole moments in the molecule, α_d, and possibly the time scale associated with some interfacial processes, α_i, are comparable to microwave frequencies. Consequently, it is these effects that require further explanation.

Dipolar Polarisation, α_d

Dipolar polarisation, such as that excited in liquid water, is due to its dipole moment which in turn results from the differing electronegativities of the oxygen and hydrogen atoms. At low frequencies the time taken by the electric field to change direction is longer than the response time of the dipoles, and the dielectric polarisation keeps in phase with the electric field. The field provides the energy necessary to make the molecules rotate into alignment. Some of the energy is transferred to the random motion each time a dipole is knocked out of alignment and then realigned. The transfer of energy is so small, however, that the temperature hardly rises. If the electric field oscillates rapidly, it changes direction faster than the response time of the dipoles. Since the dipoles do not rotate, no energy is absorbed and the water does not heat up.

In the microwave range of frequencies the time in which the field changes is about the same as the response time of the dipoles. They rotate because of the torques they experience, but the resulting polarisation lags behind the changes of the electric field. When the field is at a maximum strength, say in the upward direction, polarisation may still be low. It keeps rising as the field weakens. The lag indicates that the water absorbs energy from the field and is heated.

Two parameters define the dielectric properties of materials and are used extensively in this review. The first, e', the dielectric constant describes the ability of the molecule to be polarised by the electric field. At low frequencies this value will reach a maximum as the maximum amount of energy can be stored in the material. The dielectric loss, e'', measures the efficiency with which the energy of the electromagnetic radiation can be converted into heat. The manner in which e' and e'' vary with frequency are shown schematically in Figure 2. The dielectric loss goes through a maximum as the dielectric constant falls. The ratio of the dielectric loss and the dielectric constant define the (dielectric) loss tangent = $e''/e' = \tan \delta$, which defines the ability of a material to convert electromagnetic energy into heat energy at a given frequency and temperature.

The time taken for the dipoles in water to become polarised and then depolarised is governed by a relaxation time constant. Only when this time constant approaches the inverse excitation frequency will a particular

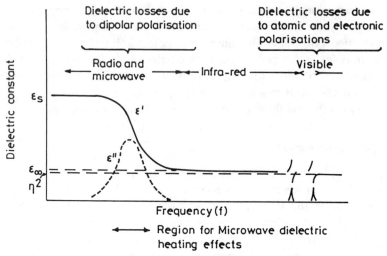

Figure 2. Schematic illustration of the real and imaginary components of the dielectric constant, e′ and e″, as a function of frequency.

polarisation mechanism become important. Note that since the time constants for electronic and atomic polarisation are much faster than 10^{-9} s these mechanisms do not contribute to microwave dielectric heating effects.

In Figure 3 the dielectric properties of distilled water are plotted as a function of frequency at 25 °C (12). It is apparent that appreciable values of the dielectric loss exist over a wide frequency range. Note that the greatest heating as measured by e″ reaches its maximum at around 20 GHz while domestic microwave ovens operate at a much lower frequency, 2.45 GHz.

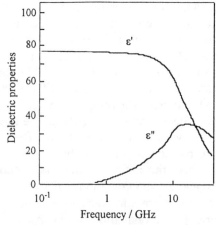

Figure 3. Dielectric properties of water as a function of frequency.

The practical reason for the lower frequency is that it is necessary to heat food efficiently throughout its interior. If the frequency is optimal for maximum heating rate, the microwaves are absorbed in the outer regions of the food, and penetrate only a short distance. Thus the penetration depth, that is the depth into a material where the power falls to one half its value on the surface, is another important parameter in the design of a microwave experiment. An approximate relationship for penetration depth D_p, when e" is small, is given by:

$$D_p \infty \lambda_o \sqrt{(\varepsilon'/\varepsilon'')} \tag{3}$$

where λ_o is the wavelength of the microwave radiation.

The theoretical examination of the frequency dependence of e' and e" began with the derivation of the Debye equations (5, 13):

$$\varepsilon'_d = \varepsilon'_\infty + \frac{(\varepsilon'_0 - \varepsilon'_\infty)}{(1 + \omega^2 \tau^2)} \tag{4}$$

$$\varepsilon''_d = \frac{(\varepsilon'_0 - \varepsilon'_\infty)\omega\tau}{(1 + \omega^2 \tau^2)} \tag{5}$$

where e'_∞ and e'_0 are defined as the high frequency and static dielectric constants, and ω and τ are the frequency and relaxation times which characterise the rate of build up and decay of polarisation. The above results apply to both liquids and solids though different models are used to derive it. It is an interesting feature of the Debye equations that the values of e'_d and e''_d at the frequency at which the dielectric loss is a maximum is independent of this frequency and of the relaxation time:

$$\varepsilon''_{max} = (\varepsilon'_0 - \varepsilon'_\infty)/2 \text{ where } \varepsilon'_d = (\varepsilon'_0 + \varepsilon'_\infty)/2 \tag{6}$$

In liquids it is generally assumed that the dipoles can point in any direction and are continually changing due to thermal agitation (6). Debye's interpretation of the relaxation is given in terms of the frictional forces in the medium. Using Stokes' theorem (14) he derived the following expression for the relaxation time of spherical dipoles:

$$\tau = 4\pi r^3 \eta / kT \tag{7}$$

where η is the viscosity of the medium, r the radius of the dipolar molecule, and k is Boltzmann's constant.

In solids due to the variable interaction of a molecule with its neighbours a dipole has a number of equilibrium positions. They are separated by potential barriers over which the dipole must pass in turning from one direction to another. In the simplest case only two equilibrium positions with opposite dipole directions exist and they are separated by an energy barrier U_a, as shown in Figure 4. From Boltzmann statistics it follows that the number of transitions from one state to another is proportional to $(1 - e^{t/\tau})$ where t is the time and τ the relaxation time constant. This leads to the following relationship between time constant and dielectric constant:

$$\tau = \frac{e^{U_a/k}T(\varepsilon_0' + 2)}{n(\varepsilon_\infty' + 2)} \tag{8}$$

where $1/n$ is the time for a single oscillation in the potential well. The magnitude of the absorption for this model has been discussed by Frohlich who concluded that the expressions should be equivalent to those followed by liquids (6). The following form of the Onsager equation (15) has been shown to apply to many liquids and solids (16–18):

$$e_0' - e_\infty' = \frac{4\pi Nm^2\varepsilon_0'(\varepsilon_\infty' + 2)^2}{9kT(2\varepsilon_0' + \varepsilon_\infty')} \tag{9}$$

where N is the number of molecules and m their mass. Note that in such a system the magnitude of the dielectric absorption decreases with increasing temperature.

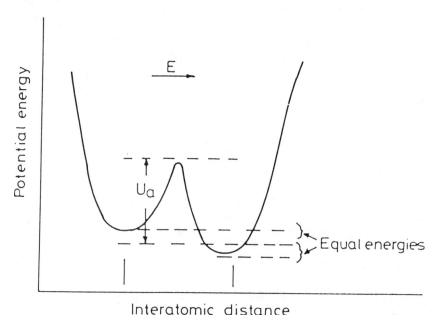

Figure 4. Potential energy diagram for two alternative positions of a dipole relative to an electric field.

In practice it is found that many polar solids, for example, aliphatic long chain ketones and most ethers, have small dielectric losses. This arises because in the solid state the energy differences between the equilibrium positions of the dipoles are very large. In most other cases, for example, many long-chain esters and some ethers, there is significant dielectric absorption at room temperatures which decreases with increasing temperature in agreement with the Onsager equation. The dielectric properties of some common solvents are given in Table I (*12*). The high dielectric losses of the alcohols are particularly noteworthy.

Many liquids and solids give dielectric loss curves much wider than a Debye curve, due to the presence of a range of relaxation times. Frequently this variation can be represented by the Fuoss and Kirkwood empirical expression (*19*):

$$\varepsilon'' = \varepsilon''_{max} \, \text{sech}\{b[\ln(f/f_{max})]\} \tag{10}$$

The value of b varies between one for a single relaxation and zero for an infinite number, f is the frequency and f_{max}, the frequency of maximum loss.

Interfacial Polarisation, α_i

A suspension of conducting particles in a non-conducting medium is an inhomogeneous material whose dielectric constant is frequency dependent. The loss relates to the build up of charges between the interfaces and is known as the Maxwell–Wagner effect. Its importance in the microwave region has not been well defined. Absorption centred around a frequency of 10^7 Hz may tail into this frequency range. Wagner (*20*) has shown that for the simplest model featuring this type of polarisation, consisting of conducting spheres distributed through a non-conducting medium, the dielectric loss factor, of volume fraction v of material, is given by:

$$\varepsilon''_i = \left(\frac{9v\varepsilon' f_{max}}{1.8 \times 10^{10}\sigma}\right)\left(\frac{\omega\tau}{1 + \omega^2\tau^2}\right) \tag{11}$$

where, σ is the conductivity (in S m^{-1}) of the conductive phase and e' its dielectric constant. The frequency variation of the loss factor is similar to that of dipolar relaxation. An experimental system approximating to Wagner's model has been made by incorporating up to 3% of roughly spherical particles of semiconducting copper phthalocyanine into paraffin wax (*21*). This gave good agreement with the theory as shown in Figure 5.

Conduction Losses

As the concentration of the conducting phase dispersed in a non-conducting phase is increased there comes a point where some account

Table I. Dielectric Properties at Three Frequencies of Common Solvents

Solvent	3×10^8 Hz ϵ'	ϵ''	3×10^9 Hz ϵ'	ϵ''	1×10^{10} Hz ϵ'	ϵ''
H_2O	77.5	1.2	76.7	12.0	55.0	29.7
NaCl, 0.1M	76	59	75.5	18.1	54	30
Heptane	1.97		1.97	2×10^{-4}	1.97	3×10^{-3}
MeOH	30.9	2.5	23.9	15.3	8.9	7.2
EtOH	22.3	6.0	6.5	1.6	1.7	1.1×10^{-1}
n-Propanol	16.0	6.7	3.7	2.5	2.3	2.0×10^{-1}
n-Butanol	11.5	6.3	3.5	1.6	0.2	
Ethylene glycol	39	6.2	12	12	7	5.5
Carbon tetrachloride	2.2		2.2	9×10^{-4}	2.2	3×10^{-3}

has to be made of the interactions between individual conducting areas. This can be viewed as an extension of simple Wagner theory (10, 22–24). A general approach to interfacial polarisation has been considered by Maxwell and Wagner in their two layer capacitor model as shown in Figure 6.

Thus the total Maxwell–Wagner effect is seen to be a composite produced by areas of differing dielectric constants and conductivity. The complex permittivity is found to be equal to:

$$\varepsilon_i^* = \varepsilon_\infty' + \frac{\left(\varepsilon_0' - \varepsilon_\infty'\right)}{(1 + j\omega t)} - \frac{j\sigma}{\omega\varepsilon_0} \tag{12}$$

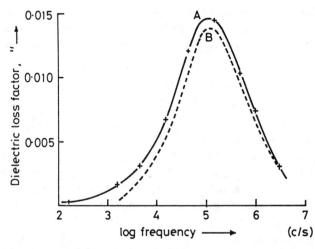

Figure 5. Dielectric absorption for a Maxwell–Wagner system consisting of copper phthalocyanine in paraffin wax. Curve A: 0.62% (volume) of copper phthalocyanine spheres; curve B: theoretical (21).

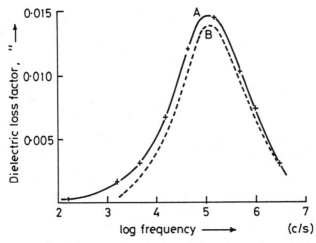

Figure 6. The Maxwell–Wagner two layer capacitor (ε_{ax} and σ_x refer to the dielectric constant and conductivity, d_x is the distance).

The real part is precisely that given by the simple Debye theory. The loss term not only gives a single relaxation time response but also a term due to dc conductivity. The contribution to the total loss due to this extra conductive part depends on the dc conductivity itself. For highly conductive liquids and solids, particularly those containing large amounts of salts, there comes a point where the conductive loss effects are larger than the dipolar relaxation effects.

In Figure 7 the dielectric loss for sintered alumina (25) as a function of temperature is illustrated. At room temperature the conduction loss effects are important only at lower frequencies. Losses in the microwave region are due to dipolar relaxation. However, as the temperature increases conduction losses increase rapidly and become as important as dipolar relaxation loss in the microwave region.

The increase of conduction with temperature in alumina is associated with the thermal activation of the electrons which pass from the oxygen 2p valence band to the 3s3p conduction band. In addition, electrical conduction is generally enhanced by material defects which sharply decrease the energy gap between the valency and conduction bands.

Attempts have been made to represent the conduction losses through the expression:

$$\varepsilon_c'' = \sigma/\omega\varepsilon_o \tag{13}$$

and Hamon (21) has again simulated this behaviour with rather higher concentrations of copper phthalocyanine in paraffin wax (*see* Figure 8). At high frequencies the loss is often greater than that suggested by a linear relationship, and expressions of the type:

$$\varepsilon_c'' = \text{constant} \times f^{-k} \tag{14}$$

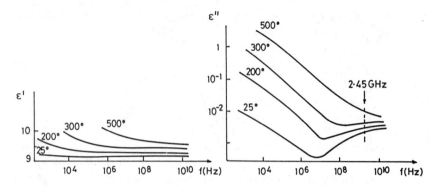

Figure 7. Dielectric properties as a function of frequency and temperature for sintered alumina (25).

where k approaches unity for high concentrations of conducting material, are found to be more appropriate (26).

Microwave Heating of Liquids and Solids

The rate of rise of temperature due to the electric field of microwave radiation is determined by the following equation (27):

$$\frac{\delta T}{\delta t} = \text{constant} \times \frac{\varepsilon'' f E_{rms}^2}{\rho C_p} \tag{15}$$

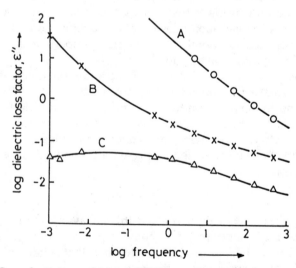

Figure 8. DC conductivity and Maxwell–Wagner absorption due to a conducting powder in paraffin wax. Curves A, B, and C: 1.9, 0.62, 0.19% (volume), respectively, copper phthalocyanine (21).

where E^2_{rms} is the r.m.s. field intensity, ρ the density, and C_p, the specific heat capacity. The losses due to radiation (28) are governed by:

$$\frac{\delta T}{\delta t} = - \frac{e\alpha}{\rho C_p} \left(\frac{A}{V}\right)_{sample} T^4 \tag{16}$$

where e is the sample emissivity and α the Stefan–Boltzmann constant. The resulting temperature rise is thus determined by the dielectric loss, specific heat capacity, and emissivity of the sample as well as the strength of the applied field. These physical properties of the liquid or solid samples are all temperature dependent, making the complete theoretical analysis of dielectric heating mathematically very complex. The dielectric constant and dielectric loss values have been established for a number of materials at room temperature. The largest collection of data is that due to von Hippel (12, 29) though many of these data apply to foodstuffs and are of limited use to the chemist. The study of dielectric constant and dielectric loss as a function of temperature is less well developed, particularly in solids, where data are needed up to sintering temperatures (1000–2200 °C). The data in Figure 9 on the dielectric properties of CuO as a function of temperature (30) comprise one of the few examples of a detailed and complete study. Several workers have, however, reported the temperatures reached by liquids and solids when placed in conventional microwave ovens. Some of the available data for liquids and solids are given in Tables II and III respectively (31, 32).

The strength of the applied field E^2_{rms}, in equation 1 is related to the applied power. The effect of applied power on the rate of microwave heating is illustrated in Figure 10 for samples of 1-propanol (31). As expected, a larger power leads to a larger E^2_{rms}, and hence a greater heating rate.

Temperature measurement inside the microwave cavity is complicated by the presence of an intense electric field (33). Thus thermocouples

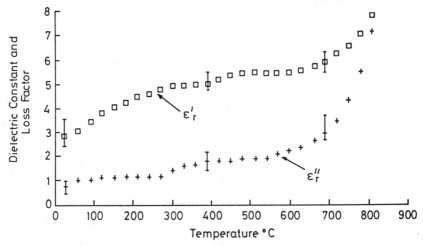

Figure 9. The dielectric properties of CuO as a function of temperature (30).

Table II. Temperature of 50 mL of Several Solvents After Heating from Room Temperature for 1 Min at 560 W and 2.45 GHz

Solvent	T (C)	Bp (C)
Water	81	100
Methanol	65	65
Ethanol	78	78
1-Propanol	97	97
1-Butanol	109	117
1-Pentanol	106	137
1-Hexanol	92	158
1-Chlorobutane	76	78
1-Bromobutane	95	101
Acetic acid	110	119
Ethyl acetate	73	77
Chloroform	49	61
Acetone	56	56
Dimethylformamide	131	153
Dimethyl ether	32	35
Hexane	25	68
Heptane	26	98
CCl_4	28	77

must be efficiently shielded and grounded to avoid sparking. These considerations are particularly important when volatile and inflammable solvents are being heated in the cavity. For temperatures up to 450 °C fibre optic thermometers have been successfully applied (34). Surface temperatures up to 3000 °C can be measured remotely using infrared pyrometry, though substantial differences exist between surface and internal temperatures, particularly at the higher temperatures.

The Microwave Oven

The principal features of a modern microwave oven are illustrated in the cutaway diagram shown in Figure 11. The important features of the magnetron in which the microwaves are generated, are shown in Figure 12. A magnetron is a thermionic diode having an anode and a directly heated cathode. As the cathode is heated, electrons are released and are attracted toward the anode. The anode is made up of an even number of small cavities, each of which acts as a tuned circuit. The gap across the end of each cavity behaves as a capacitance. The anode is therefore a series of circuits which are tuned to oscillate at a specific frequency or its overtones.

A very strong magnetic field is induced axially through the anode assembly and has the effect of bending the path of the electrons as they travel from the cathode to the anode. As the deflected electrons pass through the cavity gaps they induce a small charge into the tuned circuit

Table III. Effect of Microwave Heating on Temperature of Solids

Chemical	T (C)	Time (min)
Al	577	6
C	1283	1
Co_2O_3	1290	3
CuCl	619	13
$FeCl_3$	41	4
$MnCl_2$	53	1.75
NaCl	83	7
Ni	384	1
NiO	1305	6.25
$SbCl_3$	224	1.75
$SnCl_2$	476	2
$SnCl_4$	49	8
$ZnCl_2$	609	7
CaO	83	30
CeO_2	99	30
CuO	701	0.5
Fe_2O_3	88	30
Fe_3O_4	510	2
La_2O_3	107	30
MnO_2	321	30
PbO_2	182	7
Pb_3O_4	122	30
SnO	102	30
TiO_2	122	30
V_2O_5	701	9
WO_3	532	0.5

Note: All experiments began at room temperature. The 25 g samples were heated in a 1 kW oven (2.45 GHz) with a 1000 mL vented water load. The 5–6 g samples were heated in a 500-W oven.

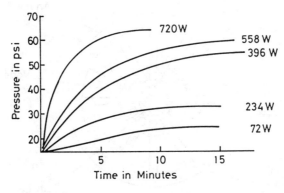

Figure 10. The effect of applied power on the heating rate of samples of 1-propanol.

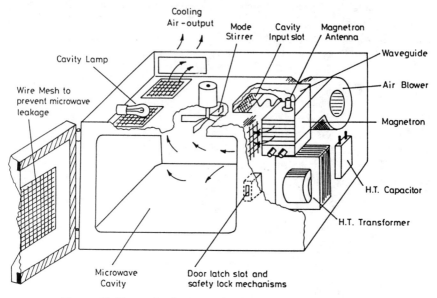

Figure 11. The major features of a domestic microwave oven.

resulting in the oscillation of the cavity. Alternate cavities are linked by two small wire straps (mode straps) which ensure the correct phase relationship. This process of oscillation continues until the oscillation has achieved a sufficiently high amplitude. It is then taken off the anode via an antenna. Of the 1200 W of electrical line power used by the magnetron around 600 W is converted into electromagnetic energy. The remainder is converted into heat that must be dissipated through air or water cooling.

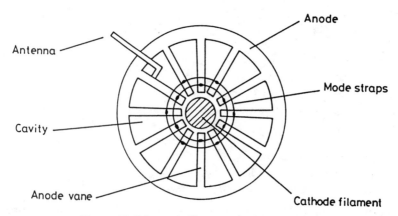

Figure 12. Schematic illustration of a magnetron.

The variable power available in domestic ovens is produced by switching the magnetron on and off according to a duty cycle. For example, a typical 600-W oven with a 30-s duty cycle can be made to deliver an average of 300 W by switching the magnetron on and off every 15 s. Large duty cycles are undesirable in chemical applications where samples may cool dramatically between switching steps and this is, therefore, one factor which must be considered in adapting an oven for chemical applications.

A waveguide is a rectangular channel made of sheet metal. Its reflective walls allow the transmission of microwaves from the magnetron to the microwave cavity or applicator. The minimum frequency which can be propagated is related to the dimensions of the rectangular cross-section through the expression $c/f = 2d$ where c is the speed of light, f the cut-off frequency, and d the larger of the dimensions of the rectangular section of the waveguide.

The reflective walls of the microwave cavity are necessary to prevent leakage of radiation and to increase the efficiency of the oven. There is rarely a perfect match between the frequency used and the resonant frequency of the load, so if the energy is reflected by the walls, absorbance is increased because the energy passes through the sample more often and can be partially absorbed on each passage. This can be particularly important if the sample is dimensionally small. If too much energy is reflected back into the waveguide the magnetron may be damaged. Most commercial ovens are protected by an automatic cut-off and there may also be protection in the form of a circulator which directs reflected energy into a dummy load. When working with small loads, poorly absorbing loads or at high powers, a dummy load, for example a beaker of water, should always be placed in the cavity along with the sample to absorb the excess energy.

In the absence of any smoothing mechanism the electric field pattern produced by the standing waves set up in the cavity may be extremely complex. Some areas may receive large amounts of energy while others may be almost neglected. To ensure that the incoming energy is smoothed out in the cavity a mode stirrer (a reflective, fan-shaped paddle) is sometimes used. Most microwave ovens are also supplied with a turntable, which ensures that the average field experienced by the sample is approximately the same in all directions.

When the liquids or solids which are being heated are poor absorbers of microwave energy and are only available in small amounts, the multimode microwave oven no longer represents the most effective system. This is particularly a problem if the samples need to be heated to a high temperature in a controlled fashion. A more expensive experimental set-up has to be used wherein a single mode resonant cavity is tuned to the characteristics of the material being heated. The main features of this type of apparatus are shown in the form of a block diagram in Figure 13. It has been used to process poorly absorbing polymers (35).

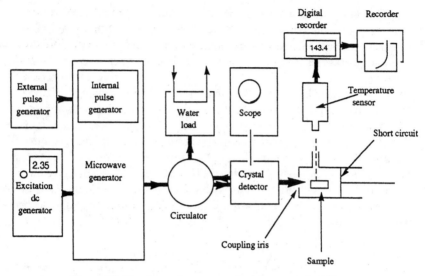

Figure 13. Single mode resonant heating system (*35*).

Briefly, a single-mode resonant heater allows a sample to be placed at positions of much higher electric field strengths than can be obtained in a multimode oven. Having chosen a particular resonant mode or electric field distribution pattern, insertion of the sample into the cavity changes the resonant frequency. By varying the position of the plunger the cavity can be made to resonate at the working frequency. Additionally the coupling iris is altered in size to ensure a good impedance match between the waveguide and the cavity so that the energy is not reflected back toward the source. The efficiency of the matching can be followed either by measuring the standing wave ratio in a slotted line or by using a network analyser. Since the dielectric properties of the material depend on temperature, the matching requirements of the cavity are continually changing and are usually controlled by a computer. The circulator protects the magnetron source from the high reflected powers that exist when there is a poor energy match. Use of a resonant cavity increases the effective cavity power by three orders of magnitude allowing the microwave heating of relatively low-loss materials such as polymers and some ceramics using small powers (<2.5 kW).

Reactions Utilising the Dielectric Loss Properties of Solvents

It is apparent from the previous discussion that any organic or inorganic solvent with a low molecular weight and a high dipole moment will couple effectively with microwaves at 2.45 GHz. The solvents which have been most commonly used for synthetic reactions are H_2O, MeOH, and

EtOH. Other solvents such as MeCN, DMF, and CH_2Cl_2 couple effectively to microwaves but have been less commonly used. Non-polar organic solvents such as C_6H_6, petroleum ethers, and CCl_4 have negligible dielectric loss and therefore do not couple efficiently with microwaves. The addition of small amounts of alcohols or water to these solvents can lead to dramatic coupling effects, however. Hence a 1:4 EtOH:toluene mixture can be heated to boiling in a few minutes in a standard domestic microwave oven. It should also be emphasised that the presence of salts in polar solvents can frequently enhance dielectric loss effects and microwave coupling.

Clearly it is unsafe to heat organic solvents with microwaves in a contained space. Sparking induced by the switching of the magnetron could lead to a violent explosion in the microwave cavity. Therefore, the problem of solvent containment is paramount in the design of safe synthetic procedures.

Low Pressure Conditions

Much important work has been performed on the synthesis and processing of polymers using microwave dielectric heating effects. Since these materials are not very volatile, containment problems are not great. For the synthesis of low molecular weight organic molecules, organometallic, and coordination compounds, containment problems are more severe. Some specific applications are given below.

Synthesis of Polymers

Typically, polymer solutions and particularly thin films are poor absorbers of microwave radiation. Accordingly resonance cavity techniques have been widely employed in the curing of polymers such as polyurethane systems (36), epoxy–glass fibre composites (37), DGEBA–DDM epoxy resins (38), and epoxy resin–aluminium particle systems (39).

Early work in this area was concerned with following the evolution of temperature in samples as a function of the applied power (40). For the cross-linking of epoxy resins a model has been developed linking the time dependence of temperature to the applied power and heat losses due to conduction. Further progress has been made by analysing the variations with time of the applied power $P_u = P_u (t)$ degraded as heat inside the samples (dielectric loss) and of the derivatives $T' = \delta T / \delta t$ and $P'_u = \delta P'_u / \delta t$ for pure resins and epoxy–glass fibre mixtures (37).

The role of solvent and polymer in the formation of films has been studied on hollowed, microwave transparent, PTFE (41). Two series of polymers were investigated: polyvinylacetate (PVAC) as a polar molecule with high dielectric constant, and polystyrene, a non-polar polymer with low dielectric losses. The solvents used were DMF, methyl ethyl ketone

(MEK), and benzene in order of decreasing polarity. The rate of polymer formation was followed by means of surface temperature measurement. Each solution of polystyrene began to heat according to the overall polarity of the mixture, then, as solvent evaporated from the mixture, the temperature decreased to approach a final steady-state value. With solutions of PVAC more complex behaviour was found, due to the non-additive combination of solvent and polymer losses. For example, MEK solutions of PVAC are less susceptible to microwaves than benzene solutions since some interaction between the two compounds probably restricts the electrical orientation of polar groups.

Improvements in film hardness and some degree of selectivity have been observed in studies using a pulsed microwave source where the on–off repetition frequency is of the order of 1×10^{-3} Hz to 20 KHz. In addition to bulk dielectric heating it is believed that specific molecular dipoles or chain segments can be influenced by using pulsed microwave cycles.

Using FTIR and ^{13}C NMR to study the cross-linking of diglycidyl ether of bisphenol-A (DGEBA) with DDS (4,4′-diaminodiphenylsulphone), it was found that the conversion rate was not a function of the highest temperature reached (42). It appears that more efficient transfer of microwave energy to dipolar sites can be achieved with specific pulses leading to high conversion rates: 2×10^{-3} s (700 W), 193 °C, 56% but 20×10^{-3} s (1500 W), 177 °C, 62%. Additionally, a pulsed microwave treatment with low repetition period (2×10^{-3} s) induces the formation of bis-aliphatic ethers by catalysing the homopolymerization reaction, while a longer pulse repetition period (50×10^{-3} s) promotes the reaction with amines only. The Persoz hardness of polyurethane films formed from 75% ethyl acetate solutions of Desmodur L75 and Desmophen 800 using pulsed irradiation compares favourably with oven cured films (43). Thus in general the films are harder (Persoz hardness 370–380 s) compared with oven cured films (340–360 s) and in the most efficient conditions (1000 W, 50×10^{-3} s; or 500 W, 2×10^{-3} s) very hard films are obtained (400 s). The cross-linking reaction of DGEBA with DDM (44–46) (4,4′′-diaminodiphenylmethane) is favoured by pulse repetition frequencies of 1000, 200, and 23.8 Hz which could correspond, probably in a rather complex fashion, to the series of segments shown in Figure 14.

The cure kinetics of polyimides have recently been studied by conventional and microwave heating (47). The polamic acid precursor (see Figure 15) was prepared by the room temperature reaction of benzophenone tetracarboxylic acid dianhydride (BDTA) and 3,3′-diaminodiphenylsulphone (DDS) in N-methylpyrrolidone (NMP). The extent of the imidization by 85 W of 2.45-GHz radiation was studied by FTIR. The equation for the kinetics of imidization can be expressed, at low initial concentrations, as:

$$\text{rate} = kf[\text{polymer}] \qquad (17)$$

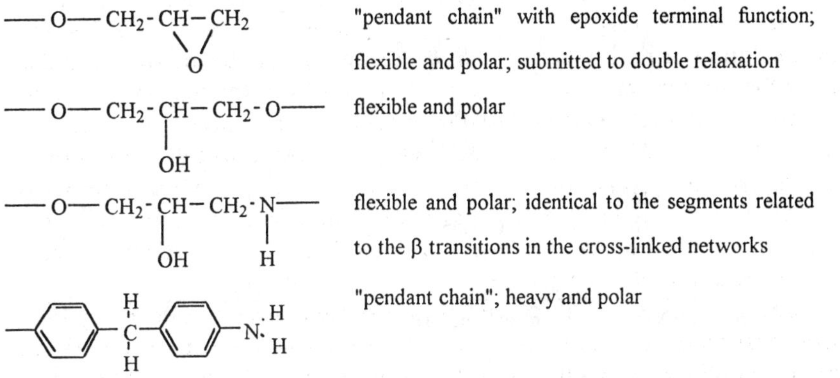

Figure 14. Possible sources of segmental relaxation in DGEBA–DDM cross-linking.

where f is the fraction of unreacted amic acid in the polymer determined directly from the FTIR by taking the ratio of the peak height (relative to a reference peak) to the maximum peak height.

The rate constants determined for the microwave solution imidization in Table IV indicate that the reaction proceeds rapidly with half lives varying from 23 to 4 min for 130 to 175 °C respectively. Furthermore, a comparison of the reaction rate constants for microwave and thermal processing at 150 and 160 °C indicate a significant enhancement in the rate of reaction when electromagnetic processing is employed. At 150 °C there is a 35-fold enhancement. However, this decreases to a 20-fold enhancement at 160 °C.

The authors have suggested two mechanisms for the observed rate acceleration. There is either some specific effect on the NMP coordinated

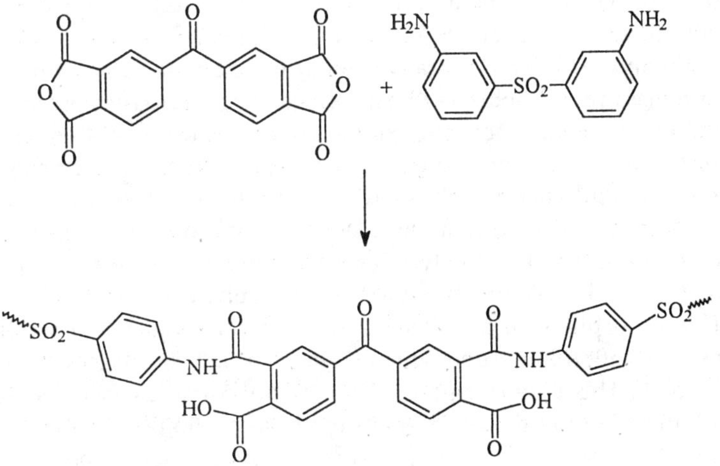

Figure 15. Synthesis of the polyamic acid precursor.

Table IV. Rate Constants for Thermal and Solution Imidization

T (C)	$k^{thermal}$ (min^{-1})	$k^{microwave}$ (min^{-1})
130		0.030
140	0.0014	
149	0.0022	
150		0.076
160		0.103
161	0.0055	
170	0.011	
175		0.169

SOURCE: Data taken from reference 47.

with the amic acid or a very localised temperature rise of say 40–60 °C due to directed absorption of the microwave radiation close to the reaction site.

Synthesis of Transition Metal Compounds

For simple chemical reactions in polar solvents, where it is desirable to limit the pressure to one atmosphere conditions, a reflux procedure has been described (48). This procedure can be introduced by modifying a conventional microwave oven as illustrated in Figure 16. The conventional chemical reflux system cannot be introduced into the microwave cavity because the circulating water coolant would absorb microwaves strongly and itself heat up rapidly. Furthermore, an air condenser would not be effective in returning the flammable solvents to the reaction flask in a safe manner. One solution to the problem would be to use a coolant that does not absorb microwaves strongly, for example, a non-polar organic or inorganic solvent. An alternative strategy is to locate the water cooled reflux condenser outside the microwave cavity. The condenser is connected to the reaction vessel by means of a port in the microwave cavity that ensures microwave leakage is kept to a safe limit.

In the modification illustrated in Figure 16 the solution is contained within a glass round-bottomed flask, which is connected via ground glass joints, to an air condenser. The condenser passes through a copper tube bolted to the side of a conventional microwave oven and is then connected to the water condenser, which is located completely outside the oven. This allows the solutions to heat rapidly when the microwave source is switched on and reflux safely without a build up of pressure and the release of volatile inflammables into the atmosphere.

The atmosphere in the round bottomed flask can be controlled by means of a Teflon inlet tube which enters through the side of the microwave cavity. This tube is connected on the outside to an inert gas supply and on the inside to the round-bottomed flask. The port through which the inlet tube passes is similarly protected by a copper tube or 'choke' which must be in good electrical contact with the body of the oven. To

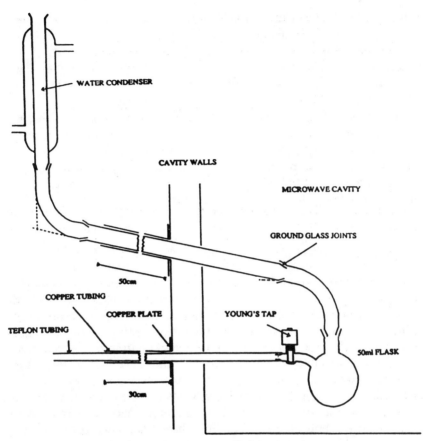

WATER CONDENSER

CAVITY WALLS

MICROWAVE CAVITY

GROUND GLASS JOINTS

50cm

COPPER TUBING

COPPER PLATE   YOUNG'S TAP

TEFLON TUBING

50ml FLASK

30cm

Modified microwave heating system applied.

Figure 16. The microwave oven modified to allow studies under refluxing conditions.

prevent leakage a choke of diameter equal to one half wavelength (i.e., 6 cm) should be at least one wavelength, 12 cm, long. The apparatus can be confirmed safe using a handheld microwave leakage detector capable of measuring power levels greater than 1 mW cm^{-2}.

When a homogeneous reaction proceeds at the normal boiling point of the solvent then the rate of reaction should be identical independent of whether the energy was introduced into the system by microwave heating or conventional conductive effects. However, for a typical unimolecular reaction with an activation energy of 100 kJ mol^{-1} and a pre-exponential factor of 4×10^{10} mol^{-1} s^{-1} it is anticipated that 10 °C superheating over the conventional reflux temperature of 77 °C will lead to a decrease in reaction time (time to reach 90% completion) by a factor of ca. 2–3.

It is well established that the rapid heating effect associated with microwave heating can cause superheating of water to about 110 °C (*49*).

The water heats so quickly that convection to the top surface and subsequent vaporisation are insufficient to dissipate the excess energy. Superheating effects can be magnified by the addition of ions (e.g., NaCl) to the water which greatly increase the conductive losses of the solution. Dramatic superheating effects have also been observed for water–oil mixtures where some interfacial polarisation effects may be important.

Experiments with pure organic solvents have demonstrated that the reflux temperature is within 5 °C of the solvent boiling point. Whan (50) has studied the rates of organic esterification reactions, under comparable microwave and conventional conditions and has shown that the rates of reaction are identical under the two methods. Therefore, there is no specific microwave effect and no superheating.

When the reaction conditions are not homogeneous or there are large quantities of mobile ions present then it is possible that localised superheating effects could lead to a small but significant increase in the reaction rate. There is some evidence, for example, that open vessel dissolution of geological and biological samples using acid solutions proceed at a faster rate in the microwave cavity than under conventional conditions (51, 52).

Table V provides some examples of organometallic syntheses which have been undertaken using the microwave reflux apparatus. The reported reaction times for these syntheses taken from the literature are also given in the table. The total synthesis times have been reduced by a factor of 6–40 times. Prior to concluding that a specific microwave acceleration effect is in action it should be emphasized that literature reaction times have a large uncertainty built into them. Nonetheless, some useful decrease in reaction time is occurring whose origin has not been well established. Contributing factors may be

1. Some superheating effects due to the presence of large number of ions present.
2. A more efficient and rapid achievement of the reaction temperature when microwave dielectric heating effects are used.
3. More efficient mixing of reactants and some localised superheating effects on the boundary between non-miscible liquids and solutions. The Maxwell–Wagner component of the dielectric loss may be particularly important in this respect.

Organic Synthesis at One Atmosphere

Some induced selectivity arising from the use of microwaves is suggested by the recent studies of the Diels–Alder reaction of 6-demethoxy-*p*-dihydrothebaine with an excess of methyl vinyl ketone to yield the two adducts shown in Scheme 1 (53). When the reaction was performed under conventional conditions extensive polymerisation of the dienophile

Table V. Microwave Synthesis of Organometallic Compounds Under Reflux Conditions

Reactants	Products	Microwave		Thermal	
		Yield (%)	Time (min)	Yield (%)	Time (h)
RhCl$_3$ xH$_2$O, C$_8$H$_{12}$, EtOH/H$_2$O (5:1)	[Rh(cod)Cl]$_2$	87	25	94	18
RuCl$_3$ xH$_2$O, EtOH, 1,3-cyclohexadiene	[Ru(C$_6$H$_6$)Cl$_2$]$_2$	89	35	95	4
RuCl$_3$ xH$_2$O, PPh$_3$, MeOH	[RuCl$_2$PPh$_3$)$_3$]	85	30	74	3
RuCl$_3$ xH$_2$O, EtOH, α-phellandrene	[Ru(η-cymene)Cl$_2$]$_2$	67	10	65	4
KReO$_4$, PPh$_3$, EtOH/H$_2$O 10:1	[ReOCl$_3$(PPh$_3$)$_2$]	94	30	95	5

NOTE: cod is 1,5-cyclooctadiene.

SOURCE: Data taken from reference 48.

resulted, which made the work-up and the isolation of the adducts cumbersome. Using a modified microwave oven 6-demethoxy-*p*-dihydrothebaine was refluxed in methyl vinyl ketone. The reaction reached completion in 24 h and substantially less polymeric material was formed.

High Pressure Conditions

The effective coupling between microwaves and polar inorganic and organic solvents can be utilised to accelerate the rates of a reaction significantly if the reactants can be contained within a closed vessel which is transparent to microwaves and can sustain the high pressures induced. The rapid heating of only the solvents within the autoclave leads to significant advantages compared with high pressure steel autoclaves used with conventional heating processes. There are a number of plastic materials which are microwave transparent and can form the basis of pressure vessels, but it is also necessary for them to be chemically inert and sufficiently strong to accommodate the pressures induced. The temperatures and pressures generated in such an autoclave depend on the level of the input

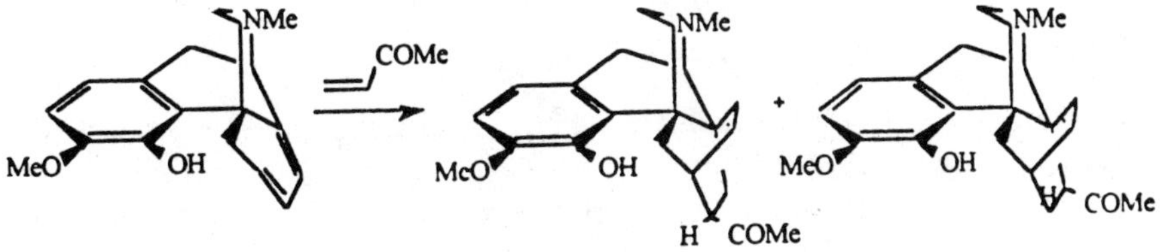

Scheme 1.

microwave power, the dielectric loss of the reacting solution, the volatility of the solvent, the volume of the container occupied by the solvent, and whether gases are generated in the reaction. Figure 17 illustrates the relationship between the temperature and pressure generated in a closed vessel for water and EtOH (54). Clearly for reactions in aqueous solutions a pressure of ca. 5 atm. is generated at 150 °C, whereas for the more volatile EtOH a pressure of 12 atm. is generated at the same temperature. These parameters largely define the pressure requirements of the vessel although it should be emphasised that it is necessary also to have a pressure release valve incorporated into the device if this methodology is to be used routinely and safely.

Parr (55) have developed a range of acid digestion vessels which can accommodate pressures up to 80 atm. and temperatures up to 250 °C. These devices, an example of which is shown in Figure 18, were originally designed for the acid digestion of geological, botanical, metallurgical, and biological samples (56), but can also be utilised for a range of inorganic and organic syntheses. The reactants are placed in a Teflon cup with a volume of ca. 23 cm^3 which is tightly sealed using an O-ring and cap, also made of Teflon. Teflon has the advantage of being chemically inert and therefore is a suitable containment vessel for acids and most commonly used organic and inorganic solvents. It does, however, suffer two disadvantages–it has a tendency to flow and creep, particularly at temperatures above 150 °C, and it is slightly porous. Therefore, repeated use of the vessel above 150 °C can lead to distortions which reduce the pressure limits of

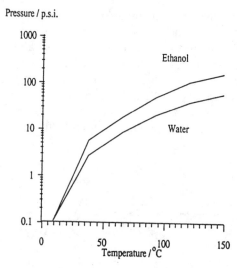

Figure 17. Temperature versus pressure curves for H_2O and EtOH (54) (1 atm. = 14.7 psi).

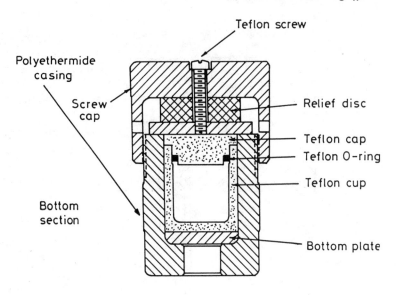

Parr. Acid Digestion Autoclave

Figure 18. Schematic illustration of the Parr acid digestion bomb (55).

the vessel from its initial 80 atm. The porosity leads to some incorporation of materials, particularly organic tars and metal powders, into the walls of the reaction vessel. The Teflon cup is contained within a larger vessel made of polyetherimide. The outer container has two sections which screw together. Pressure release is controlled by a large rubber section which presses down on the lid of the Teflon cup. When the pressure is developed within the vessel a screw rises from the top of the vessel. Each 2 mm rise corresponds to around 10 atm. pressure. Excessive pressure is released on compression of the rubber section which leads to fracture of the O-ring and release of the reactants through venting holes.

An alternative acid digestion system has been developed by CEM (57) and is shown in Figure 19. The advantage of this alternative system is that while the maximum working pressure is restricted to 14–15 atm., the pressure can be continuously monitored by a transducer outside the oven. The pressure is conveyed to the transducer via a thin Teflon tube which passes through a port in the side of the microwave oven. A feedback system has been developed by CEM to allow constant pressures to be held in the vessels over extended periods. Once again a pressure release mechanism is used to ensure pressure is kept below the specified maximum. The vessel has a much larger volume which should allow larger quantities of inorganic and organic material to be processed than has previously been possible using the Parr autoclave.

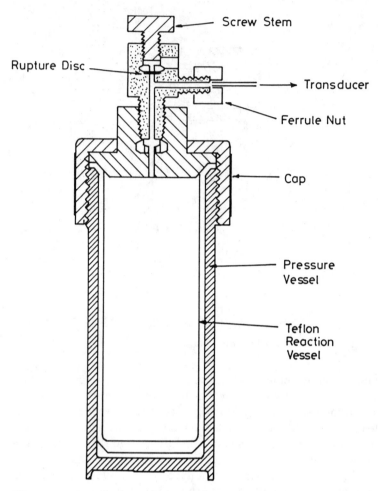

Figure 19. Schematic illustration of the CEM acid digestion vessel (57).

Both types of acid digestion apparatus have been used to shorten the time scales of a wide range of organic and inorganic reactions. Prior to a discussion of specific applications it is instructive to provide some estimate of the magnitude of the acceleration rates predicted on the basis of elementary reaction kinetics theory.

For a first order reaction which would require approximately 12–24 h at 60–80 °C to produce a yield of ca. 90% appropriate kinetic parameters are an activation energy, $E_a = 100$ kJ mol^{-1} and a pre-exponential factor of $A = 4 \times 10^{10}$ mol^{-1} s^{-1}. For these parameters the $t_{9/10}$ lifetimes (times for 90% completion) can be calculated for temperatures between 27 and 227 °C (*see* Table VI). Therefore, a reaction which commonly takes ca. 13.4 h to proceed to 90% completion in a solvent refluxing at 77 °C will achieve a

Table VI. Relationship Between Temperature and Time for a
First-Order Reaction

Temp (C)	Rate constant (k/s)	t9/10
27	1.55×10^{-7}	4126 h
77	4.76×10^{-5}	13.4 h
127	3.49×10^{-3}	11.4 min
177	9.86×10^{-2}	23.4 s
227	1.43	1.61 s

comparable yield in 23.4 s if undertaken in a closed vessel where the temperature is maintained at 177 °C. A similar time reduction factor (i.e., approximately 10^3) can be estimated for a second order reaction based on the same parameters.

Therefore this microwave methodology has the following potential: to introduce into a closed vessel which is transparent to microwaves energy sufficient to raise the temperature from ambient to ca. 200 °C in less than a minute, and thereby cause the total reaction time to be reduced by ca. 10^3. The introduction of energy so rapidly and effectively into a closed reaction vessel containing volatile organic compounds must take due regard of the likely pressure which is generated and the necessity of incorporating a safety device which allows any excess pressure to be released.

Some examples of syntheses conducted under high pressure conditions follow.

Organometallic Compounds

The rhodium(I) and iridium(I) dimers [$M_2Cl_2(diolefin)_2$] (M = Rh or Ir) are widely used as starting materials for organometallic syntheses and are conventionally synthesised from $MCl_3.xH_2O$ and the olefin in aqueous alcohol. Good yields are obtained after many hours (4–36) of refluxing in alcohol–water mixtures. Using the Parr autoclave described previously, microwave radiation of 2.45 GHz and a power level of 500 W, these dimers can be conveniently synthesised from the same reagents in good yields in less than one minute (58). In the reactions 0.2–1.0 g of the platinum metal salts were used and the total reaction volumes were approximately 15 cm³.

The range of compounds that can be synthesised in this manner, their yields, reaction mixtures and reaction times are summarised in Table VII (58). The maximum temperatures achieved in these reactions correspond approximately to 170 °C. With some olefins (e.g., cyclo-octene and cyclo-octatetraene) the dimeric complexes were not isolated and rhodium metal was the major product. In these cases the final products are unstable above 160 °C and presumably are formed and subsequently decomposed.

The same procedures were used to synthesise [$RuCl_2(\eta-C_6H_6)$]$_2$ from $RuCl_3.xH_2O$ and 1,3-cyclohexadiene in aqueous ethanol. A reaction time of

Table VII. Organometallic Syntheses in the Parr Bomb

Reactants	Products	Yield (%)	Time (s)
$RhCl_3$ xH_2O, C_8H_{12}, $EtOH/H_2$) (5:1)	$[Rh(cod)Cl]_2$	91	50
$RhCl_3$ xH_2O, $EtOH/H_2O$ (5:1), norbornadiene	$[Rh(C_7H_8)Cl]_2$	68	35
$IrCl_3$ xH_2O, C_8H_{12}, $EtOH/H_2O$ (5:1)	$[Ir(C_8H_{12})Cl]_2$	72	45
$RhCl_3$ xH_2O, $MeOH$, $C_5H_6{}^a$	$[Rh(C_5H_5)_2]^+$	62	30

NOTE: C_8H_{12} is cyclooctadiene; C_5H_6 is cyclopentadiene.

^{a}Methanolic HN_4PF_6 was added to the reaction mixture to obtain the yellow salt.

SOURCE: Data taken from reference 58.

35 s gave a yield of 89%. Interestingly, when $RhCl_3 \cdot xH_2O$ was reacted with freshly distilled cyclopentadiene and exposed to microwaves for 30 s the product was the sandwich cation $[Rh(\eta\text{-}C_5H_5)_2]^+$, which was isolated as the PF_6 salt, rather than the dimer produced conventionally. This compound is usually synthesised using the Grignard reagent C_5H_5MgBr rather than directly from cyclo-pentadiene.

Using a different type of Teflon autoclave vessel various post transition metal organometallic compounds have been synthesised (59). In the 100 cm^3 Savillex (60) Teflon bottle, with screw cap, a maximum solvent volume of 12 cm^3 was used to perform the reactions shown in Table VIII. The disadvantage of this type of vessel is that there is no pressure indication present and no safety pressure release mechanism.

In the examples of metallation of aromatic rings, ligand redistribution reactions and reactions of metallo-organic species such as $Al(OPr^1)_3$ a great reduction in reaction time has been found. It was noticed that the yields were generally of the same order as the conventional syntheses and where the separation of the products is difficult no particular advantage was evident from the microwave technique in terms of isolation.

Coordination Compounds

A wide range of coordination compounds of substitutionally inert transition metal ions have been rapidly and conveniently synthesised using microwave dielectric loss heating effects (61). The synthesis of coordination compounds of substitutionally inert transition metal ions, and particularly d^7, d^8, and low spin d^6 ions, can lead to long and tedious synthesis procedures. This is particularly true for second and third row transition metals with these electronic configurations which are exceptionally inert.

The coordination complexes summarised in Table IX represent a range of syntheses requiring from a minimum of 2 h (e.g., [IrCl3(9S3)]) to one week for [RuCl(CO)(bpy)$_2$]Cl under conventional conditions. Using the Parr autoclave the coordination compounds listed have been synthesised in good yields in a matter of minutes with 500–650-W power levels. In

Table VIII. Summary of Results of Microwave Heating

Reaction	Solvent	Microwave		Thermal	
		Time (min)	Yield (%)	Time (h)	Yield (%)
Preparation[a] of	EtOH	30	33	22	40
Preparation[a] of	EtOH	47	29	24	36
$Ph_3Bi + 3BiCl_3 \rightleftarrows 3PhBiCl_2$[b]	Propanol	6	46	3–4	30–68
$3Ph_4Sn + 2BiCl_3 \rightarrow Ph_3Bi + Ph_2SnCl_2$	EtOH–toluene	6	—	n/a	—
	EtOH	6	20	16	15
$Al(OPr^i)_3 + HOCH_2CH(OH)CH_3$[c]	None	6	—	4–6	—

[a] $Hg(OOCCH_3)_2$ used in microwave experiment. Conversion to chloro derivative via LiCl.
[b] Isolated as 2,2'-bipyridyl complex.
[c] Mixture of products; difficult to separate.
SOURCE: Data taken from reference 59.

Table IX. Coordination Complexes Synthesised by Microwave Techniques

Products	Reactants	Solvent	Yield (%)	Microwave (s)	Conventional (h)
[Cr(dpm)$_3$]	CrCl$_3$ 6H$_2$O, urea, dipivalolylmethane	EtOH (aq)	71	40	24
Ir(CO)Cl(PPh$_3$)$_2$	IrCl$_3$ xH$_2$O, PPh$_3$[a]	DMF	70	45	12
IrCl$_3$(9S3)	IrCl$_3$ xH$_2$O, 9S3	MeOH	98	16	2
[PtCl(tpy)]Cl 3H$_2$O	K$_2$PtCl$_4$, tpy	H$_2$O	47	2×30	24–100
[AuCl(tpy)]Cl 3H$_2$O	HAuCl$_4$, tpy	H$_2$O	37	2×30	24
[RuCl(CO)(bpy)$_2$]Cl	RuCl$_3$ xH$_2$O	DMF	70	3×20	168
[Ru(9S3)$_2$](PF$_6$)$_2$	RuCl$_3$ xH$_2$O, 9S3	MeOH	49	6×25	na

NOTE: dpm is 2,2',6,6'-tetramethyl-3,5-heptadionato; 9S3 is 1,4,7-trithiacyclononane; tpy is 2,2',2''-terpyridine; bpy is 2,2'-dipyridyl; DMF is 1,1'-dimethylformamide.

[a] The reaction product was at times contaminated with Ir(CO)Cl$_3$(PPh$_3$)$_2$, which was reduced to Ir(CO)Cl(PPh$_3$)$_2$ with zinc in DMF.

SOURCE: Data taken from reference 61.

some cases the pressure limitations of the vessel did not allow the whole reaction to be completed in one step and it was necessary to allow the reaction mixture to cool for approximately 15–20 min before subsequent exposures of microwaves each of 30-s duration were applied.

The synthesis of the crown thioether complex [IrCl$_3$(9S3)] required only 16 s for a virtually quantitative yield, compared to 3 h in refluxing methanol using conventional techniques. The synthesis of the related ruthenium complex [Ru(9S3)$_2$](PF$_6$)$_2$ by conventional techniques required the prior conversion of RuCl$_3$.xH$_2$O into the labile intermediate [Ru(SO$_3$CF$_3$)$_3$] or [Ru(DMSO)$_6$]$^{2+}$. Using the microwave technique it was synthesised directly from RuCl$_3$.xH$_2$O and required a total of 150-s microwave exposure (6 times 25 s) and gave the product in 49% yield.

The terpyridine complexes of Pt(II) and Au(III) listed have been extensively studied as metallo-intercalation reagents for DNA. Both compounds were obtained in good yield using H$_2$O as the solvent. The synthesis of the gold compound resulted in the formation as a byproduct of the complex salt [Au(terpy)Cl]$_2$ [AuCl$_2$]$_3$ [AuCl$_4$] which had been previously characterised.

The pentadionato- complex listed in Table IX provides an example of a substitutionally inert first row transition metal complex made by the microwave technique. It has been widely used as a spin lattice relaxation reagent in NMR. In the synthesis of [IrCl(CO)(PPh$_3$)$_2$] and [RuCl$_2$(CO)(bpy)$_2$]Cl the high dielectric loss solvent DMF was used. In these reactions DMF also acts as the source of the carbonyl ligands.

Polypyridyl complexes of Ru(II) have been the subject of widespread interest in recent years because of their unique photochemical and redox properties. The syntheses of these complexes are often time consuming and require several steps. For example, the recently reported synthesis of [Ru(bpy)$_3$](PF$_6$)$_2$ required the initial conversion of RuCl$_3$.xH$_2$O into [RuCl$_2$(CO)$_2$], by refluxing in methanoic acid followed by the substitution of the carbonyl ligands by bpy using Me$_3$NO in refluxing 2-methoxyethanol. It has been demonstrated that microwave dielectric loss heating effects can provide a general and convenient synthetic method for the polypyridyl complexes of ruthenium(II) (62), summarised in Table X. It is remarkable that yields of 60–94% were achieved using only 40 s of microwave radiation at a power level of 650 W.

Microwave techniques have recently been used in the synthesis of metal–metal bonded compounds. The preparation of [Mo$_2$(acetate)$_4$] takes between 12 and 48 h by conventional techniques and proved to be a challenge even using microwaves. The Parr autoclave containing the reactants was exposed to 325 W of 2.45 GHz microwave power for 15 min to obtain molybdenum acetate dimer in 75% yield (63):

$$Mo(Co)_6 \xrightarrow{\text{CH}_3\text{CO}_2\text{H} + \text{anhydride}} Mo_2(O_2CCH_3)_4 \qquad (18)$$

Table X. Microwave Syntheses of Ru^{II} Polypyridyl Complexes

Products	Wash Sequence	Yield (%)	Microwave Exposures
[Ru(bpy)$_3$](PF$_6$)$_2$	3 × 5 cm^3 Et$_2$O	71	40
[Ru(tpy)$_2$](PF$_6$)$_2$	3 × 5 cm^3 MeOH, 2 × 5 cm^3 Et$_2$O	70	45
[Ru(phen)$_3$](PF$_6$)$_2$	1 × 5 cm^3 MeOH, 3 × 5 cm^3 Et$_2$O	98	16
[Ru(bpy*)$_3$](PF$_6$)$_2$	10 cm^3 MeOH/H$_2$O(1:1), 2 × 5 cm^3 Et$_2$O	49	6 × 25

NOTE: bpy is 2,2'-dipyridyl; tpy is 2,2',2"-terpyridine; phen is 1,10-phenanthroline; bpy* is 4,4'-di-*t*-butyl-2,2'-dipyridyl.

SOURCE: Data from reference 62.

Organic Syntheses

Early work in this area was performed by Gedye (*31, 64, 65*) and Giguerre (*66, 67*). In the former case a variety of reactions were studied and progress has been made on defining the important experimental variables. The reactions which were performed in 150 cm^3 Savillex screw cap Teflon bottles are illustrated in Table XI. Giguerre and co-workers used 10 cm^3 sealed tubes or 15 cm^3 screw cap vials. Estimates of the reaction temperatures were obtained by using sealed capillaries containing organic compounds of known melting points either sealed within or affixed to the side of the reaction container. The results appear in Table XII. Care was taken to optimise conventional reaction conditions before comparisons were made with microwave syntheses.

Both groups suggest that the rate of heating of the solvent depends qualitatively upon the room temperature dielectric constant although, as we stressed in the section covering microwave heating of liquids and solids, the dielectric loss is the more reliable parameter. Gedye (*31*) has also shown that the rate of acceleration relative to conventional syntheses depends upon container volume, the ratio of reaction volume to container volume, and the solvent boiling point. Thus lower boiling solvents give reactions where the greatest rates of acceleration are observed. The inverse relationship between reaction flask volume and acceleration was estimated by the time required to obtain 65% yield of benzyl-4-cyanophenylether (Figure 20). These results suggest that the rate of reaction increases as the pressure increases. The effect of the ratio of the reaction volume to vessel volume is more complex. When the reaction volume is small the pressure and hence the reaction rate increases as the volume of the solvent in the reaction vessel increases. However, a point is reached where this increase is no longer observed since it takes longer to heat a large volume of solvent to the same temperature using the available microwave power.

Using this information Gedye (*31*) has increased the rate of acceleration observed in the esterification of benzoic acid in 1-propanol. The rate

Table XI. Results Obtained with a 720 W, 2.45 GHz Microwave Oven

Synthesis	Microwave	Classical	ρ
Hydrolysis of benzamide to benzoic acid	10 min, 99%	1 h, 90%	6
Oxidation of toluene to benzoic acid	5 min, 40%	25 min, 40%	5
Esterification of benzoic acid with methanol	5 min, 76%	8 h, 74%	96
Esterification of benzoic acid with propanol	18 min, 86%	7.5 h, 89%	25
Esterification of benzoic acid with n-butanol	7.5 min, 79%	1 h, 82%	8
S_N2 reaction of 4-cyanophenoxide with benzylchloride	3 min, 74%	12 h, 72%	240
	4 min, 93%	16 h, 89%	240

$$\rho = \frac{\text{rate (microwave)}}{\text{rate (classical)}}$$

enhancement was increased from 18 to 60 when the volume of the reaction mixture was altered to maximise the heating effect. Further rate enhancement to 180 times conventional was achieved by increasing the available microwave power from 560 to 630 W.

Practical application of these microwave techniques has now spread to the fields of natural product racemisation and the formation of radiochemicals.

Using sealed tubes small quantities of the natural product (–)-vincadifformine has been racemised to the synthetically more useful (+)-vincadifformine (Scheme 2) (*68*).

The optically active alkaloid is racemised through the intervention of a secodine intermediate by concurrent Diels–Alder cycloreversion and cycloaddition. In the microwave racemisation a degassed solution of (–)-vincadifformine in DMF was sealed in a glass tube under argon and heated for 20 min in a 500-W microwave oven. The (+)-enantiomer serves as the synthetic precursor of the Eburnane alkaloid, (+)-vincamine, which has been found to be a useful remedy in the treatment of cerebral insufficiency.

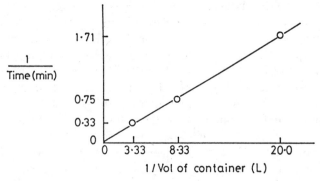

Figure 20. Effect of reaction vessel volume on the rate of formation of benzyl-1-cyanophenylether (*31*).

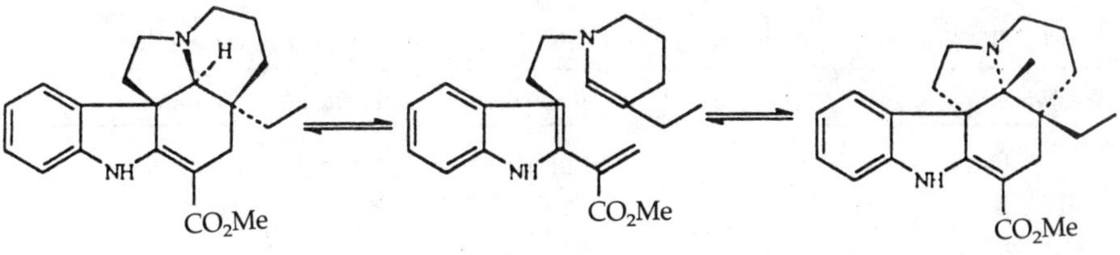

Scheme 2

(+)   (–)

The rapid and high yield synthesis of radiopharmaceuticals labelled with short lived radionuclides remains an important challenge to the synthetic chemist. With the positron emitting nuclides ^{122}I ($t_{1/2}$ 3.6 min), ^{11}C ($t_{1/2}$ 20 min), and ^{18}F ($t_{1/2}$ 110 min) this is particularly so. Using 3cm^3 Reacti-vials and a 500-W microwave oven the nucleophilic reactions of activated nitrobenzenes with ^{18}F-fluoride, and isotope exchange reactions of activated and deactivated halogenoarenes using ^{18}F-fluoride and ^{131}I-iodide have been studied (69). The microwave syntheses are much faster than comparable thermal syntheses (*see* Table XIII). The shorter microwave treatment relative to conventional heating not only leads to less decay of the radio-labelled product, but may also cause less degradation of the reagents and the generation of side-products which make isolation of the desired labelled compound difficult.

Isotope exchange of ^{131}I-iodide reached completion within 5 min in substrates both activated and deactivated towards nucleophilic substitution. Aromatic iodination yields in excess of 80% were obtained within 60 s, indicating the utility of this synthetic technique for reactions with positron-emitting ^{122}I ($t_{1/2}$ 3.6 min) and γ-emitting ^{123}I ($t_{1/2}$ 13.2 h).

As noted by Gedye, large scaling up of organic reactions in a microwave oven is not feasible using larger Teflon vessels since more microwave power is needed to process larger volumes. Instead techniques have been devised involving pumping the reactants from a reservoir through a reaction coil inside a microwave oven and on through the walls to a collection vessel. CSIRO, in Australia, has developed a continuous flow apparatus which at present can process materials at a rate of 20 cm^3/min (70). It is envisaged that their reactor will be able to process 4 litres of material a minute over many hours at temperatures up to 200 °C and pressures up to 12 atm. Recently Chen (71) has developed a similar reactor allowing more than 20 g of material to be processed. The effects of irradiation time, power level, and reaction quantity were compared with closed vessel microwave techniques. Generally the reaction conversion efficiency in the continuous flow system exceeded that found in a closed environment.

Zeolites

The synthesis of zeolites by microwave radiation is the subject of a recent patent (72). A range of zeolites have also been synthesised using a 650-W

Table XII. Microwave Acceleration of Diels–Alder, Claisen, and Ene Reactions

Reaction	Microwave	Literature	Control
(diene + alkyne diester → cyclohexadiene tetraester)	12 min, neat, 55%, 325 < 361	5 h, neat, 67%, 150	5 h, neat, 81%, 150
(furan + diethyl acetylenedicarboxylate → oxabicyclic diester)	10 min, neat, 66%, 325 < 361	4 h, neat, 95%, 100	4 h, neat, 68%, 100
(cyclohexenone + 2,3-dimethylbutadiene → octahydronaphthalenone)	15 min, neat, 25%, 400 < 425	72 h, neat, 20%, 200	a). 3 days, neat, 89%, 195 b). 2 h, neat, 75%, 195
(allyl phenyl ether → 2-allylphenol)	a). 10 min, neat, 21%, 325 < 361 b). 6 min, DMF, 92%, 325 < 361	6 h, neat, 85%, 240	6 min, neat, 17%, 320
(2-methoxyphenyl allyl ether → ortho + para allyl methoxyphenols)	a). 12 min, neat, 71%, 370 < 361 b). 5 min, DMF, 72%, 300 < 315 c). 90 s, n-methyl formamide, 87%, 276 < 300	85 min, neat, 85%, 240	a). 45 min, neat, 71%, 265 b). 12 min, neat, 92%, 320
(dienyne alcohol → cyclopentane derivative, ene reaction)	15 min, neat, 62%, 400 < 425	12 h, neat, 85%, 180	12 h, neat, 60%, 180

Table XIII. Fluorination Yields (%) for Some Aromatic Substitution Reactions
$18F^- + p\text{-}X\text{-}C_6H_4\text{-}G \rightarrow p\text{-}{}^{18}F\text{-}C_6H_4\text{-}G + X^-$

X	G	Microwave, 5 min	135 C, 5 min	135 C, 30 min
NO_2	CN	68 ± 5	52 ± 2	82 ± 2
NO_2	COMe	25 ± 3	10 ± 5	22 ± 1
NO_2	$CO[CH_2]_2CH_2$	77 ± 5	24 ± 7	80 ± 1
F	COMe	70 ± 1	12 ± 2	73 ± 2

microwave oven and a Parr autoclave (73). The long exposure times to microwave radiation quoted in Table XIV, particularly in the synthesis of Na zeolite A, mean that the autoclave is operating near its maximum temperature capability. The samples are highly uniform with particle sizes >2 μm. Under high power levels the more condensed sodalite structure tends to be obtained due to the higher pressures and temperatures involved. Some differences in sample morphology have been noted when conventional and microwave products are compared.

Additionally, zeolite exchange reactions have been performed. Replacement of Na^+ ions by Co^{2+} ions in faujasite shows comparable degrees of substitution using either 1 h of conventional heating or 60-s microwave heating. The X-ray powder patterns are identical. In the case of Co^{2+} exchange for Na^+ in sodalite, complete replacement was possible in just 4.5 min of microwave heating.

Intercalation Compounds

The intercalation of organic and organometallic compounds into layered oxide and sulphide structures has attracted considerable interest in recent years. The kinetics of the intercalation processes are often slow and consequently even after refluxing for several days these reactions do not always proceed to completion. Ultrasound has been shown to increase the rates of intercalation reactions but, as with the thermal methods, can result in loss of crystallinity of the samples. This makes structural characterisation of the intercalated compounds using X-ray powder techniques problematical (74).

Table XIV. Zeolite Syntheses Under Microwave Conditions

Zeolite	Formula	Power (W)	Time (min)
Na zeolite A	$Na_{12}[(AlO_2)_{12}(SiO_2)_{12}] \, 27H_2O$	300	45 in pulses
Faujasite	$Na_2(Al_2Si_5O_{14}) \, 10H_2O$	600	3
Analcime	$Na_{16}[(AlO_2)_{16}(SiO_2)_{32}] \, 16H_2O$	600	> 3
Hydroxysodalite	$Na_8(AlSiO_4)(OH)_2 \, 6\text{–}8H_2O$	600	2.5

SOURCE: Data taken from reference 73.

Using a Parr digestion bomb several pyridine intercalates of the layered mixed oxide α-VO(PO$_4$) have been prepared (75). In a typical experiment α-VO(PO$_4$) and the pyridine (or an equivalent volume of pyridine in xylene) were placed in the Teflon autoclave and exposed to 650 W of 2.45 GHz microwave radiation for several minutes. The length of each microwave exposure was limited to ca. 5 min by the pressure limitations of the autoclave. The maximum temperature reached was 200 °C. Table XV summarises the results of the microwave experiments, together with the results of comparable control reactions which were carried out using conventional thermal techniques. The latter were performed in glass ampoules, where 100% intercalation corresponds to a stoichiometry of VO(PO$_4$).py$_{0.85}$. Note that dramatic rate enhancements of the order 10^2–10^3 were generally observed.

An additional advantage of the microwave technique is the quality of the X-ray powder data obtained from the more crystalline microwave products. In Figure 21 the X-ray powder clearly shows the presence of two tetragonal phases. This is not apparent in the lower quality data from the conventional product. Blank experiments have established that while α-VO(PO$_4$) itself does not strongly absorb microwaves, the intercalated products do. Therefore, in the initial stages of the reaction the high temperature rise associated with the solution is primarily responsible for the rate enhancement. These efficient heating effects rapidly lead to pressures of ca. 50 atm and temperatures of 200 °C within a minute. In the latter stages of the reactions autocatalytic effects associated with the absorption of microwaves by the product may contribute to the completion of the intercalation reactions. These autocatalytic effects would have no analogues in conventional conductive heating.

Reactions Utilising the Microwave Dielectric Loss Properties of Solids

From Table III it is apparent that many solids absorb microwaves and can be heated very rapidly to high temperatures. This opens up the possibility of using the coupling to microwaves of one component in a mixture to drive a solid state chemical reaction with a second component which may be transparent to microwaves.

Chemistry of Ceramic Processing

The use of microwaves to process and sinter ceramics is thought to have the following advantages over classical heating (76):

1. *Reduced Cracking and Thermal Stress.* In microwave processing heat is generated inside the component while in furnace heating, by contrast,

heat is absorbed from the outside and can lead to thermalstresses, cracking and other problems. Hence, microwave heating can be especially useful for densifying ceramic components.

2. *Economy.* Internal heating may turn out to be more economical since the oven itself, and any microwave transparent refractory insulation material, is not heated.

3. *Increased Strength.* As a result of rapid heating by microwaves, the extent of non-isothermal processes such as segregation of impurities to the grain boundaries are minimised. Since sintering time is often reduced also, the possibility for secondary recrystallisation (exaggerated grain growth) may be reduced. By minimising impurity segregation, decreasing grain size and increasing sintered density, the mechanical properties of the ceramic are strengthened.

4. *Reduced Contamination.* The heating of the ceramic sample from the centre using microwaves limits the extent of contamination from the walls of the containment vessel.

The application of microwave energy in the field of ceramic processing and joining have been recently reviewed (77–79) and we shall restrict ourselves to those results having chemical significance.

The overall rate of a chemical reaction is governed by the speed of the slowest step or steps. For many solid state reactions this slowest step is the diffusion of the reactants towards one another through an unreactive medium. Any process which can increase the bulk diffusion coefficient could lead to a dramatic enhancement in the rate of reaction. Much of the work on the microwave processing of ceramic composite materials suggests that enhanced transport properties occur when dielectric heating effects are applied. The diffusion of various cations in Pyrex glasses and of ethylene oxide (EO) in polyvinylchloride (PVC) have been studied by both conventional and microwave heating.

Table XV. Comparison of Syntheses of $VO(PO_4) \cdot 2H_2O$ Intercalates by Using Microwave and Conventional Thermal Techniques

Guest[a]	Microwave		Conventional		Expansions[b]	
	t (min)	Stoichiometry	t (h)	Stoichiometry	c (Å)	Δc (Å)
Pyridine	5	0.84	36	0.35	9.55	5.44
4-Methylpyridine	3	0.86	12	0.60	10.56	6.45
4-Phenylpyridine	2×5	0.85	64	0.51	12.23	8.12[c]

[a]Host was $VO(PO_4) \cdot H_2O$.
[b]Coordinated water is lost during the reaction of $VO(PO_4) \cdot 2H_2O$ (c = 7.41 Å, where Δc represents the difference in length of the c axis for the intercalation compound and that of anhydrous $VOPO_4$.
[c]Another phase is also observed with a layer expansion of 9.89 Å.

SOURCE: Data taken from reference 75.

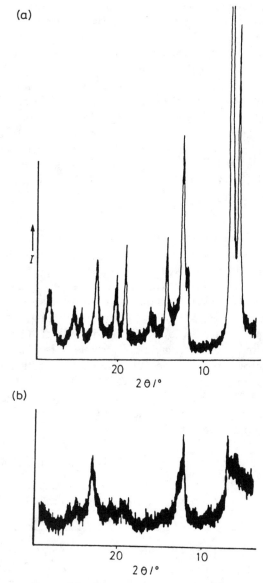

Figure 21. A comparison of the X-ray powder diffraction patterns for the intercalation compounds of α-VO(PO$_4$) with 4-phenylpyridine: (a) microwave product and (b) conventional product.

Cation diffusion in solid right circular cylinders of Pyrex glasses has been studied by back scattered electron microscopy profiling (*80*). Each cylinder was placed in a 50 cm^3 Pyrex crucible and was packed with either CoO (conventional case) or a concentrated solution of $Co(NO_3)_2 \cdot 6H_2O$ (microwave case). Diffusion occurred to a depth of 40 mm in a sample heated conventionally to 925 °C for 60 min. However, using microwave heating for just 15 min at a maximum temperature of 750 °C the diffusion depth was found to be 70 mm. Additionally several cobalt-rich phases had developed in the microwave heated sample which were absent in the conventional sample. Microwave processing at 1035 °C for just 10.5 min gave a diffusion length of 200 mm.

Microwave radiation (2.45 GHz) has been shown to greatly accelerate the diffusion of EO in polymeric materials compared to conventional heating at the same temperature (*81*). Specifically the mode of action of the microwaves was studied by examining the desorption of EO from PVC. Samples of PVC were saturated with EO in a controlled humidity apparatus prior to being heated by conventional and microwave heating. The concentration of EO remaining in the PVC as a function of time was routinely determined by heating in an enclosed vial followed by analysis by chromatography. The diffusion coefficient was found to be dependent upon the diffusant concentration which suggested that some interaction between EO and PVC was taking place.

Values of the activation energy for diffusion were obtained from Arrhenius plots of the logarithm of diffusion coefficient versus inverse temperature for both conventional and microwave-enhanced diffusion, and there was a significant reduction in activation energy for the latter as shown in Figure 22. There is agreement, to within the range of experimental error, between this value and an energy equal to the activation energy for conventional diffusion less the translational kinetic energy of the diffusant molecules. Therefore, it seems likely that the enhancement of the rate of diffusion of EO in PVC using microwaves is brought about by the active disruption of the EO–PVC hydrogen bonding resulting in a significant reduction in the proportion of immobilised diffusant molecules at any instant. This suggests that the process could be generally applicable for enhancing diffusion for polar–polarisable materials when a proportion of the diffusant molecules are immobilised in the host material.

Solid State Synthesis

Where one or more of the components of a solid state reaction mixture absorbs microwaves strongly the heat so produced can be used to drive the reaction. In Table XVI a range of syntheses achieved in this way are indicated (*82, 83*). Thus the strong absorption of V_2O_5 has been used to synthesise a sample of KVO_3 from V_2O_5 and KCO_3. The starting materials are intimately mixed and can be pelletised, prior to being placed in alumina, zirconia, or silica crucibles, as appropriate. The crucible is placed on a firebrick to protect the cavity from the high temperatures reached.

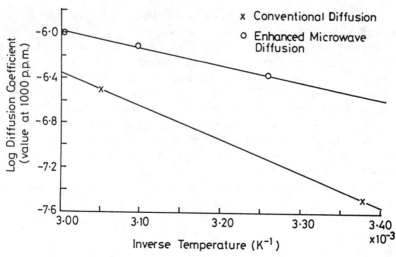

Figure 22. Arrhenius plots for conventional and enhanced desorptions (*81*).

Reaction times are shorter than those required conventionally to produce samples having comparable X-ray powder patterns. Thus, the samples appear to retain a high degree of crystallinity, and impurities and starting materials are absent from the final products. Recently the kinetics of product formation have been studied by calibrated X-ray powder techniques for microwave and thermal samples of $Sr_2LaFe_3O_{8+y}$. Using KCl as the internal standard a homogeneous microwave product was obtained after 36 min heating with around 500 W of microwave power. The sample temperature was measured at 1100 °C by a grounded sheathed thermocouple. The temperature is believed to be reliable since higher temperatures (1200 °C and higher) would be expected to produce an orthorhombic rather than a cubic X-ray pattern. A comparable thermal sample was synthesised in 26 h.

Several groups have synthesized samples of superconducting $YBa_2Cu_3O_{7-x}$ utilising the coupling of CuO to microwaves in starting reaction mixtures and eventually the product itself (*82–84*). It has been found

Table XVI. Microwave Solid-State Syntheses

Product	Starting Materials	Microwave Synthesis (min)	Conventional Synthesis (h)
KVO_3	K_2CO_3, V_2O_5	7	12
$CuFe_2O_4$	CuO, Fe_2O_3	30	23
$BaWO_4$	BaO, WO_3	30	2
$La_{1.85}Sr_{0.15}CuO_4$	La_2O_3, $SrCO_3$, CuO	35	12
$YBa_2Cu_3O_{7-x}$	Y_2O_3, $Ba(NO_3)_2$, CuO	70	24

SOURCE: Data taken from references 82 and 83.

that since the reaction time is much shorter than that required conventionally there is no need to slow cool the product in an oxygen atmosphere to obtain the orthorhombic superconducting phase. In Figure 23 the resistivity plot is shown. For a typical sample synthesised from a stoichiometric mixture of CuO, Ba(NO$_3$)$_2$, and Y$_2$O$_3$. The sample was made in a 500-W domestic microwave oven which had been modified to allow the safe removal of the various nitrogen oxides under flowing oxygen. The high onset temperature for superconductivity of 96.5 K and the narrow nature of the transition to zero resistivity are comparable to those of samples produced under conventional conditions.

SEM studies on a sample of YBa$_2$Cu$_3$O$_{7-x}$ (84), which had been synthesised conventionally but annealed in a microwave oven have shown encouraging results. The sintering occurs evenly resulting in smaller grains and lower sample porosity.

Decomposition Reactions

The rapid heating of solids by microwaves provides an alternative to conventional pyrolysis procedures. Early studies showed that the products obtained by microwave heating of cellulose were similar to those obtained thermally but the time taken to reach reaction temperature and reaction times themselves were reduced significantly (85). More recently the formation of 1,6-anhydro-β-D-glucopyranose, a useful chiral synthon, from starch or other 1 → 4 glucans has been studied (86).

An irradiation time of 5 to 15 min was required to accomplish complete reaction for 10 to 80 g of starch, respectively. The yield of (1) from different glucans (Scheme 3) varied from 0.65 to 1.7% (Table XVII). No correlation could be found between the reaction efficiency and chemical

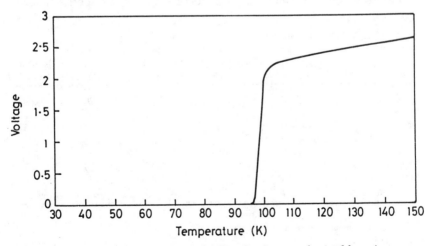

Figure 23. Resistivity plot for a sample of YBa$_2$Cu$_3$O$_{7-x}$ synthesised by microwave techniques.

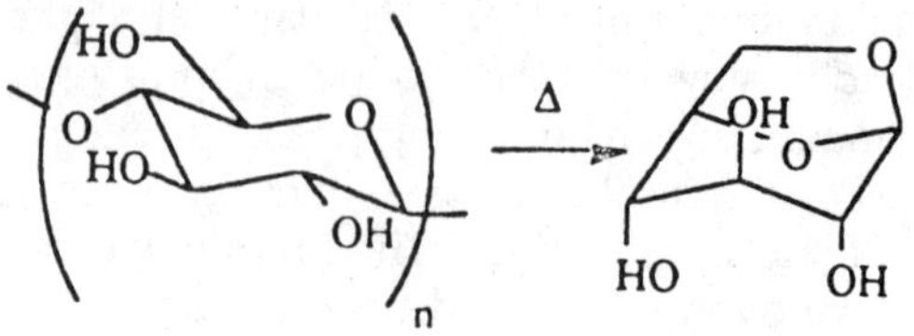

Scheme 3.

structure, degree of polymerisation or water content, though glucan density may be an important factor. For obtaining a small amount of (1) (ca. 1 g) the ease of this method outweighs its low yield.

Some evidence for a specific microwave effect has been found in the cracking of *neo*-pentane on zeolites. The stoichiometric gas phase reaction is:

$$neo - C_5H_{12} \rightarrow CH_4 + iso - C_4H_8 \qquad (19)$$

Using microwaves more methane is produced when zeolites are irradiated than when the zeolites are heated in a classical furnace and consequently less isobutene is obtained (87).

The reactions of methyl-2-pentane on a Pt–Al$_2$O$_3$ catalyst bed show some degree of selectivity too (88). The range of products produced by either cracking or isomerisation is shown in Table XVIII.

The results in Table XIX show that while most of the microwave product distribution can be produced conventionally, the production of benzene is substantially increased.

Industrially one of the most useful chemical processes is the reduction of oxide ores to yield lower oxides and metals. The strong coupling of carbon to microwave radiation has been successfully used to perform several

Table XVII. Yield of **1** from Glucans

Substrate	n^a	Water Content^b (% w/w)	Density (g/cm^3)	Yield of 1^c (mg)	Yield of 1^c (% w/w)
Cellulose	2800	6.6	0.41	220	0.94
Starch	14,000	4.2	0.95	3398	1.7
Dry starch	14,000	0.0	0.95	382	1.5
Amylopectin	2,000,000	11.0	0.33	226	1.0
Amylose	3000	13.1	0.73	262	1.2
Maltodextrin 2	50	9.5	0.46	189	0.84
Maltodextrin 10	10	7.5	0.47	150	0.65
β-Cyclodextrin	7	10.5	0.83	280	1.3

^{a}Chain average degree of polymerisation.
^{b}Water content at the start of the reaction.
^{c}According to HPLC workup.

SOURCE: Data taken from reference 86.

Table XVIII. Products Produced by Either Cracking or Isomerisation of Methyl-2-Pentane on a PtAl$_2$O$_3$ Cataylst Bed

	C1 + C5	
	C2 + C4	**cracking**
	2 C3	(Cx where x denotes the
	6 C1	number of C atoms in
	3 C2	the cracked hydrocarbon
methyl-2-pentane		
	neo-hexane (HEX)	
	methyl-3-pentane (M3P0	
	methylcyclopentane (MCP)	
	Benzene (BEN) **isomerisation**	
	cyclohexane	

reductions on a laboratory scale. For example, TiO$_2$ has been reduced to TiO and V$_2$O$_5$ to V$_2$O$_3$ after 30 min exposure to 500 W of microwave power (*89*).

The reduction of iron ores through the two reactions:

$$2Fe_2O_3 + 3C \rightarrow 4Fe + 3CO_2 \qquad \text{(reaction 1)}$$

$$Fe_3O_4 + 2C \rightarrow 3Fe + 2CO_2 \qquad \text{(reaction 2)}$$

is of particular importance in the extraction of iron from haematitic and magnetic ores. In processing a range of ores with different sources of carbon in a domestic microwave oven, it has been found that the final composition depends strongly on the type of ore used (*90*). Final products ranging from simple white cast iron to complex super hard alloy irons have been produced. The versatility of the microwave technique in the iron making process has been well demonstrated since most of the products produced conventionally have been observed as well as a few new ones.

Reactions on Supports

A novel means of accelerating a slow organic reaction is to perform the reaction on a solid support which couples effectively with microwaves.

Table XIX. Experimental Results of Reaction of Methyl-2-Pentane on PtAl$_2$O$_3$

Oven	S%	3C$_2$	C$_1$ + C$_5$	C$_2$ + C$_4$	2C$_3$	M3P	HEX	MCP	BEN
Microwave at 300 C	69.4	0.7	17.2	7.8	5.0	26.5	27.9	7.9	6.9
Classical at 300 C	67.1	0.6	17.5	9.4	5.7	21.0	28.4	17.4	0.3
Classical at 325 C	62.9	0.9	19.8	10.1	6.9	22.8	31.0	5.9	3.1

Note: S% is selectivity isomer/cracking.

Source: Data taken from reference 88.

For example, the following molecular rearrangement of pinacol to pina-colone on a phyllosilicate (M^{n+}-montmorillonite) solid support has been reported:

$$(CH_3)_2C(OH)–C(OH)(CH_3)_2 \xrightarrow{M^{n+} \text{ montmorillonite}} CH_3COC(CH_3)_3 + H_2O \quad (20)$$
$$\text{pinacol} \qquad\qquad\qquad \text{pinacolone}$$

This reaction has been studied in a conventional 650-W microwave oven using open glass or Teflon vessels (*90*). Significant rate enhancements have been found compared with the conventional supported reactions. In Table XX the percentage of pinacol transformed by 450 W of microwave power is compared with that achieved thermally at 100 °C. The reaction times were chosen so as to achieve the optimal yields: 15 h for conventional heating and 15 min by microwave techniques.

The anionic alkylation (reaction of potassium acetate with 1-bromooc-tane) carried out either on silica gel or alumina:

$$CH_3CO_2K + C_8H_{17}Br \xrightarrow{\text{silica or alumina}} CH_3CO_2C_8H_{17} + KBr \quad (21)$$

has also been studied by these methods and once again dramatic rate enhancements have been found (*90*). These are summarised in Table XXI.

Using sealed Teflon vessels similarly impressive results have been found and these are summarised in Table XXII (*92*). For example the acet-ylenic alcohol (**2**) when adsorbed onto K10 or KSF clay, acidic montmoril-lonite type phyllosilicate, at room temperature does not react. Microwave

Table XX. Conversion (%) of Pinacol to Pinacolone

M^{n+}	Conventional (100 C, 5 h)	Microwave (450 W, 15 min)
Na$^+$	5	38
Ca^{2+}	2	23
Cu^{2+}	30	94
La^{3+}	80	94
Cr^{3+}	99	98
Al^{3+}	98	99

SOURCE: Data taken from reference 91.

Table XXI. Conversion (%) of BrC_8H_{12} into $CH_3CO_2C_8H_{17}$

Substrate	Conventional (100 C, 5 h)	Microwave (600 W, 10 min)
Al$_2$O$_3$	93 ± 1	91 ± 3
SiO$_2$	69.5 ± 0.5	82 ± 12

SOURCE: Data taken from reference 91.

Table XXII. Dry Organic Reactions by Microwave Heating

Reaction	Solid Support	Microwave	Conventional
(2) → **(3)**	K10 or KSF	5 min 270 W 95%	5 min < 2%
	KSF	5 min 270 W 98%	
	KSF	5 min 270 W 98%	
	KSF	5 min 270 W 51 and 41%	
	KSF	5 min 270 W 75%	5 min no aldehyde
R—CHO + Ph—SO$_2$-CH$_2$·CN → Ph—SO$_2$	KF on Al$_2$O$_3$	20 min 55 W 95%	5 min 3%
R—CHO +	KF on Al$_2$O$_3$	20 min 55 W 90%	5 min 20%
R—CHO +	KF on Al$_2$O$_3$	20 min 55 W 62%	5 min 20%

SOURCE: Data taken from reference 92.

irradiation gave the rearranged product (3) in 92% yield. Typically, the alcohol in solution in CH_2Cl_2 was absorbed into 1 g of clay and irradiated with 270 W of 2.45 GHz microwave radiation for 5 min. A comparable thermal reaction at 170 °C gave just 2% of the rearranged product.

Summary

This review has presented the principles underlying the dielectric heating effects observed for chemical compounds in solution and in the solid state. The applications of the technique to a wide range of chemical syntheses have also been indicated. The field is in its infancy and therefore much of the effort to date has been directed towards understanding the rate enhancements of known reactions. The next few years should see the isolation of new compounds using microwave heating techniques.

Acknowledgements

The Science and Engineering Research Council is thanked for financial support to David R. Baghurst.

References

1. Jolly, W. L. *The Synthesis and Characterisation of Inorganic Compounds*; Prentice Hall: New York, 1970.
2. Harvey, A. F. *Microwave Engineering*; Academic Press: Orlando, FL, 1963.
3. Herzberg, G. *Infrared and Raman Spectra*; van Nostrand: New York, 1945.
4. Banwell, C. N. *Fundamentals of Molecular Spectroscopy*, 3rd ed.; McGraw-Hill: New York, 1983.
5. Debye, P. *Polar Molecules*; Chemical Catalog: New York, 1929.
6. Frohlich, H. *Theory of Dielectrics*, 2nd ed.; Oxford University Press: London, 1958.
7. Daniel, V. *Dielectric Relaxation*; Academic Press: Orlando, FL, 1967.
8. Cole, K. S.; Cole, R. H. *J. Chem. Phys.* **1941,** *9,* 341.
9. Hill, N.; Vaughan, W. E.; Price, A. H.; Davies, M. *Dielectric Properties and Molecular Behaviour*; von Nostrand: New York, 1969.
10. Hasted, J. B. *Aqueous Dielectrics*; Chapman and Hall: London, 1973.
11. Metaxas, A. C.; Meredith, R. J. *Industrial Microwave Heating*; Peter Perigrinus: London, 1983.
12. von Hippel, A. R. *Dielectric Materials and Applications*; Massachusetts Institute of Technology: Cambridge, MA, 1954.
13. Debye, P. *Phys. Z.* **1935,** *36,* 100.
14. Lauffer, M. A. *J. Chem. Educ.* **1981,** *58,* 250.
15. Bottcher, C. J. F. *Theory of Electric Polarisation*; Elsevier Biomedical: Amsterdam, Netherlands, 1952.
16. Meakins, R. J. *Trans. Faraday Soc.* **1955,** *51,* 953.

17. Smyth, C. P. *Dielectric Behaviour and Structure;* McGraw-Hill: New York, 1955.
18. Meakins, R. J. *Trans. Phys. Soc.* **1956**, *52*, 320.
19. Fuoss, R. M.; Kirkwood, J. G. *J. Am. Chem. Soc.* **1941**, *63*, 385.
20. Wagner, K. W. *Arch. Elektrotech.* **1914**, *2*, 371.
21. Hamon, B. V. *Aust. J. Phys.* **1953**, *6*, 304.
22. Kharadly, M. M. Z.; Jackson, W. *Proc. Inst. Electr. Eng.* **1953**, *100*, 199.
23. Anderson, J. C. *Dielectrics;* Chapman and Hall: London, 1964.
24. Coelho, R. *Physics of Dielectrics for the Engineer;* Elsevier: Amsterdam, Netherlands, 1979.
25. Berteaud, A. J.; Badot, J. C. *J. Microwave Power* **1976**, *11*, 315.
26. van Beek, K. H. *Prog. Dielectr.* **1967**, *7*, 69.
27. Meek , T. T. *J. Mater. Sci. Lett.* **1987**, *6*, 638.
28. Alberty, K. A. *Physical Chemistry*, 7th ed.; Wiley: New York, 1987; p 326.
29. Von Hippel, A. R. *Dielectrics and Waves;* Massachusetts Institute of Technology: Cambridge, MA, 1954.
30. Tinga, W. R. *Electromagn. Energy Rev.* **1988**, *1*, 1.
31. Gedye, R. N.; Smith, F. E.; Westaway, K. G. *Can. J. Chem.* **1988**, *66*, 17.
32. McGill, S. L.; Walkiewicz, J. W. *J. Microwave Power Electromagn. Energy Symp. Summ.* 1987, 175.
33. Tinga, W. R. *Soc. Manuf. Eng. Symp. Proc.* **1986**, *60*, 105.
34. Bowman, R. R. *IEEE Trans. Microwave Theory Tech.* **1976**, *24*, 43.
35. Teffal, M.; Gourdenne, A. *Eur. Polym. J.* **1983**, *19*, 543.
36. Silinski, B.; Kuzmycz, C.; Gourdenne, A. *Eur. Polym. J.* **1987**, *23*, 273.
37. Gourdenne, A.; LeVan, Q. *Polym. Prepr. (Am. Chem. Soc., Div. Polym. Chem.)* **1981**, *22*, 125.
38. LeVan, Q.; Gourdenne, A. *Eur. Polym. J.* **1987**, *23*, 777.
39. Bazaird, Y.; Gourdenne, A. *Eur. Polym. J.* **1988**, *24*, 881.
40. Gourdenne, A.; Massarani, A. H.; Monchaux, P.; Aussudre, S.; Thourel, L.; *Polym. Prepr. (Am. Chem. Soc., Div. Polym. Chem.)* **1979**, *20*, 471.
41. Julien, H.; Valot, H. *Polymer* **1983**, *24*, 810.
42. Thullier, F. M.; Jullien, H.; Grenier-Loustalot, M. F. *Polym. Commun.* **1986**, *27*, 206.
43. Jullien, H.; Valot, H. *Polymer* **1985**, *26*, 506.
44. Beldjoudi, N.; Bouazizi, A.; Doubi, D.; Gourdenne, A. *Eur. Polym. J.* **1988**, *24*, 49.
45. Beldjoudi, N.; Gourdenne, A. *Eur. Polym. J.* **1988**, *24*, 53.
46. Beldjoudi, N.; Gourdenne, A. *Eur. Polym. J.* **1988**, *24*, 265.
47. Lewis, D. A.; Ward, T. C.; Summers, J. S.; McGrath, J. E. *First Australian Symposium on Microwave Power Applications;* University of Wollongong: Wollongong, Australia, 1989.
48. Baghurst, D. R.; Mingos, D. M. P. *J. Organomet. Chem.* **1990**, *384*, C57.
49. Buffler, C. R.; Lindstrom, T. *Microwave World* **1988**, *9*, 10.
50. Whan, D., University of Hull, personal communication.
51. Mahan, K. I.; Foderaro, T. A.; Garza, T. L.; Martinez, R. M.; Maroney, G. A.; Trivisonno, M. R.; Willging, E. M. *Anal. Chem.* **1987**, *59*, 938.
52. Aysola, P.; Anderson, P. D.; Langford, C. H. *Anal. Lett.* **1988**, *21*, 2003.
53. Linders, J. T. M.; Kokje, J. P.; Overhand, M.; Lie, T. S.; Maat, L. *Recl. Trav. Chim. Pays-Bas* **1988**, *107*, 449.

54. *Chemical Engineers Handbook,* 3rd ed.; Perry, J. H., Ed.; McGraw-Hill: New York, 1950.
55. Parr Instrument Company, Moline, Il 61265.
56. *Introduction to Microwave Sample Preparation;* Kingston, H. M.; Jassie, L. B., Eds.; American Chemical Society: Washington, DC, 1988.
57. CEM Corporation, Matthews, North Carolina 28106.
58. Baghurst, D. R.; Mingos, D. M. P.; Watson, M. J. *J. Organomet. Chem.* **1989,** *368,* C43.
59. Ali, M.; Bond, S. P.; Mbogo, S. A; McWhinnie, W. R.; Watts, P. M. *J. Organomet. Chem.* **1989,** *371,* 11.
60. Savillex Corporation, Minnetoka, MN 55345.
61. Baghurst, D. R.; Cooper, S. R.; Greene, D. L.; Mingos, D. M. P.; Reynolds, S. M. *Polyhedron* **1990,** *9,* 893.
62. Greene, D. L.; Mingos, D. M. P. *Transition Met. Chem. (London)* **1991,** *16,* 71.
63. Greene, D. L., University of Oxford, unpublished results.
64. Gedye, R.; Smith, F.; Westaway, K.; Ali, H.; Balderisa, L.; Laberge, L.; Rousell, J. *Tetrahedron Lett.* **1986,** *27,* 279.
65. Gedye, R. N.; Smith, F. E.; Westaway, K. C. *Educ. Chem.* **1988,** *25,* 55.
66. Giguerre, R. J.; Bray, T. L.; Duncan, S. N.; Majetich, G. *Tetrahedron Lett.* **1986,** *28,* 4945.
67. Giguerre, R. J.; Namen, A. M.; Lopez, B. O.; Arepally, A.; Ramos, D. A.; Majetich, G.; Defauw, J. *Tetrahedron Lett.* **1987,** *28,* 6553.
68. Takano, S.; Kijima, A.; Sugihara, T.; Satah, S.; Ogasawara, K. *Chem. Lett.* **1987,** 87.
69. Hwang, D. R.; Moerlein, S. M.; Lang, L.; Welch, M. J. *J. Chem. Soc. Chem. Commun.* **1987,** 1799.
70. Peterson, C. *New Sci.* **1989,** *123,* 44.
71. Chen, S.-T.; Chiou, S.-H.; Wang, K. T. *J. Chem. Soc. Chem. Commun.* **1990,** 807.
72. Vartuli, V. C.; Chu, P.; Dwyer, F. G. Crystallisation Method Using Microwave Radiation U.S. Patent Application 4, 778, 666, 1988.
73. Gill, R. H. I.; Mingos, D. M. P., University of Oxford, unpublished results.
74. Suslick, K. S. *Science (Washington, D.C.)* **1990,** *247,* 1439.
75. Chatakondu, K.; Green, M. L. H.; Mingos, D. M. P.; Reynolds, S. M. *J. Chem. Soc. Chem. Commun.* **1989,** 1515.
76. Dagani, R. *Chem. Eng. News* **1988,** *66,* 7.
77. Sutton, W. H. *Am. Ceram. Soc. Bull.* **1989,** *68,* 376.
78. Palaith, D.; Siberglitt, R. *Am. Ceram. Soc. Bull.* **1989,** *68,* 1601.
79. *MRS Symposium Proceedings;* Sutton, W. H.; Brooks, M. H.; Chabinsky, I. J., Eds.; Society of Manufacturing Engineers: Dearborn, MI, 1988; p 124.
80. Meek, T. T.; Blake, R. D.; Katz, J. D.; Bradbury, J. R.; Brooks, M. H.; *J. Mater. Sci. Lett.* **1988,** *7,* 928.
81. Gibson, C.; Matthews, I.; Samuel, A. *J. Microwave Power Electromagn. Energy* **1988,** *23,* 17.
82. Baghurst, D. R.; Chippindale , A. M.; Mingos, D. M. P. *Nature (London)* **1988,** *332,* 311.
83. Baghurst, D. R.; Mingos, D. M. P. *J. Chem. Soc. Chem. Commun.* **1988,** 829.
84. Ahmad, I.; Chandler, G. T.; Clark , D. E. In *MRS Symposium Proceedings;* Sutton, W. H.; Brooks, M. H.; Chabinsky, I. J., Eds.; Society of Manufacturing Engineers: Dearborn, MI, 1988.

85.   Allan, G. G.; Krieger, B. B.; Work, D. W. *J. Appl. Polym.* **1980,** 25, 1839.
86.   Straathof, A. J. J.; van Bekkum, H.; Kieboom, A. P. G. *Recl. Trav. Chim. Pays-Bas* **1988,** 107, 647.
87.   Roussy, G.; Thiebaut, J. M.; Anzarmou, M.; Richard, C.; Martin, R. J. *Microwave Power Electromagn. Energy, Symp. Summ.* **1987,** 169.
88.   Thiebaut, J. M.; Roussy, G.; Maire, G.; Garin, F. *International Conference on High Frequency Microwave Processing and Heating;* N.V. Kema Laboratories: Arnhem, Netherlands, 1989.
89.   Baghurst, D. R.; Mingos, D. M. P., Oxford University, unpublished results.
90.   Barnsley, B. P.; Reilly, L.; Jones, J.; Eshman, J. *First Australian Symposium on Microwave Power Applications;* University of Wollongong: Wollongong, Australia, 1989; p 49.
91.   Gutierrez, E.; Loupy, A.; Bram, G.; Ruiz-Hitzky, E. *Tetrahedron Lett.* **1989,** 30, 945.
92.   Alloum, A.; Labaid, B.; Villemin, D. *J. Chem. Soc. Chem. Commun.* **1989,** 386.

Chapter 2

Overview of Microwave-Assisted Sample Preparation

Peter J. Walter, Stuart Chalk, and
H. M. (Skip) Kingston

Sample preparation involves numerous steps, from sample collection to sample presentation as a homogeneous solution for instrumental analysis. Sample preparation can involve combinations of the following: drying of the sample, leaching, extraction, digestion of the matrix, postdigestion chemistry, analytical separation, solvent removal, and exchange. The use of microwave technologies was shown to improve sample preparation while also reducing contamination.

This chapter and the remainder of this book will address the diverse field of microwave sample preparation and microwave chemistry. Chapter 3 addresses the use of chemistry for microwave acid dissolutions and emphasizes environmental applications. Chapter 4 addresses high-pressure acid digestion, and Chapter 5 addresses atmospheric-pressure microwave dissolution. Chapter 6 addresses flow-through microwave reactors. Chapter 11 describes the use of microwave energy in solvent extraction. Other chapters address flow-through microwave and uniquely designed microwave sample-preparation systems. Chapter 13 evaluates the use of pressure control in microwave dissolutions. This chapter reviews the literature in the field of microwave sample preparation and the chemistry of dissolution.

Historical Perspectives on Microwave Sample Preparation

Elemental analysis of nearly every matrix requires the dissolution of the sample before instrumental analysis. Despite tremendous improvements and discoveries of new analytical instruments over the past decades, few changes in dissolution methodologies have come forth. For centuries, chemists have used some variation of an open-vessel digestion or a Carius tube closed-vessel digestion. In 1975, microwaves were first used as a rapid heating source for wet open-vessel digestions (1–3). Microwaves were used to heat acids rapidly, in Erlenmeyer flasks, to digest biological matrices. Conventional sample digestion times were reduced from 1–2 h to 5–15 min by using microwave heating, a net reduction in analysis time. These papers spawned the research and development of a new sample-preparation technique.

Early microwave sample-preparation researchers used common laboratory glassware and open Teflon vessels to digest matrices at the boiling point of the acids in commercial microwave ovens. In the 1980s, researchers began using specially designed closed vessels for microwave digestions to achieve reaction temperatures above the atmospheric boiling point of the acids to increase the reaction rates and decrease reaction times. However, this development was accompanied by an increase in reaction pressures, a potential safety concern (4–7). These closed microwave-digestion vessels were fabricated from polycarbonate or Teflon and were not specifically designed for microwave (7, 8). The first closed Teflon vessels used in this transitional period were designed for leaching of nuclear waste glass samples (8).

Temperature and pressure monitors were adapted with wavelength attenuators for monitoring the reactions and evaluating the conditions in closed microwave systems (Figure 1) (4, 5). These modifications to commercial microwave systems became the foundation of the laboratory microwave

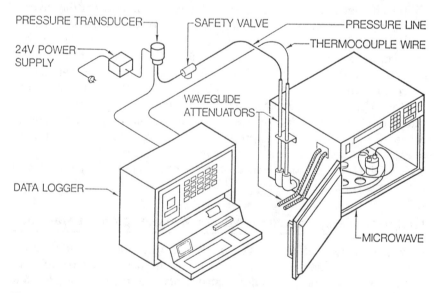

Figure 1. First research microwave sample-preparation system with temperature and pressure monitoring capabilities.

units of today. In 1987, an IR100 was awarded for the development of microwave sample preparation to the National Institutes of Standards and Technology (NIST) and CEM Corporation, led by H. M. (Skip) Kingston of NIST (9). (The IR100 is awarded by *Research and Development* magazine to the 100 most significant new technical products of the year.) Before temperature and pressure monitors were commercially available, digestion procedures were developed by a trial-and-error approach and evaluated on the basis of the recovery of an element or a suite of elements from a material, frequently a standard reference material. This approach brought about the majority of the digestion procedures and research papers that have illustrated the numerous advantages of microwave sample preparation. However, they have contributed little to the acid concentration, power, and time optimization of digestions; the understanding of the completeness of the digestion; the understanding of microwave interactions; or the understanding of the digestion mechanisms. Despite the lack of chemical knowledge that is gained by the trial-and-error approach, it was and still is the primary approach to the development of microwave-digestion procedures.

In 1985, the first laboratory multimode cavity microwave unit was introduced. Its primary improvements over home or domestic units were the added safety features. The early units, although built from domestic cavities and doors, isolated and ventilated the cavity to prevent acid fumes from attacking the electronics (Figure 2). Since the first laboratory micro-

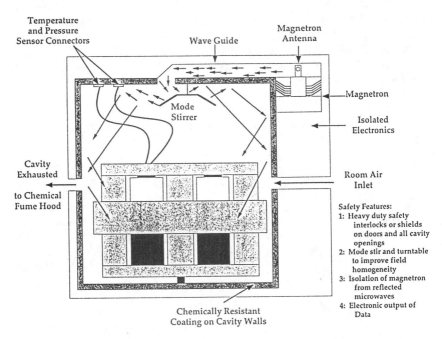

Figure 2. Typical laboratory cavity-type microwave system.

wave unit was introduced, numerous companies have continued to improve every aspect of the unit including homogeneity of the microwave field, ability to control the microwave power, and most importantly improvements in safety.

In 1986, the first completely reengineered laboratory-focused-microwave system was introduced. Contrary to microwave cavity systems, a single vessel is placed directly in a microwave waveguide (Figure 3). The vessels are constructed of either Teflon or quartz. The bottom few inches of the vessel are exposed directly to the microwaves, whereas the upper region of the vessel remains cool. This results in an effective condensing mechanism inherent to the design. While the vessels are open to the atmosphere, the refluxing action minimizes the acid and some volatile elemental losses. The vessel openings were designed to permit automated reagent addition and to restrict contamination from the atmosphere.

In the mid 1980s, a few researchers began building or modifying temperature- and pressure-monitoring equipment for use inside a microwave cavity. The primary challenges were to develop probes that were nonperturbing to the microwave field and to build wavelength attenuator cutoffs for these probes so they could enter the microwave region while preventing microwaves from leaving the microwave cavity. Monitoring temperature and/or pressure during digestions or extractions began the age of controlled digestions, the study of microwave-digestion mechanisms, and the development of transferable standard-microwave sample-preparation methods. These developments spawned an outgrowth of microwave use that is illustrated by the increase in microwave sample-preparation publications (Figure 4). Because articles on microwave sample preparation are published in all areas of chemistry, medicine, geochemistry, etc., the complete collection of many articles was only discovered years later through references.

Waveguide Type Microwave System

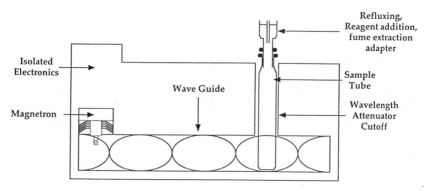

Figure 3. Typical laboratory focused-type microwave system.

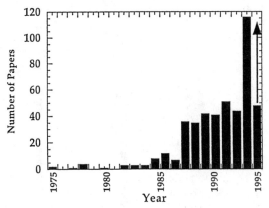

Figure 4. Growth of microwave sample preparation as seen through the growth of research papers.

The first commercial laboratory microwave unit with pressure feedback control (developed in 1989) and the first commercial laboratory microwave unit with temperature feedback control (developed in 1992) allowed a more rigorous design and control of microwave sample-preparation procedures. Concurrent with these developments, the microwave vessel evolved significantly (Figure 5). The first generation of microwave closed vessels was an all-Teflon design with low-pressure limits of approximately 7 atm. These vessels were prone to venting by exceeding their limited pressure capabilities. These pressure limits decreased as the vessel aged due to stress from previous dissolutions. The second generation microwave was the jacketed vessel. Typically these vessels were constructed of a Teflon liner and cap with a polymeric case (typically polyetherimide) that increased the pressure limits of the vessels to up to 20 atm. The third generation vessels, also lined, have been completely redesigned and are capable of extremely high pressures of 60–110 atm. The evolution of the microwave vessel and the capability to monitor the reaction conditions throughout the digestion have allowed researchers to systematically study the decomposition mechanisms of various matrices. These studies have advanced the *art* of sample preparation into *state-of-the-art* sample preparation.

In 1995, Milestone Corporation developed a completely unique approach to microwave dissolution, a microwave-heated autoclave (*10*). The ultraCLAVE uses a very high pressure (200 bar) autoclave cavity with a maximum temperature of 350 °C. The temperature and pressure capabilities far exceed all current microwave closed vessels and enable this system to be used for a wider variety of chemical applications.

As microwave sample-preparation methods have been developed, means of reproducing these methods have also been developed. One of the biggest drawbacks with the use of microwave methods has been their

All Vessels are capable of temperature and pressure connections

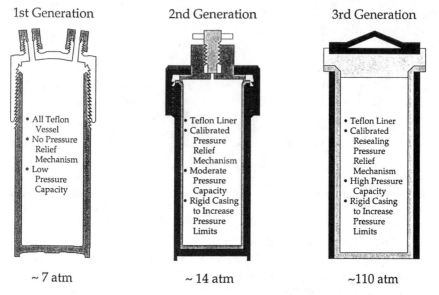

Figure 5. Evolution of closed microwave vessels.

dependence on the microwave power, because this characteristic changes significantly from unit to unit (not just manufacturer). Four approaches are used to transfer standard microwave methods: specification of microwave unit power settings, instrument power calibration, and pressure- and temperature-feedback control. A comparison of the effectiveness of these control mechanisms is shown in Figure 6. The improvements in precision and accuracy match the chronological development of these techniques.

Transferring a method by using microwave unit power settings is best illustrated by American Society for Testing and Materials (ASTM) method D4309-91 (11). This method describes two general ranges of microwave powered units and specifies the partial power setting and the time for acid leaching:

- for 575–635-W microwave units use 100% power setting for 50 min
- for 635–700-W microwave units use 100% power setting for 30 min

With the wide range in the power delivered to the cavity, the dissolution reaction conditions are never under precise control and lead to a range of maximum reaction temperatures of approximately 20 °C based on the reaction temperature profiles of the method. The variation in the reaction temperature is compounded by the variability of the reaction time. According to the specification listed previously, the reaction times can be

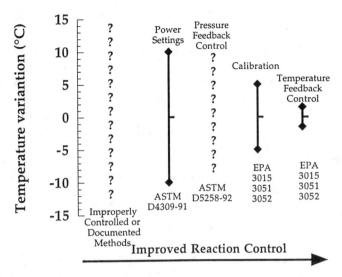

Figure 6. Reproducibility of microwave digestion temperatures for various control techniques.

either 30 or 50 min. These variations lead to a method that is incapable of accurately reproducing the method designed-reaction conditions of a maximum reaction temperature, for example 165 ± 5 °C.

Currently, methods in which calibration and feedback control are used can be documented, transferred, and reproduced. Calibration reproduces the reaction conditions through replication of the microwave field throughout the digestion procedure. Feedback control monitors temperature or pressure and makes microwave power adjustments based on the reaction conditions to reproduce the desired reaction profile.

Calibration of a cavity-type microwave involves the measurement of the microwave field versus percent power setting of the unit. Because the direct measurement of the microwave field would require expensive specialized equipment, modification of the microwave unit, and specialized training, an indirect technique of measuring the microwave field is used. The strength of the microwave field is measured by determining the amount of microwave energy absorbed by a strongly absorbing substance like water. Water is readily available and is used for power calibration almost uniformly in commercial and laboratory microwave units. By using the following thermodynamic relationship, the microwave field strength can be determined and the applied microwave field defined.

$$P = \frac{KC_p\, m\, \Delta T}{t} \tag{1}$$

where P is the apparent absorbed power (in watts), K is the conversion factor for calories/second to watts (4.184 J/cal), C_p is the heat capacity of the microwave absorbing solvent (water), m is the total mass of the microwave absorber in the cavity, ΔT is the change in temperature in the microwave absorber from the irradiation of the microwave energy, and t is the time of microwave exposure.

This equation and the method of calibration have been described in detail in several sources (5, 12, 13). Additionally, a computer program was written to guide the analyst through the collection of calibration data as well as the statistical evaluation of the data (14). This program can be downloaded from the SamplePrep Web World Wide Web (WWW) site, as described in Chapter 15. A calibrated microwave procedure can be developed and reported as watts of microwave power versus time. By simple calibration of another microwave unit and determination of the partial power settings that correspond to the watts specified in the procedure, the microwave field can be reproduced.

Calibration as a means of transferring microwave procedures is only as good as the ability to reproduce the microwave field precisely, to measure it, and to reproduce the exact loading of the microwave unit. Reproduction of the microwave field is limited by the precision with which one can measure and calibrate a microwave unit, and the ability to reproduce the exact wattage required in the procedure. The power of a microwave unit can be calculated to about ±10 watts, and the partial power settings (0–100%) of most laboratory microwave units correspond to changes of 6–12 watts. By combining these factors, the ability to reproduce the reaction conditions of a calibration-controlled method, such as U.S. Environmental Protection Agency (EPA) Method 3051, is typically ±5–10 °C or more (15, 16) (Figure 7). The temperature for a given sample procedure using the same unit can be reproduced to less than ±5 °C (15). Transfer-

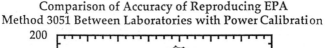

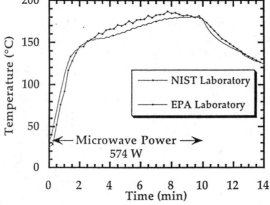

Figure 7. Reproducibility of EPA Method 3051 through calibration.

ring microwave methods through calibration, identical microwave-digestion vessels, and amounts of reagents can reproduce the reaction conditions. In some cases, the sample and the number of vessels must be the same. The most severe limitation of calibration control is that a separate determination of calibrated powers must be determined if the method is varied: that is, change in the volume or type of reagents, type of vessel, or in some cases sample size.

One of the manufacturers of focused-microwave-digestion systems performs calibration during assembly. At both 10 and 100% power settings, the magnetron power-control circuitry is adjusted so that each unit has the same power output within 5%. This result is achieved through the use of a continuous-flow calorimeter. A precision flow meter is used to ensure a constant flow rate while a thermistor measures the inlet and outlet temperature. For the focused-microwave system, the use of a flow-through calibration device forces a change on the power equation. Increasing the flow rate through the chamber decreases the residence time in the magnetic field, and thus the power absorbed decreases and a smaller temperature rise results. Therefore, an increase in the flow rate causes a decrease in the temperature at the same power as expressed in the following equation.

$$P = k Q \Delta T \tag{2}$$

where P is power (W), k is a constant (1.163 W • h/(L • °C)), Q is volumetric flow rate (L/h), and ΔT is temperature change (°C).

Equation 1 can also be used for the design of microwave procedures. If the absorption of microwave energy by an acid or combination of acids is known, eq 1 can be rearranged to predict either the time required to reach a temperature or the temperature at a given time during a digestion.

$$T_f = T_i + \frac{Pt}{KC_p m} \tag{3}$$

where i is initial and f is final temperature, and

$$t = \frac{KC_p m \Delta T}{P} \tag{4}$$

The absorption of microwave energy by numerous acids was studied and the microwave energy absorption of acids versus mass of absorbing acids plots were reported (5). By using these data, the reaction conditions during the initial stages of digestion can be predicted. The later stages of the digestion cannot be predicted because of the inability of the equation to correct for heat loss from the microwave vessels. A more complex evaluation focusing on heat loss of the vessel in a microwave field will be explained in Chapter 3.

By predicting the time to reach a temperature, eq 4 has been used to design a leaching method without the aid of using temperature feedback control. Leaching soils in concentrated nitric has been demonstrated to be nearly quantitative for soluble salts at 175 °C for 5 min (5, 17). For a specific combination of sample and reagent quantities, the final reaction temperature of 175 °C was calculated to be reached in approximately 1 min (18). Experimentally, the digestion should be complete if the digestion temperature is maintained for 5 min, which results in a total digestion time of 6 min. A plot of digestion time versus determined concentration of mercury supported the result that a quantitative recovery was accomplished in 6 min. Even without a temperature feedback-control microwave unit or temperature-measurement capability, a sophisticated microwave-digestion program can be devised with only calibration, calculations, and careful planning.

The introduction of commercial microwave digestion units in the late 1980s with first pressure and later with temperature feedback control enabled more fundamental microwave-digestion research and improved reproducibility and transferability of microwave procedures. The rate of a digestion reaction is controlled primarily by temperature and only indirectly by pressure (6). Both pressure and temperature feedback-control microwave-digestion systems are appropriate for specific types of digestions. Digestions that produce little or no gaseous products may be controlled well by pressure. If the digestion of the sample produces a minimal pressure, the primary increase in pressure is due to the acids or solvents. If the amount of the solvent is accurately dispensed into identical vessels, pressure feedback control will reproducibly repeat reaction conditions by reproducing the temperature profile through this secondary and related control.

However, when digesting materials that produce significant quantities of gas, controlling the digestion by pressure becomes more difficult. The pressure formed during the digestion is the sum of the pressures from the reagents and the pressure from the gaseous digestion products. The pressure from the gaseous digestion products depends on the sample composition and quantity. If the precise amounts of reagents and sample are not reproduced, then a specific temperature will produce different pressures and therefore different temperatures both within and between runs. An evaluation of potential problems inherent in pressure control is further discussed in Chapter 3. For some total digestions, this irreproducibility may not be serious, but if the reaction is a leach or extraction method where the reaction conditions are critical for reproducing the leaching or extraction, this type of control is critical and frequently inappropriate. In these and other cases, the best control is through the real-time adjustment and control of the reaction temperature.

Temperature feedback controls the primary factor affecting reaction rates. Regardless of the amount of sample or reagents, the temperature of the solution during the digestion can be reproduced. Methods such as EPA Method 3051 that were previously reproducible through calibration to about ±10 °C are now reproducible to ±4 °C or better. This improvement enables the development of a method for any vessel and the ability to transfer the method to any microwave unit capable of temperature feedback control by using vessels with a minimum pressure capacity. Temperature feedback control allowed for the development of EPA Method 3052, in which a wide variety of reagents and variable quantities can be used. In all variations of the method, the temperature can be accurately controlled to the method specifications. Feedback control of temperature enables improved control and reproducibility of reaction conditions over power control, pressure feedback control, and calibration as well as documentation of the digestion.

Documentation of Microwave Methods

The reproducibility of a microwave method is directly related to the detail provided in its documentation. Far too frequently, microwave methods have not been documented well enough to adequately reproduce the methods in another laboratory, as was illustrated in many of the methods listed in Appendices 1–3. Improved documentation could increase the method usefulness in analytical laboratories. This section will discuss what is critical to properly document microwave methods.

An analogy of complete documentation of a microwave method can be drawn from the documentation of a hot-plate digestion. The early procedures for hot plates probably said heat the sample until hot. This description of the procedure is lacking any information about what temperature the dissolution is to be run at or how long the digestion should be performed. Analogously, this description is similar to early microwave methods that reported heating the vessel at some power until the dissolution was complete.

Microwave units can be calibrated to produce uniform fields of energy that reproduce standardized reaction conditions. Once calibrated, the individual unit's partial power settings from a calibrated microwave unit can be converted into power in watts. Then, the method can be described as a series of power steps, each of which are described as a power in watts applied for a specific time. Despite the improvement of microwave calibration in documentation of microwave methods, this method suffers from precise reproduction of the temperature. The temperature for a specific time-reaction-conditions profile and calibration of a method are specific to the exact amount and quantities of acids and the exact type and number of microwave vessels.

The most advanced description of methods involves the complete documentation of the heating profile. For microwave dissolutions, complete documentation of a method involves the description of the entire heating profile using temperature versus time criteria. This description has been accomplished in several standard EPA methods, including 3015, 3051, and 3052.

A completely documented microwave method is independent of the specific microwave unit and vessels. A method can be reproduced by using a microwave vessel with sufficient temperature and pressure capabilities in combination with any microwave system with temperature feedback control. An example of an appropriately described method is illustrated in Figure 8.

These types of documented methods will be posted on the SamplePrep Web WWW site, as described in Chapter 15. An outline of the necessary descriptors for documenting a microwave method was described in the literature (15). The most critical descriptors are described briefly.

- description of the sample type into general classifications such as geological, metallurgical, etc.
- analytes of interest
- amount of sample per vessel (range)
- number of samples per dissolution (vessels per group)
- vessel type
- dissolution reagents specifying the precise quality of reagent (ACS grade or subboiled distilled grade, etc.) and the amount of reagent for both dissolution vessels and reagent blank vessels
- complete digestion program
- for calibration, the calibration procedure must be stated along with the specific power settings and durations of each step
- for temperature feedback control, the method should be described as a series of heating to temperature, maintaining a temperature, and cooling steps. As illustrated previously, EPA Method 3052 can be described in three stages: heat to 180 ± 5 °C in 5.5 min, maintain 180 ± 5 °C for 9.5 min, and cool for at least 5 min. The temperature profile is specified, but the tolerances to the temperatures are also stated.

This information allows scientists to reproduce methods quickly, accurately, and precisely. Some sets of these descriptors are used in the documentation of standard methods that are promulgated or suggested by various organizations worldwide. Methods also should be validated for which standard reference materials are available from many international standards organizations for this purpose.

Currently Approved International Standard Microwave Methods

As microwave sample preparation has evolved, standard microwave procedures have been developed and approved by numerous standard meth-

Sample Type: Geological, Environmental
Matrix: Clay, Silt, Organic debris
Standard Method: EPA Method 3052
Sample Size: 0.25 - 1.0 g
Control: Temperature Feedback
Stage 1 of 1:
Acid (s): 9 mL HNO₃, 3 mL HF

Program:	700 W	700 W	0 W
time	5.5 min	9.5 min	5 min
Temp.	to 180°C	at 180°C	cool
Pressure	< 15 atm	< 15 atm	< 15 atm

Unit: MLS 1200mega
Vessel: MDR 600/10
No. of Vessels: 10
Maximum Power Required: 700W
Maximum Temperature Obtained: 180°C
Maximum Expected Pressure: 12 atm

NOTE: Programming of equipment and the exact temperature and pressure profiles will depend on the microwave unit and the specific vessels.

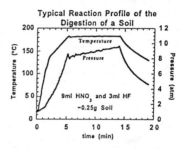

Standard Reference Material: NIST Buffalo River Sediment SRM 2704

Element	Analyzed μg/g	Certified μg/g	Technique
Cd	3.5 ± 1.2	3.45 ± 0.22	ICP-MS
Cr	132.9 ± 1.3	135 ± 5	ICP-MS
Cu	98.0 ± 4.2	98.6 ± 5.0	AA
Ni	43.6 ± 3.9	44.1 ± 3.0	ICP-MS, ET-AA
Pb	154.5 ± 9.2	161 ± 17	AA
Zn	441.9 ± 0.8	438 ± 12	AA

Comments: This analysis was used in part to write the original EPA Method 3052. It is part of a certification of EPA Method 3051 for leaching of six metals from NIST SRMs 2710 and 2711; high and moderately highly contaminated Montana soils (Report to NIST April 5, 1994, Contract # 50SBNB3C7513)

Analyst: Peter J. Walter
Location: Duquesne University

Figure 8. Example of complete documentation of a microwave decomposition method for Buffalo River sediment.

ods organizations. Currently there are 21 methods approved, or in the process of being approved, by AOAC International (*19–22*), ASTM (*11, 23–25*), the EPA (*12, 13, 26–28*), Standard Method (*29*), and French (*30*) and Chinese (*31*) national methods for either microwave drying or microwave acid dissolution. These methods are summarized in Table I.

Two microwave drying methods, ASTM Methods E1358-90 and D4643-93, have been developed to dry either wood or soil to a constant mass with the use of microwave energy. However, the bulk of the methods are acid-dissolution procedures for either total elemental analysis or

Table I. A. AOAC Official Method 977.11. Moisture in Cheese, Method IV (Microwave Oven Method)

Type	Matrix	Analytes	Detection	Control
Microwave drying	Cheese	Water	Balance	NA

Condensed Procedure
Prepare sample as in Method 955.30
Evenly Spread 10g onto Petri dish
Cover with glass paper
Place dish on weighing platform inside microwave
Determine initial mass
Microwave heat for 2:25 min at 74 %
Determine final mass and calculate % moisture

SOURCE: Summarized from reference 19.

Table I. B. AOAC Official Method 985.14. Moisture in Meat and Poultry Products, Rapid Microwave Drying Method

Type	Matrix	Analytes	Detection	Control
Microwave drying	Meat, poultry	Water	Balance	NA

Condensed Procedure
Prepare sample as in Method 983.18.
Place two glass fiber pads on balance and tare.
Rapidly and evenly spread ~4 g onto one glass pad.
Cover sample with second pad.
Determine initial mass.
Microwave heat for 3 - 5 min at 80 - 100 % depending on product type.
Determine final mass and calculate % moisture.

SOURCE: Summarized from reference 20.

acid leaching of a matrix. The reaction control of these standard acid-dissolution methods range from minimal control by using microwave unit power settings and calibration, to moderate control by using pressure feedback control, or robust control of the reaction through temperature feedback control. Some methods have been developed and documented with two types of reaction control, such as temperature feedback control and calibration. This combination allows the methods to be run by using calibration for standard microwave equipment or temperature feedback control for more advanced microwave equipment.

ASTM Method D4309-91 is an acid-leach method for water, and the reaction control is based on the power rating of the microwave unit. Basic ranges of microwave power ratings are separated into categories of digestion protocols. These power-range settings cannot accurately reproduce reaction conditions, and nonreproducible leaching of the sample results.

The Chinese method, C303.01T, and ASTM Method D5258-92 were both developed for pressure feedback control. By using pressure feedback

Table I. C. AOAC Official Method 985.15. Fat (Crude) in Meat and Poultry Products, Rapid Microwave-Solvent Extraction Method

Type	Matrix	Analytes	Detection	Control
Microwave drying with non-microwave solvent extraction	Meat, poultry	Fat	Balance	NA

<table>
<tr><td colspan="5" align="center">Condensed Procedure</td></tr>
</table>

Initial Microwave Drying:
    Prepare sample as in Method 983.18.
    Place two glass fiber pads on balance and tare.
    Rapidly and evenly spread ~4 g onto one glass pad.
    Cover sample with second pad.
    Determine initial mass.
    Microwave heat for 3 - 5 min at 80 - 100 % depending on product type.
Non-Microwave Extraction:
    Place Pads into non-microwave automated solvent extractor
    Extract with CH_2Cl_2
Final Microwave Drying
    Place Pads on balance and tare
    Microwave heat for 30 sec. at 80 - 100%
    Determine final mass and calculate % moisture.
    Repeat drying until constant weight is obtained
Determine % Fat

SOURCE: Summarized from reference 21.

Table I. D. AOAC Official Method 985.26. Solids (Total) in Processed Tomato Products, Microwave Oven Drying Method

Type	Matrix	Analytes	Detection	Control
Microwave drying	Tomato juice, tomato puree, tomato paste	Water	Balance	NA

<table>
<tr><td colspan="5" align="center">Condensed Procedure</td></tr>
</table>

Prepare samples as specified in the Method
Dry 2 glass fiber pads in microwave
Evenly spread sample onto one glass pad.
Cover sample with second pad.
Determine initial mass.
Microwave heat for 4 min at unspecified % power setting.
Determine final mass and calculate % total solids.

SOURCE: Summarized from reference 22.

Table I. E. ASTM Method D1506-94b Standard Method. Standard Test Methods for Carbon Black - Ash Content

Type	Matrix	Analytes	Detection	Control
Microwave ashing	Carbon black	Ash content	Balance	Temperature feedback control

Condensed Procedure
Pre-heat crucible and cover to 550 ± 25°C for 1 hr.
Cool to room temperature in a desiccator.
Dry sample for 1 hr. at 125°C in convection oven.
Add ~ 2 g of carbon black to crucible and determine initial mass.
Place crucibles inside microwave ashing unit.
Ashing at 550 ± 25°C for 2 hr for quartz crucibles and 6 hr. for procelain crucibles.
Cool to room temperature in a desiccator.
Determine final mass and calculate percentage of ash.

SOURCE: Summarized from reference 32.

Table I. F. ASTM Method D4309-91 Standard Method. Standard Practice for Sample Digestion Using Closed Vessel Microwave Heating Technique for the Determination of Total Recoverable Metals in Water

Type	Matrix	Analytes	Detection	Control
Microwave acid leach	Water	Al, Cd, Cr, Cu, Fe, Mn, Ni, Pb, Zn	ET-AAS, DCP-AES, FAAS, ICP-AES	Microwave unit specific power settings

Condensed Procedure
50.0 mL of sample
For ICP, DCP, FAAS - add 3 mL HNO_3 and 2 mL HCl
For ET-AAS - add 5 mL HNO_3
Prepare 12 vessels
For MW units with 575 - 635 W use 100% power for 50 min
For MW units with 635 - 700 W use 100% power for 30 min
For MW units with >700 W reduced powers are necessary to reproduce heating curves.
Both settings will allow the samples to reach a maximum temperature of 164 ± 4°C

SOURCE: Summarized from reference 11.

Table I. G. ASTM Method D4643-93 Standard Method. Standard Test Method for Determination of Water (Moisture) Content of Soil by the Microwave Oven Method

Type	Matrix	Analytes	Detection	Control
Microwave drying	Soil	Water	Balance	NA

Condensed Procedure
Place sample into a container in microwave oven
Heat for 3 min
Cool in desiccator
Reweigh
Repeat until dry mass is constant

SOURCE: Summarized from reference 25.

Table I. H. ASTM Method D5258-92 Standard Method. Standard Practice for Acid Extraction of Elements from Sediments Using Closed Vessel Microwave Heating

Type	Matrix	Analytes	Detection	Control
Microwave acid leach	Sediment, soil	As, Cd, Cu, Mg, Mn, Ni, Pb, Zn	Not specified	Pressure feedback control

Condensed Procedure
1g sample and 20 mL 1:1 HNO_3
Heat to 100 psi and maintain for 30 min

SOURCE: Summarized from reference 24.

control, these methods may more accurately reproduce the designed reaction conditions with one limitation. If the digestion produces variable amounts of gaseous by-products, variable temperatures will result.

To improve the digestion reaction control, EPA Methods 3015 and 3051 and Standard Method 3030K (a reproduction of EPA Method 3015) were designed with both calibration and temperature feedback control. The methods were developed by using temperature-measurement capabilities, and the calibration power settings for a specific microwave vessel and the specific temperature reaction profiles were provided. These methods enabled the microwave units at that time, which were primarily unable to perform temperature feedback control, and future microwave units with temperature feedback control the ability to use these methods. Because these methods have temperature reaction criteria, these methods can be very accurately and precisely run batch after batch.

The most advanced microwave standard method is EPA Method 3052 and is designed to accommodate the variation in the acid chemistry. This method uses temperature feedback control with third-generation moderate- to high-pressure vessels. One should keep in mind that many of these

Table I. I. ASTM Method D5513-94 Standard Method. Standard Practice for Microwave Digestion of Industrial Furnace Feedstreams for Trace Element Analysis

Type	Matrix	Analytes	Detection	Control
Microwave acid digestion	Coal, coke, cement raw feed materials, waste derived fuels	Ag, As, Ba, Be, Cd, Cr, Hg, Pb, Sb, Tl	Not specified	Calibration and pressure feedback control

Condensed Procedure

0.5 g sample
Prepare either 4 - 6 vessels or 8 - 10 vessels
Stage 1:
 Slowly add 8 mL of HNO_3.
 Microwave heat according to settings specified in either appendix A.1 or A.2
 Cool vessels.
Stage 2:
 Add 4 mL HF and 2 mL HCl
 Microwave heat according to settings specified in either appendix A.1 or A.2
 Cool vessels.
Stage 3:
 Add 35.5 mL of H_3BO_3.
 Microwave heat according to settings specified in either appendix A.1 or A.2
 Cool vessels.

SOURCE: Summarized from reference 33.

Table I. J. ASTM Method E1358-90 Standard Method. Standard Test Method for Determination of Moisture Content of Particulate Wood Fuels Using a Microwave Oven

Type	Matrix	Analytes	Detection	Control
Microwave drying	Wood	Water	Balance	NA

Condensed Procedure

Place 50 g sample onto paper towel
Determine initial mass
Microwave heat at full power for times as recommended in Table 1 p1096
Reweigh
Repeat until variation in successive weights are below 0.5 g

SOURCE: Summarized from reference 23.

Table I. K. ASTM Method E1645-94 Standard Method. Standard Practice for the Preparation of Dried Paint Samples for Subsequent Lead Analysis by Atomic Spectrometry

Type	Matrix	Analytes	Detection	Control
Microwave acid leach	Paint	Pb	Not specified	Calibration

Condensed Procedure

Place 0.1 - 0.2 g of sample into 30 mL polysulfone centrifuge tube.
Add 10 mL of extraction solution to each tube and tightly seal.
Pipet 31 mL of water into 120 mL PFA microwave digestion vessel
Place centrifuge tube inside digestion vessel containing water.
Fill the microwave carousel with microwave vessels (un-specified number of vessels).
Heat at 522 W for 23 min

SOURCE: Summarized from reference 34.

Table I. L. US-EPA Method 3015 Standard Method. Microwave Assisted Acid Digestion of Aqueous Samples and Extracts

Type	Matrix	Analytes	Detection	Control
Microwave acid leach	Water	Al, Ag, As, Ba, Be, Ca, Cd, Co, Cu, Cr, Fe, K, Mg, Mn, Mo, Na, Ni, Pb, Sb, Se, Tl, V, Zn	FAAS, ET-AAS, ICP-AES, ICP-MS	Temperature feedback control or calibration

Condensed Procedure

45 mL sample & 5 mL HNO_3

Performance Criteria:
Heat to 160°C ± 4°C in 10 min
Slow rise to 165-170°C ± 4°C in additional 10 min

Prescription Settings:
For five 120 mL PFA vessels 545 W for 10 min, 344 W for 10 min

Typical Reaction Profile of Method 3015

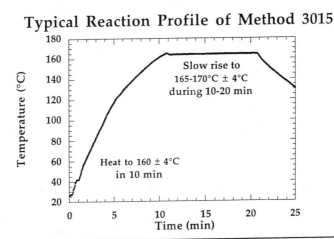

SOURCE: Summarized from reference 13.

Table I. M. US-EPA Method 3031 Standard Method. Acid Digestion of Oils for Metals Analysis by FLAA or ICP Spectroscopy

Type	Matrix	Analytes	Detection	Control
Microwave digestion	Oil	Ag, As, Ba, Be, Cd, Co, Cr, Cu, Mo, Ni, Pb, Sb, Se, Tl, V, Zn	FAAS, ICP-AES	Calibration

Condensed Procedure

Method 1: 0.5 g sample and 0.5 g $KMnO_4$
   20% for 2 min
   add 2 mL H_2SO_4, 20% for 3 min (to near dryness)
   add 10 mL HNO_3 and 2 mL HCl, 15% for 5 min,
   20% for 3 min (to near dryness), cool for 5 min
   add 5 mL HCl, 15% for 5 min
Method 2: 0.5 g sample and 10 mL H_2SO_4
   20% for 5 min, 25% for 5 min, 30% for 5 min, 35% for 5 min, cool for 5 min
   add 8 mL H_2O_2, 50% for 5 min, cool for 5 min
   add 8 mL H_2O_2, 60% for 5 min

SOURCE: Summarized from reference 26.

older standard methods were developed with limited instrumental capabilities when laboratory microwave units were little more that converted domestic microwaves systems. Many of these methods have limitations imposed by early low-pressure vessels and the lack of temperature feedback control. Newer methods, such as EPA Method 3052, were developed with and use third-generation microwave vessels and temperature feedback control.

EPA Methods 3031 and 3050B and the French standard method V03-100, a Kjeldahl nitrogen method, are examples of traditional methods that are being expanded to allow for microwave heating as an alternative heating source. Other standard methods have been proposed such as EPA Methods 351 and 365. By using atmospheric-pressure vessels and temperature feedback control, these methods are capable of improved reaction temperature control that results in improved precision. Improvements in precision and accuracy for several of the environmental standard methods are described in Chapter 3. As new methods are developed and older methods are reevaluated, more standard hot-plate acid-dissolution methods will become available as microwave methods. A complete list of current and developing standard methods will be updated continuously on the SamplePrep Web: *see* Chapter 15 for details.

Robotic Automation of Microwave Procedures

The first fundamental rule of automation of chemical processes is reactions that are not understood or in control cannot be automated successfully.

Table I. N. US-EPA Method 3050B Standard Method. Acid Digestion of Sediments, Sludges, and Soils

Type	Matrix	Analytes	Detection	Control
Microwave acid leach	Sediment, sludge, soil	Al, As, Ba, Be, Ca, Cd, Co, Cr, Cu, Fe, K, Mg, Mn, Mo, Na, Ni, Pb, Se, Tl, V, Zn	FAAS, ET-AAS, ICP-AES	Temperature feedback control or calibration

Condensed Procedure

1 g sample and 10 mL 1:1 HNO_3

Performance Criteria:
  2 min to 95°C, 5 min at 95°C, Cool for 5 min
  5 mL HNO_3, 2 min to 95°C, 5 min at 95°C, cool for 5 min
  5 mL HNO_3, 2 min to 95°C, 5 min at 95°C, cool for 5 min
  Slowly add 10 mL H_2O_2, 6 min to 95°C, 5 min at 95°C, cool for 5 min
  Add 5 mL HCl and 10 mL H_2O, 2 min to 95°C, 5 min at 95°C

Prescription Settings:
  80 W for 2 min, 30 W for 5 min, cool
  add 5 mL HNO_3, 80 W for 2 min, 30 W for 5 min, cool
  add 5 mL HNO_3, 80 W for 2 min, 30 W for 5 min, cool
  add 3 mL H_2O_2, 40 W for 5 min
  add 3 mL H_2O_2, 40 W for 5 min
  add 4 mL H_2O_2, 40 W for 5 min, 5 min cool
  add 5 mL HCl and 10 mL H_2O, 80 W for 2 min, 30 W for 5 min

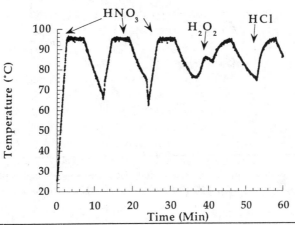

SOURCE: Summarized from reference 27.

Table I. O. US-EPA Method 3051 Standard Method. Microwave Assisted Acid Digestion of Sediments, Sludges, Soils, and Oils

Type	Matrix	Analytes	Detection	Control
Microwave acid leach	Oil, sediment, sludge, soil	Al, Ag, As, B, Ba, Be, Ca, Cd, Co, Cu, Cr, Fe, Hg, K, Mg, Mn, Mo, Na, Ni, Pb, Sb, Se, Sr, Tl, V, Zn	FAAS, ET-AAS, ICP-AES, ICP-MS	Temperature feedback control or calibration

Condensed Procedure

0.25 - 0.5 g sample and 10 mL HNO_3 (max 0.25 g for Oil)

Performance Criteria:
  Heat to 175°C in less than 5.5 min
  Maintain 175-180°C for remainder of 10 min

Prescription Settings:
  For six 120 mL PFA vessels 574 W for 10 min
  For two 120 mL PFA vessels 344 W for 10 min

Typical Reaction Profile of Method 3051

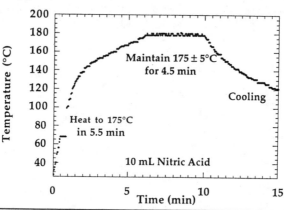

NOTE: Nitric acid alone has been reported to be insufficient to stabilize antimony (Sb) (35-38). Currently this method is being extended to include HCl to more accurately replicate the leaching of EPA Method 3050B which uses HNO_3, HCl, H_2O_2, and H_2O.

SOURCE: Summarized from reference 12.

Table I. P. US-EPA Method 3052 Standard Method. Microwave Assisted Acid Digestion of Siliceous and Organically Based Matrices

Type	Matrix	Analytes	Detection	Control
Microwave total digestion	Fly ash, oil, sediment, sludge, soil	Al, Ag, As, B, Ba, Be, Ca, Cd, Co, Cu, Cr, Fe, Hg, K, Mg, Mn, Mo, Na, Ni, Pb, Sb, Se, Sr, Tl, V, Zn	FAAS, ET-AAS, ICP-AES, ICP-MS	Temperature feedback control

Condensed Procedure

Up to 0.5 g sample and 9 ± 0.1 mL HNO_3 and 3 ± 0.1 mL HF (variable up to ± 2 mL) and
alternatively 2 ± 2 mL HCl (can not be analyzed by ET-AA)

Performance Criteria:
Heat to 180 ±5°C in less than 5.5 min
Maintain 180 ±5°C for remainder of 15 min

Typical Reaction Profile of Method 3052

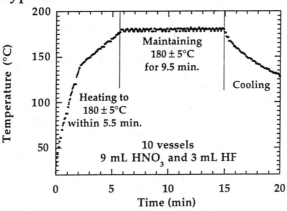

SOURCE: Summarized from reference 28.

Table I. Q. US-EPA Method 3051 Standard Method. Microwave Assisted Acid Extraction and Dissolution of Soils, Sediments, Sludges, and Oils

Type	Matrix	Analytes	Detection	Control
Microwave acid leach	Oil, sediment, sludge, soil	Al, Ag, As, B, Ba, Be, Ca, Cd, Co, Cu, Cr, Fe, Hg, K, Mg, Mn, Mo, Na, Ni, Pb, Sb, Se, Sr, Tl, V, Zn	FAAS, ET-AAS, ICP-AES, ICP-MS	Temperature feedback control or calibration

Condensed Procedure

Up to 0.5 g sample and 9 mL HNO_3 and 3 mL HCl (max 0.25 g for Oil)

Performance Criteria:
  Heat to 180 ± 5°C in less than 5.5 ± 0.25 min
  Maintain 180 ± 5°C for remainder of 10 min
Prescription Settings:
  User determinable

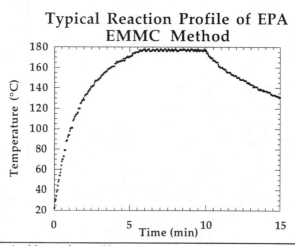

Typical Reaction Profile of EPA EMMC Method

SOURCE: Summarized from reference 39.

Table I. R. US-EPA Method NPDES Standard Method. Closed Vessel Microwave Digestion of Wastewater Samples for Metals Determination

Type	Matrix	Analytes	Detection	Control
Microwave acid leach	Domestic and industrial wastewater	Al, As, Ba, Cd, Cr, Cu, Fe, Mn, Ni, Pb, Sb, Se, Zn	FAAS, ICP-AES, DCP-AES	Microwave unit specific power settings

Condensed Procedure

50.0 mL sample, 3 mL HNO_3, and 2 mL HCl

For 1 - 6 samples:

For MW units with 364 - 420 W use 75% power for 30 min to a maximum temperature of 164 ± 4°C

For MW units with > 364 - 420 W at 75%, adjust power setting to deliver 364 - 420 W for 30 min

For 7 - 12 samples:

For MW units with 575 - 635 W use 100% power for 50 min to a maximum temperature of 165 ± 5°C.

For MW units with 635 - 700 W use 100% power for 30 min to a maximum temperature of 165 ± 5°C.

For MW units with > 700 W adjust power settings to reproduce temperature versus time profiles illustrated in the method.

SOURCE: Summarized from references 40 and 41.

Now that there are approved high-volume standard microwave-dissolution tests and methods, it makes sense to examine their automation. Two major factors make procedures using closed-vessel microwave dissolution of acid samples well-suited for intelligent automation. First, the ability to accurately transfer precise amounts of energy to the acid and the ability to control and reproduce reactions by controlling the temperature and mechanisms of the reaction provide reproducible digestion conditions. Second, the rapid heating due to the direct coupling of microwave energy into the solution reduces digestion times from hours to minutes. Automation of microwave dissolution procedures should maintain the accuracy and precision of sample preparation and increase sample throughput.

Several efforts have tried to robotically automate microwave sample preparation. The first was developed at Kidd Creak Mines in Canada for the analysis of minerals (42). The second system was developed at the National Institute of Standards and Technology (15, 16, 43, 44) with cooperation from Zymark Corporation, Hopkinton, MA, and CEM Corporation, Matthews, NC. This system was designed primarily to implement EPA microwave methods; however, it had the flexibility to perform nearly any microwave-digestion procedure because of its modular programming. Commercial versions of this system are available from Zymark Corporation. Another system was built for the dissolution of titanium dioxide by Norris et al. (45).

Table I. S. Standard Methods 3030K Standard Method. Microwave Assisted Digestion

Type	Matrix	Analytes	Detection	Control
Microwave acid leach	Water	Ag, Al, As, Au, Ba, Be, Bi, Ca, Cd, Ce, Co, Cr, Cu, Hg, Ir, K, Li, Mg, Mn, Mo, Na, Ni, Os, Pb, Pd, Pt, Rh, Sb, Se, Si, Sn, Sr, Th, Ti, Tl, V, Zn	FAAS, ET-AAS, CV-FAAS	Temperature feedback control

Condensed Procedure

45 mL sample & 5 mL HNO_3
Performance Criteria:
 Heat to $160 \pm 4°C$ in 10 min
 Slow rise to $165 - 170 \pm 4°C$ in additional 10 min
Prescription Settings:
 For five 120 mL PFA vessels 545 W for 10 min, 344 W for 10 min
Reaction Conditions are identical to EPA Method 3015

SOURCE: Summarized from reference 29.

Table I. T. Republic of China NIEA C303.01T Standard Method. Acid Digestion of Fish and Shellfish

Type	Matrix	Analytes	Detection	Control
Microwave digestion	Fish, shellfish	Al, As, Ba, Be, Ca, Cd, Co, Cr, Cu, Fe, K, Mg, Mn, Mo, Na, Ni, Os, Pb, Se, Th, V, Zn	FAAS, ET-AAS, ICP-AES	Pressure feedback control

Condensed Procedure

0.5 g sample and 5 mL HNO_3
Performance Settings:
 315 W for 10 min at 20 psi
 378 W for 10 min at 40 psi
 441 W for 10 min at 80 psi
 441 W for 10 min at 135 psi

SOURCE: Summarized from reference 31.

Table I. U. France V 03-100 Standard Method. Kjeldahl Nitrogen

Type	Matrix	Analytes	Detection	Control
Focused microwave digestion	Milk, meat products, animal food, starch & starchy foods	N	Distillation or calorimetric	?

Condensed Procedure
Specifics of the Method are unavailable at the present time

SOURCE: Summarized from reference 30.

Despite the diversity of samples and the approaches to automation, these systems have common goals and components. All three of these systems were built to increase the throughput of routine dissolution procedures while maintaining or possibly improving the quality over manual manipulations. To accomplish these goals, automated components had to be designed to perform many simple manual tasks. Some tasks had to be performed in a completely different manner with a fully automated system.

Even though the three systems were developed for different purposes and with different equipment, the overall robotic layouts are similar. Figure 9 illustrates one robotic table (43). The system uses computer software and hardware to guide the analyst through the log-in process and weighing of each sample, whereas other systems have automated powder sample dispensers (45). Once the samples have been added to the dissolution vessels, the digestion acid(s) are added by automated reagent dispensing stations. The acid-dispensing station, shown in Figure 9, is enclosed in a Class 100 Cleanhood to maintain sample integrity by protecting samples from the major source of contamination in trace elemental analysis, the environment. Once the vessel is uncapped, a reagent-dispensing arm moves over the vessel and the appropriate amounts of acids are dispensed. The vessel is then capped and torqued to the vessel specifications. The NIST system was built as automated modules as described in the Consortium on Automated Analytical Laboratory Systems (CAALS) and was documented in a government report (46–49). This system was developed to use in a standard method encapsulated for automated transfer and implementation (8, 16, 48).

Automation of microwave sample preparation relied on standard laboratory units that were modified for automated door opening, carousel indexing, and remote programming. Recently, a unit was developed with automated control integrated into its basic design (10). As with any dedicated automation instrument, it no longer bares a resemblance to a conventional microwave oven but is designed around the concept of automated operation.

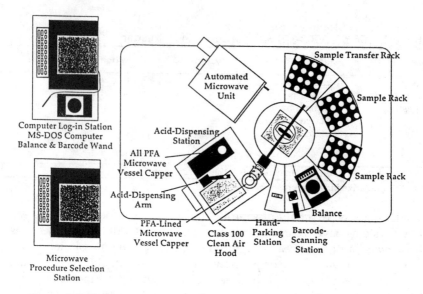

Figure 9. Fully automated robotic microwave sample-preparation system.

Because chemistry is much faster and more reliable, it leads itself to automation more readily than modifying traditional sample-preparation or chemical-reaction apparatus. This characteristic makes microwave chemistry an attractive tool for increasing efficiency in the chemical laboratory. Many of the microwave applications in subsequent chapters lend themselves to automation and will be implemented as automated systems in the next few years.

Resources for Method Development

General Guide to Literature Tables

In writing a review of any topic, incorporating a lot of information and making it easy to read as a usable resource is difficult. For this reason, in this overview of microwave sample preparation we have organized the wealth of information into tables of various formats. We hope that this will make the chapter a valuable reference for any scientist needing to develop microwave-based procedures.

In deciding how to organize the information relevant to sample dissolution, we looked at the process of developing new methods. Within each part of this process, we tried to optimize the format for each table to convey the information in a concise yet accessible manner. Because this field is still growing at a breakneck pace, we do not want this review to be a static entity. We are committed to expanding this resource as the field grows and so we welcome any suggestions that readers have about how we can improve the format, content, and areas of the interest. Not only do we plan to include updated versions of these tables in future editions of this book, but we also will be making them available on the SamplePrep Web WWW site (see Chapter 15). We hope you find the tables useful.

Review Papers

Table II includes brief descriptions of review papers in the area of microwave sample dissolution so that the reader can quickly find a more specific review on their area of interest. Not included in this table are the several chapters in the rest of this book that review current areas of microwave chemical research.

Microwave Drying

Sample drying is not often considered in the development of microwave sample-preparation procedures. However, the residual moisture content of samples can play a major part in reproducible sampling of the matrix, as well as the interaction of the microwave energy with the sample. Thus, including references on the use of microwaves for drying is relevant because this procedure can and should be included in the preparation procedure for certain matrices. In Table III, we describe the application to convey the perspective of the cited paper to the removal of free water and other free action polar solvents from the matrix. Particular importance should be placed on the indicated drying platform because several authors have commented how important the form of the sample is on the drying characteristics.

Microwave Ashing

Although the majority of work on microwave sample dissolution concerns wet acid digestions, use of an aqueous phase is not necessary while heating with microwaves. Table IV lists papers that address the use of microwave-based ashing as an alternative to muffle-furnace ashing. This approach may be more productive if the matrix is easily combustible and volatile species are not to be determined.

Microwave Dissolution

This area of research has fueled the growth of microwave technologies because of its superiority over hot-plate and block-digestor wet acid-digestion methods. Because of the large amount of work on this subject in the literature and the limited space in this book, we have chosen to include a subset of the literature. Thus, the following tables contain papers on

Table II. Microwave Sample Preparation Review Articles

Subject	Number of Papers Cited	Language	Ref.
Review and discussion on the microwave ashing technique for the determination of minerals in food.	2	English	50
Review and discussion on the advantages of microwave compared to conventional sample preparation methods.	17	German	51
Review of the use of microwave digestions for environmental analysis in atomic spectrometry	997	English	52
Review on the fundamentals of microwave-oven digestion, of modifications necessary for laboratory use particularly in flow-injection analysis.	41	English	53
Review and comparison of methods for the digestion of fat rich foods.	9	English	54
Review and optimization of variables in microwave digestions.	8	English	55
Review of the use of microwaves in sample preparation for analysis by atomic spectrometry.	6	English	56
Review of acid pressure decomposition for trace analysis.	61	English	57
Review of microwave techniques in sample dissolution, moisture determination, AES and chromatography.	69	Chinese	58
Review of the application of microwave to the digestion of geological and metallurgical samples.	47	Chinese	59
Review of leaching procedures and their application to standard reference materials.	25	English	60
Review on the use of microwave ovens to aid acid digestion of samples for analysis.	28	English	61
General review of microwave acid decomposition until 1988	45	English	62
Review of mechanized methods of sample decomposition for trace analysis.	17, 30	English	63, 64

Description		Language	
Review of microwave-generating apparatus, acids and containers, and applications to tissue samples, plant samples, ore samples and alloy samples.	25	Japanese	65
Review of critical parameters in microwave digestion procedures.	32	English	66
Review of microwave sample preparation for inorganic elemental analysis.	54	Russian	67
Review of the principles and instrumentation of microwave sample preparation.	6	German	68
Review and discussion of equipment used for microwave digestion and the application of the method to a variety of matrices.	181	English	69
Review of the analysis of biological and geological samples using microwave digestion.	21	Chinese	70
Review of high pressure closed vessel microwave digestions.	1	English	71
Review of the area of sample preparation.	40	English	72
Review of microwave sample preparation for atomic spectrometry.	14	English	73
Review of microwave sample dissolution and decomposition for elemental analysis. (with tables)	72	English	74
Review of gas phase digestions including the use of microwave irradiation.	49	English	75
Review of the Kjeldahl nitrogen method and the application of microwave techniques.	19	English	76
Review of the application of microwave sample preparation to improved speciation analysis.	11	English	77
Review of the mechanism of microwave heating, the interactions of microwaves with materials, commercial microwave sources and their application to mineralization and to protein hydrolysis.	54	French	78
Review and comparison of five different sample decomposition procedures.	13	English	79
Review of procedures for trace analysis.	59	English	80
Review of general applications of microwave heating in analytical chemistry and in sample decomposition.	38	Chinese	81
Review of microwave dissolution procedures for inorganic analysis.	9	English	82

Table III. Microwave Drying References

Matrix	Application	Drying Platform	Conditions	Comments	Ref.
Aqueous soln	Evaporation	140 mL, 8 cm tall beakers	7-20 min at 1200 W	Only mentioned in a table	71
Barium carbonate	Drying	Porclain/glass beaker with ribbed watchglass	15 min at 700 W	Can handle 1-6 20 g samples with up to 68% water content	83
Barium sulfate	Drying	Porclain/glass beaker with ribbed watchglass	15 min at 700 W	Can handle 1-6 20 g samples with up to 68% water content	83
Barium sulfate analysis precipitate	Drying	Crucible	2 min at 90 W	Analysed barium sulfate SRMs 205 and 210 (Thorn Smith, Troy MI)	84
Beef lard	Moisture content	Fiber glass pads	Not reported	Sample put between drying pads	85
Beef meat	Moisture content	Fiber glass pads	Not reported	Sample put between drying pads	85
Beef muscle	Moisture content	Fiber glass pads	Not reported	Sample put between drying pads	85
Beef, low fat	Moisture	Fiber glass pads	600 W to dryness	Collaborative study	86
Beef, medium fat	Moisture	Fiber glass pads	600 W to dryness	Collaborative study	86
Bologna	Moisture	Fiber glass pads	600 W to dryness	Collaborative study	86
Brine shrimp	Drying	Not reported	2 h at 350 W	Not focus of paper	87
Calcium carbonate	Drying	Not reported	2, 6, 10 min at 500 W	Does drying affect analytical results?	88
Calcium sulfate	Drying	Porclain/glass beaker with ribbed watchglass	15 min at 700 W	Can handle 1-6 20 g samples with up to 68% water content	83
Canned ham	Moisture	Fiber glass pads	600 W to dryness	Collaborative study	86
Carp	Drying	Petri dish	650 W to constant weight (~15 min)	Sample homogenized before drying, however intact analysis OK if less than 5 mm thick	89
Cheddar cheese	Solids content	Fiber glass pads	2-6 min at 450-600 W	Round robin study - variability seen	90
Chicken franks	Moisture	Fiber glass pads	600 W to dryness	Collaborative study	86
Coal	Inert drying	140 mL, 8 cm tall beakers	60 min at 1200 W	Only mentioned in a table	71
Coal	Total moisture	Not reported	20 min at 420 W	Compared to conventional oven drying	91
Coal	Moisture	Fiber glass pads	480 W to constant weight (~11 min)	No degradation seen after 66 min heating	92
Cobalt carbonate	Drying	Porclain/glass beaker with ribbed watchglass	15 min at 700 W	Can handle 1-6 20 g samples with up to 68% water content	83

Sample	Analysis	Container	Conditions	Comments	Ref.
Copper sulfate (pentahydrate)	Drying	Fiber glass pads or Fiber glass pad + Thermapad	25 min at 600 W	Compared with oven drying	93
Cottage cheese	Solids content	Fiber glass pads	2-6 min at 450-600 W	Round robin study - variability seen	90
Dairy products	Water content	140 mL, 8 cm tall beakers	3-4 min at 1200 W	Only mentioned in a table	71
Detergents	Solids content	Pyrex petri dish	2 min at ? W	Equivalent to oven drying	94
Drierite	Drying	Large porclain evaporating dish	20 min at 500 W or 10 min at 1000 W		95
Fibers	Moisture content	140 mL, 8 cm tall beakers	6-10 min at 1200 W	Only mentioned in a table	71
Flour	Moisture	Fiber glass pads	8 min at 415 W	Sample size determined by type of flour	96
Franks	Moisture	Fiber glass pads	600 W to dryness	Collaborative study	86
Iron (III) carbonate	Drying	Not reported	2, 6, 10 min at 500 W	Does drying affect analytical results?	88
Iron oxide ore slurry	Drying	Fluted filter paper on pyrex dish	15-60 sec at 735 W	For slurry preparation	97
Lead carbonate	Drying	Porclain/glass beaker with ribbed watchglass	15 min at 700 W	Can handle 1-6 20 g samples with up to 68% water content	83
Lead sulfate	Drying	Porclain/glass beaker with ribbed watchglass	15 min at 700 W	Can handle 1-6 20 g samples with up to 68% water content	83
Liver	Drying	Petri dish	650 W to constant weight (~15 min)	Sample homogenized before drying, however intact analysis OK if less than 5 mm thick	89
Lucerne	Drying	Petri dish	650 W to constant weight (~15 min)	Sample homogenized before drying, however intact analysis OK if less than 5 mm thick	89
Maganese carbonate	Drying	Porclain/glass beaker with ribbed watchglass	15 min at 700 W	Can handle 1-6 20 g samples with up to 68% water content	83
Magnesium carbonate	Drying	Porclain/glass beaker with ribbed watchglass	15 min at 700 W	Can handle 1-6 20 g samples with up to 68% water content	83
Mechanically deboned chicken	Moisture	Fiber glass pads	600 W to dryness	Collaborative study	86

Continued on next page.

Table III. Continued

Matrix	Application	Drying Platform	Conditions	Comments	Ref.
Milk	Solids content	Fiber glass pads	2-6 min at 450-600 W	Round robin study - variability seen	90
Mussels	Drying	120 mL PTFE vessels	3 min at 385 W, 5 min at 555 W	Compared to oven drying at 110°C for 24 hours	98
Nickel carbonate	Drying	Porclain/glass beaker with ribbed watchglass	15 min at 700 W	Can handle 1-6 20 g samples with up to 68% water content	83
NIST Milk powder 1549	Drying	Fiber glass pads or Fiber glass pad + Thermapad	8 min at 600 W	Compared with oven drying	93
NIST Coal 1632	Drying	Fiber glass pads or Fiber glass pad + Thermapad	8 min at 600 W	Compared with oven drying	93
NIST Brick Clay 679	Drying	Fiber glass pads or Fiber glass pad + Thermapad	2 min at 600 W	Compared with oven drying	93
NIST Limestone 88	Drying	Fiber glass pads or Fiber glass pad + Thermapad	5 min at 600 W	Compared with oven drying	93
NIST Flint Clay 97	Drying	Fiber glass pads or Fiber glass pad + Thermapad	5 min at 540 W	Compared with oven drying	93
NIST Plastic Clay 98	Drying	Fiber glass pads or Fiber glass pad + Thermapad	5 min at 540 W	Compared with oven drying	93
NIST Rice flour 1568	Drying	Fiber glass pads or Fiber glass pad + Thermapad	2 min at 120 W	Compared with oven drying	93
NIST Wheat flour 1567	Drying	Fiber glass pads or Fiber glass pad + Thermapad	2 min at 120 W	Compared with oven drying	93
NIST River sediment 2704	Drying	Fiber glass pads or Fiber glass pad + Thermapad	4 min at 480 W	Compared with oven drying	93

Sample	Analysis	Container	Conditions	Notes	Ref
Octocoral	Drying	Glass beaker with watchglass	20-50 min at 240 W	Followed by oven drying	99
Pork lard	Moisture content	Fiber glass pads	Not reported	Sample put between drying pads	85
Pork meat	Moisture content	Fiber glass pads	Not reported	Sample put between drying pads	85
Pork, high fat	Moisture	Fiber glass pads	600 W to dryness	Collaborative study	86
Pork, low fat	Moisture	Fiber glass pads	600 W to dryness	Collaborative study	86
Processed tomato products	Total solids	Fiber glass pads	4 min at 600 W	Collaborative study report	100
Sausage	Moisture	Fiber glass pads	600 W to dryness	Collaborative study	86
Sausages	Moisture content	Fiber glass pads	Not reported	Sample put between drying pads	85
Seafood	Drying	Not reported	? min at 90 W	Found to be equivalent to freeze drying	101
Silica gel	Drying	Large porcelain evaporating dish	20 min at 500 W or 10 min at 1000 W	Only mentioned in a table	95
Sludge	Drying	140 mL, 8 cm tall beakers	20 min at 1200 W	General discussion paper	71
Sludge	Moisture	Fiber glass pads	100 min at 600 W	General discussion paper	102
Smoked ham	Moisture	Fiber glass pads	600 W to dryness	Collaborative study	86
Sodium hydrogen phosphate (heptahydrate)	Drying	Fiber glass pads or Fiber glass pad + Thermapad	6 min at 600 W	Compared with oven drying	93
Sweet corn seed	Moisture	Not reported	600 W to constant weight	Estimation of maturities of seeds	103
Tilapia	Drying	Watch glass	750 W to dryness	Used a combined microwave convection oven	104
Titanium dioxide slurry	Moisture	Fiber glass pads	4 min at 600 W	General discussion paper	102
Tomatoes	Total solids	Not reported	AOAC method	Results depend on microwave type. Need certification	105
Waste water	Solids and water content	Porcelain, high silica glass or polymethylene beakers, 250 mL, 8 cm tall	600 W till boiling, evaporate at just below boiling, 10 min at 600 W		106
Zinc carbonate	Drying	Porcelain/glass beaker with ribbed watchglass	15 min at 700 W	Can handle 1-6 20 g samples with up to 68% water content	83

Table IV. Microwave Ashing References

Matrix	Application	Ashing Vessel	Conditions	Detection	Ref.
Alfalfa	B, Ca, Cu, Fe, K, Mg, Mn, P, Zn	Porcelain crucibles	20 min at 600 W or 40 min at 600 W	ICP-OES	107
Artichoke leaves	Ca, Fe, K, Mg, Mn, Zn	Silicon ring	30 min at 650 W, 10 min at 0 W	F-AAS	108
Bean	B, Ca, Cu, Fe, K, Mg, Mn, P, Zn	Porcelain crucibles	20 min at 600 W or 40 min at 600 W	ICP-OES	107
Blood	Pb	Quartz test tubes	30 min at 700 W	ASV, DPP	109
Citrus	B, Ca, Cu, Fe, K, Mg, Mn, P, Zn	Porcelain crucibles	20 min at 600 W or 40 min at 600 W	ICP-OES	107
Coal	Residual ash	Quartz fiber crucibles	30 min at 750 °C	Weight	110
Coal	Residual ash	Porcelain crucibles or quartz fiber crucibles	45 min at 750 °C	Weight	111
Coke from coal	Residual ash	Porcelain crucibles or quartz fiber crucibles	65 min at 750 °C	Weight	111
Corn	B, Ca, Cu, Fe, K, Mg, Mn, P, Zn	Porcelain crucibles	20 min at 600 W or 40 min at 600 W	ICP-OES	107
Dry pet food	Residual ash	Porcelain crucibles or quartz fiber crucibles	30 min at 600 °C	Weight	111
Fine paper	Residual ash	Porcelain crucibles or quartz fiber crucibles	35 min at 800 °C	Weight	111
High density polyethylene	TiO2	Porcelain crucible	3-4 min at 800 °C	Weight	112
High density polypropylene/ titanium dioxide	Residual ash	Porcelain crucibles or quartz fiber crucibles	15 min at 800 °C	Weight	111
High density polypropylene/ carbon black filler	Residual ash	Porcelain crucibles or quartz fiber crucibles	12 min at 600 °C	Weight	111
Human blood	Sb, Te	Oxygen plasma asher	20 min at 100 W	MIES (microwave induced emission spectrometry)	113

Sample	Elements	Device	Conditions	Technique	Ref.
IAEA Milk powder 11	Cu, Fe, Ni, Zn	Oxygen plasma asher	6-20 h at 350 W	F-AAS, ETV-AAS, HG-AAS	54
Linseed	Cu, Fe, Ni, Zn	Oxygen plasma asher	6-20 h at 350 W	F-AAS, ETV-AAS, HG-AAS	54
Low density polyethylene	TiO_2	Porcelain crucible	1.5-5 min at 800 °C	Weight	112
Low density polyethylene	SiO_2	Porcelain crucible	3-4 min at 800°C	Weight	112
Milk	Bi, Cd, Cu, Pb, Sb, Zn	Oxygen plasma asher	10-15 h at 200 W	DP-ASV	114
Milk chocolate	Cu, Fe, Ni, Zn	Oxygen plasma asher	6-20 h at 350 W	F-AAS, ETV-AAS, HG-AAS	54
NIST Citrus leaves 1572	Ca, Fe, K, Mg, Mn, Zn	Silicon ring	30 min at 650 W, 10 min at 0 W	F-AAS	108
NIST Tomato leaves 1573	Ca, Fe, K, Mg, Mn, Zn	Silicon ring	30 min at 650 W, 10 min at 0 W	F-AAS	108
Nutella	Cu, Fe, Ni, Zn	Oxygen plasma asher	6-20 h at 350 W	F-AAS, ETV-AAS, HG-AAS	54
Peanut butter	Cu, Fe, Ni, Zn	Oxygen plasma asher	6-20 h at 350 W	F-AAS, ETV-AAS, HG-AAS	54
Petrochemical sludge	Residual ash	Porcelain crucibles or quartz fiber crucibles	60 min at 900 °C	Weight	111
Plankton	Bi, Cd, Cu, Pb, Sb, Zn	Oxygen plasma asher	10-15 h at 200 W	DP-ASV	115
Polypropylene	TiO_2	Porcelain crucible	3-4 min at 800°C	Weight	112
Potato	B, Ca, Cu, Fe, K, Mg, Mn, P, Zn	Porcelain crucibles	20 min at 600 W or 40 min at 600 W	ICP-OES	107
Soybean flour	Cu, Fe, Ni, Zn	Oxygen plasma asher	6-20 h at 350 W	F-AAS, ETV-AAS, HG-AAS	54
Suspended particulate	Bi, Cd, Cu, Pb, Sb, Zn	Oxygen plasma asher	10-15 h at 200 W	DP-ASV	115
Tomato	B, Ca, Cu, Fe, K, Mg, Mn, P, Zn	Porcelain crucibles	20 min at 600 W or 40 min at 600 W	ICP-OES	107
Wet pet food	Residual ash	Porcelain crucibles or quartz fiber crucibles	30 min at 800 °C	Weight	111

microwave sample dissolution that have been applied to reference materials to validate the procedure used. Through this approach, we have provided known matrix and analyte data that may be critically evaluated for relevance to the analyst. This method seems to be appropriate for grouping the literature for rapid, useful evaluation. The references in this table make up about 50% of the total papers on microwave sample dissolution. The exclusion of the other 50% of papers should not imply that they are *bad science*, but they do not provide a basis by which any judgment can be made about the proposed procedures. Of the papers cited, we have not distinguished between *good* and *bad* science, rather we have provided relevant information that researchers will find useful in developing microwave methodology.

In the compilation of the tables, only those articles we have been able to obtain in hardcopy have been included. Thus, articles in journals unavailable to us have not been included. We would be delighted to obtain copies of any articles that should be included in the tables, but are not. We would also like to receive any comments regarding the accuracy of the information contained in the tables and any articles for publication updating this information.

By comparing table entries for the same reference material (a good example is NIST 1577, Bovine Liver, in Appendix 1.1), we see that the procedures used are numerous and varied in both reagents used and microwave conditions. This range is largely due to the infancy of the technique, and researchers are applying anything and everything to find procedures that will provide respectable results on a particular matrix. This approach has lead to the general uses of reagents outlined in the section, *Dissolution Reagents*, and the realization that performance-based methods (the use of temperature and pressure to control reactions) will provide more robust methods for a wider variety of matrices. To adapt these procedures for a specific sample matrix, more emphasis should be placed on the reagent compositions reported than the microwave conditions because of the wide disparity in microwave power outputs and scant information on reaction temperatures.

General Information

Appendices 1–3 have been broadly grouped into three categories based on the reference sample matrix types. Appendix 1 includes all biological reference materials, Appendix 2 contains geological and metallurgical reference materials, and Appendix 3 contains other reference materials that could not be categorized within either of the first two appendices. Appendices 1 and 2 have been subdivided for easier use.

Matrix

The institutions that supplied the reference material, the name designation, and the number are included in that order. Within each table, entries are listed in alphabetical order by supplier and sub-listed by reference material number in ascending order. Explanation of the supplier acronyms can be found in Table V. No attempt has been made to distinguish between different batches of the same reference material (e.g., 1577a and 1577b are considered an equivalent matrix).

Analytes

Elements determined in the indicated matrix are listed alphabetically by symbol. Other analytes are listed after. Speciation information has been included where possible.

Reagents and Microwave Conditions

To give a general sense of the procedure used, the information in these two columns has been organized as follows:

1. Different procedures within a paper are separated by letter designations (e.g., E., F.).
2. Different steps in a procedure are separated by number designations (e.g., 1., 2.).
3. If, within a step, more than one reagent is used, each reagent is separated by a comma.
4. Repeat steps within a profile are indicated by (x?) where ? is a numeric.
5. If the procedure involves criteria other than microwave power, such as pressure, these criteria have been included as well.
6. Other pertinent information from the procedure has been included as necessary.

Microwave Cavity and Reaction Vessel Descriptions

Considering the variety of modes in which microwave sample preparation can be performed, we have classified methods only by general descriptors (*see* Tables VI and VII). Information on the material of the digestion vessel has not been included.

Detection

Instrumental detection methods are listed. Acronyms for the techniques can be found in Table VIII.

Dissolution Reagents

As mentioned in the Microwave Dissolution section, good dissolution procedures depend on the choice of reagents that are used and the specific temperature profile. These two parameters establish the mechanisms and

Table V. Sources of reference materials

Acronym	Supplier
AIS	Academy of Iron and Steel, Pan Zhi Hua Ministry of Metallurgy, Japan
ANRT	Association Nationale de al Recherche Technique, France
AR	Alpha Resources Inc., USA
ASCRM	Australian Standard Coal Reference Material, Standard Association of Australia, Standards House, 80 Artur St., North Sydney N.S.W., Austrialia
BAM	Bundensanstalt für Materialforschung und -Prüfurg, BAM, Unter Den Eichen, 87, D-12205, Berlin, Germany
BCR	Community Bureau of Reference, Commission of the European Community, 200 Rue de la Lol, B-1046 Brussels, Belgium
BCS	British Chemical Standard, Bureau of Analysed Samples, Middlesborough,UK
BI	Behring Institute, PO Box 1140, D-3550 Marburg ,Germany
BRAMMER	Not reported
BRL	Bio-Rad Laboratories, Munich, Germany
CANMET	Canada Centre for Mineral and Energy Technology, 555 Booth St., Ottawa, Ontario Canada, K1A OG1
CCRMP	Canadian Certified Reference Material Project, Energy mines and Resources, 555 Booth St., Ottawa, Ontario Canada, K1A OG1
CRPG	Centre de Recherches Petrographiques et Geochimiques, 15 rue Notre Dame des Pauvres, 54501 Vandoeuvre, France
CSAN	Czechoslovakian Analytical Normal, Prague, Czechoslavakia
EPA	Environmental Protection Agency, Washington DC, USA
GSJ	Geological Survey of Japan, Higashi 1-1-3, Tsukuba City, Japan 305
IAEA	International Atomic Energy Agency, Analytical Quality Control Services Laboratory Seibersdorf, PO Box 100, A-1400 Vienna, Austria
IFM	Institutet Far Metallforskoring, Drottnig Kristinas Vag 48, Stockholm, Sweden
IGGE	Institute of Geophysical and Geochemical Exploration, China

kinetics of the reaction. This information is not only important for complete dissolution but also for reproducible extraction (leaching), analyte solubility, analyte volatility, species stability, and importantly, safety.

The process of developing a new microwave-digestion procedure involves identifying the primary components of the samples, understanding the chemistry required to decompose these components, and considering safety factors applied to a specific digestion and/or equipment.

To propose a digestion procedure, the reactivity of the common digestion acids and the reactivity of many matrices must be considered. If the primary components are known, procedures from the literature can be consulted to determine guidelines for decomposing the matrix. However, the specific elements of interest and the reactivity of the elements with the acids must also be considered. Therefore, consideration of reference digestion procedures and the reactivity of the acids for the analytes of interest are critical to a successful procedure.

Included here is a review of the reactivity of the primary decomposition reagents. A comprehensive review of the chemistry of each acid and combinations of acids is beyond the scope of this section. Many specific

Table V. Continued

Acronym	Supplier
IRSID	Not reported
ISS	Instituto Superiore di Santa Roma, Italy
IWG	International Working Group
KL	Kaulson Laboratories, 691 Bloomfield Avenue, West Caldwell NJ, 07006-7539, USA
MOE	The Ontario Ministry of Environment and Energy, 125 Resources Rd., Etobicoke, Ontario, Canada M9T 3Z6
NAGRA	Swiss National Cooperative for Radioactive Waste Storage, EPFL, Institut de Radiophysique appliquee, Centre Universitaire, Rue des Alambics/Entree B, CH-1015 Lausanne, Switzerland
NIES	National Institute for Environmental Studies, Japan Environmental Agency, PO Yatabe, Tsukuba Ibaraki 300-21, Japan
NIM	National Institute of Metallurgy, South African Bureau of Standards, Groenkloof Pretoria, South Africa
NIST	National Institutes of Standards and Technology, Room B311 Chemistry Building, Office of Standard Reference Materials, Gaithersburg, MD 20899, USA
NRCC	National Research Council of Canada, Institute for National Measurement Standards, Ottawa K1A OR6, Canada
RMA	Rocky Mountain Arsenal, Program Managers Office, Commerce City, CO 80022, USA
SERONORM	Nycomed Diagnostics, Oslo, Norway
SRS	Fisher Scientific, 1 Reagent lane, Fair Lawn, NJ, 07410-2802 USA
USGS	United States Geological Survey, Branch of Geochemistry, Box 25046, MS 973, Denver, CO, 80225, USA
XIGMR	Xian Institue of Geology and Mineral Resources, Chinese Academey of Geological Sciences, China
ZGI	Zentrales Geologisches Institut, Germany

NOTE: Some of these organizations have discontinued production and distribution of standard reference materials

reagent- and method-dependent mechanisms are described in a discussion of Method 3052, in Chapter 3. However, a review of the basic chemistry of several acids and common combinations of acids is discussed. The reactivities of six common dissolution reagents with the elements are presented in terms of their volatility, reactivity, solubility, complexation, stability, and catalytic effects. Additionally, when the information in the literature was available, the specific elemental species or oxidation state was noted. This information is presented in periodic-table graphical form as well as in tabular form with comprehensive citations (*see* Figures 10–15 and Tables IX–XIV). We would be delighted to obtain copies of any articles that should be included in these tables but are not, as well as receive any comments improving the accuracy and completeness of the information contained in these tables.

The most commonly used digestion reagents—nitric acid (Figure 10 and Table IX), hydrochloric acid (Figure 11 and Table X), hydrofluoric acid (Figure 12 and Table XI), sulfuric acid (Figure 13 and Table XII), perchloric acid (Figure 14 and Table XIII), and hydrogen peroxide (Figure 15 and Table XIV)—are discussed in their interactions with the elements. Even

Table VI. Microwave Type Designations

Microwave Type	Description
Multimode	Sample exposed to a multimode field (modestirred cavity system)
Single mode	Sample exposed to a single-mode field (focused waveguide system)

Table VII. Vessel Type Designations

Vessel Type	Description
HP closed	High pressure closed vessel (> 80 atm)
MP closed	Medium pressure closed vessel (>10 atm, $\leq$ 80 atm)
LP closed	Low pressure closed vessel ($\leq$ 10 atm)
Open	Atmospheric pressure vessel (1 atm)
Flow through	Flowing stream of reaction mixture passing through a microwave field
Stopped flow	Flowing stream of reaction mixture that is stopped when it is contained within a microwave field

Table VIII. Acronyms Used for Analytical Techniques

Acronym	Description
F-AAS	Flame atomic absorption spectrometry
ETV-AAS	Graphite furnace, electrothermal vaporization (atomization) AAS
CV-AAS	Cold vapor (mercury vapor) atomic absorption spectrometry
HG-AAS	Hydride generation atomic absorption spectrometry
FI-F-AAS	Flow injection flame atomic absorption spectrometry
SIMAAC	Simultaneous atomic absorption
ICP-OES	Inductively coupled plasma optical (atomic) emission spectrometry
FI-ICP-OES	Flow injection inductively coupled plasma optical (atomic) emission spectrometry
ETV-ICP	Electrothermal vaporization inductively coupled plasma
ICP/MS	Inductively coupled plasma mass spectrometry
MS	Mass spectrometry
DCP-OES	Direct current plasma optical (atomic) emission spectrometry
CV-AFS	Cold vapor atomic fluorescence spectrometry
CV-FANES	Cold vapor furnace atomic non-thermal emission spectrometry
F-OES	Flame photometry, flame emission
LC	Liquid chromatography
HPLC	High performance liquid chromatography
IC	Ion chromatography
SPC-IC	Solid phase chelation ion chromatography
UV-Vis	Ultraviolet visible spectrophotometry
ASV	Anodic stripping voltammetry
DP-ASV	Differential pulse anodic stripping voltammetry
DPP	Differential pulse polarography
NAA	Neutron activation analysis
ID	Isotope dilution
TXRF	Thermal X-ray fluorescence

NOTE: Also used fluorimetry, temperature, pressure, voltammetry, total carbon analyzer, carbon dioxide coulometry, radiochemical, α-spectrometry.

though these chemical reactivities are the basis for an understanding of a digestion procedure for a matrix, they are not absolute. The analyst will have to evaluate the reactivities of the acids with each specific matrix because every matrix will have unique chemical interactions. Many references for each acid are located in the accompanying tables and figures, and related aspects of each reagent have been provided there.

Nitric Acid

Nitric acid is an oxidizing acid that will dissolve most metals to form soluble metal nitrates. It has poor oxidizing strength below 2 M, but it is a powerful oxidizing acid in the concentrated form. Its oxidizing strength can be enhanced by the addition of chlorate, permanganate, hydrogen peroxide, or bromine or by increasing temperature and pressure. Most metals and alloys are oxidized by nitric acid; however, gold and platinum are not oxidized and some metals are passivated when attacked by concentrated nitric acid. These metals can be dissolved by use of a combination of acids or by dilute nitric acid.

Nitric acid is the most common acid for the oxidation of organic matrices. Nitric acid is a more powerful acid when used in combination with a complexing acid such as hydrochloric acid.

Hydrochloric Acid

HCl is a nonoxidizing acid that exhibits weak reducing properties during dissolution. Many metal carbonates, peroxides, and alkali hydroxides are readily dissolved by hydrochloric acid. Some metals, including Au, Cd, Fe, and Sn, can be dissolved by hydrochloric acid, but dissolution is accelerated by the addition of another acid. Most metals form soluble metal chlorides with several notable exceptions: AgCl, HgCl, and TiCl, which are insoluble, and $PbCl_2$, which is only slightly soluble. The complexing nature of HCl allows for complete dissolution of numerous metals, such as Fe(II) and Fe(III) complexing to form $(FeCl_4)^{-2}$ and $(FeCl_4)^-$, respectively (*121, 139*).

Hydrochloric acid is often used in combination with other acids for dissolution. It is frequently combined with nitric acid. Aqua regia is a 3:1 mixture of hydrochloric acid to nitric acid.

Hydrofluoric Acid

Hydrofluoric acid is a nonoxidizing acid whose reactivity is based on its strong complexing nature. It is most commonly used in inorganic analysis because it is one of the few acids that can dissolve silicates. Its strong complexation capabilities prevent the formation of sparingly soluble products of several metals and increase the solubility and stability of those elements. Dissolution with hydrofluoric acid produces primarily soluble fluorides, with the exception of insoluble or sparingly soluble fluorides of the

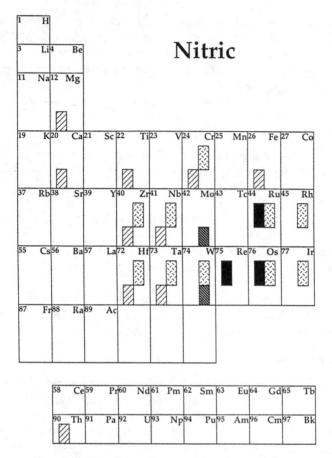

Figure 10. Reactivity of nitric acid with the elements.

alkaline earth, lanthanide, and actinide elements. To improve dissolutions, hydrofluoric acid is routinely combined with another acid, such as nitric acid. Insoluble fluorides may frequently be resolubilized by removing the hydrofluoric acid after digestion.

Sulfuric Acid

Dilute sulfuric acid does not exhibit any oxidizing properties, but the concentrated acid is capable of oxidizing many substances (121). Concentrated sulfuric acid (98.7%) has a boiling point of 339 °C, which is greater than the working ranges of all Teflons. Therefore, careful microwave sample-preparation method development must be implemented through the measurement of the reaction temperature to prevent exceeding the thermally stable temperature of Teflon vessels. Quartz vessels are the material of choice for sulfuric acid dissolutions. Sulfuric acid can corrode the surfaces of Teflon during prolonged evaporations (123).

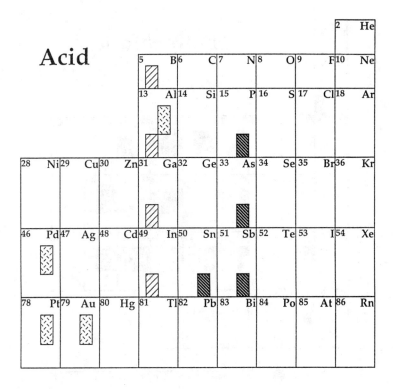

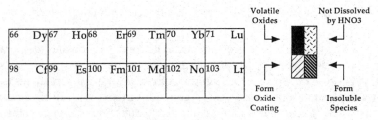

Sulfuric acid is commonly used with other acids and reagents. One of the most common combinations is with perchloric acid or hydrogen peroxide. Sulfuric acid will act as a dehydrating agent that will dramatically increase the oxidizing power of perchloric acid, but this mixture may react violently with organic matrices in closed vessels or if heated rapidly.

Perchloric Acid

Dilute aqueous perchloric acid, either warm or cold, is not an oxidizing acid. Concentrated perchloric acid (60–72%) is not an oxidizing acid cold, but becomes a powerful oxidizing acid when warm. Therefore, the oxidizing

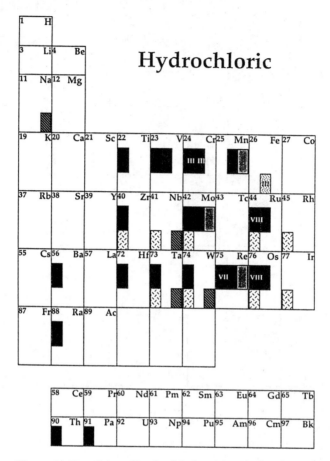

Figure 11. Reactivity of hydrochloric acid with the elements.

power of perchloric acid is proportional to its concentration and temperature. Warm perchloric acid will readily decompose organic matter, sometimes violently. Because of its extremely rapid reactivity with organic matrices (sometimes explosive), perchloric acid is generally mixed with nitric acid. This combination of acids allows for a controllable digestion of organics; the nitric acid will attack the easily oxidizable matter at lower temperature while diluting the perchloric acid. However, as the temperature rises, the perchloric acid will completely digest matter undigested by the nitric acid. Perchloric acid has been an acid of choice for the destruction of organics by using traditional heating systems because it has been shown to decompose nearly any organic matrix, and nearly all perchlorate salts are soluble. However, dry perchlorate salts of many metals are explosive!

Because of the explosive potential of perchloric acid with organic matrices and the fact that microwave sample preparation uses an extremely rapid heating source, the use of perchloric acid should be con-

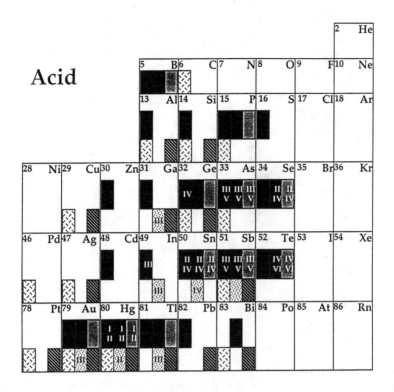

Acid

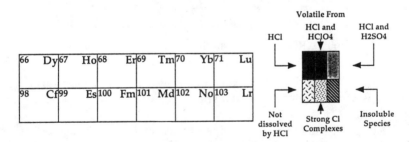

sidered a potential safety hazard. Perchloric acid has been shown to decompose at 245 °C in a microwave closed vessel and to develop dangerous amounts of gaseous by-products and tremendous excess pressure (5). General use of perchloric acid in closed or atmospheric-pressure microwave systems is not recommended because of safety concerns. If a digestion requires the use of perchloric acid then the following rules should always be obeyed:

1. Read all pertinent information about safe handling of perchloric acid. An excellent source of information about perchloric acid is reference 128.

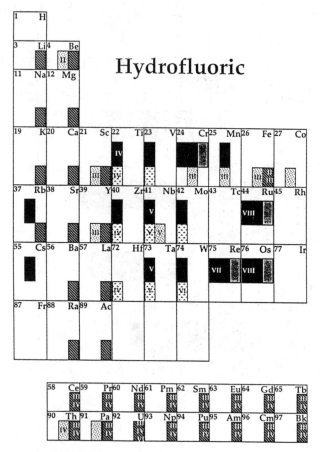

Figure 12. Reactivity of hydrofluoric acid with the elements.

2.   Always start with dilute perchloric acid. Dilute perchloric acid will slowly react with organic matter and prevent possible explosive reactions.

3.   Always use another acid to digest the easily oxidizable matter before perchloric acid can begin reacting with the matter. Similar to Rule 2, dilute the perchloric acid with an acid like nitric acid. Then, slowly heat the sample to enable the nitric acid to destroy as much of the organic matter as possible. Slowly continue heating until the perchloric acid has reached a temperature at which it can oxidize the remainder of the organic matter.

4.   The safest method of following Rules 2 and 3 is to perform a two-stage digestion. In the first stage of the digestion use only nitric acid and digest the sample as completely as possible. Once the digestion vessels are cool, open the vessels and add a minimal quantity of perchloric acid. Then, heat the digestion vessels for a second time. This

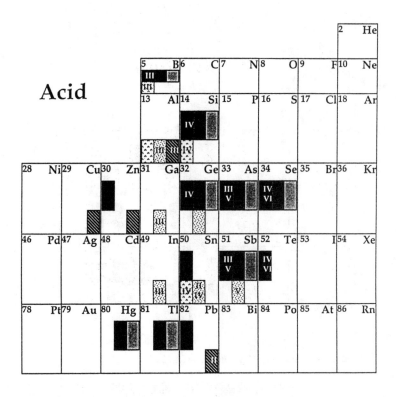

Acid

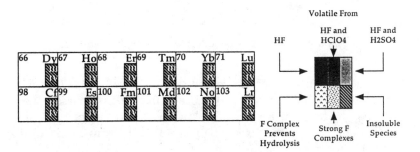

process will allow the perchloric acid to only decompose a small quantity of the toughest organic matter.

5.   Follow established safety rules regarding the disposal of perchlorates and avoid the uncontrolled collection of perchlorate residues on apparatus.

Hydrogen Peroxide

Typically concentrations of about 30% hydrogen peroxide are used in digestions, but more recently 50% concentrations are available. Hydrogen peroxide alone can react explosively with many organics, especially in the

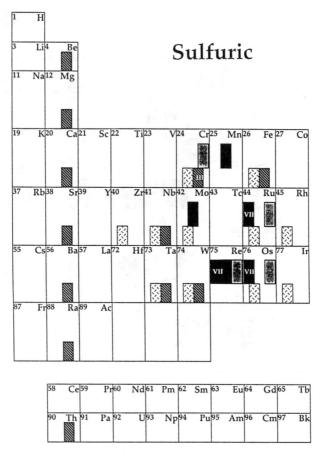

Figure 13. Reactivity of sulfuric acid with the elements.

more concentrated form. Hydrogen peroxide is usually combined with an acid because its oxidizing power increases as the acidity increases. The combination of hydrogen peroxide and sulfuric acid forms monoperoxosulfuric acid (H_2SO_5), a very strong oxidizing reagent (*121*). Because of its oxidizing power, hydrogen peroxide is frequently added after the primary acid has completed a predigestion of the matrix. The hydrogen peroxide can complete the digestion, and the potential safety hazards previously described are minimized. In this regard, hydrogen peroxide is used similarly to perchloric acid. Using these acids after the primary digestion of organic matter is completed is one way to avoid potentially violent reactions.

Physical Properties of Common Dissolution Reagents

Some general parameters for common dissolution reagents are given in Table XV. Note that the dielectric constant, an important measure of the

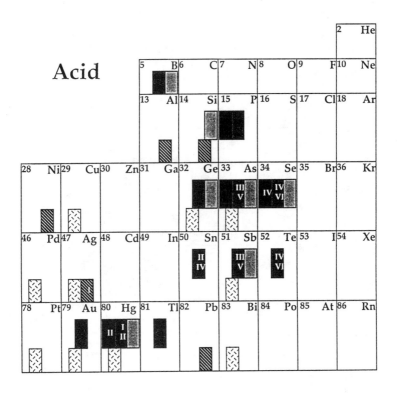

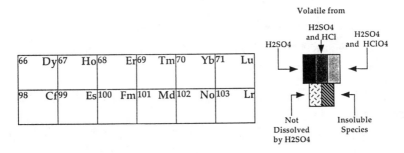

absorptivity of microwave radiation, is listed for the pure substances at specific temperatures. The values for these reagents as they are used in the laboratory will be slightly different because of the addition of water.

Guidelines for Using Chemical Reactivity Tables and Figures

Tables IX–XIV and Figures 10–15 contain the important chemical reactivities of the acids with the elements. They address the volatility of the elements, the stability either as stable soluble complexes or precipitates, the reactivity of the pure elements with the acids, and catalytic effects on the acids. Each of these categories will be addressed separately.

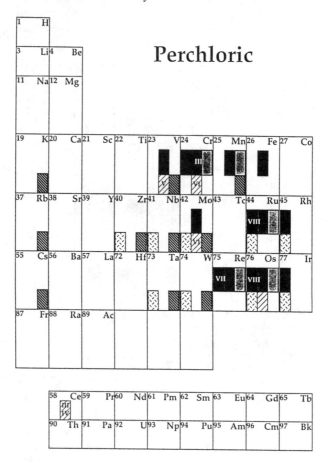

Figure 14. Reactivity of perchloric acid with the elements.

In the compilation of the information, we became aware of a considerable controversy in the literature about several of these categories. Unfortunately, the majority of the primary sources lacks the citations to the original literature eliminating the ability to resolve the literature discrepancies. As a result, some of the information in these tables and figures may have less confirmation than the authors would like, but each individual piece of information has been referenced to enable the reader further investigation. We intend to keep updating these tables and figures with new information. These updates will be included in any future updates of this book and relevant WWW pages on the SamplePrep Web site discussed in Chapter 15. We would be delighted to obtain any comments regarding the accuracy of the information contained in the tables and figures or the submission of information for inclusion into the tables.

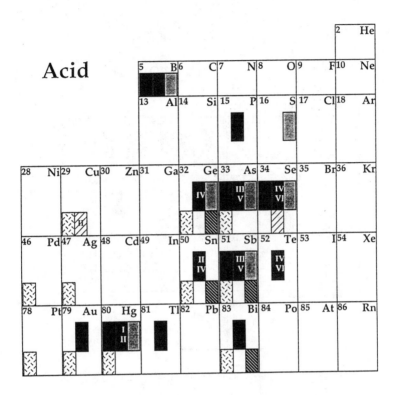

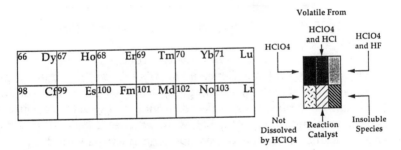

Volatility of an Elemental Species

Volatility of elemental species is the most controversial of all the information in these tables. Even though one source may state that a significant percentage of an element is volatilized under specific conditions, another source may state that it is not volatile under the same conditions. This problem led to the inclusion of the data in tabular format that allows for the complete referencing of all papers that state a particular point of view.

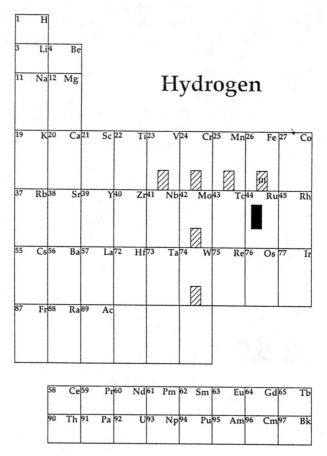

Figure 15. Reactivity of hydrogen peroxide with the elements.

At the heart of the controversy are the experimental designs that some of the works used to evaluate elemental volatility. Even though the original works were not identified, several papers addressed the same controversy and stated that numerous works were published with the conclusion that certain elements were volatile. However, the reviewers felt that the loss of some elements may have been due to interactions with the vessel, precipitation, or experimental error. Therefore, it would be wise to use these tables as a general guide and always evaluate the volatility of the analytes of interest in ones own apparatus, reagent chemistry, and specific sample.

Volatile species can be classified as either oxides, oxyhalides, halides, elements, or hydrides. The primary volatile oxides are OsO_4, RuO_4, and Re_2O_7, which are volatile in nearly any oxidizing solution. The volatile oxyhalide and halide species are generally volatile in a heated solution either with the halide acid present or a sufficient amount of the halide

Peroxide

present in the sample. The most volatile of all elements is mercury; it is volatile as the element or as numerous other forms and commonly classified as an intrinsically volatile element. Under reducing conditions, numerous elements can be reduced to hydrides that are very volatile, such as AsH_3 and SbH_3. The reagent reactivity tables list the elements, oxidation states, and volatile species with the appropriate citations. The tables are not intended to suggest that these oxidation states and species forms are the only volatile forms; they are simply a compilation of known information about these elements. A compilation of boiling points for many potentially volatile species is found in Table XVI.

Table IX. Nitric Acid

Volatile Oxides	Not Chemically Attacked	Form Oxide Coating (passivated)	Form Insoluble Acids or Hydrated Oxides
Os (OsO$_4$) (116-119,122)	Al (116)	Al (117-121)	As (122)
Re (Re$_2$O$_7$) (117-119, 122)	Au (116-121, 123)	B (121)	Mo (117, 119)
Ru (RuO$_4$) (117-119)	Cr (116, 124)	Ca (121)	P (122)
	Hf (117, 119)	Cr (117-121)	Sb (117, 119-126) (Sb$_2$O$_3$ & Sb$_2$O$_5$) (124)
	Ir (116, 117, 119, 121, 123)	Fe (121, 124)	Sn (117, 119-123, 125, 126)
	Nb (117, 119)	Ga (117, 119, 121)	W (117, 119, 125)
	Os (116-119, 121)	Hf (121)	
	Pd (116, 118, 121)	In (117, 119, 121)	
	Pt (116-119, 121, 123)	Mg (121)	
	Rh (116-119, 121, 123)	Nb (118, 121)	
	Ru (116, 118, 121)	Ta (118, 121)	
	Ta (117, 119)	Th (117, 119)	
	W (124)	Ti (120, 121)	
	Zr (117, 119)	Zr (121)	

Reactions of Elements with a Reagent

These listings, in general, are based on the reactivity of the pure element. Even though these data reflect the reactivity of the element with the acid under a specific condition, they may or may not be applicable to an impure matrix that contains the element of interest. For example, although an acid may not react with a pure element, it may react with its alloy. Nitric acid frequently forms an oxide coating with many pure metals preventing further digestion. This reaction can easily be avoided by reacting the metals with dilute acid or by the inclusion of a complexing agent.

Solubility of Elements

The solubility of the element is critical for the development of digestion procedures, but several additional factors may play a very important role in the solubility in a specific matrix. Even though a digestion method may not involve hydrochloric acid, it is quite common for samples to contain chlorides that can be solubilized during digestion and potentially precipitate the element of interest. Solubilities are not always as simple as the solubility product: for example, calcium is precipitated as the sulfate, as predicted by the solubility product. However, when calcium precipitates it will frequently cause the coprecipitation of several other elements that are not predicted to precipitate based solely on their solubility product. Sometimes, to completely decompose the matrix, the elements of interest are precipitated, such as the digestion of a silicate containing rare-earth elements with hydrofluoric acid. Despite the precipitation of the rare-earth elements during the digestion, these elements can be easily resolubilized with a postdigestion reaction and removal of the fluoride. The addition of a complexing agent such as chloride or fluoride can greatly improve the digestion of the matrix as well as the stabilization of the elements after digestion.

Stability (Complexation or Precipitation)

The stability of the element is crucial for accurate determination. Once an element is digested, it may be stable in solution as the hydrated metal ion, complexed with a complexing ligand, or precipitated from solution. To further complicate these issues, the concentration of reagents and the elements will play a very important role in the stability of the elements.

Some reagents have little or no complexation capabilities: these include perchloric acid and hydrogen peroxide. Both hydrofluoric and hydrochloric acids are very good complexing agents that can stabilize many elements in solution. For example, a number of elements will hydrolyze in an acid solution to form a precipitate, but the fluoride ion will complex a number of these elements. Nitric acid will passivate numerous metals through the formation of an oxide coating that is insoluble by nitric acid alone. In several cases, the use of dilute nitric acid alone will digest these metals, but hydrochloric acid is often added to enhance the dissolution. The hydrochloric acid will attack the oxide and complex the metal with the chloride, thus preventing the formation of an oxide coating. Silver will precipitate with hydrochloric acid to form silver chloride, but if the chloride ion concentration is increased to greater than 3 M, the silver will redissolve as a stable chloride complex.

Catalytic Effects

The reactivity of both perchloric acid and hydrogen peroxide are known to increase with the addition of specific catalysts. These catalysts, especially in the case of perchloric acid, may increase the reactivity of the reagent to a point where it may be potentially explosive.

Catalytic effects are encountered with the mixing of acids. It is common to digest an organic material with perchloric acid and a secondary acid such as nitric or sulfuric acids. The mixture of acids is designed to allow either the nitric or sulfuric acids to primarily decompose the organic matrix at low temperatures before the temperature is sufficient for the perchloric acid to begin oxidizing the remainder of the matrix (approximately >160 °C) (*121*). This concept works well with nitric acid, but sulfuric acid is dangerous. The sulfuric acid dehydrates the starting perchloric acid (~72%) to >85%, which is a strong oxidizer even at low temperatures (*121, 126*). This mixture is considered explosive and should never be used with organic material because of the postulated formation of Cl_2O_7 (*121*). Perchloric acid is potentially explosive with organics, especially hydroxyl

Table X. Hydrochloric Acid

Volatile from HCl	Volatile from HCl & HClO₄	Volatile from HCl & H₂SO₄	Not Dissolved by Dilute Acid	Strong Chloride Complexes	Insoluble Species
Al (127)	As(III) 30% (128, 129) $As(V)$ 5% (128, 129)	As(III) 100% (128, 129) As(V) 5% (128, 129)	Ag (117, 119)	Au(III) (118, 121)	AgCl (116, 118, 120, 121, 123, 124, 126, 130) > 3M HCl resoluble (124, 126)
As(III,V) (AsCl₃,AsCl₅) (117-127, 130-132)	Au 1% (128, 129)	Au 0.5% (128, 129)	As (117, 119)	Fe(III) (118, 121)	Al (121)
Au (120)	B 20% (128, 129)	B 50% (128, 129)	Au (117, 119)	Ga(III) (118, 121)	AuCl (116)
B (120, 125)	Bi 0.1% (128, 129)	Ge 90% (128, 129)	Bi (117, 119, 124)	Hg(II) (118, 121)	BiOCl (116, 124) BiCl₃ (126)
Ba (122)	Cr(III) (CrO₂Cl₂) 99.7% (127-129)	Hg(I) 75% (128, 129) Hg(II) 75% (129)	Cu (117, 119)	In(III) (118, 121)	Cu₂Cl₂ (116)
C (122)	Ge 50% (127-129)	Mn 0.02% (128, 129)	Ge (117, 119)	Sb (120)	Hg₂Cl₂ (116, 118, 120, 121)
					HgCl (124) Hg₂Cl₂O (116)
Cd (127)	Hg(I) 75% (128, 129) Hg(II) 75% (129)	Mo 5% (128, 129)	Hg (117, 119)	Sn(IV) (118, 120, 121)	Na (121)
Cr(III) (120, 125, 127, 132, 133)	Mn 0.1% (128, 129)	P 1% (128, 129)	Ir (117, 119)	Tl(III) (118, 121)	Nb (122)
Ga (127)	Mo 3% (128, 129)	Re 90% (128, 129)	Nb (117, 119)		PbCl₂ (116, 118, 124, 126, 130) > 8M HCL resoluble (126)
Ge (GeCl₄) (117-126, 130-133)	Os 100% (128, 129)	Sb(III) 33% (128, 129) Sb(V) 3% (129)	Os (117, 119)		PtCl₂ (116)
halides (122)	P 1% (128, 129)	Se(IV) 30% (129) Se(VI) 20% (128, 129)	Pd (117, 119)		SbOCl (116)
Hf (122)	Re 100% (128, 129)	Sn(II) 1% (128, 129) Sn(IV) 30% (128, 129)	Pt (117, 119)		Si (122)

Hg(I,II) (HgCl$_2$) (117-127)	Ru 99.5% (128, 129)	Te(IV) 0.1% (128, 129)	Rh (117, 119)	Ta (122)
In(III) (123, 126)	Sb(III) 2% (128, 129)	Te(VI) 0.1% (129)	Ru (117, 119)	TlCl (116, 118, 120, 121, 123, 126)
Mo (120, 127)	Sb(V) 2% (129)	Tl 0.1% (128, 129)	Sb (117, 119, 124)	W (122)
OsO$_4$ (118, 120)	Se(IV) 4% (128, 129)		Ta (117, 119)	
P (122, 132)	Se(VI) 4% (129)		W (117, 119, 124)	
Pa (122)	Sn(II) 99.8% (128, 129)		Zr (117, 119)	
Pb (120, 122, 125, 131)	Sn(IV) 100% (128, 129)			
Ra (122)	Te(IV) 0.5% (129)			
Re(VII) (118, 120, 121, 125)	Te(VI) 0.1% (128, 129)			
RuO$_4$ (118, 120)	Tl 1% (128, 129)			
S (122)	V 0.5% (128, 129)			
Sb(III,V) (SbCl$_3$,SbCl$_5$) (117-127, 132)				
Se(II,IV) (SeCl$_4$) (124)				
Si (132)				
Sn(II, IV) (SnCl$_4$) (117-119, 121-127, 132)				
Ta (127)				
TeO$_2$·2HCl (124, 125)				
Te chloride (118, 120-126, 132)				
Th (122)				
Ti (122, 125, 127)				
Tl (122)				
V (127)				
W (127)				
Zn (120, 125, 127, 131)				
Zr (122, 125, 127)				

Table XI. Hydrofluoric Acid

Volatile from HF	Volatile from HF and HClO₄	Volatile from HF and H₂SO₄	Insoluble	Strong Fluoride Complex	Fluoride Complex Prevents Hydrolysis
As(III,V) (118, 121, 123, 125, 126, 134)	As 100% (118, 120, 128, 134, 135)	As (118, 125)	Ac (123)	Al(III) (134)	Al (118)
B (BF₃) (118, 121, 125-127, 134)	B 100% (118, 120, 128, 134, 135)	B (118, 125)	Actinides (III, IV) (123, 134)	Be (II) (134)	B(III) (134)
Cr (125, 126)	Cr varies (118, 120, 128, 134, 135)	Cr (118, 125)	Al(III) (123)	Co(III) (134)	Hf(IV) (118, 120, 123, 134)
Ge(IV) (121, 123, 125, 126)	Cs 0.2% closed, 1.3% open (126)	Ge (118, 125)	Ba (116, 121, 123, 124, 134)	Cr(III) (134)	Mo (118)
Mo (127)	Ge up to 10% (118, 120, 128, 134, 135)	Hg (118, 125)	Be (116, 123)	Fe(III) (134)	Nb(V) (117-119, 126, 134)
Nb(V) (118, 121, 125, 127)	Hg (118, 120)	Os (118, 125)	Ca (116, 120, 121, 123-125, 134)	Ga(III) (134)	Si(IV) (118) (123, 134)
OsO₄ (118)	Mn up to 3% (120, 128, 134, 135)	Re (118, 125)	Cu (116)	Ge (134)	Sn(IV) (120, 134)
Pb (125)	Os (118)	Ru (118, 125)	Fe(II, III) (123)	In(III) (134)	Ta(V) (117-120, 123, 126, 134)
Re₂O₇ (118, 126)	Rb 0.3% closed, 1.4% open (126)	Sb (118, 125)	K (123)	Mn(III) (134)	Ti(IV) (118, 120, 123, 126, 134)
RuO₄ (118)	Re varies (118, 120, 128, 134, 135)	Se (118, 125)	La (123, 125)	Nb(V) (120, 123, 126, 134)	V (118, 134)
Sb(III,V) (121, 123, 125, 126)	Ru (118)	Si (118, 125)	Lanthanides (III,IV) (120, 121, 123, 134)	Pa (123)	W (VI) (118, 126, 134)

Se(IV) (121, 123, 125, 126, 134)	Sb upto 10% (118, 128, 134, 135)	Tl (126)	Li (134)	Sb (120)	Zr(IV) (117-120, 123, 134)
Se(VI) (123, 126)	Se varies (118, 120, 128, 134, 135)		Mg (116, 120, 121, 123, 134)	Sb(V) (120)	
Si (SiF4) (118, 121, 125-127)	Si 100% (118, 128, 134, 135)		Na (123)	Sc(III) (134)	
Sn (125)	Tl (126)		Pb(II) (116, 120, 123, 134)	Sn(II,V) (120, 123, 134)	
Ta(V) (118, 121, 125, 127)			Ra (116, 123)	Th(IV) (134)	
Te(IV) (121, 125, 126)			Sc (123, 134)	Y(III) (134)	
Te(VI) (126)			Sr (116, 120, 121, 123, 134)		
Ti(IV) (118, 121, 125, 127)			Th(IV) (126)		
V (127)			U(IV,VI) (121, 126)		
W (127)			Y (123, 125, 134)		
Zn (125)			Zn (116)		
Zr (125)					

Table XII. Sulfuric Acid

Volatile from H_2SO_4	Volatile from H_2SO_4 and HCl	Volatile from H_2SO_4 and HF	Not Dissolved by Dilute Acid	Insoluble
As (121)	As(III) 100% (128, 129), As(V) 5% (128, 129)	As (118)	Ag (117, 119)	Ag (Ag_2SO_4) (124, 130)
H_3BO_3 (118)	Au 0.5% (128, 129), B 50% (128, 129)	B (118)	As (117, 119)	Al (120, 122)
H_3PO_4 (118, 125)		Cr (118)	Au (117, 119)	Ba ($BaSO_4$) (116, 118, 120-124, 126, 130, 133), 0.00022g/100ml @ 18°C (133)
HBr (118)	Ge 90% (128, 129)	Ge (118)	Bi (117, 119, 124)	Be (126, 133)
HCl (118)	Hg(I) 75% (128, 129), Hg(II) 75% (129)	Hg (118)	Cr (124)	Ca ($CaSO_4$) (116, 118, 120-122, 124, 126, 130, 133, 136), 0.21g/100ml @ 30°C (133) co-ppt other metals(120, 122, 127, 130, 131)
Hg(II) (121)	Mn 0.02% (128, 129)	Os (118)	Cu (117, 119)	Cr(III) (120, 122)
HI (118), HNO₃ (118)	Mo 5% (128, 129), P 1% (128, 129)	Re (118), Ru (118)	Fe (124), Ge (117, 119)	Fe (120, 122), Mg (126, 133)

Os (OsO₄) (117-119, 122, 124)	Re 90% (128, 129)	Sb (118)	Hg (117, 119)	Nb (122)
Re (Re₂O₇) (117-119, 121, 122, 125)	Sb(III) 33% (128, 129)	Se (118)	Ir (117, 119)	Ni (122)
Ru (RuO₄) (117-119)	Sb(V) 3% (129)	Si (118)	Mo (117, 119, 124)	Pb (PbSO₄) (116, 118, 120-124, 126, 130, 133, 136, 137)
	Se(IV) 30% (129)		Nb (117, 119)	Ra (120, 122, 126, 133), 0.000002g/100ml @ 25°C (133)
Se(IV) (121, 126)	Se(VI) 20% (128, 129)		Os (117, 119)	Si (122)
	Sn(II) 1% (128, 129)		Pd (117, 119)	Sr (SrSO₄) (116, 118, 120-122, 126, 133), 0.011g/100ml @ 0°C (133)
	Sn(IV) 30% (128, 129)		Pt (117, 119)	Ta (122)
	Te(IV) 0.1% (128, 129)		Rh (117, 119)	Th (133)
	Te(VI) 0.1% (129)		Ru (117, 119)	W (122)
	Tl 0.1% (128, 129)		Sb (117, 119, 124)	
			Ta (117, 119)	
			W (117, 119, 124)	
			Zr (117, 119)	

Table XIII. Perchloric Acid

Volatile	Volatile $HClO_4$ & HCl	Volatile $HClO_4$ & HF	Not Dissolved in Dilute Acid	Insoluble Compounds	Decomposition Catalyst
As (121)	As(III) 30% (128, 129) As(V) 5% (128, 129)	As 100% (118, 120, 128, 135)	Ag (117, 119)	Ammonium (116, 122)	Ce(III) (128) Ce(IV) (121)
B (121)	Au 1% (121, 128, 129) B 20% (128, 129)	B 100% (118, 120, 128, 135) Br (120)	As (117, 119) Au (117-119)	Bi may ppt if in large amounts (122) Cs ($CsClO_4$)(116, 118, 120, 122, 123, 126)	Cr(VI) (119, 121, 123, 138) Cu(II) (121, 128)
Cr (123)	Bi 0.1% (118, 128, 129)	Cr varies (118, 120, 128, 135)	Bi (117, 119)	Ge may ppt if in large amounts (122)	Mo(VI) (121, 123)
F, Cl, Br, I, At (121)	Cr(III) (CrO_2Cl_2) 99.7% (118, 121, 128, 129)	Ge up to 10% (118, 120, 128, 135)	Cu (117, 119)	K ($KClO_4$)(116, 118, 120, 122, 123, 126)	OsO_4 (121)
H_3PO_4 (118)	Fe (121)	Hg (118, 120)	Ge (117, 119)	Mn may ppt if in large amounts (122)	SeO_2 (121)
HBr (118)	Ge(IV) 50% (121, 128, 129)	I (120)	Hg (117, 119)	Mo may ppt if in large amounts (122)	V(V) (119-121, 123, 128, 138)
HCl (118)	Hg(I) 75% (128, 129) Hg(II) 75% (129)	Mn up to 3% (120, 128, 135)	Ir (117, 119)	Nb ppt as acids. (122)	
Hg (121, 126)	Mn 0.1% (118, 121, 128, 129)	Os (118)	Mo (117, 119)	Rb ($RbClO_4$)(116, 118, 120, 122, 123, 126)	
HI (118)					

HNO3 (118)	Mo 3% (118, 121, 128, 129)	Re varies (118, 120, 128, 135)	Nb (117, 119)	Sb ppt as acids or Oxides. (122, 123)
Ir (126)	Os 100% (128, 129)	Ru (118)	Os (117–119)	Sn ppt as acids or Oxides. (122, 123)
Os OsO4 (118, 121, 123, 126)	P 1% (128, 129)	Sb up to 10% (118, 128, 135)	Pd (117–119)	Ta ppt as acids. (122)
Re Re2O7 (118, 123, 126)	Re 100% (128, 129)	Se varies (118, 120, 128, 135)	Pt (117–119)	V may ppt if in large amounts (122)
Rh (121)	Ru 99.5% (128, 129)	Si 100% (118, 128, 135)	Rh (117–119)	W (WO3)(121, 122)
Ru RuO4 (118, 121, 123, 126)	Sb(III) 2% (128, 129)	Silicium (120)	Ru (117–119)	
	Sb(V) 2% (129)			
Sb (121)	Se(IV) 4% (128, 129)		Sb (117, 119)	
	Se(VI) 4% (129)			
Se (121, 126)	Sn(II) 99.8% (128, 129)		Ta (117, 119)	
	Sn(IV) 100% (121, 128, 129)			
	Te(IV) 0.5% (129)		W (117, 119)	
	Te(VI) 0.1% (121, 128, 129)			
	Tl 1% (118, 121, 128, 129)		Zr (117, 119)	
	V 0.5% (118, 128, 129)			

Table XIV. Hydrogen Peroxide

Volatile in H_2O_2	Volatile in H_2O_2 & Cl⁻	Catalyzes Reactions
As (121)	As (121)	CrO_4^{-2} (121)
Ge (121)	Ge (121)	Cu(II) (121)
Hg (121)	Se (121)	Fe(III) (121)
Ru (121)		MnO_2 (121)
Se (121)		MoO_4^{-2} (121)
		VO_4^{-3} (121)
		WO_4^{-2} (121)

Table XV. Common Dissolution Reagent Parameters

Reagent	bp (°C)	Density at 25°C (g mL⁻¹)	Molarity (mol L⁻¹)	% reagent (w/w)	pKa (aq) (mol L⁻¹)	Dielectric Constant of Pure Substance
Nitric acid (conc)	122 (121, 123, 140)	1.410 (121)	15.5 (121)	69.2 (121)	-1.4 (140)	50 ± 10 at 14°C (140)
Hydrochloric acid (conc)	109 (123, 140)	1.18 - 1.19 (121)	11.73 (121)	36 - 38% (123)	-6.1 (141)	4.60 at 28°C (141)
Sulfuric acid (conc)	339 (121, 123)	1.84 (121, 123)	18.3M (121)	96 - 98% (121, 123)	1. ~ -3 (141) 2. 1.92 (141, 142)	100 at 25°C (140, 141)
Perchloric acid (68-72%)	203 (121, 123, 140)	1.72 (121)	11.7 (141)	72.4 (121)	~ -10 (140)	---
Hydrofluoric acid (conc)	112 (38.3 %) (121, 123, 140)	1.138 (121, 140)	21.5 - 29 (121, 123)	38 - 48 (121, 123)	3.1 - 3.45 (123, 142)	83.6 at 0°C (141)
Phosphoric acid (conc)	158 (85%) 261 (98%) (121)	1.71 (85%) 1.86 (98%) (121)	14.7 (85%) (142)	85 - 98% (121)	1. 2.15 2. 7.20 3. 12.37 (140-142)	---
Hydrogen peroxide (30%)	107 (30%) 115 (50%) (121)	1.111 (30%) 1.196 (50%) (121)	(9.7)	30% - 50% (121)	11.6 (141, 142)	70.7 (100%) 121 (35%) (140)
Water (18 MOhm)	100	1.00	(55.51)	---	---	78.4 (140, 141)

Table XVI. Boiling Points of Potentially Volatile Species

Chloride	Fluoride	Hydride	Oxide	Mixed Oxide
$AlCl_3$ 182.7 °C	AsF_3 -63 °C	AsH_3 -55 °C	Mn_2O_7 ca 25 °C	$AsFO$ 25.6°C
	AsF_5 -53 °C			
$AsCl_3$ 130.2 °C	BF_3 -99.9 °C	B_2H_6 -92.5 °C	OsO_4 130 °C	$Ba(ClO_4)_2$
				162 °C d
$AuCl_3$ 229 °C	BiF_5 230 °C	BiH_3 16.8 °C	RuO_4 40 °C	Cr_2Cl_2 117 °C
BCl_3 12.5 °C	CrF_5 117 °C	Ga_2H_6 139 °C d		$POCl_3$ 105.3 °C
				POF_3 -39.8 °C
$BaCl_2\cdot2H_2O$	GeF_4 subl	GeH_4 -88.5 °C		$ReOCl_4$ 29.3 °C
35.7 °C	-36.5°C	Ge_2H_6 29 °C		$ReOF_4$ 39.7 °C
$GaCl_3$ 201.2 °C	IrF_6 53.6 °C	PH_3 -87.5 °C		$SeOCl_2$ 176.4 °C
				$SeOF_2$ 124 °C
$GeCl_4$ 84 °C	MoF_5 214 °C	SbH_3 -18.4 °C		$VOCl$ 127 °C
	MoF_6 33.89 °C			
$NbCl_5$ 250 °C	NbF_5 225 °C	H_2Se -41.5 °C		WOF_4 187.5 °C
PCl_2 180 °C	NpF_6 76.8 °C	SiH_4 -111.8 °C		
PCl_3 75 °C				
PCl_5 166 °C d				
$SbCl_3$ 220 °C	OsF_5 225.7 °C	SnH_4 -52 °C		
$SbCl_5$ 79 °C	OsF_6 45.9 °C			
$SiCl_4$ 57.6 °C	PF_3 -101.38 °C	H_2Te -2.2 °C		
	PF_5 -84.6 °C			
$SnCl_4$ 115 °C	PtF_6 57.6 °C			
$TaCl_5$ 242 °C	PuF_6 62.16 °C			
$TiCl_4$ 136.4 °C	ReF_6 33.8 °C			
	ReF_7 77.7 °C			
VCl_4 152 °C	RuF_5 227 °C			
	SbF_5 150 °C			
	SeF_4 107.8 °C			
	SeF_6 -34.5 °C			
	SiF_4 -86 °C			
	Si_2F_6 -19 °C			
	TaF_5 229.5 °C			
	TcF_5 d60 °C			
	TeF_4 >97 °C			
	TeF_6 35.5 °C			
	UF_6 subl 56.5 °C			
	VF_5 111.2 °C			
	WF_6 17.2 °C			
	XeF_2 subl 114 °C			
	XeF_4 subl 116 °C			
	XeF_6 75.6 °C			

d is decomposed, subl is sublimation
Compiled from references 141 and 142.

compounds and fatty materials (133). Perchloric acid has been implicated in explosions with bismuth, its alloys, and Sb_2O_3 when treated with concentrated hot perchloric acid.

Hydrogen peroxide is commonly used as an oxidant that is added to an acid such as sulfuric or nitric. The combination of sulfuric acid and hydrogen peroxide has been postulated to produce permonosulfuric acid in situ, a reagent that is known to oxygenate many kinds of organic molecules (133).

Figures 10–15 are meant to be used individually when a single reagent is being used or collectively with several acids. By referring to these figures when developing a sample-preparation acid digestion, many of the relevant interactions should be apparent.

References

1. Abu-Samra, A.; Morris, J. S.; Koirtyohann, S. R *Anal. Chem.* **1975,** *47,* 1475–1477.
2. Abu-Samra, A.; Morris, J. S.; Koirtyohann, S. R. *Trace Subst. Environ. Health* **1975,** *9,* 297–301.
3. Abu-Samra, A.; Koirtyohann, S. R.; Morris, J. S., U.S. Patent 4, 080, 168, 1978.
4. Kingston, H. M.; Jassie, L. B. *Anal. Chem.* **1986,** *58,* 2534–2541.
5. Kingston, H. M.; Jassie, L. B *Introduction to Microwave Sample Preparation: Theory and Practice;* Jassie, L. B.; Kingston, H. M., Eds.; American Chemical Society: Washington, DC, 1988; pp 93–154.
6. Kingston, H. M.; Walter, P. J. *Spectroscopy* **1992,** *7,* 20–27.
7. Matthes, S. A.; Farrell, R. F.; Mackie, A. J. Technical Progress Report 120, Bureau of Mines: Washington, DC, April 1983.
8. Kingston, H. M.; Cronin, D. J.; Epstein, M. S. *Nucl. Chem. Waste Manage.* **1984,** *5,* 3–15.
9. "R&D Magazine Presents The 1987 R&D 100 Award Winners", *Res. Develop.* **1987,** *29(10),* 55–94.
10. Lautenschläger, W. U.S. Patent 5,382,414, 1995.
11. ASTM Method D4309–91: *Standard Practice for Sample Digestion Using Closed Vessel Microwave Heating Technique for the Determination of Total Recoverable Metals in Water;* The American Society for Testing and Materials: Philadelphia, PA, 1991.
12. SW–846 E.P.A. Method 3051: "Microwave Assisted Acid Digestion of Sediments, Sludges, Soils, and Oils", *Test Methods for Evaluating Solid Waste,* 3rd ed., 3rd update; U.S. Environmental Protection Agency, U.S. Government Printing Office: Washington, DC, 1995.
13. SW–846 E.P.A. Method 3015: "Microwave Assisted Acid Digestion of Aqueous Sample and Extracts", *Test Methods for Evaluating Solid Waste,* 3rd ed., 3rd update; U.S. Environmental Protection Agency, U.S. Government Printing Office: Washington, DC, 1995.
14. Walter, P. J. *Special Publication IR4718: Microwave Calibration Program,* 2. 0 ed.; National Institute of Standards and Technology: Gaithersburg, MD, 1991.

15. Settle, F. A.; Walter, P. J.; Kingston, H. M.; Pleva, M. A.; Snider, T.; Boute, W. *J. Chem. Inf. Comput. Sci.* **1992,** *32,* 349–353.

16. Kingston, H. M.; Walter, P. J.; Settle, F. A., Jr.; Pleva, M. A. In *Advances in Laboratory Automation Robotics;* Strimaitis, J. R.; Little, J. N., Eds.; 1991; Vol. 8, pp 619–629.

17. Kingston, H. M.; Jassie, L. B. *J. Res. Natl. Bur. Stand. (U.S.)* **1988,** *93,* 269–274.

18. Pagano, S. T.; Smith, B. W.; Winefordner, J. D. *Talanta* **1994,** *41,* 2073–2078.

19. AOAC Official Method 977. 11: "Moisture in Cheese", In *Official Methods of Analysis of AOAC International;* Cunniff, P., Ed.; Association of Official Analytical Chemists International: Arlington, VA, 1995.

20. AOAC Official Method 985. 14: "Moisture in Meat and Poultry Products", In *Official Methods of Analysis of AOAC International;* Cunniff, P., Ed.; Association of Official Analytical Chemists International: Arlington, VA, 1995.

21. AOAC Official Method 985. 15: "Fat (Crude) in Meat and Poultry Products", In *Official Methods of Analysis of AOAC International;* Cunniff, P., Ed.; Association of Official Analytical Chemists International: Arlington, VA, 1995.

22. AOAC Official Method 985. 26: "Solids (Total) in Processed Tomato Products", In *Official Methods of Analysis of AOAC International;* Cunniff, P., Ed.; Association of Official Analytical Chemists International: Arlington, VA, 1995.

23. ASTM Method E1358–90: *Standard Test Method for Determination of Moisture Content of Particulate Wood Fuels Using a Microwave Oven;* The American Society for Testing and Materials: Philadelphia, PA, 1990.

24. ASTM Method D5258–92: *Standard Practice for Acid Extraction of Elements from Sediments Using Closed Vessel Microwave Heating;* The American Society for Testing and Materials: Philadelphia, PA, 1992.

25. ASTM Method D4643–93: *Standard Test Method for Determination of Water (Moisture) Content by Soil by the Microwav Oven Method;* The American Society for Testing and Materials: Philadelphia, PA, 1993.

26. SW–846 EPA Method 3031: "Acid Digestion of Oils for Metals, Analysis by FLAA or ICP Spectroscopy", In *Test Methods for Evaluating Solid Waste,* 3rd ed., 3rd update; U.S. Environmental Protection Agency: Washington, DC, 1995.

27. SW–846 EPA Method 3050B: "Acid Digestion of Sediments, Sludges, and Soils", In *Test Methods for Evaluating Solid Waste,* 3rd ed., 3rd update; U.S. Environmental Protection Agency: Washington, DC, 1995.

28. SW–846 EPA Method 3052: "Microwave Assisted Acid Digestion of Siliceous and Organically Based Matricies", In *Test Methods for Evaluating Solid Waste,* 3rd ed., 3rd update; U.S. Environmental Protection Agency: Washington, DC, 1995.

29. Method 3030K: "Microwave Assisted Digestion", In *Standard Methods for the Examination of Water and Wastewater;* Greenberg, A. E., Clesceri, L. S., Eaton, A. D., Eds.; American Public Health Association: Washington, DC, 1992.

30. Method V 03–100: Kjeldahl Nitrogen; French Standard.

31. NIEA Method C303. 01T: *Acid Digestion of Fish and Shellfish;* NIEA: Republic of China, 1994.

32. ASTM Method D1506-94b: Standard Test Methods for Carbon Black–Ash Content; The American Society for Testing and Materials: Philadelphia, PA, 1994.

33. ASTM Method D5513-94: Standard Practice for Microwave Digestion of Industrial Furnace Feedstreams for Trace Element Analysis; The American Society for Testing and Materials: Philadelphia, PA, 1994.

34. ASTM Method E1645-94: Standard Practice for the Preparation of Dried Paint Samples for Subsequent Lead Analysis by Atomic Spectrometry; The American Society for Testing and Materials: Philadelphia, 1994.

35. Hewitt, A. D.; Reynolds, C. M. In "Microwave Digestion of Soils and Sediments for Assessing Contamination by Hazardous Waste Metals"; Special Report 90-19, U.S. Army Corps of Engineers Cold Regions Research & Engineering Laboratory: Hanover, NH, June 1990.

36. Hewitt, A. D.; Cragin, J. H. "Comment on "Acid Digestion for Sediments, Sludges, Soils, and Solid Wastes. A Proposed Alternataive to EPA SW 846 Method 3050"" *Environ. Sci. Technol.* **1991**, *25*, 985-986.

37. Hewitt, A. D.; Cragin, J. H. "Comment on "A Study of the Linear Ranges of Several Acid Digestion Procedures"" *Environ. Sci. Technol.* **1992**, *26*, 1848.

38. Kimbrough, D. E.; Wakakuwa, J. "A Study of the Linear Ranges of Several Acid Digestion Procedures" *Environ. Sci. Technol.* **1992**, *26*, 173-178.

39. Fordham, O., U.S. EPA Program Office, personal communication, 1996.

40. "40 CFR Part 136: Guidelines Establishing Test Procedures for the Analysis of Pollutants; Microwave Digestion"*Fed. Regist.* **1992**, *57*, 41830-41834.

41. In "Closed Vessel Microwave Digestion of Wastewater Samples for Determination of Metals"; Technical Report RD42; CEM Corporation: Matthews, NC, 1992.

42. Labrecque, J. M. "Manual and Robotically Controlled Microwave Pressure Dissolution of Minerals", In *Introduction to Microwave Sample Preparation: Theory and Practice*; Jassie, L. B.; Kingston, H. M., Eds.; American Chemical Society: Washington, DC, 1988; Chapter 10, pp 203–230.

43. Walter, P. J.; Kingston, H. M.; Settle, F. A.; Pleva, M. A.; Buote, W.; Christo, J. "Automated Intelligent Control of Microwave Sample Preparation", In *Advances in Laboratory Automation Robotics*; Strimaitis, J. R.; Little, J. N., Eds.; Hopkinton, MA, 1990; Vol. 7, pp 405–416.

44. Settle, F. A. J.; Diamondstone, B. I.; Kingston, H. M.; Pleva, M. A. *J. Chem. Inf. Comput. Sci.* **1989**, *29*, 11–17.

45. Norris, J. D.; Preston, B.; Ross, L. M. *Analyst* **1992**, *117*, 3–7.

46. Salit, M. L.; Guenther, F. R.; Kramer, G. W.; Griesmeyer, J. M. *Anal. Chem.* **1994**, *66*, 361A–367A.

47. Kingston, H. M. *Anal. Chem.* **1989**, *61*, 1381A–1384A.

48. Walter, P. J. Ph.D. Thesis "The Development and Validation of Advanced Reaction Control Techniques for Microwave Sample Preparation", Duquesne University, 1996.

49. Petersen, J. V. In "Documentation for the Microwave-Assisted Dissolution Station"; Consortium on Automated Analytical Laboratory Sytems Report; National Institute of Standards and Technology: Gaithersburg, MD, May 18, 1992.

50. Barreto, S. *Chorleywood Dig.* **1991**, *111*, 39–40.

51. Bitsch, R.; Merck, E. *Labor Praxis* **1994**, *18*, 76–81.

52. Cresser, M. S.; Armstrong, J.; Cook, J.; Dean, J. R.; Watkins, P.; Cave, M. *J. Anal. At. Spectrom.* **1993**, *8*, 1r–77r.

53. De la Guardia, M.; Salvador, A.; Burguera, J. L.; Burguera, M. *J. Flow Injection Anal.* **1988**, *5*, 121–131.
54. Dunemann, L.; Meinerling, M. *Fresenius J. Anal. Chem.* **1992**, *342*, 714–718.
55. Feinberg, M. H. *Analusis* **1991**, *19*, 47–55.
56. Gilman, L. B.; Engelhart, W. G. *Spectroscopy* **1989**, *4*, 14, 16, 18, 21.
57. Jackwerth, E.; Gomiscek, S. *Pure Appl. Chem.* **1984**, *56*, 479–489.
58. Jin, Q. *Fenxi Huaxue* **1988**, *16*, 668.
59. Jin, Q. H.; Zhang, H. Q.; Wang, D.; Liu, L.; Zou, M. Q.; Xie, S. J. *Yankuang Ceshi* **1992**, *11*, 87–94.
60. Kane, J. S.; Wilson, S. A.; Lipinski, J.; Butler, L. *Am. Environ. Lab.* **1993**, *6*, 14–15.
61. Kimber, G. M.; Kokot, S. *Trends Anal. Chem.* **1990**, *9*, 203–207.
62. Kingston, H. M.; Jassie, L. B. Introduction to Microwave Acid Decomposition, In *Introduction to Microwave Sample Preparation: Theory and Practice*; Kingston, H. M.; Jassie, L. B., Eds.; American Chemical Society: Washington, DC, 1988; pp 1–6.
63. Knapp, G. *Anal. Proc.* **1990**, *27*, 112–114.
64. Knapp, G. *Mikrochim. Acta* **1991**, *II*, 445–455.
65. Kojima, I. *Bunseki* **1992**, *1*, 14–19.
66. Kokot, S.; King, G.; Keller, H. R.; Massart, D. L. *Anal. Chim. Acta* **1992**, *268*, 81–94.
67. Kubrakova, I. V.; Kuz' min, N. M. *Zavod. Lab.* **1992**, *58*, 1–5.
68. Kuss, H. M. *Chem. Labor. Biotech.* **1991**, *42*, 11–14, 17.
69. Kuss, H. M. *Fresenius J. Anal. Chem.* **1992**, *343*, 788–793.
70. Lang, C.; Wang, M. *Lihua Jianyan: Huaxue Fence* **1989**, *25*, 315–317.
71. Lautenschläger, W. *Spectroscopy* **1989**, *4*, 16–21.
72. Majors, R. E. *LC-GC* **1995**, *13*, 82.
73. Marshall, J.; Franks, J. *Anal. Proc.* **1990**, *27*, 240–241.
74. Matusiewicz, H.; Sturgeon, R. E. *Prog. Anal. Spectrosc.* **1989**, *12*, 21–39.
75. Matusiewicz, H. *Spectrosc. Int.* **1991**, *3*, 22–28.
76. McKenzie, H. A. *Trends Anal. Chem.* **1994**, *13*, 138–144.
77. Quevauviller, P.; Maier, E. A.; Griepink, B. *Fresenius J. Anal. Chem.* **1993**, *345*, 282–286.
78. Sinquin, A.; Gorner, T.; Dellacherie, E. *Analusis* **1993**, *21*, 1–10.
79. Tatro, M. E. *Spectroscopy* **1986**, *1*, 18–21.
80. Tolg, G. *Analyst* **1987**, *112*, 365–376.
81. Wu, R.; Ding, M. *Fenxi Shiyanshi* **1991**, *10*, 51–54.
82. Zehr, B. D. *Am. Lab.* **1992**, *December*, 24–29.
83. Hesek, J. A.; Wilson, R. C. *Anal. Chem.* **1974**, *46*, 1160.
84. Tadros, S. H.; Frazier, D. O. *Analyst* **1990**, *115*, 229.
85. Crosland, A. R.; Bratchell, N. *J. Assoc. Public Anal.* **1988**, *26*, 89–95.
86. Bostian, M. L.; Fish, D. L.; Webb, N. B.; Arey, J. J. *J. Assoc. Off. Anal. Chem.* **1985**, *68*, 876–880.
87. Blust, R.; Van der Linden, A.; Verheyen, E.; Decleir, W. *J. Anal. At. Spectrom.* **1988**, *3*, 387–393.
88. Worner, H. K.; Standish, N. *Analyst* **1989**, *114*, 115–116.
89. Koh, T. S. *Anal. Chem.* **1980**, *52*, 1978–1979.
90. Barbano, D. M.; Della Valle, M. E. *J. Food Prot.* **1984**, *47*, 272–278.
91. Sumner, J. S.; Morrow, W. D. *J. Coal Qual.*

92. Jacobs, M. L. *J. Coal Qual.* **1984,** 12–15.

93. Beary, E. S. *Anal. Chem.* **1988,** *60,* 742–746.

94. Benz, C. *Soap Cosmet. Chem. Spec.* **1978,** September.

95. Latawiec, A. P.; Macbeth, S. E. *Aldrichim. Acta* **1988,** *21(2),* 30.

96. Davis, A. B.; Lai, C. S. *Cereal Chem.* **1984,** *61,* 1–4.

97. Kuehn, D. G.; Brandvig, R. L.; Lundeen, D. C.; Jefferson, R. H. *Int. Lab.* **1986,** *16,* 30–41.

98. Ybanez, N.; Cervera, M. L.; Montoro, R.; Guardia, M. *J. Anal. At. Spectrom.* **1991,** *6,* 379–384.

99. Jaffe, R.; Fernandez, C. A.; Alvarado, J. *Talanta* **1992,** *39,* 113–117.

100. Chin, H. B.; Kimball, J. R., Jr.; Hung, J.; Allen, B. *J. Assoc. Off. Anal. Chem.* **1985,** *68,* 1081–1083.

101. Cabrera, C.; Lorenzo, M. L.; Gallego, C.; Lopez Martinez, M. C.; Lillo, E. *J. Agric. Food Chem.* **1994,** *42,* 126–128.

102. Collins, M. J.; Gilman, L. B. *Powder Bulk Eng.* **1988,** February.

103. Borowski, A. M.; Fritz, V. A. *Hortic. Sci.* **1990,** *25,* 361.

104. Lamleung, S. Y.; Cheng, K. W.; Lam, Y. W. *Analyst* **1991,** *116,* 957–959.

105. Wang, S. L. *J. Assoc. Off. Anal. Chem.* **1987,** *70,* 758–759.

106. Krofta, M.; Wang, L. K. In "Development of a Microwave Method for Rapid and Accurate Analysis of Solids and Water Content"; LIR/04-81/5; U.S. Department of Commerce: Washington, DC, April 30, 1981.

107. Zhang, H.; Dotson, P. In "The Use of Microwave Muffle Furnace for Dry Ashing Plant Tissue Samples"; Agricultural Testing and Research Laboratory: Farmington, NM, 1994.

108. Morales Rubio, A.; Salvador, A.; De La Guardia, M. *Fresenius J. Anal. Chem.* **1992,** *342,* 452–456.

109. Matsumura, S.; Karai, I.; Takise, S.; Kiyota, I.; Shinagawa, K.; Horiguchi, S. *Osaka City Med. J.* **1982,** *28,* 145–148.

110. Carr, S. E.; Moser, C. R.; Matthews, N. C. *Proceedings of the International Coal Testing Conference;* 1990, 52–56.

111. Carr, S. E.; Moser, C. R. *Abstracts of Papers,* Pittsburgh Conference, Chicago, IL, 1991.

112. Budke, C. C.; McFadden, D. G. *Plast. Compd.* **1990,** March/April.

113. Van Montfort, P. F. E.; Agterdenbos, J.; Jutte, B. A. H. G. *Anal. Chem.* **1979,** *51,* 1553–1557.

114. Gillain, G.; Rutagengwa, J. *Analusis* **1985,** *13,* 471–473.

115. Gillain, G. *Talanta* **1982,** *29,* 651–654.

116. Treadwell, F. P.; Hall, W. T. *Analytical Chemistry,* 9th English ed.; John Wiley and Sons: New York, 1937.

117. Bogen, D. C. In *Treatise on Analytical Chemistry;* Kolthoff, I. M.; Elving, P. J., Eds.; Interscience: New York, 1982; Vol. 5, Part I, pp 1–22.

118. Anderson, R. *Sample Pretreatment and Separation;* John Wiley and Sons: New York, 1987.

119. Willard, H. H.; Rulfs, C. L. In *Treatise on Analytical Chemistry;* Kolthoff, I. M.; Elving, P. J., Eds.; Interscience: New York, 1961; Vol. 2: Part I, pp 1027–1050.

120. Bajo, S. In *Preconcentration Techniques for Trace Elements;* Alfassi, Z. B.; Wai, C. M., Eds.; CRC: Boca Raton, FL, 1992; pp 3–31.

121. Bock, R. *A Handbook of Decomposition Methods in Analytical Chemistry*, 1st ed.; John Wiley and Sons: New York, 1979.

122. Lundell, G. E. F.; Hoffman, J. I. *Outlines of Methods of Chemical Analysis*, 6th ed.; John Wiley and Sons: New York, 1951.

123. Sulcek, Z.; Povondra, P. *Methods of Decomposition in Inorganic Analysis*; CRC: Boca Raton, FL, 1989.

124. Scott, W. W. *Scott's Standard Methods of Chemical Analysis*, 5th ed.; D. Van Nostrand: New York, 1939.

125. Zehr, B. D. *Am. Lab.* **1992,** December, 24–29.

126. Sulcek, Z.; Povondra, P.; Dolezal, J. *CRC Crit. Rev. Anal. Chem.* **1977,** 255–323.

127. Mizuike, A. *Enrichment Techniques for Inorganic Trace Analysis*; Springer-Verlag: New York, 1983.

128. Schilt, A. A. *Perchloric Acid and Perchlorates*; G. Frederick Smith Chemical Company: Columbus, OH, 1979.

129. Hoffman, J. J.; Lundell, G. E. F. *J. Res. Natl. Bur. Stand.* **1939,** 22, 465–470.

130. O'Haver, T. C. In *Trace Analysis: Spectroscopic Methods for Elements*; Elving, P. J.; Winefordner, J. D.; Kolthoff, I. M., Eds.; John Wiley and Sons: New York, 1976; Vol. 46, pp 63–78.

131. Gorsuch, T. T. In *Accuracy in Trace Analysis: Sampling, Sample Handling, and Analysis*; National Bureau of Standards Special Publication 422; National Institute of Standards and Technology: Gaithersburg, MD, 1976; Vol. I, pp 491–507.

132. Vandecasteele, C.; Block, C. B. In *Modern Methods for Trace Element Determination*; John Wiley and Sons: New York, 1993, pp 9–53.

133. Gorsuch, T. T. *The Destruction of Organic Matter*, 1st ed.; Pergamon: New York, 1970.

134. Headridge, J. B. *CRC Crit. Rev. Anal. Chem.* **1972,** 461–490.

135. Chapman, F. W.; Marvin, G. G.; Tyree, S. Y., Jr. *Anal. Chem.* **1949,** 21, 700–701.

136. Van Loon, J. C. In *Selected Methods of Trace Metal Analysis: Biological and Environmental Samples*; John Wiley and Sons: New York, 1985; Vol. 80, pp 77–111.

137. Hillebrand, W. F.; Lundell, G. E. F.; Bright, H. A.; Hoffman, J. I. *Applied Inorganic Analysis*, 2 ed.; John Wiley and Sons: New York, 1953.

138. Dunlop, E. C.; Ginnard, C. R. In *Treatise on Analytical Chemistry*; Kolthoff, I. M.; Elving, P. J., Eds.; Interscience: New York, 1982; Vol. 5, Part I, pp 23–61.

139. Cotton, F. A.; Wilkinson, G. *Advanced Inorganic Chemistry*, 5th ed.; John Wiley and Sons: New York, 1988.

140. Greenwood, N. N.; Earnshaw, A. *Chemistry of the Elements*; Pergamon: Oxford, England, 1984; pp 1064–1068.

141. Dean, J. A. *Lange's Handbook of Chemistry*, 12th ed.; McGraw-Hill: New York, 1979.

142. Weast, R. C. *CRC Handbook of Chemistry and Physics*, 54th ed.; CRC: Cleveland, OH, 1973.

Appendices

Appendix 1. Biological Reference Materials

Appendix 1.1 Animal Tissue and Fluid

Matrix	Analytes	Reagents	Cavity/Vessel	Conditions	Detection	Ref.
BCR Milk powder 63R	Cd, Cu, Pb	HNO_3, $HClO_4$, H_2SO_4	Multimode/ HP closed	10 min at 85 bar	DP-ASV	A1
BCR Milk powder 150	Hg	HNO_3	Multimode/ LP closed	8 min at 140 W, 8 min at 280 W, 4 min at 420 W	CV-AAS	A2
	Cu, Pb	HNO_3	Multimode/ LP closed	3 min at 650 W	ETV-AAS	A3
	Cd	HNO_3, HCl	Multimode/ HP closed	4 min at 360 W, 10 min at 180 W	ETV-AAS	A4
	Cd, Cu, Fe, Pb	1. HNO_3, H_2O, H_2O_2 2. Aqua regia, HF (optional)	Multimode/ open	1. 15 min at 70 W, 15 min at 150 W, 15 min at 180 W, 1 min at 250 W, 1 min at 600 W 2. 15 min at 70 W, 1 min at 600 W	ETV-AAS	A5
BCR Skim milk powder 151	Hg	HNO_3	Multimode/ LP closed	8 min at 140 W, 8 min at 280 W, 4 min at 420 W	CV-AAS	A2
	Cd	HNO_3, HCl	Multimode/ HP closed	4 min at 360 W, 10 min at 180 W	ETV-AAS	A4
	Cd, Cu, Fe	A. HNO_3, H_2O_2	Multimode/ HP closed	75 sec at 665 W	ETV-AAS	A6
		B. HNO_3, H_2O_2	Multimode/ HP closed	1 min at 250 W, 2 min at 0 W, 2 min at 250 W, 2 min at 400 W, 2 min at 600 W		

C.

Sample	Element	Reagents	Mode/Pressure	Conditions	Detection	Ref.
	Hg	HNO_3, $HClO_4$	Multimode/HP closed	150 sec at 950 W, 60 sec at 0 W, 90 sec at 300 W, 90 sec at 500 W, 90 sec at 700 W, 90 sec at 850 W	CV-AFS	A7
	Pb	HNO_3	Multimode/MP closed	8 min at 140 W, 8 min at 280 W, 4 min at 420 W	ETV-AAS	A8
BCR Bovine muscle 184	Cu, Pb	HNO_3, HCl	Multimode/flow through	18 sec at 700 W (0.4 mL coil, 1.5 mL min^{-1})	DP-ASV	A1
	Ni	HNO_3, HCl	Multimode/HP closed	4 min at 350 W, 10 min at 140 W	ETV-AAS	A9
	Cd, Cu, Fe, Mn, Pb	1. HNO_3, H_2O, H_2O_2 2. Aqua regia, HF (optional)	Multimode/open	1. 15 min at 70 W, 15 min at 150 W, 15 min at 180 W, 1 min at 250 W, 1 min at 600 W 2. 15 min at 70 W, 1 min at 600 W	ETV-AAS	A5
	Cd	HNO_3, HCl	Multimode/HP closed	4 min at 360 W, 10 min at 180 W	ETV-AAS	A4
	Ni	HNO_3, HCl	Multimode/HP closed	4 min at 360 W, 10 min at 180 W	ICP-OES	A10, A11
	Pb	HNO_3	Multimode/LP closed	3 min at 650 W	ETV-AAS	A3
BCR Bovine liver 185	Cd, Pb	HNO_3, $HClO_4$, H_2SO_4	Multimode/HP closed	10 min at 85 bar	DP-ASV	A1
BCR Pig kidney 186	Pb	HNO_3, HCl	Multimode/flow through	18 sec at 700 W (0.4 mL coil, 1.5 mL min^{-1})	ETV-AAS	A8
	Hg	HNO_3	Multimode/LP closed	8 min at 140 W, 8 min at 280 W, 4 min at 420 W	CV-AFS	A7

Appendix 1.1 Animal Tissue and Fluid

Matrix	Analytes	Reagents	Cavity/Vessel	Conditions	Detection	Ref.
BCR Pig kidney 186 (contd.)	Ni	HNO_3, HCl	Multimode/ HP closed	4 min at 350 W, 10 min at 140 W	ETV-AAS	A9
	Cd, Cu, Fe, Mn, Pb	1. HNO_3, H_2O, H_2O_2 2. Aqua regia, HF (optional)	Multimode/ open	1. 15 min at 70 W, 15 min at 150 W, 15 min at 180 W, 1 min at 250 W, 1 min at 600 W 2. 15 min at 70 W, 1 min at 600 W	ETV-AAS	A5
	Cd	HNO_3, HCl	Multimode/ HP closed	4 min at 360 W, 10 min at 180 W	ETV-AAS	A4
	Ni	HNO_3, HCl	Multimode/ HP closed	4 min at 360 W, 10 min at 180 W	ICP-OES	A10, A11
BCR Whole blood 194	Pb	HNO_3	Multimode/ LP closed	15 min at 540 W, 20 min at 420 W	TXRF	A12
BCR Whole blood 195	Pb	HNO_3	Multimode/ LP closed	15 min at 540 W, 20 min at 420 W	TXRF	A12
BCR Whole blood 196	Pb	HNO_3	Multimode/ LP closed	15 min at 540 W, 20 min at 420 W	TXRF	A12
BCR Human hair 397	Hg	HNO_3	Multimode/ LP closed	8 min at 140 W, 8 min at 280 W, 4 min at 420 W	CV-AFS	A7
	Cd, Hg, Pb, Zn	A. 1. HNO_3 2. HNO_3 3. H_2O_2 4. H_2O	Single mode/ open	1. 5 min at 10 W, 10 min at 30 W, 10 min at 60 W 2. 10 min 60 W 3. 5 min at 60 W 4. 5 min at 50 W	ICP-OES ICP-MS HPLC-ICP-MS	A13

Sample	Analyte	Reagents	Mode	Conditions	Method	Ref.
BI Control blood for metals 620401	Hg(II), methylmercury, mercurescein, phenylmercury	B. 1. HNO$_3$, HCl 2. HNO$_3$, HCl 3. H$_2$O$_2$ 4. H$_2$O 1. HCl, KBr, KBrO$_3$ 2. KMnO$_4$ 3. NaBH$_4$, NaOH	Single mode/ open Single mode/ flow through	1. 5 min at 40 W, 10 min at 50 W 2. 20 min 54 W 3. 5 min at 40 W 4. 5 min at 50 W 1. 25.4 sec at 70 W (14–16 mL min^{-1})	CV-AAS	A14
BI Control metals in urine Lanonorm M1	Cd	HNO$_3$	Multimode/ LP closed	3 min at 650 W	ETV-AAS	A3
BI Control metals in urine Lanonorm M1 (contd.)	Hg	1. K$_2$Cr$_2$O$_7$, HNO$_3$, KBr, KBrO$_3$ 2. HCl	Single mode/ flow through	1. Offline 2. 65 sec at 90 W (8.5 mL min^{-1})	CV-AAS	A15
BI Control metals in urine Lanonorm M2	Hg	1. K$_2$Cr$_2$O$_7$, HNO$_3$, KBr, KBrO$_3$ 2. HCl	Single mode/ flow through	1. Offline 2. 65 sec at 90 W (8.5 mL min^{-1})	CV-AAS	A15
BI Control metals in urine Lanonorm M3	Cd, Pb	HNO$_3$	Multimode/ LP closed	3 min at 650 W	ETV-AAS	A3
BI Control blood for metals OSSD-20/21	Hg	HNO$_3$	Multimode/ HP closed	70 sec at 600 W	CV-AAS	A16
BRL Urine metals level II 39202	Hg	1. K$_2$Cr$_2$O$_7$, HNO$_3$, KBr, KBrO$_3$ 2. HCl	Single mode/ flow through	1. Offline 2. 65 sec at 90 W (8.5 mL min^{-1})	CV-AAS	A15

Appendix 1.1 Animal Tissue and Fluid

Matrix	Analytes	Reagents	Cavity/Vessel	Conditions	Detection	Ref.
IAEA Milk powder A11	I	HNO_3, N_2H_4	Multimode/ HP closed	35 sec at 675 W	NAA	A17
	Se	1. HNO_3 2. H_2SO_4, $HClO_4$ 3. H_2O	Single mode/ open	1. 15 min at 45 W, 10 min at 75 W 2. 10 min at 90 W, 35 min at 120 W 3. 8 min at 120 W	ETV-AAS	A18
IAEA Milk powder A11 (contd.)	Cu, Fe, Ni, Zn	A. 1. HNO_3 2. H_2O_2	Multimode/ LP closed	1. 3 min at 280 W, 2-5 min at 0 W, 3 min at 420 W, 2-5 min at 0 W, 2 min at 560 W 2. 2 min at 260 W	F-AAS ETV-AAS HG-AAS	A19
		B. 1. HNO_3, H_2SO_4 2. HNO_3 3. HNO_3	Single mode/ open	1. 5 min at 80 W 2. 5 min at 60 W, 5 min at 80 W 3. 5 min at 80 W (for fat rich foods)		
		C. HNO_3	Single mode/ open	10 min at 80 W (for lower fat foods)		
IAEA Animal blood A13	Al	HNO_3	Multimode/ HP closed	1 min at 150 W, 30 min at 0 W, 1 min at 450 W	ICP-OES	A20
	Ca, Cu, Fe, K, Ni, P, Pb, Rb, S, Se, Sr, Zn	HNO_3	Multimode/ LP closed	15 min at 540 W, 20 min at 420 W	TXRF	A12
IAEA Whey powder A155	Al	HNO_3	Multimode/ HP closed	1 min at 150 W, 30 min at 0 W, 1 min at 450 W	ICP-OES	A20

Sample	Analytes	Reagents	Mode/Vessel	Conditions	Method	Ref.
IAEA Animal muscle H4	I	HNO$_3$, N$_2$H$_4$	Multimode/HP closed	35 sec at 675 W	NAA	A17
	Se	1. HNO$_3$ 2. H$_2$SO$_4$, HClO$_4$ 3. Water	Single mode/open	1. 15 min at 45 W, 10 min at 75 W 2. 10 min at 90 W, 35 min at 120 W 3. 8 min at 120 W	ETV-AAS	A18
	Al, As, Ba, Ca, Cd, Ce, Cr, Cs, Cu, Fe, La, Li, Mg, Mn, Mo, Ni, Rb, Sb, Se, Sr, Tl, V, Y, Zn	HNO$_3$	Multimode/HP closed	75 sec at 350 W	ICP/MS	A21
	Al, Ca, Cu, Fe, K, Mg, Mn, Na, P, Sr, Zn	HNO$_3$	Multimode/LP closed	10 min at 70 psi 15 min at 450 W	ICP-OES	A22
	Al, Cd, Co, Cu, Fe, Mg, Mn, Mo, Rb, Sr, Zn	HNO$_3$	Multimode/LP closed	2 min at 600 W	ICP/MS	A23
IAEA Horse kidney H8	Al, Cd, Co, Cu, Fe, Mg, Mn, Mo, Rb, Sr, Zn	HNO$_3$	Multimode/LP closed	2 min at 600 W	ICP/MS	A23
	Ca, Cd, Cu, Fe, K, Mg, Mn, Na, P, Sr, Zn, Residual carbon	A. 1. HNO$_3$, H$_2$SO$_4$ 2. H$_2$O$_2$(x16) 3. HNO$_3$ or HNO$_3$, HCl or NH$_3$ (aq) 4. NH$_4$EDTA B. HNO$_3$	Single mode/open Multimode/MP closed	1. 4 min at 30 W, 4 min at 120 W 2. 1 min at 120 W(x16) 3. 4 min at 30 W 2 min at 20 psi, 5 min at 40 psi, 2 min at 60 psi, 2 min at 80 psi, 2 min at 100 psi, 2 min at 120 psi, 2 min at 140 psi, 15 min at 160 psi	ICP-OES	A24

Appendix 1.1 Animal Tissue and Fluid

Matrix	Analytes	Reagents	Cavity/Vessel	Conditions	Detection	Ref.
IAEA Horse kidney H8 (contd.)	Ca, Cd, Fe, Mg, Zn	A. HNO_3	Multimode/stopped flow	2 min at 300 W, 1 min at 0 W, 3 min at 150 W	F-AAS ICP-OES	A25
		B. HNO_3	Multimode/LP closed	30 min at 420 W		
KL Heavy metal urine control 0140	Hg	HNO_3	Multimode/HP closed	70 sec at 600 W	CV-AAS	A16
ISS Green algae MMM-2	Cr, Co, Mn, Ni, Pb	HNO_3, H_2O_2	Multimode/LP closed	6 min at 700 W	F-AAS ICP-OES	A26
NIES Human hair 5	Ca, Cu, Fe, Mg, Mn, Zn	HNO_3, $HClO_4$, HCl, HF	Multimode/MP closed	14 min intermittent 200 W	F-AAS	A27
	Cd, Co, Cu, Ni, Pb	HNO_3, HCl, $HClO_4$, HF	Multimode/MP closed	5 min at 200 W (water load), 3.5 min at 200 W (no load)	ICP-OES ETV-ICP	A28
	Cu, Pb, Residual carbon	A. HNO_3	Multimode/LP closed	0.5-5 min at 550 W (with water ballast)	ETV-AAS DP-ASV Coulometry	A29
		B. HNO_3, H_2O_2	Multimode/LP closed	0.5-5 min at 550 W (with water ballast)		
		C. HNO_3, H_2O_2	Multimode/MP closed	15-60 sec at 550 W (with water ballast)		
		D. 1. HNO_3 or H_2O_2 2. HNO_3 vapor	Multimode/MP closed	10 min at 550 W (with water ballast)		
	Cr, Hg, Se	1. HNO_3 2. H_2O_2	Multimode/HP closed	1.3 min at 450 W	Radiochemical	A30

Sample	Elements	Reagents	Mode/Vessel	Conditions	Technique	Ref.
NIST Human serum 909	Cr, Cu, Fe, K, Li, Mg, Mn, Mo, Na, P, Pb, S, Sr, Zn	1. HNO_3 2. $HClO_4$	Multimode/ MP closed	1. 1 min at 90 W, 1 min at 0 W (x5)	ICP-OES	A31
	Ni	HNO_3, HCl	Multimode/ HP closed	4 min at 350 W, 10 min at 140 W	ETV-AAS	A9
	Ca, Cr, Cu, Fe, K, Mn, Ni, P, Pb, Rb, S, Se, Sr, Zn	HNO_3	Multimode/ LP closed	15 min at 540 W, 20 min at 420 W	TXRF	A12
NIST Albumin 926	No analysis	HNO_3	Multimode/ LP closed	Varied	Temperature Pressure	A32
NIST Milk powder 1549	I	HNO_3, N_2H_4	Multimode/ HP closed	35 sec at 675 W	NAA	A17
	Ca, Mg	1. HNO_3, H_2SO_4, H_2O_2, H_2O 2. H_2O_2 3. HNO_3, H_2O 4. HNO_3, H_2O	Single mode/ stopped flow	1. 3 min at 20 W (condenser on), 2 min at 56 W (condenser on), 10 min at 80 W (condenser off) 2. 3 min at 116 W (condenser on) 3. 2 min at 80 W (condenser on) 4. 2 min at 80 W (condenser on)	ICP-OES	A33
	Cr, Cu, Ni, Pb, Sn, Zn	HNO_3, H_2O_2	Multimode/ LP closed	3 min at ? W	ETV-AAS F-AAS	A34
	Al, Ba, Ca, K, Mg, Mn, Na, P, S, Si, Zn	A. 1. HNO_3 2. H_2O_2 3. H_2O_2 B. 1. HNO_3 2. H_2O_2, HF 3. H_3BO_3	Multimode/ open (with reflux top)	1. 5 min at 300 W, 600 W until 1 mL remains 2. 30-40 min at 600 W until 1 mL remains 1. 5 min at 300 W, 20-30 min at 600 W reflux not on 2. 30 min at 180 W reflux on 3. 180 W reflux on	FI-ICP-OES	A35

Appendix 1.1 Animal Tissue and Fluid

Matrix	Analytes	Reagents	Cavity/Vessel	Conditions	Detection	Ref.
NIST Milk powder 1549 (contd.)	Ca, Cd, Cu, Fe, K, Mg, Mn, Na, P, Sr, Zn, Residual carbon	**A.** 1. HNO_3, H_2SO_4 2. H_2O_2(x16) 3. HNO_3 or HNO_3, HCl or NH_3 (aq) 4. NH_4EDTA **B.** HNO_3	Single mode/ open Multimode/ MP closed	1. 4 min at 30 W, 4 min at 120 W 2. 1 min at 120 W(x16) 3. 4 min at 30 W 2 min at 20 psi, 5 min at 40 psi, 2 min at 60 psi, 2 min at 80 psi, 2 min at 100 psi, 2 min at 120 psi, 2 min at 140 psi, 15 min at 160 psi	ICP-OES	A24
	Zn, Residual carbon	**A.** HNO_3, H_2SO_4 **B.** HNO_3 or HNO_3, H_2SO_4 or HNO_3, H_2O_2 or HNO_3, H_2SO_4, H_2O_2 **C.** HNO_3, H_2O_2, H_2SO_4	Multimode/ LP closed Multimode/ MP closed Single mode/ open	2 min at 130 W, 4 min at 260 W, 2 min at 390 W, 2 min at 520 W, 12, 24 or 72 min at 650 W 2 min at 20 psi, 5 min at 40 psi, 2 min at 60 psi, 2 min at 80 psi, 2 min at 100 psi, 2 min at 120 psi, 2 min at 140 psi, 15 min at 160 psi 4 min at 30 W, 6 9 or 16 min at 120 W, 10 20 or 30 min at 205 W	ICP-OES	A36, A37

	Elements	Reagents	Mode	Conditions	Technique	Ref.
NIST Brewers yeast 1569	Al, Ba, Ca, K, Mg, Mn, Na, P, S, Si, Zn	A. 1. HNO$_3$ 2. H$_2$O$_2$ 3. H$_2$O$_2$ B. 1. HNO$_3$ 2. H$_2$O$_2$, HF 3. H$_3$BO$_3$	Multimode/ open Multimode/ open (with reflux top)	1. 5 min at 300 W, 600 W until 1 mL remains 2. 30-40 min at 600 W until 1 mL remains 1. 5 min at 300 W, 20-30 min at 600 W reflux not on 2. 30 min at 600 W reflux on 3. 180 W reflux on	FI-ICP-OES	A35
NIST Bovine liver 1577	Zn, Cd	HNO$_3$	Multimode/ open	8 min at 200 W	FI-F-AAS	A38
	Ca, Fe, Mg, Zn	HNO$_3$	Multimode/ flow through	100 sec at 525 W (10 mL coil, 6 mL min^{-1})	F-AAS	A39
	Ag, Al, As, Ba, Ca, Cd, Co, Cr, Cu, Fe, Hg, K, Li, Mg, Mn, Mo, Na, P, Pb, S, Sb, Se, Si, Sr, Ti, Tl, V, Zn	1. HNO$_3$ 2. HClO$_4$	Multimode/ MP closed	1. 1 min at 90 W, 1 min at 0 W (x5)	ICP-OES	A31
	Pb	HNO$_3$, HCl	Multimode/ flow through	18 sec at 700 W (0.4 mL coil, 1.5 mL min^{-1})	ETV-AAS	A8
	Se	1. HNO$_3$ 2. H$_2$SO$_4$, HClO$_4$ 3. H$_2$O	Single mode/ open	1. 15 min at 45 W, 10 min at 75 W 2. 10 min at 90 W, 35 min at 120 W 3. 8 min at 120 W	ETV-AAS	A18
	I	1. HNO$_3$ 2. N$_2$H$_4$	Multimode/ HP closed	1. 35 sec at 675 W	NAA	A17
	Cd	HNO$_3$	Multimode/ LP closed	2 min on defrost three times	ETV-AAS	A40

Appendix 1.1 Animal Tissue and Fluid

Matrix	Analytes	Reagents	Cavity/Vessel	Conditions	Detection	Ref.
NIST Bovine liver 1577 (contd.)	Hg	HNO_3	Multimode/ LP closed	8 min at 140 W, 8 min at 280 W, 4 min at 420 W	CV-AAS	A2
	Na	1. HNO_3 2. HNO_3, H_2SO_4 3. H_2O_2	Multimode/ LP closed	1. 1 min at 630 W, 2. 5 min at 273 W, 20 min at 189 W (2-5 samples) or 4 min at 630 W, 8 min at 378 W (6-12 samples)	F-AAS	A41
	Ba, Ca, K, Mg, Mn, Na, P, S, Zn	1. HNO_3 2. H_2O_2	Multimode/ open	1. 30 min at 540 W 2. 30 min at 540 W	ICP-OES	A42
	Ca, Cu, Fe, Mg, Mn, Zn	HNO_3, $HClO_4$, HCl, HF	Multimode/ MP closed	14 min at 200 W	F-AAS	A27
	Residual amino acids	HNO_3	Multimode/ LP closed	Varied	Fluorimetry	A43
	—	HNO_3	Multimode/ LP closed	Varied	Temperature Pressure	A32
	Al, Ba, Ca, K, Mg, Mn, Na, P, S, Si, Zn	A. 1. HNO_3 2. H_2O_2 3. H_2O_2 B. 1. HNO_3 2. H_2O_2, HF 3. H_3BO_3	Multimode/ open Multimode/ open (with reflux top)	1. 5 min at 300 W, 600 W until 1 mL remains 2. 30-40 min at 600 W until 1 mL remains 1. 5 min at 300 W, 20-30 min at 600 W reflux not on 2. 30 min at 180 W reflux on 3. 180 W reflux on	FI-ICP-OES	A35

Analytes	Reagents	Mode/vessel	Conditions	Detection	Ref.
Se	HNO_3	Multimode/LP closed	(1.5 h at 60°C on hotplate) 4 min at 287 W	ETV-AAS	A44
Residual organic species, Cu, Zn	HNO_3	Multimode/LP closed	5 min at 132 W, 5 min at 207 W (additional 3 min at 230 W if temp not at 180°C for 100 sec)	Voltammetry LC UV-Vis	A45
As, Co, Cu, Se, Zn	HNO_3, $HClO_4$	Multimode/open	600 W until first signs of $HClO_4$ fumes	F-AAS NAA	A46-A48
Ca, Cd, Cu, Fe, K, Mg, Mn, Na, Zn	HNO_3	Multimode/HP closed	30 sec at 665 W	F-OES F-AAS ICP-OES	A49
Cd, Cr, Cu, Fe, Pb, Zn	HNO_3, $H2SO_4$	Multimode/LP closed	60 sec at 700 W	F-AAS	A50
Cd, Cr, Cu, Fe, Pb, Zn	HNO_3, H_2SO_4	Multimode/open	10 sec at 70 W, 180 sec at 0 W (x6)	F-AAS	A51
Al, As, Ba, Ca, Cd, Ce, Cr, Cs, Cu, Fe, La, Li, Mg, Mn, Mo, Ni, Rb, Sb, Se, Sr, Tl, V, Y, Zn	HNO_3	Multimode/HP closed	75 sec at 350 W	ICP/MS	A21
B, Ca, Fe, K, Mg, Mn, Na, Ni, P	HNO_3, $HClO_4$	Multimode/open	20 min at 600 W	ICP-OES	A52
Fe	HNO_3, H_2O_2	Multimode/LP closed	15 min at 150 W	F-AAS	A53
Cu, Fe, Mn, Zn	HNO_3	Multimode/LP closed	8 min at 150 W	F-AAS ETV-AAS SIMAAC	A54
Ca, Cu, Fe, K, Mg, Mn, Na, P, S, Zn	HNO_3	Multimode/LP closed	10 min at 70 psi, 15 min at 450 W	NAA ICP-OES	A22
Al, Ba, Ca, Cu, Fe, K, Mg, Mn, Na, P, Zn	1. Aqua regia 2. HBO_3	Multimode/MP closed (evacuated slightly at start)	1. 3 min at 625 W	ICP-OES	A55
As, Be, Ca, Cd, Co, Cr, Cu, Fe, K, Mg, Mn, Na, Ni, Pb, Sb, Se, V, Zn	Aqua regia	Multimode/LP closed	10 min at 300 W, 5 min at 600 W, 10 min at 480 W	ICP-OES ETV-AAS	A56

Appendix 1.1 Animal Tissue and Fluid

Matrix	Analytes	Reagents	Cavity/Vessel	Conditions	Detection	Ref.
NIST Bovine liver 1577 (contd.)	Cu, Mn, Zn	1. HNO₃ 2. HClO₄	Multimode/ LP closed	Varied	SPC-IC	A57
	Al, Ba, Ca, Cd, Cu, Fe, Mg, Pb, Zn	A. Aqua regia	Multimode/ stopped flow	2 min at 720 W	ICP-OES	A58
		B. Aqua regia	Multimode/ open	2 min at 720 W, 2 min at 0 W, 2 min at 720 W, 2 min at 0 W, 2 min at 720 W (with water load)		
	Al, Cd, Co, Cu, Fe, Mg, Mn, Mo, Rb, Sr, Zn	HNO₃	Multimode/ LP closed	2 min at 600 W	ICP/MS	A23
	Cu, Fe	HNO₃	Multimode/ LP closed	3 min at 650 W	ETV-AAS	A3
	Residual carbon	HNO₃	Multimode/ HP closed	5 min at 500 W (200 bar)	Total carbon analyzer Carbon dioxide coulometer	A59
	As	1. HNO₃, H₂SO₄ 2. H₂O₂ 3. H₂O₂	Multimode/ LP closed	1. 4 min at 100 W, cool to RT, 6 min at 325 W, cool to RT 2. 6 min at 325 W, cool to RT, 6 min at 650 W, cool to RT 3. 1 h at 90°C (waterbath, open)	HG-AAS	A60
	Residual carbon	HNO₃	Multimode/ HP closed	4 min at 90 W	Total carbon analyzer Carbon dioxide coulometer	A61

Analytes	Reagents	Mode	Conditions	Technique	Reference
Ca, Cd, Cu, Fe, K, Mg, Mn, Na, P, Sr, Zn, Residual carbon	A. 1. HNO_3, H_2SO_4 2. H_2O_2(x16) 3. HNO_3 or HNO_3, HCl or NH_3 (aq) 4. NH_4EDTA B. HNO_3	Single mode/ open	1. 4 min at 30 W, 4 min at 120 W, 2. 1 min at 120 W(x16), 3. 4 min at 30 W	ICP-OES	A24
		Multimode/ MP closed	2 min at 20 psi, 5 min at 40 psi, 2 min at 60 psi, 2 min at 80 psi, 2 min at 100 psi, 2 min at 120 psi, 2 min at 140 psi, 15 min at 160 psi		
Ag, Al, As, Cd, Co, Cu, Fe, Hg, Mn, Mo, Pb, Rb, Sb, Se, Sr, Tl, V, Zn	HNO_3	Multimode/ HP closed	2 min at 300 W	ICP/MS	A62
Al	HNO_3	Multimode/ LP closed	40 sec at 700 W, 10 min at 0 W (vent), 60 sec at 700 W, 10 min at 0 W (vent), 90 sec at 700 W, 5 min at 70 W	ETV-AAS	A63
Cd, Cu, Fe, Mn, Pb	1. HNO_3, H_2O, H_2O_2 2. Aqua regia, HF (optional)	Multimode/ open	1. 15 min at 70 W, 15 min at 150 W, 15 min at 180 W, 1 min at 250 W, 1 min at 600 W 2. 15 min at 70 W, 1 min at 600 W	ETV-AAS	A5

Appendix 1.1 Animal Tissue and Fluid

Matrix	Analytes	Reagents	Cavity/Vessel	Conditions	Detection	Ref.
NIST Bovine liver 1577 (contd.)	Cd	HNO$_3$, HCl	Multimode/ HP closed	4 min at 360 W, 10 min at 180 W	ETV-AAS	A4
	Se	HNO$_3$, HClO$_4$	Multimode/ open	Not reported	Fluorimetry	A65
	Fe, Mn, Zn	HNO$_3$	Multimode/ flow through	2 min at ? W	F-AAS ICP-OES ICP/MS	A66
	Ca, Cu, Fe, Mg, Mn, K, Na, Zn	1. HNO$_3$ 2. HNO$_3$ 3. H$_2$O$_2$	Single mode/ open	1. 10 min at 60 W, 5 min at 80 W 2. 5 min at 70 W 3. 3 min at 60 W	F-AAS	A67
	Se	HNO$_3$	Multimode/ open	2 min on MedHi, 2 min at 0 W (x3)	ETV-AAS	A68
NIST Bovine serum 1598	Al, As, Cd, Cr, Cu, Mn, Mo, Sb, V	HNO$_3$	Multimode/ LP closed	8 min at 164 W	NAA	A69
NIST Freeze dried urine 2670	Residual amino acids	HNO$_3$	Multimode/ LP closed	Varied	Fluorimetry	A43
	Hg	1. K$_2$Cr$_2$O$_7$, HNO$_3$, Kbr, KBrO$_3$ 2. HCl	Single mode/ flow through	1. Offline 2. 65 sec at 90 W (8.5 mL min^{-1})	CV-AAS	A15
	B	HNO$_3$, H$_2$O$_2$	Multimode/ HP closed	2 min at 100 W, 2 min at 0 W, 5 min at 250 W, 3 min at 0 W, 5 min at 500 W, 15 min at 250 W	UV-Vis Fluorimetry ICP-OES ICP/MS	A64

Analyte	Reagents	Mode	Conditions	Technique	Ref.
Ni	HNO_3, HCl	Multimode/HP closed	4 min at 350 W, 10 min at 140 W	ETV-AAS	A9
Cd	HNO_3, HCl	Multimode/HP closed	4 min at 360 W, 10 min at 180 W	ETV-AAS	A4
Ca, Cd, Cu, K, Mg, Mn, Na, Residual carbon	A. 1. HNO_3, H_2SO_4 2. H_2O_2 (x16) 3. HNO_3 or HNO_3, HCl or NH_3 (aq) 4. NH_4EDTA B. HNO_3	Single mode/open Multimode/MP closed	1. 4 min at 30 W, 4 min at 120 W 2. 1 min at 120 W (x16) 3. 4 min at 30 W 2 min at 20 psi, 5 min at 40 psi, 2 min at 60 psi, 2 min at 80 psi, 2 min at 100 psi, 2 min at 120 psi, 2 min at 140 psi, 15 min at 160 psi	ICP-OES	A24
Ca, Mg	1. HNO_3, H_2SO_4, H_2O_2, H_2O 2. H_2O_2 3. HNO_3, H_2O 4. HNO_3, H_2O	Single mode/stopped flow	1. 3 min at 20 W (condenser on), 2 min at 56 W (condenser on), 10 min at 80 W (condenser off) 2. 3 min at 116 W (condenser on) 3. 2 min at 80 W (condenser on) 4. 2 min at 80 W (condenser on)	ICP-OES	A33
Ni	HNO_3, HCl	Multimode/HP closed	4 min at 360 W, 10 min at 180 W	ICP-OES	A11
NIST Bovine serum 8419	Ca, Cu, Fe, K, Mg, Na	HNO_3	Multimode/HP closed	30 sec at 665 W	F-OES F-AAS ICP-OES	A49
SERONORM Trace elements in blood B115	Hg	1. $K_2Cr_2O_7$, HNO_3, KBr, $KBrO_3$ 2. HCl	Single mode/flow through	1. Offline 2. 65 sec at 90 W (8.5 mL min^{-1})	CV-AAS	A15

Appendix 1.1 Animal Tissue and Fluid

Matrix	Analytes	Reagents	Cavity/Vessel	Conditions	Detection	Ref.
SERONORM Trace elements in urine 116	As	1. HNO_3, H_2SO_4 2. H_2O_2 3. H_2O_2	Multimode/ LP closed	1. 4 min at 100 W, cool to RT, 6 min at 325 W, cool to RT 2. 6 min at 325 W, cool to RT, 6 min at 650 W, cool to RT 3. 1 h at 90°C (waterbath, open)	HG-AAS	A60
SERONORM Trace elements in blood #904	Pb	HNO_3, HCl	Multimode/ flow through	18 sec at 700 W (0.4 mL coil, 1.5 mL min^{-1})	ETV-AAS	A8
SERONORM Trace elements in blood #905	Hg(II), methylmercury, mercurescein, phenylmercury	1. HCl, KBr, $KBrO_3$ 2. $KMnO_4$ 3. $NaBH_4$, NaOH	Single mode/ flow through	1. 25.4 sec at 70 W (14-16 mL min^{-1})	CV-AAS	A14
SERONORM Trace elements in blood #906	Hg(II), methylmercury, mercurescein, phenylmercury	1. HCl, KBr, $KBrO_3$ 2. $KMnO_4$ 3. $NaBH_4$, NaOH	Single mode/ flow through	1. 25.4 sec at 70 W (14-16 mL min^{-1})	CV-AAS	A14
SERONORM Trace elements in urine # 009024	Bi, Hg	1. HCl, KBr, $KBrO_3$ 2. $NaBH_4$, NaOH	Single mode/ flow through	1. 65 sec at 90-120 W (8.5 mL min^{-1})	CV-AAS HG-AAS	A70, A71
SERONORM Trace elements in blood #010010	Hg(II), methylmercury, mercurescein, phenylmercury	1. HCl, KBr, $KBrO_3$ 2. $KMnO_4$ 3. $NaBH_4$, NaOH	Single mode/ flow through	1. 25.4 sec at 70 W (14-16 mL min^{-1})	CV-AAS	A14
SERONORM Trace elements in blood #010011	Hg(II), methylmercury, mercurescein, phenylmercury	1. HCl, KBr, $KBrO_3$ 2. $KMnO_4$ 3. $NaBH_4$, NaOH	Single mode/ flow through	1. 25.4 sec at 70 W (14-16 mL min^{-1})	CV-AAS	A14

SERONORM Trace elements in blood #010012	Hg(II), methylmercury, mercurescein, phenylmercury	1. HCl, KBr, KBrO$_3$ 2. KMnO$_4$ 3. NaBH$_4$, NaOH	Single mode/ flow through	1. 25.4 sec at 70 W (14-16 mL min^{-1})	CV-AAS	A14
SERONORM Trace elements in blood #205052	Hg(II), methylmercury, mercurescein, phenylmercury	1. HCl, KBr, KBrO$_3$ 2. KMnO$_4$ 3. NaBH$_4$, NaOH	Single mode/ flow through	1. 25.4 sec at 70 W (14-16 mL min^{-1})	CV-AAS	A14
SERONORM Trace elements in blood #203056	Hg(II), methylmercury, mercurescein, phenylmercury	1. HCl, KBr, KBrO$_3$ 2. KMnO$_4$ 3. NaBH$_4$, NaOH	Single mode/ flow through	1. 25.4 sec at 70 W (14-16 mL min^{-1})	CV-AAS	A14

Appendix 1.2 Botanical

Matrix	Analytes	Reagents	Cavity/Vessel	Conditions	Detection	Ref.
BCR Aquatic plant 60	Ca, Cd, Cu, Fe, K, Mg, Mn, Na, Pb, Zn	HNO$_3$, HCl	Multimode/ LP closed	15 min at 300 W, 5 min at 0 W, 20 min at 510 W	F-AAS ETV-AAS F-OES	A72
	Ni	HNO$_3$, HCl	Multimode/ HP closed	4 min at 350 W, 10 min at 140 W	ETV-AAS	A9
	Fe	A. HNO$_3$	Multimode/ HP closed	30 sec at 700 W	ETV-AAS	A6
		B. 1. HNO$_3$ 2. HF	Multimode/ HP closed	Same as A		
	Cd, Cu, Pb	1. HNO$_3$, H$_2$O, H$_2$O$_2$ (optional) 2. Aqua regia, HF	Multimode/ open	1. 15 min at 70 W, 15 min at 150 W, 15 min at 180 W, 1 min at 250 W, 1 min at 600 W 2. 15 min at 70 W, 1 min at 600 W	ETV-AAS	A5

Appendix 1.2 Botanical

Matrix	Analytes	Reagents	Cavity/Vessel	Conditions	Detection	Ref.
BCR Aquatic plant 60 (contd.)	Cd	HNO₃, HCl	Multimode/HP closed	4 min at 360 W, 10 min at 180 W	ETV-AAS	A4
	Cu, Cd, Mn, Pb, Zn	HNO₃	Multimode/HP closed	30 sec at 665 W	F-OES F-AAS ICP-OES	A49
	Ni	HNO₃, HCl	Multimode/HP closed	4 min at 360 W, 10 min at 180 W	ICP-OES	A10, A11
BCR Aquatic moss 61	Cu, Cd, Mn, Pb, Zn	HNO₃	Multimode/HP closed	30 sec at 665 W	F-OES F-AAS ICP-OES	A49
	Fe	A. HNO₃	Multimode/HP closed	30 sec at 700 W	ETV-AAS	A6
		B. 1. HNO₃ 2. HF	Multimode/HP closed	Same as A		
	Cd	HNO₃, HCl	Multimode/HP closed	4 min at 360 W, 10 min at 180 W	ETV-AAS	A4
	Ni	HNO₃, HCl	Multimode/HP closed	4 min at 350 W, 10 min at 140 W	ETV-AAS	A9
	Ni	HNO₃, HCl	Multimode/HP closed	4 min at 360 W, 10 min at 180 W	ICP-OES	A10, A11
BCR Olive blossom 62	Cu, Cd, Mn, Pb, Zn	HNO₃	Multimode/HP closed	30 sec at 665 W	F-OES F-AAS ICP-OES	A49
	Ni	HNO₃, HCl	Multimode/HP closed	4 min at 350 W, 10 min at 140 W	ETV-AAS	A9
	Ni	HNO₃, HCl	Multimode/HP closed	4 min at 360 W, 10 min at 180 W	ICP-OES	A10, A11

Sample	Elements	Reagents	Mode	Power/time program	Detection	Ref.
	Ca, Cd, Cu, Fe, K, Mg, Mn, Na, Pb, Zn	HNO_3, HCl	Multimode/LP closed	15 min at 300 W, 5 min at 0 W, 20 min at 510 W	F-AAS, ETV-AAS, F-OES	A72
	Cd, Cu, Mn, Pb	1. HNO_3, H_2O, H_2O_2 2. Aqua regia, HF (optional)	Multimode/open	1. 15 min at 70 W, 15 min at 150 W, 15 min at 180 W, 1 min at 250 W, 1 min at 600 W 2. 15 min at 70 W, 1 min at 600 W	ETV-AAS	A5
	Pb	HNO_3, HCl	Multimode/flow through	18 sec at 700 W (0.4 mL coil, 1.5 mL min^{-1})	ETV-AAS	A8
	Cu, Mg, Pb	HNO_3	Multimode/LP closed	3 min at 650 W	ETV-AAS	A3
BCR Spruce needles 101	Cd, Cu, Pb	HNO_3, $HClO_4$, H_2SO_4	Multimode/HP closed.	10 min at 85 bar	DP-ASV	A1
	Al, Ca, Mg, Mn, P, Zn	1. HNO_3 2. HNO_3 3. H_2O_2 4. H_2O	Single mode/open	1. 5 min at 10 W, 10 min at 30 W, 10 min at 60 W 2. 10 min 60 W 3. 5 min at 60 W 4. 5 min at 50 W	ICP-OES ICP-MS HPLC-ICP-MS	A13
BCR Wholemeal flour 189	Cd, Cu, Fe, Mn, Zn	A. HNO_3, H_2O_2	Multimode/HP closed	75 sec at 665 W	ETV-AAS	A6
		B. HNO_3, H_2O_2	Multimode/HP closed	1 min at 250 W, 2 min at 0 W, 2 min at 250 W, 2 min at 400 W, 2 min at 600 W		

Appendix 1.2 Botanical

Matrix	Analytes	Reagents	Cavity/Vessel	Conditions	Detection	Ref.
BCR Wholemeal flour 189 (contd.)	Cd, Cu, Fe, Mn, Zn	C. HNO_3, $HClO_4$	Multimode/ HP closed	150 sec at 950 W, 60 sec at 0 W, 90 sec at 300 W, 90 sec at 500 W, 90 sec at 700 W, 90 sec at 850 W		
BCR Ulva lactuca 279	Ca, Cd, Cu, Fe, K, Mg, Mn, Na, Pb, Zn	HNO_3	Multimode/ LP closed	5 min at 300 W, 5 min at 0 W, 5 min at 300 W, 5 min at 450 W, 5 min at 0 W, 5 min at 450 W	F-AAS ETV-AAS F-OES	A72
BCR White clover 402	Co, Mo	A. 1. HNO_3 2. HNO_3 3. H_2O_2 4. H_2O B. 1. HNO_3, HCl 2. HNO_3, HCl 3. H_2O_2 4. H_2O	Single mode/ open Single mode/ open	1. 5 min at 10 W, 10 min at 30 W, 10 min at 60 W 2. 10 min 60 W 3. 5 min at 60 W 4. 5 min at 50 W 1. 5 min at 40 W, 10 min at 50 W 2. 20 min 54 W 3. 5 min at 40 W 4. 5 min at 50 W	ICP-OES ICP-MS HPLC-ICP-MS	A13
IAEA Hay powder V10	Mo	1. HNO_3 2. Perhydrol	Multimode/ open then Multimode/ LP closed	1. 30 min at 180W (open), 10 min at 420W, 10 min at 600W 2. 10 min at 600W	MS	A73
	I	HNO_3, N_2H_4	Multimode/ HP closed	35 sec at 675 W	NAA	A17

Material	Elements	Reagent	Mode/system	Power/time program	Detection	Ref.
	Ca, Cd, Cu, Fe, K, Mg, Mn, Na, Pb, Zn	HNO₃, HCl	Multimode/ LP closed	15 min at 300 W, 5 min at 0 W, 20 min at 510 W	F-AAS ETV-AAS F-OES	A72
	Hg	1. HNO₃ 2. H₂O₂	Multimode/ HP closed	3 min at 450 W	Radiochemical	A30
	Ag, Ba, Bi, Cd, Co, Cr, Cs, Cu, Ga, Hg, Li, Mo, Ni, Pb, Rb, Sn, Sr, Th, U, Zn, Zr	HNO₃	Multimode/ HP closed	1 min at 300 W (x3-5), 7 min at 300 W, 2 min at 600 W	ICP/MS	A74
MOE Vegetation V85-1	Al, Ba, Ca, Cu, Fe, Mg, Mn, Zn	Aqua regia	Multimode/ stopped flow	2 min at 720 W or 32 min at 720 W	ICP-OES	A58
MOE Norway maple	Al, Ba, Ca, Cu, Fe, Mg, Mn, Pb, Zn	A. Aqua regia	A. Multimode/ stopped flow	2 min at 720 W	ICP-OES	A58
		B. Aqua regia	B. Multimode/ open	2 min at 720 W, 2 min at 0 W, 2 min at 720 W, 2 min at 0 W, 2 min at 720 W (with water load)		
MOE White birch	Al, Ba, Ca, Cd, Cu, Fe, Mg, Mn, Zn	A. Aqua regia	A. Multimode/ stopped flow	2 min at 720 W	ICP-OES	A58
		B. Aqua regia	B. Multimode/ open	2 min at 720 W, 2 min at 0 W, 2 min at 720 W, 2 min at 0 W, 2 min at 720 W (with water load)		
NIES Pepperbush	Ca, Fe, Mg, Zn	HNO₃	Multimode/ flow through	100 sec at 525 W (10 mL coil, 6 mL min⁻¹)	F-AAS	A39
1	Ca, Cu, Fe, Mg, Mn, Zn	HNO₃, HClO₄, HCl, HF	Multimode/ MP closed	14 min intermittent 200 W	F-AAS	A27

Appendix 1.2 Botanical

Matrix	Analytes	Reagents	Cavity/Vessel	Conditions	Detection	Ref.
NIES Pepperbush 1 (contd.)	Co	HNO$_3$, HCl	Multimode/ MP closed	5 min at 600 W (50 mL water load), 5 min at 600 W (no load)	UV-Vis	A75
	Cd, Co, Cu, Ni, Pb	HNO$_3$, HCl, HClO$_4$, HF	Multimode/ HP closed	5 min at 200 W (water load), 3.5 min at 200 W (no load)	ICP-OES ETV-ICP	A28
	Cd, Cu, Fe, Mn, Zn	A. HNO$_3$, H$_2$O$_2$	Multimode/ HP closed	75 sec at 665 W	ETV-AAS	A6
		B. HNO$_3$, H$_2$O$_2$	Multimode/ HP closed	1 min at 250 W, 2 min at 0 W, 2 min at 250 W, 2 min at 400 W, 2 min at 600 W		
		C. HNO$_3$, HClO$_4$	Multimode/ HP closed	150 sec at 950 W, 60 sec at 0 W, 90 sec at 300 W, 90 sec at 500 W, 90 sec at 700 W, 90 sec at 850 W		
	Fe, Mn, Zn	HNO$_3$	Multimode/ flow through	-	F-AAS ICP-OES ICP/MS	A66
	Cd	1. HNO$_3$, HClO$_4$, HCl, HF 2. HClO$_4$	Multimode/ MP closed	1. 5 min at 200 W (using water load), 4 min at 200 W (no load)	F-AAS	A76
NIES Tea leaves 7	Ca, Cu, Fe, Mg, Mn, Zn	HNO$_3$, HClO$_4$, HCl, HF	Multimode/ MP closed	14 min intermittent 200 W	F-AAS	A27

Sample	Elements	Reagents	Mode	Power/Time	Detection	Ref.
Co	HNO₃, HCl		Multimode/ LP closed	5 min at 600 W (50 mL water load), 5 min at 600 W (no load)	UV-Vis	A75
	Cd, Co, Cu, Ni, Pb	HNO₃, HCl, HClO₄, HF	Multimode/ MP closed	5 min at 200 W (water load), 3.5 min at 200 W (no load)	ICP-OES ETV-ICP	A28
NIES Sargasso 9	Ca, Fe, Mg, Zn	HNO₃	Multimode/ flow through	100 sec at 525 W (10 mL coil, 6 mL min⁻¹)	F-AAS	A39
	Fe, Mn, Zn	HNO₃	Multimode/ flow through	-	F-AAS ICP-OES ICP/MS	A66
	As, Cr, V	HNO₃, H₂O₂	Multimode/ LP closed	3 min at 300 W	ICP/MS	A77
	Ca, Cu, Fe, Mg, Mn, Zn	HNO₃, HClO₄, HCl, HF	Multimode/ MP closed	14 min intermittent 200 W	F-AAS	A27
NIES Rice flour 10	Mo	1. HNO₃ 2. Perhydrol	Multimode/ open *then* Multimode/ LP closed	1. 30 min at 180W (open), 10 min at 420W, 10 min at 600W 2. 10 min at 600W	MS	A73
	Cd	1. HNO₃, HClO₄, HCl, HF 2. HClO₄	Multimode/ MP closed	1. 5 min at 200 W (using load), 4 min at 200 W (no load)	F-AAS	A76
NIST Apple leaves 1515	Ba, Ce, Co, Cr, Eu, Fe, Hf, K, La, Na, Sb, Sc, Sm, Sr, Tb, Zn	1. HNO₃, HF 2. HClO₄	Multimode/ LP closed	1. 20 min at 373 W (9 vessels) 2. 20 min at 545 W (6 vessels)	NAA	A78
NIST Peach leaves 1547	Ba, Ce, Co, Cr, Eu, Fe, Hf, K, La, Na, Sb, Sc, Sm, Sr, Tb, Zn	1. HNO₃, HF 2. HClO₄	Multimode/ LP closed	1. 20 min at 373 W (9 vessels) 2. 20 min at 545 W (6 vessels)	NAA	A78
NIST Wheat flour 1567	Mo	1. HNO₃ 2. Perhydrol	Multimode/ open *then* Multimode/ LP closed	1. 30 min at 180 W (open), 10 min at 420 W, 10 min at 600 W 2. 10 min at 600W	MS	A73

Appendix 1.2 Botanical

Matrix	Analytes	Reagents	Cavity/Vessel	Conditions	Detection	Ref.
NIST Wheat flour 1567 (contd.)	Se	1. HNO$_3$ 2. H$_2$SO$_4$, HClO$_4$ 3. H$_2$O	Single mode/open	1. 15 min at 45 W, 10 min at 75 W, 2. 10 min at 90 W, 35 min at 120 W 3. 8 min at 120 W	ETV-AAS	A18
	Cd	HNO$_3$	Multimode/LP closed	2 min on defrost three times	ETV-AAS	A40
	Hg	HNO$_3$	Multimode/LP closed	8 min at 140 W, 8 min at 280 W, 4 min at 420 W	CV-AAS	A2
	Ba, Ca, K, Mg, Mn, Na, P, S, Zn	1. HNO$_3$ 2. H$_2$O$_2$	Multimode/open	1. 30 min at 540 W 2. 30 min at 540 W	ICP-OES	A42
	Residual amino acids	HNO$_3$	Multimode/LP closed	Varied	Fluorimetric	A43
	Al, Ba, Ca, K, Mg, Mn, Na, P, S, Si, Zn	A. 1. HNO$_3$ 2. H$_2$O$_2$ 3. H$_2$O$_2$ B. 1. HNO$_3$ 2. H$_2$O$_2$, HF 3. H$_3$BO$_3$	Multimode/open Multimode/open (with reflux top)	1. 5 min at 300 W, 600 W until 1 mL remains 2. 30-40 min at 600 W until 1 mL remains 1. 5 min at 300 W, 20-30 min at 600 W reflux not on 2. 30 min at 180 W reflux on 3. 180 W reflux on	FI-ICP-OES	A35
	Se	HNO$_3$	Multimode/LP closed	(1.5 h at 60°C on hotplate) 4 min at 287 W	ETV-AAS	A44
	Ca, Cu, Fe, K, Mn, Zn	HNO$_3$	Multimode/HP closed	30 sec at 665 W	F-OES F-AAS ICP-OES	A49

As, Ca, Cu, Fe, Mg, Mn, Mo, P, S, Se, U, V, Zn	HNO_3, H_2O_2	Multimode/MP closed (Max 350 psi - power cutoff at 150 psi)	4 min at 296 W, 8 min at 360 W	F-AAS, ICP/MS, ICP-OES	A79
Cu, Fe, K, Mn, Zn	HNO_3	Multimode/LP closed	8 min at 150 W	F-AAS, ETV-AAS, SIMAAC	A54
Ca, Cu, Fe, K, Mg, Mn, P, Zn	HNO_3, HCl	Multimode/LP closed	5 min at 300 W, 5 min at 0 W, 5 min at 300 W, 5 min at 450 W, 5 min at 0 W, 5 min at 450 W	ICP-OES	A80
Cd, Cu, Fe, Mn	1. HNO_3, H_2O, H_2O_2 2. Aqua regia, HF (optional)	Multimode/open	1. 15 min at 70 W, 15 min at 150 W, 15 min at 180 W, 1 min at 250 W, 1 min at 600 W 2. 15 min at 70 W, 1 min at 600 W	ETV-AAS	A5
Ca, Fe, K, Mg, Mn, Na, P, S	1. HNO_3 2. H_2O_2 3. HCl	Multimode/LP closed	1. 30 min at 540 W, 5 min at 0 W 2. 5 min at 0 W, 15 min at 540 W, 5 min at 0 W 3. 10 min at 180 W	ICP-OES	A81
Ba, Ca, K, Mg, Mn, Na, P, S, Zn	1. HNO_3 2. H_2O_2	Multimode/open	1. 30 min at 540 W 2. 30 min at 540 W	ICP-OES	A42
Residual amino acids	HNO_3	Multimode/LP closed	Varied	Fluorimetry	A43
NIST Rice flour 1568	Al, Ba, Ca, K, Mg, Mn, Na, P, S, Si, Zn	A. 1. HNO_3 2. H_2O_2 3. H_2O_2	Multimode/open	1. 5 min at 300 W, 600 W until 1 mL remains 2. 30–40 min at 600 W until 1 mL remains	FI-ICP-OES	A35

Appendix 1.2 Botanical

Matrix	Analytes	Reagents	Cavity/Vessel	Conditions	Detection	Ref.
NIST Rice flour 1568 (contd.)	Al, Ba, Ca, K, Mg, Mn, Na, P, S, Si, Zn	B. 1. HNO_3 2. H_2O_2, HF 3. H_3BO_3	Multimode/ open (with reflux top)	1. 5 min at 300 W, 20-30 min at 600 W reflux not on 2. 30 min at 180 W reflux on 3. 180 W reflux on		
	B	HNO_3, H_2O_2	Multimode/ HP closed	2 min at 100W, 2 min at 0 W, 5 min at 250 W, 3 min at 0 W, 5 min at 500 W, 15 min at 250 W	UV-Vis Fluorimetry ICP-OES ICP/MS	A64
	As, Ca, Cu, Fe, Mg, Mn, Mo, P, S, Se, U, V, Zn	HNO_3, H_2O_2	Multimode/ MP closed (Max 350 psi - power cutoff at 150 psi)	4 min at 296 W, 8 min at 360 W	F-AAS ICP/MS ICP-OES	A79
	Cr, Hg, Se	1. HNO_3 2. H_2O_2	Multimode/ HP closed	1. 3 min at 450 W	Radiochemical	A30
NIST Spinach 1570	I	HNO_3, N_2H_4	Multimode/ HP closed	35 sec at 675 W	NAA	A17
	Ba, Ca, K, Mg, Mn, Na, P, S, Zn	1. HNO_3 2. H_2O_2	Multimode/ open	1. 30 min at 540 W 2. 30 min at 540 W	ICP-OES	A42
	Al, Ba, Ca, K, Mg, Mn, Na, P, S, Si, Zn	A. 1. HNO_3 2. H_2O_2 3. H_2O_2	Multimode/ open	1. 5 min at 300 W, 600 W until 1 mL remains 2. 30-40 min at 600 W until 1 mL remains	FI-ICP-OES	A35

Material	Elements	Reagents	Microwave system	Conditions	Method	Ref.
NIST Orchard leaves 1571	Al, Ba, Ca, Cr, Cu, Fe, K, Li, Mg, Mn, Na, P, Si, Sr, Ti, Zn	B. 1. HNO_3 2. H_2O_2, HF 3. H_3BO_3 1. Aqua regia 2. HBO_3	Multimode/open (with reflux top) Multimode/MP closed (evacuated slightly at start)	1. 5 min at 300 W, 20-30 min at 600 W reflux not on 2. 30 min at 180 W reflux on 3. 180 W reflux on 1. 3 min at 625 W	ICP-OES	A55
	I	HNO_3, N_2H_4	Multimode/HP closed	35 sec at 675 W	NAA	A17
	Cd	HNO_3	Multimode/LP closed	2 min on defrost three times	ETV-AAS	A40
	As	HNO_3, $HClO_4$, H_2SO_4	Multimode/HP closed	10 min at 85 bar	DP-ASV	A1
	Al, Ba, Ca, K, Mg, Mn, Na, P, S, Si, Zn	A. 1. HNO_3 2. H_2O_2 3. H_2O_2 B. 1. HNO_3 2. H_2O_2, HF 3. H_3BO_3	Multimode/open Multimode/open (with reflux top)	1. 5 min at 300 W, 600 W until 1 mL remains 2. 30-40 min at 600 W until 1 mL remains 1. 5 min at 300 W, 20-30 min at 600 W reflux not on 2. 30 min at 180 W reflux on 3. 180 W reflux on	FI-ICP-OES	A35
	As, Co, Cr, Cu, Ni, Pb, Se, Zn	HNO_3, $HClO_4$	Multimode/open	600 W until first signs of $HClO_4$ fumes	F-AAS NAA	A46-A48
	Cd	1. HNO_3, $HClO_4$, HCl, HF 2. $HClO_4$	Multimode/MP closed	1. 5 min at 200 W (using water load), 4 min at 200 W (no load)	F-AAS	A76
	Al, Ba, Ca, Cr, Co, Cu, Fe, K, Mg, Mn, Na, P, S, Zn	HNO_3	Multimode/LP closed	10 min at 70 psi, 15 min at 450 W	NAA ICP-OES	A22
	Po	HNO_3	Multimode/MP closed	60 min at 60 W, cool 60 min at 60 W	α-spectrometry	A82

Appendix 1.2 Botanical

Matrix	Analytes	Reagents	Cavity/Vessel	Conditions	Detection	Ref.
NIST Orchard leaves 1571 (contd.)	Cd, Co, Fe, Ni, Pb	HNO_3, HF	Multimode/LP closed	1 min at 240 W, cool (x15)	ETV-AAS X-ray Fluorescence	A83
	Al, Ba, Ca, Cu, Fe, K, Mg, Mn, Na, P, Zn	1. Aqua regia 2. HBO_3	Multimode/MP closed (evacuated slightly at start)	3 min at 625 W	ICP-OES	A55
	Se	HNO_3, $HClO_4$	Not reported	-	Fluorimetry	A65
	Cu, Pb, Zn	HNO_3, $HClO_4$	Multimode/open	3 min at 600 W	ETV-AAS	A84
	Al, Ba, Ca, Cu, Fe, Mg, Pb, Zn	A. Aqua regia	Multimode/stopped flow	2 min at 720 W	ICP-OES	A58
		B. Aqua regia	Multimode/open	2 min at 720 W, 2 min at 0 W, 2 min at 720 W, 2 min at 0 W, 2 min at 720 W (with water load)		
	Al	HNO_3	Multimode/HP closed	1 min at 150 W, 30 min at 0 W, 1 min at 450 W	ICP-OES	A20
NIST Citrus leaves 1572	Al, Ag, As, Ba, Ca, Cl, Cr, Cu, Fe, K, Li, Mg, Mn, Mo, N, Na, Ni, P, Pb, Rb, S, Sb, Si, Sr, Ti, Zn	HNO_3, $HClO_4$	Multimode/MP closed	15 min at 0 W, 10 min at 150 W, 10 min at 450 W, 15 min at 0 W	ICP-OES	A31
	I	HNO_3, N_2H_4	Multimode/HP closed	35 sec at 675 W	NAA	A17
	Cd	HNO_3	Multimode/LP closed	2 min on defrost three times	ETV-AAS	A40
	Ba, Ca, K, Mg, Mn, Na, P, S, Zn	1. HNO_3 2. H_2O_2	Multimode/open	1. 30 min at 540 W 2. 30 min at 540 W	ICP-OES	A42

Elements	Reagents	Mode/System	Power Program	Method	Ref.
Al, Ba, Ca, K, Mg, Mn, Na, P, S, Si, Zn	**A.** 1. HNO₃ 2. H₂O₂ 3. H₂O₂ **B.** 1. HNO₃ 2. H₂O₂, HF 3. H₃BO₃	Multimode/ open Multimode/ open (with reflux top)	1. 5 min at 300 W, 600 W until 1 mL remains until 1 mL remains 2. 30-40 min at 600 W until 1 mL remains 1. 5 min at 300 W, 20-30 min at 600 W reflux not on 2. 30 min at 180 W reflux on 3. 180 W reflux on	FI-ICP-OES	A35
-	HNO₃	Multimode/ LP closed	10 min at 360 W, 5 min at 480 W	-	A85
Cd	1. HNO₃, HClO₄, HCl, HF 2. HClO₄	Multimode/ MP closed	1. 5 min at 200 W (using water load), 4 min at 200 W (no load)	F-AAS	A76
Hg	HNO₃	Multimode/ LP closed	800 W until 75°C, 3 min at 75°C	CV-FANES	A86
Hg	HNO₃	Multimode/ HP closed	90 sec at 600 W (estimated power)	CV-AAS	A87
Al	HNO₃	Multimode/ MP closed	4 min at 200 W, cool, 4 min at 350 W, cool, 8 min at 250 W, cool, 10 min at 400 W (x4)	ETV-AAS	A88
Sr	HNO₃	Multimode/ open	Heat until dryness	ID (stable isotope dilution activation analysis)	A89
As, Ca, Cu, Fe, Mg, Mn, Mo, P, S, Se, U, V, Zn	HNO₃, H₂O₂	Multimode/ MP closed (Max 350 psi - power cutoff at 150 psi)	4 min at 296 W, 8 min at 360 W	F-AAS ICP/MS ICP-OES	A79
Ca, Cu, Fe, K, Mg, Mn, P, Zn	HNO₃, HCl	Multimode/ LP closed	5 min at 300 W, 5 min at 0 W, 5 min at 300 W, 5 min at 450 W, 5 min at 0 W, 5 min at 450 W	ICP-OES	A80

Appendix 1.2 Botanical

Matrix	Analytes	Reagents	Cavity/Vessel	Conditions	Detection	Ref.
NIST Citrus leaves 1572 (contd.)	As, Cr, Hg, Sb	1. HNO_3 2. H_2O_2	Multimode/HP closed	3 min at 450 W	Radiochemical	A30
	Ba, Ca, Cu, Mg, Mn, Zn	1. HNO_3, HF 2. HBO_3	Multimode/LP closed	2 min at 240 W, 2 min 360 W, 16 min at 160°C (600W)	ICP-OES	A90
	Ni	HNO_3, HCl	Multimode/HP closed	4 min at 350 W, 10 min at 140 W	ETV-AAS	A9
	Ni	HNO_3, HCl	Multimode/HP closed	4 min at 360 W, 10 min at 180 W	ICP-OES	A10, A11
	Al, Ca, Fe, K, Mg, Mn, Na, P, S	1. HNO_3 2. H_2O_2 3. HCl	Multimode/LP closed	1. 30 min at 540 W, 5 min at 0 W 2. 5 min at 0 W, 15 min at 540 W, 5 min at 0 W 3. 10 min at 180 W	ICP-OES	A81
	P	1. HNO_3, $HClO_4$ 2. $HClO_4$	Multimode/open (with scrubber)	1. 15 min at 750 W	ICP-OES	A91
	Al	HNO_3, H_2SO_4	Multimode/MP closed	5 min at 1170 W	IC	A92
	As	HNO_3, H_2O_2	Multimode/HP closed	-	ICP/MS	A93
	Ni	HNO_3, HCl	Multimode/HP closed	4 min at 460 W, 10 min at 180 W (10 mL water load)	ICP-OES	A94
	Cu, Fe, Mn, Pb	1. HNO_3, H_2O, H_2O_2 2. Aqua regia, HF (optional)	Multimode/open	1. 15 min at 70 W, 15 min at 150 W, 15 min at 180 W, 1 min at 250 W, 1 min at 600 W 2. 15 min at 70 W, 1 min at 600 W	ETV-AAS	A5

Sample	Elements	Reagents	Mode	Power program	Method	Ref.
	Ca, Fe, K, Mg, Mn, Zn	-	Multimode/open	30 min at 650 W	F-AAS	A95
	Ca, Cu, Fe, Mg, Mn, K, Na, Zn	1. HNO_3 2. HNO_3 3. H_2O_2	Single mode/open	1. 10 min at 60 W, 5 min at 80 W 2. 5 min at 70 W 3. 3 min at 60 W	F-AAS	A67
	Al	HNO_3	Multimode/HP closed	1 min at 150 W, 30 min at 0 W, 1 min at 450 W	ICP-OES	A20
	Al, As, Ba, Cr, Cu, Fe, K, Mg, Mn, Ni, P, Pb, Rb, Sr, Zn	1. HNO_3 2. HF 3. H_2O_2	Multimode/HP closed	1. 5 min at 300 W, 2 min at 600 W, 2 min at 0 W, 3 min at 300 W 2. 5 min at 300 W, 2 min at 600 W, 2 min at 0 W 3. 5 min at 300 W, 2 min at 600 W, 2 min at 0 W, 3 min at 300 W	DCP-OES F-AAS	A96
NIST Tomato leaves 1573	Cu, Mn	HNO_3, H_2O_2	Multimode/flow through	2 min at 650 W (digestate recirculated through oven)	F-AAS	A97
	Ba, Ca, K, Mg, Mn, Na, P, S, Zn	1. HNO_3 2. H_2O_2	Multimode/open	1. 30 min at 540 W 2. 30 min at 540 W	ICP-OES	A42
	Al, Ba, Ca, K, Mg, Mn, Na, P, S, Si, Zn	A. 1. HNO_3 2. H_2O_2 3. H_2O_2 B. 1. HNO_3 2. H_2O_2, HF 3. H_3BO_3	Multimode/open — Multimode/open (with reflux top)	1. 5 min at 300 W, 600 W until 1 mL remains 2. 30–40 min at 600 W until 1 mL remains — 1. 5 min at 300 W, 20-30 min at 600 W reflux not on 2. 30 min at 180 W reflux on 3. 180 W reflux on	FI-ICP-OES	A35

Appendix 1.2 Botanical

Matrix	Analytes	Reagents	Cavity/Vessel	Conditions	Detection	Ref.
NIST Tomato leaves 1573 (contd.)	As, Cr, Cu, Fe, Mn, Pb, Rb, Sr, Th, U, Zn	1. HNO_3 2. HF	Multimode/ MP closed	1. Leave overnight, 3 min at 130 W, 6 min at 195 W, 6 min at 260 W, 6 min at 325 W, cool, vent 2. 6 min at 325 W	ICP/MS	A98
	Cd	1. HNO_3, $HClO_4$, HCl, HF 2. $HClO_4$	Multimode/ MP closed	1. 5 min at 200 W (using water load), 4 min at 200 W (no load)	F-AAS	A76
	Ca, Cu, Fe, K, Mn, P, Pb, Zn	HNO_3	Multimode/ HP closed	30 sec at 665 W	F-OES F-AAS ICP-OES	A49
	Sr	HNO_3	Multimode/ open	Heat until dryness	ID (stable isotope dilution activation analysis)	A89
	Al, Ba, Ca, Cd, Cr, Fe, K, Mg, Mn, Na, P	1. Aqua regia 2. HBO_3	Multimode/ MP closed (evacuated slightly at start)	3 min at 625 W	ICP-OES	A55
	Cd, Co, Fe, Ni, Pb	HNO_3, HF	Multimode/ LP closed	1 min at 240 W, cool (x15)	ETV-AAS X-ray Fluorescence	A83
	Cr, Hg, Se	1. HNO_3 2. H_2O_2	Multimode/ HP closed	3 min at 450 W	Radiochemical	A30
	Fe	A. HNO_3 B. 1. HNO_3 2. HF	Multimode/ HP closed Multimode/ HP closed	30 sec at 700 W Same as A	ETV-AAS	A6
	Ca, Fe, K, Mg, Mn, Zn	-	Multimode/ open	30 min at 650 W	F-AAS	A95

Elements	Reagents	Mode/Vessel	Conditions	Detection	Reference
B	HNO_3, H_2O_2	Multimode/HP closed	2 min at 100W, 2 min at 0 W, 5 min at 250 W, 3 min at 0 W, 5 min at 500 W, 15 min at 250 W	UV-Vis Fluorimetry ICP-OES ICP/MS	A64
Mn	HNO_3	Multimode/flow through	2 sec at 650 W (0.5 mL coil, 15.4 mL min^{-1})	F-AAS	A99
Al, As, Be, Ca, Cd, Co, Cr, Cu, Fe, K, Mg, Mn, Na, Ni, Pb, Sb, Se, V, Zn	Aqua regia	Multimode/LP closed	10 min at 300 W, 5 min at 600 W, 10 min at 480 W	ICP-OES ETV-AAS	A56
Ca, Fe, K, Mg, Mn, Na, P, S	1. HNO_3 2. H_2O_2 3. HCl	Multimode/LP closed	1. 30 min at 540 W, 5 min at 0 W, 2. 5 min at 0 W, 15 min at 540 W, 5 min at 0 W, 3. 10 min at 180 W	ICP-OES	A81
NIST Pine needles 1575 Pb	HNO_3, HCl	Multimode/flow through	18 sec at 700 W (0.4 mL coil, 1.5 mL min^{-1})	ETV-AAS	A8
Ba, Ca, K, Mg, Mn, Na, P, S, Zn	1. HNO_3 2. H_2O_2	Multimode/open	1. 30 min at 540 W 2. 30 min at 540 W	ICP-OES	A42
Al, Ba, Ca, K, Mg, Mn, Na, P, S, Si, Zn	A. 1. HNO_3 2. H_2O_2 3. H_2O_2 B. 1. HNO_3 2. H_2O_2, HF 3. H_3BO_3	Multimode/open (with reflux top)	1. 5 min at 300 W, 600 W until 1 mL remains, 2. 30-40 min at 600 W until 1 mL remains; 1. 5 min at 300 W, 20-30 min at 600 W reflux not on, 2. 30 min at 180 W reflux on, 3. 180 W reflux on	FI-ICP-OES	A35

Appendix 1.2 Botanical

Matrix	Analytes	Reagents	Cavity/Vessel	Conditions	Detection	Ref.
NIST Pine needles 1575 (contd.)	Al, As, Cr, Cu, Fe, Hg, Mn, Pb, Rb, Sr, Th, U	1. HNO_3 2. HF	Multimode/ MP closed	1. Leave overnight, 3 min at 130 W, 6 min at 195 W, 6 min at 260 W, 6 min at 325 W, cool, vent 2. 6 min at 325 W	ICP/MS	A98
	Ca, Cu, Fe, K, Mn, P, Pb	HNO_3	Multimode/ HP closed	30 sec at 665 W	F-OES F-AAS ICP-OES	A49
	Hg	HNO_3	Multimode/ LP closed	800 W until 75°C, 3 min at 75°C	CV-FANES	A86
	As, Ca, Cu, Fe, Mg, Mn, Mo, P, S, Se, U, V, Zn	HNO_3, H_2O_2	Multimode/ MP closed (Max 350 psi - power cutoff at 150 psi)	4 min at 296 W, 8 min at 360 W	F-AAS ICP/MS ICP-OES	A79
	Al, Ba, Ca, Cr, Fe, K, Mg, Mn, Na, P, Zn	1. Aqua regia 2. HBO_3	Multimode/ MP closed (evacuated slightly at start)	1. 3 min at 625 W	ICP-OES	A55
	Al, As, Be, Ca, Cd, Co, Cr, Cu, Fe, K, Mg, Mn, Na, Ni, Pb, Sb, Se, V, Zn	Aqua regia	Multimode/ LP closed	10 min at 300 W, 5 min at 600 W, 10 min at 480 W	ICP-OES ETV-AAS	A56
	Cd, Co, Fe, Ni, Pb	HNO_3, HF	Multimode/ LP closed	1 min at 240 W, cool (x15)	ETV-AAS XRF	A83
	Ca, Fe, K, Mg, Mn, Na, P, S	1. HNO_3 2. H_2O_2 3. HCl	Multimode/ LP closed	1. 30 min at 540 W, 5 min at 0 W 2. 5 min at 0 W, 15 min at 540 W, 5 min at 0 W 3. 10 min at 180 W	ICP-OES	A81

Elements	Reagents	Mode / Vessel	Power / Time program	Technique	Ref.
Ca, K, Mg, P	1. HNO_3, $HClO_4$ 2. $HClO_4$	Multimode/ open (with scrubber)	15 min at 750 W	ICP-OES	A91
Pb	HNO_3	Multimode/ LP closed	3 min at 650 W	ETV-AAS	A3
Al	HNO_3	Multimode/ LP closed	40 sec at 700 W, 10 min at 0 W (vent), 60 sec at 700 W, 10 min at 0 W (vent), 90 sec at 700 W, 5 min at 70 W	ETV-AAS	A63
Cu	1. HNO_3 2. H_2O_2	Multimode/ LP closed	9.5 min at 540 W	F-AAS	A100
Cu, Fe, Pb	1. HNO_3, H_2O, H_2O_2 2. Aqua regia, HF (optional)	Multimode/ open	1. 15 min at 70 W, 15 min at 150 W, 15 min at 180 W, 1 min at 250 W, 1 min at 600 W 2. 15 min at 70 W, 1 min at 600 W	ETV-AAS	A5
B	HNO_3, H_2O_2	Multimode/ HP closed	2 min at 100W, 2 min at 0 W, 5 min at 250 W, 3 min at 0 W, 5 min at 500 W, 15 min at 250 W	UV-Vis Fluorimetry ICP-OES ICP/MS	A64
Ca, Fe	HNO_3, HCl, HF	Multimode/ MP closed	15 min at 540 W	DCP-OES	A101
Al, As, Ba, Cr, Cu, Fe, K, Mg, Mn, Ni, P, Pb, Rb, Sr, Zn	1. HNO_3 2. HF 3. H_2O_2	Multimode/ HP closed	1. 5 min at 300 W, 2 min at 600 W, 2 min at 0 W, 3 min at 300 W 2. 5 min at 300 W, 2 min at 600 W, 2 min at 0 W 3. 5 min at 300 W, 2 min at 600 W, 2 min at 0 W, 3 min at 300 W	DCP-OES F-AAS	A96

Appendix 1.2 Botanical

Matrix	Analytes	Reagents	Cavity/Vessel	Conditions	Detection	Ref.
NIST Corn stalks 8412	Al, Ba, Ca, K, Mg, Mn, Na, P, S, Si, Zn	A. 1. HNO_3 2. H_2O_2 3. H_2O_2	Multimode/ open	1. 5 min at 300 W, 600 W until 1 mL remains 2. 30-40 min at 600 W until 1 mL remains	FI-ICP-OES	A35
		B. 1. HNO_3 2. H_2O_2, HF 3. H_3BO_3	Multimode/ open (with reflux top)	1. 5 min at 300 W, 20-30 min at 600 W reflux off 2. 30 min at 180 W reflux on 3. 180 W reflux on		
NIST Corn kernel 8413	Al, Ba, Ca, K, Mg, Mn, Na, P, S, Si, Zn	A. 1. HNO_3 2. H_2O_2 3. H_2O_2	Multimode/ open	1. 5 min at 300 W, 600 W until 1 mL remains 2. 30-40 min at 600 W until 1 mL remains	FI-ICP-OES	A35
		B. 1. HNO_3 2. H_2O_2, HF 3. H_3BO_3	Multimode/ open (with reflux top)	1. 5 min at 300 W, 20-30 min at 600 W reflux not on 2. 30 min at 180 W reflux on 3. 180 W reflux on		

Appendix 1.3 Marine

Matrix	Analytes	Reagents	Cavity/Vessel	Conditions	Detection	Ref.
BCR Mussel tissue 278	As	HNO$_3$, H$_2$O$_2$	Multimode/ LP closed	1 min at 555 W, 4 min at 300 W	ETV-AAS	A102
	Cu, Pb	HNO$_3$, HClO$_4$, H$_2$SO$_4$	Multimode/ HP closed	10 min at 85 bar	DP-ASV	A1
	Cd	HNO$_3$, V$_2$O$_5$	Multimode/ HP closed	90 sec at 600 W	ETV-AAS	A103
	Pb	HNO$_3$, V$_2$O$_5$	Multimode/ HP closed	90 sec at 460 W	ETV-AAS	A104
	Hg	HNO$_3$	Multimode/ HP closed	90 sec at 600 W	CV-AAS	A105
BCR Plankton 414	Cd, Cr, Cu, Hg, Mn, Ni, Pb, V, Zn	A. 1. HNO$_3$ 2. HNO$_3$ 3. H$_2$O$_2$ 4. H$_2$O	Single mode/ open	1. 5 min at 10 W, 10 min at 30 W, 10 min at 60 W 2. 10 min 60 W 3. 5 min at 60 W 4. 5 min at 50 W	ICP-OES ICP-MS HPLC-ICP-MS	A13
		B. 1. HNO$_3$, HCl 2. HNO$_3$, HCl 3. H$_2$O$_2$ 4. H$_2$O	Single mode/ open	1. 5 min at 40 W, 10 min at 50 W 2. 20 min 54 W 3. 5 min at 40 W 4. 5 min at 50 W		
BCR Cod muscle 422	Cu	HNO$_3$, HClO$_4$, H$_2$SO$_4$	Multimode/ HP closed	10 min at 85 bar	DP-ASV	A1
	As, Cu, Fe, Hg, Mn, Zn	A. 1. HNO$_3$ 2. HNO$_3$ 3. H$_2$O$_2$ 4. H$_2$O	Single mode/ open	1. 5 min at 10 W, 10 min at 30 W, 10 min at 60 W 2. 10 min 60 W 3. 5 min at 60 W 4. 5 min at 50 W	ICP-OES ICP-MS HPLC-ICP-MS	A13

Appendix 1.3 Marine

Matrix	Analytes	Reagents	Cavity/Vessel	Conditions	Detection	Ref.
BCR Cod muscle 422 (contd.)		B. 1. HNO_3, HCl 2. HNO_3, HCl 3. H_2O_2 4. H_2O	Single mode/ open	1. 5 min at 40 W, 10 min at 50 W 2. 20 min 54 W 3. 5 min at 40 W 4. 5 min at 50 W		
		C. 1. HNO_3, H_2SO_4 2. HNO_3 3. H_2O_2 4. H_2O_2 5. H_2O	Single mode/ open	1. 5 min at 20 W, 10 min at 40 W, 10 min at 100 W 2. 10 min at 100 W 3. 5 min at 100 W 4. 5 min at 100 W 5. 5 min at 80 W		
IAEA Copepod MAA-1	Cr, Hg, Se	1. HNO_3 2. H_2O_2	Multimode/ HP closed	1. 3 min at 450 W	Radiochemical	A30
IAEA Fish flesh MAA-2	Se	1. HNO_3 2. H_2SO_4, $HClO_4$ 3. H_2O	Single mode/ open	1. 15 min at 45 W, 10 min at 75 W 2. 10 min at 90 W, 35 min at 120 W 3. 8 min at 120 W	ETV-AAS	A18
	As	HNO_3, $HClO_4$, H_2SO_4	Multimode/ HP closed.	10 min at 85 bar	DP-ASV	A1
	Cr, Hg, Se	1. HNO_3 2. H_2O_2	Multimode/ HP closed	1. 3 min at 450 W	Radiochemical	A30
IAEA Mussel tissue MAM-2	Cr, Hg, Se	1. HNO_3 2. H_2O_2	Multimode/ HP closed	1. 3 min at 450 W	Radiochemical	A30
NIES Chlorella 3	Ca, Fe, Mg, Zn	HNO_3	Multimode/ flow through	100 sec at 525 W (10 mL coil, 6 mL min^{-1})	F-AAS	A39

Analyte	Reagent	Mode/Vessel	Conditions	Detection	Ref.
Fe, Mn, Zn	HNO_3	Multimode/ flow through	-	F-AAS ICP-OES ICP/MS	A66
Cd, Co, Cu, Ni, Pb	HNO_3, HCl, $HClO_4$, HF	Multimode/ HP closed	5 min at 200 W (water load), 3.5 min at 200 W (no load)	ICP-OES ETV-ICP	A28
I	1. HNO_3 2. N_2H_4	Multimode/ MP closed	35 sec at 675 W	NAA	A17
Ca, Fe, Mg, Zn	HNO_3	Multimode/ flow through	100 sec at 525 W (10 mL coil, 6 mL min^{-1})	F-AAS	A39
Ca, Cd, Cu, Fe, K, Mg, Mn, Na, P, Sr, Zn, Residual carbon	**A.** 1. HNO_3, H_2SO_4 2. H_2O_2(x16) 3. HNO_3 or HNO_3, HCl or NH_3 (aq) 4. NH_4EDTA	Single mode/ open	1. 4 min at 30 W, 4 min at 120 W 2. 1 min at 120 W(x16) 3. 4 min at 30 W	ICP-OES	A24
	B. HNO_3	Multimode/ MP closed	2 min at 20 psi, 5 min at 40 psi, 2 min at 60 psi, 2 min at 80 psi, 2 min at 100 psi, 2 min at 120 psi, 2 min at 140 psi, 15 min at 160 psi		
NIES Mussel 6 Cd, Co, Cu, Ni, Pb	HNO_3, HCl, $HClO_4$, HF	Multimode/ MP closed	5 min at 200 W (water load), 3.5 min at 200 W (no load)	ICP-OES ETV-ICP	A28
Cd, Cu, Fe, Mn, Zn	**A.** HNO_3, H_2O_2	Multimode/ HP closed	75 sec at 665 W	ETV-AAS	A6
	B. HNO_3, H_2O_2	Multimode/ HP closed	1 min at 250 W, 2 min at 0 W, 2 min at 250 W, 2 min at 400 W, 2 min at 600 W		
	C. HNO_3, $HClO_4$	Multimode/ HP closed	150 sec at 950 W, 60 sec at 0 W, 90 sec at 300 W, 90 sec at 500 W, 90 sec at 700 W, 90 sec at 850 W		

Appendix 1.3 Marine

Matrix	Analytes	Reagents	Cavity/Vessel	Conditions	Detection	Ref.
NIES Mussel 6	Se	1. HNO_3, H_2SO_4 2. H_2O_2 3. H_2O_2	Multimode/ LP closed	1. 6 min at 330 W, cool 2. 4 min at 450 W, cool 3. 4 min at 600 W	DPP	A106
(contd.)	Se	A. 1. HNO_3 2. H_2O_2 3. H_2O_2	Multimode/ LP closed	1. 6 min at 330 W, cool 2. 4 min at 450 W, cool 3. 4 min at 600 W	HG-AAS	A107
		B. 1. HNO_3, H_2SO_4 2. H_2O_2 3. H_2O_2	Multimode/ LP closed	Same as A		
		C. 1. HNO_3, H_3PO_4 2. H_2O_2 3. H_2O_2	Multimode/ LP closed	Same as A		
		D. 1. HNO_3, $K_2S_2O_8$ 2. H_2O_2 3. H_2O_2	Multimode/ LP closed	Same as A		
	Fe, Mn, Zn	HNO_3	Multimode/ flow through	-	F-AAS ICP-OES ICP/MS	A66
	Ca, Cu, Fe, Mg, Mn, Zn	HNO_3, $HClO_4$, HCl, HF	Multimode/ MP closed	14 min intermittent 200 W	F-AAS	A27
	Al	HNO_3	Multimode/ MP closed	4 min at 200 W, cool, 4 min at 350 W, cool, 8 min at 250 W, cool, 10 min at 400 W (x4)	ETV-AAS	A88
NIST Albacore tuna 50	Hg	HNO_3	Multimode/ HP closed	70 sec at 600 W	CV-AAS	A16

Sample	Analytes	Reagents	Mode/vessel	Microwave program	Detection	Ref.
NIST Oyster tissue 1566	Zn, Cd	HNO_3	Multimode/open	8 min at 200 W	FI-F-AAS	A38
	I	1. HNO_3 2. N_2H_4	Multimode/HP closed	1. 35 sec at 675 W	NAA	A17
	Hg	HNO_3	Multimode/LP closed	8 min at 140 W, 8 min at 280 W, 4 min at 420 W	CV-AAS	A2
	Hg	HNO_3	Multimode/LP closed	8 min at 140 W, 8 min at 280 W, 4 min at 420 W	CV-AFS	A7
	Ba, Ca, K, Mg, Mn, Na, P, S, Zn	1. HNO_3 2. H_2O_2	Multimode/open	1. 30 min at 540 W 2. 30 min at 540 W	ICP-OES	A42
	As	HNO_3, H_2O_2	Multimode/LP closed	1 min at 555 W, 4 min at 300 W	ETV-AAS	A102
	Residual amino acids	HNO_3	Multimode/LP closed	Varied	Fluorimetry	A43
	Al, Ba, Ca, K, Mg, Mn, Na, P, S, Si, Zn	A. 1. HNO_3 2. H_2O_2 3. H_2O_2 B. 1. HNO_3 2. H_2O_2, HF 3. H_3BO_3	Multimode/open Multimode/open (with reflux top)	1. 5 min at 300 W, 600 W until 1 mL remains 2. 30-40 min at 600 W until 1 mL remains 1. 5 min at 300 W, 20-30 min at 600 W reflux not on 2. 30 min at 180 W reflux on 3. 180 W reflux on	FI-ICP-OES	A35
	As, Ca, Cd, Cu, Fe, K, Mg, Mn, Na, Pb, Zn	HNO_3	Multimode/HP closed	30 sec at 665 W	F-OES F-AAS ICP-OES	A49
	Al, As, Ba, Ca, Cd, Ce, Cr, Cs, Cu, Fe, La, Li, Mg, Mn, Mo, Ni, Rb, Sb, Se, Sr, Tl, V, Y, Zn	HNO_3	Multimode/MP closed	75 sec at 350 W	ICP/MS	A21
	Al	HNO_3	Multimode/MP closed	4 min at 200 W, cool, 4 min at 350 W, cool, 8 min at 250 W, cool, 10 min at 400 W (x4)	ETV-AAS	A88

Appendix 1.3 Marine

Matrix	Analytes	Reagents	Cavity/Vessel	Conditions	Detection	Ref.
NIST Oyster tissue 1566 (contd.)	Al, As, Ba, Ca, Cd, Cu, Fe, K, Mg, Mn, Na, P, Si, Sr, Ti, Zn	1. Aqua regia 2. HBO_3	Multimode/ MP closed (evacuated slightly at start)	1. 3 min at 625 W	ICP-OES	A55
	Hg, Se	1. HNO_3 2. H_2O_2	Multimode/ HP closed	1. 3 min at 450 W	Radiochemical	A30
	Cu, Fe, Zn	1. HNO_3 2. $HClO_4$, HF	Multimode/ LP closed	Variable	SPC-IC	A57
	Residual carbon	HNO_3	Multimode/ HP closed	5 min at 500 W (200 bar)	Total carbon analyzer Carbon dioxide coulometer	A59
	Ca, Cd, Cu, Fe, K, Mg, Mn, Na, P, Sr, Zn, Residual carbon	A. 1. HNO_3, H_2SO_4 2. H_2O_2(x16) 3. HNO_3 or HNO_3, HCl or NH_3 (aq) 4. NH_4EDTA B. HNO_3	Single mode/ open Multimode/ MP closed	1. 4 min at 30 W, 4 min at 120 W 2. 1 min at 120 W(x16) 3. 4 min at 30 W 2 min at 20 psi, 5 min at 40 psi, 2 min at 60 psi, 2 min at 80 psi, 2 min at 100 psi, 2 min at 120 psi, 2 min at 140 psi, 15 min at 160 psi	ICP-OES	A24
	As, arsenocholine, arsenobetaine, tetramethylarsonium iodide	HNO_3	Multimode/ HP closed	90 sec at 500 W	ETV-AAS	A108

Sample	Analyte	Reagent	Mode/Vessel	Conditions	Detection	Ref.
	Cd, Cr, Cu, Mn, Mo, Pb	HNO_3	Multimode/LP closed	3 min at 650 W	ETV-AAS	A3
NRCC Dogfish liver DOLT-1	Cd, Cu, Mn, Pb, Zn	HNO_3, H_2SO_4	Multimode/open	1000 W until nitric acid boils off	F-AAS	A109
	Cu, Fe, Zn	HNO_3	Multimode/HP closed	3 min at 418 W	F-AAS	A110
	Residual carbon	HNO_3	Multimode/HP closed	5 min at 500 W (200 bar)	Total carbon analyzer Carbon dioxide coulometer	A59
NRCC Dogfish liver DOLT-1 (contd.)	Ni	HNO_3, HCl	Multimode/HP closed	4 min at 350 W, 10 min at 140 W	ETV-AAS	A9
	As, Se	HNO_3, H_2O_2	Multimode/LP closed	3 min at 300 W	ICP/MS	A77
	Cd	HNO_3, HCl	Multimode/HP closed	4 min at 360 W, 10 min at 180 W	ETV-AAS	A4
	Cu, Fe, Zn	HNO_3	Multimode/HP closed	-	ICP-OES F-AAS	A111
	Cu, Fe, Zn	HNO_3	Multimode/HP closed	3 min at 418 W	F-AAS	A112
	Hg, Se	HNO_3	Multimode/LP closed	8 min at 75 W, 8 min at 225 W, 4 min at 375 W	CV-AAS HG-AAS	A113
	Ni	HNO_3, HCl	Multimode/HP closed	4 min at 360 W, 10 min at 180 W	ICP-OES	A10, A11
NRCC Dogfish muscle	Hg, Se	HNO_3	Multimode/LP closed	8 min at 75 W, 8 min at 225 W, 4 min at 375 W	CV-AAS HG-AAS	A113
DORM-1	Ni	HNO_3, HCl	Multimode/HP closed	4 min at 350 W, 10 min at 140 W	ETV-AAS	A9
	As, dimethylarsenic acid, arsenite, monomethylarsonic acid, arsenate, methanearsonate	HNO_3	Multimode/LP closed	-	HPLC-ICP/MS	A114

Appendix 1.3 Marine

Matrix	Analytes	Reagents	Cavity/Vessel	Conditions	Detection	Ref.
NRCC Dogfish muscle DORM-1 (contd.)	As, Cr	HNO_3, H_2O_2	Multimode/ LP closed	3 min at 300 W	ICP/MS	A77
	Ni	HNO_3, HCl	Multimode/ HP closed	4 min at 460 W, 10 min at 180 W (10 mL water load)	ICP-OES	A94
	Cd	HNO_3, HCl	Multimode/ HP closed	4 min at 360 W, 10 min at 180 W	ETV-AAS	A4
	Residual carbon	HNO_3	Multimode/ HP closed	5 min at 500 W (200 bar)	Total carbon analyzer Carbon dioxide coulometer	A59
	Ni	HNO_3, HCl	Multimode/ HP closed	4 min at 360 W, 10 min at 180 W	ICP-OES	A10, A11
NRCC Non-defatted lobster hepato-pancreas LUTS-1	Residual carbon	HNO_3	Multimode/ HP closed	5 min at 500 W (200 bar)	Total carbon analyzer Carbon dioxide coulometer	A59
	Cd, Co, Cu, Hg, Mn, Ni, Pb, Sr, Zn	1. HNO_3 2. H_2O_2	Multimode/ LP closed	1. 10 min at 3.5 atm 2. 5 min at 4 atm	ICP/MS	A115
	Cd	HNO_3	Multimode/ LP closed	2 min at 600 W, 10 min at 90 W	ASV	A116
NRCC Cod liver tissue NOAA-K	Ag, As, Cd, Cr, Cu, Fe, Hg, Mn, Ni, Pb, Sn, Zn	A. HNO_3, $HClO_4$	Multimode/ LP closed	25 min at 70 psi	ICP/MS	A117
		B. HNO_3	Multimode/ LP closed	Same as A		
NRCC Shellfish tissue NOAA-L	Ag, As, Cd, Cr, Cu, Fe, Hg, Mn, Ni, Pb, Sn, Zn	A. HNO_3, $HClO_4$	Multimode/ LP closed	25 min at 70 psi	ICP/MS	A117
		B. HNO_3	Multimode/ LP closed	Same as A		

Sample	Elements	Reagents	Vessel	Conditions	Method	Ref.
NRCC Lobster hepato-pancreas	As, Cd, Co, Cr, Cu, Fe, Mn, Ni, Pb, Se, Zn	HNO_3, $HClO_4$	Multimode/ LP closed	20 min at 60-65 psi	ETV-AAS F-AAS	A118
TORT-1	As, Cd, Co, Cr, Cu, Fe, Mn, Mo, Ni, Pb, Se, Sr, V, Zn	A. HNO_3, H_2O_2 B. HNO_3, H_2O_2 C. HNO_3, H_2O_2	Multimode/ LP closed Multimode/ LP closed Multimode/ HP closed	3 min at 600 W (Pressure release) 1 min at 600 W 1 min at 600 W	ETV-AAS F-AAS	A119
	Cd, Cr, Pb	HNO_3	Multimode/ LP closed	15 min at 180 W, 10 min at 0 W, 15 min at 180 W	F-AAS	A120
	As, Ni, Co	HNO_3, $HClO_4$, H_2SO_4	Multimode/ HP closed	10 min at 85 bar	DP-ASV	A1
	Residual carbon	HNO_3	Multimode/ HP closed	5 min at 500 W (200 bar)	Total carbon analyzer Carbon dioxide coulometer	A59
	As, Cd, Co, Cr, Cu, Mn, Ni, Pb, Sn, V, Zn	HNO_3	Multimode/ MP closed	5 min at 50W, 5 min at 90 W, 15 min at 150 W	ETV-AAS F-AAS	A121
	As	HNO_3, H_2O_2	Multimode/ HP closed	-	ICP/MS	A93
	Al	HNO_3	Multimode/ LP closed	40 sec at 700 W, 10 min at 0 W (vent), 60 sec at 700 W, 10 min at 0 W (vent), 90 sec at 700 W, 5 min at 70 W	ETV-AAS	A63
	Ni	HNO_3, HCl	Multimode/ HP closed	4 min at 350 W, 10 min at 140 W	ETV-AAS	A9

Appendix 1.3 Marine

Matrix	Analytes	Reagents	Cavity/Vessel	Conditions	Detection	Ref.
NRCC Lobster hepato-pancreas TORT-1 (contd.)	As, Cr, Se, V	HNO_3, H_2O_2	Multimode/ LP closed	3 min at 300 W	ICP/MS	A77
	Ni	HNO_3, HCl	Multimode/ HP closed	4 min at 460 W, 10 min at 180 W (10 mL water load)	ICP-OES	A94
	Cd	HNO_3, HCl	Multimode/ HP closed	4 min at 360 W, 10 min at 180 W	ETV-AAS	A4
	Ni	HNO_3, HCl	Multimode/ HP closed	4 min at 360 W, 10 min at 180 W	ICP-OES	A10, A11
	Ca, Cu, Fe, Zn	HNO_3, HCl, HF	Multimode/ MP closed	15 min at 540 W	DCP-OES	A101

A1. Schramel, P.; Hasse, S. "Destruction of organic materials by pressurized microwave digestion" *Fresenius' J. Anal. Chem.* 1993, *346*, 794-799.

A2. Vermeir, G.; Vandecasteele, C.; Dams, R. "Microwave dissolution for the determination of mercury in biological samples" *Anal. Chim. Acta* 1989, *220*, 257-261.

A3. Littlejohn, D.; Egila, J. N.; Gosland, R. M.; Kunwar, U. K.; Smith, C.; Shan, X. "Graphite furnace analysis - getting easier and achieving more?" *Anal. Chim. Acta* 1991, *250*, 71-84.

A4. Espinosa Almendro, J. M.; Bosch Ojeda, C.; Garcia de Torres, A.; Cano Pavon, J. M. "Solvent extraction of cadmium as a previous step for its determination in biological samples by electrothermal atomization atomic-absorption spectrometry" *Talanta* 1993, *40*, 1643-1648.

A5. Chakraborti, D.; Burguera, M.; Burguera, J. L. "Analysis of standard reference materials after microwave-oven digestion in open vessels using graphite-furnace atomic-absorption spectrophotometry and Zeeman-effect background correction" *Fresenius' J. Anal. Chem.* 1993, *347*, 233-237.

A6. Mingorance, M. D.; Perez Vazquez, M. L.; Lachica, M. "Microwave digestion methods for the atomic-spectrometric determination of some elements in biological samples" *J. Anal. At. Spectrom.* 1993, *8*, 853-858.

A7. Vermeir, G.; Vandecasteele, C.; Dams, R. "Atomic fluorescence spectrometry combined with reduced aeration for the determination of mercury in biological samples" *Anal. Chim. Acta* 1991, *242*, 203-208.

A8. Burguera, J. L.; Burguera, M. "Determination of lead in biological materials by microwave-assisted mineralization and flow injection electrothermal atomic absorption spectrometry" *J. Anal. At. Spectrom.* 1993, *8*, 235-241.

A9.   Vereda Alonso, E.; Cano Pavon, J. M.; Garcia de Torres, A.; Siles Cordero, M. T. "Determination of nickel in biological samples prepared by microwave dissolution using electrothermal atomic-absorption spectrometry after extraction with 1,5-bis[phenyl-(2-pyridy)methylene]thiocarbonhydrazide" *Anal. Chim. Acta* 1993, *283*, 224-229.

A10. Vereda Alonso, E.; Garcia de Torres, A.; Cano Pavon, J. M. "Determination of nickel in biological samples by ETA-AAS and ICP-OES after acidic dissolution (with microwave heating and pre-concentration by liquid - liquid extraction)" *Mikrochim. Acta* 1993, *110*, 41-45.

A11. Vereda Alonso, E.; Garcia de Torres, A.; Cano Pavon, J. M. "Determination of nickel in biological materials after microwave dissolution using inductively coupled plasma atomic emission spectrometry with prior extraction into butan-1-ol" *Analyst* 1992, *117*, 1157-1160.

A12. Prange, A.; Boeddeker, H.; Michaelis, W. "Multi-element determination of trace elements in whole blood and blood serum by TXRF" *Fresenius' J. Anal. Chem.* 1989, *335*, 914-918.

A13. Quevauviller, P.; Imbert, J. L.; Olle, M. "Evaluation of the use of microwave oven systems for the digestion of environmental samples" *Mikrochim. Acta* 1993, *112*, 147-154.

A14. Guo, T.; Baasner, J. "On-line microwave sample pretreatment for the determination of mercury in blood by flow injection cold vapor atomic absorption spectrometry" *Talanta* 1993, *40*, 1927-1936.

A15. Welz, B.; Tsalev, D. L.; Sperling, M. "On-line microwave sample pre-treatment for the determination of mercury in water and urine by flow-injection cold-vapor atomic-absorption spectrometry" *Anal. Chim. Acta* 1992, *261*, 91-103.

A16. Tahan, J. E.; Granadillo, V. A.; Sanchez, J. M.; Cubillan, H. S.; Romero, R. A. "Mineralization of biological materials prior to determination of total mercury by cold-vapor atomic-absorption spectrometry" *J. Anal. At. Spectrom.* 1993, *8*, 1005-1010.

A17. Rao, R. R.; Chatt, A. "Microwave acid digestion and preconcentration neutron activation analysis of biological and diet samples for iodine" *Anal. Chem.* 1991, *63*, 1298-1303.

A18. Hocquellet, P.; Candillier, M. P. "Evaluation of microwave digestion and solvent extraction for the determination of trace amounts of selenium in feeds and plant and animal tissues by electrothermal atomic absorption spectrometry" *Analyst* 1991, *116*, 505-509.

A19. Dunemann, L.; Meinerling, M. "Comparison of different microwave-based digestion techniques in view of their application to fat-rich foods" *Fresenius' J. Anal. Chem.* 1992, *342*, 714-718.

A20. Schelenz, R.; Zeiller, E. "Influence of digestion methods on the determination of total aluminum in food samples by ICP-OES" *Fresenius' J. Anal. Chem.* 1993, *345*, 68-71.

A21. Friel, J. K.; Skinner, C. S.; Jackson, S. E.; Longerich, H. P. "Analysis of biological reference materials, prepared by microwave dissolution, using inductively coupled plasma mass spectrometry" *Analyst* 1990, *115*, 269-273.

A22. Andrasi, E.; Dozsa, A.; Bezur, L.; Ernyei, L.; Molnar, Z. "Characterization of biological RMs [reference materials] for potential use in human brain analysis" *Fresenius' J. Anal. Chem.* 1993, *345*, 340-342.

A23. Lyon, T. D. B.; Fell, G. S.; McKay, K.; Scott, R. D. "Accuracy of multi-element analysis of human tissue obtained at autopsy using inductively coupled plasma mass spectrometry" *J. Anal. At. Spectrom.* 1991, 6, 559-564.

A24. Krushevska, A.; Barnes, R. M.; Amarasiriwaradena, C. "Decomposition of biological samples for inductively coupled plasma atomic-emission spectrometry using an open focused microwave digestion system" *Analyst* 1993, 118, 1175-1181.

A25. Gluodenis, T. J.; Tyson, J. F. "Flow-injection systems for directly coupling on-line digestions with analytical atomic spectrometry. II. Reactions in a microwave field" *J. Anal. At. Spectrom.* 1993, 8, 697-704.

A26. Heltai, G.; Percisch, K. "Moderated pressure microwave digestion system for preparation of biological samples" *Talanta* 1994, 41, 1067-1072.

A27. Kojima, I.; Kato, A.; Iida, C. "Microwave digestion of biological samples with acid mixture in a closed double PTFE vessel for metal determination by "one-drop" flame atomic absorption spectrometry" *Anal. Chim. Acta* 1992, 264, 101-106.

A28. Uchida, T.; Isoyama, H.; Oda, H.; Wada, H.; Uenoyama, H. "Determination of ultra-trace metals in biological standards by inductively coupled plasma atomic-emission spectrometry with ultrasonic nebulization" *Anal. Chim. Acta* 1993, 283, 881-886.

A29. Matusiewicz, H.; Suszka, A.; Ciszewski, A. "Efficiency of wet oxidation with pressurized sample digestion for trace analysis of human hair material" *Acta Chim. Hung.* 1991, 128, 849-859.

A30. Vasconcellas, M. B. A.; Maihara, V. A.; Favaro, D. I. T.; Armelin, M. J. A.; Toro, E. C.; Ogris, R. "Radiochemical separation methods for the determination of some toxic elements in biological reference materials" *J. Radioanal. Nucl. Chem.* 1991, 153, 185-199.

A31. Que Hee, S. S.; Boyle, J. R. "Simultaneous multielemental analysis of some environmental and biological samples by inductively coupled plasma atomic emission spectrometry" *Anal. Chem.* 1988, 60, 1033-1042.

A32. Kingston, H. M.; Jassie, L. B. "Microwave acid sample decomposition for elemental analysis" *J. Res. Nat. Bur. Stand.* 1988, 93, 269-274.

A33. Martines Stewart, L. J.; Barnes, R. M. "Flow-through, microwave-heated digestion chamber for automated sample preparation prior to inductively coupled plasma spectrochemical analysis" *Analyst* 1994, 119, 1003-1010.

A34. Munro, J. L. "Determination of six elements in infant formula by flame and furnace atomic absorption spectrometry" *At. Spectrosc.* 1987, 8, 92.

A35. White, R. T.; Open Reflex Vessels for Microwave Digestion: Botanical, Biological, and Food Samples for Elemental Analysis In *Introduction to Microwave Sample Preparation: Theory and Practice*; Jassie, L. B., Kingston, H. M., Eds.; ACS: Washington DC, 1988, pp 53-78.

A36.   Krushevska, A.; Barnes, R. M.; Amarasiriwaradena, C. J.; Foner, H. A.; Martines, L. "Determination of the residual carbon content by inductively coupled plasma atomic emission spectrometry after decomposition of biological samples" *J. Anal. At. Spectrom.* 1992, 7, 845 - 850.

A37.   Krushevska, A.; Barnes, R. M.; Amarasiriwaradena, C. J.; Foner, H. A.; Martines, L. "Comparison of sample decomposition procedures for the determination of zinc in milk by inductively coupled plasma atomic emission spectrometry" *J. Anal. At. Spectrom.* 1992, 7, 851 - 858.

A38.   Burguera, M.; Burguera, J. L.; Alarcon, O. M. "Determination of zinc and cadmium in small amounts of biological tissues by microwave-assisted digestion and flow-injection atomic-absorption spectrometry" *Anal. Chim. Acta* 1988, 214, 421-427.

A39.   Haswell, S. J.; Barclay, D. A. "On-line microwave digestion of slurry samples with direct flame atomic absorption spectrometric elemental detection" *Analyst* 1992, 117, 117-120.

A40.   Manca, D.; Lefebvre, M.; Trottier, B.; Lapare, S.; Ricard, A. C.; Van Tra, H.; Chevalier, G. "Micro method for determination of cadmium in tissues and slurried samples by use of flameless atomic-absorption spectrometry" *Microchem. J.* 1992, 46, 249-258.

A41.   Sapp, R. E.; Davidson, S. D. "Microwave digestion of multi-component foods for sodium analysis by atomic absorption spectrometry" *J. Food Sci.* 1991, 56, 1412-1414.

A42.   White, R. T.; Doughit, G. E. "Use of microwave oven and nitric acid-hydrogen peroxide digestion to prepare botanical materials for elemental analysis by inductively coupled argon plasma emission spectroscopy" *J. Assoc. Off. Anal. Chem.* 1985, 68, 766-769.

A43.   Kingston, H. M.; Jassie, L. B. "Microwave energy for acid decomposition at elevated temperatures and pressures using biological and botanical samples" *Anal. Chem.* 1986, 58, 2534-2541.

A44.   Patterson, K. Y.; Veillon, C.; Kingston, H. M.; Microwave Digestion of Biological Samples: Selenium Analysis by Electrothermal Atomic Absorption Spectrometry In *Introduction to Microwave Sample Preparation: Theory and Practice*; Kingston, H. M., Jassie, L. B., Eds.; ACS: Washington DC, 1988, pp 155-166.

A45.   Pratt, K. W.; Kingston, H. M.; MacCrehan, W. A.; Koch, W. F. "Voltammetric and liquid chromatographic identification of organic products of microwave-assisted wet ashing of biological samples" *Anal. Chem.* 1988, 60, 2024-2027.

A46.   Abu Samra, A.; Koirtyohann, S. R.; Morris, J. S.: "Method and apparatus for the wet digestion of organic and biological samples", USA Patent, 1978,

A47.   Abu Samra, A.; Morris, J. S.; Koirtyohann, S. R. "Wet ashing of some biological samples in a microwave oven" *Anal. Chem.* 1975, 47, 1475-1477.

A48.   Abu Samra, A.; Morris, J. S.; Koirtyohann, S. R. "Wet ashing of some biological samples in a microwave oven" *Trace Subst. Environ. Health* 1975, 9, 297-301.

A49. Lachica, M. "Use of microwave oven for the determination of mineral elements in biological material" *Analusis* 1990, *18*, 331-333.

A50. Aysola, P.; Anderson, P. D.; Langford, C. H. "Wet ashing in biological samples in a microwave oven under pressure using poly(tetrafluoroethylene) vessels" *Anal. Chem.* 1987, *59*, 1582-1583.

A51. Aysola, P.; Anderson, P. D.; Langford, C. H. "An open-vessel pulse-microwave technique for wet ashing of metal-contaminated animal tissues" *Anal. Lett.* 1988, *21*, 2003-2010.

A52. De Boer, J. L. M.; Maessen, F. J. M. J. "A comparative examination of sample treatment procedures for ICP-OES analysis of biological tissue" *Spectrochim. Acta* 1983, *38B*, 739-746.

A53. Van Wyck, D. B.; Schifman, R. B.; Stivelman, J. C.; Rulz, J.; Martin, D. "Rapid sample preparation for determination of iron in tissue by closed-vessel digestion and microwave energy" *Clin. Chem.* 1988, *34*, 1128-1130.

A54. Miller Ihli, N. J. "Trace-element determinations in biologicals using atomic-absorption spectrometry" *J. Res. Nat. Bur. Stand.* 1988, *93*, 350-354.

A55. Nadkarni, R. A. "Applications of microwave oven sample dissolution in analysis" *Anal. Chem.* 1984, *56*, 2233-2237.

A56. Bettinelli, M.; Baroni, U.; Pastorelli, N. J. "Microwave oven sample dissolution for the analysis of environmental and biological materials" *Anal. Chim. Acta* 1989, *225*, 159-174.

A57. Siriraks, A.; Kingston, H. M.; Riviello, J. M. "Chelation ion chromatography as a method for trace elemental analysis in complex environmental and biological samples" *Anal. Chem.* 1990, *62*, 1185-1193.

A58. Karanassios, V.; Li, F. H.; Liu, B.; Salin, E. D. "Rapid stopped-flow microwave digestion system" *J. Anal. At. Spectrom.* 1991, *6*, 457-463.

A59. Matusiewicz, H.; Sturgeon, R. E. "Comparison of the efficiencies of on-line and high-pressure closed vessel approaches to microwave heated sample decomposition" *Fresenius' J. Anal. Chem.* 1994, *349*, 428-433.

A60. Mayer, D.; Haubenwallner, S.; Kosmus, W. "Modified electrical heating system for hydride generation atomic absorption spectrometry and elaboration of a digestion method for the determination of arsenic and selenium in biological materials" *Anal. Chim. Acta* 1992, *268*, 315-321.

A61. Matusiewicz, H. "Development of a high pressure/temperature focused microwave heated teflon bomb for sample preparation" *Anal. Chem.* 1994, *66*, 751-755.

A62. Schmit, J. P.; Youla, M.; Gelinas, Y. "Multi-element analysis of biological tissues by inductively coupled plasma mass spectrometry" *Anal. Chim. Acta* 1991, *12*, 495-501.

A63. Motkosky, N.; Kratochvil, B. G. "Characterization of trace amounts of aluminum in biological reference materials by electrothermal atomic-absorption spectrometry" *Analyst* 1993, *118*, 1313-1316.

A64. Evans, S.; Krahenbuhl, U. "Boron analysis in biological material: microwave digestion procedure and determination by different methods" *Fresenius' J. Anal. Chem.* 1994, *349*, 454-459.

A65. Koh, T. S. "Microwave drying of biological tissues for trace-element determination" *Anal. Chem.* 1980, *52*, 1978-1979.

A66. Lofty, D. "Faster microwave digestion by flow-injection" *Lab. Equip. Dig.* 1992, *30*, 13,15.

A67. Oles, P. J.; Graham, W. M. "Microwave acid-digestion of various food matrices for nutrient determination by atomic-absorption spectrophotometry" *J. Assoc. Off. Anal. Chem.* 1991, *74*, 812-814.

A68. Martin, C. K.; Williams, J. C. "Determination of selenium in bovine liver by Zeeman-effect atomic-absorption spectrometry using a palladium - copper chemical modifier" *J. Anal. At. Spectrom.* 1989, *4*, 691-695.

A69. Greenberg, R. R.; Zeisler, R.; Kingston, H. M.; Sullivan, T. M. "Neutron activation analysis of the NIST bovine serum standard reference material using chemical separations" *Fresenius' J. Anal. Chem.* 1988, *332*, 652-656.

A70. Tsalev, D. L.; Sperling, M.; Welz, B. "On-line microwave sample pre-treatment for hydride-generation and cold-vapor atomic-absorption spectrometry. II Chemistry and applications" *Analyst* 1992, *117*, 1735-1741.

A71. Tsalev, D. L.; Sperling, M.; Welz, B. "On-line microwave sample pre-treatment for hydride generation and cold vapor atomic-absorption spectrometry. I. The manifold" *Analyst* 1992, *117*, 1729-1733.

A72. Nieuwenhuize, J.; Poley Vos, C. H. "A rapid microwave dissolution method for determination of trace and minor elements in lyophilized [freeze-dried] plant material" *At. Spectrosc.* 1989, *10*, 148-153.

A73. Saumer, M.; Gantner, E.; Reinhardt, J.; Ache, H. J. "Determination of molybdenum in plant reference material by thermal-ionization isotope-dilution mass spectrometry" *Fresenius' J. Anal. Chem.* 1992, *344*, 109-113.

A74. Noltner, T.; Maisenbacher, P.; Puchelt, H. "Microwave acid digestion of geological and biological standard reference materials for trace element analysis by inductively coupled plasma-mass spectrometry" *Spectroscopy* 1990, *5*, 49-53.

A75. Yamane, T.; Koshino, K. "Flow-injection spectrophotometric determination of trace cobalt with 2-(5-bromo-2-pyridylazo)-5-(Nitrogen-propyl-Nitrogen-sulfopropylamino)aniline: use of the rate-enhancing effect of copper(II) on the complex formation reaction of cobalt" *Anal. Chim. Acta* 1992, *261*, 205-211.

A76. Kojima, I.; Kondo, S. "Sensitive "one drop" flame atomic-absorptiometric determination of cadmium in botanical samples using direct nebulization of chloroform extract" *J. Anal. At. Spectrom.* 1993, *8*, 115-118.

A77. Ebdon, L.; Fisher, A. S.; Worsfold, P. J. "Determination of arsenic, chromium, selenium and vanadium in biological samples by inductively coupled plasma mass spectrometry using on-line elimination of interference and pre-concentration by flow injection" *J. Anal. At. Spectrom.* 1994, *9*, 611-614.

A78. Greenberg, R. R.; Kingston, H. M.; Watters, R. L.; Pratt, K. W. "Dissolution problems with botanical reference materials" *Fresenius' J. Anal. Chem.* 1990, *338*, 394-398.

A79. Miller, R. O. "Principles and practices of microwave dissolution/digestion of plant material", International Soil Testing and Plant Analysis Symposium, Orlando, FL 1991.

A80. Schelkoph, G. M.; Milne, D. B. "Wet microwave digestion of diet and fecal samples for inductively coupled plasma analysis" *Anal. Chem.* 1988, *60*, 2060-2062.

A81. Kalra, Y. P.; Maynard, D. G.; Radford, F. G. "Microwave digestion of tree foliage for multi-element analysis" *Can. J. For. Res.* 1989, *19*, 981-985.

A82. Towler, P. H.; Smith, J. D. "Recovery of polonium from microwave bomb digestions" *Anal. Chim. Acta* 1994, *292*, 209-212.

A83. Alvarado, J.; Cristiano, A. R. "Determination of cadmium, cobalt, iron, nickel and lead in Venezuelan cigarettes by electrothermal atomic-absorption spectrometry" *J. Anal. At. Spectrom.* 1993, *8*, 253-259.

A84. Krasowski, J. A.; Copeland, T. R. "Matrix interferences in furnace atomic absorption spectrometry" *Anal. Chem.* 1979, *51*, 1843-1849.

A85. Kimber, G. M.; Kokot, S. "Practical options for microwave digestions" *Trends in Anal. Chem.* 1990, *9*, 203-207.

A86. Baxter, D. C.; Nichol, R.; Littlejohn, D. "Evaluation of cold-vapor furnace atomic non-thermal excitation spectrometry for the determination of mercury in environmental samples" *Spectrochim. Acta* 1992, *47B*, 1155-1163.

A87. Navarro Alarcon, M.; Lopez Martinez, M. C.; Sanchez Vinas, M.; Lopez Garcia De La Serrana, H. "Determination of mercury in crops by cold-vapor atomic-absorption spectrometry after microwave dissolution" *J. Agric. Food Chem.* 1991, *39*, 2223-2225.

A88. Xu, N.; Majidi, V.; Ehmann, W. D.; Markesbery, W. R. "Determination of aluminum in human brain tissue by electrothermal atomic-absorption spectrometry" *J. Anal. At. Spectrom.* 1992, *7*, 749-751.

A89. Yagi, M.; Masumoto, K. "Stable-isotope dilution activation analysis for special samples in which the self-shielding effect is negligible. Determination of strontium in biological materials by means of photon activation" *J. Radioanal. Nucl. Chem.* 1985, *90*, 91-103.

A90. Mincey, D. W.; Williams, R. C.; Giglio, J. J.; Graves, G. A.; Pacella, A. J. "Temperature controlled microwave oven digestion system" *Anal. Chim. Acta* 1992, *264*, 97-100.

A91. Mateo, M. A.; Sabate, S. "Wet digestion of vegetable tissue using a domestic microwave oven" *Anal. Chim. Acta* 1993, *279*, 273-279.

A92. Dean, J. R. "Ion-chromatographic determination of aluminum with ultra-violet spectrophotometric detection" *Analyst* 1989, *114*, 165-168.

A93. Ebdon, L.; Ford, M. J.; Hutton, R. C.; Hill, S. J. "Evaluation of ethene addition to the nebulizer gas in inductively coupled plasma mass spectrometry for the removal of matrix-, solvent-, and support-gas-derived polyatomic-ion interferences" *Appl. Spectrosc.* 1994, *48*, 507-516.

A94. Vereda Alonso, E.; Garcia de Torres, A.; Cano Pavon, J. M. "Determination of nickel in biological samples by inductively coupled plasma atomic-emission spectrometry after extraction with 1,5-bis[phenyl-(2-pyridyl)methylene]thiocarbonohydrazide" *J. Anal. At. Spectrom.* 1993, 8, 843-846.

A95. Morales Rubio, A.; Salvador, A.; De La Guardia, M. "Microwave muffle-furnace-assisted decomposition of vegetable samples for flame atomic-spectrometric determination of calcium, magnesium, potassium, iron, manganese and zinc" *Fresenius' J. Anal. Chem.* 1992, 342, 452-456.

A96. Zunk, B. "Microwave digestion of plant material for trace element determination" *Anal. Chim. Acta* 1990, 236, 337-343.

A97. Carbonell, V.; Morales Rubio, A.; Salvador, A.; De La Guardia, M.; Burguera, J. L.; Burguera, M. "Atomic-absorption spectrometric analysis of solids with on-line microwave-assisted digestion" *J. Anal. At. Spectrom.* 1992, 7, 1085-1089.

A98. Amarasiriwaradena, C. J.; Gercken, B.; Argentine, M. D.; Barnes, R. M. "Semi-quantitative analysis by inductively coupled plasma mass spectrometry" *J. Anal. At. Spectrom.* 1990, 5, 457-462.

A99. De La Guardia, M.; Carbonell, V.; Morales Rubio, A.; Salvador, A. "On-line microwave-assisted digestion of solid samples for their flame-atomic-spectrometric analysis" *Talanta* 1993, 40, 1609-1617.

A100. Stock, D. J. "Flame atomic absorption spectrometric determination of copper-8-quinilinolate anti-stain on treated lumber surfaces" *J. Anal. At. Spectrom.* 1990, 5, 631-634.

A101. Mohd, A. A.; Dean, J. R.; Tomlinson, W. R. "Factorial design approach to microwave dissolution" *Analyst* 1992, 117, 1743-1748.

A102. Ybanez, N.; Cervera, M. L.; Montoro, R.; Guardia, M. "Comparison of dry mineralization and microwave-oven digestion for the determination of arsenic in mussel products by platform in furnace Zeeman-effect atomic absorption spectrometry" *J. Anal. At. Spectrom.* 1991, 6, 379-384.

A103. Cabrera, C.; Lorenzo, M. L.; Gallego, C.; Lopez Martinez, M. C.; Lillo, E. "Cadmium contamination levels in seafood determined by electrothermal atomic-absorption spectrometry after microwave dissolution" *J. Agric. Food Chem.* 1994, 42, 126-128.

A104. Cabrera, C.; Lorenzo, M. L.; Gallego, C.; Lopez Martinez, M. C. "Determination of lead in fish by electrothermal atomic absorption spectrometry" *Anal. Chim. Acta* 1991, 246, 375-378.

A105. Navarro Alarcon, M.; Lopez Martinez, M. C.; Lopez, H.; Sanchez, M. "Microwave dissolution for the determination of mercury in fish by cold vapor atomic absorption spectrometry" *Anal. Chim. Acta* 1992, 257, 155-158.

A106. Lan, W. G.; Wong, M. K.; Sin, Y. M. "Microwave digestion of fish tissue for selenium determination by differential pulse polarography" *Talanta* 1994, 41, 53-58.

A107. Lan, W. G.; Wong, M. K.; Sin, Y. M. "Comparison of four microwave digestion methods for the determination of selenium in fish tissue by using hydride generation atomic-absorption spectrometry" *Talanta* 1994, 41, 195-200.

A108. Pergantis, S. A.; Cullen, W. R.; Wade, A. P. "Simplex optimization of conditions for the determination of arsenic in environmental samples by using electrothermal atomic-absorption spectrometry" *Talanta* 1994, *41*, 205-209.

A109. Demura, R.; Tsukada, S.; Yamamoto, I. "Rapid determination of trace metals in foods by using the microwave oven-digestion method II. Determination of zinc, copper, manganese, lead and cadmium" *Eisei Kagaku* 1985, *31*, 405-409.

A110. Soto Ferreiro, R. M.; Bermejo Barrera, P. "High pressure acid digestion using microwave heating for the determination of zinc, iron and copper in mussels by flame atomic-absorption spectrometry" *Analusis* 1993, *21*, 197-199.

A111. Copa Rodriguez, F. J.; Basadre Pampin, M. I. "Determination of iron, copper and zinc in tinned mussels by inductively coupled plasma atomic-emission spectrometry (ICP-OES)" *Fresenius' J. Anal. Chem.* 1994, *348*, 390-395.

A112. Soto Ferreiro, R. M.; Laino, C. C.; Bermejo Barrera, P. "Comparative study of sample preparation methods for zinc, iron and copper determination in mussels by flame atomic absorption spectrometry" *Anal. Lett.* 1991, *24*, 2277-2292.

A113. Lamleung, S. Y.; Cheng, K. W.; Lam, Y. W. "Application of a microwave oven for drying and nitric acid extraction of mercury and selenium from fish tissue" *Analyst* 1991, *116*, 957-959.

A114. Branch, S.; Ebdon, L.; O'Neill, P. "Determination of arsenic species in fish by directly coupled high-performance liquid chromatography-inductively coupled plasma mass spectrometry" *J. Anal. At. Spectrom.* 1994, *9*, 33-37.

A115. McLaren, J. W.; Siu, K. W. M.; Lam, J. W.; Willie, S. N.; Maxwell, P. S.; Palepu, A.; Koether, M.; Berman, S. S. "Applications of ICP-MS in marine analytical chemistry" *Fresenius' J. Anal. Chem.* 1990, *337*, 721-728.

A116. Fernando, A. R.; Kratochvil, B. G. "Internal standards in differential pulse anodic stripping voltammetry" *Can. J. Chem.* 1991, *69*, 755-758.

A117. Beauchemin, D.; McLaren, J. W.; Berman, S. S. "Use of external calibration for the determination of trace metals in biological materials by inductively coupled plasma mass spectrometry" *J. Anal. At. Spectrom.* 1988, *3*, 775-780.

A118. Nakashima, S.; Sturgeon, R. E.; Willie, S. N.; Berman, S. S. "Acid digestion of marine samples for trace element analysis using microwave heating" *Analyst* 1988, *113*, 159-163.

A119. Matusiewicz, H.; Sturgeon, R. E.; Berman, S. S. "Trace element analysis of biological material following pressure digestion with nitric acid - hydrogen peroxide and microwave heating" *J. Anal. At. Spectrom.* 1989, *4*, 323-327.

A120. McCarthy, H. T.; Ellis, P. C. "Comparison of microwave digestion with conventional wet ashing and dry ashing digestion for analysis of lead, cadmium, chromium, copper, and zinc in shellfish by flame atomic absorption spectroscopy" *J. Assoc. Off. Anal. Chem.* 1991, *74*, 566-569.

A121. Matusiewicz, H.; Sturgeon, R. E.; Berman, S. S. "Vapor-phase acid digestion of inorganic and organic matrices for trace element analysis using a microwave heated bomb" *J. Anal. At. Spectrom.* 1991, *6*, 283-287.

Appendix 2. Geological and Metallurgical Reference Materials

Appendix 2.1 Rocks and Minerals

Matrix	Analytes	Reagents	Cavity/Vessel	MW conditions	Detection	Ref.
AIS Vanadium titanium ore BH-102	Al, Si, Ca, Co, Cu, Mg, Fe, Ti, Mn, Ni	HNO_3, HF	Multimode/ LP closed	10 min at 480 W, 8 min at 360 W (for ICP) 10 min at 480 W, 5 min at 360 W (for AA)	ICP-OES F-AAS	B1
AIS Vanadium titanium ore BH-104	Al, Si, Ca, Co, Cu, Mg, Fe, Ti, Mn, Ni	HNO_3, HF	Multimode/ LP closed	10 min at 480 W, 8 min at 360 W (for ICP) 10 min at 480 W, 5 min at 360 W (for AA)	ICP-OES F-AAS	B1
ANRT Bauxite BX-N	Al, Ba, K, Li, Ni, Si, Ti, Y, Zr	1. HNO_3, HCl, HF 2. H_3BO_3	Multimode/ LP closed	1. 2.5 min at 650 W 2. 10 min at 650 W	ICP-OES	B2
ANRT Serpentine UB-N	Al, Cr, K, Li, Si, V, Zn	1. HNO_3, HCl, HF 2. H_3BO_3	Multimode/ LP closed	1. 2.5 min at 650 W 2. 10 min at 650 W	ICP-OES	B2
AR Trace metals in coal 1800	Ni, V	HNO_3, HCl	Multimode/ LP closed	1 min at 600 W, cool (x2)	ETV-AAS	B3, B4
AR Trace metals in coal 1801	Ni, V	HNO_3, HCl	Multimode/ LP closed	1 min at 600 W, cool (x2)	ETV-AAS	B3
AR Trace metals in coal 1802	Ni, V	HNO_3, HCl	Multimode/ LP closed	1 min at 600 W, cool (x2)	ETV-AAS	B3
AR Trace metals in coal 1803	Ni, V	HNO_3, HCl	Multimode/ LP closed	1 min at 600 W, cool (x2)	ETV-AAS	B3
ASCRM Coal 009	P	1. HCl, HF 2. H_3BO_3	Multimode/ LP closed	1. 45 sec at 500 W 2. 60 sec at 500 W	UV-Vis	B5

Appendix 2.1 Rocks and Minerals

Matrix	Analytes	Reagents	Cavity/Vessel	MW conditions	Detection	Ref.
BCR Gas coal 180	Al, As, Ba, Ca, Cd, Cr, Co, Cu, Fe, K, Mg, Mn, Na, Ni, Pb, Sb, Sc, Se, Si, Sn, Ti, Tl, V, Zn, Zr	1. HF, Aqua regia 2. H_3BO_3	Multimode/ LP closed	1. 10 min at 300 W, 5 min at 600 W, 10 min at 480 W 2. 6 min at 300 W	ICP-OES F-AAS ETV-AAS	B6
BCR Coking coal 181	Al, As, Ba, Ca, Cd, Cr, Co, Cu, Fe, K, Mg, Mn, Na, Ni, Pb, Sb, Sc, Se, Si, Sn, Ti, Tl, V, Zn, Zr	1. HF, Aqua regia 2. H_3BO_3	Multimode/ LP closed	1. 10 min at 300 W, 5 min at 600 W, 10 min at 480 W 2. 6 min at 300 W	ICP-OES F-AAS ETV-AAS	B6
BCR Steam coal 182	Al, As, Ba, Ca, Cd, Cr, Co, Cu, Fe, K, Mg, Mn, Na, Ni, Pb, Sb, Sc, Se, Si, Sn, Ti, Tl, V, Zn, Zr	1. HF, Aqua regia 2. H_3BO_3	Multimode/ LP closed	1. 10 min at 300 W, 5 min at 600 W, 10 min at 480 W 2. 6 min at 300 W	ICP-OES F-AAS ETV-AAS	B6
CANMET Uranium rich sandstone DL - 1	Th, U	1. HCl, HNO₃, HF 2. Aqua regia	Multimode/ HP closed	1. 4 - 8 min at 600 W fume off acid (x3) 2. If residue 3 min at 600 W	α spectrometry	B7
CANMET Ore SY-2	Al, Ca, Fe, K, Mg, Mn, Na, Ti, Si	**A.** 1. HCl, HF 2. HBO₃ **B.** 1. HCl, HF, H_2SO_4 2. HBO₃ **C.** 1. Aqua regia, HF 2. HBO₃ **D.** 1. HCl, HF, H_2O_2 2. HBO₃	Multimode/ LP closed Multimode/ LP closed Multirode/ LP closed Multimode/ LP closed	1. 10-13 sec at 1470 W 1. Same as A 1. Same as A 1. Same as A	F-AAS	B8
CCRMP Copper concentrate CCU-1	Al, Ca, Cu, Fe, K, Li, Pb, Si	1. HNO₃, HCl, HF 2. H_3BO_3	Multimode/ LP closed	1. 2.5 min at 650 W 2. 10 min at 650 W	ICP-OES	B2

Sample	Elements	Reagents	Mode/Vessel	Power program	Technique	Ref.
CCRMP Lead concentrate CPB-1	Al, K, Li, Mg, Si, Sn	1. HNO_3, HCl, HF 2. H_3BO_3	Multimode/LP closed	1. 2.5 min at 650 W 2. 10 min at 650 W	ICP-OES	B2
CCRMP Zinc concentrate CZN-1	Al, As, K, Li, Si, Sn	1. HNO_3, HCl, HF 2. H_3BO_3	Multimode/LP closed	1. 2.5 min at 650 W 2. 10 min at 650 W	ICP-OES	B2
CCRMP Zinc lead tin silver ore KC-1	Al, Ag, K, Li, Si, Sn	1. HNO_3, HCl, HF 2. H_3BO_3	Multimode/LP closed	1. 2.5 min at 650 W 2. 10 min at 650 W	ICP-OES	B2
CCRMP Gabbro MRG-1	Al, Cr, Co, K, Li, Si, Zn	1. HNO_3, HCl, HF 2. H_3BO_3	Multimode/LP closed	1. 2.5 min at 650 W 2. 10 min at 650 W	ICP-OES	B2
CCRMP Zinc tin copper lead ore MP-1	Al, K, Li, Si, Sn	1. HNO_3, HCl, HF 2. H_3BO_3	Multimode/LP closed	1. 2.5 min at 650 W 2. 10 min at 650 W	ICP-OES	B2
CCRMP SO-2	Nb, Mo, Ta, W	1. HNO_3 2. HF, H_2O	Multimode/MP closed	1. 30 min at 200 W 2. 1 min at 1000 W, 30 min at 450 W	ID-ICP-MS	B9
CCRMP Syenite SY-3	Al, Co, K, Li, Si, Zr	1. HNO_3, HCl, HF 2. H_3BO_3	Multimode/LP closed	1. 2.5 min at 650 W 2. 10 min at 650 W	ICP-OES	B2
	Nb, Mo, Ta, W	1. HNO_3 2. HF, H_2O	Multimode/MP closed	1. 30 min at 200 W 2. 1 min at 1000 W, 30 min at 450 W	ID-ICP-MS	B9
CRPG Biotite Mica-Fe	Al, Ce, K, La, Li, Si, Y, Zr	1. HNO_3, HCl, HF 2. H_3BO_3	Multimode/LP closed	1. 2.5 min at 650 W 2. 10 min at 650 W	ICP-OES	B2
	Nb, Mo, Ta, W	1. HNO_3 2. HF, H_2O	Multimode/MP closed	1. 30 min at 200 W 2. 1 min at 1000 W, 30 min at 450 W	ID-ICP-MS	B9
CRPG Mica-Mg	Nb, Mo, Ta, W	1. HNO_3 2. HF, H_2O	Multimode/MP closed	1. 30 min at 200 W 2. 1 min at 1000 W, 30 min at 450 W	ID-ICP-MS	B9

Appendix 2.1 Rocks and Minerals

Matrix	Analytes	Reagents	Cavity/Vessel	MW conditions	Detection	Ref.
GSJ Andesite JA-1	Fe, K, Mg, Mn, Na, Si	A. HNO_3, HF	Multimode/MP closed	0.5 - 2 min to 68 atm (500 W)	F-AAS	B10
		B. HNO_3, HCl, HF	Multimode/MP closed	Same as A		
GSJ Andesite JA-1 (contd.)	Fe	H_2SO_4, HF	Multimode/LP closed	1 min at 500 W, 5 min at 0 W (x3)	UV-Vis	B11
GSJ Andesite JA-3	Fe	H_2SO_4, HF	Multimode/LP closed	1 min at 500 W, 5 min at 0 W (x3)	UV-Vis	B11
GSJ Basalt JB-1	Cu	HNO_3, HF	Multimode/LP closed	1 min at full power (x2)	FIA	B12
	Fe	H_2SO_4, HF	Multimode/LP closed	1 min at 500 W, 5 min at 0 W (x3)	UV-Vis	B11
	Nb, Mo, Ta, W	1. HNO_3 2. HF, H_2O	Multimode/MP closed	1. 30 min at 200 W 2. 1 min at 1000 W, 30 min at 450 W	ID-ICP-MS	B9
GSJ Basalt JB-2	Fe	H_2SO_4, HF	Multimode/LP closed	1 min at 500 W, 5 min at 0 W (x3)	UV-Vis	B11
GSJ Basalt JB-3	Fe	H_2SO_4, HF	Multimode/LP closed	1 min at 500 W, 5 min at 0 W (x3)	UV-Vis	B11
GSJ Gabbro JG-1	Fe	H_2SO_4, HF	Multimode/LP closed	1 min at 500 W, 5 min at 0 W (x3)	UV-Vis	B11
	Nb, Mo, Ta, W	1. HNO_3 2. HF, H_2O	Multimode/MP closed	1. 30 min at 200 W 2. 1 min at 1000 W, 30 min at 450 W	ID-ICP-MS	B9

Sample	Elements	Reagents	Mode/Vessel	Program	Detection	Ref.
IWG AC-E	Nb, Mo, Ta, W	1. HNO_3 2. HF, H_2O	Multimode/ MP closed	1. 30 min at 200 W 2. 1 min at 1000 W, 30 min at 450 W	ID-ICP-MS	B9
IWG AL-I	Nb, Mo, Ta, W	1. HNO_3 2. HF, H_2O	Multimode/ MP closed	1. 30 min at 200 W 2. 1 min at 1000 W, 30 min at 450 W	ID-ICP-MS	B9
NAGRA Granite BOE-7	Th, U	1. HCl, HNO_3, HF 2. Aqua regia	Multimode/ HP closed	1. 4 - 8 min at 600 W fume off acid (x3) 2. If residue 3 min at 600 W	α spectrometry	B7
NAGRA Granite BOE-8	Th, U	1. HCl, HNO_3, HF 2. Aqua regia	Multimode/ HP closed	1. 4 - 8 min at 600 W fume off acid (x3) 2. If residue 3 min at 600 W	α spectrometry	B7
NIM Granite G	Ce, Dy, Er, Eu, Gd, Ho, La, Lu, Nd, Pr, Sm, Tb, Tm, Yb	1. HF 2. HNO_3, HCl, HF	Multimode/ MP closed	1. Fume to dryness (hotplate) 2. 10 min at 30 bar	γ-spectrometry, X-ray ICP-OES	B13
NIM Lajaurite L	Ce, Dy, Er, Eu, Gd, Ho, La, Lu, Nd, Pr, Sm, Tb, Tm, Yb	1. HF 2. HNO_3, HCl, HF	Multimode/ MP closed	1. Fume to dryness (hotplate) 2. 10 min at 30 bar	γ-spectrometry, X-ray ICP-OES	B13
NIM Norite N	Ce, Dy, Er, Eu, Gd, Ho, La, Lu, Nd, Pr, Sm, Tb, Tm, Yb	1. HF 2. HNO_3, HCl, HF	Multimode/ MP closed	1. Fume to dryness (hotplate) 2. 10 min at 30 bar	γ-spectrometry, X-ray ICP-OES	B13
NIM Syenite S	Ce, Dy, Er, Eu, Gd, Ho, La, Lu, Nd, Pr, Sm, Tb, Tm, Yb	1. HF 2. HNO_3, HCl, HF	Multimode/ MP closed	1. Fume to dryness (hotplate) 2. 10 min at 30 bar	γ-spectrometry, X-ray ICP-OES	B13
NIST Limestone 1	Al, K, Li, Si, Ti	1. HNO_3, HCl, HF 2. H_3BO_3	Multimode/ LP closed	1. 2.5 min at 650 W 2. 10 min at 650 W	ICP-OES	B2
NIST Mn ore 25	Al, K, Li, Si	1. HNO_3, HCl, HF 2. H_3BO_3	Multimode/ LP closed	1. 2.5 min at 650 W 2. 10 min at 650 W	ICP-OES	B2
NIST Fe ore 27	Al, Fe, K, Li, Si	1. HNO_3, HCl, HF 2. H_3BO_3	Multimode/ LP closed	1. 2.5 min at 650 W 2. 10 min at 650 W	ICP-OES	B2
NIST Phosphate rock 56	Al, K, Li, Si	1. HNO_3, HCl, HF 2. H_3BO_3	Multimode/ LP closed	1. 2.5 min at 650 W 2. 10 min at 650 W	ICP-OES	B2

Appendix 2.1 Rocks and Minerals

Matrix	Analytes	Reagents	Cavity/Vessel	MW conditions	Detection	Ref.
NIST Bauxite 69	Al, K, Li, Mg, Si, Ti, Zr	1. HNO₃, HCl, HF 2. H₃BO₃	Multimode/ LP closed	1. 2.5 min at 650 W 2. 10 min at 650 W	ICP-OES	B2
NIST Feldspar potash 70	Al, Ca. K, Na, Si	HCl, HF, HNO₃	Multimode/ LP closed	1-3 min at 650 W	F-AAS	B14
NIST Fluorspar 79	Al, K, Li, Si	1. HNO₃, HCl, HF 2. H₃BO₃	Multimode/ LP closed	1. 2.5 min at 650 W 2. 10 min at 650 W	ICP-OES	B2
NIST Limestone, dolomitic 88	Al, K, Li, Si, Ti	1. HNO₃, HCl, HF 2. H₃BO₃	Multimode/ LP closed	1. 2.5 min at 650 W 2. 10 min at 650 W	ICP-OES	B2
	Al, Fe, Ca, K, Mg, Na	1. Ethanoic acid 2. HNO₃, HF 3. H₃BO₃	Multimode/ LP closed	1. 150°C to near dryness (hotplate) 2. 3 min at 630 W, 30 min at 567 W 3. 10 min at 378 W	F-AAS	B15
NIST Flint clay 97	Al, Ca, Cr, K, Li, Mg, Si, Ti, Zr	1. HNO₃, HCl, HF 2. H₃BO₃	Multimode/ LP closed	1. 2.5 min at 650 W 2. 10 min at 650 W	ICP-OES	B2
	Cd, Cr	1. HNO₃, HCl, HF 2. HBO₃	Multimode/ LP closed	1. 30 sec at 600 W, 1 min at 0 W (x10) 2. 30 sec at 600 W, 1 min at 0 W (x6)	ETV-AAS	B16
NIST Feldspar soda 99	Al, K, Li, Si, Ti	1. HNO₃, HCl, HF 2. H₃BO₃	Multimode/ LP closed	1. 2.5 min at 650 W 2. 10 min at 650 W	ICP-OES	B2
NIST Chrome refractory 103	Al, Cr, Fe, K, Li, Mg, Mn, Si	1. HNO₃, HCl, HF 2. H₃BO₃	Multimode/ LP closed	1. 2.5 min at 650 W 2. 10 min at 650 W	ICP-OES	B2
NIST Zinc ore 113	Al, K, Li, Si	1. HNO₃, HCl, HF 2. H₃BO₃	Multimode/ LP closed	1. 2.5 min at 650 W 2. 10 min at 650 W	ICP-OES	B2

Sample	Analytes	Reagents	Mode/Vessel	Program	Detection	Ref.
NIST Phosphate rock 120	Al, K, Li, Si, Ti	1. HNO_3, HCl, HF 2. H_3BO_3	Multimode/LP closed	1. 2.5 min at 650 W 2. 10 min at 650 W	ICP-OES	B2
	-	HBF_4, HNO_3	Multimode/LP closed	2 min at 600 W		B17
	Cd, Cr	1. HNO_3, HCl, HF 2. HBO_3	Multimode/LP closed	1. 30 sec at 600 W, 1 min at 0 W (x10) 2. 30 sec at 600 W, 1 min at 0 W (x6)	ETV-AAS	B16
	Al, Ba, Ca, Cd, Cr, Fe, K, Li, Mg, Mn, Na, P, S, Si, Sr, Ti, V, Zn	A. 1. HNO_3 2. HF 3. HBO_3	Multimode/MP closed	1. 15 min at 0 W 2. 8 min at 60 W, 6 min at 90 W, 9 min at 60 W	ICP-OES	B18
		B. HNO_3, $HClO_4$	Multimode/open	15 min at 0 W, 10 min at 10 W, 10 min at 40 W, 15 min at 0 W (repeat procedure)		
NIST Sn ore 138	Al, K, Li, Si, Sn	1. HNO_3, HCl, HF 2. H_3BO_3	Multimode/LP closed	1. 2.5 min at 650 W 2. 10 min at 650 W	ICP-OES	B2
NIST Spodumene lithium ore 181	Al, K, Li, Si	1. HNO_3, HCl, HF 2. H_3BO_3	Multimode/LP closed	1. 2.5 min at 650 W 2. 10 min at 650 W	ICP-OES	B2
NIST Petalite lithium ore 182	Al, K, Li, Si	1. HNO_3, HCl, HF 2. H_3BO_3	Multimode/LP closed	1. 2.5 min at 650 W 2. 10 min at 650 W	ICP-OES	B2
NIST Lepidolite lithium ore 183	Al, K, Li, Si	1. HNO_3, HCl, HF 2. H_3BO_3	Multimode/LP closed	1. 2.5 min at 650 W 2. 10 min at 650 W	ICP-OES	B2

Appendix 2.1 Rocks and Minerals

Matrix	Analytes	Reagents	Cavity/Vessel	MW conditions	Detection	Ref.
NIST Obsidian rock 278	Pb, U	HNO_3, HF, $HClO_4$	Multimode/ LP closed	5 min at 90 W, 15 min at 138 W, cool, 1 - 5 hr at 138 W	ID-MS	B19
	Al, Ba, Ca, Co, Cr, Cu, Fe, K, Mg, Mn, Na, Ni, Pb, Sc, Si, Ti, V, Zn, Zr	1. HF, Aqua regia 2. H_3BO_3	Multimode/ LP closed	1. 8 min at 300 W, 4 min at 600 W, 7 min at 480 W 2. 6 min at 600 W	ICP-OES	B20
	Al, As, Ba, Ca, Cd, Cr, Cu, Co, Fe, K, Mg, Mn, Na, Ni, Pb, Sb, Sc, Se, Si, Sn, Ti, Tl, V, Zn, Zr	1. HF, Aqua regia 2. H_3BO_3	Multimode/ LP closed	1. 8 min at 300 W, 11 min at 600 W 2. 6 min at 300 W	ICP-OES F-AAS ETV-AAS	B6
NIST Mo ore 333	Al, K, Li, Mo, Si	1. HNO_3, HCl, HF 2. H_3BO_3	Multimode/ LP closed	1. 2.5 min at 650 W 2. 10 min at 650 W	ICP-OES	B2
NIST Potassium feldspar 607	Pb, U	HNO_3, HF, $HClO_4$	Multimode/ LP closed	5 min at 90 W, 15 min at 138 W, cool, 1-5 hr at 138 W	ID-MS	B19
NIST Basalt rock 688	Al, Ba, Ca, Co, Cr, Cu, Fe, K, Mg, Mn, Na, Ni, Pb, Sc, Si, Ti, V, Zn, Zr	1. HF, Aqua regia 2. H_3BO_3	Multimode/ LP closed	1. 8 min at 300 W, 4 min at 600 W, 7 min at 480 W 2. 6 min at 600 W	ICP-OES	B20
	Al, As, Ba, Ca, Cd, Cr, Cu, Co, Fe, K, Mg, Mn, Na, Ni, Pb, Sb, Sc, Se, Si, Sn, Ti, Tl, V, Zn, Zr	1. HF, Aqua regia 2. H_3BO_3	Multimode/ LP closed	1. 8 min at 300 W, 11 min at 600 W 2. 6 min at 300 W	ICP-OES F-AAS ETV-AAS	B6
NIST Fe ore 692	Al, Fe, K, Li, Mg, Si	1. HNO_3, HCl, HF 2. H_3BO_3	Multimode/ LP closed	1. 2.5 min at 650 W 2. 10 min at 650 W	ICP-OES	B2

Sample	Elements	Reagents	Microwave mode	Program	Detection	Ref.
NIST Phosphate rock 694	Al, Ba, Ca, Cd, Cr, Fe, K, Li, Mg, Mn, Na, P, S, Si, Sr, Ti, V, Zn	A. 1. HNO_3 2. HF 3. HBO_3	Multimode/MP closed	1. 15 min at 0 W, 2. 8 min at 60 W, 6 min at 90 W, 9 min at 60 W	ICP-OES	B18
		B. HNO_3, $HClO_4$	Multimode/open	15 min at 0 W, 10 min at 10 W, 10 min at 40 W, 15 min at 0 W (repeat procedure)		
NIST Coal Bituminous 1632	Al, Ba, Ca, Cr, Cu, Fe, K, Li, Mg, Mn, Na, Ni, P, Si, Sr, Ti, V, Zr	Aqua regia, HF	Multimode/open	3 min at 625 W	ICP-OES	B21
	Fe, Ni, V	HNO_3, HCl	Multimode/LP closed	1 min at 600 W, cool (x2)	ETV-AAS	B3, B4
	Al, As, Ba, Ca, Cd, Cr, Cu, Co, Fe, K, Mg, Mn, Na, Ni, Pb, Sb, Sc, Se, Si, Sn, Ti, Tl, V, Zn, Zr	1. HF, Aqua regia 2. H_3BO_3	Multimode/LP closed	1. 10 min at 300 W, 5 min at 600 W, 10 min at 480 W, 2. 6 min at 300 W	ICP-OES F-AAS ETV-AAS	B6
NIST Coal fly ash 1633	Al, As, Ba, Be, Ca, Co, Cr, Cu, Fe, K, Li, Mg, Mn, Na, Ni, P, Pb, Si, Sr, Ti, V, Zr	Aqua regia, HF	Multimode/open	3 min at 625 W	ICP-OES	B21
	As, Cd, Pb, Sb, Se, Tl	A. 1. HF, Aqua regia 2. H_3BO_3	Multimode/LP closed	1. 8 min at 300 W, 4 min at 600 W, 7 min at 480 W, 2. 6 min at 600 W	ETV-AAS NAA	B22
		B. HNO_3	Multimode/LP closed	8 min at 300 W, 4 min at 600 W, 7 min at 480 W		
	Al, As, Ba, Be, Ca, Co, Cr, Cu, Fe, Mg, Mn, Mo, Ni, Pb, S, Sb, Ti, V, Zn	A. Aqua regia	Multimode/LP closed	2 min at 600 W, 5 min at 360 W, 15 min at 180 W	ICP-OES NAA	B23

Appendix 2.1 Rocks and Minerals

Matrix	Analytes	Reagents	Cavity/Vessel	MW conditions	Detection	Ref.
NIST Coal fly ash 1633 (contd.)	Al, As, Ba, Be, Ca, Co, Cr, Cu, Fe, Mg, Mn, Mo, Ni, Pb, S, Sb, Ti, V, Zn	**B.** 1. Aqua regia, HF 2. H_3BO_3	Multimode/LP closed	1. 1 min at 600 W, 5 min at 360 W, 15 min at 180 W 2. 15 min at 330 W	ETV-AAS X-ray Fluorescence	B24
	Cd, Co, Fe, Ni, Pb	1. HNO_3, HCl 2. HF	Multimode/LP closed	1. 7 min at 240 W, cool 2. 3 min at 240 W, cool	ICP-OES	B20
	Al, Ba, Ca, Co, Cr, Cu, Fe, K, Mg, Mn, Na, Ni, Pb, Sc, Si, Ti, V, Zn, Zr	1. HF, Aqua regia 2. H_3BO_3	Multimode/LP closed	1. 8 min at 300 W, 4 min at 600 W, 7 min at 480 W 2. 6 min at 600 W	ICP-OES	
	Al, As, Ba, Ca, Cd, Cr, Cu, Co, Fe, K, Mg, Mn, Na, Ni, Pb, Sb, Sc, Se, Si, Sn, Ti, Tl, V, Zn, Zr	1. HF, Aqua regia 2. H_3BO_3	Multimode/LP closed	1. 10 min at 300 W, 5 min at 600 W, 10 min at 480 W 2. 6 min at 300 W	ICP-OES F-AAS ETV-AAS	B6
	Al, As, Ba, Be, Ca, Co, Cr, Fe, Mg, Mn, Ni, P, Pb, S, Ti, V, Zn	**A.** HF, Aqua regia **B.** $HNO3$	Multimode/LP closed Multimode/LP closed	5 min at 200 W, 5 min at 400 W Same as A	ICP-OES	B25
NIST Subbituminous coal 1635	Al, Ba, Ca, Cr, Cu, Fe, K, Li, Mg, Mn, Na, P, Si, Sr, Ti, V, Zn	Aqua regia, HF	Multimode/open	3 min at 625 W	ICP-OES	B21
	Al, As, Ba, Ca, Cd, Cr, Cu, Co, Fe, K, Mg, Mn, Na, Ni, Pb, Sb, Sc, Se, Si, Sn, Ti, Tl, V, Zn, Zr	1. HF, Aqua regia 2. H_3BO_3	Multimode/LP closed	1. 10 min at 300 W, 5 min at 600 W, 10 min at 480 W 2. 6 min at 300 W	ICP-OES F-AAS ETV-AAS	B6
USGS Manganese nodule A-1	Al, K, Li, Si	1. HNO_3, HCl, HF 2. H_3BO_3	Multimode/LP closed	1. 2.5 min at 650 W 2. 10 min at 650 W	ICP-OES	B2

Sample	Elements	Reagents	Mode	Conditions	Technique	Ref.
USGS Andesite AGV-1	Al, K, Li, Si, Zr	1. HNO_3, HCl, HF 2. H_3BO_3	Multimode/ LP closed	1. 2.5 min at 650 W 2. 10 min at 650 W	ICP-OES	B2
	Ce, Dy, Er, Eu, Gd, Ho, La, Lu, Nd, Pr, Tb, Tm, Sm, Yb	1. HF 2. HNO_3, HCl, HF	Multimode/ MP closed	1. Fume to dryness (hotplate) 2. 10 min at 30 bar	γ-spectrometry X-ray ICP-OES	B13
USGS Basalt BCR-1	Al, Ba, Ca, Cr, Cu, Fe, K, Li, Mg, Mn, Na, P, Si, Sr, Ti, V, Zr	Aqua regia, HF	Multimode/ open	3 min at 625 W	ICP-OES	B21
	Pb, U	HNO_3, HF, $HClO_4$	Multimode/ LP closed	5 min at 90 W, 15 min at 138 W, cool, 1-5 hr at 138 W	ID-MS	B19
	Al, K, Li, Si	1. HNO_3, HCl, HF 2. H_3BO_3	Multimode/ LP closed	1. 2.5 min at 650 W 2. 10 min at 650 W	ICP-OES	B2
USGS Basalt BHVO-1	Ag, As, B, Ba, Be, Co, Cr, Cs, Cu, Ga, Hf, Li, Mn, Mo, Nb, Ni, Pb, Rb, Sb, Sc, Sn, Sr, Ta, Th, U, V, Y, Zn, Zr	HNO_3, HF	Multimode/ MP closed	5 min at 300 W, cool, vent, 20 min using increasing power up to 600 W	ICP-MS	B26
	Nb, Mo, Ta, W	1. HNO_3 2. HF, H_2O	Multimode/ MP closed	1. 30 min at 200 W 2. 1 min at 1000 W, 30 min at 450 W	ID-ICP-MS	B9
USGS Dunite DTS-1	Al, Ca, Cr, K, Li, Si	1. HNO_3, HCl, HF 2. H_3BO_3	Multimode/ LP closed	1. 2.5 min at 650 W 2. 10 min at 650 W	ICP-OES	B2
USGS Granite G-2	Pb, U	HNO_3, HF, $HClO_4$	Multimode/ LP closed	5 min at 90 W, 15 min at 138 W, cool, 1 - 5 hr at 138 W	ID-MS	B19
	Al, K, Li, Si, Ti, Zr	1. HNO_3, HCl, HF 2. H_3BO_3	Multimode/ LP closed	1. 2.5 min at 650 W 2. 10 min at 650 W	ICP-OES	B2
	Si, Al, Zn, Cr	HCl, HNO_3, HF	Multimode/ LP closed	50 - 60 sec at 700 W	F-AAS	B27
	Nb, Mo, Ta, W	1. HNO_3 2. HF, H_2O	Multimode/ MP closed	1. 30 min at 200 W 2. 1 min at 1000 W, 30 min at 450 W	ID-ICP-MS	B9

Appendix 2.1 Rocks and Minerals

Matrix	Analytes	Reagents	Cavity/Vessel	MW conditions	Detection	Ref.
USGS Granodiorite GSP-1	Al, K, La, Li, Si, Ti, Y, Zr	1. HNO_3, HCl, HF 2. H_3BO_3	Multimode/ LP closed	1. 2.5 min at 650 W 2. 10 min at 650 W	ICP-OES	B2
	Pb, U	HNO_3, HF, $HClO_4$	Multimode/ LP closed	5 min at 90 W, 15 min at 138 W, cool, 1 - 5 hr at 138 W	ID	B19
USGS Jasperoid GXR-1	Al, Cr, K, Li, Mg, Si, Sb, Ti	1. HNO_3, HCl, HF 2. H_3BO_3	Multimode/ LP closed	1. 2.5 min at 650 W 2. 10 min at 650 W	ICP-OES	B2
USGS GXR-4	Nb, Mo, Ta, W	1. HNO_3 2. HF, H_2O	Multimode/ MP closed	1. 30 min at 200 W 2. 1 min at 1000 W, 30 min at 450 W	ID-ICP-MS	B9
USGS GXR-6	Nb, Mo, Ta, W	1. HNO_3 2. HF, H_2O	Multimode/ MP closed	1. 30 min at 200 W 2. 1 min at 1000 W, 30 min at 450 W	ID-ICP-MS	B9
USGS Gabbro JG-1	Nb, Mo, Ta, W	1. HNO_3 2. HF, H_2O	Multimode/ MP closed	1. 30 min at 200 W 2. 1 min at 1000 W, 30 min at 450 W	ID-ICP-MS	B9
USGS Manganese nodule P-1	Al, K, Li, Si	1. HNO_3, HCl, HF 2. H_3BO_3	Multimode/ LP closed	1. 2.5 min at 650 W 2. 10 min at 650 W	ICP-OES	B2
USGS Peridotite PCC-1	Al, Ca, Cr, K, Li, Mg, Si, Ti, V	1. HNO_3, HCl, HF 2. H_3BO_3	Multimode/ LP closed	1. 2.5 min at 650 W 2. 10 min at 650 W	ICP-OES	B2
	Nb, Mo, Ta, W	1. HNO_3 2. HF, H_2O	Multimode/ MP closed	1. 30 min at 200 W 2. 1 min at 1000 W, 30 min at 450 W	ID-ICP-MS	B9
USGS QLO-1	Nb, Mo, Ta, W	1. HNO_3 2. HF, H_2O	Multimode/ MP closed	1. 30 min at 200 W 2. 1 min at 1000 W, 30 min at 450 W	ID-ICP-MS	B9
USGS Rhyolite RGM-1	Al, Cu, K, Li, Si	1. HNO_3, HCl, HF 2. H_3BO_3	Multimode/ LP closed	1. 2.5 min at 650 W 2. 10 min at 650 W	ICP-OES	B2

Sample	Elements	Reagents	Mode/Vessel	Microwave program	Technique	Ref.
USGS SDC-1	Nb, Mo, Ta, W	1. HNO_3 2. HF, H_2O	Multimode/MP closed	1. 30 min at 200 W, 2. 1 min at 1000 W, 30 min at 450 W	ID-ICP-MS	B9
USGS Devonian Ohio shale SDO-1	Ag, As, B, Ba, Be, Co, Cr, Cs, Cu, Ga, Hf, Li, Mn, Mo, Nb, Ni, Pb, Rb, Sb, Sc, Sn, Ta, Th, U, V, Y, Zn, Zr	1. HNO_3, H_2O_2 2. HF, $HClO_4$	Multimode/MP closed	1. 4 × 7 min at 300 W, 1.5 min at 600 W, cool, vent 2. 3 × 7 min at 300 W, 2 min at 600 W, 0.5 min at 900 W	ICP-MS	B26
USGS Green river shale SGR-1	Al, Co, K, Li, Si, Ti, Zr	1. HNO_3, HCl, HF 2. H_3BO_3	Multimode/LP closed	1. 2.5 min at 650 W 2. 10 min at 650 W	ICP-OES	B2
USGS Syenite STM-1	Al, Ca, Fe, K, Li, Mg, Na, Si, Ti, Zr	1. HNO_3, HCl, HF 2. H_3BO_3	Multimode/LP closed	1. 2.5 min at 650 W 2. 10 min at 650 W	ICP-OES	B2
	Nb, Mo, Ta, W	1. HNO_3 2. HF, H_2O	Multimode/MP closed	1. 30 min at 200 W 2. 1 min at 1000 W 30 min at 450 W	ID-ICP-MS	B9
USGS Diabase W-1	Al, Ba, Ca, Co, Cr, Cu, Fe, K, Li, Mg, Mn, Na, Ni, P, Si, Sr, Ti, V, Zr	Aqua regia, HF	Multimode/open	3 min at 625 W	ICP-OES	B21
	Al, K, Li, Si, Ti, Zr	1. HNO_3, HCl, HF 2. H_3BO_3	Multimode/LP closed	1. 2.5 min at 650 W 2. 10 min at 650 W	ICP-OES	B2
USGS Diabase W-2	Ag, As, B, Ba, Be, Co, Cr, Cs, Cu, Ga, Hf, Li, Mn, Mo, Nb, Ni, Pb, Rb, Sb, Sc, Sn, Ta, Th, U, V, Y, Zn, Zr	HNO_3, HF	Multimode/MP closed	5 min at 300 W, cool, vent, 20 min using increasing power up to 600 W	ICP-MS	B26
XIGMR DZE-1	Nb, Mo, Ta, W	1. HNO_3 2. HF, H_2O	Multimode/MP closed	1. 30 min at 200 W, 2. 1 min at 1000 W, 30 min at 450 W	ID-ICP-MS	B9
XIGMR DZE-2	Nb, Mo, Ta, W	1. HNO_3 2. HF, H_2O	Multimode/MP closed	1. 30 min at 200 W 2. 1 min at 1000 W, 30 min at 450 W	ID-ICP-MS	B9
ZGI Slate TB	Al, Ce, K, Li, Si, Sr, Ti, Y, Zr	1. HNO_3, HCl, HF 2. H_3BO_3	Multimode/LP closed	1. 2.5 min at 650 W 2. 10 min at 650 W	ICP-OES	B2

Appendix 2.2 Steels and Alloys

Matrix	Analytes	Reagents	Cavity/Vessel	MW conditions	Detection	Ref.
BAM High-alloy steel 228-1	Cr	1. HNO_3, HCl, HF 2. HNO_3	Multimode/ LP closed	1. 15 min at 55% 2. 15 min at 55% (if incompletely digested)	Titration with Fe(II)	B28
BAM High-alloy steel 277-1	Cr	1. HNO_3, HCl, HF 2. HNO_3	Multimode/ LP closed	1. 15 min at 55% 2. 15 min at 55% (if incompletely digested)	Titration with Fe(II)	B28
BAM High-alloy steel 278-1	Cr	1. HNO_3, HCl, HF 2. HNO_3	Multimode/ LP closed	1. 15 min at 55% 2. 15 min at 55% (if incompletely digested)	Titration with Fe(II)	B28
BAM High-alloy steel 328-1	Cr	1. HNO_3, HCl, HF 2. HNO_3	Multimode/ LP closed	1. 15 min at 55% 2. 15 min at 55% (if incompletely digested)	Titration with Fe(II)	B28
BAM Ferro-chromium 530-1	Al, Co, Cr, Cu, Fe, Mn, Mo, Ni, Si, Ti, V	H_2SO_4, H_2O, H_3PO_4	Multimode/ LP closed	10 min at 330 W	ICP-OES	B29
BAM Ferro-chromium 533-1	Al, Co, Cr, Cu, Fe, Mn, Mo, Ni, Si, Ti, V	H_2SO_4, H_2O, H_3PO_4	Multimode/ LP closed	10 min at 330 W	ICP-OES	B29
BAM High-alloy steel CrMnMoNiTi1	Cr	1. HNO_3, HCl, HF 2. HNO_3	Multimode/ LP closed	1. 15 min at 55% 2. 15 min at 55% (if incompletely digested)	Titration with Fe(II)	B28
BAM High-alloy steel CrNiSiMn1	Cr	1. HNO_3, HCl, HF 2. HNO_3	Multimode/ LP closed	1. 15 min at 55% 2. 15 min at 55% (if incompletely digested)	Titration with Fe(II)	B28

Sample	Elements	Reagents	Mode	Conditions	Method	Ref.
BCS Ferro-chromium 203/2	Al, Co, Cr, Cu, Fe, Mn, Mo, Ni, Si, Ti, V	H_2SO_4, H_2O, H_3PO_4	Multimode/LP closed	10 min at 330 W	ICP-OES	B29
BCS Ferro-chromium 204/1	Al, Co, Cr, Cu, Fe, Mn, Mo, Ni, Si, Ti, V	H_2SO_4, H_2O, H_3PO_4	Multimode/LP closed	10 min at 330 W	ICP-OES	B29
BCS Ferro-manganese 208/1	Al, Co, Cr, Cu, Fe, Mn, Mo, Ni, Si, Ti, V	H_2SO_4, H_2O, H_3PO_4, HNO_3	Multimode/LP closed	10 min at 330 W	ICP-OES	B29
BCS High-alloy steel 211/1	Cr	1. HNO_3, HCl, HF 2. HNO_3	Multimode/LP closed	1. 15 min at 55% 2. 15 min at 55% (if incompletely digested)	Titration with Fe(II)	B28
BCS Ferro-manganese 280	Al, Co, Cr, Cu, Fe, Mn, Mo, Ni, Si, Ti, V	H_2SO_4, H_2O, H_3PO_4, HNO_3	Multimode/LP closed	10 min at 330 W	ICP-OES	B29
BCS Fe-Si alloy 305	Al, Ca, Cr, Cu, Fe, Hf, Mn, Ni, Ru, Si, Ti, Zr	H_3PO_4	Multimode/open	10-15 min to 150-200°C	ICP-OES	B30
BCS High-alloy steel 338	Cr	1. HNO_3, HCl, HF 2. HNO_3	Multimode/LP closed	1. 15 min at 55% 2. 15 min at 55% (if incompletely digested)	Titration with Fe(II)	B28
BCS High-alloy steel 339	Cr	1. HNO_3, HCl, HF 2. HNO_3	Multimode/LP closed	1. 15 min at 55% 2. 15 min at 55% (if incompletely digested)	Titration with Fe(II)	B28
BCS High-alloy steel 340	Cr	1. HNO_3, HCl, HF 2. HNO_3	Multimode/LP closed	1. 15 min at 55% 2. 15 min at 55% (if incompletely digested)	Titration with Fe(II)	B28

Appendix 2.2   Steels and Alloys

Matrix	Analytes	Reagents	Cavity/Vessel	MW conditions	Detection	Ref.
BCS High-alloy steel 341	Cr	1. HNO_3, HCl, HF 2. HNO_3	Multimode/ LP closed	1. 15 min at 55% 2. 15 min at 55% (if incompletely digested)	Titration with Fe(II)	B28
BCS High-alloy steel 342	Cr	1. HNO_3, HCl, HF 2. HNO_3	Multimode/ LP closed	1. 15 min at 55% 2. 15 min at 55% (if incompletely digested)	Titration with Fe(II)	B28
BCS Nickel alloy 345	As	HNO_3, HF	Multimode/ LP closed	15 min at 200 W, 15 min at 325 W, 30 min at 260 W	HG-AAS	B31
BCS Nickel alloy 346	As	HNO_3, HF	Multimode/ LP closed	15 min at 200 W, 15 min at 325 W, 30 min at 260 W	HG-AAS	B31
	As	HNO_3, HF	Multimode/ LP closed	15 min at 200 W, 15 min at 325 W, 30 min at 260 W	HG-AAS	B32
BCS Steel 452	As, Cr, Cu, Mn, Ni, P, Si, Sn, Ti, W	HNO_3, HCl, HF	Multimode/ LP closed	80 sec at 625 W	DCP-OES	B33
BCS High-alloy steel JK 8F	Cr	1. HNO_3, HCl, HF 2. HNO_3	Multimode/ LP closed	1. 15 min at 55% 2. 15 min at 55% (if incompletely digested)	Titration with Fe(II)	B28
BRAMMER Slag ST-100	Ca, Fe, Mg, Mn, Si	HCl, HF, HNO_3	Multimode/ LP closed	2 min at 650 W	F-AAS	B14
CSAN Ferro-silicon 4-1-01	Al, Ca, Cr, Cu, Fe, Hf, Mn, Ni, Ru, Si, Ti, Zr	H_3PO_4	Multimode/ open	10-15 min at 150-200°C	ICP-OES	A30

Sample	Elements	Reagents	Mode	Conditions	Method	Ref
CSAN Ferro-silicon 4-1-02	Al, Ca, Cr, Cu, Fe, Hf, Mn, Ni, Ru, Si, Ti, Zr	H_3PO_4	Multimode/ open	10-15 min at 150-200°C	ICP-OES	B30
CSAN Ferro-chromium 4-2-01	Al, Co, Cr, Cu, Fe, Mn, Mo, Ni, Si, Ti, V	H_2SO_4, H_2O, H_3PO_4	Multimode/ LP closed	10 min at 330 W	ICP-OES	B29
CSAN Ferro-chromium 4-2-02	Al, Co, Cr, Cu, Fe, Mn, Mo, Ni, Si, Ti, V	H_2SO_4, H_2O, H_3PO_4	Multimode/ LP closed	10 min at 330 W	ICP-OES	B29
CSAN Ferro-chromium 4-2-03	Al, Co, Cr, Cu, Fe, Mn, Mo, Ni, Si, Ti, V	H_2SO_4, H_2O, H_3PO_4	Multimode/ LP closed	10 min at 330 W	ICP-OES	B29
CSAN Ferro-chromium 4-2-04	Al, Co, Cr, Cu, Fe, Mn, Mo, Ni, Si, Ti, V	H_2SO_4, H_2O, H_3PO_4	Multimode/ LP closed	10 min at 330 W	ICP-OES	B29
CSAN Ferro-manganese 4-3-01	Al, Co, Cr, Cu, Fe, Mn, Mo, Ni, Si, Ti, V	H_2SO_4, H_2O, H_3PO_4	Multimode/ LP closed	10 min at 330 W	ICP-OES	B29
CSAN Ferro-manganese 4-3-02	Al, Co, Cr, Cu, Fe, Mn, Mo, Ni, Si, Ti, V	H_2SO_4, H_2O, H_3PO_4	Multimode/ LP closed	10 min at 330 W	ICP-OES	B29
CSAN Ferro-manganese silicon 4-5-02	Al, Ca, Cr, Cu, Fe, Hf, Mn, Ni, Ru, Si, Ti, Zr	H_3PO_4	Multimode/ open	10-15 min at 150 - 200 °C	ICP-OES	B30

Appendix 2.2   Steels and Alloys

Matrix	Analytes	Reagents	Cavity/Vessel	MW conditions	Detection	Ref.
CSAN Ferro-chromium silicon 4-5-03	Al, Ca, Cr, Cu, Fe, Hf, Mn, Ni, Ru, Si, Ti, Zr	H_3PO_4	Multimode/ open	10-15 min at 150 - 200 °C	ICP-OES	B30
IFM Slag 7	Al, Ca, Fe, K, Mg, Mn, Na, Ti, Si	A. 1. HCl, HF 2. HBO_3 B. 1. HCl, HF, H_2SO_4 2. HBO_3 C. 1. Aqua regia, HF 2. HBO_3 D. 1. HCl, HF, H_2O_2 2. HBO_3	Multimode/ LP closed Multimode/ LP closed Multimode/ LP closed Multimode/ LP closed	1. 10-13 sec at 1470 W 1. Same as A 1. Same as A 1. Same as A	F-AAS	B8
IRSID High-alloy steel 204-1	Cr	1. HNO_3, HCl, HF 2. HNO_3	Multimode/ LP closed	1. 15 min at 55% 2. 15 min at 55% (if incompletely digested)	Titration with Fe(II)	B28
NIST High-alloy stainless steel 73	Cr	1. HNO_3, HCl, HF 2. HNO_3	Multimode/ LP closed	1. 15 min at 55% 2. 15 min at 55% (if incompletely digested)	Titration with Fe(II)	B28
NIST High-alloy stainless steel 101	Cr	1. HNO_3, HCl, HF 2. HNO_3	Multimode/ LP closed	1. 15 min at 55% 2. 15 min at 55% (if incompletely digested)	Titration with Fe(II)	B28
NIST Stainless steel Cr-Ni-Ti 121	-	HCl, HNO_3, HF	Multimode/ LP closed	30 sec at 300 W (x2)	-	B17

Sample	Analyte	Reagents	Mode	Conditions	Method	Ref.
NIST Solder 40 Sn - 60 Pb 127		HBF_4, H_2O, HNO_3	Multimode/ LP closed	Varying time at 600 W		B17
NIST High-alloy stainless steel 133	Cr	1. HNO_3, HCl, HF 2. HNO_3	Multimode/ LP closed	1. 15 min at 55% 2. 15 min at 55% (if incompletely digested)	Titration with Fe(II)	B28
NIST High-alloy stainless steel 160	Cr	1. HNO_3, HCl, HF 2. HNO_3	Multimode/ LP closed	1. 15 min at 55% 2. 15 min at 55% (if incompletely digested)	Titration with Fe(II)	B28
NIST High-alloy steel 161	Cr	1. HNO_3, HCl, HF 2. HNO_3	Multimode/ LP closed	1. 15 min at 55% 2. 15 min at 55% (if incompletely digested)	Titration with Fe(II)	B28
NIST Nickel copper alloy 162	Al, Cu, Fe, Mn, Ni, Si	HCl, HF, HNO_3	Multimode/ LP closed	2 min at 650 W	F-AAS	B14
NIST High-alloy steel 168	Cr	1. HNO_3, HCl, HF 2. HNO_3	Multimode/ LP closed	1. 15 min at 55% 2. 15 min at 55% (if incompletely digested)	Titration with Fe(II)	B28
	-	HF, HNO_3, H_2O	Multimode/ LP closed	5 min at 600 W	-	B17
NIST LA steel high silicon 179	Al, Cr, Cu, Mn, Mo, Ni, P, Si, Sn, Ti, V	HNO_3, HCl, HF	Multimode/ LP closed	80 sec at 625 W	DCP-OES	B33
NIST High-alloy stainless steel 133	Cr	1. HNO_3, HCl, HF 2. HNO_3	Multimode/ LP closed	1. 15 min at 55% 2. 15 min at 55% (if incompletely digested)	Titration with Fe(II)	B28

Appendix 2.2   Steels and Alloys

Matrix	Analytes	Reagents	Cavity/Vessel	MW conditions	Detection	Ref.
NIST High-alloy valve steel 346	Cr	1. HNO_3, HCl, HF 2. HNO_3	Multimode/ LP closed	1. 15 min at 55% 2. 15 min at 55% (if incompletely digested)	Titration with Fe(II)	B28
NIST Waspalloy Ni-Co-Cr 349	Co, Cr, Fe, Mo, Ni, W	HCl, HF, HNO_3	Multimode/ LP closed	30 sec at 325 W	ICP-OES	B34
NIST Cast iron steel 890	Cr	1. HNO_3, HCl, HF 2. HNO_3	Multimode/ LP closed	1. 15 min at 55% 2. 15 min at 55% (if incompletely digested)	Titration with Fe(II)	B28
NIST Cast iron steel 892	Cr	1. HNO_3, HCl, HF 2. HNO_3	Multimode/ LP closed	1. 15 min at 55% 2. 15 min at 55% (if incompletely digested)	Titration with Fe(II)	B28

Appendix 2.3   Soils and Sediments

Matrix	Analytes	Reagents	Cavity/Vessel	MW conditions	Detection	Ref.
BCR Calcareous loam 141	Cd, Cr, Cu, Fe, Mn, Pb, Zn	HCl, HNO_3	Multimode/ LP closed	1 min at 180 W, 4 min at 480 W, 60 min at 600 W	F-AAS ETV-AAS	B35
	Cd, Cu, Pb	1. HNO_3, H_2O, H_2O_2 2. Aqua regia, HF (optional)	Multimode/ open	1. 15 min at 70 W, 15 min at 150 W, 15 min at 180 W, 1 min at 250 W, 1 min at 600 W 2. 15 min at 70 W, 1 min at 600 W	ETV-AAS	B36
BCR Soil - light sandy 142	Cd, Cr, Cu, Fe, Mn, Pb, Zr	HCl, HNO_3	Multimode/ LP closed	1 min at 180 W, 4 min at 480 W, 60 min at 600 W	F-AAS ETV-AAS	B35

Sample	Elements	Reagents	System	Program	Detection	Ref.
BCR Estuarine sediment 277	As, Cd, Cr, Cu, Hg, Ni, Pb, Zn	A. 1. HCl, HNO_3 2. HCl, HNO_3 3. H_2O_2 4. H_2O	Single mode/ open	1. 5 min at 40 W, 10 min at 50 W 2. 10 min at 54 W 3. 5 min at 40 W 4. 5 min at 50 W	ICP-OES ICP-MS CV-AAS HPLC-ICP-MS	B37
		B. HCl, HNO_3, HF	Multimode/ LP closed	10 min at 120 W, 20 min at 240 W, 20 min at 300 W		
CANMET Regosolic C horizon soil SO-1	As	1. HNO_3 2. HNO_3, $HClO_4$, H_2SO_4	Multimode/ LP closed	1. 2.5 min at 639 W, cool, vent 2. 2.5 min at 639 W, 2 min 0 W (x3)	F-AAS	B38
	Pb	HNO_3, HF	Multimode/ LP closed	7 min at 400 W	DP-ASV	B39
CANMET Podzolic B horizon soil SO-2	As	1. HNO_3 2. HNO_3, $HClO_4$, H_2SO_4	Multimode/ LP closed	1. 2.5 min at 639 W, cool, vent 2. 2.5 min at 639 W, 2 min 0 W (x3)	F-AAS	B38
	Pb	HNO_3, HF	Multimode/ LP closed	7 min at 400 W	DP-ASV	B39
CANMET Calcareous C horizon soil SO-3	As	1. HNO_3 2. HNO_3, $HClO_4$, H_2SO_4	Multimode/ LP closed	1. 2.5 min at 639 W, cool, vent 2. 2.5 min at 639 W, 2 min 0 W (x3)	F-AAS	B38
	Pb	HNO_3, HF	Multimode/ LP closed	7 min at 400 W	DP-ASV	B39
CANMET Black Chernozemic A horizon soil SO-4	As	1. HNO_3 2. HNO_3, $HClO_4$, H_2SO_4	Multimode/ LP closed	1. 2.5 min at 639 W, cool, vent 2. 2.5 min at 639 W, 2 min 0 W (x3)	F-AAS	B38
	Pb	HNO_3, HF	Multimode/ LP closed	7 min at 400 W	DP-ASV	B39

Appendix 2.3   Soils and Sediments

Matrix	Analytes	Reagents	Cavity/Vessel	MW conditions	Detection	Ref.
IAEA Lake sediment SL-1	Al, Ba, Ca, Cr, Cu, Fe, K, Li, Mg, Mn, Na, P, Pb, Si, Sr, Ti, V, Zn	Aqua regia, HF	Multimode/ open	3 min at 625 W	ICP-OES	B21
	Cd, Cr, Cu, Fe, Mn, Pb, Zn	HCl, HNO$_3$	Multimode/ LP closed	1 min at 180 W, 4 min at 480 W, 60 min at 600 W	F-AAS ETV-AAS	B35
IAEA Soil 5	Hg	HCl, HNO$_3$, H$_2$O	Multimode/ LP closed	1 min at 180 W, 4 min at 480 W, 10 min at 600 W	CV-AAS NAA	B40
IAEA Soil 7	Cd, Cr, Cu, Fe, Mn, Pb, Zn	HCl, HNO$_3$	Multimode/ LP closed	1 min at 180 W, 4 min at 480 W, 60 min at 600 W	F-AAS ETV-AAS	B35
IGGE Stream sediment GSD-2	Al, Be, Cr, K, La, Li, Nb, Si, Sn, Ti, Y, Zr	1. HNO$_3$, HCl, HF 2. H$_3$BO$_3$	Multimode/ LP closed	1. 2.5 min at 650 W 2. 10 min at 650 W	ICP-OES	B2
IGGE Stream sediment GSD-3	Al, Cr, K, La, Li, Si, Ti, V, Zr	1. HNO$_3$, HCl, HF 2. H$_3$BO$_3$	Multimode/ LP closed	1. 2.5 min at 650 W 2. 10 min at 650 W	ICP-OES	B2
IGGE Stream sediment GSD-8	Al, K, Li, Nb, Si, Sn, Ti, Y, Zr	1. HNO$_3$, HCl, HF 2. H$_3$BO$_3$	Multimode/ LP closed	1. 2.5 min at 650 W 2. 10 min at 650 W	ICP-OES	B2
NIES Pond sediment 2	Hg	HNO$_3$	Multimode/ HP closed	70 sec at 600 W	CV-AAS	B41
NIST River sediment 1645	Ca, Cr, Fe, Mn, Pb, Zn	Exchangeable: MgCl$_2$	Multimode/ open	4 min at ? W	F-AAS	B42
		Carbonate Bound: Sodium Acetate	Multimode/ open	5 min at ? W		

Sample	Elements	Reagents	Mode	Conditions	Detection	Ref
		Fe-Mn Oxide: Hydroxylamine hydrochloride	Multimode/ open	4 min at ? W (x3)		
		Organic: H$_2$O$_2$, HNO$_3$	Multimode/ open	4 min at ? W		
		Residual: HNO$_3$, HCl, HF	Multimode/ open	5 min at ? W		
	Al, As, Ba, Ca, Cr, Cu, Fe, K, Mg, Mn, Na, P, Pb, Si, Sr, Tl, Zn	Aqua regia, HF	Multimode/ open	3 min at 625 W	ICP-OES	B21
	Al, K, Li, Si	1. HNO$_3$, HCl, HF 2. H$_3$BO$_3$	Multimode/ LP closed	1. 2.5 min at 650 W 2. 10 min at 650 W	ICP-OES	B2
	Th, U	1. HCl, HNO$_3$, HF 2. Aqua regia	Multimode/ HP closed	1. 4-8 min at 600 W, fume off acid (x3) 2. If residue 3 min at 600 W	α spectrometry	B7
	Hg	HNO$_3$	Multimode/ LP closed	Heat to 75°C (800 W), 2 min at 75°C	CV-FANES	B43
	As, Cd, Cu, Mg, Mn, Ni, Pb, Zn	HNO$_3$, H$_2$O	Multimode/ LP closed	Heat to 100 psi, 55 min at 100 psi	-	B44
	Al, Ba, Ca, Co, Cr, Cu, Fe, K, Mg, Mn, Na, Ni, Pb, Sc, Si, Ti, V, Zn, Zr	1. HF, Aqua regia 2. H$_3$BO$_3$	Multimode/ LP closed	1. 8 min at 300 W, 4 min at 600 W, 7 min at 480 W 2. 6 min at 600 W	ICP-OES	B20
	Al, As, Ba, Ca, Cd, Cr, Cu, Co, Fe, K, Mg, Mn, Na, Ni, Pb, Sb, Sc, Se, Si, Sn, Tl, V, Zn, Zr	1. HF, Aqua regia 2. H$_3$BO$_3$	Multimode/ LP closed	1. 8 min at 300 W, 11 min at 600 W 2. 6 min at 300 W	ICP-OES F-AAS ETV-AAS	B6
NIST Estuarine sediment 1646	Al, As, Ba, Ca, Cd, Cr, Cu, Co, Fe, K, Mg, Mn, Na, Ni, Pb, Sb, Sc, Se, Si, Sn, Ti, Tl, V, Zn, Zr	1. HF, Aqua regia 2. H$_3$BO$_3$	Multimode/ LP closed	1. 8 min at 300 W, 11 min at 600 W 2. 6 min at 300 W	ICP-OES F-AAS ETV-AAS	B6

Appendix 2.3   Soils and Sediments

Matrix	Analytes	Reagents	Cavity/Vessel	MW conditions	Detection	Ref.
NIST Estuarine sediment 1646 (contd.)	Ca, Cr, Fe, Mn, Pb, Zn	**Exchangeable:** MgCl$_2$	Multimode/open	4 min at ? W	F-AAS	B42
		Carbonate Bound: Sodium Acetate	Multimode/open	5 min at ? W		
		Fe-Mn Oxide: Hydroxylamine hydrochloride	Multimode/open	4 min at ? W (x3)		
		Organic: H$_2$O$_2$, HNO$_3$	Multimode/open	4 min at ? W		
		Residual: HNO$_3$, HCl, HF	Multimode/open	5 min at ? W		
	Al, As, B, Ba, Be, Ca, Cd, Co, Cr, Cu, Fe, K, Li, Mg, Mn, Na, Ni, Pb, Ti, V, Zn	HNO$_3$	Multimode/LP closed	10 min at 400 W; 50 min at 330 W	ICP-OES	B45
	Al, Ba, Ca, Co, Cr, Cu, Fe, K, Mg, Mn, Na, Ni, Pb, Sc, Si, Ti, V, Zn, Zr	1. HF, Aqua regia 2. H$_3$BO$_3$	Multimode/LP closed	1. 8 min at 300 W, 4 min at 600 W, 7 min at 480 W 2. 6 min at 600 W	ICP-OES	B20
NIST Buffalo river sediment 2704	As, Cr, Cu, Mn, Ni, Pb, Se, Zn	HNO$_3$, H$_2$O	Multimode/LP closed	12 min to 100 psi (651 W), 30 min at 100 psi (651 W, 181°C)	F-AAS ETV-AAS	B46
	Ag, Al, B, Ba, Be, Ca, Cd, Co, Cr, Cu, Fe, Mg, Mn, Mo, Ni, Pb, Sr, V, Zn	HNO$_3$	Multimode/LP closed	5.5 min to 175°C (574 W), 4.5 min at 175-180°C (574 W)	ICP-OES	B47, B48

Elements	Reagents	Mode	Conditions	Technique	Ref.
Ag, As, Ba, Cd, Cu, Cr, Hg, Ni, Pb, Se, Tl, Zn	A. HNO_3 B. HNO_3, HF	Multimode/LP closed Multimode/LP closed	2:20 min at 600 W, 9:25 min at 480 W Same as A	ETV-AAS CV-AAS	B49
Ag, As, Ba, Cd, Cu, Cr, Hg, Ni, Pb, Sb, Se, Tl, Zn	Extraction: HNO_3 Total: HNO_3, HF	Multimode/LP closed Multimode/LP closed	2:20 min at 604 W, 9:25 min at 570 W Same as extraction		B50
Ba, Cd, Cr, Co, Cu, Pb, Ni, Sr, V, Zn	HNO_3	Multimode/LP closed	5.5 min to 175 °C (574 W), 4.5 min at 175–180 °C (574 W)		B51
Co, Cr, Cu, Ni, Pb, Zn	A. 1. HF, HNO_3, HCl 2. H_3BO_3 B. HNO_3 C. Same as A D. 1. HF, HNO_3 2. H_3BO_3 E. HNO_3, HCl F. Same as A G. Same as E	Multimode/LP closed (all same)	8 min at 630 W, 15 min at 315 W Same as A 10 min at 630 W, 20 min at 473 W, 5 min at 158 W Same as C Same as C 10 min at 630 W, 20 min at 410 W, 5 min at 189 W Same as A 10 min at 630 W, 20 min at 441 W, 5 min at 158 W	F-AAS ETV-AAS	B52, B53

Appendix 2.3   Soils and Sediments

Matrix	Analytes	Reagents	Cavity/Vessel	MW conditions	Detection	Ref.
NIST Buffalo river sediment 2704 (contd.)	Co, Cr, Cu, Ni, Pb, Zn	**H.** HF, HNO$_3$, HCl		15 min at 630 W		
		I. HNO$_3$, H$_2$O$_2$		Same as G		
		J. HNO$_3$, H$_2$O$_2$, HF		Same as G		
		K. 1. HNO$_3$, H$_2$O$_2$, HF 2. H$_3$BO$_3$		Same as G		
		L. Acetic acid		Same as G		
		M. Same as E		7 min at 441 W, 5 min at 630 W		
		N. Same as H		Same as M		
	Al, As, Ba, Be, Ca, Co, Cr, Cu, Fe, Hg, Mg, Mn, Ni, P, Pb, S, Se, Tl, U, V, Zn	HNO$_3$	Multimode/ LP closed	Heat to 175°C within 5.5 min, 4.5 min at 175-180°C	ICP-MS ETV-AAS	B54
	Cd, Cr, Cu, Ni, Pb, Zn	**A.** HNO$_3$	Multimode/ LP closed	Heat to 175 °C within 5.5 min, 4.5 min at 175-180 °C	F-AAS ETV-AAS ICP-MS	B55
		B. HNO$_3$, HF	Multimode/ LP closed	Heat to 175 °C within 5.5 min, 9.5 min at 175-180 °C		
NIST Montana soil highly elevated traces 2710	Cd, Cr, Cu, Ni, Pb, Zn	**A.** HNO$_3$	Multimode/ LP closed	Heat to 175 °C within 5.5 min, 4.5 min at 175-180°C	F-AAS ETV-AAS ICP-MS	B55
		B. HNO$_3$, HF	Multimode/ LP closed	Heat to 175 °C within 5.5 min, 9.5 min at 175-180°C		

Sample	Elements	Reagents	Mode	Program	Method	Ref.
NIST Montana soil moderately elevated traces 2711	Cd, Cr, Cu, Ni, Pb, Zn	A. HNO_3 B. HNO_3, HF	Multimode/ LP closed Multimode/ LP closed	Heat to 175 °C within 5.5 min, 4.5 min at 175-180 °C Heat to 175 °C within 5.5 min, 9.5 min at 175-180 °C	F-AAS ETV-AAS ICP-MS	B55
NIST Peruvian soil 4355	As, Cr, Cu, Mn, Ni, Pb, Se, Zn	HNO_3, H_2O	Multimode/ LP closed	12 min to 100 psi (651 W), 30 min at 100 psi (651 W/181 °C)	F-AAS ETV-AAS	B46
	Ag, Al, B, Ba, Be, Ca, Cd, Co, Cr, Cu, Fe, Mg, Mn, Mo, Ni, Pb, Sr, V, Zn	HNO_3	Multimode/ LP closed	5.5 min to 175 °C (574 W), 4.5 min at 175-180 °C (574 W)	ICP-OES	B47, B48
NRCC Marine sediment BCSS-1	Al, Ba, Be, Ca, Cd, Cr, Cu, Fe, Hg, Mg, Mn, Na, Ni, P, Pb, Sr, Ti, V, Zn	HNO_3, HCl	Multimode/ LP closed	1-20 min at 720 W	ICP-OES ETV-AAS	B56
	Al, As, Ba, Be, Ca, Co, Cr, Cu, Fe, Mg, Mn, Mo, Ni, Pb, S, Sb, Ti, V, Zn	A. Aqua regia B. 1. Aqua regia, HF 2. H_3BO_3	Multimode/ LP closed Multimode/ LP closed	2 min at 600 W, 5 min at 360 W, 15 min at 180 W 1. 1 min at 600 W, 5 min at 360 W, 15 min at 180 W, 15 min at 180 W 2. 15 min at 330 W	ICP-OES NAA	B23
NRCC Marine sediment MESS-1	Al, Cr, K, Li, Si, Ti, Zr	1. HNO_3, HCl, HF 2. H_3BO_3	Multimode/ LP closed	1. 2.5 min at 650 W 2. 10 min at 650 W	ICP-OES	B2
	Si, Al, Zn, Cr	HCl, HNO_3, HF	Multimode/ LP closed	50 - 60 sec at 700 W	F-AAS	B27
	Al, Ba, Be, Ca, Cd, Cr, Cu, Fe, Hg, Mg, Mn, Na, Ni, P, Pb, Sr, Ti, V, Zn	HNO_3, HCl	Multimode/ LP closed	1-20 min at 720 W	ICP-OES ETV-AAS	B56
	Al, As, Ba, Be, Ca, Co, Cr, Cu, Fe, Mg, Mn, Mo, Ni, Pb, S, Sb, Ti, V, Zn	A. Aqua regia	Multimode/ LP closed	2 min at 600 W, 5 min at 360 W, 15 min at 180 W	ICP-OES NAA	B23

Appendix 2.3   Soils and Sediments

Matrix	Analytes	Reagents	Cavity/Vessel	MW conditions	Detection	Ref.
NRCC Marine sediment MESS-1 (contd.)	Si, Al, Zn, Cr	B. 1. Aqua regia, HF 2. H_3BO_3	Multimode/ LP closed	1. 1 min at 600 W, 5 min at 360 W, 15 min at 180 W 2. 15 min at 330 W		
	As, Cd, Co, Cr, Cu, Mn, Ni, Pb, Sn, V, Zn	HNO_3, HF	Multimode/ MP closed	25 min at 210 W	ETV-AAS F-AAS	B57
	As, Cd, Co, Cr, Cu, Fe, Mn, Ni, Pb, Se, Zn	A. HNO_3, HF, $HClO_4$ B. Same as A	Multimode/ LP closed Multimode/ LP closed	100% power to 60-65 psi, 20 min at 65 psi Heat as in A, vent, repeat A	F-AAS ETV-AAS ICP-OES	B58
NRCC Marine Sediment PACS-1	As, Cr, Cu, Mn, Ni, Pb, Se, Zn	HNO_3, H_2O	Multimode/ LP closed	12 min to 100 psi (651 W), 30 min at 100 psi (651 W/181°C)	F-AAS ETV-AAS	B46
	Al, As, Ba, Be, Ca, Co, Cr, Cu, Fe, Mg, Mn, Mo, Ni, Pb, S, Sb, Ti, V, Zn	A. Aqua regia B. 1. Aqua regia, HF 2. H_3BO_3	Multimode/ LP closed Multimode/ LP closed	2 min at 600 W, 5 min at 360 W, 15 min at 180 W 1. 1 min at 600 W, 5 min at 360 W, 15 min at 180 W 2. 15 min at 330 W	ICP-OES NAA	B23
NRCC Soil SO-2	Cu, Zn	HNO_3, HF	Multimode/ LP closed	15 min at 105 W, vent, 7 min at 350 W	F-AAS	B59
NRCC Soil SO-3	Cu, Zn	HNO_3, HF	Multimode/ LP closed	15 min at 105 W, vent, 7 min at 350 W	F-AAS	B59

Sample	Elements	Acids	Mode	Conditions	Method	Ref
NRCC Soil SO-4	Cu, Zn	HNO_3, HF	Multimode/ LP closed	15 min at 105 W, vent, 7 min at 350 W	F-AAS	B59
RMA Standard soil	Ag, As, Ba, Cd, Cu, Cr, Hg, Ni, Pb, Se, Tl, Zn	A. HNO_3	Multimode/ LP closed	2:20 min at 600 W, 9:25 min at 480 W	ETV-AAS CV-AAS	B49
		B. HNO_3, HF	Multimode/ LP closed	Same as A		B50
	Ag, As, Ba, Cd, Cu, Cr, Hg, Ni, Pb, Sb, Se, Tl, Zn	Extraction: HNO_3	Multimode/ LP closed	2:20 min at 604 W, 9:25 min at 570 W		
		Total: HNO_3, HF	Multimode/ LP closed	Same as extraction		
USGS Soil	Al, K, Li, Si, Ti, Zr	1. HNO_3, HCl, HF 2. H_3BO_3	Multimode/ LP closed	1. 2.5 min at 650 W 2. 10 min at 650 W	ICP-OES	B2
GXR-2	Nb, Mo, Ta, W	1. HNO_3 2. HF, H_2O	Multimode/ MP closed	1. 30 min at 200 W 2. 1 min at 1000 W, 30 min at 450 W	ID-ICP-MS	B9
USGS Fe-Mn-W rich hot spring deposit GXR-3	Al, K, Li, Mg, Si, Ti	1. HNO_3, HCl, HF 2. H_3BO_3	Multimode/ LP closed	1. 2.5 min at 650 W 2. 10 min at 650 W	ICP-OES	B2
USGS Soil GXR-5	Al, Cr, K, Li, Si, Ti, Zr	1. HNO_3, HCl, HF 2. H_3BO_3	Multimode/ LP closed	1. 2.5 min at 650 W 2. 10 min at 650 W	ICP-OES	B2
	Nb, Mo, Ta, W	1. HNO_3 2. HF, H_2O	Multimode/ MP closed	1. 30 min at 200 W 2. 1 min at 1000 W, 30 min at 450 W	ID-ICP-MS	B9
USGS Marine sediment	Nb, Mo, Ta, W	1. HNO_3 2. HF, H_2O	Multimode/ MP closed	1. 30 min at 200 W 2. 1 min at 1000 W, 30 min at 450 W	ID-ICP-MS	B̄9
MAG-1	Al, K, Li, Si, Ti, Y, Zr	1. HNO_3, HCl, HF 2. H_3BO_3	Multimode/ LP closed	1. 2.5 min at 650 W 2. 10 min at 650 W	ICP-OES	B2

B1.　Bao hou, L.; Zhong quan, Y.; Kai, H. "Determination of silicon, aluminum, calcium, magnesium, iron, titanium, manganese, copper, cobalt, and nickel in vanadium - titanium - iron ore by microwave oven digestion, ICP, AA and chemical analysis methods", Pittsburgh Conference and Exposition, Atlanta, GA 1989; Paper 1375.

B2.   Lamothe, P. J.; Fries, T. L.; Consul, J. "Evaluation of a microwave oven system for the dissolution of geological samples" *Anal. Chem.* 1986, *58*, 1881-1886.

B3.   Alvarado, J.; Leon, L. E.; Lopez, F.; Lima, C. "Comparison of conventional and microwave wet acid digestion procedures for the determination of iron, nickel, and vanadium in coal by electrothermal atomization atomic absorption spectrometry" *J. Anal. At. Spectrom.* 1988, *3*, 135-138.

B4.   Alvarado, J.; Alvarez, M.; Cristiano, A. R.; Marco, L. M. "Extraction of vanadium from petroleum coke samples by means of microwave wet acid digestion" *Fuel* 1990, *69*, 128-130.

B5.   Riley, K. W.; Schafer, H. N. S.; Orban, H. "Rapid acid extraction of bituminous coal for the determination of phosphorus" *Analyst* 1990, *115*, 1405-1406.

B6.   Bettinelli, M.; Baroni, U.; Pastorelli, N. J. "Microwave oven sample dissolution for the analysis of environmental and biological materials" *Anal. Chim. Acta* 1989, *225*, 159-174.

B7.   Alexander, W. R.; Shimmield, T. M. "Microwave oven dissolution of geological samples: Novel application in the determination of natural decay series radionuclides" *J. Radioanal. Nucl. Chem.* 1990, *145*, 301-310.

B8.   Westbrook, W. T.; Jefferson, R. H. "Dissolution of sample by heating with a microwave oven in a teflon vessel for instrumental analysis" *J. Micro. Power* 1986, *21*, 25-32.

B9.   Goguel, R. "Group separation by solvent extraction from silicate rock matrix of niobium, molybdenum, tantalum and tungsten at low levels for ICP-OES" *Fresenius' J. Anal. Chem.* 1992, *344*, 326-333.

B10. Suzuki, T.; Sensui, M. "Application of the microwave acid digestion method to the decomposition of rock samples" *Anal. Chim. Acta* 1991, *245*, 43-48.

B11. Endo, M.; Sasaki, I.; Abe, S. "Kinetic spectrophotometric determination of iron oxidation states in geological materials" *Fresenius' J. Anal. Chem.* 1992, *343*, 366-369.

B12. Yoshimura, K.; Matsuoka, S.; Inokura, Y.; Hase, U. "Flow analysis for trace amounts of copper by ion-exchanger phase absorptiometry with 4,7-diphenyl-2,9-dimethyl-1,10-phenanthroline disulfonate and its application to the study of karst groundwater storm runoff" *Anal. Chim. Acta* 1992, *268*, 225-233.

B13. Bauer Wolf, E.; Wegscheider, W.; Posch, S.; Knapp, G.; Kolmer, H.; Panholzer, F. "Determination of traces of rare-earth elements in geological samples" *Talanta* 1993, *40*, 9-15.

B14. Matthes, S. A.; Farrell, R. F.; Mackie, A. J. in "A microwave system for the acid dissolution of metal and mineral samples"; Technical Progress Report: 120, National Bureau of Mines, Albany, OR, USA, April, 1983

B15. Kemp, A. J. "Microwave digestion of carbonate rock samples for chemical analysis" *Analyst* 1990, *115*, 1197-1199.

B16. Alvarado, J.; Petrola, A. "Determination of cadmium, chromium, lead, silver and gold in Venezuelan red mud by atomic-absorption spectrometry" *J. Anal. At. Spectrom.* 1989, *4*, 411-414.

B17. Matthes, S. A.; Guidelines for Developing Microwave Dissolution Methods for Geological and Metallurgical Samples In *Introduction to Microwave Sample Preparation: Theory and Practice*; Jassie, L. B., Kingston, H. M., Eds.; ACS: Washington DC, 1988, pp 33-52.

B18. Que Hee, S. S.; Boyle, J. R. "Simultaneous multielemental analysis of some environmental and biological samples by inductively coupled plasma atomic emission spectrometry" *Anal. Chem.* 1988, 60, 1033-1042.

B19. Fischer, L. B. "Microwave dissolution of geologic material: Application to isotope dilution analysis" *Anal. Chem.* 1986, 58, 261-263.

B20. Bettinelli, M.; Baroni, U.; Pastorelli, N. "Analysis of coal fly ash and environmental materials by inductively coupled plasma atomic emission spectrometry: Comparison of different decomposition procedures" *J. Anal. At. Spectrom.* 1987, 2, 485-489.

B21. Nadkarni, R. A. "Applications of microwave oven sample dissolution in analysis" *Anal. Chem.* 1984, 56, 2233-2237.

B22. Bettinelli, M.; Baroni, U.; Pastorelli, N. J. "Determination of arsenic, cadmium, lead, antimony, selenium, and thallium in coal fly ash using the stabilized temperature platform furnace and Zeeman-effect background correction" *J. Anal. At. Spectrom.* 1988, 3, 1005-1011.

B23. Paudyn, A. M.; Smith, R. G. "Microwave decomposition of dusts, ashes, and sediments for the determination of elements by ICP-OES" *Can. J. Appl. Spectrosc.* 1992, 37, 94-99.

B24. Alvarado, J.; Cristiano, A. R. "Determination of cadmium, cobalt, iron, nickel and lead in Venezuelan cigarettes by electrothermal atomic-absorption spectrometry" *J. Anal. At. Spectrom.* 1993, 8, 253-259.

B25. Paudyn, A. M.; Smith, R. G. "Applications of inductively coupled plasma atomic emission spectrometry in occupational health" *J. Anal. At. Spectrom.* 1990, 5, 523-529.

B26. Noltner, T.; Maisenbacher, P.; Puchelt, H. "Microwave acid digestion of geological and biological standard reference materials for trace element analysis by inductively coupled plasma-mass spectrometry" *Spectroscopy* 1990, 5, 49-53.

B27. Rantala, R. T. T.; Loring, D. H. "Teflon bomb decomposition of silicate materials in a microwave oven" *Anal. Chim. Acta* 1989, 220, 263-267.

B28. Berglund, B.; Wichardt, C. "Accurate and precise reference method for the determination of chromium in high-alloy steel" *Anal. Chim. Acta* 1990, 236, 399-410.

B29. Hlavacek, I.; Hlavackova, I. "Determination of minor and trace elements in ferrochromium and ferromanganese by inductively coupled plasma atomic emission spectrometry" *J. Anal. At. Spectrom.* 1991, 6, 535-540.

B30. Hlavackova, I.; Hlavacek, I. "Multi-element analysis of some high-silicon-content ferro-alloys by inductively coupled plasma atomic-emission spectrometry" *J. Anal. At. Spectrom.* 1994, 9, 251-255.

B31.   Riby, P. G.; Haswell, S. J.; Grzeskowiak, R. "Determination of arsenic in a nickel-based alloy using a microwave digestion procedure and a continuous-flow hydride-generation atomic-absorption system incorporating on-line matrix removal" *J. Anal. At. Spectrom.* 1989, *4*, 181-184.

B32.   Tyson, J. F.; Offley, S. G.; Seare, N. J.; Kibble, H. A. B.; Fellows, C. "Determination of arsenic in a nickel-based alloy by flow-injection hydride-generation atomic-absorption spectrometry incorporating continuous-flow matrix isolation and stopped-flow pre-reduction procedures" *J. Anal. At. Spectrom.* 1992, *7*, 315-322.

B33.   Fernando, L. A.; Heavner, W. D.; Gabrielli, C. C. "Closed-vessel microwave dissolution and comprehensive analysis of steel by direct current plasma atomic emission spectrometry" *Anal. Chem.* 1986, *58*, 511-512.

B34.   Whitten, C. W. "ICP analysis of recycled superalloy scrap" *At. Spectrosc.* 1987, *8*, 81-83.

B35.   Nieuwenhuize, J.; Poley Vos, C. H.; Van Den Akker, A. H.; Van Delft, W. "Comparison of microwave and conventional extraction techniques for the determination of metals in soil, sediment and sludge samples by atomic spectrometry" *Analyst* 1991, *116*, 347-351.

B36.   Chakraborti, D.; Burguera, M.; Burguera, J. L. "Analysis of standard reference materials after microwave-oven digestion in open vessels using graphite-furnace atomic-absorption spectrophotometry and Zeeman-effect background correction" *Fresenius' J. Anal. Chem.* 1993, *347*, 233-237.

B37.   Quevauviller, P.; Imbert, J. L.; Olle, M. "Evaluation of the use of microwave oven systems for the digestion of environmental samples" *Mikrochim. Acta* 1993, *112*, 147-154.

B38.   Huang, J.; Goltz, D.; Smith, F. "Microwave dissolution technique for the determination of arsenic in soils" *Talanta* 1988, *35*, 907-908.

B39.   Fernando, A. R.; Plambeck, J. A. "Digestion of soil samples for the determination of trace amounts of lead by differential-pulse anodic stripping voltammetry" *Analyst* 1992, *117*, 39-42.

B40.   Van Delft, W.; Vos, G. "Comparison of digestion procedures for the determination of mercury in soils by cold-vapor atomic absorption spectrometry" *Anal. Chim. Acta* 1988, *209*, 147-156.

B41.   Tahan, J. E.; Granadillo, V. A.; Sanchez, J. M.; Cubillan, H. S.; Romero, R. A. "Mineralization of biological materials prior to determination of total mercury by cold-vapor atomic-absorption spectrometry" *J. Anal. At. Spectrom.* 1993, *8*, 1005-1010.

B42.   Mahan, K. I.; Foderaro, T. A.; Garza, T. L.; Martinez, R. M.; Maroney, G. A.; Trivisonno, M. R.; Willging, E. M. "Microwave digestion techniques in the sequential extraction of calcium, iron, chromium, manganese, lead, and zinc in sediments" *Anal. Chem.* 1987, *59*, 938-945.

B43.   Baxter, D. C.; Nichol, R.; Littlejohn, D. "Evaluation of cold-vapor furnace atomic non-thermal excitation spectrometry for the determination of mercury in environmental samples" *Spectrochim. Acta* 1992, *47B*, 1155-1163.

B44.   Gilman, L. B.; Engelhart, W. G. "Recent advances in microwave sample preparation" *Spectroscopy* 1989, *4*, 14, 16, 18, 21.

B45. Kammin, W. R.; Brandt, M. J. "ICP-OES evaluation of microwave digestion" *Spectroscopy* 1989, *4*, 49-50, 52, 55.

B46. Revesz, R.; Hasty, E. "Microwave digestion of soils, sediments, and waste water for analysis of environmentally significant elements", International Conference on Metals in Soils, Waters, Plants, and Animals 1990; Paper FR-54.

B47. Binstock, D. A.; Grohse, P. M.; Gaskill, A.; Sellers, C.; Kingston, H. M.; Jassie, L. B. "Development and validation of a method for determining elements in solid waste using microwave digestion" *J. Assoc. Off. Anal. Chem.* 1991, *74*, 360-366.

B48. Binstock, D. A.; Yeager, W. M.; Groshe, P. M.; Gaskill, A. in "Validation of a method for determining elements in solid waste by microwave digestion"; Technical Report Draft: Research Triangle Park, NC, November, 1989

B49. Hewitt, A. D.; Reynolds, C. M. "Dissolution of metals from soils and sediments with a microwave-nitric acid digestion technique" *At. Spectrosc.* 1990, *11*, 187-192.

B50. Hewitt, A. D.; Reynolds, C. M. in "Microwave digestion of soils and sediments for assessing contamination by hazardous waste metals"; Special Report 90-19, U.S. Army Corps of Engineers Cold Regions Research & Engineering Laboratory, June 1990, 1990

B51. Kane, J. S.; Wilson, S. A.; Lipinski, J; Butler, L. "Leaching procedures: A brief review of their varied uses and their application to selected standard reference materials" *Am. Environ. Lab.* 1993, *6*, 14-15.

B52. Kokot, S.; King, G.; Keller, H. R.; Massart, D. L. "Application of chemometrics for the selection of microwave digestion procedures" *Anal. Chim. Acta* 1992, *268*, 81-94.

B53. Kokot, S.; King, G.; Keller, H. R.; Massart, D. L. "Microwave digestion: Analysis of procedures" *Anal. Chim. Acta* 1992, *259*, 267-279.

B54. Kingston, H. M.; Walter, P. J. "Comparison of microwave versus conventional dissolution for environmental applications" *Spectroscopy* 1992, *7*, 20-27.

B55. Kingston, H. M.; Walter, P. J.; Lorentzen, E. M. L.; Lusnak, G. P. in "The performance of leaching studies on soil SRMs 2710 and 2711"; Report of Analysis: Duquesne University, April 5, 1994

B56. Millward, C. G.; Kluckner, P. D. "Microwave digestion technique for the extraction of minerals from environmental marine sediments for analysis by inductively coupled plasma atomic emission spectrometry and atomic absorption spectrometry" *J. Anal. At. Spectrom.* 1989, *4*, 709-713.

B57. Matusiewicz, H.; Sturgeon, R. E.; Berman, S. S. "Vapor-phase acid digestion of inorganic and organic matrices for trace element analysis using a microwave heated bomb" *J. Anal. At. Spectrom.* 1991, *6*, 283-287.

B58. Nakashima, S.; Sturgeon, R. E.; Willie, S. N.; Berman, S. S. "Acid digestion of marine samples for trace element analysis using microwave heating" *Analyst* 1988, *113*, 159-163.

B59. Kratochvil, B. G.; Mamba, S. "Microwave acid dissolution of soil samples for elemental analysis" *Can. J. Chem.* 1990, *68*, 360-362.

Appendix 3. Miscellaneous Reference Materials

Appendix 3. Miscellaneous

Matrix	Analytes	Reagents	Cavity/Vessel	Conditions	Detection	Ref.
BCR Sewage amended soil 143	Cd, Cr, Cu, Fe, Mn, Pb, Zn	HCl, HNO$_3$	Multimode/LP closed	1 min at 180 W, 4 min at 480 W, 60 min at 600 W	F-AAS ETV-AAS	C1
	Cd, Cu, Ni, Pb, Zn	1. HNO$_3$ 2. HNO$_3$ 3. H$_2$O$_2$ 4. H$_2$O	Single mode/open	1. 5 min at 10 W, 10 min at 30 W, 10 min at 60 W 2. 10 min at 60 W 3. 5 min at 60 W 4. 5 min at 50 W	ICP-OES ICP-MS	C2
BCR Domestic sewage sludge 144	Pb	HNO$_3$	Multimode/flow through	5 min at 650 W	F-AAS	C3
	Cu, Mn	HNO$_3$, H$_2$O$_2$	Multimode/flow through	3-5 min at 650W (digestate recirculated through oven)	F-AAS	C4
	Cu, Mn	HNO$_3$, Isoamyl alcohol	Multimode/LP closed	3 min at 520 W	F-AAS	C5
	Cd, Cu, Fe, Mn, Pb, Zn	HNO$_3$, 2-ethyl-hexan-1-ol	Multimode/LP closed	3 min at 520 W	F-AAS	C6
	Cu, Mn, Pb, Zn	HNO$_3$	Multimode/flow through	2 sec at 650 W (0.5 mL coil, 15.4 mL min^{-1})	F-AAS	C7
	Al, As, Ba, Ca, Cd, Cr, Cu, Co, Fe, K, Mg, Mn, Na, Ni, Pb, Sb, Sc, Se, Si, Sn, Ti, Tl, V, Zn, Zr	1. HF, Aqua regia 2. H$_3$BO$_3$	Multimode/LP closed	1. 8 min at 300 W, 11 min at 600 W 2. 6 min at 300 W	ICP-OES F-AAS ETV-AAS	C8
BCR Sewage sludge 145	Cd, Cr, Cu, Fe, Mn, Pb, Zn	HCl, HNO$_3$	Multimode/LP closed	1 min at 180 W, 4 min at 480 W, 60 min at 600 W	F-AAS ETV-AAS	C1
	Al, As, Ba, Ca, Cd, Cr, Cu, Co, Fe, K, Mg, Mn, Na, Ni, Pb, Sb, Sc, Se, Si, Sn, Ti, Tl, V, Zn, Zr	1. HF, Aqua regia 2. H$_3$BO$_3$	Multimode/LP closed	1. 8 min at 300 W, 11 min at 600 W 2. 6 min at 300 W	ICP-OES F-AAS ETV-AAS	C8

Sample	Elements	Reagents	Mode	Conditions	Detection	Ref.
BCR Industrial sewage sludge 146	Pb	HNO_3	Multimode/flow through	5 min at 650 W	F-AAS	C3
	Cu, Mn	HNO_3, H_2O_2	Multimode/flow through	3–5 min at 650 W (digestate recirculated through oven)	F-AAS	C4
	Cu, Mn, Pb	HNO_3	Multimode/LP closed	3 min at 90°C (650 W)	ETV-AAS	C9
	Cu, Mn	HNO_3, Isoamyl alcohol	Multimode/LP closed	3 min at 520 W	F-AAS	C5
	Cu, Mn, Pb, Zn	HNO_3	Multimode/flow through	2 sec at 650 W (0.5 mL coil, 15.4 mL min^{-1})	F-AAS	C7
	Cd, Cu, Fe, Mn, Pb, Zn	HNO_3, 2-ethyl-hexan-1-ol	Multimode/LP closed	3 min at 520 W	F-AAS	C6
	Al, As, Ba, Ca, Cd, Cr, Cu, Co, Fe, K, Mg, Mn, Na, Ni, Pb, Sb, Sc, Sc, Si, Sn, Ti, Tl, V, Zn, Zr	1. HF, Aqua regia 2. H_3BO_3	Multimode/LP closed	1. 8 min at 300 W, 11 min at 600 W 2. 6 min at 300 W	ICP-OES F-AAS ETV-AAS	C8
BCR Brown bread 191	Cd, Cu, Pb	HNO_3, $HClO_4$, H_2SO_4	Multimode/HP closed	10 min at 85 bar	DP-ASV	C10
EPA Solid LCS	Al, As, Be, Cd, Co, Cr, Cu, Mg, Mn, Ni, Pb, Sb, V, Zn	HNO_3, HCl	Multimode/MP closed	1 min at 600 W	ICP-OES	C11
IAEA Human diet H9	I	HNO_3, N_2H_4	Multimode/HP closed	35 sec at 675 W	NAA	C12
	Cr, Hg, Se	1. HNO_3 2. H_2O_2	Multimode/HP closed	1. 3 min at 450 W	Radiochemical	C13
	Cd, Cu, Fe, Mn	HNO_3	Multimode/LP closed	3 min at 650 W (90°C)	ETV-AAS	C9
	Al	HNO_3	Multimode/HP closed	1 min at 150 W, 30 min 0 W, 1 min at 450 W	ICP-OES	C14
IAEA Fresh water W4	Cu, Mn, Pb	HNO_3	Multimode/LP closed	3 min at 650 W (90°C)	ETV-AAS'	C9

Appendix 3. Miscellaneous

Matrix	Analytes	Reagents	Cavity/Vessel	Conditions	Detection	Ref.
NIST Trace elements in glass 613	Pb, U	HNO_3, HF, $HClO_4$	Multimode/ LP closed	5 min at 90 W, 15 min at 138 W	ID	C15
NIST Trace elements in glass 615	Pb, U	HNO_3, HF, $HClO_4$	Multimode/ LP closed	5 min at 90 W, 15 min at 138 W	ID	C15
NIST Portland Cement 636	NA	HBF_4, HNO_3	Multimode/ LP closed	2 min at 600 W	NA	C16
NIST Doped Platinum 681	Au, Pd, Pt	Aqua regia	Multimode/ LP closed	1-3 h at 860 KPa	ETV-AAS	C17
NIST Alumina 699	Ca, Fe, Mn, Ti	H_2SO_4	Multimode/ MP closed	3 min at 630 W, 5-15 min at 315 W (170°C)	ICP-OES	C18
NIST Wear metals in oil 1085	Ag, Al, B, Ba, Be, Ca, Cd, Co, Cr, Cu, Fe, Mg, Mn, Mo, Ni, Pb, Sr, V, Zn	HNO_3	Multimode/ LP closed	5.5 min to 175°C (574 W) 4.5 min at 175-180°C (574 W)	ICP-OES	C19, C20
NIST Total diet 1548	Cu, Fe, Mn, Zn	A. HNO_3, H_2O_2	Multimode/ HP closed	75 sec at 665 W	ETV-AAS	C21
		B. HNO_3, H_2O_2	Multimode/ HP closed	1 min at 250 W, 2 min at 0 W, 2 min at 250 W, 2 min at 400 W, 2 min at 600 W		

Sample	Elements	Reagents	Mode/vessel	Conditions	Method	Ref.
NIST Powdered lead base paint 1579	Pb	C. HNO$_3$, HClO$_4$	Multimode/HP closed	150 sec at 950 W, 60 sec at 0 W, 90 sec at 300 W, 90 sec at 500 W, 90 sec at 700 W, 90 sec at 850 W	F-AAS	C22
		A. HNO$_3$	Multimode/LP closed	7 min at 675 W (90 psi max)		
		B. Same as A	HP closed vessel	3 min at 270 W		
	Ag, Al, Ba, Ca, Cd, Co, Cr, Cu, Fe, K, Li, Mg, Mn, Na, P, Pb, S, Sb, Si, Sr, Ti, Zn	1. HNO$_3$ 2. HF 3. HBO$_3$	Multimode/MP closed	1. 15 min at 0 W, 2. 8 min at 60 W, 6 min at 90 W, 9 min at 60 W	ICP-OES	C23
	Al, Ba, Ca, Cu, Mg, Mn, Pb, S, Sb, Ti, Zn	A. Aqua regia, H$_2$O	Multimode/MP closed	20 min at 180 psi	ICP-OES NAA	C24
		B. HNO$_3$, HF	Multimode/LP closed	20 min at 190 W, 15 min at 0 W		
NIST Trace elements in fuel oil 1634	Ag, Al, B, Ba, Be, Ca, Cd, Co, Cr, Cu, Fe, Mg, Mn, Mo, Ni, Pb, Sr, V, Zn	HNO$_3$	Multimode/LP closed	5.5 min to 175 °C (574 W) 4.5 min at 175-180 °C (574 W)	ICP-OES	C19, C20
NIST Trace elements in water 1643	Bi, Hg	1. HCl, KBr, KBrO$_3$ 2. NaBH$_4$, NaOH	Single mode/flow through	1. 65 sec at 90-120 W (8.5 mL min^{-1})	CV-AAS	C25, C26
NIST Urban particulate 1648	Cd, Cu, Fe, Pb	1. HNO$_3$, H$_2$O, H$_2$O$_2$ 2. Aqua regia, HF (optional)	Multimode/open	1. 15 min at 70 W, 15 min at 150 W, 15 min at 180 W, 1 min at 250 W, 1 min at 600 W 2. 15 min at 70 W, 1 min at 600 W	ETV-AAS	C27

Appendix 3. Miscellaneous

Matrix	Analytes	Reagents	Cavity/Vessel	Conditions	Detection	Ref.
NIST Metals on filter media 2676	Al, As, Ba, Be, Ca, Co, Cr, Fe, Mg, Mn, Ni, P, Pb, S, Ti, V, Zn	A. HF, Aqua regia B. HNO_3	Multimode/ LP closed Multimode/ LP closed	5 min at 200 W, 5 min at 400 W Same as A	ICP-OES	C28
NIST Multi element mix A 3171	Al	HNO_3	Multimode/ MP closed	4 min at 200 W, cool, 4 min at 350 W, cool, 8 min at 250 W, cool, 10 min at 400 W (x4)	ETV-AAS	C29
NIST Mixed diet 8431	Ca, Cu, Fe, K, Mg, Mn, P, Zn	HNO_3, HCl	Multimode/ LP closed	5 min at 300 W, 5 min at 0 W, 5 min at 300 W, 5 min at 450 W, 5 min at 0 W, 5 min at 450 W	ICP-OES	C30
	Cr, Hg, Se	1. HNO_3 2. H_2O_2	Multimode/ HP closed	1.3 min at 450 W	Radiochemical	C13
	Al	HNO_3	Multimode/ HP closed	1 min at 150 W, 30 min at 0 W, 1 min at 450 W	ICP-OES	C14
NRCC Seawater NASS-2	As, Fe, Se, V	HNO_3, H_2O_2	Multimode/ HP closed	-	ICP-MS	C31
NRCC Riverine water SLRS-2	As, Fe, Se, V	HNO_3, H_2O_2	Multimode/ HP closed	-	ICP- MS	C31

C1. Nieuwenhuize, J.; Poley Vos, C. H.; Van Den Akker, A. H.; Van Delft, W. "Comparison of microwave and conventional extraction techniques for the determination of metals in soil, sediment and sludge samples by atomic spectrometry" *Analyst* **1991**, *116*, 347-351.

C2. Quevauviller, P.; Imbert, J. L.; Olle, M. "Evaluation of the use of microwave oven systems for the digestion of environmental samples" *Mikrochim. Acta* **1993**, *112*, 147-154.

C3.   Carbonell, V.; De La Guardia, M.; Salvador, A.; Burguera, J. L.; Burguera, M. "On-line microwave oven digestion flame atomic absorption analysis of solid samples" *Anal. Chim. Acta* **1990**, *238*, 417-421.

C4.   Carbonell, V.; Morales Rubio, A.; Salvador, A.; De La Guardia, M.; Burguera, J. L.; Burguera, M. "Atomic-absorption spectrometric analysis of solids with on-line microwave-assisted digestion" *J. Anal. At. Spectrom.* **1992**, *7*, 1085-1089.

C5.   Martinez Avila, R.; Carbonell, V.; De La Guardia, M.; Salvador, A. "Slurries introduction in flow-injection atomic-absorption spectroscopic analysis of sewage sludges" *J. Assoc. Off. Anal. Chem.* **1990**, *73*, 389-393.

C6.   Morales, A.; Pomares, F.; De La Guardia, M.; Salvador, A. "Determination of cadmium, copper, iron, manganese, lead and zinc in sewage sludges with prior acid digestion in a microwave oven and slurry introduction" *J. Anal. At. Spectrom.* **1989**, *4*, 329-332.

C7.   De La Guardia, M.; Carbonell, V.; Morales Rubio, A.; Salvador, A. "On-line microwave-assisted digestion of solid samples for their flame-atomic-spectrometric analysis" *Talanta* **1993**, *40*, 1609-1617.

C8.   Bettinelli, M.; Baroni, U.; Pastorelli, N. J. "Microwave oven sample dissolution for the analysis of environmental and biological materials" *Anal. Chim. Acta* **1989**, *225*, 159-174.

C9.   Littlejohn, D.; Egila, J. N.; Gosland, R. M.; Kunwar, U. K.; Smith, C.; Shan, X. "Graphite furnace analysis - getting easier and achieving more?" *Anal. Chim. Acta* **1991**, *250*, 71-84.

C10.   Schramel, P.; Hasse, S. "Destruction of organic materials by pressurized microwave digestion" *Fresenius' J. Anal. Chem.* **1993**, *346*, 794-799.

C11.   Kammin, W. R.; Brandt, M. J. "Simulation of EPA method 3050 using a high-temperature and high-pressure microwave bomb" *Spectroscopy* **1989**, *4*, 22-24.

C12.   Rao, R. R.; Chatt, A. "Microwave acid digestion and preconcentration neutron activation analysis of biological and diet samples for iodine" *Anal. Chem.* **1991**, *63*, 1298-1303.

C13.   Vasconcellas, M. B. A.; Maihara, V. A.; Favaro, D. I. T.; Armelin, M. J. A.; Toro, E. C.; Ogris, R. "Radiochemical separation methods for the determination of some toxic elements in biological reference materials" *J. Radioanal. Nucl. Chem.* **1991**, *153*, 185-199.

C14.   Schelenz, R.; Zeiller, E. "Influence of digestion methods on the determination of total aluminum in food samples by ICP-OES" *Fresenius' J. Anal. Chem.* **1993**, *345*, 68-71.

C15.   Fischer, L. B. "Microwave dissolution of geologic material: Application to isotope dilution analysis" *Anal. Chem.* **1986**, *58*, 261-263.

C16.   Matthes, S. A.; Guidelines for Developing Microwave Dissolution Methods for Geological and Metallurgical Samples In *Introduction to Microwave Sample Preparation: Theory and Practice*; Jassie, L. B., Kingston, H. M., Eds.; ACS: Washington DC, 1988, pp 33-52.

C17. Hinds, M. W.; Littau, S.; Moulinie, P. "Determination of trace metals in platinum by electrothermal atomic-absorption spectrometry following a closed-vessel microwave dissolution procedure" *Analyst* **1992**, *117*, 1473-1475.

C18. Tatar, E.; Varga, I.; Zaray, G. "Microwave assisted dissolution of aluminum oxide samples" *Mikrochim. Acta* **1993**, *111*, 45-54.

C19. Binstock, D. A.; Grohse, P. M.; Gaskill, A.; Sellers, C.; Kingston, H. M.; Jassie, L. B. "Development and validation of a method for determining elements in solid waste using microwave digestion" *J. Assoc. Off. Anal. Chem.* **1991**, *74*, 360-366.

C20. Binstock, D. A.; Yeager, W. M.; Groshe, P. M.; Gaskill, A. in "Validation of a method for determining elements in solid waste by microwave digestion"; Technical Report Draft: Research Triangle Park, NC, November, 1989

C21. Mingorance, M. D.; Perez Vazquez, M. L.; Lachica, M. "Microwave digestion methods for the atomic-spectrometric determination of some elements in biological samples" *J. Anal. At. Spectrom.* **1993**, *8*, 853-858.

C22. Corl, W. E. "Comparison of microwave versus hot-plate dissolution techniques: Determining lead in paint chips" *Spectroscopy* **1991**, *6*, 40-43.

C23. Que Hee, S. S.; Boyle, J. R. "Simultaneous multielemental analysis of some environmental and biological samples by inductively coupled plasma atomic emission spectrometry" *Anal. Chem.* **1988**, *60*, 1033-1042.

C24. Paudyn, A. M.; Smith, R. G. "Determination of elements in paints and paint scrapings by inductively coupled plasma atomic-emission spectrometry using microwave-assisted digestion" *Fresenius' J. Anal. Chem.* **1993**, *345*, 695-700.

C25. Tsalev, D. L.; Sperling, M.; Welz, B. "On-line microwave sample pre-treatment for hydride-generation and cold-vapor atomic-absorption spectrometry. II. Chemistry and applications" *Analyst* **1992**, *117*, 1735-1741.

C26. Tsalev, D. L.; Sperling, M.; Welz, B. "On-line microwave sample pre-treatment for hydride generation and cold vapor atomic-absorption spectrometry. I. The manifold" *Analyst* **1992**, *117*, 1729-1733.

C27. Chakraborti, D.; Burguera, M.; Burguera, J. L. "Analysis of standard reference materials after microwave-oven digestion in open vessels using graphite-furnace atomic-absorption spectrophotometry and Zeeman-effect background correction" *Fresenius' J. Anal. Chem.* **1993**, *347*, 233-237.

C28. Paudyn, A. M.; Smith, R. G. "Applications of inductively coupled plasma atomic emission spectrometry in occupational health" *J. Anal. At. Spectrom.* **1990**, *5*, 523-529.

C29. Xu, N.; Majidi, V.; Ehmann, W. D.; Markesbery, W. R. "Determination of aluminum in human brain tissue by electrothermal atomic-absorption spectrometry" *J. Anal. At. Spectrom.* **1992**, *7*, 749-751.

C30. Schelkoph, G. M.; Milne, D. B. "Wet microwave digestion of diet and fecal samples for inductively coupled plasma analysis" *Anal. Chem.* **1988**, *60*, 2060-2062.

C31. Ebdon, L.; Ford, M. J.; Hutton, R. C.; Hill, S. J. "Evaluation of ethene addition to the nebulizer gas in inductively coupled plasma mass spectrometry for the removal of matrix-, solvent-, and support-gas-derived polyatomic-ion interferences" *Appl. Spectrosc.* **1994**, *48*, 507-516.

Chapter 3

Environmental Microwave Sample Preparation: Fundamentals, Methods, and Applications

H. M. (Skip) Kingston, Peter J. Walter, Stuart Chalk, Elke Lorentzen, and Dirk Link

Microwave sample preparation is now a standard analytical tool employing a variety of microwave equipment, including both low-to-high pressure closed vessels and atmospheric-pressure open vessels. Schemes for decomposition of environmental samples in preparation for elemental analysis are evaluated and individual procedures for a variety of sample types are examined. Theoretical and practical considerations in the development and implementation of microwave sample preparation are discussed including the use of clean chemistry principles, synergistic techniques, and automation. Methods for the total decomposition or leaching of trace elements in soils, sludges, sediments, oils, biological materials, water, and wastewater are described. The Environmental Protection Agency, Office of Solid Waste microwave Methods 3050B, 3051, and 3052 are discussed in detail. Standardized, specific, and alternative chemistries used and proposed for these microwave sample preparation methods are provided and evaluated. Supporting reference data and documentation are presented to demonstrate specific methods and alternative capabilities.

For centuries, chemists performing analytical procedures have searched for methods that are accurate, precise, and as independent of the analyst as possible. This goal was described by Berzelius (1) in 1814, and is translated as "seek to find the method of analysis which depends least on the skill of the operating chemist".

This goal is certainly applicable to sample preparation, one of the analytical chemist's most time-consuming, error-prone, and difficult steps in an analysis. Today, many chemists continue to use century-old beaker and hot plate acid dissolution methods to liberate elements for determination by new detection methods. Because of a number of constraints and ambiguities inherent in hot plate methods, they fall short of the goal of being accurate, precise, and analyst-independent. More recently, newer sample preparation techniques based on microwave heating have revolutionized sample preparation and provide a viable and preferable alternative.

Microwave sample preparation is becoming established as the sample preparation technique of choice due to its efficiency, relatively low cost, and robustness. A new understanding of reaction-enhancing mechanisms and the development of sophisticated procedures in the field of environmental elemental analysis mark the broad applicability of this technique. As will be seen through the applications described in this chapter, microwave-enhanced reaction methods of many kinds have been established and standardized throughout the world. One of the many reasons for the popularity of microwave sample preparation is that it often augments the analytical detection technique and the judgment of the analyst, providing a more robust analytical approach.

Strong mineral acids used in sample dissolution have been known since the 1300s when sulfuric and nitric acid were first reported. Mineral acids have been the preferred medium for dissolving samples, as they liberate elements of interest in a form that can be readily analyzed. In his publication *A Short History of Chemistry*, Isaac Asimov hailed the discovery of mineral acids for their contribution to human kinds' technological advance. He stated (2):

> The discovery of mineral acids was the most important chemical advance after that of the successful production of iron from its ore some three thousand years before. Many chemical reactions could be carried through, and many substances dissolved, by Europeans with the aid of the strong mineral acids, which the earlier Greeks and Arabs could not have brought about with vinegar, the strongest acid at their disposal.

Mineral acid dissolution is not usually acknowledged as the important chemical process that it actually is. The reason why mineral acids are so important to instrumental analysis is that they liberate element ions into a homogeneous solution (at the molecular level) that can be directly introduced into the analytical instrument. For quantitative analysis, most instruments require a solution so they may determine a phenomenon, such as wavelength energy intensity or mass. These properties are integrated over some time interval. Thus, the quantitation relationship (and molecular homogeneity) must remain constant over this period. Even other solubilizing procedures frequently require the addition, at later steps, of acid to further dissolve solids (e.g., fusions) or to stabilize the solution.

Historically, sample preparation procedures were performed in open vessels until 1860, when Carius developed pressurized acid dissolution in closed glass tubes with steel casings. The 20th century has seen further advances, such as steel-jacketed Teflon bombs, direct energy coupling through open-vessel microwave systems and finally, in the early 1980s, closed-vessel microwave digestion (3–5).

Early use of microwave energy for acid dissolution reactions was carried out in domestic microwave ovens and can be best described as trial-and-error development. The measurement or regulation of either energy or temperature was not performed (6). This approach persisted until the mid 1980s when the theoretical basis of controlling input energy for the control of the reaction temperature was described (3). Temperature control was achieved by isolating the reaction in closed, pressurized Teflon vessels during microwave acid decomposition and balancing microwave energy absorption against heat loss. In 1988, the first book on microwave sample preparation detailed the theoretical basis and predictive capabilities of microwave energy transfer for these specific purposes, and it described numerous applications of closed- and open-vessel microwave-assisted acid dissolutions (7). In 1987, the technique was considered to be one of the top 100 developments in research and industry (8).

The number of publications dealing with microwave sample preparation continues to increase annually at a rate appropriate for a developing technology. Today, microwave dissolution is used for elemental analysis of almost all types of samples (9), including biological (10–12), botanical (13–15), geological (16–21), metallurgical (19, 22, 23) matrices, and for many other unique applications. Refer to Chapter 2 for an organized review of microwave sample preparation literature including matrices, analytes, analytical detection methods, microwave equipment types, reagent combinations, and decomposition for analysis of standard materials.

Through this research a number of standard methods were developed. Table I lists the current approved standard methods that are based on microwave irradiation. With the exception of the drying methods, these standard methods are categorized as sample preparation methods for acid decomposition or leaching of specific matrices. They represent some of the most robust and accurate methods for sample preparation and are among the most time and reagent efficient.

Microwave methods, while being robust and versatile in their implementation, are not governed by intuitively obvious mechanisms. Classical

Table I. Standard Microwave Sample Preparation Methods

Standards Organization and Method Number	Brief Method Description
ASTM D 4309–91	Standard practice for sample digestion using closed-vessel microwave heating technique for the determination of total recoverable metals in water
ASTM D 4643–93	Standard test method for determination of water (moisture) content of soil by the microwave oven method
ASTM D 5258–92	Standard practice for acid extraction of elements from sediments using closed-vessel microwave heating
ASTM E 1358–90	Standard test method for determination of moisture content of particulate wood fuels using a microwave oven
US–EPA 3015 SW–846 Update II	Microwave-assisted acid digestion of aqueous samples and extracts
US–EPA Draft 3031 SW–846 Update III	Acid digestion of oils for metals analysis by FAAS or ICP spectroscopy—proposed microwave adaptation
US–EPA 3050B SW–846 Update III	Acid digestion of sediments, sludges, and soils—proposed microwave adaptation
US–EPA 3051 SW–846 Update II	Microwave-assisted acid digestion of sediments, sludges, soils, and oils
US–EPA 3052 SW–846 Update III	Microwave-assisted acid digestion of siliceous and organically based matrices
NPDES method	Closed-vessel microwave digestion of wastewater samples for metals determination
Standard Methods of Water & Waste Water Analysis 3030 K	Microwave-assisted leach/digestion for metals in water
French Standard AFNOR NF V 03–100	Kjeldahl nitrogen
Republic of China Environmental Protection Agency NIEA C303.01T	Acid digestion of fish and shellfish

education and experience with convection- and conduction-based heating systems does not equip the chemist to predict conditions in microwave systems. This situation has been the origin of many accidents, near accidents, and unexpected equipment failures when using this newer technology (see Chapter 16). Many subtle, but distinct, differences exist between the standard heating of acids in open or sealed containers and their microwave energy equivalents. Many of these little understood phenomena are described and illustrated in this chapter. The current standard environmental procedures take advantage of these unique relationships. A detailed understanding of their influence on vessel conditions will aid the practitioner in implementing microwave methods.

Closed-vessel microwave decomposition at low pressure (≤10 atm) and moderate pressure (>10 and ≤80 atm) was originally developed for elemental certification of Standard Reference Materials (SRMs) at the National Institute of Standards and Technology (NIST). It was used there for many years to certify elemental constituents in a variety of SRMs covering a broad spectrum of elements and sample matrices. The original impetus for the development of closed-Teflon-vessel microwave sample preparation was the need to break the perceived barrier of part-per-billion analysis and create a reliable, ultra-trace, and total analysis method for material certifications (*8*). Soon after these total decomposition methods were developed, they were tested and standardized for both the leaching and the total decomposition of environmental samples. These procedures resulted in a suite of methods further developed in conjunction with the Environmental Protection Agency (EPA). They include methods 3015, 3051, 3052, and the alternative procedures in 3031 and 3050B (Table I), which target the matrix categories most frequently analyzed (*see* Chapter 2).

This chapter includes both fundamental descriptions of the nonintuitive characteristics of microwaves and a detailed examination of the major environmental microwave methods, which provides a thorough discussion of acid decomposition of samples using microwave energy.

Fundamentals of Microwave Interactions

Energy Absorption Mechanisms and Reaction Conditions

The speed, efficiency, and completeness of a dissolution procedure is a function of the reaction conditions and the reagent chemistry. Experimental conditions are an inherent limitation to the speed of reaction in dissolution techniques. In conventional dissolution procedures, hot plates, heating mantles, heating blocks, and laboratory ovens conduct heat to the reaction vessel, which transfers heat only to the solution in direct contact with the vessel walls. The heat is distributed via convection currents throughout the reagent–sample digestion mixture. Because of the nature of this process, these heating methods are relatively slow.

Hot plate and conventional open-vessel dissolution procedures are also limited by the boiling points of the acids or azeotropic mixtures. In addition, there can be variations between laboratories as a result of differences in acid boiling points due to changes in atmospheric pressure, colligative properties (convection), and conductive heating properties of conventional apparatus.

In contrast, the use of microwave irradiation has a more direct, and thus controllable, heating mechanism and is less susceptible to the variables mentioned previously. Both open-vessel and closed-vessel microwave systems use direct absorption of microwave radiation through essentially microwave transparent vessel materials. Atmospheric pressure microwave systems can generate more stable temperature conditions, are not limited by heating mechanisms of convection or conduction, and exhibit superheating effects resulting in temperatures above the boiling points of the solvents (24–27). In comparison, closed-vessel microwave dissolution systems are limited only by the temperature and pressure safety tolerances of the reaction vessel and the microwave absorption characteristics of the solution. Another unique difference is in the energy absorption by the liquid phase versus the gas phase, which produces an inherent solvent refluxing that may be exploited over conventional heating systems (detailed later).

Solution absorption of microwave radiation is controlled by fundamental thermodynamic properties and is totally different from conventional heating mechanisms. An informative discussion of the advantages of conventional high temperature and pressure acid dissolution was addressed in an International Union of Pure and Applied Chemistry monograph published prior to the adoption of microwave methodology (28). Both atmospheric-pressure and closed-vessel microwave systems demonstrate additional efficiency gains over their conventional counterparts through the generation of unique reaction conditions.

In microwave heating, energy is directly transferred to absorbing molecules from both the sample and reagents as detailed in Chapter 1. An examination of the role played by each of the absorption mechanisms will assist the analyst in understanding why unique conditions are achieved in microwave systems. Recall the two microwave absorbing mechanisms: ionic conductance and dipole rotation (29, 30). In simplified terms, ionic conductance refers to the phenomenon by which ions in solution migrate when an electromagnetic field is applied. The solution resistance to the free flow of ions results in friction that heats the solution. This mechanism is much less dependent on microwave frequency than is the dipole rotation mechanism. Dipole rotation is the alignment of a molecule dipole with the applied field. Molecular "friction" results from the very rapid forced molecular movement caused by the oscillation of the applied field. At 2450 MHz, the dipoles align (lose entropy) and then randomize (gain entropy) 4.9×10^9 times per second, resulting in fast, efficient, and thermodynamically predictable power absorption.

Because mineral acids in aqueous solution are all ionized, polar, and charged, microwave absorption by these reagents is very high. The power absorption values, and the ability to predict them at 2450 MHz for mineral acids, was developed in the first publication dealing with closed Teflon vessels, and in Chapter 6 of the first book on microwave sample preparation (3, 31). Microwave energy absorption is dependent on the specific acid, azeotropic mixture, diluted acid, the total mass of acid in the cavity, and the available microwave power. The control of acid decomposition methods benefits from this consistent, predictable, and reproducible energy absorption (3, 29–31). This is not the case when using conventional heating apparatus.

Fundamental Relationships in Microwave Energy Absorption

At first, predicting reaction temperature and pressure conditions appears to involve simple relationships and to be straightforward. As we will see, microwave closed-vessel systems have counter-intuitive phenomena. Microwave energy absorption can be expressed in a fundamental thermodynamic relationship that relates energy absorption to the specific heat capacity, mass in the microwave field, temperature increase, and time of sample exposure as follows:

$$P_{abs} = \frac{KC_p m\Delta T}{t} \tag{1}$$

where P is the apparent power absorbed by the sample in watts (W = joule/s); K is the conversion factor for calories/s to W (4.184); C_p is the heat capacity, thermal capacity, or specific heat (cal g^{-1} °C^{-1}); m is the mass of the sample in grams; ΔT is the final temperature, T_f, minus the initial temperature T_i (°C); and t is the time in seconds. This equation, governing power absorption, can be rearranged and used to predict the temperature (T_f) of the reagent at some point in time (eq 2), or the time (t) necessary to reach a particular temperature (eq 3). These predictions work well for most vessels over the 20–250 °C temperature range and for extended periods of time dependent on the microwave vessel system (31).

$$T_f = T_i + \frac{P_{abs}t}{KC_p m} \tag{2}$$

$$t = \frac{KC_p m\Delta T}{P_{abs}} \tag{3}$$

In addition, a quartic model (eq 4) has been developed to predict the power absorbed in closed microwave vessels by a quantity of mineral acid used in decomposition (31). This equation is valid within the mass range of 25–1000 g of reagent. Coefficients A through E are reported for specific concentrations of nitric, hydrochloric, hydrofluoric, sulfuric acids, water, and for specific acid concentrations. The specific power absorption can be determined for other acid concentrations and mixtures by using this method.

$$\ln(P_{abs}) = A + B\ln(m) + C[\ln(m)]^2 + D[\ln(m)]^3 + E[\ln(m)]^4 \tag{4}$$

By using this equation, the calculation of the power absorption can be made within 4–10% at the 95% confidence level, depending on the mineral acid and aqueous dilution used.

Microwave Heating at Atmospheric Pressure

Replacement of traditional sample preparation methods and temperature-dependent chemical reactions with microwave-induced heating is occurring at a rapid rate. There are many reasons for this shift; most focus on control and efficiency improvements. Open-vessel procedures allow for the direct adaptation of existing hot plate and heating mantle methods with greatly improved control of reaction conditions.

In an atmospheric pressure system, the temperature can be calculated by eq 2 up to the solvent boiling point, including any superheating of the acid, dilution, or azeotropic effect. This equation is not valid after a phase change is reached in atmospheric-pressure systems. However, the boiling point does limit the oxidation potential (the ability of the reagents to destroy the matrix) in open-vessel procedures.

Due to the fundamental limitations of working at atmospheric pressure, researchers have turned to other methods to increase the oxidation potential of digest solutions. In addition to using azeotropic mixtures and taking advantage of any superheating, analysts have used acids with higher boiling points, such as sulfuric acid, to produce more rigorous digestion–leaching conditions. Hydrogen peroxide can be added safely in an open-vessel system. It increases the oxidation potential and also improves the conversion of microwave energy into heat because of its high dielectric constant (121) when diluted appropriately in water (35% (v/v)) (32).

Also, a significant advantage with an atmospheric pressure system should be mentioned. When digesting large samples (3–10 g), especially those with a high carbon content, large amounts of gas and vapors are generated. In open vessels, these are released from the reaction mixture and are continuously swept from the headspace above the sample. Thus, in contrast to closed vessels, reactions producing gases are favored to go to completion according to Le Chatelier's principle.

Temperature and Pressure Relationships in Closed Microwave Vessels

Normally, vessels used in microwave systems are relatively transparent to microwave radiation and are inert to the reagents used. The materials in contact with the sample are usually of fluoropolymer, quartz, or glass composition. Closed, pressurized vessels are frequently multilayered with a structural outer casing of a microwave transparent polymer, such as polyetherimide. This polymer is commonly used for tensile strength, rigidity, and the containment of accumulated pressure.

Two distinct types of closed vessels exist. One type of vessel is characterized as an uninsulated, relatively thin, single-walled fluoropolymer vessel, or a fluoropolymer vessel with a thin outer liner made of polyetherimide or rigid composite plastic. These vessels have minimal insulating characteristics and permit large quantities of heat to escape. The other major type can be described as a well-insulated vessel, usually of a very thick-walled fluoropolymer, or a vessel with a very thick outer layer or casing, or both. These vessels retain heat in a very efficient manner and therefore do not permit rapid cooling by the ambient air forced over the vessels within the microwave cavity.

Unlike open-vessel systems, the temperature predictions from eq 2 are also applicable for temperatures above the boiling point for closed-vessel microwave containers. The equations provide an excellent means of estimating the temperature in insulated closed vessels up to several hundred degrees over several minutes. However, because the majority of microwave vessels are noninsulating, and thus lose heat readily, the predictions get more inaccurate as temperature rises. A significant amount of energy is lost as heat through the walls of the vessel, to the cooler air of the microwave cavity, or to the cooling system. Consequently, gas pressures within closed vessels are usually not what would be predicted from the temperature of the liquid phase. In fact, they are dramatically lower in most cases. This unique feature of microwave closed-vessel systems is one of the key reasons that microwave systems are revolutionizing sample preparation.

Evidence of this phenomenon can be seen by performing some simple experiments. The power absorption relationship and the predictive model can be used to describe the heating temperature profile for a concentrated acid (e.g., nitric acid) in a totally insulating microwave closed vessel. Figure 1a is a plot of temperature data for concentrated nitric acid from eqs 1 and 3 (the theoretical temperature rise), and actual measurements of the temperature for one of six all-Teflon perfluoroalkoxy (PFA) vessels, each containing 10 mL of acid in a 574 ± 5 W, 2450 MHz microwave energy field (the measured temperature rise). The acid temperature agrees well with the predicted temperature for the first 2 min, and then begins to plateau in the Teflon PFA vessel as heat is lost through the vessel walls. Thus, for the majority of closed vessels, eq 2 should be modified to include a term that takes this into consideration (eq 5).

$$T_f = T_i + \frac{P_{abs}t}{KC_p m} - \text{Heat Loss} \tag{5}$$

Measurement of the outer wall temperature of a heavily insulated vessel (Paar Instruments Company, Moline, IL), when the acid inside was heated to over 200 °C, showed only an ~10 °C increase above ambient temperature after 5 min when the solution inside the vessel was 95 °C, measurement at the bottom outer wall of an all-Teflon PFA vessel (Savillex Corporation, Minnetonka, MN) produced a temperature of 90 °C, and the temperature at the top outer wall was 75 °C.

At the present time, no convenient method of estimating this heat loss in the fundamental equation predicting temperature or time to temperature has been effective. The amount of heat lost varies with a number of parameters such as the temperature, vessel geometry, liquid volume, duration of heating, and heat capacity of the liquid phase. Thus, this amount of heat loss has not been accurately modeled at this time.

As a consequence of the lower vapor-phase temperature, the measured pressure in the same all-Teflon PFA vessel (Figure 1b) is very different from the predicted pressure. In a completely insulating vessel, the gas phase will be the same temperature as the liquid phase after equilibration (Figure 2). On the basis of the measured liquid phase temperature, the pressures would have to be at least those corresponding to the equivalent vapor pressure of nitric acid (33). From Figure 1b, it is evident that the measured pressure is considerably less. This result is because in a microwave field, the assumption that all the components of the system (liquid, gas, and vessel) are in thermal equilibrium is not true. Even though the whole vessel is within the microwave field, the gas phase is not heated as efficiently as the liquid phase. The ionic conduction mechanism previously discussed is absent in the gas phase as free ions are left behind in solution. The lack of discrete ions deprives the gas phase of the ionic conduction heating mechanism and leaves only the molecular rotation mechanism. In addition, the efficiency of heat transfer by rotation is decreased significantly because of the statistically smaller number of molecule–molecule collisions.

By examining what is occurring inside the vessel while being heated by microwave energy, Figure 3 illustrates that the temperature of the gas phase does not reach that of the liquid phase. No thermal equilibrium of the gas and liquid phases exists, as would be the case in a convection oven where the gas, liquid, and vessel walls are all at the same temperature. If the gas phase is not at the same temperature, the pressure will not be described by the standard table, but rather by the ideal gas law, and the temperature, volume, and composition of the gas phase.

In the gas phase, molecules transfer heat upon collision to the walls of the vessel. Because the walls are transparent to microwave energy, the walls are cool and absorb heat from the gas and transfer heat upon colliding with the vessel walls. A temperature gradient develops from the warmest part of the vessel, in contact with the liquid phase, to the coolest region near the top of the vessel. The resulting temperature gradient depicted in Figure 3 causes the acid vapors to condense and effectively creates a refluxing action inside the vessel. This *sustained, dynamic, thermal nonequilibrium* between the phases is a direct result of the heating mechanisms and physical interactions found only in a microwave field (34).

This surprising finding is the reason that pressure control for microwave procedures is not generically applicable for sample preparation methods. Vessels of different design can produce the same liquid-phase temperature but have different gas-phase pressures. Demonstration of this phenomenon was shown for the standard EPA microwave leach method 3051, where the same temperature profile (that must be maintained to

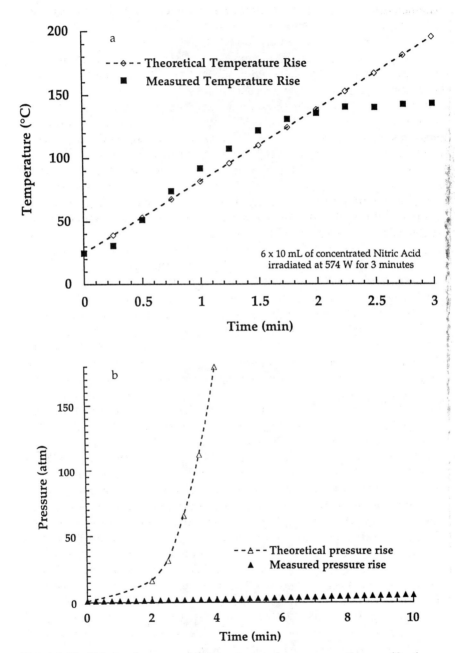

Figure 1. Predicted and measured temperature and pressure vs. time profiles for a closed all-Teflon PFA microwave vessel.

Insulating Microwave Vessel

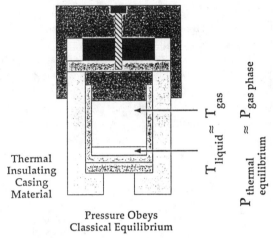

Thermal Insulating Casing Material

Pressure Obeys Classical Equilibrium

$T_{liquid} \approx T_{gas}$

$P_{thermal\ equilibrium} \approx P_{gas\ phase}$

Figure 2. Thermal equilibrium of liquid and gas phases in an insulating microwave vessel.

Non-insulating Microwave Vessel

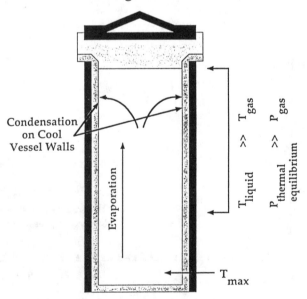

Condensation on Cool Vessel Walls

Evaporation

T_{max}

$T_{liquid} >> T_{gas}$

$P_{thermal\ equilibrium} >> P_{gas}$

Figure 3. Refluxing inside a noninsulating microwave vessel.

meet the performance criteria) could be generated in a variety of commercial vessels (Figure 4). Although each nitric acid solution reaches the same approximate temperature, the vessels are at different pressures. Figure 5 shows a plot of the pressures of each of these vessels during identical runs of Method 3051. It can be seen that the pressures differ significantly due to different rates of heat loss. Some manufacturer's vessels resulted in similar pressures, which is to be expected because, for example, the CEM (Matthews, NC) and O. I. Analytical (College Station, TX) vessels tested were very similarly constructed.

The phenomenon of lower internal pressure at relatively high temperatures is one of the main advantages of microwave closed-vessel decomposition (assuming there are no reactions that are kinetically limited by low pressures). In these systems, it is desirable to work against a thermal heat loss from the gas phase in the vessel. In fact, internal vessel pressures may be further lowered by cooling the gas phase in the vessel. Some designs of closed microwave vessels take advantage of heat loss to improve the safety and robustness of digestion procedures.

In the literature, discussions, and through courses taught on the subject of microwave chemistry, indications are that this concept is often misunderstood. It is, however, a pivotal concept affecting the usefulness of closed-vessel microwave acid dissolution, organic extraction, leaching, synthesis, and many other closed-vessel microwave chemical reactions, including some that have not become popular. Frequently, many analysts depend on this reduced pressure at a particular temperature without

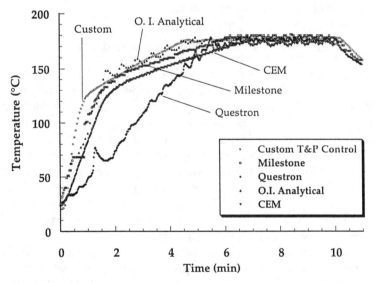

Figure 4. Temperature vs. time profiles for research and commercial microwave equipment using EPA Method 3051.

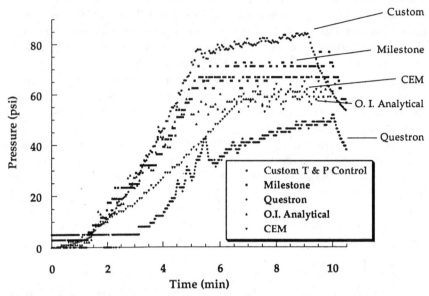

Figure 5. Pressure vs. time profiles for research and commercial microwave equipment using EPA Method 3051.

realizing this situation exists. This phenomenon has been only briefly described in previous communications (*31*). Understanding the cause, importance, and consequences of this nonintuitive effect is necessary for development of closed-vessel procedures.

Sample Preparation and Microwave Chemistry

The art of preparing a sample for analysis has come a long way. We now understand more fully what is required to efficiently extract analytes from complicated matrices. Not surprisingly, the primary area of understanding of this process lies in the chemistry involved. This is not just a matter of the interaction of the dissolution reagents and the sample, but it also involves the control of the sample environment and the reaction conditions. The temperature of the reaction mixture, in particular, has a significant impact on the rates, extents, reproducibilities, and types of reactions that occur. With this understanding, it may be possible to design dissolution procedures for specific matrices and analytes, such as those necessary to determine speciated elemental concentrations.

Microwave-based acid dissolution procedures, particularly those carried out in closed vessels, seem synergistic with the previously mentioned factors for producing representative sample preparation. The move to directly transfer energy through microwave heating allows the analyst to control the environment of the digestion and the reaction conditions, essentially isolating the sample processing from the laboratory environment. This feature also allows the analyst to gain more information about the reaction processes because the system is more easily controlled.

Evolution of Microwave Environmental Sample Preparation

A historical evaluation of microwave use in sample preparation shows that developments have been driven by a need for standardized methods in environmental and analytical chemistry. If it were not for the large analytical instrument market, manufacturers would probably not have developed specialized laboratory microwave equipment. Analytical chemists, and those in other areas of chemistry, would still be converting domestic microwave ovens for use in their research.

Environmental applications were the impetus for the development of the closed-vessel microwave method. Two U.S. government laboratories, the Bureau of Mines and the National Bureau of Standards (now NIST), both began microwave closed-vessel decomposition development by using domestic microwave systems in 1979. At the Bureau of Mines, sealed polycarbonate bottles were used, and at NIST, Teflon PFA vessels were evaluated. These efforts resulted in the first two papers published on sealed microwave applications in 1983 and 1984 (4, 5). In 1981, two commercial companies were requested to make laboratory microwave equipment for research at NIST. Essentially, both used commercial microwave equipment with minor modifications and thus the units were not viable. CEM produced the first viable commercial laboratory unit, and although much of it was recycled from domestic microwave systems, it was modified to an extent that produced a usable first generation laboratory instrument. Vessels originally designed for the nuclear waste leaching studies (for NIST research) and manufactured by Savillex Corporation were employed by CEM and are still in use in a modified form today. These first systems paved the way for development of modern microwave sample preparation.

Almost immediately after these developments, the demonstration of efficiency, precision, and accuracy for the dissolution and analysis of environmental samples launched development of standard microwave methods. More recently, many of the limitations of early commercial equipment have been overcome and more capable laboratory microwave instrumentation has emerged and provided chemists with unique capabilities not possible with traditional laboratory reaction apparatus. Temperature feedback control and robust reaction vessels have opened up many new applications for microwave-enhanced chemistry. These new applications are so much more efficient than traditional methods that the field of microwave sample preparation and microwave-enhanced chemistry is rapidly advancing in laboratories around the world. All of this improvement has evolved in just over one decade and has established a new field of chemical research. The expansion of the field from 1984 to 1997 can be measured by its growth from a small niche market in microwave drying equipment, with approximately $1 million in sales per year, to an instrument market of over $100 million in sales per year. Some of the key developments in research, applications, and equipment that have led to the development, expansion, and standardization of this technique have been compiled in Table II.

Digestion Reaction Temperatures

The question of what temperature and which mineral acid is needed to decompose a sample is a difficult one that does not have a single answer.

Table II. Laboratory Microwave Equipment and Application Evolution

Year	Equipment	Ref.
1975	Domestic microwave oven open-beaker heating of acids and samples	6
1979–1984	Closed polycarbonate tube and closed Teflon PFA vessel acid digestions	4, 5
1984	Temperature and pressure measurement in situ demonstrated	3
1985	Commercial closed-vessel microwave system introduced	8
1986	Commercial focused atmospheric pressure microwave introduced	35
1986	Flow-through microwave digestion demonstrated	36
1986	Microwave extraction demonstrated	37
1986	Fundamental closed-vessel sample decomposition paper	3
1987	Measurement of oxidation temperatures of organics in microwave systems	31
1987	Sample Preparation IR-100 award for microwave instrumentation	8
1988	First book on microwave sample preparation	7
1988	Draft Methods 3051 and 3015 presented to EPA	38
1989	Commercial pressure feedback control introduced	39
1989–1990	Validation of EPA Method 3051, adoption by CERCLA	17
1989–1992	Automation of EPA microwave methods	40–43
1990	Simultaneous temperature and pressure feedback control first introduced	40
1991	Demonstration of solvent superheating in microwave systems	26, 27
1993	Commercial temperature feedback control introduced	39
1992	Commercial flow-through microwave instrumentation introduced	44
1994	Development of environmental microwave leach standards	45
1995	Total microwave sample processing	46
1995	Approval of EPA RCRA Methods 3051 and 3015	47
1995	Development and proposal of EPA Method 3052	48, 49
1995–1996	Conversion of EPA Methods 3050B, 3031, 3060A to microwave heating	49–52

A logical way exists to evaluate the problem. Samples are composed of various classes of compounds that collectively make up the matrix. One class is silicates, typified by a silicon–oxygen backbone, which are known to require hydrofluoric acid to break the Si–O bond. Organics are another, more complicated class, containing various types of carbon–carbon bonds. Even further definition can be given to each of these classes of compounds. Organics may be further described as oils, plants, or biological tissues. Biological tissue can be subdivided into classes such as carbohydrate, cellulose, protein, fat, etc.

A fundamental premise adopted here and by many other analysts is that if the decomposition of each of the subclasses of the matrix is known, then decomposition of the combined matrix as a discrete sample can be understood and predicted. This premise has been proven useful for many well-characterized SRMs. The advantage of using SRMs is that frequently the matrix is well-characterized and its subcomponents are identified. An example is NIST Buffalo River Sediment SRM 2704. During its certification, microscopic analysis of the subcomponents identified small bits of coal, rock, organic matter, clay, iron oxide, and other subcomponents that make up this sample. Organic and inorganic components in this and other SRMs (e.g., 1571, orchard leaves) have been identified over the years. Schemes for total decomposition were based on accounting for the chemistry of each subcomponent. Even difficult combinations of subcomponents that present challenging matrices have eventually been decomposed and quantitatively analyzed after liberating the elements from each subcomponent.

How do you determine the most appropriate acid, appropriate temperature, and appropriate reaction time? These are questions the analyst must predict correctly to attain quantitative results. Some examples of how appropriate temperatures are set, and how long these reactions take have been developed in fundamental studies. Other examples are from work that used conditions and evaluated the elemental result but did not actually characterize the remaining matrix. Both sets of information are useful although fundamental studies are more satisfying and more generally applicable. Examples of sample types, reagent combinations, elements determined, equipment configurations, and methods of detection have been compiled in Chapter 2, and are grouped by reference materials of similar matrix types. These examples provide the advantage of being able to duplicate the matrix in the laboratory. They serve as good examples of typical conditions and starting points.

Other studies have provided guidance into the fundamental questions posed earlier of specific temperatures and times. For example, fundamental studies of the analysis of biological food samples for Kjeldahl nitrogen by sulfuric acid–hydrogen peroxide digestions have been reported (53). The authors showed that a triangle of fundamental components of biological tissues—carbohydrates, proteins, and fats—can be used to calculate digestion conditions. Times and power settings for open-vessel microwave decompositions are suggested on the basis of the amount of each of these components. This concept also uses the premise that each of the sample components must be addressed in an optimal manner by altering the conditions of the procedure or the reagent mixture. General classes of matrices along with the subcomponent charts were used in a similar manner for the analysis of biological, mineral, and synthetic products (54). Again, acid combinations and microwave power settings were recommended for open-vessel microwave systems. In open-vessel systems, heating of the acid is more efficient, but the temperature is limited to the boiling point of the acid or azetrope plus any superheating.

In closed-vessel microwave sample preparation, a slightly different approach has been taken. Because of the change in oxidation potential with temperature, oxidizing reagents such as nitric acid were studied to determine optimal decomposition temperatures for different matrix components (31).

Oxidation of organic sample components by using nitric acid is a fundamental component of many standard methods. Even if the sample is an environmental soil, organic matter may still be present and chelate metals and prevent them from being determined in solution. Because the oxidizing power of nitric acid is temperature dependent, minimum temperatures are required to achieve rapid, efficient oxidation of specific organic bonds. Components of biological tissues were used to represent organic molecule types with the traditional subcomponents of carbohydrates, proteins, and fats–fatty acids being tested independently and in combination. Temperatures of 140 °C, 150 °C, and 160–165 °C were needed to efficiently decompose carbohydrates, proteins, and fats and oils, respectively, as exemplified by the rise in pressure at a specific temperature (31). Thus, instead of programming power and time, a specific temperature was identified that achieves efficient and rapid decomposition.

These temperatures were used as minimum conditions in standard methods to consistently reproduce decompositio n conditions. They were also used to set the temperature criterion for nitric acid digestion in the standard environmental methods such as EPA Resource Conservation and Recovery Act (RCRA) 3015, 3051, and 3052. In both leach procedures (Methods 3015 and 3051), the temperature profile and acid combination reproducibility were important to achieve good precision for the leach. In a total decomposition method such as 3052, a certain minimum temperature is necessary to achieve efficient dissolution. Higher temperatures and longer digestion times generally achieve similar results and can produce greater efficiencies.

When the closed-vessel nitric acid decomposition of a complete tissue sample, Bovine Liver SRM 1577a, was compared with that of the biological constituents, the tissue decomposed readily at 160–170 °C. Because SRM 1577a contains carbohydrates, proteins, and lipids, destruction of the matrix required a temperature that would decompose all of these molecular species (*55, 56*). Rather than evaluate just the elemental recovery, analysis of the residual organic molecules provided a complete description of the final sample matrix. Some researchers have approximated this description by using total residual carbon analysis (*57, 58*).

Chromatographic separation of the tissue digestate was undertaken to determine what organic molecules remain after a 10 min nitric acid decomposition that sustained temperatures of >175 °C for 5 min. The predominant organic molecules remaining are *ortho-*, *meta-*, and *para-*nitrobenzoic acid. The one organic bond not decomposed under these conditions is the *pi*-bond of the benzene ring (*57, 58*). The aromatic amino acids are the origin of these ring structures. No significant traces of carbohydrates, proteins, or fatty acids were detected by high-performance liquid chromatography (HPLC), a result suggesting that for a 10 min digestion, decomposition may be sufficiently complete for most instrumental techniques (*56, 59*).

To test this hypothesis, an albumin sample was processed as described previously and the temperature and pressure were recorded (Figure 6 predigestion curve). It was then removed from the microwave and left to stand until temperature and pressure had equilibrated with the laboratory conditions. The sample was then returned to the microwave and a higher power digestion program was used to complete the destruction of the matrix (Figure 6, second digestion). Figure 7 shows an overlay plot of the second temperature–pressure curve and a curve of nitric acid alone, which indicates that the sample is completely decomposed in the first digestion procedure.

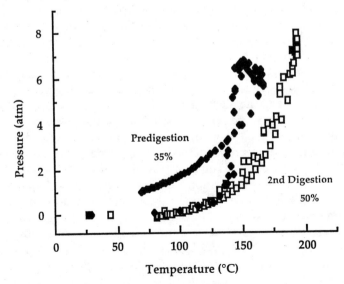

Figure 6. Temperature and pressure curves for the nitric acid digestion of albumin.

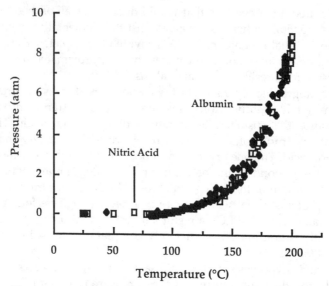

Figure 7. Temperature and pressure curves for nitric acid and the second nitric acid digestion of albumin.

Most modern inorganic analytical instruments such as inductively coupled plasma–mass spectrometry (ICP–MS), inductively coupled plasma–optical emission spectrometry (ICP–OES), and flame atomic absorption spectrometry (F–AAS) are not seriously affected by trace quantities of benzoic acids. Other analytical instrument methods such as polarography and graphite furnace–atomic absorption spectrometry (GF–AAS) may be significantly affected (56, 60). Additional treatment at higher temperatures may be necessary to ensure that these interferences are minimized.

Such experiments demonstrate that knowledge about the specific makeup of a sample matrix is essential to determine the most efficient temperatures necessary for decomposition. Understanding the characteristics of a digestion and the interaction of components with specific reagents allows the analyst to more readily control the sample digestion process. Efficiently designed microwave digestions require reaching and maintaining minimum temperatures that rapidly decompose the major organic components in the matrix. Previously, quality control in sample dissolution has only been assessed empirically by comparing the results of elemental analysis without identifying the specific characteristics of the digestion. Complete decomposition of organic matrices is rarely achieved by using special reagents such as perchloric acid. Because the use of this strong oxidizer should be avoided, higher pressure and temperature decompositions using nitric acid are the acceptable alternative.

The extent of complete dissolution of organic samples by using nitric acid in steel-jacketed Teflon bombs at high temperatures and pressures has also been studied (*28, 61, 62*). These results show that the decomposition of these samples was incomplete, even after heating at 180–200 °C for 3 h. When the residual organics from these dissolutions of biological tissue and model compounds were analyzed, essentially the same molecules were found as those determined in the 10–15 min (175–180 °C) microwave decompositions. Similar ratios of *ortho*-, *meta*-, and *para*-nitrobenzoic acids were found, as well as small amounts of dinitrobenzoic acid and other aromatic compounds (*62*). This result indicated that temperature, and not the microwave energy, was the major influence on the decomposition. Only recently have microwave systems become capable of achieving temperatures over 300 °C and pressures in excess of 100 atm. Studies will evaluate these new microwave systems where complete decompositions may be achieved for certain analytical applications.

Elemental analysis using techniques that do not require complete decomposition of organic materials should be evaluated for their tolerance to these molecules. The closed-vessel microwave nitric acid decomposition of organic tissue gives comparable results to traditional steel-jacketed Teflon bomb decompositions. Because of direct transfer of the microwave energy, the same conditions are achieved in a much shorter time and with greater reproducibility. The usefulness of high temperature and pressure decompositions in conjunction with a variety of instrumental trace element analyses has been demonstrated in the literature (*28, 31*) and is reviewed in Chapter 2.

Despite the effectiveness of high temperature and pressure decomposition, the specific chemistry of an acid's interaction with a particular matrix must be understood to optimize the dissolution. The use of microwave technology should not overshadow the need to understand the chemical reactions taking place, many of which are accelerated by elevated temperatures. It is necessary to have knowledge of acids appropriate for the matrix, elemental analytes, and suitable temperature and pressure conditions to effect the optimum dissolution procedures. Microwave technology should be implemented only after these conditions have been determined.

Minerals, and possibly some alloys, have similar optimal conditions under which they efficiently and reproducibly decompose. These decomposition conditions indicate that the majority of these reactions are characterized as opening out reactions of fully oxidized materials. They proceed under similar temperature conditions with other acids such as hydrofluoric and hydrochloric acids. Many of these components are not as sensitive to temperature as oxidation reactions and the combination of these two fundamental types of decomposition reactions is feasible at relatively low oxidation temperatures.

Standard microwave sample preparation methods are more appropriate than methods less controllable and descriptive. The controlled reaction conditions reproduce temperature profiles in standard methods very accurately. Closed-vessel microwave decomposition provides a method of standardizing sample preparation decomposition procedures. These procedures are more robust than previous methods because they are duplicated in each laboratory through instrumental control of the microwave energy. Studies to standardize microwave procedure by software encapsulation have demonstrated the feasibility of describing these reactions (40–42). These capabilities are the basis for the current standard microwave closed-vessel environmental methods and will permit their electronic transfer to future instrumentation.

Microwave Reaction Monitoring Techniques

From the previous conclusions that stated that pressures developed in closed microwave vessels do not correlate with solution temperatures, it seems more appropriate to measure temperature to control reactions. This result is compounded by the existence of the direct leaching–temperature–reaction relationship. Thus, the area of microwave sample preparation has moved toward using temperature feedback as the only way to gain reproducible and accurate methods.

Temperature Measurement in a Microwave Field

Currently, four types of temperature measurement devices are used commercially in laboratory microwave systems for the direct measurement of the reaction temperature: thermocouples, fiber optics, IR, and gas-filled bulbs. Each of these temperature measurement systems have overcome the inherent problems and limitations of working in a radio frequency (RF) electromagnetic field.

Thermocouple measurement devices have been used for measuring temperatures for many years. In an RF field, this measurement can occasionally lead to problems when sensors are not correctly engineered. When thermocouples are exposed to microwaves, the microwaves couple with the thermocouple measurement circuitry (63). Errors are produced when the metal surface interacts with the RF field. A combination of shielding, grounding the shielding to the microwave cavity wall, and electroplating the thermocouple with gold either eliminates or minimizes these errors (3, 31). The gold-plated shielding essentially removes the thermocouple from the RF field. After calibration, thermocouples are useful in the general range of 0–400 °C in microwave systems, and typically are accurate to ±0.1%. They can measure well beyond this range (may exceed 1000 °C) depending on the specific bi-metallics and signal processing algorithms being used.

A special safety consideration is the use of unshielded or ungrounded thermocouples in single or multimode microwave environments. Placing any metallic device or wire, such as a thermocouple, into a microwave environment can be hazardous. Without grounding the shielding to the cavity wall or wavelength attenuator prior to leaving the microwave environment, microwave energy will be transmitted along the surface of the shielding and into the laboratory. Prevention of this phenomenon by proper shielding and grounding of the thermocouple to prevent any leakage of microwaves has been described (3).

Fiber-optic systems are microwave transparent and nonperturbing to the microwave field. Fiber optics have been used in phosphorus-based and remote IR temperature sensors. Phosphorus fiber-optic sensors use a light source outside the microwave cavity that emits an excitation wavelength that travels through the fiber-optic cabling to a phosphor tip. Once excited, the phosphor tip fluoresces, with a 1/e decay profile, and its emission travels back through the fiber-optic cabling. The measurement system converts the temperature-dependent fluorescent decay signal into a temperature measurement by comparing the decay rate with calibration values (64–69). After calibration, phosphorus fiber-optic thermometers are linear from 0–250 °C in microwave systems (dependent on the specific phosphor and fiber-optic cabling). They typically have an accuracy of ±2 °C, down to ±0.2 °C with calibration near the measurement temperature, and a precision of ±0.5 °C, down to ±0.1 °C with the integration of several temperature measurements (65, 68).

IR temperature sensors have been used for direct and indirect temperature measurements. Direct IR sensors use fiber optics to directly measure the digestion solution temperature. Indirect IR sensors measure the temperature of the digestion vessel, usually the bottom (51), by the IR emission. Because the temperature of the digestion vessel may lag behind the temperature of the solution, this type of sensor may suffer from hysteresis.

IR sensors have also been used to measure the temperature of a multilayer digestion vessel (70). In this case, the sensor is not currently used for reaction control, but for safety. An IR sensor is aimed at the microwave vessel outer casings. As the turntable revolves, the IR sensor measures the temperature of each vessel. If the vessel outer casing has reached an unsafe temperature, the microwave power is shut off until the temperature has dropped to a safe level, preventing overheating and extending vessel life.

Gas-filled bulb thermometers have recently been introduced by two companies (51, 71). The measurement of temperature with these gas-filled bulbs is based on the gas law principle; the temperature is proportional to the internal gas pressure. After calibration, glass-bulb thermometers are linear from 0–250 °C and have an accuracy of 2% (72).

These temperature measurement devices should also be protected from the harsh acid environment of the decomposition. Most devices are protected by either temporary or permanent layers of glass or Teflon. Protecting with glass (optimally with quartz for controlling contamination) is mechanically the strongest method, but it is inappropriate for any digestion with hydrofluoric acid. Teflon is chemically inert, but neutral molecules, such as HCl and HF, penetrate and begin to attack the temperature measurement device. Chemical attack of the device can be minimized by the prompt washing and chemical neutralization of the temperature measurement probe after digestions. To assure accurate temperature measurements, all types of temperature sensors need to be periodically recalibrated.

Pressure Measurement in a Microwave Field

All commercial pressure measurement systems use pressure transducers. The only variation is whether the pressure transducer is shielded and installed directly into the vessel cap, or whether the cap has a fitting that connects the vessel to a pressure transducer mounted outside the microwave cavity. The pressure transducer should be made with a Teflon coating to prevent chemical attack of the transducer by any acid vapors. These systems have remained remarkably constant over the past decade, having been originally described in closed-vessel decomposition in 1986 (3).

Pressure Feedback Control

Pressure feedback control was commercially introduced in 1989, three years before the commercial introduction of temperature feedback control. Because of its head start, pressure feedback control has probably been used in the development of more methods than temperature feedback control. Unfortunately, pressure feedback control relies on secondary effects in a digestion. The production of vapors from acids and other reagents, byproducts of the digestion, and the vessel type all contribute to the resultant vessel pressure. It is inappropriate for the accurate, controlled digestion of matrices that produce gases and for standardized procedures. Despite this severe limitation, all too often uninformed analysts believe that they are safely and accurately controlling the microwave reaction when in fact the reaction may be completely out of control. This lack of control has resulted in numerous vessel ventings and even explosions.

To illustrate the inability of pressure feedback control to accurately control the digestion of a matrix that contains components that decompose into gaseous products, consider the digestion of motor oil. Motor oil, either new or used, is primarily composed of organics that are decomposed by nitric acid, as described elsewhere in this chapter.

A decomposition program that would allow for the safe digestion of motor oil was performed with 10 mL nitric acid and a range of motor oil sample size (0.111–0.389 g). This method was designed to be safe because it uses very low microwave powers to slowly heat the nitric acid, thus minimizing the rapid decomposition of the organics that could produce a pressure wave that could cause the vessels to vent. The microwave procedure is listed below:

- 2 min at 250 W up to a controlling pressure of 15 bar
- 4 min at 350 W up to a controlling pressure of 15 bar
- 15 min at 450 W up to a controlling pressure of 15 bar

The intentional low power during the first two stages, 250 and 350 W, limited the pressure in all cases below the controlling pressure. This power limitation resulted in a very slow heating to prevent over-pressurization of the vessels. During the third stage, the increased microwave power of 450 W allowed the samples to reach the controlling pressure. The pressure feedback control algorithm then adjusted the microwave power to maintain the control pressure for the duration of the 15 min.

Despite the exact same controlling program, Figures 8a–8d illustrates the dramatically different reaction profiles that occurred with four differ-

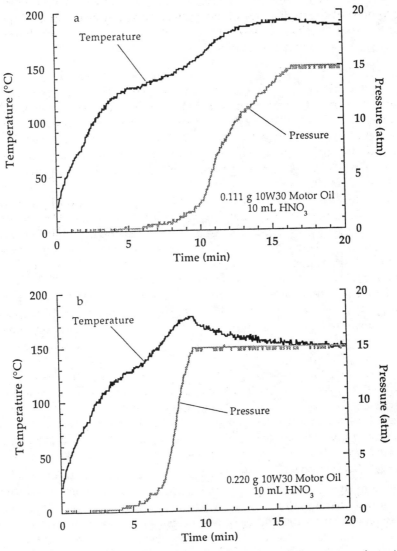

Figure 8. Temperature and pressure vs. time profiles for the digestion of varying amounts of motor oil (a–d). (Continued on next page.)

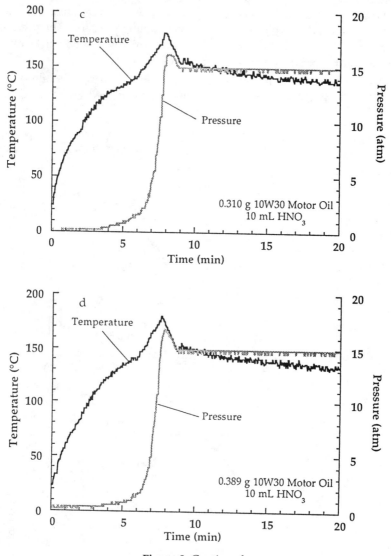

Figure 8. Continued.

ent sample sizes. The smallest sample size, 0.111 g of motor oil, reached the controlling pressure of 15 bar in 16.4 min. As the sample size was increased, the time to reach 15 bar was dramatically reduced. This difference resulted from the exothermic decomposition of the motor oil. Clearly, with pressure feedback control, the decomposition reaction condition profiles were not reproduced. To further illustrate the inability of pressure feedback control to safely decompose the samples, consider what happened when the reaction pressure reached the controlling pressure of 15 bar. In the first two cases (Figures 8a and 8b), the reaction was accurately controlled. In the second two cases (Figures 8c and 8d), 0.310 and 0.389 g of oil, the pressure feedback control was not capable of accurately and safely stopping the reaction at 15 bar. Because these digestions are decomposing rapidly, the pressure feedback control system stopped the microwave energy and the sample continued to decompose. The highest pressures inside these vessels were 16.1 and 16.8 bar, respectively; this result is an over-pressurization of 12%. Because pressure feedback control does not monitor temperature, it was possible to quickly heat the sample to the point that the sample decomposed too rapidly for the control algorithm to stop the reaction. Several digestions of organics using pressure feedback control that lead to runaway reactions by the scenario just discussed have been reported to the authors. For this reason, the authors suggest that when any digestion method is being developed, it should never be designed to exceed 80% of the microwave vessel temperature and pressure capabilities, providing a 20% safety buffer.

Varying the temperature heating profiles by controlling with pressure affects the safety of the digestion, and the completeness of the digestion. For the smallest sample, as the pressure is maintained, the temperature inside the vessel continues to decrease. The continual decrease in the reaction temperature while maintaining a constant pressure is a function of the insulating nature of the particular microwave vessels used. Throughout the microwave digestion, the microwave vessel is being heated. Depending on its thermal characteristics, the vessel will retain varying amounts of heat. This heating results in an increased pressure at the same temperature by using temperature feedback control. Conversely, if the vessel is kept at the same pressure, the temperature of the reaction must be decreased.

The smallest sample, 0.111 g reached 15 bar at a temperature of 192 °C. The temperature only decreased 6 °C while the pressure was controlled at 15 bar. As discussed earlier, the final temperature of 186 °C is capable of digesting most organic molecules, and therefore the sample has been digested. The second sample, 0.220 g, reached 15 bar in only 9.3 min and maintained that pressure for 11.5 min. During this time, the temperature inside the vessel decreased from 178 to 148 °C, a reduction of 30 °C. Because the final temperature was high enough to decompose some types of organics and the temperature did not rapidly decrease below sufficient reaction temperatures, this sample was considered marginally decomposed. The third sample, 0.310 g, reached 15 bar in 7.8 min. While the pressure was maintained for 13 min, the temperature dropped by 45 °C. Figure 8c illustrates the rapid drop in temperature when the microwave power was turned off, which was needed to decrease the pressure and maintain 15 bar. Because of this rapid decrease in the reaction temperature, this sample was not decomposed. These results indicate that published microwave methods based on pressure feedback control must be written to be vessel specific, or with defined vessel design criteria.

Despite the technical capability of pressure feedback control, its usefulness is hampered by several key limitations. The variation of sample size can lead to a severe variation in the time to pressure, resulting in a tremendous variation in the reaction conditions observed during an experiment. The pressure observed for a specific temperature in a particular reaction will vary for different types of microwave vessels. This means that a method developed in one type of microwave vessel is not transferable or reproducible between vessel types. Although laboratory microwave manufacturers introduce new microwave vessels yearly, they may also obsolete previous models. This process means that a standard method developed by using a particular vessel would be obsolete as soon as the microwave vessel is obsolete.

The most severe limitation is the variable reaction temperatures that occur at a specified control pressure. As illustrated previously, the same pressure control method can lead to some samples that are completely digested, some samples marginally digested, and other samples that are not digested. The conditions vary with sample size, sample matrix type, and vessel construction. This type of control is not appropriate for standard method specifications.

Power Calibration Control

Calibration control was developed to accurately transfer microwave methods based on the reproduction of microwave field strength and resulting temperature. Calibration control was extensively used with the development and validation of early EPA Methods 3015 and 3051 (73, 74). Section 7.1 describes the calibration of a cavity or multimode microwave unit in EPA Method 3051.

7.1. Calibration of Microwave Equipment

7.1.1. Measurement of the available power for heating is evaluated so that absolute power in watts may be transferred from one microwave unit to another. For cavity type microwave equipment, this is accomplished by measuring the temperature rise in 1 kg of water exposed to microwave radiation for a fixed period of time. The analyst can relate power in watts to the partial power setting of the unit. The calibration format required for laboratory microwave units depends on the type of electronic system used by the manufacturer to provide partial microwave power. Few units have an accurate and precise linear relationship between percent power settings and absorbed power. Where linear circuits have been utilized, the calibration curve can be determined by a three-point calibration method (7.1.3), otherwise, the analyst must use the multiple point calibration method (7.1.2).

Many cavity microwave units do not have completely linear calibrations. Most, if not all, manufacturers' units are linear over a very wide range of partial power settings. Figures 9 and 10 illustrate the calibration of two commercial laboratory microwave cavity units. Figure 9 illustrates a microwave unit whose calibration is linear from approximately 50 to 100% power. The partial power settings below 50% are essentially nonmeasurable. Figure 10 illustrates a commercial laboratory microwave unit whose calibration is apparently linear over the entire range, with the exception of the partial power settings from 90% to 99%. These figures illustrate the importance of a comprehensive calibration rather than a simple one-point calibration, where the calibration curve is assumed to intersect the origin.

Section 7.1.4 of EPA Method 3051 describes the measurement of a single calibration point. A calibration program was developed for the federal government (RCRA) to guide the analyst through the collection of a complete calibration data set (73). The program performs a statistical analysis of the data and reports the linear portion of the calibration curve. Once calibrated, the software can convert power settings in watts into a unit-specific partial power setting and report the accuracy of the calculated partial power setting. This calibration program is available to the analyst at the SamplePrep Web site discussed in Chapter 15.

7.1.4. Equilibrate a large volume of water to room temperature (23 ± 2 °C). One kg of reagent water is weighed (1000.0 g ± 0.1 g) into a fluorocarbon beaker or a beaker made of some other material that does not significantly absorb microwave energy (glass absorbs microwave energy and is not recommended). The initial temperature of the water should be 23 ± 2 °C measured to ±0.05 °C. The covered beaker is circulated continuously (in the normal sample path) through the microwave field for 2 min at the desired partial power setting with the unit exhaust fan on maximum (as it will be during normal operation). The beaker is removed and the water vigorously stirred. Use a magnetic stirring bar inserted immediately after microwave irradiation and record the maximum temperature within the first 30 s to ±0.05 °C. Use a new sample for each additional measurement. If the water is reused both the water and the beaker must have returned to 23 ± 2 °C. Three measurements at each power setting should be made. The absorbed power is determined by the following relationship:

$$P = \frac{KC_pM\Delta T}{t}$$

where P is the apparent power absorbed by the sample in watts (W/(joule · s); K is the conversion factor for calories/s to watts (4.184), Cp is the heat capacity, thermal capacity, or specific heat (cal/(g · °C), m is the mass of the water sample in grams, ΔT is the final temperature minus the initial temperature (°C), and t is the time in seconds.

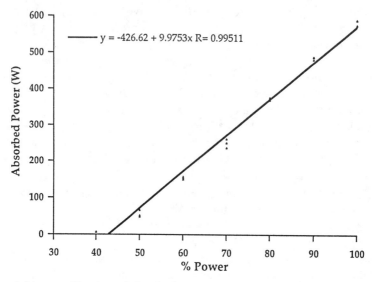

Figure 9. Linear calibration of absorbed power vs. partial power setting for a commercial laboratory microwave unit.

The limitations of the calibration of a microwave unit are included in the following section of EPA Method 3051.

NOTE: Stable line voltage is necessary for accurate and reproducible calibration and operation. The line voltage should be within manufacturer specification, and during measurement and operation should not vary by more than ±2 V. A constant power supply may be necessary for microwave use if the source of the line voltage is unstable.

Electronic components in most microwave units are matched to the units' function and output. When any part of the high voltage circuit, power source, or control components in the unit have been serviced or replaced, it will be necessary to recheck the units' calibration. If the power output has changed significantly (±10 W), then the entire calibration should be reevaluated.

Calibration control has been used by many laboratories for EPA Methods 3015 and 3051. Calibration can result in accurate reproduction of EPA methods, including Method 3051. As a test of the accuracy of calibration control, EPA Method 3051 was evaluated in two separate laboratories (*see* Figure 11). This example should be considered the optimum calibration control of a microwave method because both microwave units were calibrated and experiments were run by the same analyst using the same

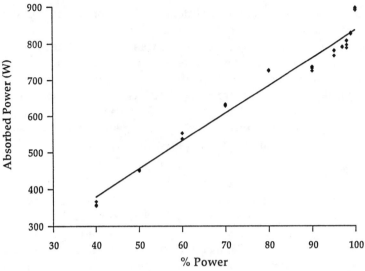

Figure 10. Nonlinear calibration of absorbed power vs. partial power setting for a commercial laboratory microwave unit.

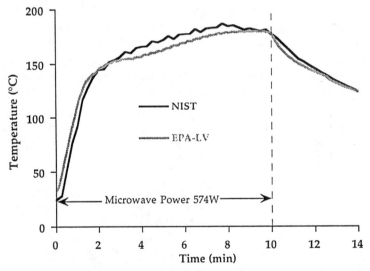

Figure 11. Comparison of the reaction conditions, temperature vs. time, using calibration to transfer EPA Method 3051.

measurement equipment. Throughout our research, the difference in the reaction temperatures, controlled through properly calibrated systems, has always been less than 15 °C.

Calibration control has a number of limitations. A calibrated microwave method can reproduce the desired reaction conditions only if the same type and number of microwave vessels, the exact combination and quantity of acids, and, to a lesser extent, similar samples are digested. If a different number or type of vessels is used, or the quantity or choice of reagents is adjusted, a new set of power versus time settings will have to be determined by using a microwave equipped with a temperature measurement system. Because of the wide variety of sample types, most laboratories will prefer to use temperature feedback control as the control technique of choice if it is available on their equipment. However, calibration control is a viable procedure, especially for repetitive standardized methods, and can dramatically reduce equipment costs.

Temperature Feedback Control

The limitations of both pressure feedback control and calibration control can be overcome through temperature feedback control. Whether a soil or an oil is being digested, the reaction temperature can, in most cases, be accurately controlled and the method performance criteria reproduced, *see* Figures 4 and 12. Temperature feedback control overcomes the limitations of calibration and allows the digestion of a variable number of vessels. Figure 13 illustrates the reproduction of EPA Method 3051 through 2- or 6-vessel calibration, or temperature feedback control. Temperature feedback control more accurately reproduces the method, and is capable of digesting odd numbers of vessels whose calibration settings have not been determined.

Temperature feedback control is considerably more accurate than either pressure feedback or calibration control because it controls the rate

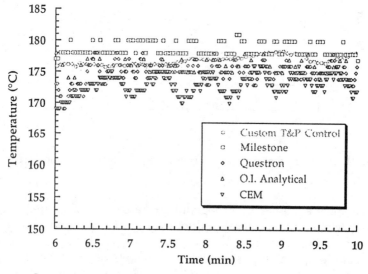

Figure 12. Comparison of commercial and research microwave equipment reproduction of EPA Method 3051 maintaining a constant reaction temperature.

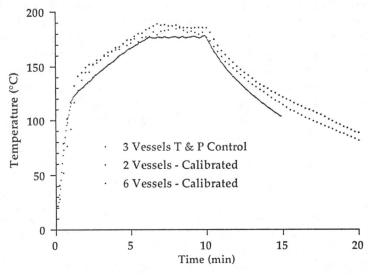

Figure 13. Comparison of the reaction conditions, temperature vs. time, by using calibration for 2 or 6 vessels or temperature feedback control of EPA Method 3051.

of the reaction by the temperature, thus reproducing the primary reaction conditions in standard methods. It also is capable of compensating for the addition of energy from the reaction by reducing the microwave power when exothermic reaction energy becomes significant, such as in the digestion of organics and oils. Because specific reactions have been shown to be temperature dependent, temperature feedback control can be used to heat a sample to specific component digestion temperatures.

Simultaneous Temperature and Pressure Feedback Control

Prior to the commercial introduction of temperature feedback control, a simultaneous temperature and pressure feedback control system was developed (75–77) in our laboratory. Temperature was identified as the primary controlling factor, but the vessel pressure was monitored secondarily for both safety and reaction progress.

In this system, the software primarily controls temperature, but also evaluates pressure and relative rises in pressure. The control algorithm allows the temperature of the acids to reach the boiling point prior to full temperature control. It only monitors for unsafe increases in temperature or pressure. Once the boiling point is exceeded, the system computes the slope required to achieve the goal temperature in the goal time. It then computes the temperature that should be achieved at the next measurement time point. At the next measurement time point, the system determines if the predicted temperature has been achieved or whether correction to the microwave power is necessary. This process is repeated until the desired temperature has been reached. The software decisions are based on a proportional integrating derivative (PID) algorithm (77).

Once the reaction temperature reaches the goal temperature, a second subroutine in the software begins the maintenance of a temperature for a given time. This subroutine is also based on a PID algorithm. Both subroutines for heating to a temperature and maintaining a temperature have been individually tuned for optimal control by using EPA Methods 3051 and 3015. Because of the algorithm flexibility, the control software can be adjusted for unique or specialized reaction control.

EPA Method 3051 was tested several times to determine the reproducibility of reaction conditions as illustrated in Figure 14. During three independent batch runs, the temperature deviations were minute. More importantly, during Run 2, a single vessel vented at approximately 5 min. If this method were run by using calibration control, no corrections to the microwave power would occur and the remaining acid would absorb more energy. Also with calibration control, the temperature of the acids in the remaining vessels would exceed the prescribed temperature and potentially exceed the pressure limitations of the vessels, resulting in a potential safety hazard. In contrast, the software detected an excessive decrease in the microwave power needed to maintain the prescribed temperature, and notified the operator by a warning message. With the PID software control, the temperature was properly maintained, preventing a safety hazard and the set of samples was completed within specifications.

At the end of each run, the software automatically stored the average percent power required to maintain a temperature and the accumulated

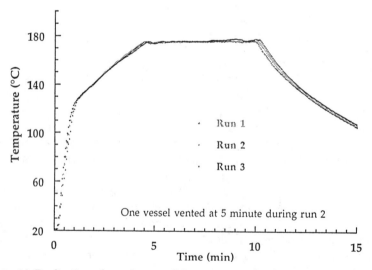

Figure 14. Replication of reaction conditions, temperature vs. time, using temperature feedback control for EPA Method 3051.

integration value. These parameters could be used in future runs to help teach the algorithm about controlling any specific digestion method. A detailed description of how to construct a PID control system has been published (77).

A comparison of four commercial temperature feedback control microwave systems was performed by using EPA Method 3051 specifications. The differences between manufacturer compliance to the overall conditions of Method 3051 are minor when comparing the complete digestion profile, as was illustrated in Figure 4. A more detailed comparison identified a significant difference between the control systems.

Comparing a 4-min period during the digestions in which the same temperature should be achieved shows the precision of temperature control (*see* Figure 12). Commercial control systems are limited to temperature control range of 6–8 °C, whereas the custom PID controls within better than 2 °C. Because the reaction rate is directly related to the temperature of the digestion, improved control of the digestion temperature will improve the reproducibility of a leach test, such as EPA Method 3051 or 3015. As more sophisticated control algorithms such as the PID become commercially available, the use of microwave energy for highly temperature-dependent chemical reaction control will mature.

Clean Chemistry and the Analytical Blank

The improvements in reproducibility gained by the use of microwave-based sample preparation has lead to analysts being able to uncover significant sources of error, such as the analytical blank, that previously were obscured by the error of the preparation procedures. The significance of the role of the analytical blank in chemical analysis of trace metals cannot be overstressed. It is a primary source of error in trace element analysis. Trace analysis is as much dependent on the control of the analytical blank as it is on the accuracy and precision of the instrument making the measurement. Inability to control contamination that is external to the sample, or those contributions of the analyte coming from all sources other than the sample, is frequently the limiting factor in trace (parts per million (ppm) to parts per billion (ppb)) and ultra-trace analysis (ppb to parts per trillion (ppt)). Analytical blank contributions occur from four major sources:

- the atmosphere in which the sample preparation and analysis are conducted
- the purity of the reagents used in sample preparation, including all reagents and the quantities added directly to the sample
- the materials used in digestion or extraction vessels that come in contact with the sample during the sample preparation and analysis

- the analyst's technique and skill in preparing the samples and performing the analyses (78–80)

Each of these sources of contamination will be discussed and solutions provided to assist the analyst in understanding how they may be controlled. Implementing microwave chemistry can significantly assist with controlling this analytical blank challenge and is integrated into these discussions. The way in which uncertainty is handled when measurement data is combined provides an understanding of why the analytical blank has been described as the "Achilles' heel" of trace analysis (78).

The authors originally began to develop closed-vessel microwave sample preparation methods to reduce the analytical blank and to permit certification–quality measurements by a variety of instrumental analysis methods following sample preparation. This development of microwave sample preparation, coupled with clean chemistry techniques, broke some of the traditional barriers to blank limitations and permitted measurements of this quality to be made.

Only under very few circumstances can the analyst ignore the contribution of the uncertainty of the blank in calculating the overall measurement uncertainty. One condition would be if the calculation is insignificant compared to the data. For example, when the blank value is less than 10^3–10^4 times smaller than the sample measurement and the error in the sample measurement is in the percent range, the uncertainty of the blank measurement is insignificant compared with the uncertainty of the measurement. This result only occurs when the blank is extremely low compared to the measurement, which is rarely the case when trace and ultra-trace analyses are being conducted. It is more likely that the blank is significant and must be subtracted from the measurement. Because the blank measurement is closer to the detection limit of the instrument, the imprecision of the blank is larger than the measurement itself. This relationship causes the analytical blank to frequently become the limiting factor in the overall measurement precision.

To compute the overall standard deviation for a final measurement, several sources of error and imprecision must be combined. Each standard deviation for the computation of y must be considered to obtain the overall measurement uncertainty.

$$y(\pm s_y) = a(\pm s_a) - b(\pm s_b) + c(\pm s_c) \tag{6}$$

The standard deviation of the result s_y is given by combining the standard deviations of the measurements. In this case, the a term represents the standard deviation of the measurement, b represents the standard deviation of the blank that must be subtracted, and c represents the uncertainty associated with the sampling error (81). This example will only consider the uncertainty of the measurement (*a*) and the blank (*b*), because the sampling uncertainty (*c*) is beyond the scope of this discussion and is covered in other literature (*82*).

$$s_y = \sqrt{s_a^2 + s_b^2 + s_c^2} \qquad (7)$$

The following example illustrates a common relationship that causes the imprecision of the blank to be limiting. A set of samples is analyzed and has a mean value of 55.5 ± 0.3 and the analytical blank is 11 ± 5 (which is too large to be ignored). The uncertainties on these numbers are the standard deviations of the replicate measurements. The analytical blank becomes the dominant uncertainty in calculating the uncertainty of the final result. Here $y = 55.5 - 11 = 44.5$ and $s_y = \sqrt{s_a^2 + s_b^2}$ is $s = \sqrt{0.3^2 + 5^2}$; therefore, the result is 44.5 ± 5. Essentially, the entire uncertainty is due to the uncertainty of the analytical blank.

It has been suggested by some environmental laboratories that having the blank concentration just below the detection limit while the measurement is detectable provides a more convenient measurement. This approach, however, is not valid. A blank value below the limit of detection does not remove its influence. Just because it is not detectable does not mean it is not influencing the measurement. An accurate measurement of the blank value with high precision actually provides the most accurate overall analytical estimate of the concentration.

There have been many examples of the difficulty of performing sample preparation in concentration ranges considered trace and ultra-trace. Sample preparation is the largest source of error, frequently due to contamination of the sample during decomposition. An excellent demonstration of this problem was an interlaboratory study of lead in blood. Figure 15 illustrates the two samples analyzed collaboratively by seven laboratories (*78*). The first sample was an acidified aqueous standard of 140 ng Pb/g, which needed no sample preparation. The range of the measurements was between −30 and +75% of the real value and was apparently randomly distributed in both positive and negative bias. The second sample, analyzed by the same laboratories, was whole blood and required acid decomposition. Its concentration was only 30 ng Pb/g and the range was +30 to +450% of the value. The data clearly illustrated that when sample preparation was required, all of the reported values were biased high. At lower concentration levels, the blank became a more significant part of the analysis and sample preparation became the most significant source of this contamination. This result illustrates a typical bias when sample preparation is required in addition to instrumental analysis.

To reach lower levels of analysis, the blank must be lower and more precise. An example of the trend that must be continued in trace and ultra-trace analysis is provided by reviewing the progress over a decade, of certifications of organic and biological NIST SRMs for vanadium. Table III

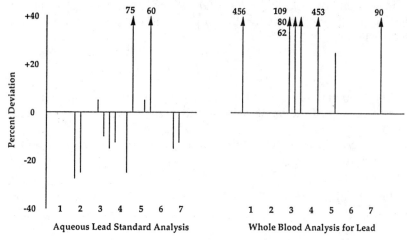

Figure 15. Comparison of interlaboratory results for direct analysis of aqueous standards vs. whole blood requiring sample preparation.

Table III. Progression of Vanadium and Blank Concentrations in Biological SRMs

NIST SRMs	Ref.	Concentration (µg/g)	RSD (%)	Blank (pg)
8505 Crude Oil	84	390	1.06	4200
1572 Citrus Leaves	85	245	0.245	3160
1566 Oyster Tissue	85	2316	0.28	1330
1577 Bovine Liver	85	98.7	1.6	410
1634b Fuel Oil	86	55.70	0.27	270
909 Human Serum	85	2.63	11.6	150
1598 Bovine Serum	83	0.055[a]	40	<1

[a]Information value only.

lists six organically based SRMs that were certified for vanadium. Bovine serum could not be certified for vanadium because a second independent method of analysis at such low levels could not be developed (83). As the concentration decreased, methods were also required to decrease the blank value. The relative standard deviation of the concentration, however, increased as the blank became a more significant factor in the overall mean value.

As stated previously, the analytical blank is derived from four primary sources. We will briefly examine each source, integrate them with microwave digestion, and discuss technological influences that assist in reducing the analytical blank.

Sample Preparation and Analysis Atmosphere

The environment in which the sample is prepared is the major source of contamination in the laboratory for most elements. Some rare elements may be an exception, but for the majority of elements of interest, contamination from airborne sources is the most significant of the four main sources. Table IV illustrates concentrations of lead in air.

Any laboratory air that comes in contact with the sample will deposit some portion of its elemental content into the sample. The sample is especially vulnerable to this transfer when it is being decomposed in acid. The acid will digest most particles it comes in contact with transforming the particles into ions in solution, combining them with those of the sample.

The microwave sample preparation system minimizes continuous sustained transfer of elements from the air to the sample solution by providing a controlled, closed, or restricted sample container during decomposition. It prevents the entry of the laboratory atmosphere and thus prevents the majority of environmental airborne contamination. This result leaves only the air within the vessel, captured during sample transfer, and the digestion–extraction process.

To prevent this volume of air from contaminating the sample, the sample must be transferred in a clean environment. This task is much easier than it might at first appear. These precautions are becoming state-of-the-art in many analytical and environmental laboratories. The prevention of airborne contamination is most frequently dealt with by employing a laminar flow clean bench or a clean laboratory facility. The authors constructed both from component parts and found the process to be inexpensive and uncomplicated, once the concepts are understood.

Many sources of airborne contamination exist. Several of the sources have been described and their particle size ranges are provided in Figure 16. These diverse sources primarily provide particulates in discrete size ranges. Depending on whether the laboratory is located in an industrial, urban, or rural area, or near the sea, the distribution of these source particles will be different, as will their composition. The vertical dashed line in Figure 16 indicates the cutoff for the high efficiency particulate air (HEPA) filter used to prevent particulate contamination, usually 0.5 μm. Particles above this size cannot pass through a HEPA filter that is in good

Table IV. Examples of Lead Concentrations in Air

Site	Lead Concentration ($\mu g/m^3$)	Ref.
Downtown Air, St. Louis, MO	18.84	87
Rural Park Air, Southeastern MO	0.77	88
Laboratory Air, NIST, MD	0.4	78

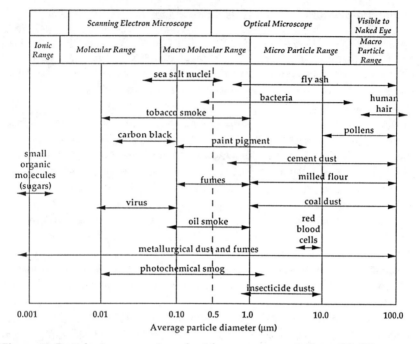

Figure 16. Particle size comparison chart for common particulates (*90, 91*).

working order. These filters were developed jointly by the Massachusetts Institute of Technology and Arthur D. Little & Company, Inc. for the Manhattan Project (*89*).

Before continuing with descriptions of the clean laboratory and clean room, definitions of *class* and *clean* are needed. The definition of clean air is derived from Federal Standard 209a that defines cleanliness levels. Table V lists these conditions.

Laminar flow is directed, coherent air movement that does not contain any turbulence. The idea behind its use is akin to standing in a rapidly flowing stream and facing the shore. The upstream leg cannot be contaminated by sunscreen lotion on the downstream leg, because the water is carrying it away from the upstream leg. The principle is then to keep the sample that you do not want to become contaminated by something else in the room or on the bench "upstream" of other objects. The air coming into the room, or recirculated within it, is free of dust and contaminants 0.5 μm or larger. The analyst and other equipment in the room now become the most significant contaminating sources. Laminar flow hoods are almost always directed down in the analytical laboratory and are shielded from crossflow air turbulence, where feasible.

A clean bench is the most simple solution to processing samples in a relatively dust free and clean environment. The authors have produced portable clean benches (or clean carts) to allow for addition of sample and reagents, and to open or close vessels of both the atmospheric and closed-vessel design. An example of the clean cart is shown in Figure 17. These carts can be constructed for $1000–$1400, depending on whether they need to be exhausted or are just islands of clean laboratory air. The cart in Figure 17 is a stand alone, exhausted, class-100 clean air hood that can also exhaust a second clean air hood when connected via a flexible port. Some commercial units can also be purchased fully assembled, but they are generally not designed to be portable and are more expensive.

We constructed a clean laboratory 18' × 15' for approximately $30,000 by using readily available materials and specialized HEPA clean hood components. Figure 18 provides a diagram of this room with clean working benches and clean areas for instrumental analysis. The walls were sealed with epoxy paint and a drop ceiling of Plexiglas was installed with six class-100 clean air handling filter units. Similar and more expensive air quality clean facilities have been described in the literature (79). This clean facility is used for both microwave vessel sample preparation and ICP–MS analysis. The ICP–MS is located in the clean laboratory, so its extremely low detection capability may be used.

Examples of the laminar flow air are indicated in Figure 18 and illustrate that the room has two separated clean areas. The area under the bench against the wall is made from hoods that refilter already clean room air and isolate the chemistry done in these areas from individuals and equipment in the room. The main area in the clean laboratory has equipment that is vented out of the clean laboratory because of the presence of the main source of particulates—the analyst.

The clean laboratory was constructed to comply with two of the standard clean-laboratory definitions. This clean-laboratory configuration has enabled analysis of very low level samples with a minimum of contamination from atmospheric sources (92). Figures 19a and 19b are photographs

Table V. Cleanliness Levels in Federal Standard 209a

Class[a]	Maximum Contamination in Work Area (particles/ft^{-3})
100	100 particles >0.5 μm
	0 particles >5.0 μm
10,000	10,000 particles >0.5 μm
	65 particles >5.0 μm
100,000	100,000 particles >0.5 μm
	700 particles >5.0 μm

[a]The standard requires laminar-flow equipment to attain this level of cleanliness. Because measurement of dust particles smaller than 0.5 μm introduces substantial errors, 0.5 μm has been adopted as the criterion of measurement.
SOURCE: Data from reference 89.

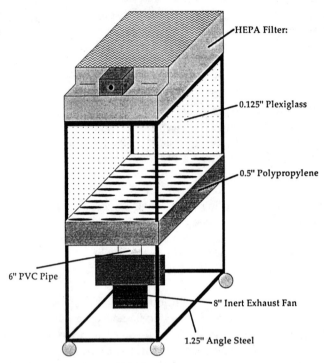

Figure 17. Schematic of a portable clean room with exhaust system for microwave vessel preparation and assembly.

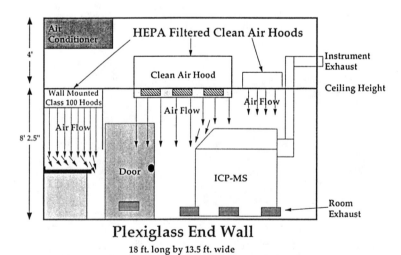

Figure 18. Schematic of Duquesne University clean room laboratory for sample preparation and analysis.

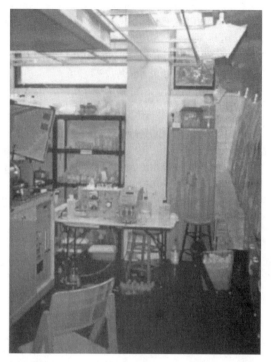

Figure 19a. Duquesne University environmental analysis clean room.

Figure 19b. Microwave vessel being prepared in a clean room.

of the laboratory and microwave vessels being prepared on the class-100 clean bench. A dramatic reduction from airborne contaminants can be expected by using HEPA-filtered air in laminar flow clean hoods, or entire clean laboratories. Table VI demonstrates the dramatic differences in airborne contaminant concentrations in an ordinary laboratory, a clean laboratory, and a clean hood in a clean laboratory.

Microwave vessels prepared in a clean environment and sealed in clean air are essentially free from any contamination due to trapped particles. They may be transported and microwaved in a normal laboratory environment without risk of further contamination and can then be returned to the clean environment for subsequent processing.

After construction, the clean facility was tested to verify its effectiveness. The test data in Figure 20 demonstrate the advantage of having some type of clean laboratory environment in which to prepare samples and its potential to reduce the analytical blank. Coulter Counter data certifying

Table VI. Particulate Concentrations ($\mu g/m^3$) in Laboratory Air

Air Source	Iron	Copper	Lead	Cadmium
Ordinary laboratory	0.2	0.02	0.4	0.002
Clean room	0.001	0.002	0.0002	ND[a]
Clean hood	0.0009	0.007	0.0003	0.0002

[a]ND is not detected.
SOURCE: Data from reference 88.

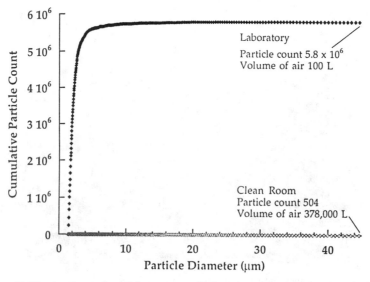

Figure 20. Comparison of particle count analysis, counts vs. particle size, of a clean room and a standard laboratory.

the multistage clean facility after construction are presented in Figure 20 and compare the normal laboratory air just outside the clean laboratory with the clean laboratory air. Obviously, the air inside this facility is significantly cleaner than normal laboratory air, by a factor of 10^7. Further validation of the multistage clean room provided evidence that the clean benches exceed class-100 specifications and the clean laboratory exceeds class-10,000 specifications. Additional pictures, diagrams, and materials specifications for construction of the portable clean hoods and clean laboratories are available on the SamplePrep Web site described in Chapter 15.

Reagent Purity

For acid decomposition, leaching, and extraction, the purity of the reagents used is extremely important to the overall level of the blank. Reagents have very different purities depending on their processing grade and purpose. Frequently, the analyst must purchase special reagents or purify reagents prior to use to minimize the analytical blank.

In addition to the purity of the reagents, the quantity that must be added is also significant. When reagents are added, they bring with them elemental components that exist as contaminants. The more reagent that must be used due to reasons other than the stoichiometric reaction, the higher the blank. Closed microwave vessels, as described in Chapter 2, reduce the amount of reagent required because it is not boiled away. In atmospheric reactions, microwave vessels are designed to reflux and return reagents to the reaction, also preventing evaporation losses. Therefore, lower amounts of reagent are necessary to complete reactions using microwave vessel designs. Other apparatus such as conventional closed fluoropolymer steel-jacketed pressure digestion devices and Carius tubes have these same characteristics.

Reagents of high purity must either be purchased or produced in the laboratory. There are good reasons to do both. First, the production of some reagents is difficult and purchasing them from commercial sources is the most viable alternative. Also, if the amount of reagent used in a laboratory is minimal and does not warrant establishing a continuous supply of high purity reagent, then purchasing again is the most logical alternative. However, other reagents used in significant quantities can be easily purified at a relatively low cost in a laboratory setting. Reagents are also difficult to store in high purity, so another rationale for continuous production of high purity reagents is to maintain their purity. For example, water and acids accumulate both organic and inorganic constituents from a variety of sources when stored (*79, 93, 94*). High purity water is best made fresh from one of the commercially available 18 MΩ water deionization systems. High purity reagents are not generally stored for long periods of time because their purity degrades with storage (*80*).

High purity reagents once required elaborate facilities and expensive equipment and maintenance to produce. This fact has changed dramatically over the past several years with the introduction of commercially available reagent preparation apparatus (*95*).

In the preparation of high purity reagents there is only one choice for the method of purification. Subboiling distillation (*96, 97*), different from normal distillation, uses an IR radiation source so as not to boil the liquid reagent. Preventing boiling avoids the Brownian movement of solution droplets produced when bubbles burst at the surface of the liquid. These aerosolized solution particles are carried everywhere in the apparatus and physically transport metal ions and contaminants that should have been left in solution. Subboiling distillation is a slower but very reliable method of purifying all of the common mineral acids and many organic reagents used in analytical methods. It relies exclusively on the vapor pressure of the material, reagent, and contaminant, and is therefore specifically optimized for purification of the mineral acids if the object is to remove metal ions. Of all acids, nitric acid, for a variety of reasons, can be purified to excellent quality.

Figure 21 illustrates the level of metal contamination in commercial subboiling distilled acids as supplied by Fisher Scientific (Pittsburgh, PA) in acid-leached Teflon bottles (*98–101*). This figure shows that nitric acid is almost always more contaminant-free than other mineral acids, such as hydrofluoric or perchloric acid. It is a good indication as to whether the purity supplied is sufficient to make many of the acid-stabilized storage

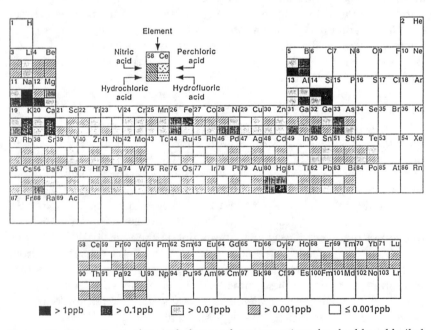

Figure 21. Comparison of typical elemental concentrations for double subboiled distilled nitric, perchloric, hydrochloric, and hydrofluoric acids (*98–101*).

and decomposition decisions necessary in sample preparation. Table VII compares subboiling distilled reagent grade and commercially available subboiled nitric acid obtained from Fisher Scientific. The two orders of magnitude difference between subboiled distilled nitric acid and ACS standard reagent grade is frequently the main source of metals added to the analytical blank. Tables of metal impurity concentrations for subboiling distilled acids have been published with measurements of high purity acid, reagent, and other grades (*78, 96, 97*).

The preparation of subboiling distilled reagents has several alternative equipment designs. Older references list using a Teflon block, threaded to accept two Teflon bottles. One of the methods that has been successful in the authors laboratory employs a modern version of this approach by using Teflon PFA molded segments (Savillex Corporation) that screw together (*104*). Four of these Teflon segments are placed in one class-100 clean hood for subboiling acid purification. This arrangement can accommodate four stills that produce high purity acids at a relatively low, but a reliable rate of 100–150 mL in 24 h for the standard mineral acids, such as nitric and hydrochloric. Acids with higher vapor pressures, such as acetic acid, may reach as much as 300 mL in a 24-h period. Figure 22a is a drawing of the Teflon PFA stills; Figure 22b is a photograph of four units mounted in the class-100 clean hood. Several configurations of these stills

Table VII. Impurity Concentrations in Subboiled and ACS Reagent Grade Nitric Acid

Element	ACS Grade[a] (ppb)	Subboiled[b] (ppb)	Double Subboiled[c] (ppb)
Arsenic	≤4	<0.3	<0.1
Iron	≤200	<2	≤0.3
Aluminum	≤200	<3	≤0.3
Boron	≤100	<8	<0.03
Calcium	≤200	<8	<0.2
Chromium	≤100	<3	0.04
Copper	≤50	<1	0.019
Lead	≤100	<0.3	0.002
Magnesium	≤300	<2	0.014
Nickel	≤50	<1	≤0.1
Potassium	≤300	<4	<0.2
Sodium	≤300	<4	<0.5
Tin	≤300	<0.5	<0.05
Titanium	≤300	<0.5	≤0.008
Zinc	≤300	<1	≤0.3
Totals	≤2800	<39	≤2.2

[a]Data from reference 102.
[b]Data from reference 103.
[c]Data from reference 99.

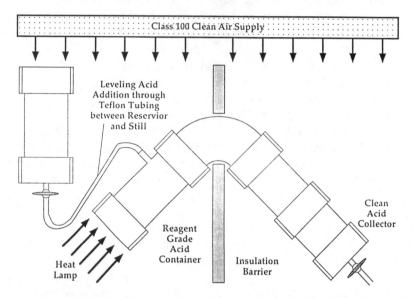

Figure 22a. Teflon IR heated subboiling still in class-100 clean air hood with leveling reservoir constructed from modular Teflon PFA segments (*105*).

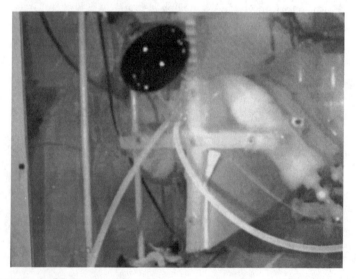

Figure 22b. Picture of Teflon PFA stills in a clean hood.

were produced by putting together different segments to make the still, feedstock, and condensing segments, and this design was optimal for consistently producing high quality and inexpensive acids.

Our laboratory uses such large quantities of nitric, hydrochloric, and acetic acids that a commercial quartz tandem subboiling still was purchased from Milestone S. R. L. (Bergamo, Italy). These stills are capable of producing 300 mL/h, or up to 7 L in a 24-h period. They have been specifically designed to optimize the subboiling preparation of mineral acids, water, acetic acid, and other reagents. Figure 23a is a diagram of these stills, and Figure 23b is a photograph of the two stills mounted in the control housing without a clean hood, to permit examination of the construction. The units are completely automated and produce consistently high purity reagents. Because these reagents sometimes cost hundreds of dollars per liter, the expense of the unit was quickly recovered.

Table VIII describes a total blank audit that was prepared during the study of a procedure that included both sample preparation and analysis. The analysis table demonstrates the additive nature of reagent blank contribution and the difficulty of eliminating that contribution. It also illustrates that when control of the atmosphere, reagents, material, and technique are considered, a consistent blank at very low levels can be achieved and stabilized.

Table VIII demonstrates that statistically significant control of the analytical blank is possible, even at very low levels, with appropriate clean chemistry sample preparation procedures. Practice, skill level, apparatus, and methods all contribute to bringing the analytical blank into a manageable range.

Materials for Sample Preparation, Storage, and Analysis

For elemental analysis, specific, preferred materials are used for the construction of microwave vessels. Over the past two decades, materials identified as being noncontaminating have become the top choices for bottles, beakers, reaction vessels, and storage containers in trace and ultra-trace analysis. These materials are the same as those being used in microwave vessels that are typically transparent to microwave radiation, thermally durable, chemically resistant or inert, noncontaminating, and have appropriate compression and tensile strength. Table IX lists the specific types of materials, in order of preference, for their noncontaminating nature and chemical inertness to most acid reactions. These materials have been evaluated by others and tested extensively for their elemental contamination characteristics (*78, 93, 106*).

With the exception of polyethylene, these are the most common materials used for microwave vessels, both atmospheric pressure vessels and closed-vessel liners that come in contact with the sample. These materials are the most stable to acid reactions (with the exception of quartz and

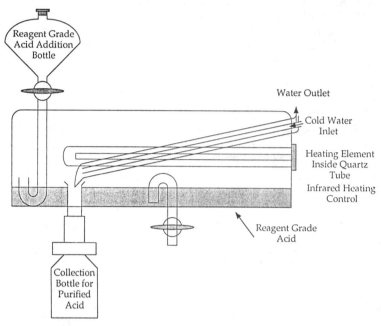

Figure 23a. Commercially available IR quartz self-contained subboiling still system with leveling reservoir and onboard controller (95).

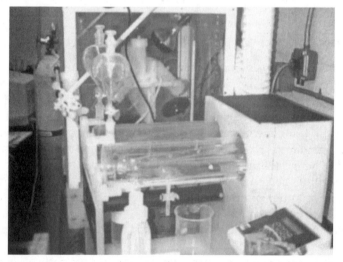

Figure 23b. Picture of Milestone quartz stills.

Table VIII. Sources of Vanadium Blank in Human Serum (SRM 909) During a Blank Contribution Audit Including Sample Preparation and Analysis

Reagent	Vanadium Concentration (pg/g)	Amount Used (g)	Total Vanadium Contribution (pg)
HNO_3	1.8	13	23
$HClO_4$	2.0	7	14
CH_3COONH_4	1.0	22	22
H_2O	0.5	10	5
NH_4OH	0.7	13	9

SOURCE: Data from reference 85.

Total calculated reagent blank	73 pg
Measured loading blank	Range 2–41 pg, Average 19 pg
(sub-sampling and instrument introduction)	
Total known blank	73 + 19 = 92 pg
Measured analysis blank	Range 100–170 pg (n = 5)
	Average—150 pg*
Unknown blank level	58 pg

NOTE: All samples prepared using subboiled distilled reagents, Teflon PFA containers, in a class-100 clean laboratory.

Table IX. Non-Contaminating Materials and Specifications for Use in Ultra-Trace Analysis and as Decomposition Vessels and Sample Containers (*78, 93, 106, 107*)

- Fluoropolymers: PFA[a], TFM, TFE[a], FEP[a], Tefzel[a]
- Quartz, synthetic
- Polyethylene (suitable for storage only; not for acid digestion)
- Quartz, natural
- Borosilicate glass

NOTE: Listed from highest to lowest preference for use in containing samples.
[a]Various forms of Teflon.

glass if hydrofluoric acid is used). Fluoropolymers are the most common and were adapted from other chemical uses for application in microwave systems. Of the fluoropolymers, perfluoropropylvinylether (TFM), perfluoroalkoxy (PFA), and tetrafluoroethylene (TFE) (or PTFE) have the highest use temperatures, ranging from 270 to 300 °C. They are also chemically inert to the majority of mineral acids and combinations thereof. Sulfuric acid has a boiling point of approximately 330 °C and can damage all fluoropolymers by melting them. Quartz and glass can thermally handle sulfuric acid, but borosilicate glass is not appropriate for ultra-trace elemental analysis (*78, 106*). Glass actually forms a gel layer that hydrates and leaches, transferring elemental components to the sample solution. Even though these are minute quantities, there are many low level analyses where these contributions would be detected in the blank.

Structurally stabilizing outer casings for closed pressurized microwave vessels are made from various transparent polymers, with and without fiber reinforcement, such as polyetherimide. These materials do not come in contact with samples and are only used for structural integrity during pressurized chemical reactions. They also have advantages over the metal composite casings used in Carius tubes and steel-jacketed bombs. The presence of rust or corroded metal can provide significant amounts of particulate that are avoided with the use of polymers.

Polyethylene is suitable for storage of diluted samples after decomposition, but it does not have a thermal-use temperature appropriate for decomposition (71–93 °C, mp 120–135 °C). It is also not inert enough to be useful as a decomposition vessel or vessel liner, as are polycarbonate and polypropylene. Polyethylene's low cost and relative inertness to cool, weak acid solutions make it an excellent storage container for trace element solutions (93, 106).

A unique feature of microwave-induced heating is that the container is cooler than the liquid. The heat in the container is generally transferred from the acid and water in the reagent and sample, and the vessel removes heat from the solution. This function assists by reducing the leaching of the container materials because they are at lower temperatures. This difference could also play a role in leaching from fluoropolymers, but has not yet been studied with direct comparison of identical materials and heating sources. One additional improvement that has not yet been implemented by vessel manufacturers is the molding of these liner materials in clean air environments. This molding would prevent entrainment of dust in the polymer of the vessel and thus its leaching during decomposition with strong acids.

Analytical Technique

The final significant source of analytical blank contamination is the skill of the analyst and the appropriateness of the technique that is being performed. Analytical blank control has been explained as the combination of atmosphere, reagent, material, and the protocol being used correctly. Microwave sample preparation assists each of these parameters in synergistic ways to lower the analytical blank, increase blank precision, and improve quality control.

- If a closed or controlled atmospheric microwave vessel is prepared in a clean hood and sealed before leaving the clean environment, the sample will not be affected by atmospheric contamination during the reaction, because it has not been removed from a clean environment.
- The vessel materials would normally not be used by many laboratories, so the advantages of the fluoropolymers would not be realized if they were not required in many microwave reaction vessels.

- The time the sample spends in decomposition, leaching, or extraction is typically reduced from hours to minutes, thus reducing the potential leaching from the container walls and evaporation of the reagent before it reacts productively.

The blank is both reduced in size and is more consistent due to limiting the exposure variables. An example of these components working together has been provided in the literature when analysis under different conditions has verified these conclusions (108, 109). For example, Table X illustrates and isolates the blank optimization areas: environment, reagents, materials, and analysis skills. The skill of the analyst was kept constant as the same individual changed environments, reagents, and combinations of these parameters in the analysis (78).

The principles are also illustrated for chromium in Figure 24. The blanks associated with samples analyzed by isotope dilution mass spec-

Table X. Examples of the Analytical Blank Influence on Trace Analysis of Elements in Glass

Conditions	Pb (ng)	Ag (ng)
Initial analysis of TEG standard	330 ± 250	970 ± 500
Analysis using subboiled distilled acids	260 ± 200	—
Analysis in class-100 hood	20 ± 8	207 ± 200
Analysis using subboiled acids in class-100 hood	2 ± 1	3 ± 2

NOTE: TEG is trace element in glass, SRMs 610–619.
SOURCE: Data from reference 78.

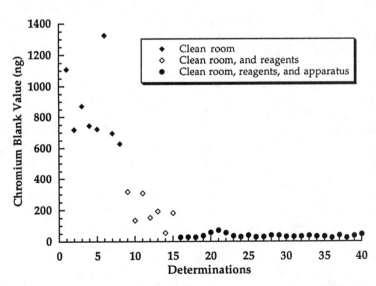

Figure 24. Progression of analytical blanks for the determination of chromium.

trometry were sequentially evaluated in a similar manner to the progression in Table XI. In the first group of data (far left, determinations 1–8), the analyses were performed in a class-100 clean air environment by using subboiling purified reagents (except for unpurified ceric sulfate; the major source of the chromium blank) and standard laboratory glassware. The next group (determinations 9–15) were performed by using a class-100 clean room environment, high purity reagents, and standard laboratory glassware (the major source of contamination). The final group (determinations 16–40) was achieved by using a class-100 clean room and hood environment, all subboiled purified reagents, and apparatus constructed from Teflon and quartz. Under these final conditions the chromium analytical blank was in statistical control with variability of less than 10 ng. The adherence to each of the principles that have been discussed permits such quality control and high-quality data (*110*). This setup was the state of the art in clean chemistry procedures prior to closed-vessel microwave sample decomposition methods. It is an excellent example of the "part-per-billion (ppb) barrier" that prompted the development of this tool to continue driving down the analytical blank.

It was not possible to evaluate microwave sample preparation in Table X because these studies were done prior to integrating and implementating microwave sample preparation with these analysis steps. However, Table XI summarizes studies spanning from 1980 to 1990 and makes incremental improvements in the certification of vanadium and the blank associated with these measurements. It shows the implementation of closed-vessel and microwave methods using a clean room and high purity reagents for sample preparation and analysis. A distinct improvement can be observed that is directly attributable to the synergistic addition of microwave sample preparation added to these other key parameters. This table is an extension of Tables III, VIII, and X, where, after implementing a clean environment, high purity subboiled reagents and fluoropolymer vessels, microwave technology was added to the analyst's tools, furthering the improvement of the blank. This addition resulted in noticeable synergistic relationships, and specific improvements were observed in the blank level and its precision.

Table XI shows the blank values relating to the analyses in Table III. The analyses are listed in approximate chronological order of their completion. The technology applied for the first five SRMs was a class-100 clean laboratory, subboiled reagents, Teflon FEP or PFA beakers, and a hot plate in a class-100 clean laboratory exhausted clean hood. A major change was implemented in the analysis of human serum. The samples were processed as previously stated, but with closed Teflon PFA beakers instead of open Teflon beakers. This change reduced the amount of acid and time necessary for decomposition. Finally, the decomposition of bovine serum used Teflon PFA microwave closed-vessel heating, which further reduced the amount of acid and time necessary for decomposition (*83, 111*).

Table XI. Progression of Chemical Blanks for Vanadium (ng)

No.	Oyster Tissue (SRM 1566)	Crude Oil (RM 8505)	Citrus Leaves (SRM 1572)	Bovine Liver (SRM 1577a)	Fuel Oil (SRM 1634b)	Human Serum (SRM 909)	Bovine Serum (SRM 1598)
1	2.35	3.6	2.32	0.39	0.24	0.10	—
2	1.30	7.1	3.07	0.54	0.25	0.17	—
3	0.34	2.0	4.09	0.30	0.28	0.14	—
4					0.24	0.16	
5					0.32	0.16	
X	1.33	4.2	3.16	0.41	0.27	0.15	<0.001[a]
SD (%)	1.01	2.6	0.89	0.12	0.034	0.028	—[a]
RSD (%)	75.6	61.6	28.1	30	13	19	—[a]

[a]Blanks were so low they were at the detection limit of the instrument.
SOURCE: Data from reference 85.

Precision improvements were associated with using closed-vessel higher pressure digestions that took much less time. The same clean facilities were used for each analysis and the same personnel produced reagents by using the same quartz subboiling stills. This progression of achieving lower blanks was only accomplished by enhanced skills in clean chemistry, laboratory cleanliness, and the use of a microwave sample preparation system.

The blank was reduced to a point that it became indeterminable at that low level by using the sophisticated neutron activation analysis techniques being applied. These levels are some of the lowest that were achieved at NIST. Unfortunately, a second independent method could not be found to duplicate the results of this ultra-trace analysis, and information numbers were produced instead of certified numbers for vanadium. In the other standard reference materials, a second method verified these measurements and certified measurements were produced. These blank values for vanadium in Bovine Serum were added as less than picogram levels and are plotted in Figure 25. The values extend blank evaluation beyond Table X and demonstrate that additional blank improvements may be observed when microwave sample preparation is added to the traditional clean chemistry precautions.

As requirements for even lower analysis are developed, these blank tools may need to be developed further. For the present, this sample preparation technique has reduced the blank below the detection limits available on current analytical instrumentation.

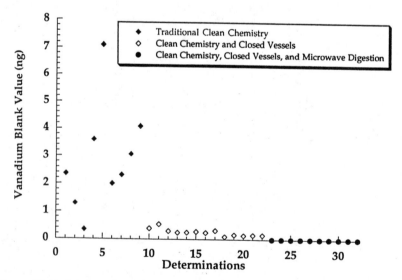

Figure 25. Reduction of vanadium blanks through the use of clean chemistry and closed-vessel microwave digestion.

Synergy of Microwave Sample Preparation and Ultra-Trace Analysis

Reducing and stabilizing the analytical blank requires the capability to control a combination of the aforementioned contamination sources. This control produced the dramatic improvement in the specific example of vanadium analytical blanks. The use of microwave decomposition, in conjunction with the other main areas of blank reduction, permitted even further improvements. Microwave sample preparation, considered as part of the fourth criteria, the analyst's skill, offers an improved, general, applicable tool that is more capable than previous techniques. With the experience of two decades of ultra-trace analysis striving to minimize each of the four areas of principle blank contribution, adding microwave sample preparation to the tools available improves all four areas. Microwave sample preparation reduces the environmental exposure as the sample retains the clean environment, excludes additional environmental sources of contamination, and makes this isolation more feasible. This technique reduces reagent use to stoichiometric quantities and prevents losses from evaporation. It also reduces the time of decomposition by increasing the temperature and coupling the energy directly to the reagents.

Microwave sample preparation is a more reproducible method of duplicating ultra-trace sample preparation methods, both within a laboratory, or, as has been seen in environmental analyses, between laboratories. It improves the tools and reduces the overall skill level necessary for the analyst. This method goes a long way to help Berzelius' 1814 statement of making a method of analysis that "depends least on the skill of the operating chemist...". This statement implies that the method is capable of permitting many analysts to achieve these desired analytical goals with a minimum of variation without loosing control of the method. Microwave sample preparation qualifies as the most appropriate tool in controlling the quality of the analysis through its capability and reproducible contribution to the overall sample preparation and analysis process.

Environmental Microwave Sample Preparation Standard Methods

Current Status

Microwave-enhanced chemical sample preparation methods have become approved standard methods and are among a new generation of applied environmental regulatory analysis techniques. Currently, there are three EPA RCRA methods that are microwave-specific. Two are leach methods and the other is a comprehensive and adaptable dissolution method for total metal analysis. Several other methods are capable of being implemented through microwave instrumentation. The National Pollutant Discharge Elimination Systems (NPDES) also has an alternative leach method.

The RCRA Program in the Office of Solid Waste (OSW) and the Superfund Program (CERCLA) in the Office of Emergency and Remedial Response (OERR) are cooperating and approving shared microwave acid dissolution methods. Method 3051 is a key acid-assisted leaching sample preparation method using microwave energy for the preparation of 26 elements in soils, sediments, sludges, and oils. Method 3015 is for acid leaching and digestion of 23 elements from medium and heavily sedimented aqueous samples and extracts, and is being used cooperatively across these EPA programs. Both Methods 3051 and 3015, draft methods approved in 1990, were approved for Agency use and were published in SW-846, Update II on January 13, 1995 (47). These methods are capable of providing equivalent results for more than a dozen previous methods based on older technology. Method 3051 results are usually compared to results from Method 3050. In September 1992, the alternate microwave leaching procedure for wastewater was approved in the NPDES program for the testing of specific metals in water matrices (112).

In addition to the distinctly designated microwave methods, a number of additional methods have been modified to permit alternative microwave implementation. One of the most environmentally used leaching methods is 3050B (50), the "Acid Digestion of Sediments, Sludges, and Soils", which can be implemented by using atmospheric-pressure microwave systems as an alternative to traditional hot plate, heating mantle, block digestor, or other heating devices (47). Its unique aspects will be discussed in detail later in this chapter to clarify both the conventional and microwave alternatives. Experimental evidence will also be provided to technically support the microwave implementation of Method 3050B. Method 3031, "Acid Digestion of Oils for Metals Analysis by FLAA or ICP Spectroscopy", is another method that may alternately be implemented by using microwave technology. This method is a recent one that first appeared in proposed SW-846 Update III and is not discussed in this chapter.

Method 3052, proposed for inclusion in SW-846 Update III, is a new and more comprehensive method. It is a total sample decomposition as compared to more typical environmental leaching procedures. It is designed to produce a total analytical composition analysis as opposed to a consistent leach recovery of a range of elements (49). Method 3052, "Microwave Assisted Acid Digestion of Siliceous and Organically Based Matrices", is designed to be applicable to all common environmental matrix types and has many performance-based alternative chemistries. It is approved for 26 elements and is capable of preparing samples for any analytical detection method or other elements that may be required by the RCRA program. Method 3052 is a performance-based sample preparation method that has extensive alternatives. It is capable of being adapted to a variety of media and matrix applications; more than previously promulgated methods. It represents the first of a new generation of methods aimed at producing accurate total analytical composition through performance-based criteria.

Method 3052: A New Moderate Pressure Total Decomposition Method

Introduction to Method 3052

Method 3052 was created to answer the need for a general total analysis method for inorganic constituents. Total sample decomposition is interpreted to mean the sufficient destruction of the matrix to permit the total analysis to be performed for the elements of interest. Method 3052 is validated for 26 elements and extends the sample preparation to most of the metals and many nonmetal elements in the periodic table. It permits optimization for specific elements, matrices, and detection techniques.

The original closed-vessel microwave protocol that became EPA Method 3052 was developed for, and evolved from, elemental certification of SRMs at NIST. This technology was developed for the elemental sample preparation procedure for elemental analysis by many instrumental techniques, including isotope dilution mass spectrometry (IDMS), ICP–MS, ICP–OES, AAS, GF–AAS, neutron activation analysis (NAA), chelation ion chromatography, polarography, and laser enhanced ionization (LEI) (*31, 34, 56, 111, 113, 114*). An extensive list of SRMs that have been analyzed after using microwave decomposition for sample preparation is found in Chapter 2. Examples of elements that have been certified, or provided for reference, in specific NIST SRMs during the development of what is now the SW-846 Method 3052 are listed in Table XII.

Many additional examples of microwave sample preparation by using acid dissolution of reference materials and applied instrumental analysis techniques, elements, combinations of elements, and a wider variety of matrices are presented in Chapter 2 of this text. EPA Office of Solid Waste publication SW-846 Method 3052 is designed for the total analysis of many difficult and challenging matrices and is therefore applicable to many industrial and common laboratory samples. The total elemental nature and complete decomposition design of this method make it more applicable than many other environmental methods. Also, the method's adaptive

Table XII. Examples of Standard Reference Materials the Elements Certified After Method 3052 Microwave Dissolution Sample Preparation

SRM Number	SRM Name	Elements Certified in SRM
1515	Apple Leaves	Mn, Ni
1547	Peach Leaves	Mn, Ni
1548	Total Diet	Cu, Mn, Ni, Pb
1567a	Wheat Flour	P
1598	Bovine Serum	Al, Cu, Mn, Ni, V
1646	Estuarine Sediment	V
2704	Buffalo River Sediment	As, Cd, Hg, Mn, Ni, Pb, S, Se, U, Tl

and alternative procedures enable optimization for specific matrix types and elemental chemistries demonstrating the comprehensive nature of the method. This method is in contrast to other methods (3051, 3015, and 3050), which are leaching methods and therefore not designed to provide a total analysis, but rather were constructed to provide a worst-case scenario of leaching of hazardous materials and mobility of constituents. Although both types of methods have their place, the total elemental analysis method permits broad application both within EPA matrix types and programs and for general analytical applications.

To facilitate discussion of Method 3052, we have duplicated selected portions of the method. This duplication permits a detailed explanation of the method basis and highlights the alternatives and flexibility built into the method. The scope and application of Method 3052 as stated in section 1.1 of the method will be examined first (48).

Scope

1.0 SCOPE AND APPLICATION

1.1 This method is applicable to the microwave-assisted acid digestion of siliceous matrices, organic matrices and other complex matrices. Ashes, biological tissues, oils, oil contaminated soils, sediments, sludges, and soils may be digested using this method for total decomposition (relative to the target analyte list) and if analysis is required. This method is applicable for the following elements:

Al, Sb, As, B, Ba, Be, Cd, Ca, Cr, Co, Cu, Fe, Pb, Mg, Mn,
Hg, Mo, Ni, K, Se, Ag, Na, Sr, Tl, V, and Zn.

Other elements and matrices may be analyzed by this method if performance is demonstrated for the analyte of interest, in the matrices of interest, at the concentration levels of interest (see Sec.8.0)*.

---

*Please refer to Method 3052 for other sections, graphics, and tables.

The method is applicable to all elements (listed previously) necessary for assessment under the RCRA program, but is also generically applicable to others that may be necessary in the future or for specific programs. Other elements that are not specified in the RCRA list are included with the caveat of performance demonstration. This approach is practical to permit the expansion and general use of this method. Sufficient alternative chemistries and reagent choices are provided to accommodate practical and adequate protocols for most matrices. Many elements that are not on the RCRA list, such as phosphorous, sulfur, uranium, and others, have been analyzed by using this sample preparation procedure. Chapter 2 has many examples of elements that have been satisfactorily analyzed in matrices using similar and modified chemical reagents. Table XIII provides reference conditions for several matrices, represented by SRMs, that have been provided to show general guidelines and examples of more common matrix types. Although some crystalline mineral forms and refractories are very acid-resistant, the method is generally applicable for most specific matrices that incorporate the usual environmental samples, and many that represent the most common matrix types. Adjustments in the reagent composition are also required to optimize solubility and stability of specific analytes. Several notable examples are described.

Expandability

> **1.2** This method is provided as a rapid multi-element, microwave-assisted acid digestion prior to analysis protocol so that decisions can be made about the site or material. Digests and alternative procedures produced by the method are suitable for analysis by flame atomic absorption spectroscopy (FLAA), cold vapor atomic absorption spectroscopy (CV–AAS), graphite furnace atomic absorption spectroscopy (GF–AAS), inductively coupled plasma optical emission spectroscopy (ICP–OES), inductively coupled plasma mass spectrometry (ICP–MS), and other analytical elemental analysis techniques where applicable. Due to the rapid advances in microwave technology, consult your manufacturer's recommended instructions for guidance on their microwave digestion system and refer to the SW-846 "DISCLAIMER" when conducting analyses using Method 3052.

The method has been expanded for potential use with many analytical techniques. This expansion is appropriate and prudent because of many years of testing Method 3052, the robustness of its performance, and numerous examples of similar applications. Many examples of similar applicable methods are referenced in Chapter 2. The majority of elemental analysis techniques have been used to analyze samples prepared by using Method 3052, or modified versions. The conditions of these techniques are not as well-defined, but the reagent chemistries provide guidance and a reference point. Because Method 3052 produces a matrix of known characteristics in most cases, it permits a generic application for many analytical instruments. However, one notable caution is the use of polarography, which is sensitive to trace residual organics, for the determination of specific elements (*56*). Several specific suitable instrumental methods are mentioned for clarity, but most inorganic analytical instrumental methods of analysis are compatible.

Total elemental sample preparation methods are inherently compatible with many analytical detection systems. New and unique instrumentation may be validated through the use of SRM analysis, which is a traditional method of evaluating accuracy and appropriateness.

As an example of a very different elemental analysis system, mercury was determined in two soil and sediment standards prepared with Method 3052. Aliquots of the digest solution were analyzed with an automated mercury analyzer (*115*). Table XIV illustrates the 95% confidence intervals of the certified SRMs and the analysis. The statistical agreement between the SRM and the Method 3052-prepared digest validates the analysis of mercury with this new instrument.

Table XIII. Recommended Digestion Parameters for the Analysis of Matrices by EPA Method 3052

| Matrix | Standard | Volume of Acid (mL) | | |
		HNO_3	HF	HCl
Soil	NIST SRM 2710	9	3	0–2[a]
	Highly Contaminated Montana Soil			
	NIST SRM 2711	9	3	0–2[a]
	Moderately Contaminated Montana Soil			
Sediment	NIST SRM 2704	9	3	0–2[a]
	Buffalo River Sediment			
Biological	NIST SRM 1566a	9	0	0
	Oyster Tissue			
	NIST SRM 1577a	9	0	0
	Bovine Liver			
Botanical	NIST SRM 1515	9	0	0
	Apple Leaves			
	NIST SRM 1547	9	0	0
	Peach Leaves			
	NIST SRM 1572	9	0.5	0
	Citrus Leaves			
Oil	NIST SRM 1084a	9	0.5	0–2[a]
	Wear-Metals in Lubricating Oil			

[a]HCl is added to stabilize elements such as Ag and Sb when they are analyzed.

Table XIV. Mercury Analysis in Soils by Using an Automated Mercury Analyzer

SRM	Certified (µg/g)	Analyzed (µg/g)
2704	1.44 ± 0.07	1.45 ± 0.03
2710	32.6 ± 1.8	33.4 ± 0.2

NOTE: Values are 95% confidence intervals.

Performance-Based Method

1.3 The goal of this method is *total* sample decomposition and with judicious choice of acid combinations this is achievable for most matrices (see Sec. 3.2). Selection of reagents which give the highest recoveries for the target is considered the optimum method condition.

As stated in sections 1.1 and 1.3, Method 3052 is a performance-based protocol that permits tremendous latitude in element, matrix, instrumentation, and conditions to achieve the method goal. Total sample decomposition is interpreted to mean the sufficient destruction of the matrix to permit the total analysis to be performed for the elements of interest. As has been discussed, the composition of the solution is in many cases very well defined and characterized (56, 116). Decomposition is, in the case of nitric acid oxidation of organic based samples, not complete with known specific organic moieties remaining (62, 116, 117). In opening out reactions such as oxide dissolution, the reactivity is less dependent on reagent temperature and more dependent on the specific crystal structure and elemental species involved.

The highest recovery of an element can be defined as the highest found concentration with the appropriate correction for the blank applied. It is usually the case that low recoveries are encountered when a component of the sample matrix sequesters some portion of an element, thus preventing it from participating in the detection mechanism specific to the instrumental approach used. A classic example of this is the analysis of vanadium in NIST SRM 1571 Orchard Leaves (118). A study was conducted in which the laboratories that used combinations of acids did not include hydrofluoric acid. This procedure resulted in low recoveries of vanadium by 33%, regardless of the detection method used. It was discovered that 66% of the vanadium was derived from organic constituents, and 33% from silacial constituents. Decomposition procedures that included hydrofluoric acid recovered the total concentration of vanadium in the SRM, whereas all acid combinations that did not include hydrofluoric acid obtained results approximately 33% lower due to omission of the concentration from the silacial component. The suggestion that the highest recoveries (maximum concentration) are the optimum method conditions (more correct concentration) is thus based on this matrix-segmented elemental distribution. Each matrix segment must be sufficiently decomposed to free all the elemental concentrations for analysis.

Microwave Instrumentation

4.0 APPARATUS AND MATERIALS

4.1 Microwave apparatus requirements.

4.1.1 The temperature performance requirements necessitate the microwave decomposition system sense the temperature to within ±2.5 °C and automatically adjust the microwave field output power within 2 seconds of sensing. Temperature sensors should be accurate to ±2 °C (including the final reaction temperature of 180 °C); verification at two points >50 °C apart should be determined periodically. Temperature feedback control provides the primary control performance mechanism for the method. Due to the flexibility in the reagents used to achieve total analysis, temperature feedback control is necessary for reproducible microwave heating.

Alternatively, for a specific set of reagent(s) combination(s), quantity, and specific vessel type, a calibration control mechanism can be developed similar to previous microwave methods (*see* Method 3051 in Reference 5). Through calibration of the microwave power, vessel load and heat loss, the reaction temperature profile described in section 7.6 can be reproduced. The calibration settings are specific for the number and type of vessel used and for the microwave system in addition to the variation in reagent combinations. Therefore, no specific calibration settings are provided in this method. These settings may be developed by using temperature monitoring equipment for each specific set of equipment and reagent combination. They may only be used if not altered as previously described in other methods such as 3051 and 3015. In this circumstance, the microwave system provides programmable power with a minimum of 600 W, which can be programmed to within ±12 W of the required power. Typical systems provide a nominal 600 W to 1200 W of power (References 1, 2, 6). Calibration control provides backward compatibility with older laboratory microwave systems without temperature monitoring or feedback control and with lower cost microwave systems for some repetitive analyses. Older lower pressure vessels may not be compatible.

While drafting Method 3052, a survey of manufacturer microwave laboratory instrumentation was performed. The specifications were provided to manufacturers to permit compliance. Many manufacturers' equipment did not need to be modified because they had already incorporated provisions to contain the vessel pressures produced in Method 3052. The vessel pressure limit is one critical parameter of microwave equipment that could prevent a manufacturer from meeting the specifications. Both temperature feedback control and calibration control of the microwave and sample vessel systems are provided as alternatives to permit the largest number of instrument suppliers to be able to provide appropriate equipment. Many temperature control sensors and electronic feedback control mechanisms are appropriate to control the decomposition method. An attempt was made to eliminate any proprietary technology requirements that may limit the choice of equipment. Rather, the merits of the equipment permit the analyst to choose from a variety of vendors whose equipment meets the specifications and is capable of performing the method in a safe manner. Several microwave vendors have recently improved their vessel pressure performance, increasing them to permit application of Method 3052.

The use of microwave power calibration permits inexpensive method compliance when a repetitive analysis is to be performed. The microwave vessel and exact reagent load must be evaluated and a set of microwave field strengths for this specific configuration identified. These specific and reproducible microwave loads and conditions must be developed by the user or the manufacturer and stringently adhered to for calibration control. Methods 3051 and 3015 had these parameters evaluated for a specific vessel type, but these vessels are now essentially obsolete. This obsolescence, based on energy balancing, demonstrates the need to provide a protocol that specifies reaction conditions rather than the equipment. This strategy has been followed in this section of the method. Specific understanding of variable pressures that are obtained in microwave sample vessels at the same cavity load and field strength are described in the earlier energy absorption and reaction conditions section of this chapter. These significant, but subtle interactions require the vessel heat loss to be considered when using calibration control.

Method

7.0 PROCEDURE

7.1 Temperature control of closed-vessel microwave instruments provides the main feedback control performance mechanism for the method. Control requires a temperature sensor in one or more vessels during the entire decomposition. The microwave decomposition system should sense the temperature to within ±2.5 °C and permit adjustment of the microwave output power within 2 seconds.

This discussion assumes temperature feedback control of the microwave system and considers a variety of alternative chemistries that require the ability to neglect the microwave cavity loading. The temperature sensing and control requirement reflects the ability of thermocouples, fiber optics, gas bulbs, and infrared thermometry to sense the temperature inside the microwave sample vessels. The temperature is then used to adjust the microwave field strength to achieve temperature conformation. This adjustment is handled in a variety of ways by different manufacturers and equipment. To permit the flexibility for decomposition of a wide variety of matrix types and the use of many of the alternative procedures, feedback control automatically compensates for variables such as absolute acid power absorption, a different number of vessels in the microwave cavity, or different mass loading in the microwave field. Once a specific procedure has been chosen, control can be accomplished with calibration if the acid load and specific vessel type are kept constant. However, calibration control would not permit development of the method parameters without having access to temperature measurement equipment to adjust the microwave field intensity and verify conditions.

Digestion

> **7.3 Sample Digestion**
>
> **7.3.1** Weigh a well-mixed sample to the nearest 0.001 g into an appropriate vessel equipped with a pressure relief mechanism. For soils, ash, sediments, sludges, and siliceous wastes, initially use no more than 0.5 g. For oil or oil contaminated soils, initially use no more than 0.25 g.

The difference in sample weight, based on sample type, is primarily due to matrix decomposition products that build up in the closed microwave vessels. Figures 26 and 27 demonstrate the difference in pressure for these two types of decompositions. A 0.25-g sample of soil, ash, or silicate-based rock will typically produce pressures of 8–15 atm, depending on the vessel heat loss characteristics and the combination of acids used. In highly insulated vessels, this pressure can be much higher. In this same vessel and with the same nitric acid mass, a 0.25-g sample of organic material will produce a pressure of 25 atm. The difference between these two matrix types is the reaction gases that are produced. The important reactions (eqs 8 and 9) have been reiterated below to demonstrate the origin of the reaction products. The large quantity of carbon dioxide produced by organic materials makes the largest difference between the two samples.

$$SiO_2 + 4HF \xrightarrow{\Delta} SiF_4 + 2H_2O \tag{8}$$

$$(CH_2)_x + 2xHNO_3 \xrightarrow{\Delta} xCH_2 + 2xNO + 2xH_2O \tag{9}$$

Figure 28 shows the temperature profile for nitric acid without a sample present using recommended temperature specifications of Method 3052. A comparison of Figures 26 and 27 indicates how much of the vessel pressure is due to the partial pressure of the acid and the amount of pressure that is due to the residual reaction products. A sample of oil, or highly organic-contaminated soil, will combine these two reactions and the pressure would be expected to be somewhere between those in Figures 26 and 27.

Sample size scaleup provisions have been anticipated and are discussed in section 7.3.12 of Method 3052. Any sample size scaleup should be done in small increments to prevent exceeding vessel pressure capabilities.

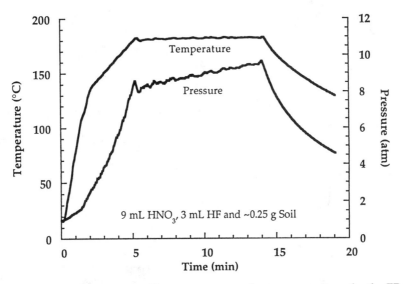

Figure 26. Typical reaction profile, temperature and pressure vs. time, for the EPA Method 3052 digestion of soil.

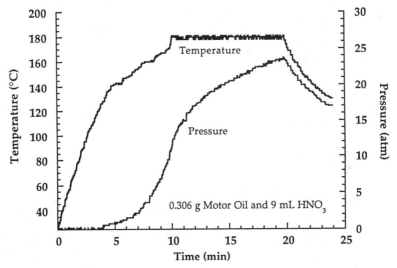

Figure 27. Typical reaction profile, temperature and pressure vs. time, for the EPA Method 3052 digestion of oil.

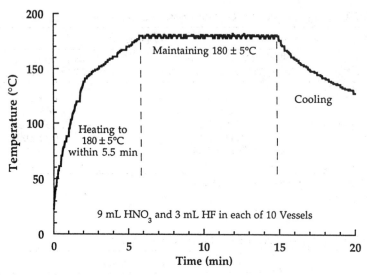

Figure 28. Typical reaction profile, temperature vs. time, for the EPA Method 3052.

Nitric Acid

7.3.2 Add 9 ± 0.1 mL concentrated nitric acid and 3 ± 0.1 mL concentrated hydrofluoric acid to the vessel in a fume hood. If the approximate silicon dioxide content of the sample is known, the quantity of hydrofluoric acid may be varied from 0 to 5 mL for stoichiometric reasons. Samples with higher concentrations of silicon dioxide (>70%) may require higher concentrations of hydrofluoric acid (>3 mL). Alternatively samples with lower concentrations of silicon dioxide (<10% to 0%) may require much less hydrofluoric acid (0.5 mL to 0 mL). Examples are presented in Tables 1, 2, 3, and 6.

Nitric acid is the primary acid used because of several of its attributes. Most nitrates are soluble and the final solution will therefore be appropriate from the standpoint of solubility. Because of the solubility of nitrates, an excess of nitric acid does not precipitate metals, but concentrated solutions may become saturated and the addition of water may help with solubility. Nitric acid is an excellent oxidizing acid, especially under elevated temperature conditions because its oxidation potential increases with temperature. At 180 °C for a period of 5–10 min almost complete oxidation of organic-based matrices will occur. For example, biologicals such as oyster tissue and bovine liver are reduced to acid solutions with free metal ions appropriate for most spectroscopic measurements (Tables XV and XVI). The predominant organic species remaining are simple nitrobenzoic acids

Table XV. Analysis of NIST SRM 1566 Oyster Tissue

	Concentration (µg/g)	
Element	Certified Value ±95% Confidence Limits	Determined Concentration ±95% Confidence Limits
Cadmium	3.5 ± 0.4	4.39 ± 0.11
Copper	63 ± 3.5	61 ± 2
Manganese	17.5 ± 1.2	11.69 ± 0.41
Nickel	1.03 ± 0.19	1.22 ± 0.91
Lead	0.48 ± 0.04	—[a]
Zinc	852 ± 14	882 ± 17

NOTE: Tissue was digested with 9 mL HNO_3.
[a]Not detected.

Table XVI. Analysis of NIST SRM 1577a Bovine Liver

	Concentration (µg/g)	
Element	Certified Value ±95% Confidence Limits	Determined Concentration ±95% Confidence Limits
Cadmium	0.44 ± 0.06	0.43 ± 0.03
Copper	158 ± 7	165 ± 3
Manganese	9.9 ± 0.8	10.51 ± 0.11
Lead	0.14 ± 0.02	—[a]
Zinc	123 ± 8	123.7 ± 1.5

NOTE: Tissue was digested with 9 mL HNO_3.
[a]Not detected.

that do not complex metals, or interfere with plasma or highly energetic ionization sources (*see* the section on "Digestion Reaction Temperatures"). However, these trace organics will interfere with specific elements, such as Ni, in polarographic measurements and may cause matrix effects in analyses by graphite furnace (*56, 62*).

Under these conditions, nitric acid is applicable for the oxidation of most organic matrices. It will decompose most organic molecules, converting hydrocarbons into carbon dioxide and water, as previously described. Other reagents enhance the oxidizing potential of nitric acid, such as hydrogen peroxide, potassium iodide, perchlorate, permanganate, bromine, and sulfuric acid (*119, 120*). It also is effective for digesting most metals and alloys, with the exception of the noble metals and metals that passivate as oxides or hydroxides in pure form, such as high purity Al, Cr, Fe, and others (*see* Chapter 2). Volatile oxides that are formed in nitric acid include OsO_4, RuO_4, and Re_2O_7. Osmium tetroxide, as a neutral molecule, diffuses rapidly through Teflon vessels during decomposition (*121*). This compound can pass through Teflon and would not be totally recovered in this type of vessel. On reactions with nitric and other acids, other volatile neutral molecules can diffuse through Teflon such as the acid molecules themselves, carbon dioxide, and water. However, charged metal ions do not enter Teflon materials on digestion, even under pressure. Diagrams of nitric acid solubility, volatility, passivation, and other reactions are included in Chapter 2, and will be updated on SamplePrep Web.

At room temperature, nitric acid will cleave most organic covalent bonds. However, these reactions are greatly accelerated with temperatures above 120 °C, the boiling point of nitric acid. Temperatures significantly above the boiling point are only developed in closed pressurized containers and have been used in various configurations since the development of the sealed Carius tube nitric acid digestions in 1860 (*28, 34, 122, 123*). The completeness of the degradation of organic matter depends on the temperature of the decomposition. The optimum temperature conditions for nitric acid oxidation have been previously determined and discussed for both microwave and traditional pressurized closed vessels (*28, 30, 62, 116*). Nitric acid becomes a better oxidizer as it becomes hotter, and it completely oxidizes organic tissue bonds at 300 °C. At 180 °C, nitric acid will oxidize most organic molecular bonds within 10 min. The only well-documented exceptions are the specific bonds in aromatic rings (*56, 62*). As higher temperatures can be safely sustained, increasing the temperature will only improve the completeness of the decomposition and is an appropriate adaptation for nitric acid closed-vessel organic oxidation methods. The sustained reaction temperature of 180 °C in Method 3052 should be viewed as a minimum. It can be increased as matrix and analytes require, and as apparatus permits.

Some inorganic molecules are not attacked by nitric acid. For example, silicon dioxide, as well as many other metal oxides, is essentially inert in nitric acid at most concentrations. This limitation distinguishes Method 3051 from Method 3052, or a leach method from a total decomposition sample preparation method.

Hydrofluoric Acid

Hydrofluoric acid is primarily included in Method 3052 for the decomposition of constituents containing silicates. It was first applied for this purpose by Berzelius (*1*). Most soils are based on oxides, and the highest percentage are silicon-based. Silicon oxides are taken to silicon tetrafluoride by the action of hydrofluoric acid, as has been well documented (*124, 125*). A few specific silicates (Coesite, Stishovite, polymorphous silicon dioxide, and high aluminum silicates, especially topaz) are not dissolved by hydrofluoric acid (*119*). Twenty-one of the 28 common mineral silicates are decomposed by hydrofluoric acid at 105–110 °C. The remaining silicates can be decomposed by using elevated temperature and pressure bomb techniques (*126, 127*).

Hydrofluoric acid is a weak, nonoxidizing acid whose characteristics depend primarily on its complexing properties. Fluoride is a powerful complexing anion that forms many soluble metal complexes, especially with the difficult-to-stabilize refractory elements such as Cr, Zr, Hf, V, etc. (for a comprehensive list of elements *see* Chapter 2). It is also a very hazardous acid to work with because it is odorless, colorless, does not produce pain or stain on contact with skin, and passes easily through the skin to readily attack bone. Thus, it should be handled very carefully. Boric acid solution should be available at all times for washing hands after using hydrofluoric acid.

Hydrofluoric acid does not take part in redox reactions, but fluoride ion complexation of elements can dramatically change their redox potentials. The use of hydrofluoric acid in combination with other acids therefore enhances the effect of these oxidizing acid capabilities (*119*). Its action can lead to a variety of insoluble transition and rare earth element compounds. The lanthanide and actinide elements form relatively insoluble fluorides that can precipitate. Removal of fluoride by complexation with boric acid is the most common method of resolubilizing these precipitates. On evaporation of the acid and concentration of the solution, precipitates containing Al^{3+}, Ca^{2+}, Fe^{3+}, Fe^{2+}, Mg^{2+}, Na^+, and K^+ may form. Mineral formation may cause the precipitation of insoluble mineral fluorides that are difficult to redissolve, such as $NaAlF_4$, $MgAlF_5$, $Fe(II)(Al,Fe(III))F_5$, $Na(Mg,Al)_6(F,OH)_{18} \cdot 3H_2O$ that have been previously identified (*120*). The composition of the mineral form depends on the composition of the soil, rock, or sediment. These compositions cannot easily be predicted or generalized (*120*). Elimination of excess hydrofluoric acid and use of boric acid complexation to resolubilize fluoride is discussed in a later section.

Hydrofluoric acid contribution to dissolution is not as temperature dependent as nitric acid. Its reactions are sometimes called "opening out" reactions for rocks and minerals, because the decomposition is usually fully oxidized materials and mineral forms. Many mineral forms seem to respond by solubilization in a vigorous combination of hot nitric and hydrofluoric acids, a result demonstrating a beneficial temperature preference. Closing the system and heating the acids with microwave energy increases the temperature of hydrofluoric and nitric acids to well above their boiling points, thus accelerating the reaction of hydrofluoric acid.

The combination of nitric and hydrofluoric acids yields total elemental recoveries of numerous elements from mixed origin sediment, SRM 2704 (Buffalo River Sediment), and soil, SRMs 2710 and 2711 (Highly and Moderately Contaminated Montana Soil) (*see* Tables XVII–XIX). Many of the most difficult elements to prepare for analysis are available from this single sample preparation procedure, including As, Cr, Hg, P, S, Se, Tl, and U. Specific soil and sediment reference materials have been provided as examples of plausible starting decomposition mixtures in Table XIII.

In addition to discrete matrix types (soil), other sample types (oils and botanicals, Tables XX and XXI) can be prepared using this single sample

Table XVII. Analysis of NIST SRM 2704 Buffalo River Sediment After Sample Preparation by Using EPA Method 3052

Element	Analyzed (µg/g ± SD)	Certified (µg/g ± SD)
Arsenic	23.4 ± 2.6	23.4 ± 0.8
Cadmium	3.5 ± 1.2	3.45 ± 0.22
Chromium	132.9 ± 1.3	135 ± 5
Copper	98.0 ± 4.2	98.6 ± 5.0
Lead	155 ± 9.2	161 ± 17
Mercury	1.49 ± 0.14	1.44 ± 0.07
Nickel	43.6 ± 3.9	44.1 ± 3.0
Phosphorus	1.016 ± 0.016 (mg/g)	0.998 ± 0.028 (mg/g)
Selenium	1.13 ± 0.9	(1.1)
Sulfur	3.56 ± 0.16	—
Thallium	1.15 ± 0.22	1.2 ± 0.2
Uranium	2.97 ± 0.04	3.13 ± 0.13
Zinc	441.9 ± 0.8	438 ± 12

NOTE: Samples digested with 9 mL HNO_3 and 4 mL HF.

Table XVIII. Analysis of NIST SRM 2710 Montana Soil After Sample Preparation by Using EPA Method 3052

Element	Analyzed (µg/g ± SD)	Certified (µg/g ± SD)
Antimony	39.3 ± 0.9	38.4 ± 3.0
Cadmium	21.9 ± 0.7	21.8 ± 0.2
Chromium	34.0 ± 3.2	(39)
Copper	2902 ± 83	2950 ± 130
Lead	5425 ± 251	5532 ± 80
Nickel	13.5 ± 1.0	14.3 ± 1.0
Silver	36.6 ± 0.5	35.3 ± 1.5
Zinc	7007 ± 111	6952 ± 91

NOTE: Samples digested with either 9 mL HNO_3 and 4 mL HF or 9 mL HNO_3, 3 mL HF, and 2 mL HCl.

Table XIX. Analysis of NIST SRM 2711 Montana Soil After Sample Preparation by Using EPA Method 3052

Element	Analyzed (µg/g ± SD)	Certified (µg/g ± SD)
Cadmium	40.5 ± 1.0	41.70 ± 0.25
Chromium	45.5 ± 1.0	(47)
Copper	106.8 ± 3.4	114 ± 2
Lead	1161 ± 49	1162 ± 31
Nickel	19.6 ± 0.9	20.6 ± 1.1
Silver	4.3 ± 1.0	4.63 ± 0.39
Zinc	342 ± 9.4	350.4 ± 4.8

NOTE: Samples digested with 9 mL HNO_3 and 4 mL HF.

Table XX. Analysis of NIST SRM 1084a Wear Metals in Oil (100 ppm)

Element	Analyzed (µg/g ± SD)	Certified µg/g ± SD)
Chromium	98.1 ± 1.1	98.3 ± 0.8
Copper	102.4 ± 2.4	100.0 ± 1.9
Lead	99.2 ± 2.3	101.1 ± 1.3
Nickel	99.2 ± 2.4	99.7 ± 1.6
Silver	102.7 ± 2.2	101.4 ± 1.5

NOTE: Samples digested with 9 mL HNO_3 and 0.5 mL HF.

Table XXI. Analysis of NIST SRM 1572 Citrus Leaves

	Concentration (µg/g)	
Element	Certified Value ±95% Confidence Limits	Determined Concentration ±95% Confidence Limits
Cadmium	0.03 ± 0.01	0.019 ± 0.004
Copper	16.5 ± 1.0	14.8 ± 1.0
Manganese	23 ± 2	23.4 ± 0.5
Nickel	0.6 ± 0.3	1.41 ± 0.15
Lead	13.3 ± 2.4	14.0 ± 0.3
Zinc	29 ± 2	23.8 ± 1.4

NOTE: Leaves were digested with 9 mL HNO_3 and 0.5 mL of HF.

preparation procedure, as can any combination of these three. As previously described, small quantities of hydrofluoric acid are required for minor silacial components for preparation of some plant materials (*88*).

Excessive dilution of nitric acid, the principle oxidizing acid, with hydrofluoric acid could result in incomplete digestion of the organic fraction of the matrix. Because of the insolubility of fluorides, a large excess of hydrofluoric acid is not desirable. The minimum amount of hydrofluoric acid should be used. The quantity of reagent is governed by stoichiometry and the quantity of silicon dioxide and other oxides requiring hydrofluoric acid. Four hydrofluoric molecules are required for each silicon atom with a suggested excess of 50–100%. The reaction with silicon dioxide produces fluorosilicic acid (or silicon fluoride), which dissociates into volatile silicon fluoride and hydrofluoric acid in the following reactions.

$$SiO_2 + 6HF \longrightarrow H_2SiF_6 + 2H_2O \qquad (10)$$

$$H_2SiF_6 \xrightarrow{\Delta} SiF_4(g) + 2HF \qquad (11)$$

The reaction is thought to be a two step mechanism. A mechanism for the HF attack of glass surfaces involves water, where hydration of the surface is the first step:

$$SiO_2 + 2H_2O \longrightarrow Si(OH)_4 \qquad (12)$$

Followed by the formation of a compound that is soluble in HF:

$$Si(OH)_4 + 6HF \longrightarrow H_2SiF_6 + 4H_2O \qquad (13)$$

These mechanisms are influenced by surface area and temperature because they are limited by the rate of the transport of the $Si(OH)_4$ from the sand grains and glass surfaces. They provide stoichiometric examples that aid in determining the minimum amount of hydrofluoric acid to add, depending on the silicon dioxide or glass content of the matrix. HF–H_2SiF_6–H_2O form a constant boiling ternary mixture containing 10% HF, 36% H_2SiF_6, and 54% H_2O that has a boiling point of 112 °C (*128*).

Excess nitric and hydrofluoric acids may be eliminated later in a procedure. Hydrofluoric acid is poisonous; fumes and liquid must not come in contact with laboratory personnel. HF has some unique and very serious consequences if it comes in contact with skin or pulmonary tissue, which should be absolutely avoided. Washing with boric acid after handling containers that are contaminated with hydrofluoric acid, or its fumes, is a common precaution practiced in many laboratories. Boric acid complexes fluoride and allows it to be washed from the skin.

Hydrochloric Acid

7.3.3 (Continued)

The addition of 2 ± 2 mL concentrated hydrochloric acid to the nitric and hydrofluoric acids is appropriate for the stabilization of Ag, Ba, and Sb and high concentrations of Fe and Al in solution. The amount of HCl needed will vary depending on the matrix and the concentration of the analytes. The addition of hydrochloric acid may, however, limit the techniques or increase the difficulties of analysis. Examples are presented in Table 4.

Hydrochloric acid is necessary to attack certain elements, for the stability of specific elements (such as Sb and Ag), and to assist in the solubilization of other elements (such as Fe and Al) by complexation. Silver, with minimal quantities of chlorides, will precipitate with typical concentrations of chlorides in soil and tissue matrices. However, with greater than 3 M HCl, a complex of silver chloride forms that resolubilizes and stabilizes silver (*119, 129*).

$$Ag^+_{(aq)} + Cl^-_{(aq)} \longrightarrow AgCl_{(s)} \qquad (14)$$

$$AgCl_{(s)} + HCl_{(aq)} \longrightarrow AgCl^-_{2(aq)} + H^+ \qquad (15)$$

Up to 70% antimony can be lost in single oxidizing acids, such as nitric acid. This is due to the formation of antimony oxides and mixed oxides which adsorb onto vessel and silicate surfaces (*130*). The proposed compound, Sb_2O_5, forms during the nitric acid leach of soils and adsorbs on to siliceous material and precipitates from oxidizing acid solutions. However, the addition of HCl forms the complexed $SbCl_6^-$ anion that does not adsorb and is stable in solution. Hydrochloric acid, up to a 50:50 ratio, has been proposed to stabilize Sb in soil solutions (*131, 132*). Acid mixtures of nitric, hydrofluoric, and hydrochloric acid have been shown to be effective at recovering Sb in soil samples. Experiments described in Table XXII showed low recovery of Sb and Ag when using the current suggested acid starting conditions of nitric and hydrofluoric acids. As can be seen, the addition of hydrochloric acid solubilized and stabilized both Sb and Ag in solution, and provided total recoveries for both standard soils tested (SRMs 2710 and 2711). These examples of stabilization by complexation, and resolubilization of a precipitate through complexation, demonstrate the effects of the addition of a complexing acid, such as hydrochloric acid.

In testing the method, the addition of hydrochloric acid to nitric and hydrofluoric acids did not result in the loss of any of the elements tested while closed-vessel microwave technology was employed. Hydrochloric acid forms strong complexes with many ions, stabilizes them in solution, and prevents them from adsorbing onto particulate matter or vessel walls. The following ions are stabilized as strong chloro complexes: Au(III), Fe(III), Ga(III), Hg(II), In(III), Sb, Sn(IV), Tl(III). Compilation of these and other effects for the mineral acids is presented in Chapter 2.

Some metallic elements, such as Au, Pd, Pt, Pd, and Rh, are only attacked by a combination of nitric and hydrochloric acids. This mechanism involves the formation of nitrosyl chloride (NOCl), a third reagent. The ratio of nitric to hydrochloric acid will determine the extent of formation of this reagent. Nitrosyl chloride is the main ingredient in aqua regia and is formed in the following manner:

$$3HCl + HNO_3 \longrightarrow NOCl + Cl_2 + 2H_2O \tag{16}$$

Table XXII. Stabilization and Recovery of Elements with HCl from NIST SRM 2710 Montana Soil

Element	HNO_3 and HF ($\mu g/g$)	HNO_3, HF and HCl ($\mu g/g$)	Certified ($\mu g/g$)
Antimony	33.1 ± 2.1	39.3 ± 0.9	38.4 ± 3.0
Silver	10.6 ± 4.5	36.6 ± 0.5	35.3 ± 1.5

NOTES: HNO_3 and HF digestion was performed by using 9 mL and 3 mL, respectively. HNO_3, HF, and HCl digestion was performed by using 9 mL, 3 mL, and 2 mL, respectively.

This reaction is sufficiently slow to permit the use of aqua regia in the microwave closed vessels if the pure reagents of nitric and hydrochloric acids are added directly to the microwave vessel and not premixed. Premixing the reagents significantly increases the quantities of chlorine gas and could create over-pressurization of the vessel because the gas is liberated from solution during heating. The partial pressure of chlorine gas is added to those of the acids and digestion products; excess chlorine gas should be avoided.

Mixtures of nitric and hydrochloric acids and the specific ratio producing classic aqua regia are good solvents for many ores, alloys, and special steels (*120, 133, 134*). A strict definition of aqua regia is a 3:1 ratio of hydrochloric and nitric acids, but sufficient quantities of NOCl are formed to permit the solubilization of precious metals and some species in other ratios.

Even though a vast number of elements form relatively volatile species, closed-vessel microwave technology minimizes or eliminates loss through volatilization. The decomposition occurs at relatively high temperatures, but the vessels are opened at lower temperature. Vessels are opened when solutions are below 100 °C and compounds are not volatile.

Other Reagents

7.3.3 The addition of other reagents with the original acids prior to digestion may permit more complete oxidation of organic sample constituents, address specific decomposition chemistry requirements, or address specific elemental stability and solubility problems.

The potential to add complementary reagents that can assist with some needed or specific reactions is appropriate to permit the flexibility to handle additional matrix and analyte requirements. Although the method is robust and appropriate, it requires adaptability that this section provides. Actually, with the diversity of reagents already specified, a host of secondary reagents already exists. For example, the combination of nitric acid and hydrochloric acid produces nitrosyl chloride (NOCl) in situ. Adding hydrogen peroxide in the presence of sulfuric acid produces monoperoxosulphuric acid (H_2SO_5) (*122*). Free radicals, free oxygen, chlorine, and other species all play complex roles in decomposition. Other reagents may be beneficial and necessary to produce specific chemistry and aid in decomposition. Other reagents that are added should be evaluated for compatibility with the current reagents and with the intended analytical technique. Because the method is a total decomposition method, the performance criteria is that the method should produce a stable solution for analysis of the total analytes in the matrix of interest. Additional reagents are appropriate as long as they are safe and assist with this goal. Reagents that have already been identified will be evaluated for their specific use and potential in this method.

7.3.3 continued

The addition of hydrogen peroxide (30%) in small or catalytic quantities (such as 0.1 to 2 mL) may aid in the complete oxidation of organic matter.

Hydrogen peroxide, H_2O_2, is an excellent complement to acid oxidation because its oxidizing potential, or redox potential, increases with acidity. It is available in relatively adequate purity as a 30% w/w solution and is also obtainable as a 50% w/w solution. It decomposes autocatalytically, producing heat, water, and molecular oxygen. Its decomposition is also catalyzed by heavy metal ions and metal oxides, such as Fe^{3+}, Cu^{2+}, MoO_4^{2-}, Ce^{4+}, CrO_4^{2-}, WO_4^{2-}, VO_4^{3-}, MnO_2, and others (*122*).

The oxidation of organic matter using hydrogen peroxide has been well-represented in the literature for over a century. Two of the most popular peroxide combinations are with sulfuric acid, or as Fenton's reagent. Sulfuric acid has typically been in an open glass vessel with increments of hydrogen peroxide added slowly and carefully. The formation of H_2SO_5 is thought to be an effective component in this combination. Fenton's reagent, discovered in 1894, is a mixture of hydrogen peroxide and Fe^{2+} (catalytic). Both combinations, with sulfuric acid or iron catalyst, are used for digestion of large quantities of organic material and tissue, but are done in open-vessel systems with frequent addition of fresh hydrogen peroxide. In closed-vessel decompositions, smaller quantities of these reagents must be used because of the tremendous quantities of molecular oxygen produced. Hydrogen peroxide is not as effective in these small quantities as in the traditional open vessels mentioned in the literature where large quantities of hydrogen peroxide are continuously added. It is, however, effective in assisting nitric, hydrochloric, and hydrofluoric acid to decompose many components in complex matrices. Hydrogen peroxide, in combination with mineral acids such as nitric, hydrochloric, and hydrofluoric for the decomposition of organic matter, should be permitted to prereact in the vessel before final sealing and digestion. This prereaction will permit excess molecular oxygen and carbon dioxide to escape prior to the acceleration of the reaction on heating (*122*).

Many metals and metal alloys are soluble in hydrogen peroxide, but most metals are dissolved in an acid with this reagent. Steel, brazing metals, nonferrous metals, metallic molybdenum, tungsten, and rhenium are dissolved at moderate temperatures with hydrogen peroxide (*122*). Combinations of nitric acid and peroxide are effective at dissolving steel, Cu, Cd, Pb, Zn, calcium tungstate, and pyrites. Hydrogen peroxide with complexing acids such as hydrochloric and hydrofluoric acids is also effective in the dissolution of iron, steel, ferrotungsten, and silicon alloys.

A number of acid and hydrogen peroxide combinations in Teflon-lined pressure digestion bombs are reported in the literature (*122*). These acids are hydrogen peroxide combinations that provide effective solvents for metals and molecular species. Hydrogen peroxide is an effective reagent and may assist in providing a reagent solvent that is more effective than acids alone in the solubilization of many metals, alloys, and molecular species. Caution must be taken to prevent detrimental effects from molecular oxygen released by this reagent.

7.3.3 continued

The addition of 0 to 5 mL water (double deionized) may improve the solubility of minerals and prevent temperature spikes due to exothermic reactions.

The addition of water to the reaction mixture has several benefits. It assists in solubilizing ions and molecules that are solubility limited because of the concentration that has been reached in the solution after release from the matrix. This result may also be observed on dilution following initial dissolution. Solubility should be evaluated depending on the matrix components. Frequently, the insolubility of a compound may be solved by adding water after dissolution. During dissolution, however, water may also assist with solubilizing and preventing mineral formation from concentrated solutions.

Adding water also prevents temperature and pressure spiking in rapid closed-vessel decompositions (*135*). As documented, some reactions are temperature dependent and the decomposition is sudden at specific temperatures. The sudden temperature spike increases the efficiency of the decomposition and, if it is a gaseous digestion, may also generate a pressure spike. The addition of water moderates these reactions and reduces these spikes in both pressure and temperature. Water acts as a thermal and pressure shock absorber and is very effective at reducing sudden over-pressurization in closed-vessel acid decomposition of organic material.

Reaction Conditions, Control, and Environment

7.3.6 This method is a performance-based method, designed to achieve or approach total decomposition of the sample through achieving specific reaction conditions. The temperature of each sample should rise to 180 ± 5 °C in approximately 5.5 min and remain at 180 ± 5 °C for 9.5 min. The temperature–time and pressure–time profiles are given for a standard soil sample in Figure 1. The number of samples simultaneously digested is dependent on the analyst. The number may range from 1 to the maximum number of vessels that the microwave unit's magnetron can heat according to the manufacturer or literature specifications (the number will depend on the power of the unit, the quantity and combination of reagents, and the heat loss from the vessels).

7.3.6.1 For reactive substances, the heating profile may be altered for safety purposes. The decomposition is primarily controlled by maintaining the reagents at 180 ± 5 °C for 9.5 min; therefore, the time it takes to heat the samples to 180 ± 5 °C is not critical. The samples may be heated at a slower rate to prevent potential uncontrollable exothermic reactions. The time to reach 180 ± 5 °C may be increased to 10 min provided that 180 ± 5 °C is subsequently maintained for 9.5 min. Decomposition profiles are presented in Figures 1 & 2. The extreme difference in pressure is due to the gaseous digestion products.

Temperature is an effective tool for the acceleration of acid decomposition of most matrix components. Certain organic bond oxidation reactions occur in nitric acid at specific temperatures (*3, 116*). Additional examples are also presented in this chapter. The efficient and rapid reaction of nitric acid at specific temperatures indicates that temperature is necessary to rapidly decompose matrix components, such as organic material. A temperature of 180 °C, or higher, must be maintained to allow enough time for these reactions to occur and be completed.

The amount of time necessary to reach these temperatures is not critical, but has been suggested to permit efficient average microwave equipment compliance and relative efficiency and safety. The performance of the decomposition is the most significant factor and it may be necessary to alter the time necessary to reach this temperature. For example, when decomposing an automotive oil, it is appropriate to extend the temperature ramp time to approximately 10 min. This extension prevents pressure and temperature spikes resulting from reactions that are very exothermic and pressure intensive. Figure 27 demonstrates this alteration to optimize both safety and efficiency in this difficult reaction. It also demonstrates the high pressures generated from the decomposition of the oil, producing carbon dioxide, which must be contained in such reactions. As can be seen, the vessel is pressurized to over 80% of its required minimum pressure capacity. On the basis of their construction, different vessels have varying capacities to contain pressure and will generate different temperatures depending on their heat loss characteristics.

At present, two polymer classes are being used to construct microwave vessels: fluoropolymers for reaction chamber liners, and more rigid polymers and composites for casing materials and associated outer containment structures. In general, the fluoropolymers have upper-use temperatures in the 250 °C range, whereas the casement materials, such as polyetherimide, have lower use temperatures. These thermal constraints are also a reason to provide minimum and target temperatures for material compatibility and safety. Whereas higher temperatures are more efficient at oxidizing organic material and performing decomposition reactions, practical temperature and pressure limits exist based on materials currently used for vessel construction which must have both solvent resistance and certain microwave absorption characteristics. Temperatures and pressures of 300 °C and 200 atm, respectively, are now available in laboratory microwave vessels. A table of current manufacturer specifications can be accessed on the SamplePrep Web site (*see* Chapter 15).

These temperature profiles suggest a practical and experimentally tested optimum/minimum for reaction performance and material stability. These temperatures and the time required to reach them are not as restrictive as in leaching methods because of the objective of total decomposition versus reproducible leaching of sample materials. These temperatures are important in that certain reactions cannot occur below specific temperatures. The oxidizing potential of nitric acid and hydrogen peroxide is increased to critical values at specific temperatures that are necessary to perform oxidation of specific organic bonds, or produce specific elemental oxidation states. The reactions will not progress appropriately if the specified minimum reaction temperature, 180 °C, is not reached (*116*). The maintenance of this specified temperature is necessary to render the matrix decomposition complete to the point that the analytes of interest exist as free ions in solution. Even though some decomposition will occur at lower temperatures, it will be significantly slower and often causes matrix components to interfere with some instrumental detection of various analytes. For most organic and inorganic compounds, specific molecular components will remain in solution but will not significantly affect instrument performance (*56, 62, 111*).

Beyond the theoretical considerations, it appears from many experiments and evaluations on a wide variety of samples that the recommended temperature profile, mineral acids, and reagents represent a practical optimum for general decomposition. This protocol and specific conditions accomplish the decomposition of the widest variety of environmental and generally desired materials.

7.3.6.2 Calibration control is applicable in reproducing this method, provided the power in watts versus time parameters are determined to reproduce the specifications listed in 7.3.6. The calibration settings will be specific to the quantity and combination of reagents, quantity of vessels, and heat loss characteristics of the vessels (Reference 1). If calibration control is being used, any vessels containing acids for analytical blank purposes are counted as sample vessels; when fewer than the recommended number of samples are to be digested, the remaining vessels should be filled with the same acid mixture to achieve the full complement of vessels. This provides an energy balance, since the microwave power absorbed is proportional to the total absorbed mass in the cavity (Reference 1). Irradiate each group of vessels using the predetermined calibration settings. (Different vessel types should not be mixed.)

Even though temperature feedback control is the primary reaction control mechanism, it is appropriate to permit calibration control that has been used previously to continue to provide a low cost option for some repetitive analyses. Calibration control incorporates the calibration of the microwave field intensity, microwave energy absorption for the reagents, and the heat loss characteristics of the particular vessel used. Although the microwave field intensity is the only parameter calibrated by the analyst for each microwave unit, the other parameters are necessary to achieve reproducible conditions for a digestion protocol. These components permit conditions to be predicted, allowing the field, vessel, and reagent combination that will produce the temperature profile required in the method to be used in combination. These calibration procedures, methods, and philosophies have previously been described to control this particular type of sample preparation reaction (3, 31, 34, 73). Calibration will normalize the microwave field. This microwave field can be reproduced in any microwave cavity by using 2450 MHz frequency and the identical vessel and reagent configuration.

The microwave power settings should be set by using temperature measurement equipment and the same type and model vessels being used in the calibration-controlled system. This process may involve another microwave system with temperature measurement capability in the same laboratory, or may be provided by the manufacturer or an independent laboratory by using the published vessel specifications. The microwave unit must be calibrated to evaluate the correlation between the power output settings and the power delivered to the cavity. Multimode or cavity-type microwave units are typically nominally power rated by the manufacturer, and may be 100–200 W different in actual power delivered to the cavity compared with their nominal power rating. Microwave energy exposes all of the mass of reagent and absorbing material in the cavity simultaneously; it is the total amount of absorber and the total amount of heat loss from all vessels that must be measured before calibration control is effective. Therefore, the vessels should contain as similar a combination of absorbing solvent and sample as possible. If these conditions are met, reliable and reproducible analytical sample preparation can be performed by using this method of control.

Once established, this method of control is the least expensive and requires less equipment than any method established for the use of microwave equipment on a routine basis. It will not, however, readily permit development of new acid combinations or a varying number of vessels until measurement of energy transfer and temperature conditions are evaluated and new field strength conditions are provided for the specific reagent, volume, vessel type, and number of vessels.

An advantage of calibration control is that it does permit the use of microwave sample preparation for decompositions in many laboratories with older equipment that may perform repetitive decompositions of soils, sludges, oils, or limited combinations of defined and characterized matrix samples. For organic-based matrices, such as oils, calibration control may require evaluation of the exothermic contributions of the reaction to the reaction conditions to accurately reproduce the method's performance criteria. Without this provision, new equipment would be required for many of the existing laboratories that made an early investment in microwave technology, because calibration was the primary control method of first-generation commercial laboratory microwave equipment.

Digest Matrix Modification

7.3.10 The removal or reduction of the quantity of the hydrochloric and hydrofluoric acids prior to analysis may be desirable. The chemistry and volatility of the analytes of interest should be considered and evaluated when using this alternative. Evaporation to near dryness in a controlled environment with controlled pure gas and neutralizing and collection of exhaust interactions is an alternative where appropriate. This manipulation may be performed in the microwave system, if the system is capable of this function, or external to the microwave system in more common apparatus. This option must be tested and validated to determine analyte retention and loss and should be accompanied by equipment validation possibly using the standard addition method and standard reference materials. This alternative may be used to alter either the acid concentration or acid composition. Note: The final solution typically requires nitric acid to maintain appropriate sample solution acidity and stability of the elements. Commonly, a 2% (v/v) nitric acid concentration is desirable. Examples of analysis performed with and without removal of the hydrofluoric acid are presented in Table 5. Waste minimization techniques should be used to capture reagent fumes. This procedure should be tested and validated in the apparatus and on standards before using on unknown samples.

Both hydrofluoric and hydrochloric acids may interfere with analytical methods of analysis and detection, or equipment with quartz or glass components. Both of these acids may be removed by mild heating to drive off these reagents. In testing this option with microwave-assisted reagent evaporation, we experienced no losses among the elements tested. Chapter 2 describes the literature references of volatile species where solutions were taken to dryness, or near dryness, by using conventional heating.

The absence of volatilization problems may also be partially due to the relatively cool and controlled conditions created in the microwave evaporation system. Conventional evaporation has been known to cause losses due to volatilization of halide salts and other volatile species by using conventional heating methods that overheat the vessel and solution. Other physical losses may occur when boiling produces micro-fine aerosol particles that are randomly transported by Brownian movement and lost from the vessel. Active boiling is prevented in microwave evaporation; the temperature is controlled and vapor-phase transport is increased through engineered vessel porting. When these solutions are heated by microwave energy, only the acid vapor is removed by molecular excitation, and a gas sweep or vacuum assist is used.

In these tests, the microwave vessel liner was also used as the evaporation container. By using the microwave system, sample drying, decomposition, evaporation, and reconstitution were accomplished in the same vessel liner (*see* Figure 29). The samples were microwave heated under mild temperature conditions (below the boiling point), while the reagents were recovered and trapped in a neutralization system to minimize reagent loss to the atmosphere. The acids were automatically neutralized

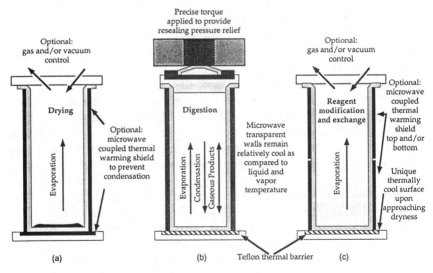

Figure 29. Total microwave sample processing in a single vessel.

to sodium nitrate, chloride, or fluoride solutions, thus lowering their polluting effects in the environment and minimizing contaminated waste. Figure 29 also illustrates the total microwave chemistry processing concept.

Recall that a unique situation is created under microwave-induced heating where the vessel remains cooler than the solution. Direct molecular motion is induced while physical transport out of the vessel is controlled. Overheating the digest solution is avoided during evaporation. Microwave drying selectively evaporates free water while leaving water of hydration still bound to sample species.

In experimentation with optional thermal induction at controlled rate and location, unique advantages of microwave solution manipulation are again dependent on the direct induced and controlled thermal interactions. Deliberate manipulation creating specific cool vessel floors while evaporating directly excited reagent molecules is a unique physical phenomenon related to microwave-induced heating and molecular transport from thermally transparent polymers. Precise thermal induction used in conjunction with these fluoropolymer vessels and controlled energy absorption permit evaporation and solvent exchange unique for analytical chemical manipulation. This phenomenon again takes advantage of the unique conditions that are created in the microwave vessel, as described earlier in this chapter. As a new tool, it may significantly reduce many traditional elemental losses due to hot vessel surfaces, overheating of residues, and thermally induced molecular losses found with conventional heating.

The analytical approach that uses microwave energy for multiple preparation steps, including drying, digestion or leaching, and evaporation is called total microwave processing (46). To reduce contamination while improving reproducibility and control of the sample preparation steps, microwave energy is used. By using a single TFM vessel liner, the undried sample is weighed. The sample is heated while moisture and vapors are evacuated from the vessel. The samples are dry with only moderate microwave heating in 10–20 min. Once cool, the vessels are reweighed to determine the percent moisture lost. Then, under a clean hood, the digestion acids are added and the vessels sealed and digested. Once digested, the vessels are vented and placed into the evaporation rotor. By using microwave energy, the samples are gently heated while a vacuum (or gas) removes the vapors to a neutralization system. No significant difference in analyte concentrations are seen (Table XXIII). This series of analytical sample preparation steps illustrates an entire sample preparation using a single vessel to minimize contamination and a controlled clean environment throughout the process, ensuring clean chemistry.

During these procedures, the samples were under controlled atmospheric conditions and not exposed to normal laboratory air. Class-100 clean chemistry conditions were used to prepare samples, add reagents, and control gaseous input and output from the system. A small class-100

Table XXIII. NIST SRM 2710 Montana Soil: Highly Elevated
Trace Element Concentrations

Element	Direct Analysis ($\mu g/g \pm SD$)	After Fuming ($\mu g/g \pm SD$)	Certified ($\mu g/g \pm SD$)
Antimony	39.3 ± 0.9	39.4 ± 0.9	38.4 ± 3.0
Cadmium	21.9 ± 0.7	23.3 ± 1.6	21.8 ± 0.2
Chromium	34.0 ± 3.2	32.4 ± 0.4	(39)
Copper	2902 ± 83	2870 ± 150	2950 ± 130
Lead	5425 ± 251	5502 ± 106	5532 ± 80
Nickel	13.5 ± 1.0	13.5 ± 0.8	14.3 ± 1.0
Silver	36.6 ± 0.5	38.9 ± 1.1	35.3 ± 1.5
Zinc	7007 ± 111	6992 ± 132	6952 ± 91

NOTE: For direct analysis, samples were digested by using 9 mL HNO_3
and 3 mL HCl or 9 mL HNO_3, 3 mL HF, and 2 mL HCl. For fumed analysis, samples were digested by using 9 mL HNO_3 and 3 mL HCl followed by the removal of the HF. The digest solution was fumed in a
microwave system under vacuum to ~1 mL and 3 mL HCl were added.
The digest solution was fumed to ~1 mL and 3 mL HNO_3 added. The
solution was fumed for a final step to ~1 mL and quantitatively transferred and diluted to final volume.

clean air unit is sufficient for preventing contamination during vessel
transfer, reagent addition, and manipulation during sample preparation.

With removal of specific acids, such as hydrofluoric acid, the insoluble
salts that are created during digestion may be resolubilized. By avoiding the
addition of boric acid, boron may be analyzed. Additionally, the acid composition of the digest solution may be adjusted for optimal solubility and complexation of analytes. These manipulations are all acceptable alternatives,
including the addition of boric acid which is discussed later in this section.

Sample Size Adaptation

7.3.12 Sample size may be scaled up from 0.1, 0.25, or 0.5 g to 1.0 g
through a series of 0.2 g sample size increments. Scaleup can produce different reaction conditions and/or produce increasing gaseous reaction
products. Increases in sample size may not require alteration of the acid
quantity or combination, but other reagents may be added to permit a
more complete decomposition and oxidation of organic and other sample
constituents where necessary (such as increasing the HF for the complete
destruction of silicates). Each step of the scaleup must demonstrate safe
operation before continuing.

The method has been tested with a wide range of sample sizes and
has produced excellent results for elements by using the basic reagent volumes of nitric and hydrofluoric acids for 0.25 g to 2 g of soil. An example

of sample size performance is presented in Figure 30 where many sample sizes for SRM 2710 have been used to illustrate the robustness of the conditions for a single element. For soil, sludge, sediment, ash, and dust samples, similar results may be expected. If the concentration of metal ions becomes too high, mineral formation may be initiated due to supersaturation and high concentrations of elements relative to the reagents. In general, stoichiometry or reagent limits, mineral formation, and residual carbon dioxide pressure from organic oxidation will be limiting factors. For example, either exceeding the 6:1 ratio for HF to silicon dioxide required, or exceeding the 30 atm of residual pressure containment will impose a limit on the size sample that can be decomposed in systems that meet the minimum requirements.

Scaling up should be done in small, controlled increments. Increasing the sample size by 0.2 g increments is suggested and will permit the sample to be evaluated without over-stressing the equipment and generally exhausting a limiting reagent in a single stage. Section 7.3.1 also limits the suggested sample size for oil and organic samples to 0.25 g. These quantities will cause most vessels to approach the 30 atm pressure limit as the carbon in organic samples is converted almost entirely to carbon dioxide. This result does not limit the number of discrete organic samples that may be combined from independent digestions to achieve a gram or more from individual 0.25 g increments. This combination of samples would be one way of increasing sample size beyond the 0.25 g specification if detection limits are of concern due to the level of analyte or the instrumental method used for analysis. As can be seen when using ICP–MS analysis, a 0.25 g sample is sufficient to provide detection of most of the elemental

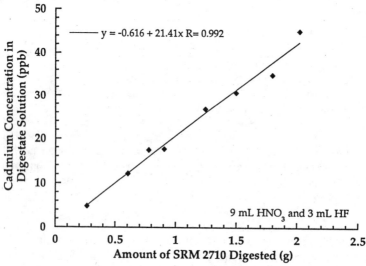

Figure 30. Increasing the size of sample.

constituents from even low levels in tissue SRMs. An alternative for large organic samples would be the use of high-pressure vessels, typically with >100 atm capacity. These vessels may be capable of up to 1 g of organic material. Additionally, for carbonate-based or easily digested organic samples, it may be advisable to let them react at room temperature and outgas carbon dioxide before they are sealed. This procedure can greatly reduce the digestion pressure, but should be checked to see if additional loss of volatile elements or sample contamination has occurred.

Scaling up the reaction should also be performed cautiously because exothermic reactions may actually increase the temperature beyond $180 \pm 5\,°C$ without additional microwave heating. Exothermic reactions that accelerate the reaction beyond the control of the equipment have been measured and reported, causing runaway reactions. The excess energy produced by the reaction is responsible for these reactions. This problem can be directly related to the sample size necessary to initiate such a reaction (3). Opening out reactions of geological materials are less prone to produce these conditions and are more easily scaled up. However, if unknown sample matrices are being handled, caution should be taken, understanding that an exothermic reaction may occur.

Refractory Components

7.3.8 Complete the preparation of the sample by carefully uncapping and venting each vessel in a fume hood. Vent the vessels using the procedure recommended by the vessel manufacturer. Transfer the sample to an acid-cleaned bottle. If the digested sample contains particulates which may clog nebulizers or interfere with injection of the sample into the instrument, the sample may be centrifuged, allowed to settle, or filtered.

7.3.8.1 Centrifugation: Centrifugation at 2,000–3,000 rpm for 10 min is usually sufficient to clear the supernatant.

7.3.8.2 Settling: If undissolved material remains such as TiO_2, or other refractory oxides, allow the sample to stand until the supernatant is clear. Allowing a sample to stand overnight will usually accomplish this. If it does not, centrifuge or filter the sample.

7.3.8.3 Filtering: If necessary, the filtering apparatus must be thoroughly cleaned and prerinsed with dilute (approximately 10% v/v) nitric acid. Filter the sample through qualitative filter paper into a second acid-cleaned container.

As section 7.3.8.2 indicates, there are refractory elements that may be present in samples in high concentration that cannot be solubilized. A refractory is defined in the American Heritage Dictionary as "...obstinately resistant to authority or control...difficult to melt or work; resistant to heat. . .". Over many years of experience certifying the concentration of elements in real matrices at NIST, we have seen refractory materials continually appear in samples, such as SRM 1648 Urban Particulate (dust which is several percent titanium dioxide). Even though this method comes the closest to being a robust universal method for most matrices, certain matrix components still pose problems. Additional manipulation may be required to isolate and solubilize these compounds, but it is not required by the method.

Frequently, the refractory is a single high purity oxide or element, such as TiO_2 (rutile), Al_2O_3 (corundum), or C (graphite), and the crystalline form of the refractory is important. If these compounds are pure and do not adsorb analyte material, they may be dealt with by centrifugation or filtration. The remaining solution may be analyzed and these elemental compounds will not interfere with the detection of most of the trace elements in solution.

Boric Acid

7.3.9 If the hydrofluoric acid concentration is a consideration in the analysis technique such as with ICP methods, boric acid may be added to permit the complexation of fluoride to protect the quartz plasma torch. The amount of acid added may be varied, depending on the equipment and the analysis procedure. If this option is used, alterations in the measurement procedure to adjust for the boric acid and any bias it may cause are necessary. This addition will prevent the measurement of boron as one of the elemental constituents in the sample. Alternatively, a hydrofluoric acid resistant ICP torch may be used and the addition of boric acid would be unnecessary for this analytical configuration. All major manufacturers have hydrofluoric resistant components available for the analysis of solutions containing hydrofluoric acid.

Most analytical instrument manufacturers, when contacted about the tolerance of their equipment to hydrofluoric acid concentrations, suggested that their standard equipment would handle a substantial amount, or that they have, for example, an alternative hydrofluoric acid-resistant (1–10% HF) configuration of ICP torch and nebulizer. Therefore, it may not be necessary to complex or remove the hydrofluoric acid to accommodate equipment. Solutions containing substantial quantities of this acid may be analyzed directly.

Boric acid reacts with HF according to the following equations:

$$H_3BO_3 + HF \longrightarrow HBF_3(OH) + 2H_2O \tag{17}$$

$$HBF_3(OH) + HF \longrightarrow HBF_4 + H_2O \tag{18}$$

The rate-determining step is the formation of fluoroboric acid (*136*). To overcome the reaction's slow kinetics, the solution is chilled in an ice bath, or, in most cases, the HF solution is treated with a vast excess (10 times) of boric acid.

Boric acid complexation of hydrofluoric acid is also a very traditional way of dealing with excess hydrofluoric acid (*4*). When we asked many prominent analytical chemists about specific ratios of boric acid to hydrofluoric, we received suggestions varying by over an order of magnitude. This method has a wide range of favorite working possibilities and personal preferences. Because a scientific rationale for restricting the ratio of boric acid could not be found to satisfy all conditions, analysts are left to choose their favorite because many different ratios seem to work well.

An example of how boric acid can be used to resolubilize many rare earth elements and complex excess hydrofluoric acid in a complex matrix has been presented in an early version of this method (*137*). The application was for the dissolution and solubilization of simulated and real nuclear waste material, for the analysis of half of the periodic table. The method used nitric, hydrofluoric, hydrochloric, and boric acids in separate steps as described. The method protocol is similar to an earlier procedure developed by NIST for the Department of Energy.

Summary

Method 3052 is one of the most robust and practical sample preparation methods to be developed. It has been tested for over a decade on SRM certification, nuclear waste sample preparation development, and general laboratory use. Analysts from all over the world have tested various iterations of this protocol. While not perfect, it is one of the most advanced and generally applicable elemental sample preparation methods developed to date, and represents an appropriate balance of efficiency, ease of use, safety, and robustness. The current version of Method 3052 will be maintained on the SamplePrep Web site, described in Chapter 15 of this text. New adaptations and specific suggestions will be maintained there, along with suggestions for specific analyte matrix problems, solutions, and safety information. The method will continue to evolve with many modifications and alternative implementations to come. Please contribute your solutions to specific problems using this method by contacting SamplePrep Web and the authors of the method at the site.

Environmental Microwave Acid Leaching of Metals from Solids

Comparing EPA Methods 3051 and 3050

Methods 3051 and 3050 are leach methods, primarily for solids as documented in SW-846 Update II. Method 3051, a closed-vessel microwave sample preparation method, was originally developed as an alternative to the conventional hot plate Method 3050. Method 3050 is approved for the analysis of 21 elements in sediments, sludges, and soils (Al, As, Ba, Be, Ca, Cd, Co, Cr, Cu, Fe, K, Mg, Mn, Mo, Na, Ni, Pb, Se, Tl, V, and Zn). However, to analyze for all 21 elements, two digestions of the same material are required. Each digest is suitable for the analysis of selected elements by certain atomic spectrometry techniques. Method 3051 is approved for the analysis of 26 environmentally important elements in sediments, sludges, soils, and oils (Al, Ag, As, B, Ba, Be, Ca, Cd, Co, Cr, Cu, Fe, Hg, K, Mg, Mn, Mo, Na, Ni, Pb, Sb, Se, Sr, Tl, V, and Zn) (17). A single digest is suitable for the analysis of all 26 elements by several atomic spectrometry techniques.

Neither of these methods is a total decomposition procedure, but rather an acid leaching protocol. They leach elements from the material in an attempt to simulate the amount of each element that could be leached or become labile in the environment under worst-case scenarios. Because the methods are leach tests, the choice of reaction conditions and reagents will dramatically affect the elements available for analysis. As opposed to total decomposition sample preparation methods where there is a single total elemental concentration and a single correct answer, the leach data are dependent on the reaction conditions and the rate of these reactions to provide the given composition of the leachate solution.

Methods 3051 and 3050 have several key variations that are based on the difference between open and closed systems. Method 3051, using a closed vessel at elevated temperatures, has demonstrated retention of ions during dissolution. Because mineral acids convert most of the elements into ionic forms, they do not behave as neutral volatile molecules and are retained by the closed vessel in solution. Several elements analyzed by both methods could potentially form volatile compounds, such as molecular forms of arsenic, antimony, tin, selenium, and mercury (122). In addition, chromium may be volatile under oxidizing conditions in the presence of chloride (122). The sealed vessel used in Method 3051 prevents both volatilization and atmospheric contamination during dissolution. Many errors from sample losses and from contamination are encountered in open systems and have been previously documented (78). The closed nature of the microwave system is inherently more amenable to protect against these types of errors. Cooperative laboratory studies confirmed these conclusions and were reported in the literature (17). In addition, the loss of acid fumes during open-vessel decomposition is prevented in the Method 3051 closed system, and excess reagents are not used and reagent waste and reagent blank are minimized.

The methods are very different in their specifications. The microwave digestion Method 3051 provides performance criteria requiring that the sample and concentrated nitric acid be heated to 175 °C within the first 5.5 min and be maintained at 175 ± 5 °C for the remainder of 10 min. These temperatures were optimized for the destruction of organic matrix material as described previously (*31, 34, 116*). The temperature of 175 °C is required to be maintained for sufficient time to permit the completion of the reactions and the destruction of organic material. As has been described for many silicate decompositions, the temperature is not as critical; the inclusion of key reagents, such as hydrofluoric acid, is the dominant requirement. In these leaching reactions, the ability to repeat the conditions is the most important parameter. The reaction specifications, including the temperature profile and reagents, primarily determine the chemical reactions that are allowed to proceed.

Method 3050 consists of a series of steps that involves the addition of various reagents refluxed at 95 °C, and includes variable options for the analyst. Table XXIV provides a brief summary of the two methods. These methods are very different from a reagent perspective in that they do not duplicate the same chemistry. Method 3050 uses an additional strong oxidizer, 30% hydrogen peroxide, and a strong complexer, hydrochloric acid, both of which were required to be excluded from the original version of Method 3051. The restriction in Method 3051 to only use a single reagent, nitric acid, was required by EPA to assure that the four instrumental anal-

Table XXIV. Comparative Procedural Outline of Methods 3051 and 3050 for Protocol Steps and Reagents

Parameter	Method 3051	Method 3050
Vessel	Closed microwave vessel capable of 7.5 ± 0.7 atm	Usually a covered beaker or flask (atmospheric pressure)
Sample Size	0.25–0.50 g	1–2 g
Reagents	10 mL concentrated HNO_3	20 mL concentrated HNO_3, 17 mL H_2O, 5 mL HCl, 10 mL 30% H_2O_2
Time	10 min	2–6 h
Procedure	Weigh sample into vessel, add 10 mL concentrated HNO_3. Place in microwave unit. Heat to 175 °C within 5.5 min. Maintain 175 °C for remaining part of 10 min.	Weigh sample into vessel. Add 10 mL 1:1 HNO_3. Reflux at 95 °C for 10–15 min. Cool, then add 5 mL concentrated HNO_3. Reflux at 95 °C for 30 min. Cool, then add 5 mL concentrated HNO_3. Reflux at 95 °C for 30 min. Condense down to 5 mL. Add 2 mL H_2O and 3 mL 30% H_2O_2. Heat until effervescence subsides, cool, add 1 mL 30% H_2O_2 until no change. Add 5 mL concentrated HCl and 10 mL H_2O. Reflux at 95 °C for 15 min.

SOURCE: Summarized from reference 34.

ysis methods specified for subsequent analysis were applicable: ICP–MS, ICP–OES, GF–AAS, and F–AAS. The difference in chemistries caused by the addition of different reagents is the reason for elemental biases between Methods 3050 and 3051 and must be discussed in detail.

Increased Efficiency as a Result of Higher Temperature

An explanation as to why the microwave digestion is more efficient than hot plate digestion can be described by kinetic reaction theory (34). The comparison of the method kinetics will illuminate the fundamental differences between hot plate and closed-vessel microwave sample digestion. Arrhenius defined the following empirical relationship:

$$\frac{d\ln K}{dT} = \frac{E_a}{RT^2} \tag{19}$$

The integrated form is:

$$\ln \frac{K_2}{K_1} = \frac{E_a}{R} \left(\frac{1}{T_1} - \frac{1}{T_2} \right) \tag{20}$$

where k_1 and k_2 are the rate constants at temperatures T_1 and T_2, E_a is the Arrhenius activation energy, and R is the gas constant. In most cases, the reaction rate increases exponentially with increased temperature (138). Assuming a nominal energy of activation of 80 KJ/mol that is invariant over the temperature range, the rate constant at 175 °C is 107 times greater than at 95 °C. This difference increases with higher activation energies. Applying this result to the comparison of Methods 3050 and 3051, if the microwave method takes 4.5 min at 175 °C, then the hot plate digestion method should take at least 107 times 4.5 min, or 480 min (8 h) at 95 °C. Although this is not a detailed kinetic comparison, it illustrates the magnitude of the reactivity, and hence efficiency, differences that can exist between the methods.

Significance of Reagents

This previous comparison is not entirely appropriate because the two methods share only nitric acid, and Method 3050 uses several additional reagents. The microwave method uses only nitric acid, whereas the hot plate method (prior to Method 3050B, which is discussed in the next section) uses nitric acid, 30% hydrogen peroxide, and hydrochloric acid. As a result of the different reagents, these reaction mixtures contain several different species during dissolution, resulting in variations in the leaching of several specific analyte elements. The reaction species are compared in Table XXV and in most cases these differences can be understood from the

Table XXV. Comparison of Chemical Species Possible During Dissolution by Methods 3051 and 3050

Parameter	Microwave Method 3051	Hot Plate Method 3050
Reagents	HNO_3	HNO_3
		H_2O_2
		HCl
Species generated in situ	H_2O	H_2O
	H_3O^+	H_3O^+
	NO_3^-	NO_3^-
	NO_2^+	NO_2^+
	NO_2	NO_2
	O_2	O_2
	Digest products	Cl^-
		NOCl
		Cl_2
		HO_2^-
		O_2^{2-}
		O_2
		Digest products

SOURCE: Data are from reference 34.

chemistry. The differences in chemistry are not due to the differences in temperature. Temperature is mainly involved in the reaction rate and not the specific equilibrium or reaction mechanisms that are reached.

A study to compare the leachable concentrations produced by the two methods was performed by using a well-characterized homogeneous material from the NIST Buffalo River Sediment SRM (SRM 2704). A typical temperature versus time profile of Method 3051 with a cooling stage is illustrated in Figure 31. Of the 15 elements studied, Method 3051 was statistically significantly lower in the analysis of seven of the elements (*see* Table XXVI (*17, 38, 139*)). This difference is primarily the result of the difference in reagents. The use of up to 10 mL of 30% hydrogen peroxide significantly increases the rate of digestion to the point that Method 3050 can be completed within a few hours. Because of the increased efficiency of the microwave method, the two methods yield similar results for most of the elements tested. Many of the chemistry differences, such as hydrogen peroxide, which complexes vanadium to increase its analyzed quantity, are well known (*85, 140*).

The addition of large amounts of chloride in Method 3050 has a stabilizing effect on aluminum, iron, antimony, and silver by forming chloro complexes that tend to draw additional concentrations of these elements into solution. Nitric acid is not a stable medium for the determination of silver because trace quantities of chloride, found in the acid (and in most environmental samples), will cause precipitation of silver chloride. This results in unpredictably low analysis of silver by Method 3051, whereas higher concentrations of chloride resolubilize the silver (*120*).

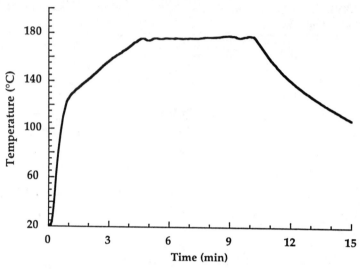

Figure 31. Typical reaction profile (temperature and pressure vs. time) of EPA Method 3051.

Because the goal of these leach procedures is to simulate the potential leaching of materials in the environment and to evaluate their pollution hazard, the very harsh conditions imposed by both methods give worst-case scenarios. High hydronium ion concentrations are a reality in the environment in acid mine drainage, to a lesser extent as acid rain, and under other naturally occurring conditions. These tests were appropriate because they were under low pH conditions that liberate and stabilize large metal concentrations in environmental systems.

If the reagent chemistries of Methods 3050 and 3051 are made similar, the composition of the leachate is projected to become similar. Alternative reagent mixtures containing hydrochloric acid are being developed for Method 3051 to alleviate the reagent-induced bias. Hydrochloric acid and peroxide are compatible reagents for the microwave and can be easily controlled by using newer and more modern control mechanisms for microwave sample preparation systems. The alternative Method 3051 procedure, using hydrochloric acid, is anticipated to show similar results to EPA Method 3050.

Control of Method 3051 was originally based on calibration criteria with specific temperature requirements provided. This method was developed and accepted for use by CERCLA, although the approval process started in 1989 for its use in the RCRA program. During this time, instrumentation advanced and calibration criteria become less significant in laboratory microwave instrument systems. Table XXVII describes the microwave cavity power required to balance the energy to control Methods 3051 and 3015 for two types of vessels, with slightly different heat loss characteristics. The calibration must be done for a specific vessel construction

Table XXVI. Comparison of Elemental Concentrations in Acid Leaching of SRM 2704 Buffalo River Sediment by Methods 3501 and 3050

Element	3051 Mean	σ	3050 Mean	σ	Statistical Difference α = 0.01	% Bias
Al	1.25	0.08%	2.5	0.19	Significant	−50
As	11.6	0.5	12.8	1.1		
Ba	79.3	3.4	132	10	Significant	−40
Be	0.689	0.11	1.05	0.05	Significant	−34
Ca	1.87	1.04%	1.88	0.01		
Cr	69.4	3.4	78.9	2.9	Significant	−9.5
Co	9.42	1.26	10.8	0.5		
Cu	89.1	3.6	88.5	1.7		
Fe	2.91	0.11%	3.29	0.07	Significant	−12
Pb	153	19	169	8		
Mg	7990	240	9080	150	Significant	−12
Mn	465	15	486	4		
Ni	37.8	3.2	41.8	0.6		
V	25.1	1.5	49.4	2.8	Significant	−49
Zn	392	19	403	11		

Note: Method 3050 digestions for all elements involved the use of HCl. Values are µg/g unless noted.

Table XXVII. Reaction Conditions and Reagents for Methods 3051 and 3015 by Using Calibration Control

Method	No. vessels	Sample Size	HNO₃ (mL)	Temperature (C)	Time (min)	Vessel Type PFA Power (W)	LD Power (W)
3051 (soils)	6	0.5 g	10	175	5.5	574	445
				175 ± 5	4.5	574	236
3015 (aqueous)	5	45 mL	5	160 ± 4	10	545	473
				165–170	10	344	237

Note: PFA is 120-mL all Teflon PFA vessel; LD is 120-mL Teflon PFA lined vessel. Temperature and time are performance criteria. Time and vessel type are prescription settings.

and for a specific number of vessels. A detailed description of the reasons for this was previously described and showed variation of pressure for different vessels with the same nominal temperature. These specifications of power, number of vessels, required reagents, and rigid times are described as prescription criteria.

Contrast this result with the requirement of reaching 175 ± 5 °C in 4.5 min and maintaining these conditions for 5.5 min. As was seen in Figure 4, when using temperature feedback control, all equipment evaluated met Method 3051 specifications by using differing numbers of vessels of various construction and materials and differing temperature measurement systems. Method 3051 is a generic, nonproprietary environmental method that is capable of being executed successfully by most instrument manufacturers. It has lower vessel pressure requirements than the newer 3052 Method, and was developed with first-generation microwave technology. Method 3051a may well be upgraded to take advantage of the new, more capable laboratory microwave systems. The addition of hydrochloric acid will offer an alternative Method 3051 procedure that will more accurately reproduce the chemistry of Method 3050 in the same 10 min time frame. It will also require increasing the pressure capabilities of the vessels because added gaseous reaction products are produced by hydrochloric acid.

Figure 12 demonstrated control of Method 3051 by using both calibration and a noncommercial research feedback control microwave system. Both methods of control are adequate for standard methods. Method 3051 originally specified conditions for calibration control of either two or six vessels. Any number of vessels can be controlled by using temperature feedback control, and samples that contribute thermal energy or gaseous digestion products are more applicable to temperature feedback control as described previously. In addition, the failure of a vessel seal in a single vessel in Figure 13 demonstrated the robustness of temperature feedback control because the instrument is dynamically reprogrammed by using temperature data to complete the specified conditions without altering the conditions of any of the remaining reaction vessels. The compromised vessel can easily be identified by its weight loss on removal from the microwave system.

Because 3051 and 3050 are leach methods, accuracy in these methods is based on the precision of replicating the reaction conditions. With the use of a hot plate, the variation in the reaction temperature in Method 3050 may exceed 25 °C. Temperature variability in seven beakers on a hot plate at separate locations during Method 3050 was measured. A representative diagram is displayed in Figure 32. Through the use of temperature feedback control, Method 3051 can be precisely reproduced to better than ±4 °C. Improvements in precision of Method 3051 are primarily due to more precise parameter/reaction control. Of these parameters, temperature has been proven to be one of the most important in determining leaching precision. When evaluating element leaching parameters and precision for simulated nuclear waste, temperature was the key to reducing data precision from over 50% to 3% for all nine elements tested (5).

The EPA has also compared several environmental acid leaching methods including Method 3050 and calibration-controlled Method 3051 (141). The interlaboratory comparison report of several EPA acid leaching methods of a homogenized soil found median relative standard deviations for Method 3051 to be 11%, compared with the nonmicrowave Methods 3050, 3055, and Environmental Methods Management Committee (EMMC), which were 15, 21, and 19%, respectively. The report concluded that, even though there were biases between the leaching of Methods 3051

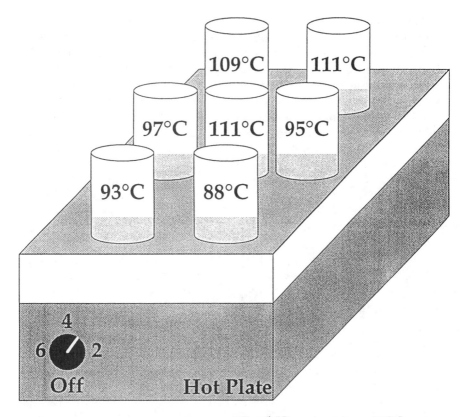

• Goal Temperature 95°C
• Mean Temperature 101 ± 9°C

Figure 32. Temperature measurements of Method 3050 done by using beakers on a hot plate.

and 3050, Method 3051 resulted in excellent interlaboratory precision. These biases are due to the different chemical leaching mechanisms resulting from reagent differences.

Leach methods are prone to error due to the incomplete leaching of the elements, irreproducible or uncontrollable leach conditions, and the lack of suitable leach standards. Although temperature feedback control improves the reproducibility of a method's reaction conditions, any further improvements in the accuracy of these methods will necessitate the development of leach method-specific standard reference leach materials.

As with total elemental analysis, the comparison of leach analysis data with certified leach measurements from specific SRMs provides a valuable quality control tool. To date, two NIST SRMs have been studied to evaluate their appropriateness as standard leach materials for EPA Method 3051 with respect to six of the more environmentally important elements: Cd, Cr, Cu, Ni, Pb, and Zn. The NIST SRMs 2710 and 2711 (highly contaminated and moderately contaminated Montana soils) were selected for their appropriateness to meet the environmental laboratory need. Total decomposition by using microwave Method 3052 was also performed as a quality control evaluation, as described previously (34) and in this chapter.

The overall ranges using both Methods 3051 and 3050 show similar mean leaching values for both methods illustrated in Figures 33 and 34.

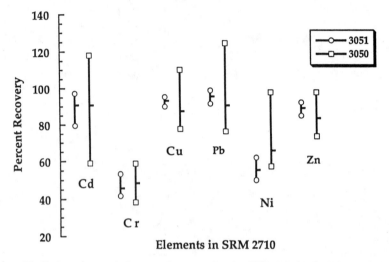

Figure 33. Comparison of elemental recoveries for EPA Methods 3050 and 3051 from SRM 2710.

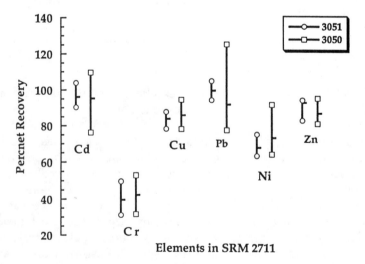

Figure 34. Comparison of elemental recoveries for EPA Methods 3050 and 3051 from SRM 2711.

However, the leach ranges and precision for Method 3050 is, in all cases, larger than for Method 3051. This result illustrates the precise reproducible leaching of the matrix by using the microwave Method 3051, approaching that of total analysis. This result also illustrates the poorer precision typical of all hot plate methods, including Method 3050. In addition, classical laboratory airborne contamination is indicated in the hot plate method evidenced by the range frequently extending above 100% recovery of the standard reference sample. Traditional contamination is absent from closed-vessel microwave-prepared leach solutions of these standard materials. These trends are representative of the superiority of microwave-assisted sample preparation methods over older technology. We believe this is a trend that again speaks to Berzelius' desire for more robust technological solutions to sample preparation and analysis needs.

Microwave Acid Methods for Leaching of Water and Wastes for Metals

Currently, four microwave-assisted acid leaching methods have been approved for water and wastewater: ASTM D4309-91, NPDES, EPA 3015, and Standard Method 3030K (*see* Table XXVIII). ASTM D4309-91 and NPDES are essentially the same methods, as are EPA 3015 and 3030K. These methods were developed with two entirely different reaction control mechanisms: microwave unit power settings versus both microwave calibration control and temperature feedback control.

ASTM Method D4309-91 and NPDES methods were both developed without the advanced reaction control techniques used with the development of the other methods. These methods have suffered from their inability to accurately reproduce the leaching conditions that are prescribed in the methods. ASTM Method D4309-91 states that the acid and sample must be heated to a maximum temperature of either 164 ± 4 or $165 \pm 5\,°C$, without a hold period. However, the heating curves that are included in the method for different microwave unit power ratings show deviations in temperatures inside the vessels of nearly 20 °C between vessels heated with 575 W and those heated with 635 W. The apparent maximum temperature in both cases is approximately $165 \pm 5\,°C$. This result leads to a very imprecise control mechanism, and because the accuracy of a leach method is directly related to the leaching conditions, these methods are inherently inaccurate.

These methods were also developed to take advantage of every vessel position in the microwave carousel. Development involved a 12-position carousel. Twelve vessels that were very poor thermal insulators, each containing 55 mL of solution, were heated in a limited power 550–700 W microwave unit system. The method states that in units ranging from 575 to 635 W, the heating takes 50 min to reach temperature, but for units

Table XXVIII. Standard Water and Wastewater Methods

ASTM D4309-91	NPDES	U.S. EPA 3015	Standard Methods 3030K
50 mL Sample	50 mL Sample	45 mL Sample	45 mL Sample
3 mL HNO₃ and 2 mL HCl or 5 mL HNO₃	3 mL HNO₃ and 2 mL HCl	5 mL HNO₃	5 mL HNO₃
Microwave unit power setting control	Microwave unit power setting control	Temperature feedback or calibration control	Temperature feedback or calibration control
For 1–6 120 mL all PFA vessels	For 1–6 120 mL all PFA vessels	For variable number of vessels	For variable number of vessels
575–635 W microwave units use 75% power for 30 min.	364–420 W microwave units use 75% power for 30 min.	Heat to 160 ± 4 °C in 10 min.	Heat to 160 ± 4 °C in 10 min.
635–700 W microwave units use 75% power for 25 min to a maximum temperature of 164 ± 4 °C.	To a maximum temperature of 164 ± 4 °C.	Slow rise to 165–170 ± 4 °C in additional 10 min. or for five 120 mL all PFA vessels 545 W for 10 min and 344 W for additional 10 min.	Slow rise to 165–170 ± 4 °C in additional 10 min or for five 120 mL all PFA vessels 545 W for 10 min and 344 W for additional 10 min.
For 12 120 mL all PFA vessels	For 7–12 120 mL all PFA vessels		
575–635 W microwave units use 100% power for 50 min.	575–635 W microwave units use 100% power for 50 min.		
635–700 W microwave units use 100% power for 30 min to a maximum temperature of 165 ± 5 °C.	635–700 W microwave units use 100% power for 30 min to a maximum temperature of 165 ± 5 °C.		

ranging from 635 to 700 W, the heating takes only 30 min to reach temperature. As the power of the microwave unit increases, the reaction time decreases. If either fewer vessels or higher powered microwave units are used (typically greater than 1000 W), these methods should be able to reach temperature with a variable number of vessels in less than 10 min to better use the speed of microwave heating and optimize throughput.

The microwave power settings of these methods are also dependent on the specific type of microwave vessel used. If another type of microwave vessel is used, the thermal insulating properties may vary from those in the method design, and the microwave unit power settings that will reproduce the specifications of this method will have to be redetermined. Alternatively, temperature feedback control could be employed to reproduce these temperature profiles. The specified control mechanism and slow, power-limited temperature profiles limit the usefulness of these methods to first-generation microwave vessels that were prone to venting (*see* Chapter 2), and antiquated reaction control techniques.

EPA Method 3015 and a replicate method, Standard Method 3030K (for water and wastewater), were developed with an understanding of the importance of the reaction temperature for destruction of organics and reproducible reaction control. These methods were developed with two types of control, power calibration and temperature feedback. Even though it was known that calibration could improve the reproducibility of the reaction conditions occurring in a leach method as compared with microwave unit power settings, the performance criteria for temperature feedback control were included (*31*).

This result has several ramifications. With calibration control, a given set of power settings, in watts versus time, is only capable of replicating the method's reaction conditions if the precise amount of acids and sample are placed into the identical type of vessel and the prescribed number of vessels are placed into the microwave unit. Calibration is only as good as the capability to reproduce the prescribed method. Any changes, even minor, may dramatically affect the reaction conditions, causing lower recoveries of analytes or poorer reproducibility. Calibration, like microwave unit power settings, is dependent on the use of the identical type of microwave vessels. However, by controlling the actual temperature inside the microwave vessels, the reaction conditions can be reproduced accurately, even when the number of vessels or the type of vessel is changed.

By using temperature feedback control, these methods can be implemented with any microwave vessel capable of maintaining a modest amount of pressure. Therefore, these methods are not instrument- or vessel-dependent and will be able to be used with all of the current microwave equipment. These methods will not be obsolete by changing vessel configuration, or the ongoing development and improvement in laboratory microwave equipment.

Method 3050B: Evaluation of an Atmospheric Pressure Microwave Leach Procedure

Because the control of temperature is paramount to reproducible digestions or extractions of elements, we have investigated using microwave digestions with temperature feedback control for EPA Method 3050. Previously, most leach methods like Method 3050 were accomplished on a hot plate in a beaker, flask, or sealed container. These methods are very traditional, time consuming, fairly inefficient, and in general, imprecise. By substituting microwave control, we gain a number of advantages including efficiency, higher accuracy, reduced waste, and better repeatability. This improvement is due to direct microwave energy coupling and controlled energy input into digest solutions.

To date, microwave oven technology has been coupled with closed vessels that have become a standard for EPA Methods 3051, 3015, and 3052, discussed previously in this chapter. The control of temperature in leaching is responsible for much of the precision of these procedures. During the early 1980s, fundamental research established temperature control as the most significant contributor to errors in leach test (5). These tests focused on nuclear waste glass materials. Temperature was found to be the dominant parameter in leaching uncertainty and imprecision. The control of temperature to within ±0.04%, instead of ±1% over a 28-day leach period, reduced the leaching measurement uncertainty from approximately 50% to approximately 3% for the elements evaluated.

Method 3050, a method for evaluating solid waste (as part of SW-846 Update II), has traditionally been a leach test performed on a hot plate. In recent years, the EPA has proposed to revise certain testing methods that are used to comply with the requirements of subtitle C of the RCRA act of 1976. Method 3050B is one of several methods included in the list of draft revised methods proposed in SW-846 Update III in the *Federal Register* (49). This method has been made more broadly applicable by adapting the prescription-based method to create a performance-based method. By adding the words "or equivalent" to the electric hot plate designation, and specifying that the heating device be "adjustable and capable of maintaining a temperature of 90–95 °C", the method permits the use of heating blocks and microwave energy as acceptable heating alternatives within the structure of the method. In addition, with the incorporation of feedback control of the leachate temperature, it is easy to increase the reproducibility of the measurement while automating the process.

In comparison to indirect heating by convection and conduction in hot plate digestions, microwave irradiation is directly absorbed by the mechanisms of ionic conductance and dipole rotation (*see* Chapter 1). Because the sample is heated directly, equilibrium conditions are obtained much more rapidly. In Method 3050B, the sample preparation time can be reduced from over 2.5 h to 1 h by using microwave technology rather than conventional means. As well as allowing greater sample throughput, this may also improve reproducibility because the sample is exposed to reagents and the atmosphere for a shorter period of time. The reaction conditions and reagents influence the results of the leach analysis, and a large number of different species are involved (*see* Table XXV).

Validation Study of Atmospheric-Pressure Microwave Implementation

The procedural comparison of hot plate Method 3050B to a microwave-assisted Method 3050B is shown in Table XXIX. The method is validated for the analysis of 21 elements (Al, As, Ba, Be, Ca, Cd, Co, Cr, Cu, Fe, K, Mg, Mn, Mo, Na, Ni, Pb, Se, Tl, V, and Zn) in sediments, sludges, and soils. Two elements, silver and antimony, have been added to the revised version of 3050B. Two digestions of the same material are required to analyze all elements. For the digestion of samples to be analyzed by F–AAS or ICP–OES, the sample is leached with nitric acid, hydrogen peroxide, and hydrochloric acid. The addition of hydrochloric acid helps the solubility, and thus recovery of Sb, Bi, Pb, and Ag by complexation with chloride ion. Thus, both antimony and silver are included in the method. This inclusion complements the 3050B procedure for the determination by ICP–MS or GF–AAS, which uses only nitric acid and hydrogen peroxide.

Table XXIX. Comparative Procedural Outlines of EPA Method 3050B and Modified Microwave-Assisted 3050B

Parameter	Microwave-Assisted Method 3050B	Hot Plate Method 3050B
Vessel	Open microwave vessel at atmospheric pressure	Usually a covered beaker or flask at atmospheric pressure
Sample size	1–2 g	1–2 g
Reagents	15 mL concentrated HNO_3, 17 mL H_2O, 5 mL concentrated HCl, 10 mL 30% H_2O_2	15 mL concentrated HNO_3, 10 mL H_2O, 10 mL concentrated HCl, 10 mL 30% H_2O_2
Time	60 min.	2–6 h
Procedure	Weigh sample into vessel and add 10 mL 1:1 HNO_3. Reflux at 95 °C for 5 min. Cool, then add 5 mL concentrated HNO_3. Reflux at 95 °C for 5 min. Cool, then add 5 mL concentrated HNO_3. Reflux at 95 °C for 5–10 min. Cool, add max. 10 mL 30% H_2O_2. Heat until effervescence is minimal. Reflux at 95 °C for 5–10 min. Cool, add 15 mL 1:2 HCl, reflux at 95 °C for 5 min.	Weigh sample into vessel and add 10 mL 1:1 HNO_3. Reflux at 95 °C for 10–15 min. Cool, then add 5 mL concentrated HNO_3. Reflux at 95 °C for 30 min. Cool, then add 5 mL concentrated HNO_3. Reflux at 95 °C for 30 min, evaporate to 5 mL, or heat for 2 h. Cool, add max. 10 mL 30% H_2O_2. Heat until effervescence is minimal. Evaporate to 5 mL or heat for 2 h. Cool, add 10 mL concentrated HCl, reflux at 95 °C for 15 min.

Our evaluation involved studying standard Method 3050B on a hot plate in comparison to three configurations of microwave equipment. To directly compare the sample preparation procedures, all analyses for metal content were determined by using the same analysis instrumentation and solutions. The adaptation of Method 3050B to a modified microwave-assisted method used the following equipment and is summarized in Table XXX. Microwave leaching using power control was accomplished with an automatic version of a Microdigestor A301 (Prolabo, Paris, France) consisting of a TX32 programmer, an exhaust system (ASPIVAP) to evaporate and neutralize the acid fumes, a pump unit for the automatic addition of reagents, and a sample carousel. The Microdigestor A301 was set up in an independent hood with a ventilation fan connected to a hood exhaust.

The temperature feedback control units, Megal 500 (gas bulb thermometer, Prolabo) with a Microdigest 401 (Prolabo), and a Maxidigest MX350 (Prolabo) with an IR temperature sensor M402 (Prolabo) were used for temperature-controlled microwave-assisted leaching. A fluoroptic temperature probe was used for temperature acquisition in the microwave digestor with power control and for calibration of the digestate temperature (Luxtron 750, Santa Clara, CA; MFW2-probe). The conventional procedure was accomplished on a hot plate. SRM 2704 Buffalo River Sediment, SRM 2710 Montana Soils, Highly Elevated Trace Element Concentrations, and SRM 2711 Montana Soils, Moderately Elevated Trace Element Concentrations were obtained from NIST.

For the conventional hot plate EPA Method 3050B, a sample size of approximately 1.00 ± 0.01 g was weighted in a 250-mL borosilicate vessel, and 10 mL HNO_3 1:1 (v/v) was added. The sample was heated for 15 min on a hot plate at ~95 °C without boiling. After cooling, 5 mL HNO_3 were added and the sample was refluxed for 30 min at ~95 °C without boiling. This step was repeated again. The sample was evaporated to ~5 mL without boiling. After cooling, 2 mL 18 MΩ water was added followed by the slow addition of 10 mL H_2O_2 (30%). After cooling, 5 mL concentrated HCl and 10 mL 18 MΩ water were added and the sample was refluxed for 15 min without boiling. After cooling, the sample was filtered and diluted to 100 mL by using 18 MΩ water.

For power-controlled microwave implementation of Method 3050B (Figure 35), approximately 1.00 ± 0.01 g sample were weighed into a borosilicate vessel (100 mL) and 10 mL HNO_3 1:1 (v/v) were added. A microwave digestion program consisting of 80 W for 2 min and 30 W for 5 min was applied. After cooling for 5 min, 5 mL concentrated HNO_3 was automatically added and a second power program of 80 W for 2 min, and 30 W for 5 min. This step was repeated a second time. After cooling for 5 min, 3 mL H_2O_2 (30% (v/v)) were added slowly and a third power program of 40 W for 5 min was applied. This step was repeated three times.

Table XXX. Instrumentation for Conventional and Microwave-Assisted Method 3050B

Use	Atmospheric-Pressure Microwave-Assisted Methods			Method 3050B by Hot Plate
	Power Control	Temperature Control 1	Temperature Control 2	
Heating device	Microwave system with power control	Microwave system with temperature feedback control	Microwave system with temperature feedback control	Hot plate with power control
Temperature measurement	Fiber optic sensor	Gas bulb thermometer	IR sensor	Thermocouple, thermometer
Vessel	Borosilicate glass, 100 mL	Borosilicate glass, 250 mL	Quartz glass, 250 mL	Borosilicate glass, 250 mL

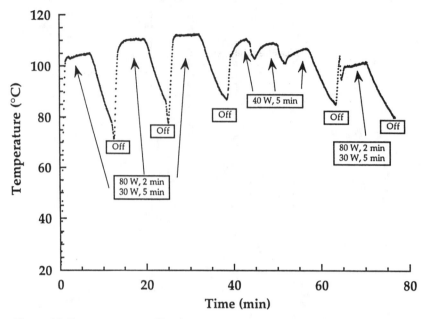

Figure 35. Temperature profile of microwave-assisted 3050B with power control.

After cooling for 5 min, 5 mL concentrated HCl and 10 mL 18 MΩ water were added and heated with a fourth power program of 80 W for 2 min and 30 W for 5 min was applied. After cooling, the sample was filtered and diluted to 100 mL by using 18 MΩ water.

For the temperature feedback controlled microwave implementation of Method 3050B (Figures 36 and 37), approximately 1.00 ± 0.01 g sample were weighed into a borosilicate glass vessel or quartz glass vessel (*see* Table XXX), and 10 mL HNO_3 1:1 (v/v) were added. A microwave temperature feedback control program consisting of heating the solution to 95 °C in 2 min and maintaining the temperature for 5 min was applied. After cooling for 5 min, 5 mL concentrated HNO_3 were automatically added and a second temperature program of heating the solution to 95 °C in 2 min and maintaining the temperature for 5 min was applied. This step was repeated a second time. After cooling for 5 min, 10 mL H_2O_2 (30%) were added slowly, the solution was heated to 95 °C in 6 min, and the temperature was maintained for 5 min. After cooling for 5 min, 5 mL concentrated HCl and 10 mL 18 MΩ water were added, the solution was heated to 95 °C in 2 min, and the temperature was maintained for 5 min. After cooling, the sample was filtered and diluted to 100 mL by using 18 MΩ water.

Power and Temperature Feedback Control

For EPA Method 3050B, the digestion requires heating the sample to 95 °C and holding this approximate temperature. This process is easier to repro-

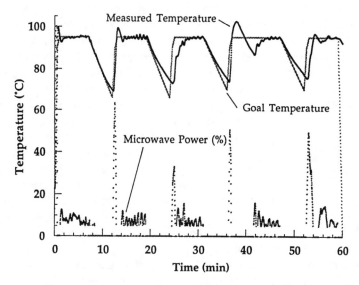

Figure 36. Temperature profile of microwave-assisted 3050B with temperature feedback control (gas bulb thermometer).

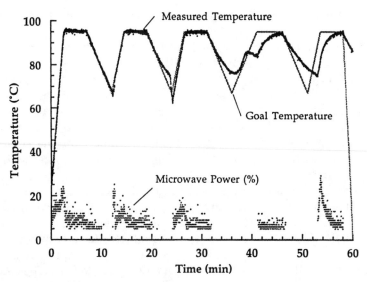

Figure 37. Temperature profile of microwave-assisted 3050B with temperature feedback control (IR sensor).

duce and maintain by using microwave irradiation than on a hot plate. The temperature profile for the microwave-assisted modified Method 3050B with power control was obtained by simultaneous temperature measurement with the Luxtron 750 unit and fluoroptic temperature probe MFW2 (Figure 35). The resultant temperature was above the required 95 °C. Control was limited because of the power increments and time settings of power-controlled microwave-assisted digestions. In addition, heat loss was considered because the digestions were performed in glass flasks.

The alternative to this method is to control the temperature rather than power during microwave digestions. Temperature measurement in a microwave field is limited to devices that are transparent to the field. Classical temperature measurement devices, such as mercury thermometers or unshielded thermocouples, cannot be used because of possible leakage of microwave radiation. As discussed previously, fiber-optic thermometry is commonly used in microwave cavity systems because of the absence of interaction with microwave energy and transparency in the field (69). Drawbacks of this approach are the high cost of the equipment and the fragility of the fiber-optic probes.

Microwave digestors with temperature feedback control units based on two different temperature measurement methods were used:

- measuring the pressure difference of heated air in a gas bulb inside the sample flask
- measuring the intensity of IR radiation emitted from the vessel base

The gas bulb thermometer method is based on the measurement of the change in the bulb gas pressure with the change in temperature. The pressure transducer signal is sent to an RS232 C serial port. The gas bulb, made of glass, is fragile and must be handled very carefully. After calibration by the manufacturer, the thermometer is suitable for a broad temperature range from 0 to 500 °C. This device measures the actual temperature inside the digestion flask. The accuracy of the temperature measurement was checked by simultaneous acquisition of temperature data with the fiber-optic probe.

For the noninvasive temperature control, an IR detector, a thermopile Pt-sensor suitable for a temperature range from 20 to 450 °C, was used. An energy transducer converts the amount of heat energy emitted from the vessel, as IR radiation, into an electrical signal. The detector measures the spectral range between 8 and 14 µm. This range is detected through a hole in the bottom of the microwave cavity under the sample flask. The emitted IR radiation is reflected by a mirror, at 90°, toward the IR detector. The different emissivities of objects emitting IR radiation are corrected for by an emissivity factor (0.1–1.0), controlled within the M402 software. Calibration of the IR sensor was accomplished by simultaneous acquisition of the temperature data with the fiber-optic sensor or gas bulb thermometer.

Only quartz glass vessels with flat bottoms were used. A major advantage of the IR temperature sensor is that there is less chance of cross contamination due to contact of the temperature probe with the sample. Therefore, no cleaning step of the temperature device after each digestion is required, as compared to the gas bulb thermometer.

Both temperature measurement devices were volume-dependent. The gas bulb has to be fully covered with sample solution to make an accurate temperature measurement. The IR temperature sensor also measures energy emitted or reflected by any surface directly viewed by the IR sensor. Temperature profiles obtained with the gas bulb thermometer and the IR temperature sensor are shown in Figures 36 and 37.

Results and Discussion

The regulation of a certain temperature on a hot plate was much more difficult than using the microwave instruments. Tests on a hot plate showed that temperatures differed between 85 °C and 118 °C, depending on the location of the flask on the hot plate (Figure 32). Similar temperature variations have been reported by Kane (*142*). The advantage of preparing several samples at the same time diminishes the problem of regulating and controlling a certain temperature for all sample flasks. In comparison, the IR sensor is able to regulate and control 95 °C for 5 min with a mean temperature of 95.4 ± 1.1 °C (4 replicates, 300 data points), but only one sample can be processed at a time.

Spikes, as a result of overshooting when the goal temperature is reached, were mainly observed with the Megal 500 unit (gas bulb thermometer) and are probably due to the algorithms used for data acquisition. Only minor spikes were observed with the noninvasive temperature feedback control unit M402 (IR sensor). Except for these spikes, which occur at the transition from temperature ramping to plateau, the temperature is controlled to a high accuracy.

Results of analysis using microwave-assisted Method 3050B (power control and temperature control) and conventional Method 3050B are shown in Tables XXXI–XXXIII for three SRMs (2704, 2710, and 2711). The results are compared with certified values for total digestion, and NIST "leachable concentrations" using Method 3050 (*143*). All the results are in the range of the NIST reference data.

The leachable concentrations of six elements, as percentage of certified total concentrations of SRM 2710 and 2711, are shown in Figures 38 and 39. The majority of the values are in the same range or higher as the leach reference data published by NIST. The percentage recoveries attained with temperature control (gas bulb and IR sensor) are slightly lower than the corresponding values of power control microwave and hot plate. This result can be explained by more accurately maintained, but relatively lower, temperature during the leaching procedures because the temperature with power control and on the hot plate is higher than the required 95 °C.

Table XXXI. Results of Analysis of NIST Standard Reference Material 2704, by Using Method 3050B

Element	Atmospheric-Pressure Microwave-Assisted Methods			Method 3050B by Hot Plate ($\mu g/g \pm SD$)	NIST Certified Values for Total Digestion ($\mu g/g \pm 95\%$ CI)
	Power Control ($\mu g/g \pm SD$)	Temperature Control (gas bulb) ($\mu g/g \pm SD$)	Temperature Control (IR sensor) ($\mu g/g \pm SD$)		
Cu	101 ± 7	89 ± 1	98 ± 1.4	100 ± 2	98.6 ± 5.0
Pb	160 ± 2	145 ± 6	145 ± 7	146 ± 1	161 ± 17
Zn	427 ± 2	411 ± 3	405 ± 14	427 ± 5	438 ± 12
Cd	NA	3.5 ± 0.66	3.7 ± 0.9	NA	3.45 ± 0.22
Cr	82 ± 3	79 ± 2	85 ± 4	89 ± 1	135 ± 5
Ni	42 ± 1	36 ± 1	38 ± 4	44 ± 2	44.1 ± 30

NOTE: NA is not available.

Table XXXII. Results of Analysis of NIST Standard Reference Material 2710 by Using Method 3050B

Element	Atmospheric-Pressure Microwave-Assisted Methods			Method 3050B by Hot Plate ($\mu g/g \pm SD$)	NIST Leachable Concentrations Using Method 3050B[a] ($\mu g/g$)	NIST Certified Values for Total Digestion ($\mu g/g \pm 95\%$ CI)
	Power Control ($\mu g/g \pm SD$)	Temperature Control (gas bulb) ($\mu g/g \pm SD$)	Temperature Control (IR sensor) ($\mu g/g \pm SD$)			
Cu	2640 ± 60	2790 ± 41	2480 ± 33	2910 ± 59	2700	2950 ± 130
Pb	5640 ± 117	5430 ± 72	5170 ± 34	5720 ± 280	5100	5532 ± 80
Zn	6410 ± 74	5810 ± 34	6130 ± 27	6230 ± 115	5900	6952 ± 91
Cd	NA	20.3 ± 1.4	20.2 ± 0.4	NA	20	21.8 ± 0.2
Cr	20 ± 1.6	19 ± 2	18 ± 2.4	23 ± 0.5	19	39[b]
Ni	7.8 ± 0.29	10 ± 1	9.1 ± 1.1	7 ± 0.44	10.1	14.3 ± 1.0

NOTE: NA is not available.
[a] Values are NIST published leachable concentrations (143).
[b] Noncertified values for information only.

Table XXXIII. Results of Analysis of NIST Standard Reference Material 2711 by Using Method 3050B

| Element | Atmospheric-Pressure Microwave-Assisted Methods | | | Method 3050B by Hot Plate (μg/g ± SD) | NIST Leachable Concentrations Using Method 3050[a] (μg/g) | NIST Certified Values for Total Digestion (μg/g ± 95% CI) |
	Power Control (μg/g ± SD)	Temperature Control (gas bulb) (μg/g ± SD)	Temperature Control (IR sensor) (μg/g ± SD)			
Cu	107 ± 4.6	98 ± 5	98 ± 3.8	111 ± 6.4	100	114 ± 2
Pb	1240 ± 68	1130 ± 20	1120 ± 29	1240 ± 38	1100	1162 ± 31
Zn	330 ± 17	312 ± 2	307 ± 12	340 ± 13	310	350.4 ± 4.8
Cd	NA	39.6 ± 3.9	40.9 ± 1.9	NA	40	41.7 ± 0.25
Cr	22 ± 0.35	21 ± 1	15 ± 1.1	23 ± 0.9	20	47[b]
Ni	15 ± 0.2	17 ± 2	15 ± 1.6	16 ± 0.4	16	20.6 ± 1.1

NOTE: NA is not available.

[a] Values are NIST published leachable concentrations (143).

[b] Noncertified values for information only.

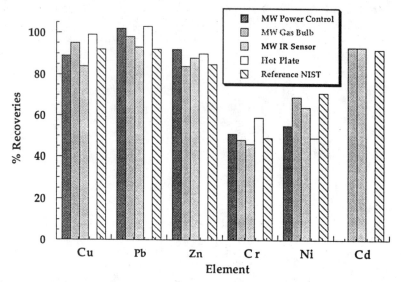

Figure 38. Recoveries of leachable concentrations of the analytes as percentage of certified values for total digestion of SRM 2710.

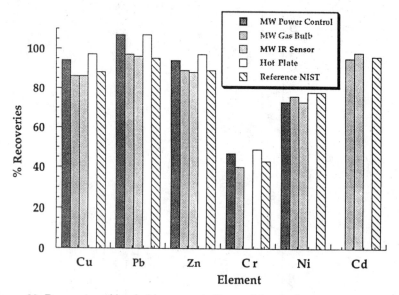

Figure 39. Recoveries of leachable concentrations of the analytes as percentage of certified values for total digestion of SRM 2711.

Figures 40a–40d show examples of 95% confidence limit ranges for acid-leached elements (SRM 2710) obtained with a microwave-assisted modified Method 3050, power-controlled and temperature-controlled, in comparison to conventional Method 3050B and reference leach data by NIST. The 95% confidence limits ranges for the six analyzed elements obtained with the microwave-assisted Method 3050B are frequently better than those obtained with the conventional Method 3050B (Tables XXXI–XXXIII).

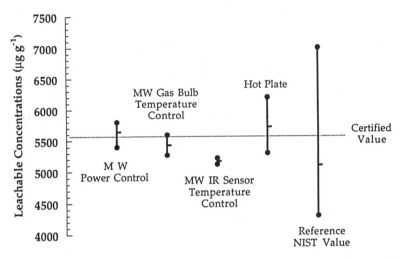

Figure 40a. Comparison of 95% confidence limit ranges with NIST reference data for acid-leached lead (SRM 2710).

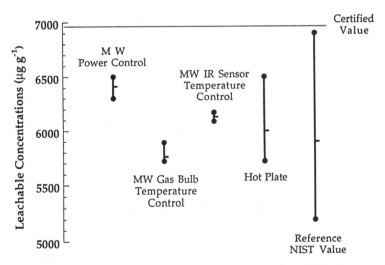

Figure 40b. Comparison of 95% confidence limit ranges with NIST reference data for acid-leached zinc (SRM 2710).

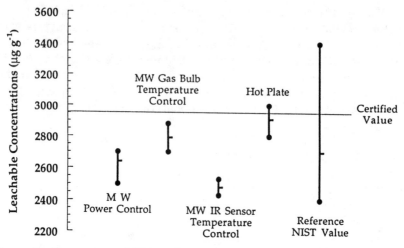

Figure 40c. Comparison of 95% confidence limit ranges with NIST reference data for acid-leached copper (SRM 2710).

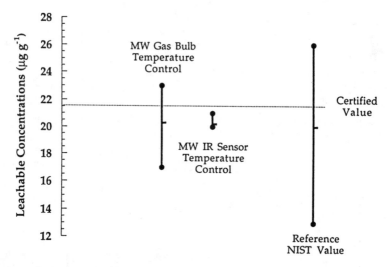

Figure 40d. Comparison of 95% confidence limit ranges with NIST reference data for acid-leached cadmium (SRM 2710).

Conclusion

Sample preparation, usually the time-consuming step in analytical chemistry, can be performed more than twice as fast by using microwave irradiation with temperature feedback control. More importantly, precision and accuracy are achieved faster. This study shows that controlling the temperature rather than the power is the better option for microwave digestions. In addition, several unique capabilities of atmospheric-pressure microwave digestions exist, including effective handling of gas-forming matrices, and sequential and incremental addition of reagents during digestion. Microwave-assisted leaching also offers increased precision over heat control by convection and conduction. The instrument is easy to use and can also be automated or semiautomated to reduce attendance by the chemist.

It is likely that the number of applications using microwave sample preparation under atmospheric pressure will increase, because many other methods can easily be converted into microwave-assisted atmospheric pressure methods (i.e., EPA Methods 3060, 3031, and the future Method 3000). Many other traditional sample preparation methods can be converted to microwave methods with reduced reagent loss, higher precision, and the possibility of automation. Several configurations of atmospheric pressure microwave (including CEM Star System) are appropriate and more traditional sample preparation methods such as these environmental leach methods may achieve similar advantages by conversion from the use of hot plates and heating mantles to atmospheric pressure microwave implementation.

Total Kjeldahl Nitrogen and Total Phosphorus Microwave Leach Method

Nitrogen and phosphorus are major nutrients that can become pollutants in the environment. This characteristic is due to the impact on eutrofication of nitrates and phosphates from sources such as fertilizer and sewage. In an effort to keep pace with the demand for analysis of both elements, a single, simple, fast, and precise microwave sample preparation method for the determination of both total Kjeldahl nitrogen and total phosphorus has been developed (*144*). Unlike several of the established EPA water and wastewater methods, a catalyst is not required for the microwave digestion, reducing both the possibility of reagent contamination and the production of harmful waste. Because the method quantitatively releases all Kjeldahl nitrogen and phosphorus from the sample matrix, it is possible for the same final sample digestate solution to be analyzed for both analytes. This method was tested for feasibility by using four well-characterized SRMs. The accuracy of the results for the determination of total Kjeldahl nitrogen and total phosphorus is excellent, and the precision is equal to or better than existing standard methods.

Phosphorus and Nitrogen Analysis

A significant amount of research has been done in the development of procedures for both nitrogen (*145*) and phosphorus (*146*), a testament to the importance of both these analytes. This work has lead to the generation of several standard methods. For Kjeldahl nitrogen, EPA Method 351 (*147*) and the French National Association of Standardization (AFNOR)

(*148*) Method NF V 03-100 are currently used. For phosphorus, EPA Method 365 (*149*) in all its variations is used.

The digestion time of the procedure developed is 25 min, without the need for catalyst reduction. In comparison, the EPA water/wastewaters procedures for total Kjeldahl nitrogen and total phosphorus have 30–150-min sample preparation steps, and most require a catalyst. Even the 30-min digestion step outlined in one variation of the EPA procedure (351.3) requires a mercury catalyst. The AFNOR method is used in the analysis of foods for Kjeldahl nitrogen determination and takes 20–40 min depending on the matrix, and without the use of a catalyst (*53, 54, 150*).

Although no current standard method is specifically designed for the determination of both phosphorus and nitrogen, it is possible to analyze digestates from EPA Method 365.4 for Kjeldahl nitrogen. In the literature, a 45-min persulfate procedure for the preparation of water samples for nitrogen and phosphorus analysis has been reported (*151*). However, this method is prolonged somewhat due to an optimization step that is necessary for the oxidizing reagent.

Procedure and Analysis

The proposed method involves the automated, sequential addition of two oxidizing reagents, H_2SO_4 and H_2O_2, and microwave heating to ~300 °C, resulting in complete digestion in less than 30 min for a variety of solid matrices. Only minimal procedural changes are required to accommodate all samples, which simplifies the use of the method significantly.

A 0.5–1.0-g sample is accurately measured and placed in an open borosilicate glass vessel, custom-designed (Prolabo) for use in the Microdigest 401 (Prolabo). The vessel is "open" only with respect to pressure equilibration, because a refluxing head allows the flask to retain the solvent and volatile species. This system is designed to digest sample sizes up to 10 g, which is a big advantage. Other unique features of the atmospheric pressure unit include the capability for automated sequential addition of reagents during digestion and precise control of the energy due to the focused-microwave coupling mechanism.

After weighing out the sample, 20 mL of concentrated sulfuric acid (ACS certified grade) were introduced into the vessel cavity for a 10-min digestion step. Following a short cool-down period (3 min), 6–12 mL of 30% hydrogen peroxide (Fisher Scientific) were automatically added until complete oxidation was obtained. This step required 6–12 additional minutes (typically 10 min), depending on the sample matrix. A graph illustrating the sample preparation program (microwave power and reagent addition versus time) is shown in Figure 41. Minor modifications to the program, such as slow temperature ramping during the initial step, may be necessary depending on the sample matrix being digested. More complex matrices, such as bovine liver, require a slightly longer oxidation

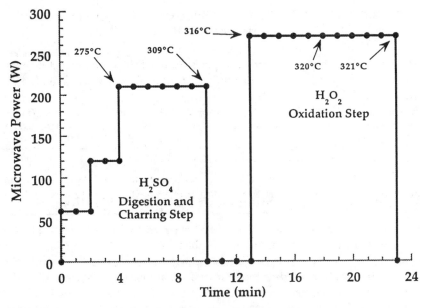

Figure 41. Programmed digestion procedure for bovine liver.

period. ACS certified tryptophan and SRMs from the NIST have been analyzed to validate the digestion procedure. The SRMs digested were SRM 1572 Citrus Leaves, SRM 1577a Bovine Liver, and SRM 1566 Oyster Tissue. All reference materials can be prepared for analysis in 25 min or less.

Analysis was performed in quadruplicate on all the samples. Total Kjeldahl nitrogen was determined by the ammonia procedure, as outlined in EPA Method 351, and total phosphorus was determined by the colorimetric procedure described in EPA Method 365. The rapid digestion is possible by using the efficient energy coupling associated with microwave energy and using a high-temperature boiling acid (sulfuric acid, bp 340 °C), which allows elevated temperatures (>300 °C) to be used in the procedure.

Results for the four sample types are shown in Table XXXIV. Excellent agreement was obtained with comparable or better precision for both elements. The results of this study verified the atmospheric pressure microwave sample preparation procedure for total Kjeldahl nitrogen and total phosphorus as an excellent alternative procedure for either or both analytes. These studies demonstrate the feasibility of microwave technology to unify classical environmental analysis procedures. It may also be possible to apply this method to water and waste samples, and thus ultimately develop one single sample preparation procedure for the analysis of nitrogen and phosphorus in both solid and liquid matrices. These adaptations also demonstrate the ability to reduce waste catalyst and improve the efficiency of classical analytical tests.

Table XXXIV. Results of Atmospheric-Pressure Microwave Digestion and
Analysis for Total Kjeldahl Nitrogen and Total Phosphorus

| Sample | Nitrogen % (w/w) ± 95% CI | | Phosphorus % (w/w) 95% CI | |
	Certified	Experimental	Certified	Experimental
Tryptophan	13.72^a	13 ± 0.62		
Bovine Liver	10.6 ± 0.6	10 ± 0.91	1.11 ± 0.04	1.1 ± 0.12
Oyster Tissue			(0.81)	0.807 ± 0.0048
Citrus Leaves			0.13 ± 0.02	0.133 ± 0.0016

^{a}Value calculated from stoichiometry (NIST noncertified values).

Future Directions of Microwave Sample Preparation

We see microwave sample preparation as a growing area that has not yet
stabilized, as have more mature techniques such as spectroscopy and chromatography. Sample preparation supports these areas and extends the
efficiency that has been developed in analytical instrumental detection. We
expect the fields of microwave sample preparation and microwave chemistry to continue to mature, as the fundamental knowledge base expands.
Continued growth requires ongoing development of the fundamental
understanding of mechanisms and interactions of direct microwave
energy coupling with solvent, sample, and their influence on reactions and
vessels. Much of the success of microwave sample preparation has been
due to the exceptional efficiency with reaction control, reducing both time
and cost of doing what has previously been done by using conduction and
convection heating mechanisms.

Future development will continue to build on the reaction control and
unique capabilities that have been discussed in this chapter. Microwave
sample preparation will become a standard laboratory tool, similar to
spectroscopy and chromatography, and will have as many varieties of specialized equipment as these earlier analytical staples. The future will see
specialized fields developing that are not just equivalent procedures to traditional heating, but that exploit the unique advantages of using microwave energy. Many new areas are starting to emerge. Some of these application areas, many of which are discussed elsewhere in this book, have
been compiled in Table XXXV.

Sample preparation equipment will also continue to evolve now that
markets have been established. Equipment will become both more specialized and more flexible. Niche markets will exist for truly unique microwave instrumentation and for specialized reaction control. Microwave cavity instruments will adopt a more versatile approach where a single
microwave field generator will functionalize many applications by replacing specialized cavity and waveguide attachments. Some of these solutions
are becoming available, providing further stimulation for innovation. A
few instrument evolutions have been projected in Table XXXVI.

Table XXXV. Future Areas of Development in Microwave Sample Preparation and Microwave-Enhanced Chemistry

- Extraction for speciation
- High pressure digestions
- Optimization of organic extraction
- Flow through microwave chemistry
- On-line derivatization integrated into analytical instrumentation
- Reagent exchange during reactions
- Ultra-trace analysis and clean chemistry
- Microwave-enhanced organic reactions
- Reliance on performance-based microwave procedures
- Standardized microwave equipment control
- Inorganic enhanced-microwave reactions
- Exchangeable software-based applications, method encapsulation
- Protein hydrolysis and biochemical microwave reactions
- World Wide Web dynamic chemical information exchange

Table XXXIV. Future Instrument Developments

- Evolution of a generic chemical reactor
- Educationally-directed microwave chemistry instrumentation
- Total microwave sample preparation and sample handling
- Continued evolution of microwave vessels
- Departure from re-manufactured home units to unique instrumentation
- Instrument-integrated microwave sample preparation
- Variable frequency alternative microwave instrumentation
- Continuously variable microwave power
- Autoclaves and sterilization equipment based on microwave energy
- Reaction vessels with reagent delivery and integrated instrument connections
- Centrifuge microwave reactors and concentrators
- Frequency-dependent reaction control
- Micro sample preparation using flow through systems and vessel inserts
- Innovative re-engineering of flowing systems
- Microwave absorbing materials used in active heating and shielding vessels
- Adjustable reaction gas phase pressure at set temperatures
- Sensors measuring reaction progress (i.e., refractive index, field strength, reflected energy)
- Next generation analytical microwave systems

We envision this field to remain very dynamic in this "information age". New innovative methods of using microwave sample preparation and microwave-enhanced chemistry will continue to be developed. The World Wide Web will most certainly be the home for much of the method and application development and information exchange. Software and computers will become invaluable conduits to the latest research and application developments.

Acknowledgments

We express our appreciation to the many professionals who have contributed and assisted us in developing the research and this contribution to microwave sample preparation.

EPA Personnel

EPA personnel were invaluable throughout the development of the environmental microwave sample preparation standard methods. We thank

- Ollie Fordham, EPA National Inorganic Methods Program Manager for RCRA, for his leadership and assistance in the development of Method 3052.
- Charles Sellers, the former EPA Inorganic Methods Program Manager, for his leadership and assistance in the development of Methods 3051 and 3015.
- Mike Hurd, Inorganic Manager in the EPA CERCLA program, for cooperation on Methods 3051 and 3015.
- David Freedman, EPA Office of Solid Waste, for his early leadership in initiating research collaborations to evaluate new microwave technology that led to Methods 3051 and 3015.
- Pat Sosinski, EPA Region 3, who assisted in the validation testing of Method 3015 and validation of the microwave calibration software.
- Ort Villa, former director of EPA Region 3 laboratory, whose leadership in innovative development assisted in validation of these Methods.
- Larry Buttler, EPA Las Vegas Laboratory, who coordinated the development and validation of the robotic implementation of Methods 3051 and 3015.

Current and Former Graduate Students and Post-Doctoral Students

- Leo Collins, for his contribution of text and data related to a paper he developed in the adaptation of EPA Methods 351 and 365.
- Lois Jassie, Dr. Kingston's first student in microwave chemistry, and currently Senior Scientist at CEM, for her contribution to the development of Methods 3051, 3015 and fundamental relationships.
- Jim Peterson, for his invaluable efforts in the advanced robotic microwave automation CAALS demonstration at NIST.
- George Lusnak, for the construction and testing of portable modular clean stations.

University and NIST Collaborators

- Frank Settle and Mike Pleva, Virginia Military Institute and Washington and Lee University, respectively, for their collaboration on expert system and automation research in microwave sample preparation.

- Gary Kramer, Project Manager of CAALS at NIST, for his assistance in the robotic automation of microwave sample preparation.
- Larry Machlin and Tom Murphy of NIST, for a decade of instruction in clean chemistry techniques, and John Moody for assistance in clean room design and construction and High purity reagent preparation.
- Laura Powell and Ernest Garner for use of the chromium lots and demonstrating the progression of blank improvement using clean chemistry.

Industrial Collaborators

The microwave manufacturers that have supported our research into fundamental microwave sample preparation:

- CEM, for their early support of fundamental microwave research and the development of EPA Methods 3051 and 3015.
- Milestone SRL, for their ongoing support of fundamental microwave interactions and applications that permit expansion of the field to both scientist and developer, especially in assisting in the validation of EPA Method 3052.
- Prolabo Corporation, for their ongoing support of fundamental atmospheric pressure microwave research and the development of microwave alternative to EPA Method 3050B.
- Russ Saville, Savillex Corporation, for early, continued, and custom development of high-quality fluoropolymer microwave vessels.
- Jim Fete, Coulter Corporation for his expertise in measuring the particulate levels in the clean laboratory.
- Frank Calovini, SAIC Corporation for assistance in preparing EPA microwave manuscripts and reviews.
- Bill Boute and Jim Christo (formerly) of Zymark Corporation, for contributions to the robotic automation of microwave closed-vessel systems.
- Ken Wickersheim, Luxtron Corporation, for the donation of a Luxtron fiber-optic unit and research interactions on microwave temperature measurements.

References

1. Berzelius, J. J. *Lehrbuch der Chemie*; Wöhler, F., Translator; Arnoldsche Buchhandlung: Dresden, Germany, 1831; p 74.
2. Asimov, I. *A Short History of Chemistry*; Doubleday and Company: Garden City, NY, 1965; pp 25–26.
3. Kingston, H. M.; Jassie, L. B. *Anal. Chem.* **1986,** *58,* 2534–2541.
4. Matthes, S. A.; Farrell, R. F.; Mackie, A. J. "A Microwave System for the Acid Dissolution of Metal and Mineral Samples," Report 120, National Bureau of Mines: Albany, OR, April 1983.

5. Kingston, H. M.; Cronin, D. J.; Epstein, M. S. *Nucl. Chem. Waste Manage.* **1984,** *5,* 3–15.

6. Abu-Samra, A.; Morris, J. S.; Koirtyohann, S. R. *Anal. Chem.* **1975,** *47,* 1475–1477.

7. Kingston, H. M.; Jassie, L. B. *Introduction to Microwave Sample Preparation: Theory and Practice;* American Chemical Society: Washington, DC, 1988.

8. "The 100 Most Significant New Products of the Year", *Ind. Res. Dev.* **1987,** *29,* 55–94.

9. Nadkarni, R. A. *Anal. Chem.* **1984,** *56,* 2233–2237.

10. Ybanez, N.; Cervera, M. L.; Montoro, R.; Guardia, M. *J. Anal. At. Spectrom.* **1991,** *6,* 379–384.

11. Lyon, T. D. B.; Fell, G. S.; McKay, K.; Scott, R. D. *J. Anal. At. Spectrom.* **1991,** *6,* 559–564.

12. McCarthy, H. T.; Ellis, P. C. *J. Assoc. Off. Anal. Chem.* **1991,** *74,* 566–569.

13. White, R. T., Jr. In *Introduction to Microwave Sample Preparation: Theory and Practice;* Jassie, L. B.; Kingston, H. M., Eds.; American Chemical Society: Washington. DC, 1988; Chapter 4, pp 53–78.

14. Friel, J. K.; Skinner, C. S.; Jackson, S. E.; Longerich, H. P. *Analyst* **1990,** *115,* 269–273.

15. Kimber, G. M.; Kokot, S. *Trends Anal. Chem.* **1990,** *9,* 203–207.

16. Nieuwenhuize, J.; Poley-Vos, C. H.; van den Akker, A. H.; van Delft, W. *Analyst* **1991,** *116,* 347–351.

17. Binstock, D. A.; Grohse, P. M.; Gaskill, A., Jr.; Sellers, C.; Kingston, H. M.; Jassie, L. B. *J. Assoc. Off. Anal. Chem.* **1991,** *74,* 360–366.

18. Noltner, T.; Maisenbacher, P.; Puchelt, H. *Spectroscopy* **1990,** *5,* 49–53.

19. Matthes, S. A. In *Introduction to Microwave Sample Preparation: Theory and Practice;* Jassie, L. B.; Kingston, H. M., Eds.; American Chemical Society: Washington, DC, 1988; Chapter 3, pp 33–52.

20. Lamothe, P. J.; Fries, T. L.; Consul, J. *Anal. Chem.* **1986,** *58,* 1881–1886.

21. Smith, F.; Cousins, B.; Bozic, J.; Flora, W. *Anal. Chim. Acta* **1985,** *177,* 243–245.

22. Fernando, L. A.; Heavner, W. D.; Gabrielli, C. C. *Anal. Chem.* **1986,** *58,* 511–512.

23. Lautenschläger, W. *Spectroscopy* **1989,** *4,* 16–21.

24. Bond, G.; Moyes, R. B.; Pollington, S. D.; Whan, D. A. *Chem. Ind.* **1991,** *September 16,* 686–687.

25. Baghurst, D. R.; Mingos, D. M. P. *J. Chem. Soc. Dalton Trans.* **1992,** 1151–1155.

26. Majetich, G.; Neas, E.; Hooper, T. *Proceedings of the First World Congress on Microwave Chemistry;* International Microwave Power Institute: Breukelen, The Netherlands, 1992.

27. Majetich, G.; Neas, E.; Hooper, T. *Abstracts of Papers,* 201st National Meeting of the American Chemical Society, Atlanta, GA; April 16, 1991; Paper ORGN 231.

28. Jackwerth, E.; Gomiscek, S. *Pure Appl. Chem.* **1984,** *56,* 479–489.

29. Metaxas, A. C.; Meredith, R. J. *Industrial Microwave Heating;* Peter Pergrinus Ltd.: London, 1983.

30. Neas, E. D.; Collins, M. J. In *Introduction to Microwave Sample Preparation: Theory and Practice;* Jassie, L. B.; Kingston, H. M., Eds.; American Chemical Society: Washington, DC, 1988; Chapter 2, pp 7–32.

31. Kingston, H. M.; Jassie, L. B. In *Introduction to Microwave Sample Preparation: Theory and Practice*; Jassie, L. B.; Kingston, H. M., Eds.; American Chemical Society: Washington, DC, 1988; Chapter 6, pp 93–154.

32. Greenwood, N. N.; Earnshaw, A. *Chemistry of the Elements*; Pergamon: Oxford, England, 1984; p 744.

33. Gordon, C. L. *J. Res. Natl. Bur. Stand.* **1943**, *30*, 107–111.

34. Kingston, H. M.; Walter, P. J. *Spectroscopy* **1992**, *7*, 20–27.

35. Commarmot, R.; Didenot, D.; Gardais, J.-F. *Apparatus for the Chemical Reaction by Wet Process of Various Products,* U.S. Patent 4,681,740; July 21, 1987.

36. Burguera, M.; Burguera, J. L.; Alarcon, O. M. *Anal. Chim. Acta* **1986**, *179*, 351–357.

37. Ganzler, K.; Bati, J.; Valko, K. *A New Method for the Extraction and High-Performance Liquid Chromatographic Determination of Vicine and Convicine in Fababeans*; Proceedings for the International Eastern European–American Symposium of Chromatography: 1984; pp 435–442.

38. Kingston, H. M. "Microwave Method Development Quarterly Report"; Report IAG# DWI-393254-01-0; National Institutes of Standards and Technology: Gaithersburg, MD, June 1988.

39. Fidler, B., CEM Corporation, personal communication, 1996.

40. Kingston, H. M.; Walter, P. J.; Settle, F. A., Jr.; Pleva, M. A. In *Advances in Laboratory Automation Robotics*; Strimaitis, J. R.; Little, J. N., Eds.; Hopkinton, MA, 1991; Vol. 8, pp 619–629.

41. Settle, F. A. J.; Diamondstone, B. I.; Kingston, H. M., Pleva, M. A. *J. Chem. Inf. Comput. Sci.* **1989**, *29*, 11–17.

42. Settle, F. A.; Walter, P. J.; Kingston, H. M.; Pleva, M. A.; Snider, T.; Boute, W. *J. Chem. Inf. Comput. Sci.* **1992**, *32*, 349–353.

43. Walter, P. J.; Kingston, H. M.; Settle, F. A.; Pleva, M. A.; Buote, W.; Christo, J. In *Advances in Laboratory Automation Robotics*; Strimaitis, J. R., Little, J. N., Eds.; Zymark Corporation: Hopkinton, MA, 1990; Vol. 7, pp 405–416.

44. Haswell, S. J.; Barclay, D. A. *Analyst* **1992**, *117*, 117–120.

45. Kingston, H. M.; Walter, P. J.; Lorentzen, E. M. L.; Lusnak, G. P. *The Performance of Leaching Studies on Soil SRMs 2710 and 2711*; Final Report to the National Institute of Standards and Technology; Duquesne University: Pittsburgh, PA, April 5 1994.

46. Walter, P. J.; Kingston, H. M. *Total Microwave Processing Using Microwave Technologies*; Federation of Analytical Chemistry and Spectroscopy Societies: Cincinnati, OH, 1995.

47. *Fed. Regist.* **1995**, *60(9)*, 3089–3095.

48. EPA Method 3052: "Microwave Assisted Acid Digestion of Silceous and Organically Based Matricies", In *Test Methods for Evaluating Solid Waste–Update (III)*; U.S. Environmental Protection Agency: Washington, DC, 1996.

49. *Fed. Regist.* **1995**, *60(142)*, 37974–37980. (3052).

50. EPA Method 3050B: "Acid Digestion of Sediments, Sludges, and Soils", In *Test Methods for Evaluating Solid Waste–Update (III)*; U.S. Environmental Protection Agency: Washington, DC, 1996.

51. Lorentzen, E. M.; Kingston, H. M. "The Advantage of Temperature Feedback Controlled Microwave Assisted Leaching Under Atmospheric Pressure", In *Total Microwave Processing Using Microwave Technologies*; Federation of Analytical Chemistry and Spectroscopy Societies: Cincinnati, OH, 1995.

52. Lorentzen, E. M.; Kingston, H. M. *Anal. Chem.* **1996**, *68*, 4316–4320.

53. Feinberg, M. H.; Ireland Ripert, J.; Mourel, R. M. *Anal. Chim. Acta* **1993**, *272*, 83–90.

54. Feinberg, M. H.; Suard, C.; Ireland Ripert, J. *Chemom. Intell. Lab. Syst.* **1994**, *22*, 37–47.

55. Kingston, H. M.; Jassie, L. B. *Symposium on Accuracy in Trace Analysis— Accomplishments, Goals, Challenges;* National Bureau of Standards: Gaithersburg, MD, September 29, 1987.

56. Pratt, K. W.; Kingston, H. M.; MacCrehan, W. A.; Koch, W. F. *Anal. Chem.* **1988**, *60*, 2024–2027.

57. Matusiewicz, H.; Suszka, A.; Ciszewski, A. *Acta Chim. Hung.* **1991**, *128*, 849–859.

58. Krushevska, A.; Barnes, R. M.; Amarasiriwaradena, C. J.; Foner, H. A.; Martines, L. *J. Anal. At. Spectrom.* **1992**, *7*, 845–850.

59. Jiang, W., M.A. Thesis; *Ozone Degradation of Residual Carbon from Microwave Digestion of Biological Samples;* Duquesne University: Pittsburgh, PA, 1996.

60. Knapp, G. *Mikrochim. Acta* **1991**, *II*, 445–455.

61. Stöppler, M.; Muller, K. P.; Backhaus, F. *Fresenius' J. Anal. Chem.* **1979**, *297*, 107–112.

62. Würfels, M.; Jackwerth, E.; Stöppler, M. *Anal. Chim. Acta* **1989**, *226*, 1–41.

63. Chakraborty, D. P.; Brezovich, I. A. *J. Micro. Power* **1982**, *17*, 17–28.

64. Wickersheim, K. A.; Sun, M. *Res. Develop. Ind.* **1985**, *November*, 114.

65. Wickersheim, K. A.; Sun, M. H. *Med. Electron.* **1987**, *February*, 84–91.

66. Wickersheim, K. A.; Sun, M.; Kamal, A. *Int. Microwave Power Inst.* **1990**, *25*, 141–148.

67. Wickersheim, K. A. *On-line Measurements in Microwave Ovens;* Proceedings for Polymers, Laminations and Coatings Conference; Orlando, FL, 1989; pp 51–58.

68. Wickersheim, K. A.; Jensen, E. M.; Sun, M. H. *Interference Technol. Eng. Master* **1989.**

69. Wickersheim, K. A.; Sun, M. H. *J. Micro. Power* **1987**, *22*, 85–93.

70. New Product Announcement: IR-TC-500, Milestone Corporation: Sorisole, Italy, July 1991.

71. Collins, L. W. Presented at the Pittsburgh Conference, Chicago, IL, 1996; paper 976.

72. Collins, L. W., O. I. Analytical, personal communication, 1995.

73. Walter, P. J. *Microwave Calibration Program v2.0;* Report IR4718, National Institute of Standards and Technology: Gaithersburg, MD, 1991.

74. Kingston, H. M.; Walter, P. J.; Jassie, L. B.; Chalk, S. J.; Bruce, R. "Calibration of Laboratory Microwave Equipment for the Accurate Transfer of Methods" in preparation, 1997.

75. Walter, P. W.; Kingston, H. M. Presented at the Pittsburgh Conference, Atlanta, GA, 1993; paper 560.

76. *CAALS Oversight Board Meeting;* National Institute of Standards and Technology: Gaithersburg, MD, November 1, 1990.

77. Walter, P. J. Ph.D. Dissertation, *The Development and Validation of Advanced Control Techniques for Microwave Sample Preparation;* Duquesne University, Pittsburgh, PA, 1996.

78. Murphy, T. J. *The Role of the Analytical Blank in Accurate Trace Analysis,* In *National Bureau of Standards Special Publication 422: Accuracy in Trace Analysis: Sampling, Sample Handling, and Analysis;* National Institute of Science and Technology: Gaithersburg, MD, 1976; pp 509–539.

79. Moody, J. R. *Anal. Chem.* **1982**, *54*, 1358A–1376A.
80. Adeloju, S. B.; Bond, A. M. *Anal. Chem.* **1985**, *57*, 1728–1733.
81. Skoog, D. A.; West, D. M.; Holler, F. J. *Fundamentals of Analytical Chemistry*, 6th ed.; Saunders College Publishing: Fort Worth, TX, 1992; pp 22–24.
82. *Principles of Environmental Sampling*; Keith, L. H., Ed.; American Chemical Society: Washington, DC, 1988.
83. Greenberg, R. R.; Zeisler, R.; Kingston, H. M.; Sullivan, T. M. *Fresenius' J. Anal. Chem.* **1988**, *332*, 652–656.
84. Kingston, H. M.; Fassett, J. D. *Report of Analysis—Vanadium in Crude Oil Reference Material (RM 8505)*; National Institute of Standards and Technology: Gaithersburg, MD, 1983.
85. Fassett, J. D.; Kingston, H. M. *Anal. Chem.* **1985**, *57*, 2474–2478.
86. Kingston, H. M.; Fassett, J. D. *Report of Analysis—Determination of Vanadium in Oil (SRM 1634b)*; National Institute of Standards and Technology: Gaithersburg, MD, 1984.
87. Rabinowitz, M. B.; Wetherill, G. W. *Environ. Sci. Technol.* **1972**, *6*, 705–709.
88. Maienthal, E. J. In *U. S. National Bureau of Standards Technical Note 545*; Taylor, J. K., Ed.; U. S. Governmental Printing Office: Washington, DC, 1970; pp 53–54.
89. Zief, M.; Mitchell, J. W. *Contamination Control in Trace Element Analysis*; Elving, P. J., Ed.; Chemical Analysis, Vol. 47; John Wiley and Sons: New York, 1976.
90. "The Filtration Spectrum"; Report P/N 17978; Osmonics, Inc.: Minnetonka, MN, 1984.
91. Miller, G. T. *Living in the Environment*; Wadsworth Inc.: Belmont, CA, 1994; p 571.
92. Taylor, D. B.; Hutton, R.; Koller, D.; Nogay, D.; Kingston, H. M. *J. Anal. At. Spectrom.* **1996**, *11*, 187–191.
93. Moody, J. R. *Philos. Trans. R. Soc. London* **1982**, *305*, 669–680.
94. Gabler, R.; Hege, R.; Hughes, D. *J. Liq. Chromatogr.* **1983**, *6*, 1565–2570.
95. Milestone Product Catalog; Milestone SLR: Sorisole, Italy, 1995.
96. Kuehner, E. C.; Alvarez, R.; Paulsen, P. J.; Murphy, T. J. *Anal. Chem.* **1972**, *44*, 2050–2056.
97. Moody, J. R.; Beary, E. S. *Talanta* **1982**, *29*, 1003–1010.
98. "Certificate of Analysis: Fisher *OPTIMA*™ Perchloric Acid", Report 2D92-1; Fisher Scientific: Pittsburgh, PA, 1995.
99. "Certificate of Analysis: Fisher *OPTIMA*™ Nitric Acid", Report A467-12503; Fisher Scientific: Pittsburgh, PA, 1995.
100. "Certificate of Analysis: Fisher *OPTIMA*™ Hydrochloric Acid", Report A466-4D93-6; Fisher Scientific: Pittsburgh, PA, 1995.
101. "Certificate of Analysis: Fisher *OPTIMA*™ Hydrofluoric Acid", Report A463-5D92-6; Fisher Scientific: Pittsburgh, PA, 1995.
102. *The Fisher Catalog*; Fisher Scientific: Pittsburgh, PA, 1995; p 192C.
103. "Certificate of Analysis: Fisher Trace Metal Grade Nitric Acid", Report A509-11507; Fisher Scientific: Pittsburgh, PA, 1995.
104. Mattinson, J. M. *Anal. Chem.* **1972**, *44*, 1715.
105. PFA Labware Catalog; Savillex Corp.: Minnetonka, MN, 1995.
106. Moody, J. R.; Lindstrom, R. M. *Anal. Chem.* **1977**, *49*, 2264–2267.
107. Kuehner, E. C.; Freeman, D. H. In *Purification of Inorganic and Organic Materials*; Zief, M., Ed.; Marcel Dekker: New York, 1969; Chapter 30, pp 297–306.

108.   Skelly, E. M.; DiStefano, F. T. *Appl. Spectrosc.* **1988,** *42,* 1302–1306.
109.   Prevatt, F. J. *Environ. Test. Anal.* **1995,** *4,* 24–27.
110.   Powell, L.; Garner, E. L., The National Institute of Standards and Technology, unpublished work, 1996.
111.   Greenberg, R. R.; Kingston, H. M.; Watters, R. L.; Pratt, K. W. *Fresenius' J. Anal. Chem.* **1990,** *338,* 394–398.
112.   *Fed. Regist.* **1992,** *57(177),* 41830–41834. (NPDES).
113.   Siriraks, A.; Kingston, H. M.; Riviello, J. M. *Anal. Chem.* **1990,** *62,* 1185–1193.
114.   Turk, G. C.; Kingston, H. M. *J. Anal. At. Spectrom.* **1990,** *5,* 595–601.
115.   Milestone Product Catalog, Milestone SLR: Sorisole, Italy, 1996.
116.   Kingston, H. M.; Jassie, L. B. *J. Res. Natl. Bur. Stand.* **1988,** *93,* 269–274.
117.   Krushevska, A.; Barnes, R. M.; Amarasiriwaradena, C. J. *Analyst* **1993,** *118,* 1175–1181.
118.   Heydorn, K.; Damsgaard, E.; Rietz, B. *Anal. Chem.* **1980,** *52,* 1045–1049.
119.   Sulcek, Z.; Povondra, P.; Dolezal, J. *Crit. Rev. Anal. Chem.* **1977,** 255–323.
120.   Sulcek, Z.; Povondra, P. *Methods of Decomposition in Inorganic Analysis;* CRC: Boca Raton, FL, 1989.
121.   Kingston, H. M., The National Institute of Standards and Technology, unpublished work, 1988.
122.   Bock, R. *A Handbook of Decomposition Methods in Analytical Chemistry;* Marr, I. L., Translator; John Wiley and Sons: New York, 1979; p 202.
123.   Lundell, G. E. F.; Bright, H. A.; Hoffman, J. I. *Applied Inorganic Analysis: With Special Reference to the Analysis of Metals, Minerals, and Rocks;* John Wiley and Sons: New York, 1953; p 212.
124.   Bernas, B. *Anal. Chem.* **1968,** *40,* 1682–1686.
125.   Langmyrh, F. J.; Paus, P. E. *Anal. Chim. Acta* **1968,** *43,* 397–408.
126.   Bogen, D. C. In *Treatise on Analytical Chemistry;* Kolthoff, I. M.; Elving, P. J., Eds.; Interscience: New York, 1982; Vol. 5, Part I, Chapter 1, pp 1–22.
127.   Willard, H. H.; Rulfs, C. L. In *Treatise on Analytical Chemistry;* Kolthoff, I. M., Elving, P. J., Eds.; Interscience: New York, 1961; Vol. 2, Part I, Chapter 24, pp 1027–1050.
128.   Sulcek, Z.; Povondra, P. *Methods of Decomposition in Inorganic Analysis;* CRC: Boca Raton, FL, 1989; p 36.
129.   Scott, W. W. *Scott's Standard Methods of Chemical Analysis,* 5th ed; D. Van Nostrand Company: New York, 1939.
130.   Hewitt, A. D.; Reynolds, C. M. "Microwave Digestion of Soils and Sediments for Assessing Contamination by Hazardous Waste Metals"; Report 90-19; U.S. Army Corps of Engineers Cold Regions Research and Engineering Laboratory: Hanover, NH, June 1990.
131.   Hewitt, A. D.; Cragin, J. H. *Environ. Sci. Technol.* **1991,** *25,* 985–986.
132.   Hewitt, A. D.; Cragin, J. H. *Environ. Sci. Technol.* **1992,** *26,* 1848.
133.   Raoot, S.; Athavale, S. V.; Rao, T. H. *Analyst* **1986,** *111,* 115.
134.   Abele, C.; Weichbrodt, G.; Wichmann, K. *Fresenius' Z. Anal. Chem.* **1985,** *322,* 11.
135.   Engelhardt, G., Milestone Inc., personal communication, 1995.
136.   Sulcek, Z.; Povondra, P. *Methods of Decomposition in Inorganic Analysis;* CRC: Boca Raton, FL, 1989; p 66.
137.   Strucken, E. F.; Floyd, T. S.; Manchester, D. P. In *Introduction to Microwave Sample Preparation: Theory and Practice;* Jassie, L. B.; Kingston, H. M., Eds.; American Chemical Society: Washington, DC, 1988; Chapter 9, pp 187–202.

138. Moore, J. W.; Pearson, R. G. *Kinetics and Mechanism,* 3rd ed.; John Wiley and Sons: New York, 1981.

139. EPA Method 3051: "Microwave Assisted Acid Digestion of Sediments, Sludges, Soils, and Oils", In *Test Methods for Evaluating Solid Waste;* U.S. Environmental Protection Agency: Washington, DC, 1986.

140. Cotton, F. A.; Wilkinson, G. *Advanced Inorganic Chemistry,* 5th ed.; John Wiley and Sons: New York, 1988.

141. Longbottom, J. E.; Edgell, K. W.; Sunderhaus, M. D. *Digestion Method Comparison Study: Methods 3050A, 3051, 3055, and EMMC;* Report RCRA Study 23; U.S. Environmental Protection Agency: Washington, DC, July 6 1993.

142. Kane, J. S. *Fresenius' J. Anal. Chem.* **1995,** *352,* 209–213.

143. Addendum to the "Certificate of Analysis for SRM's 2709, 2710, 2711"; National Institute of Standards and Technology: Gaithersburg, MD, August 23 1993.

144. Collins, L.; Chalk, S. J.; Kingston, H. M. *Anal. Chem.* **1996,** *68,* 2610–2614.

145. McKenzie, H. A. *Trends Anal. Chem.* **1994,** *13,* 138.

146. Robards, K.; McKelvie, I. D.; Benson, R. L.; Worsfold, P. J.; Blundell, N. J.; Casey, H. *Anal. Chim. Acta* **1994,** *287,* 147–190.

147. *Total Kjeldahl Nitrogen, Environmental Monitoring and Support Laboratory;* Report EPA-600/4-79-020; U.S. Environmental Protection Agency: Cincinnati, OH, March 1983.

148. *Kjeldahl Nitrogen Determination with Microwave Sample Preparation;* Report NF V 03-100; Association Française de Normalisation: Paris, France, September 6, 1992.

149. *Total Phosphorus, Environmental Monitoring and Support Laboratory;* Report EPA-600/4-79-020; U.S. Environmental Protection Agency: Cincinnati, OH, March 1983.

150. Suard, C. L.; Feinberg, M. H.; Ireland Ripert, J.; Mourel, R. M. *Analusis* **1993,** *21,* 287–291.

151. Johnes, P. J.; Heathwaite, A. L. *Water Res.* **1992,** *26,* 1281–1287.

Alternative Laboratory
Microwave Instruments

Development of High-Pressure Closed-Vessel Systems for Microwave-Assisted Sample Digestion

Henryk Matusiewicz

The historical development of high-pressure closed-vessel systems in microwave-assisted sample digestion is presented. The current state of the art, including advantages and limitations of this approach, is discussed. The construction and methodological controls for a series of special microwave-digestion bombs and a universal complex microwave-digestion system for chemical preparation for transferring the sample into a liquid state are described. Emphasis is laid on the completeness of the digestion, for which the use of high-pressure high-temperature focused-microwave-heated Teflon bomb is prerequisite. The novel prototype system uses focused microwaves, operated at 2.45 GHz, to improve digestion capability. Methodology was developed using powdered biological reference material. With this new decomposition device, organic material is totally oxidized with nitric acid in a single-step procedure.

Sample digestion is not often recognized as an important step in analytical chemistry. However, it is time consuming and an important source of errors. Therefore, the complete digestion of a sample is required to achieve reproducible and accurate elemental results with instrumental analytical methods. Two significant developments in sample preparation procedures are

- the use of sealed high-pressure bombs (vessels) to accelerate sample digestion and minimize contamination and loss of volatile elements
- the use of microwave radiation to assist in digestion

The merits of high-pressure digestion in Teflon bombs are widely recognized (1) and such techniques are attracting considerable attention. Acid-digestion bombs are generally Teflon-lined, stainless-steel containers (2) that, when sealed, are capable of digesting resistant materials with suitable solvents under elevated temperature and pressure conditions while retaining potentially volatile compounds.

In 1975, Abu-Samra et al. (3) described one of the first uses of microwave heating for the rapid wet acid digestion of biological materials. This report stimulated a long-term development of microwave technology for the preparation of all types of samples for analysis, as summarized in a popular book (4), general review articles (5–11), and in Chapter 2 of this book. These studies have led to commercial microwave-digestion systems and microwave-digestion bombs that have replaced many other classical procedures and are now quite common in the laboratory.

Microwave-Digestion Bombs

The vessels used for microwave acid digestion are either low-pressure or high-pressure bombs. With high-pressure bombs, the pressures and temperatures needed for the total digestion of the sample are reached very quickly, volatile elements are not lost from such sealed vessels, and the sensitive parts of the microwave oven are not subject to corrosive acid fumes. However, no universal recipe can be recommended for the meaning of *low* or *high* pressure related always to a *complete* destruction or dissolution of the organic and inorganic materials. However, because of the higher pressures, more safety considerations are involved.

Parr bombs (Parr Instrument Company, Moline, IL) are designed specifically for microwave heating (12). Parr 4781 and 4782 microwave-digestion bombs consist of a sealable 23-mL (4781) or 45-mL (4782) poly-(tetrafluoroethylene) (PTFE) cup with cover contained within a polymer–resin shell. The bombs can be used at temperatures up to 250 °C and pressures up to 82 atm without corrosion or loss of volatile elements. The bomb has a pressure indicator and an internal pressure-release mechanism for safety.

The Berghof all-PTFE digestion vessel (Berghof GmbH, Eningen, Germany) DAP 20 is isostatically pressed PTFE of 20-mL capacity (13, 14). A thin fluoroplastic foil is inserted between the body and the cap, which acts as a seal and rupture disc. The bomb can operate at temperatures of up to 170 °C and a maximum pressure of 25 atm.

Recently, Kürner Analysentechnik (Germany) announced availability of PTFE digestion vessels (23 mL and 45 mL) equipped with a pressure monitor. A maximum operating pressure of 84 atm at 250 °C is claimed.

GEC Alsthom International S.A./N.V. (Brussel, Belgium) introduced a new line of Teflon bombs for preparing analytical samples that are designed specifically for microwave heating. These bombs [all Teflon and

PTFE–PEEK (poly(ether ether ketone)) models] are available from 1.5-mL to 250-mL capacity and having a maximum operating pressure of about 150 atm and a working temperature of up to 250 °C.

Lorran digestion vessels (Lorran International, Porters Lake, Nova Scotia, Canada) are sealable 20-mL or 40-mL all-PTFE cups with covers (*15*). A unique system allows the pressure to reach about 17 atm at 100 °C before venting. No rupture disks are required because the venting simply occurs through a pressure-relief hole in the cap. As the pressure drops the vessel reseals. The same vessels can be used in both microwave and conventional heating procedures for comparative purposes.

Very recently, CAL Laborgeräte (Germany) announced availability of CAL–MICRO PFA (tetrafluoroethylene with a fully fluorinated alkoxy side chain) digestion vessels (100 mL) equipped with a pressure safety valve. A maximum operating pressure of 35 atm at 170 °C is claimed.

Laboratory-made all-Teflon bombs also are appropriate for microwave-heated digestion purposes (*16*). In principle, any vessel transparent to microwave radiation may be used for microwave digestions, and vessels made from material with a softening point well above 150 °C are generally preferred. The optimum digestion vessels to be used are those made of polyfluorocarbons, typically PTFE. This material has a softening point of about 250 °C and shows particularly low surface adsorption of ions.

Kojima et al. (*17*) modified a Teflon digestion bomb by using a closed double-Teflon vessel with a polypropylene jacket to permit leak-free and safe digestion of samples. In this design, the high pressure and temperature necessary for complete and rapid sample decomposition were easily attained. In my opinion, this concept of a double-Teflon vessel was subsequently implemented by San'ai Kagaku Co., Nagoya (Japan), who introduced a sealed PTFE vessel with a polypropylene jacket (*18*). A 25-mL PTFE pressure vessel with a 7-mL Teflon–PFA interior vessel is available for this purpose. Water can be introduced into the outer PTFE reactor pressure vessel, which serves to compensate for the rise in pressure in the inner vessel.

Pougnet et al. (*19*) designed a general-purpose microwave-digestion pressure vessel suitable for a wide range of applications that could be easily manufactured from available materials at a reasonable cost while providing the necessary safety features and ease of operation. The 85-mL vessel is constructed entirely from PTFE and has a body-wall thickness of 12 mm machined from solid PTFE cylinders. The protection shield is made of 20-mm polypropylene. The most simple approach for protection against excess pressure was to incorporate a rupture disc (0.2-mm thick PTFE disc of 12-mm diameter that was subjected to gamma radiation). The disc ruptured reproducibly at 10 atm and 110 °C. A torque wrench set at 15 Nm can be used to reproducibly seal the vessel.

A microwave reactor suitable for organic synthesis and kinetics studies was described by Constable et al. (*20*), and in my opinion, the microwave reaction vessel could be well-suited to microwave-heated acid digestion of samples. The 100-mL Teflon reaction vessel can be operated up to 200 °C and 10 atm. It has two ports on the lid, one of which is connected to a pressure gauge whereas the other accommodates a Luxtron optic-fiber thermometer. With such a device it is possible to stir and directly monitor the temperature and pressure in the microwave environment.

Vapor-phase sample digestion in pressurized PFA–Teflon vessels by using microwave radiation is an extremely effective method for the digestion of organic and inorganic material (21). A 100-mL lined digestion vessel (PFA–Teflon) is available for this purpose from CEM Corporation (Matthews, NC). The bomb can be used up to 250 °C and about 14 atm. A laboratory-built all-PTFE microsampling device can be employed for vapor-phase sample digestion.

Several microwave-heating configurations were presented by Pougnet et al. (22, 23) based on 500-W or 1200-W, 2.45-GHz fundamental-mode microwave waveguide cavities, which heat the cavities inside pressure vessels currently used in laboratories for sample digestion and other applications. A computer is used to accurately control time and power and to continuously monitor temperature of a sample by using standard thermocouples. Pressure measurements inside closed-vessels taken by using a pressure transducer are also performed. Feedback control allows constant temperature and pressure to be maintained at predetermined levels. However, no description of the maximum temperature and pressure limits of these devices is made anywhere. The waveguide is shown to heat samples very reproducibly.

The capsule concept was reviewed in detail by Légére and Salin (24). The sample is handled in an encapsulated form until it is in the digestion solvent. The operation of the capsule-based microwave-digestion system proceeds in several steps for which temperature and pressure are monitored. The heating in such a system, as in all microwave bomb systems, is from the solution outward, and the system performance is dictated by the same chemical and physical laws governing other microwave-assisted systems.

Commercial Microwave-Digestion Systems

To meet market preference, several microwave-digestion systems for elemental analyses have received commercial acceptance. This result has lead to many different technical constructions frequently being used for the dissolution of inorganic material and the decomposition of organic material in routine as well as in research applications.

The first commercial microwave-digestion system developed specifically for trace-metal sample preparations and for research laboratories was produced by CEM Corporation, Model MDS-81D, introduced in 1982 (25). Recently, CEM engineered the MDS-2000/2100 series microwave sample-preparation systems. They subsequently introduced a pressure monitor (controller) to allow users to observe the pressure generated in a digestion vessel while developing sample preparation methods. The built-in pressure-control system allows unattended digestion of samples with established methods programmed in the system memory. More recently, CEM introduced the thermooptic temperature control system, which provides a unique design solution to temperature measurement in a microwave field. CEM Advanced Composite Vessels combine the chemical resistance and inertness of Teflon–PFA with the mechanical strength of an Ultem polyetherimide outershell. A unique double-wall construction of lined digestion vessels (advanced composite sleeves) allows acid digestion of samples at temperatures up to 200 °C and pressures up to about 13.5 atm. In addition, CEM offers a microwave vessel designed for high-pressure digestion of samples. As many as six heavy duty vessels can be used with the MDS-2000 series microwave-digestion system to perform in-situ acid-digestion reactions at pressures to 40 atm and temperatures to 200 °C. A corrosion-protected cavity (Teflon coated) and efficient exhaust system make this oven attractive for use in a laboratory environment. A number of accessories, including a vessel-capping station and smaller digestion bombs (3 mL and 7 mL) are also available.

Floyd Corp. (Lake Wylie, SC, now marketed by O.I. Analytical, College Station, TX) markets the RMS-150 remote microwave system with integrated pressure control (*26, 27*). The digestion vessels incorporate a double-wall vessel design for operation at pressures as high as 20 atm and with internal temperatures up to 250 °C. The thread design enables the vessels to be hand tightened, thereby eliminated the need for a capping station.

The Q Wave-1000 (Questron Corporation, Princeton, NJ) is engineered for computer control (*28*). Pressure is sensed and controlled by a sealed pressure transducer and a thermocouple sensor is fully contained in the sensor head. The thermocouple sensor is designed and manufactured specifically for an acid microwave environment. The Q Wave vessel has three compartments: the outer wall is made from a high-density microwave-transparent composite; the middle wall is made from PTFE; and the inner compartment is made of PFA–Teflon. The vessels are rated in excess of 15 atm and 200 °C.

The Spex CDS 7000 closed-digestion system (Spex Industries Incorporation, Edison, NJ) is, in my opinion, exactly the same unit offered by Questron Corp., the Q Wave-1000.

Prolabo (Paris, France), in collaboration with Anton Paar K.G. (Graz, Austria), developed the Superdigest pressure-decomposition system with focused-microwave heating. This instrument uses closed quartz vessels for fast and reliable sample preparation for elemental trace analysis. Up to 4 digestion units can be operated simultaneously. Digestion vessels are made of quartz (80 mL) enclosed in plastic cylinders (quartz is universally known to be the best material for reaction vessels because it offers low contamination and adsorption effects and excellent temperature resistance). In these vessels microwave decomposition can be performed at 80 atm and more than 300 °C. The vessels are intensively cooled by an air stream to keep the temperature low. The pressure, controlled by an opto-mechanical pressure sensor, is kept constant at the maximum pressure level during the entire decomposition procedure. For safety a rupture disc and acrylic transparent shield are employed.

Anton Paar (Graz, Austria) offers the pressurized microwave decomposition system (29), which is designed for a maximum pressure of 80 atm and a corresponding temperature of approximately 300 °C. The most significant advantage of the system is that of controlling the microwave power by the pressure optical sensor built in the digestion vessel. The device works like a pressure cooker. The digestion vessels are made from high-purity quartz; only the seal at the top of the vessel is made from Teflon. During the decomposition cooling air is forced through the bomb to prevent overheating of plastic parts and to facilitate rapid cooling of the vessel (5–10 min) at the end of the process. Simultaneous digestion of a maximum of two samples can be achieved. More recently, Anton Paar introduced the Multiwave microwave digestion system (1000 W with unpulsed power control, 2.45 GHz) (30). The system consists of a rotor with 6 pressure–temperature controlled vessels. The digestion vessels are made of quartz (25-mL or 50-mL volume for a maximum pressure of 74 atm) or of TFM (100-mL volume for 30 atm).

Meditest Ltd. (Budapest, Hungary) very recently introduced the Digmed microwave-digestion system (31). A special new type of Teflon bomb was constructed in which the vapor pressure can be maintained at a moderate level (up to 5 atm) by means of an internal quartz or Teflon cooling spiral (a water cooling spiral is inserted into a closed space through the cover). During operation, the reflux of the condensed acid and water vapors continuously renews the liquid phase over the sample. The Digmed microwave system consists of a microwave oven (2.45 GHz, 27-L internal volume, 700-W power, adjustable continuous-impulse mode), a rotating carousel for 6 Teflon bombs and a water cooling system (safety control valves, distribution and collector system, and PTFE tubing).

The MLS-1200 microwave-digestion system (Milestone s.r.l., Sorisole, Italy) is a commercially available instrument operated at a maximum of 1200 W. This system features programmable time and power, a unique *unpulsed* 250-W mode, an exhaust module system for venting vapors, fume absorbing modules, and a cooling station (32–34). This system is designed for ease of implementation with a robotic interface. The newest unit, the MLS-1200 Mega, has been developed with automatic pressure control up to 30 atm and automatic temperature control up to 240 °C coupled with feedback to input microwave power, which may be used to maintain pre-set pressures and temperatures. In addition, Milestone introduced a noninvasive IR probe assembly to monitor the external temperatures of each digestion vessel shield (35). Milestone lined digestion vessels of 80-mL and 100-mL volume, manufactured from a patented polymer tetrafluoromet-oxil (i.e., Hostaflon TFM, a chemically modified PTFE), allow the use of high-boiling-point acids, such as sulfuric acid (33). The bomb can be used (according to the manufacturer) at a temperature up to 350 °C and a maximum pressure of 110 atm.

Six or 10 vessels at a time can be placed in the microwave-digestion rotor (MDR). The MDR technology system consists of a core (rotor) of high-resistance polypropylene, in which a number of niches are carved along the periphery. A high-purity TFM–Teflon vessel, protected by a high-impact shield, is lodged in every niche. The vessel cover is kept in place by a plastic dome-shaped spring with a flat top, loaded by a plastic screw. The MDR is placed in the microwave cavity of the MLS-1200 standard unit or MLS-1200 Mega unit and rotates to ensure homogeneous microwave absorption. The vessels can be tightened and opened easily by hand by using either a special wrench or torque wrench. Very recently, Milestone introduced QV.40 air-cooled 35-mL quartz vessels. The vessel can be used up to 260 °C and 100 atm. Six high-purity quartz vessels at a time can be placed in this same MDR system.

Linn High Therm GmbH (Hirschbach, Germany) markets the Lifumat Mic 1.2/2450 microwave generator (900 W) with pressurized sample chamber (4 atm) for microwave acid digestion (particularly digestion bombs). Merck (Darmstadt, Germany) recently introduced MW500 micro-wave-digestion system (500 W, 2.45 GHz). The system consists of a set of the digestion vessels made of Teflon (30 mL).

Milestone in cooperation with Andreas Hofer (Hochdrucktechnik GmbH, Germany) very recently introduced the ultraCLAVE microwave digestion system. The prototype device combines microwave heating with high-pressure vessel technology (reactions can be conducted at pressures and temperatures up to 200 atm and 350 °C, respectively). In my opinion, this concept suggested by Matusiewicz (40) to fill the gap between the sophisticated, expensive, and effective thermal high-pressure digestion system [i.e., high-pressure Asher (HPA)] and the sophisticated and expensive microwave heated digestion systems (i.e., CEM, Milestone) was subsequently implemented by Milestone who introduced the ultraCLAVE system. In addition, in my opinion, the technique developed is an extension of the pressurized sample chamber previously reported by Linn High Therm model Lifumat Mic.

Berghof Laborprodukte GmbH (Eningen, Germany) recently introduced two microwave-digestion systems developed for trace analysis. The MWS-1 system offers direct in situ temperature measurement by IR sensor technique through the TFM wall of the closed reaction vessels (30- or 80-mL). The microwave power is regulated according to the reaction temperature from up to 6 vessels. The second system, a high-pressure high-temperature microwave digestor, MDA-II, is in my opinion similar to the microwave-heating configurations presented by Pougnet et al. (*22, 23*). This system with focused-microwave heating (500 W power) is designed for a maximum pressure of 100 atm and a temperatures of 270–290 °C. The apparatus has a high standard of safety with stainless-steel vessel and rupture disc fitting and in situ temperature control with combined power regulation. Several of these commercially available microwave-digestion systems have the possibility for the simultaneous digestion of up to 10 samples with applied microwave power between 600 and 1200 W.

Perhaps the current fascination for using microwave heating for on-line digestion has led to the introduction of commercial instruments based on this hybrid technique. CEM developed the SpectroPrep continuous-flow automated microwave-digestion system. Questron is marketing the AutoPrep-Q5000, an automated, discrete-flow, high-pressure microwave-digestion system. Similarly, Perkin Elmer (Überlingen, Germany) offers an on-line flow injection microwave-digestion system. The pressure developed in these systems can be as high as in many closed bombs. Some additional details will be discussed in Chapter 6 of this book and will not be presented here.

High-Pressure High-Temperature Focused-Microwave-Heated Teflon Bomb

From the previous discussion, we see that microwave acid digestion can be easily adapted for closed-vessel digestions; hence, its application has been limited to decompositions in closed Teflon-lined vessels made of nonmetallic microwave-transparent materials operating with a maximum upper safe pressure of around 80–100 atm. For some techniques (especially voltammetric determinations of metals) it is necessary to destroy all electroactive organic matter (*36*). For complete mineralization of organic compounds in closed vessels, an obvious need exists to achieve both higher temperatures and pressures while containing the reaction products, including copious amounts of gaseous reaction by-products. Even under optimized conditions, biological samples cannot be completely oxidized within the temperature and pressure limit of Teflon vessels alone. Unfortunately, Teflon-lined reaction bombs placed inside stainless-steel jackets are unusable because of their metallic construction; therefore, various jackets (containers) made of high-strength nonmetallic microwave-transparent polymeric materials have been used (with the internal pressure not exceeding the design limits of the vessel shell). Also, the optimum temperature range for Teflon vessels has a maximum around 200 °C (the recommended maximum temperature for short-term use of Teflon bombs is 250 °C).

An additional disadvantage of microwave oven systems is that as much as a 10–20% error in temperature can result from the disparity between rated and actual power outputs, and uneven heating (nonhomogeneous distribution of microwave radiation in the oven cavity) may be a problem (37–39). Reproducibility of field strength or density for different equipment is essential if reproducible results are to be obtained. Furthermore, the ability to transfer methods or reproduce them on different equipment requires good calibration methods, which at present is not a common practice. The feedback control that is now just becoming available can also help to alleviate this problem. Finally, an additional limitation is the necessary delay in opening digestion bombs, which must first be cooled to room temperature to reduce the internal pressure to a safe level and ensure the condensation of volatile analyte species. Therefore, the capacity of the present generation of microwave ovens for sample digestion is rather limited, and significant improvements could be made to the technology currently in use.

In response to these limitations and focusing on the fact that rapid heating of solvents and samples within a polymer vessel can lead to significant advantages over high-pressure steel-jacketed Teflon bombs (which are thermally heated), research and development engineers from Enterprise for Implementation of Scientific and Technological Progress, Plazmatronika Ltd., Wrocław, Poland, and I sought to develop a focused-microwave-heated bomb that would exceed the operational capabilities of existing microwave-digestion systems and permit construction of an integrated microwave source or bomb. The combined advantage of having a high-pressure Teflon bomb (1, 2) incorporating microwave heating (4–11) has produced a focused high-pressure high-temperature microwave-heated digestion system (40, 41) capable of being water or fluid cooled in situ. Preliminary tests indicate that this arrangement works well with both low- and high-power unpulsed or continuous-mode microwave generators and permits precise control of the calories absorbed by the sample. The following section of this chapter describes the design characteristics and preliminary operating conditions established for the focused-microwave-heated digestion system.

Design Concept

The proposed device should consist of an integrated high-pressure high-temperature polymer bomb and a focused-microwave-heated digestion system operating under the very high pressure and temperature such as those typically employed with a classical Parr-type acid-digestion bomb. The design criteria for this device included obtaining a configuration that would match the dissolution and decomposition performance of conventional stainless-steel thermally heated bombs but would permit precise control of the calories absorbed by the sample. These criteria were met by the system illustrated in Figure 1 (patent pending, Poland) (40, 41). This arrangement uses a rectangular brass waveguide to direct and focus the

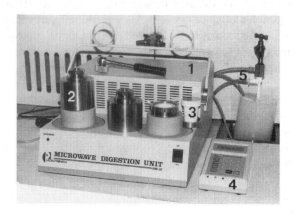

A

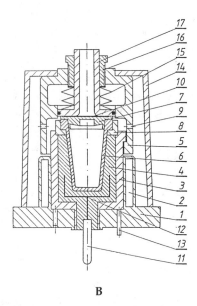

B

Figure 1. Experimental setup: (A) overview photograph (high-pressure high-temperature focused-microwave-digestion system); (1) microwave power generator; (2) microwave-heated assembled digestion bomb; (3) TFM–Teflon sample cup vessel; (4) remote control panel; (5) fluid cooling system. (B) Schematic of the TFM–Teflon high-pressure high-temperature focused-microwave-heated digestion bomb; (1) bottom outer body (stainless-steel); (2) fluid cooling system; (3) pressure bomb (stainless-steel jacket); (4) ceramic vessel liner; (5) bayonet closure cap (stainless-steel); (6) TFM–Teflon sample cup vessel; (7) TFM–Teflon lid; (8) ring (stainless steel); (9) relief port; (10) metal rupture foil; (11) external coupling antenna; (12) internal coupling antenna system; (13) antirotation post; (14) polypropylene protection jacket; (15) compressible rupture discs (safety valve); (16) hexagonal lug; (17) stainless-steel compression plate.

microwaves into the sample by means of the antenna coupler and a polymer bomb to contain the sample and acids. A special configuration of antennas is used: internal and external antenna systems are carefully sized to achieve efficient coupling of microwave energy between a magnetron and the sample and acids. The design offers very efficient (>90%) energy transfer from the microwave generator to the sample and acids and can be employed at low as well as high powers. Details of a 700-W low-power ripple, 2.45-GHz regulated microwave generator, Model MPC-01a (Plazmatronika Ltd.), were reported (42). The high-pressure high-temperature digestion vessel (inner volume, 28 mL) is constructed of a patented polymer TFM–PTFE that permits use of high-boiling-point acids (such as sulfuric acid) at temperatures up to 350 °C. The use temperature of TFM–PTFE polymer material is 260–350 °C, and the melting or decomposition temperature is 350–380 °C (43).

The TFM–Teflon sample-digestion vessel is conically shaped with a handle that can be attached to the middle of the top ring (stainless steel) allowing it to be moved up. The vessel is mounted in an autoclave (outer bomb casing) of acid-proof stainless steel, which can withstand pressures in excess of 200 atm. However, a safety disc will rupture if the internal pressure exceeds 110 atm. Two safety systems deal with overpressure conditions: the first is actuation of the helical (or disc-shaped) spring in case of a very slow rise to overpressure, and the second is release of the rupture disc foil in case of a step-wise excess in pressure. During operation, a polypropylene protection shield is used around the bomb. The hexagonal lug top must always be carefully tightened by using a torque wrench that protects the thread from forces higher than 1.5 Nm. No calculations have yet been made to determine the force on the TFM–Teflon lid, because of the difficulty in estimating the coefficient of friction in the thread. As the need for a more effective cooling method to minimize the delay in opening such pressure vessels has been recognized (31, 44), a water or fluid (up to −10 °C) is used for in-situ vessel cooling in both a pre- and postdigestion mode. The vessel can be easily tightened and opened by hand using a torque wrench. The focused-microwave-heated digestion concept is versatile, and additional arrangements (i.e., one microwave power generator and 4 or 5 separate TFM–Teflon bombs) are feasible.

The new commercially available version of the system (Model BM-1 S, BM-2 S, and BM-5 S, Plazmatronika Ltd., Wrocław, Poland) is equipped with microprocessor control and accommodates one, two, or five samples at a time. The commercial version of the bomb vessel is presently not equipped with a temperature indicator but with pressure sensors (an indirect measurement via a medial separation Teflon membrane). However, for research and development purposes it is being equipped with sensors for monitoring temperature (Teflon-immersed thermocouple submerged inside the vessel). A significant feature of the new design is the relatively easy machining of the microwave-heated bomb assembly in which the only precisely defined dimensions are the length and diameters of antennas as well as their configuration.

Application

The primary purpose of this work was to evaluate the performance of the proposed high-pressure high-temperature focused-microwave-heated Teflon bomb system for the decomposition of organic materials, indicated by the completeness of sample destruction (40). The usefulness of the digestion technique was judged from the residual carbon content, not from an optical point of view (i.e., clear and colorless solution). The amount of residual carbon was calculated as a percentage of the original carbon content in the sample. Biological reference materials were chosen as suitable samples for demonstrating the feasibility of the closed-vessel microwave-heated procedure and evaluating the effectiveness of sample destruction. The standard reference material from the National Institute of Standards and Technology, NIST SRM 1577a Bovine Liver, contains 51% C (dry weight). The results, summarized in Table I, clearly show that the organic matrix was completely oxidized during pressurized focused-microwave-heated bomb digestion.

Good agreement between the microwave-heated bomb decomposition and a traditional thermal acid-decomposition procedure in a quartz bomb was obtained. As seen in Table I (40), the microwave technique required only about 3% of the time necessary for the thermal high-pressure technique to decompose biological powdered material. In this study, the completeness of decomposition was investigated for samples of about 0.1 g because this mass is sufficient for the typical analysis of many trace elements (metals) in such samples. The amount of acid should be minimized to avoid contamination as well as for safety and economy. This result supports the validity of the present decomposition procedure and operating conditions and confirms the suitability of the focused-microwave-heated bomb decomposition system for complete oxidation of such biological material.

Table I. Total Residual Carbon Content in Samples of Bovine Liver (NIST–SRM 1577a) Digested by High-Pressure Focused-Microwave Heating

				Residual Carbon[a]		
Digestion Method	Sample Mass (g)	Final Volume (mL)	Digestion Time (min)	In Digestate ($\mu g/mL$)	In Dry Sample (mg/g)	Efficiency of Oxidation[b] (%)
Microwave heated bomb	0.1	10	4	30 ± 3	3 ± 0.3	99.4
High-pressure asher	0.1	10	120	20 ± 2	2 ± 0.2	99.6

[a]Total carbon content of undigested sample: 510 ± 10 mg/g ($n = 3$), dry-mass basis. Mean and standard deviation reported.
[b]Five measurements from a triplicate sample preparation.

The cavity and the antennas did not overheat during this digestion study, a result indicating efficient transfer of energy to the sample and solution. No microwave radiation leakage above 1 mW/cm^2 was detected at a distance of 5 cm from the stainless-steel bomb under all operating conditions. However, the limiting pressure required for the rupture of the inner TFM–Teflon vessel has not yet been determined.

Relatively few studies have been reported concerning the completeness of microwave-heated sample decomposition (*21, 36, 45–57a*); hence, monitoring the residual carbon content of test digestions proved to be a useful tool in the development of the procedure (*21, 45–49, 51–57*). To limit the complexity of the experimental work a single power absorbing component, nitric acid, was chosen as the digestion acid. The selection of samples studied was designed to offer as wide a range of initial carbon content as possible (ca. 40–60%). Different volumes of acid, heating times and power were investigated. The final optimized results are summarized in Table II (*45*) and clearly show that the organic matrices can be completely oxidized within 5 min. Beyond 5 min no further destruction occurs. In addition to time, the relative volume of HNO$_3$ and applied power play an important role. Thus, a 5-min decomposition with 2 mL of concentrated HNO$_3$ and 100 W of applied power was necessary for complete destruction of these biological materials. Therefore, a digestion time of 5 min was selected for all the samples to generalize the procedure. Complete decomposition can be achieved by using HClO$_4$ and, in fact, many such decompositions report no residual carbon. In this work, HClO$_4$ was not used simply because it is not required; high-pressure and high-temperature digestion destroys the organic carbon completely by using only HNO$_3$. Although the bomb under study was not equipped with any pressure or temperature indicators, it may be assumed that a temperature between 250 °C and 320 °C was reached along with a pressure above 100 atm (but definitively below 110 atm). These estimates are based on the fact that complete oxidation of organic matrices occurs at temperatures between 250 °C (*58*) and 320 °C (at about 100 atm) (*59, 60*).

Conclusions

In a review of acid microwave-heated sample digestion by Matusiewicz and Sturgeon (*6*) one interesting comment highlighting future study was made: "Microwave-digestion systems could also be designed specifically for use with steel-jacketed Teflon bombs utilizing a waveguide cavity design rather than an oven. This approach (energy from the magnetron focused through a waveguide directly into the sample) should attain much faster heating and higher pressures. Additionally, this should result in higher precision and sample-to-sample reproducibility compared to microwave oven digestions". In this respect, the high-pressure high-temperature

Table II. Total Residual Carbon in Samples Digested by High-Pressure Focused-Microwave Heating

Material	Original Carbon Content[a] (%)	Fat (%)	Lipids (%)	Residual Carbon Content[b] (%)	Residual Carbon[c] (%)	Efficiency of Oxidation (%)
NRCC TORT-1	43.5 ± 0.9	—[d]	—[d]	0.1	0.4	99.6 ± 2.0
NRCC DORM-1	44.4 ± 0.9	5	—[d]	0.2	0.9	99.1 ± 2.0
NIST 1566a	49.1 ± 1.0	—[d]	—[d]	0.2	0.4	99.6 ± 2.0
NIST 1577	49.5 ± 1.0	—[d]	—[d]	0.3	0.6	99.4 ± 2.0
NRCC DOLT-2	50.6 ± 1.0	24	—[d]	0.4	0.8	99.2 ± 2.0
NRCC LUTS-1	58.8 ± 1.2[e]	—[d]	55[e]	0.6[e]	1.0[e]	99.0 ± 2.0

NOTE: Digestion of samples was for 5 min at 100 W. Values are in percent and are triplicate measurements from a duplicate sample preparation. Mean and standard deviation are reported.
[a]Total carbon content of undigested samples.
[b]Sample mass 0.1 g, final volume 10 mL.
[c]Residual carbon calculated as a percentage of original carbon in the sample.
[d]No declaration.
[e]Dry-mass basis.

bomb-focused-microwave-heated digestion system described here can make a significant contribution.

Through an outline of existing commercial closed-vessel microwave-heated digestion systems, this chapter described the development of a high-pressure high-temperature focused-microwave closed-vessel device. Such a device is seen to bridge the gap between the sophisticated, expensive, and effective thermal high-pressure digestion system (i.e., HPA) and the sophisticated and expensive microwave-heated digestion systems (i.e., CEM, Milestone) that do not provide complete digestion efficiency.

The combined use of a high-pressure high-temperature TFM–Teflon decomposition vessel with a stainless-steel jacket and microwave heating permits complete destruction of the organic carbon by using only HNO_3. This method is extremely effective and simple for the digestion of organic materials. Microwave-heated digestion can be used to rapidly and completely oxidize organics in a single step procedure by using a closed system if sufficiently high pressure can be achieved. This result contrasts with the multistep procedures often required (i.e., cycles of heating and cooling to limit pressure build up and postmicrowave digestion).

However, the completeness of the reaction generally is not necessary if the trace organics are known and the instrumental system is appropriate. For example, current methodology based on inductively coupled plasma (ICP)–mass spectrometric and ICP–optical emission spectroscopic instruments often adds milligram quantities of complex organics to the final matrix and improve the analytical accuracy of the system. Indeed the many other methods of microwave decomposition should not be excluded as accurate alternatives for sample preparation.

The use of water or other fluids for in-situ vessel cooling reduces the delay in opening the vessel and processing the digests. The primary advantage of the proposed design is that the closed TFM–Teflon focused-microwave-heated bomb enables very high pressure and temperature to be reached. In this apparatus, the pressure increases with a rise in the temperature and the generated pressure is not controlled independent of the temperature. However, because there is presently no way of monitoring the pressure and temperature inside the high-pressure vessel, the effects of these parameters on the decomposition cannot be evaluated. In general, the effects of high pressure and high temperature during microwave-heated digestion remain to be evaluated, especially for safe application to an exothermic spontaneous reaction of HNO_3 and H_2O_2, which may occur during sample digestion (*61*).

Further development is still required in order to improve the design of the high-pressure high-temperature systems. For example, one assembly may be augmented by even four or five such arrangements (thereby permitting simultaneous multisample digestions). Recently, the control of instantaneous power and prototype of 5 separate TFM–Teflon bombs and one microwave power generator was proposed by Plazmatronika Ltd. This same stabilized generator could be used both to provide microwave power to the bomb assembly as well as to a microwave cavity for microwave-induced plasma atomic emission spectrometry.

Finally, this arrangement brings a new dimension to sample preparation and may combine the qualities of the Parr-type Teflon bomb in retaining volatiles and of the high-pressure Asher in completing effective decomposition within a short heating time characteristic of the microwave technique. The final goal of these studies is the development of an on-line apparatus based on the concepts described herein.

Acknowledgment

Financial support by the Poznań University of Technology, Grant No. DS 31-471/96 and cooperation with the research and development staff of Plazmatronika Ltd. is gratefully acknowledged.

References

1. Jackwerth, E.; Gomišček, S. *Pure Appl. Chem.* **1984,** *56*, 479.
2. Karpov, Y. A.; Orlova, V. A. *Vysokochist Veshchestva* **1990,** *2*, 40.
3. Abu-Samra, A.; Morris, J. S.; Koirtyohann, S. R. *Anal. Chem.* **1975,** *47*, 1475.
4. *Introduction to Microwave Sample Preparation: Theory and Practice;* Kingston, H. M.; Jassie, L. B., Eds.; American Chemical Society: Washington, DC, 1988.
5. de la Guardia, M.; Salvador, A.; Burguera, J. L.; Burguera, M. J. *Flow Injection Anal.* **1988,** *5*, 121.
6. Matusiewicz, H.; Sturgeon, R. E. *Prog. Anal. Spectrosc.* **1989,** *12*, 21.

7. Šulcek, Z.; Novak, J.; Vyskočil, J. *Chem. Listy* **1989**, *83*, 388.
8. Šulcek, Z.; Povondra, P. *Methods of Decomposition in Inorganic Analysis*; CRC Press: Boca Raton, FL, 1989; Chapter 6, pp 159–164.
9. Kuss, H. M. *Fresenius J. Anal. Chem.* **1992**, *343*, 788.
10. Smith, F. E.; Arsenault, E. A. *Talanta* **1996**, *43*, 1207.
11. Chakraborty, R.; Das, A. K.; Cervera, M. L.; de la Guardia, M. *Fresenius J. Anal. Chem.* **1996**, *355*, 99.
12. Zehr, B. D.; Fedorchak, M. A. *Am. Lab.* **1991**, *23*, 40.
13. Buresch, O.; Hönle, W.; Haid, U.; v. Schnering, H. G. *Fresenius Z. Anal. Chem.* **1987**, *328*, 82.
14. Feuerbacher, H.; Böckler, J. *6. Colloquium Atomspektrometrische Spurenanalytik* **1991**, 647.
15. Rantala, R. T. T.; Loring, D. H. *Anal. Chim. Acta* **1989**, *220*, 263.
16. Xu, L.; Shen, W. *Fresenius Z. Anal. Chem.* **1988**, *332*, 45.
17. Kojima, I.; Uchida, T.; Iida, C. *Anal. Sci.* **1988**, *4*, 211.
18. Miyahara, M.; Saito, Y. *J. Agric. Food Chem.* **1994**, *42*, 1126.
19. Pougnet, M. A. B.; Schnautz, N. G.; Walker, A. M. *S. Afr. J. Chem.* **1992**, *45*, 86.
20. Constable, D.; Raner, K.; Somlo, P.; Strauss, C. *J. Microwave Power Electromagnetic Energy* **1992**, *27*, 195.
21. Matusiewicz, H.; Sturgeon, R. E.; Berman, S. S. *J. Anal. At. Spectrom.* **1991**, *6*, 283.
22. Pougnet, M.; Michelson, S.; Downing, B. *J. Microwave Power Electromagnetic Energy* **1991**, *26*, 140.
23. Pougnet, M.; Downing, B.; Michelson, S. *J. Microwave Power Electromagnetic Energy* **1993**, *28*, 18.
24. Légére, G.; Salin, E. D. *Appl. Spectrosc.* **1995**, *49*, 14A.
25. Gilman, L. B.; Engelhart, W. G. *Spectroscopy* **1989**, *4*, 14.
26. Sturcken, E. F.; Floyd, T. S.; Manchester, D. P. *Introduction to Microwave Sample Preparation: Theory and Practice*; Kingston, H. M.; Jassie, L. B., Eds.; American Chemical Society: Washington, DC, 1988; Chapter 9, pp 187–202.
27. Grillo, A. C. *Spectroscopy* **1990**, *5*, 14.
28. Grillo, A.; Moses, M. *Am. Lab.* **1992**, *24*, 14.
29. Schramel, P.; Hasse, S. *Fresenius J. Anal. Chem.* **1993**, *346*, 794.
30. Kainrath, P.; Kettisch, P.; Schalk, A.; Zischka, M. *Labor Praxis* **1995**, *19*, 34.
31. Heltai, G.; Percsich, K. *Talanta* **1994**, *41*, 1067.
32. Lautenschlaeger, W. *Spectrosc. Int.* **1990**, *2*, 18.
33. Matusiewicz, H. *Mikrochim. Acta* **1993**, *111*, 71.
34. DeMenna, G. J.; Brown, G. J. *Int. Lab.* **1994**, *24*, 25.
35. Mincey, D. W.; Williams, R. C.; Giglio, J. J.; Graves, G. A.; Pacella, A. J. *Anal. Chim. Acta* **1992**, *264*, 97.
36. Pratt, K. W.; Kingston, H. M.; MacCrehan, W. A.; Koch, W. F. *Anal. Chem.* **1988**, *60*, 2024.
37. Morales-Rubio, A.; Cerezo, J.; Salvador, A.; de la Guardia, M. *Microchem. J.* **1993**, *47*, 270.
38. Vereda Alonso, E.; Garcia de Torres, A.; Cano Pavon, J. M. *Mikrochim. Acta* **1993**, *110*, 41.
39. Dema-Khalaf, K.; Morales-Rubio, A.; de la Guardia, M. *Anal. Chim. Acta* **1993**, *281*, 249.
40. Matusiewicz, H. *Anal. Chem.* **1994**, *66*, 751.

41. Report 12/93; Plazmatronika Ltd.: Wrocław, Poland, 1993.

42. Matusiewicz, H. *Spectrochim. Acta* **1992,** *47B,* 1221.

43. Lautenschläger, W.; Schweizer, T. *Labor Praxis* **1990,** *14,* 376.

44. Reid, H. J.; Greenfield, S.; Edmonds, T. E. *Analyst* **1993,** *118,* 443.

45. Matusiewicz, H.; Sturgeon, R. E. *Fresenius J. Anal. Chem.* **1994,** *349,* 428.

46. Matusiewicz, H.; Suszka, A.; Ciszewski, A. *Acta Chim. Hung.* **1991,** *128,* 849.

47. Kingston, H. M.; Jassie, L. B. *Anal. Chem.* **1986,** *58,* 2534.

48. Nakashima, S.; Sturgeon, R. E.; Willie, S. N.; Berman, S. S. *Analyst* **1988,** *113,* 159.

49. Matusiewicz, H.; Sturgeon, R. E.; Berman, S. S. *J. Anal. At. Spectrom.* **1989,** *4,* 323.

50. Friel, J. K.; Skinner, C. S.; Jackson, S. E.; Longerich, H. P. *Analyst* **1990,** *115,* 269.

51. Knapp, G. *Mikrochim. Acta* **1991,** *II,* 445.

52. Krushevska, A.; Barnes, R. M.; Amarasiriwaradena, C. J.; Foner, H.; Martines, L. *J. Anal. At. Spectrom.* **1992,** *7,* 845.

53. Krushevska, A.; Barnes, R. M.; Amarasiriwaradena, C. J.; Foner, H.; Martines, L. *J. Anal. At. Spectrom.* **1992,** *7,* 851.

54. Dunemann, L.; Meinerling, M. *Fresenius' J. Anal. Chem.* **1992,** *342,* 714.

55. Gluodenis, T. J., Jr.; Tyson, J. F. *J. Anal. At. Spectrom.* **1993,** *8,* 697.

56. Krushevska, A.; Barnes, R. M.; Amarasiriwaradena, C. J. *Analyst* **1993,** *118,* 1175.

57. Reid, H. J.; Greenfield, S.; Edmonds, T. E. *Analyst* **1995,** *120,* 1543. (a) Jiang, W.; Chalk, S. J.; Kingston, H. M. *Analyst* **1997,** *122,* 211.

58. White, R. T., Jr. *J. Assoc. Off. Anal. Chem.* **1989,** *72,* 387.

59. Würfels, M.; Jackwerth, E.; Stoeppler, M. *Fresenius Z. Anal. Chem.* **1987,** *329,* 459.

60. Knapp, G. *Fresenius Z. Anal. Chem.* **1984,** *317,* 213.

61. Sah, R. N.; Miller, R. O. *Anal. Chem.* **1992,** *64,* 230.

Chapter 5

Focused-Microwave-Assisted Reactions

Atmospheric-Pressure Acid Digestion, On-Line Pretreatment and Acid Digestion, Volatile Species Production, and Extraction

J. M. Mermet

Microwave-assisted acid digestion based on the use of focused-microwave, atmospheric pressure technology is described. The advantages of using open vessels over closed, pressurized vessels is discussed, particularly about the possibility of adding reagents during the digestion processes and obtaining complete dryness of the digest. Examples of applications that include the determination of volatile species are given for inorganic and organic samples. The operating conditions and the acid selection are reported for a wide variety of samples. The use of focused-microwave, atmospheric pressure technology is also described for the Kjeldahl nitrogen determination in organic materials, the chemical oxygen demand, the preparation of samples for speciation, the on-line treatment, and the solvent extraction.

An interesting application of the use of microwaves is based on the design of so-called focused microwaves. This development has been applied mainly to microwave-assisted acid digestion at atmospheric pressure (incorrectly called open-vessel digestion). However, this technology has been used more recently for the on-line heating of reagents and samples to obtain pretreatment and acid digestion. It has also been used to produce volatile species and to assist extraction.

In this chapter, we will recall the advantages of using focused-microwave-assisted reactions, the technology involved, the various commercially available systems including their operating parameters and

371

accessories, and examples of applications for elemental analysis and Kjeldahl nitrogen determination. Moreover, some recent developments such as on-line sample preprocessing and sample treatment for speciation, volatile species production, and extraction will be described.

Atmospheric-Pressure Acid Digestion

Limitation of Pressurized Microwave-Assisted Digestion

Sample digestion has been recognized as a crucial step in elemental analytical chemistry. Conventional conductive-heating hot-plate digestion is widely used but it is time-consuming and can lead to analyte loss and sample contamination. Also, no precise control of the energy deposited into the sample exists. Thus, microwave-assisted digestion has gained wide acceptance because of its ability in overcoming the limitation of the hot plate: the digestion time is shorter, the reagent consumption is lower, and accuracy and reproducibility are better. Both closed, pressurized vessels and open, atmospheric-pressure vessels are in common use. The characteristics and the advantages of closed, pressurized vessels were described in Chapters 2 and 3. Some major advantages are the use of HNO_3 under pressure for inorganic material digestion and the possibility of using several vessels simultaneously. Although closed, pressurized vessels are used extensively, they suffer from some limitations:

- safety problems because of the use of high pressure and the possibility of explosions due to the generation of hydrogen with alloys and metals
- limited amount of sample, usually less than 1 g for inorganic samples and 0.5 g for organic samples
- limited temperature of the solution due to the nature of the material that is usually used, poly(tetrafluoroethylene) (PTFE), for the vessel
- difficulty in obtaining completeness of the dissolution
- not very suitable for organic compounds
- the single-step procedure offers no possibility of adding reagents
- need for cooling before opening the vessel
- possible memory effects due to the use of porous PTFE

In any instance, PTFE is a material difficult to clean. Moreover, the vessel cannot usually be used for further addition of reagents. Thus, to complement the closed-vessel system a microwave-assisted digestion system operating at atmospheric pressure was needed.

Potential Benefits

Use of atmospheric pressure offers some significant advantages over pressurized vessels:

- safety resulting from the operation at atmospheric pressure with open vessels, for instance with gas-forming species
- possible addition of reagents at any time during the digestion, which allows sequential acid attacks
- use of various types of material to construct the vessel such as PTFE, glass, or quartz
- possibility of working at high temperature with quartz material when using H_2SO_4 near the boiling point to destroy organic compounds
- possibility of obtaining complete dryness of the digest, thus eliminating excess acid
- handling of large sample size
- no need for cooling down depressurizing period

In addition to these advantages, the following specifications would be useful:

- high-efficiency transfer and precise control of the energy deposited into the sample
- full automation
- ability to dissolve any type of materials such as metals and alloys, polymers, and organic compounds (in particular petroleum and food products) and to perform Kjeldahl digestion
- no loss of volatile species (e.g., As, Se, and Cd)
- no change in the composition of the reagents during the digestion procedure
- no risk of external pollution due to the surrounding atmosphere

So far, only one manufacturer (1) has developed a series of microwave systems fulfilling most of the previous requirements. In contrast to the closed systems where the microwave energy is dispersed throughout the oven, the commercially available atmospheric-pressure systems are based on the use of microwaves that are focused into the vessel. The purpose of this design was to develop microwave systems really dedicated to chemistry instead of using a conventional technology mainly developed for domestic applications.

Focused-Microwave Technology

In any commercially available microwave system, the electromagnetic energy is produced by a magnetron at 2450 MHz. The antenna of the magnetron is setup inside a rectangular metallic waveguide. Its location is selected according to the Rieke diagram. Two normalized International Electrotechnological Commission (IEC) waveguides can be used at 2450 MHz: R22 and R26. Their dimensions are 109.22 × 54.61 mm and 86.36 × 43.18 mm, respectively. When using a conventional design, the waveguide is connected to a microwave oven that acts as a cavity through a metallic rotating deflector system. This system is used to disperse the electromagnetic energy to the different resonant modes of the cavity so as to improve the homogeneity of the heating within the cavity. This system allows the use of several vessels in the oven. However, the energy is spread among the various vessels and it is difficult to obtain a high homogeneity of the energy.

Also, a waveguide can be used as a single-mode cavity (2). In this instance, the R26 waveguide is closed at both sides. An opening (typically 42 mm) is used to facilitate the incorporation of a flask into the waveguide. This opening is located so that the axis of the flask corresponds to an antinode of the waves transmitted by the waveguide. A precise location of the wave antinode is obtained by a careful positioning of the opening and length selection of the waveguide along with an adjustment of a tuning device, for instance two metal rods. Therefore, a high coupling between the electromagnetic energy and the sample is obtained (Figure 1), which leads to a high power density. The manufacturer claims that an increase of a factor of 10 in coupling efficiency is obtained compared with the use of a multimode cavity. Preventing the propagation of the microwaves outside the waveguide is obtained by using a tubular cylindrical stack that surrounds the portion of the flask that is outside the waveguide (Figure 1). Therefore, the top of the flask is easily accessible. This arrangement allows the user to add reagents during the process or a device such as a reflux system.

Even though the principle of focused microwaves is efficient in terms of energy transfer, it only allows for the use of one flask at time. Automation for some systems can permit the sequential use of flasks, but this characteristic can be considered a limitation. A more recent development involves the use of four flasks at the same time by symmetrically splitting the microwave energy among the four flasks at the end of the waveguide (Figure 2).

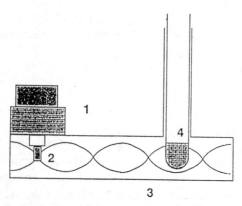

Figure 1. Schematic drawing of a single-mode, focused-microwave system: (1) magnetron, (2) antenna, (3) waveguide, (4) flask.

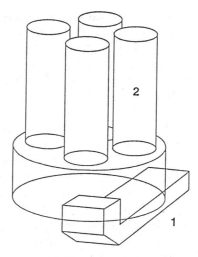

Figure 2. Schematic drawing of a microwave system using four flasks at the same time: (1) waveguide, (2) flasks.

A continuous adjustable percentage of the maximum power is used during a given period of time. This result differs from a multimode cavity where pulsed power is usually used. The maximum power is in the 200–800-W range, depending on the type of microwave system (Table I). Calibration of the power is obtained during the production of the system by measuring the increase in the temperature of a volume of water circulating at the location of the sample.

Operating Parameters and Automation

The main parameters to be considered in developing a method are the percentage of applied power and the duration time. The power and the corresponding time will depend on the type of sample and the reagents to be added. Several examples are given. Full automation includes the addition of reagents via dispensing modules and change of flasks, which are setup on a carousel. Up to 16 different or identical samples can be sequentially digested with 16 different power and time programs if necessary. Flasks are made of borosilicate glass or silica glass or PTFE. Several sizes are available (Table I). Sample size depends on the microwave system and can be up to 10 g with a single-flask system (Table I).

With the commercial system a large variety of accessories are available: reflux systems, acid-fume aspiration and trapping system, dispensing pumps (0.5–5 mL/min), vacuum pumps, etc. One accessory that is particularly useful is the gas thermometer (Megal 500), which allows the user to verify whether the selected duration for the forward power is long enough. An example of temperature measurement related by power is

Table I. Main Characteristics of the Commercially Available Systems Based on the Principle of Focused-Microwave-Assisted Atmospheric-Pressure Digestion

System	Max. Power (W)	Max. Sample Size (g)	Max. Flask Size (mL)	Full Automation
300	200	1	100	No
301	200	2	100	No
A301	200	2	100	Yes
401	300	10	250	No
MX 350	300	10	250	No
MX4350	800	4×10	250	No

SOURCE: Data are from reference 1.

shown in Figure 3. An alternative is the use of the IR emission from the borosilicate tube passing through an aperture and being measured on a calibrated pyrometer.

Use of Acids

Use of microwaves does not modify the intrinsic nature of acid reactions with samples. As mentioned, the major improvement is an efficient control and deposition of energy into the sample with the possibility of sequential additions of reagents. Atmospheric-pressure microwave-assisted digestion is based mainly on the use of acids, which do not differ significantly from those used in conventional hot plates and conductive heating. The most commonly used acids are HNO_3, HCl, and H_2SO_4. Also, HF, H_3PO_4, and aqua regia may be used. Addition of hydrogen peroxide can be carried out to enhance the oxidizing properties of the acids. Summaries of the acid properties are given in Table II.

Use of acids is not only related to their chemical properties but also to the analytical system that is used to analyze the sample. Flame atomic absorption spectrometry, graphite furnace atomic absorption spectrometry (GFAAS), inductively coupled plasma–atomic emission spectrometry (ICP–AES), and inductively coupled plasma–mass spectrometry (ICP–MS) are the most commonly used analytical methods for elemental analysis. Their use implies some requirements to the physical and chemical forms of the sample. In ICP–AES and ICP–MS, low concentration of acids is preferred because a depressive effect is observed with increasing concentrations of acids. This result is true particularly with high-viscosity acids such as H_2SO_4 and H_3PO_4, which modify both the nebulizer aerosol production and the aerosol transport. Low concentration is also preferred for acids such as HCl and HF because of the release of Cl and F. A sample-introduction system can be modified to accommodate the presence of HF. However, when the products at the exit of the torch are not properly evacuated, the torch box, the collimating system, and the sampler for ICP–AES

Temperature (°C) and % power

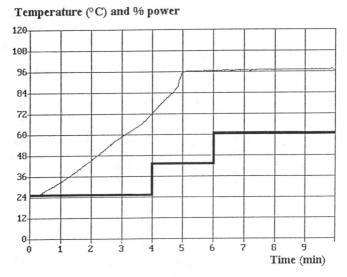

Figure 3. Example of measurement by using a gas thermometer. A 401 system was used with 15 mL of solution. Key: the thick line is power and the thin line is temperature.

and ICP–MS, respectively, can be strongly degraded by Cl or F. Moreover, the presence of Cl, P, or S leads to the formation of polyatomic species in ICP–MS, which may result in the observation of isobaric interferences. A classical isobaric interference is $^{40}Ar^{35}Cl$ on ^{75}As. The ideal acid for ICP–AES and ICP–MS is HNO_3, because its dissociation results only in N and O. ICP–MS users can refer to reference 3 for more details.

At high concentration, HNO_3 is a strong oxidizing agent. It is certainly one of the most widely used acids for digestion of metals, alloys, geological samples, environmental samples, and some organic products (e.g., biological fluids and food). With organic compounds, it is often selected for pretreatment before using $HClO_4$. The main limitation is a relatively low boiling temperature that can result in long digestion times. However, HNO_3 is easy to purify.

HCl is an efficient solvent for metals and alloys. It is often combined with other acids such as HNO_3 for complex matrices. HCl can also be used for geological materials and environmental samples. In contrast, HCl is not suitable for the digestion of organic compounds because it is a weak reducing agent. HCl may be associated to HNO_3 to form aqua regia (1 + 3 (v/v) of 16 M HNO_3 and 12 M HCl, respectively) for the digestion and solubilization of metals such as Au, Cd, In, Pd, and Pt.

$HClO_4$ is a powerful oxidizing acid and will oxidize elements to their highest oxidation states. However, $HClO_4$ can form highly unstable perchlorates with elements and explosive products with organic compounds. A pretreatment is necessary prior to the use of $HClO_4$ with organic com-

Table II. Main Properties and Fields of Applications of Acids Used for Digestion

Parameter	HNO_3	HCl	$HClO_4$	H_2SO_4	HF	H_3PO_4
Form	Azeotrope: 68%, 16 M	36%, 12 M azeotrope: 20%	Azeotrope: 72%, 12 M	Concentrated: 98%, 18 M	48%, 29 M azeotrope: 38%, 22 M	85%, 15 M
Boiling temp. C	122	110	203	338	112	213
Oxidizing ability	Strong	Reducing agent	Strong	Low		
Vessel	Any	Any	Any	Borosilicate glass or quartz	PTFE	Any
Inorganic products; metals and alloys	Highly suitable	Suitable for most metals and alloys	Alloys if W is present Suitable	Suitable	Silica-based materials	Fe-based and Al-based alloys
Inorganic products: geological materials	Suitable	Suitable		Suitable	For SiO_2	
Inorganic products: ceramics					For Si	Suitable
Environment, samples	Suitable	Suitable		Suitable	For Si	
Organic products	Suitable	Not suitable	Need prior treatment, not for fats and oils	Highly suitable		
Plastics and polymers				Highly suitable HNO_3 for organic compounds		
Acid combination	Suitable HCl for alloys, aqua regia	HNO_3 for alloys			Complexation with H_3BO_3	
Losses		Volatile metal chlorides		Some sulfates		
Unsoluble products			Unstable perchlorates	Large amount of acid for organic compounds	K and Ca	
Safety	Explosive nitrates				Corrosive and toxic	
AAS	Highly suitable	Volatile chlorides	Interferences	Not very suitable	Degradation of furnace	Unsuitable
ICP–AES	Highly suitable	Low concentration (Cl)	Low concentration (Cl)	Low concentration (viscosity)	Low concentration	Low concentration (viscosity)
ICP–MS	Highly suitable	Unsuitable (Cl)	Unsuitable (Cl)	Not very suitable (S) (viscosity)	Not very suitable (F) (F attack)	Not very suitable (P)

pounds. Even so, use of $HClO_4$ is strongly discouraged when organic compounds contain fats or oils.

Because of its high boiling temperature, H_2SO_4 can only be used in borosilicate glass or quartz vessels. It is a versatile acid that can be used for metals and alloys, geological samples, and environmental samples. However, digestion of organic products is the main field of application of H_2SO_4. It can be combined with HNO_3, and hydrogen peroxide is often used in a second step to form monopersulfic acid, H_2SO_5, to compensate for the low oxidizing ability of H_2SO_4. Sulfates of Ba, Ca, Sr, and Pb are insoluble.

HF is used for the digestion of silica-based materials (e.g., geological samples, refractory materials, and glass). Fluorine can be eliminated by formation of volatile SiF_4. Alternatively, HF may be complexed by addition of H_3BO_3. HF is often combined with an oxidizing acid such as HNO_3 or $HClO_4$ to obtain high oxidation states and complete digestion.

H_3PO_4 is not widely used with atmospheric-pressure microwave-assisted digestion. This acid is suitable for Al- and Fe-based alloys or alloys containing W.

Statistical evaluation on use of these acids in atmospheric-pressure microwave-assisted digestion was reported by Feinberg in application notes (4). Although these results cannot be considered as a general rule, the data confirm that the most commonly used acids are as follows: HCl, HNO_3, and H_2SO_4 for metals and alloys; HCl, HNO_3, HF, and H_2SO_4 for geological samples; HNO_3, HCl, HF, H_2SO_4, and aqua regia for environmental samples; H_2SO_4 and HNO_3 for food products; and H_2SO_4 for most organic compounds. However, the capability of multielement techniques such as ICP–AES and ICP–MS makes the digestion of sample a challenge as to the number of elements that can be present in a single solution. Currently, the limitation is linked to wet chemistry, that is, the number of soluble elements, pollution, and losses, than to the performance of the analytical system.

Applications for Elemental Analysis

A wide variety of materials are subject to elemental analysis. Most of the time, these materials require a pretreatment or a complete digestion to be in a form that can be introduced into the analytical system. These materials can be classified into several categories that may need different digestion procedures (5–32):

1. Inorganic materials

- metals and alloys
- geological materials (ores, rocks, and soils)
- environmental materials (water, sludges, and fly ash)
- chemical products (acids)

2. Organic materials

- chemical products
- oil plant materials
- polymers
- foodstuffs
- beverages
- biological liquids (blood, urine)
- animal and plant tissues

A special section will be devoted to volatile species because the use of an atmospheric-pressure system is often thought to lead to losses of such elements. Many elements can be lost as volatile species during acid digestion, and Hg is probably the most commonly cited. However, volatile species can be in the form of halogen compounds such as As, B, Cr, Ge, Pb, Si, Sn, Te, Ti, Zn, and Zr. Some volatile species can be formed under oxidizing conditions (Os, Rh, and Ru) or reducing conditions (Se and Te). Some examples of determination of Hg, As, Pb, and Zn will illustrate the capability of atmospheric-pressure microwave-assisted digestion systems to handle these elements.

Determination of Volatile Elements

Mercury has been determined in two certified reference materials (26): National Institute of Standards and Technology (NIST) Skim Milk Powder (NIST 151) and Pig Kidney (NIST 186). Summary of the procedure is given in Figure 4. A Microdigest 301 Prolabo system was used. A Vigreux reflux column was setup at the top of the vessel to avoid losses. Four steps were used and a maximum of 25% of the available power was used. The total digestion time was less than 30 min. For the determination of Hg, AAS cold-vapor generation was used. Results were compared with those obtained with the use of a conventional digestion method (HNO_3 and H_2SO_4, 120 min) and with the use of a closed vessel (14 min). Results were 0.105, 0.103, and 0.084 mg/kg for atmospheric-pressure vessels, pressurized vessels, and conventional method, respectively, for the Skim Milk Powder (certified value of 0.101 mg/kg). Good agreement was obtained for the microwave-assisted method but not for the conventional method. Similarly, results for the three previously listed methods were 1.98, 1.97, and 1.82 mg/kg, respectively, for Pig Kidney (certified value of 1.97 mg/kg). In this instance agreement was good for the three methods; therefore, atmospheric-pressure microwave-assisted digestion is suitable for the determination of Hg in various matrices.

Arsenic was determined in various foodstuffs by hydride generation AAS (8). A Microdigest A 300 Prolabo system was used for digestion of Pig Kidney (NIST 186), Bovine Liver (NIST 185 and 1577 A), and Oyster

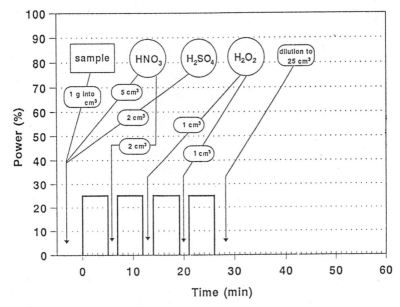

Figure 4. Summary of the procedure used for the determination of mercury in milk powder and pig kidney.

Tissue (NIST 1566 A). The procedure is summarized in Figure 5. Although the concentration of As was at the trace level, the agreement was good with certified values (Table III).

Zinc has been determined in milk samples (*19*) such as Nonfat Milk Powder (NIST 1549) and various types of milk (Similac, Ross Laboratories, Columbus). A Microdigest M300 Prolabo system was used for digestion of milk sample and determination was made by means of ICP–AES. Atmospheric-pressure microwave-assisted digestion was compared with dry ashing, a hot-plate procedure, high-pressure ashing, and low- and high-pressure microwave-assisted digestion. Several procedures were tested for the atmospheric-pressure microwave-assisted digestion. In every instance a Vigreux reflux column was used. An example of the procedure is given in Figure 6. After evaporation of HNO_3, the microwave power was increased to reach the boiling temperature of H_2SO_4 and the subsequent charring of the sample. Hydrogen peroxide was used to enhance the oxidizing properties of H_2SO_4. Regardless of the procedure, the zinc recovery was in the 97–101% range. Even when using an acid with a high boiling point, no loss occurred in the Zn content.

Lead is one of the most commonly determined elements in environmental samples. Lead can occur as a volatile species that can be readily lost during the dry-ashing procedure resulting in temperatures >500 °C. Sulfuric acid at high concentration cannot be used for sample digestion because lead sulfate is insoluble. By using a Maxidigest MX 350 Prolabo

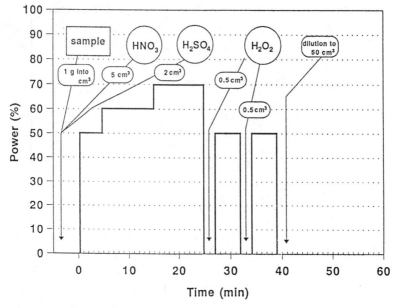

Figure 5. Summary of the procedure used for the determination of arsenic in foodstuffs.

system (Table I), lead was determined in wine and wine-related products such as grapes, leaves, soils, and dregs by GFAAS (21). The procedure was validated by using three certified reference materials—Nonfat Milk Powder (NIST 1549), Pine Needles (NIST 1575), and Sewage Sludges (BCR 145)—based on 10 replicate digestions. The digestion procedure is summarized in Figure 7. Use of the Maxidigest systems allowed the use of up to a 10-g sample. Use of HNO_3 and H_2O_2 was sufficient to obtain an efficient Pb recovery over a wide range of concentration, and results were in good agreement with certified values (Table IV).

Table III. Comparison of Experimental Values of As Concentration in Foodstuffs with Certified Values

Sample	Certified Value	Number of Digestions	Experimental Value
Pig kidney (NIST 186)	63 ± 9 µg/kg	12	63 ± 8
Bovine liver (NIST 185)	24 ± 3 µg/kg	8	25 ± 2
Bovine liver (NIST 1577A)	47 ± 6 µg/kg	6	44 ± 3
Oyster tissue (NIST 1566A)	14 ± 1.2 mg/kg	12	13.4 ± 0.6

NOTE: Focused-microwave-assisted atmospheric-pressure digestion was associated with hydride generation AAS.

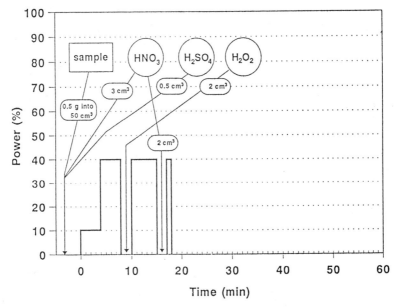

Figure 6. Summary of the procedure used for the determination of zinc in milk.

In addition to the volatiles species, the focused-microwave-assisted atmospheric-pressure digestion can be used for many of the other elements in the periodic table; for example, the determination of Cu, Pb, Zn, Cr, and Ni by using the conventional Environmental Protection Agency Method 3050B and the focused-microwave-assisted, atmospheric-pressure digestion (*30*). GFAAS and flame AAS were used for the detection. Method 3050B makes use of HNO_3 and H_2O_2 and its duration is about 2 h. Several standard reference materials (SRMs) 2704, 2710, and 2711 were used. The improvement in sample preparation time was 50%. Results are summarized in Table V. The authors claimed that better precision were obtained by using the microwave-assisted digestion method (*see* Chapter 3). These results demonstrate the capability of the methodology described.

Geological and Environmental Samples

Geological and environmental samples are usually complex and difficult to dissolve. An example is the determination of rare earth elements in oceanic suspended-matter samples (*28*). An analytical method such as the ICP–MS has the capability to determine these elements at low concentrations. In this work an A 301 system was used (Table I) and the microwave-assisted-digestion procedure was validated by using the U.S. Geological Survey (USGS) marine mud MAG-1 and USGS shale SCO-1 certified refer-

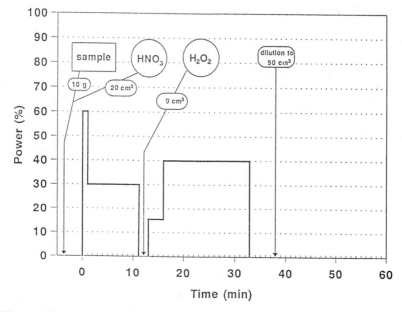

Figure 7. Summary of the procedure used for the determination of lead in environmental samples.

ence materials. Several procedures were tested based on the use of HNO_3, HF, and HCl. This procedure is summarized in Figure 8. Because of the difficulty of digesting the sample, steps 1–4 were repeated with increasing powers (25%, 35%, 85%, and 85%, respectively) until dissolution was achieved. The 85% power was used to go to dryness. Best results were obtained for a sample weight of 0.5 g and using the standard-additions method for calibration.

A further example is the determination of trace amounts of elements in a large variety of geological certified reference materials (29). Ba, Cr, Cu,

Table IV. Comparison of the Experimental Values of Pb Concentration in Reference Materials with Certified Values

Sample	Certified Value ($\mu g/g$)	Number of Digestions	Experimental Values
Nonfat milk powder (NIST 1549)	0.019 ± 0.003	10	$0.0193 \pm 0.001(\sigma)$
Pine needles (NIST 1575)	10.8 ± 0.5	10	10.9 ± 0.47
Sewage sludges (SRM 145)	349 ± 15	10	351.0 ± 0.15

NOTE: Focused-microwave-assisted atmospheric pressure digestion was associated with GF/AAS.

Table V. Comparison of Concentrations Obtained by Using the Conventional EPA 3050 and 3050B Method and the Focused-Microwave-Assisted Atmospheric-Pressure Digestion Method

Sample	Certified Reference Values for Total Digestion (μg/g)	NIST Leachable Concentration Using Method 3050 (μg/g)	Microwave-Assisted Digestion (μg/g)
SRM 2704			
Cu	98.6	100	101
Pb	161	146	160
Zn	438	427	427
Cr	135	89	82
Ni	44	44	42
SRM 2711			
Cu	100	111	107
Pb	1100	1240	1240
Zn	310	340	330
Cr	20	23	22
Ni	16	16	15
SRM 2710			
Cu	2700	2910	2640
Pb	5100	5720	5640
Zn	5900	6230	6410
Cr	19	23	20
Ni	10	7	8

Source: Data are from reference 30.

Co, Ga, Ni, Pb, Sr, V, and Zn were determined by ICP–AES, and As, Cd, and Sb were determined by GFAAS. Several procedures were tested. One procedure included the use of $HClO_4$, and the results are summarized in Figure 9. Agreement was excellent between the certified and the experimental values (Table VI).

Organic Samples

As specified previously, H_2SO_4 is mainly used for the digestion of organic materials, with or without addition of HNO_3. A general procedure can be described: the first step makes use of a power in the 20–60% range and the duration can be up to 20 min to obtain an homogeneous black solution. Then, the second step consists of an oxidation by using either HNO_3 or H_2O_2. The power can be up to 80% and the duration to about 10 min to obtain a clear solution with little precipitate. Then, during the third step, the power is applied for a few minutes to evaporate the reagent. The fourth step consists of an acid further treatment at high power. Steps 3 and 4 can be repeated until complete dissolution of any precipitate. Several examples of applications of focused-microwave-assisted atmospheric-pressure digestion for organic and inorganic matrices are summarized in Table VII.

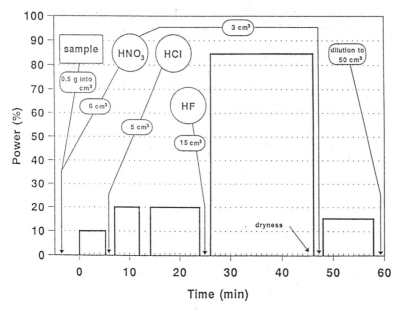

Figure 8. Summary of the procedure used for the determination of rare-earth elements in oceanic suspended matter samples.

Kjeldahl Nitrogen Determination in Organic Materials

Microwave-assisted digestion at atmospheric pressure is particularly suitable for nitrogen determination in organic compounds following the Kjeldahl method. Conventional use of the Kjeldahl method is time-consuming. Use of microwave technology leads to shorter digestion time and high recovery of nitrogen. Decomposition of the organic matrix usually can be obtained by using concentrated, pure H_2SO_4 without addition of catalyst or salt (23). Then, oxidation of the remaining compounds is obtained by addition of H_2O_2. Use of H_2O_2 avoids the possible contamination due to the presence of a catalyst. However, complete acid decomposition must be achieved and heating is stopped for a few minutes before adding H_2O_2 so as to avoid any possible explosion with undecomposed organic products. Hydrogen peroxide can be added in one or several steps. Then, NH_3 titration is performed. Usually, the Maxidigest MX 350 is used (Table I), and allows the decomposition of up to 10 g of sample to be carried out. The system is equipped with a reflux column. Power and time are selected according to the type of matrix (23): protein-rich matrix (fish, chicken, beef, liver), carbohydrate-rich matrix (rice), fat-rich matrix (butter), and mixed foods containing nearly equivalent amount of carbohydrates, fats and proteins (hazelnut and soy flour).

Experimental design can be used to optimize the procedure (23) by considering four factors for the acid decomposition and four similar factors for

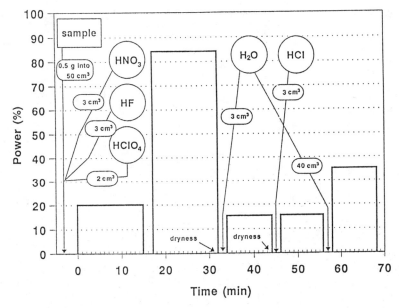

Figure 9. Summary of the procedure used for the determination of various elements in certified geological reference materials.

oxidation: that is, number of steps, maximum power, time of the decomposition or oxidation plateau, and total time. In any instance, 20 mL of H_2SO_4 are added to 0.1–2 g of sample. Then, 5 mL of H_2O_2 are added for each subsequent step. Power and time depend on the type of matrix, but some general conditions can be given. Two or three steps with increasing power (up to 70%) can be used for foods rich in proteins and carbohydrates, whereas a single step is used for oxidation (power of 80–90%). For food rich in carbohydrate, it is crucial to use several steps with increasing power to avoid the foam formation. Similarly, the same acid-decomposition procedure is used for fat-rich foods to avoid formation of foam, whereas several oxidation steps are necessary to account for the difficulty of oxidation. The number of steps can be even larger for mixed foods. Tryptophan is often used to verify the digestion conditions and the nitrogen recovery (i.e., 13.72 g per 100 g (23, 30)) because this product is known as difficult to digest. The Kjeldahl method based on the use of the focused-microwave technology has been officially accepted as an alternative method by the AFNOR (French Association of Standardization) after the determination of nitrogen in milk, wheat flour, powdered egg, casein, meat and dactyl hay in the 0.4–14 g per 100 g range by 11 laboratories (22). For samples such as amine, arginine, champagne, fish, glucose syrup, horse serum, jam, malt, motor oil, phenol products, pizza, soya peptone, tobacco, addition of 6–12 g of a 93% K_2SO_4–7% $CuSO_4$ mixture tablet to the 20 mL of H_2SO_4 increased the working temperature (application notes, Prolabo).

Table VI. Comparison Between Experimental and Certified Values for Some Elements in Geological Materials by Using the Focused-Microwave-Atmospheric-Pressure Assisted Digestion

Material	Type	Cr Certified (µg/g)	Found (µg/g)	Pb Certified (µg/g)	Found (µg/g)	As Certified (µg/g)	Found (µg/g)
BX-N (ANRT	bauxite	290	281	135	133	120	116
MA-N (GIT-IWG)	granite	3	2.9	29	30	13	14.5
SY-3 (CCRMP)	syenite	(11)	8	133	130	18.8	19
AL-1 (GIT-IWG)	albite	4	4	4.5	6		
AN-G (GIT-IWG)	anorthosite	50	47	2	/		
BE-N (GIT-IWG)	basalt	360	356	4	/		
DR-N (ANRT)	diorite	42	36	55	56		
FK-N (ANRT)	k-feldspar	5	4.85	240	237		
Mica Fe (CRPG)	biotite	90	94	13	14		
CRM 320 (BCR)	river sediment	138	130	42	47	76.7	75
GXR 6 (USGS)	soil	96	90	101	115	330	326

Source: Data are from reference 29.

Chemical Oxygen Demand (COD)

The focused-microwave technology can be used to accelerate the oxidation of the organic products and some inorganic salts in water by $K_2Cr_2O_7$ (33) so as to measure the amount of material that can be oxidized. The chemical oxygen demand is the amount of oxygen that is consumed during the oxidation reaction. From a practical point of view, boiling of the sample in acidic medium (H_2SO_4) is conducted with Ag_2SO_4 as an oxidation catalyst, $HgSO_4$ to complex the chlorides, and an excess of $K_2Cr_2O_7$. Cr(VI) is reduced in Cr(III). An on-line colorimetric detection of Cr(III) allows a permanent control of the reaction. Instead of a 2-h duration with the conventional method, only 5–15 min are necessary for the total reaction.

Sample Preparation for Speciation

Demand is growing for the knowledge of the chemical form of an element. For instance, the toxicity of As depends on its chemical form. Inorganic arsenic [As(III) and As(V)] is far more toxic, particularly As(III), than organic species such as monomethylarsenic acid (MMA), dimethylarsenic acid (DMA), arsenobetaine, or arsenocholine. The oxidation state is the crucial factor with inorganic species, whereas the substituents are the important parameters with organic species. In other words, the determination of total arsenic is meaningless with respect to its toxicity. For instance, the DORM-1 dogfish reference material contains 18 ppm of total arsenic, whereas only 0.2 ppm is toxic. It is, therefore, important to at least differentiate between inorganic and organic arsenic. Similarly, Cr(VI) is more toxic than Cr(III), and Se(IV) is more toxic than Se(VI). For other elements, it is important to determine methyl mercury, dimethylmercury, tetraethyl lead, tetramethyl lead, methylethyl lead, and tetra and triorganotin compounds. Therefore, development of analytical tools allows the determination of these species is necessary. In particular, development of sample preparation procedures that do not modify the interspecies equilibrium is strongly needed.

Focused-microwave-assisted atmospheric-pressure digestion has been applied to the speciation of As and Sn (34, 35). The principle is to use moderate microwave power. High-pressure liquid chromatography (HPLC)–ICP–MS and HPLC–hydride generation–quartz tube AAS were used for the identification and quantitation of As. Soil, sediments, and fish (fresh cod) tissues were analyzed. The sediment was spiked with four As compounds [As(III), DMA, MMA and As(V)]. Procedures for the determination of total arsenic and the various As species are summarized in Table VIII. A moderate power of 10% is used in the instance where speciation is needed (A 300 system). As(V), MMA and DMA are stable, whereas As(III) is oxidized into As(V). Therefore, the inorganic arsenic can be differentiated from DMA and MMA. A further improvement was made by using

Table VII. Examples of Applications Based on the Use of Focused-Microwave-Assisted Atmospheric-Pressure Digestion

Elements	Matrix	Microwave System	Sample Size (g)	Reagent	Amount (mL)	Power %	Time (min)	Total Time (min)	Detection	Ref.
Al, Cu, Fe, Mn	Wheat Flour NIST SRM 1567a	A300	0.5	HNO_3 HNO_3 HNO_3 H_2O_2	10 10 10 5	15 30 30 30	10 22 13 5	50	ICP–AES	6
As, Hg, Se, Pb	Spiked wheat flour	A300	0.5	HNO_3 HNO_3 HNO_3 H_2O_2	10 10 5 5	10 40 40 40 15 40	10 5 10 10 5 5	45	ICP–AES	6
Al, Cu, Fe, Mn, Ni, Pb, Zn	Citrus leaves NIST SRM 1572	A300	0.5	HNO_3 HNO_3 H_2O_2 H_2O_2	10 10 10 5	15 30 30 30	10 22 10 5	47	ICP–AES	6
Ca, K, Mg, Na, P	Pine needles NIST SRM 1575, Citrus leaves NIST SRM 1572	MX 350	1	HNO_3^+ H_2SO_4	10	45	4	15	ICP–AES	29
Al, Ba, Ca, Cu, Fe, Mg, Mn, K, Sr, Zn	Pine needles	A300	0.05	HNO_3 H_2O_2 HNO_3	5 10 4	60 90 20	4 4 15		ICP–AES	10
Ca, Cu, Fe, Mg, Mn, K, Na, Zn	Bovine Liver NIST SRM 1577a, Citrus leaves NIST SRM 1572	A300	0.5–1	HNO_3	20	30 40 35 30	10 5 5 3		AAS	12
Ca, K, Na, Mg	Diet milk BIPEA	A300	1	H_2SO_4 H_2O_2 H_2O_2	10 5 5	60 30 30	10 5 5		AAS	15
Se	Feeds, plant, animal tissues	A300	1	H_2NO_3 $H_2SO_4^+$ $HClO_4$	20 2 3	15 25 30 40	15 10 10 35	80	AAS	11

U	Sediment	A300	1	HNO_3 HNO_3 H_2O_2	2 2 2	30, 50, 60 50, 60, 30 50, 30	5, 2, 2 5, 3, 2 5, 2	60	Fluorimetry	14
Cu, Cd, Pb Al, Cr, Ni	BCR lake sediment	A301	1	HNO_3 + HF + $HClO_4$ + HCl +	10 20 5	40, 50	10, 35	65	AAS ICP–AES	32
Multielement	Glass, NIST SRM93a, BCR 126A	A300	1	H_2O + HNO_3 + HF $HClO_4$	20 5 8 8	40, 30 40, 60, 40, 60, 80, 30	10, 2 3, 5, 2, 3, 15, 10	40	ICP–AES	20
C, Mn, P, S, Si	Non-alloyed steel IRSID CECA 002-2 and 007-2	A301	0.1–1	H_2O + HCl HCl H_2O HNO_3	10 20 10 10 5	30, 20	1, 2	3	ICP–AES	27
C, Mn, P, Pb, S, Si	Sulfated steel IRSID 302–1, 304-01	A301	0.1–1	HCl HNO_3	15 10	25, 30	1, 2	3	ICP–AES	27
C, Ca, Cu, Cr, Mn, Ni, P, S, Si	Cast iron, BCS 234/8, CTIF 48501 …	A301	0.1–1	HCl HNO_3	10 5	50, 50	1, 2	3	ICP–AES	27
Co, Cr, Cu, Fe, Mn, Mo, Ni V	Steel, NIST 339	A300	0.2	HCl HNO_3 H_3PO_4	4.5 12.7 12.7	35	35		ICP–AES	16

Table VIII. Procedure for Focused-Microwave-Assisted Atmospheric-Pressure Digestion of Sample Containing Arsenic Species so as To Perform Speciation

Goal	Reagents	Amount (mL)	Power (%)	Time (min)
Total arsenic	HCl +	7	20	5
	HNO$_3$	3	25	10
	HCl +	7	30	10
	HNO$_3$	3		
	H$_2$O$_2$	1	20	5
	H$_2$O	5	25	5
Speciation	HCl +	10	10	10
	HNO$_3$	5		

H$_3$PO$_4$ under the same mild operating conditions. By using an acid concentration as low as possible (<2 N), preliminary investigations indicated that it was possible to keep the As(III) oxidation state.

Setup for On-Line Pretreatment

Some sample introduction techniques such as hydride generation (for As, Bi, Pb, Sb, Se, and Te) or cold vapor (for Hg) require a pretreatment (*see also* Chapter 6), in particular to oxidize organic materials and to obtain the appropriate oxidation state for hydride generation. This pretreatment is usually carried out off-line. Flow injection (FI) methods can be used for pretreatment and have the advantages of less reagent consumption, less analyte losses and risk of contamination, and ease of automation. Microwave ovens could be used to increase the kinetic of the pretreatment by heating almost instantaneously the liquid sample flowing through an inert plastic tube. Use of conventional microwave technology could lead to the following problems (36):

- inhomogeneity of the power field within the multimode cavity and poor coupling efficiency with a small load
- short reaction times
- incomplete digestion
- evolution of gases during digestion leading to pressure build up

These problems are why several publications have illustrated the benefits of coupling FI with the focused-microwave technology (36–40). The potential advantage of the focused-microwave technology is to permit the user to locate the reaction coil within the interaction volume (circa 50 cm^3) where the microwave power density is high. Therefore, a high heating efficiency is obtained.

The crucial part of the system is the residence time of the sample within the interaction volume. The optimized residence time depends on the forward microwave power, the internal diameter (id) of the tube used for the reaction coil, and the flow rate of the liquid sample. Even at a low power of 50 W, an aqueous sample could be brought to the boil in less than 10 s, leading to a short reaction time. A higher power results in bubble formation and flow disturbances. To improve the control of the reaction, two coils were actually used: a reaction coil and a ballast coil (*36*) made of PTFE. The reaction and ballast coils were 10.2-m and 2.3-m long with an id of 1.07 mm and 1.33 mm, respectively. The two coils were wound perpendicularly around a rectangular PTFE shaft (280 × 38 × 15 mm). This shaft was located in the chimney (Figure 1) so that only its lower part, and therefore a part of the two coils (14% for the reaction coil), was in the microwave interaction zone. Several flow sample rates in the 3.7–12.3 mL/min range were used. With a sample flow rate of 8.5 mL/min the interaction time was 0.35 s in each of the 18 rows for a total interaction time of 6.3 s, whereas the total reaction time was 46 s. Deionized water was circulated in the ballast coil to absorb 10–20% of the microwave power and reduce the power not absorbed by the reaction coil. The ballast coil acted as a thermal and load buffer.

At the exit of the reaction coil, the hot effluent was merged with a reductant flow and an argon purge gas flow in the mercury–hydride system (MHS) manifold. Then, the mixture passed through the MHS reaction coil and entered the gas–liquid phase separator (GLS). Because of the length of the reaction coil, the system was inherently subject to an increase of the dispersion in the system, and consequently to a peak broadening and a decrease in the peak height intensity. However, this decrease was partly compensated by an introduction of hot samples into the MHS manifold. The system was applied (*37, 38*) to the determination of Hg, As, Bi, Pb, and Sn in urine and environmental water samples by AAS.

A different setup was used for the on-line acid digestion and prereduction (*39*) of As for hydride generation. In the instance of As, an oxidizing acid digestion can lead to the highest (V) oxidation state, which decreases the efficiency of hydride formation. Therefore, a prereduction stage is necessary to obtain the adequate (III) oxidation state. In this setup, the sample passed through two coils. The first coil was used for acid digestion. It was made of perfluoroalkoxy polymer (PFA) and had a length of 8 m and an id of 0.8 mm. It was coiled and totally located in the microwave interaction zone. A pressure of 6 bar was obtained within the coil, which resulted in a significant reduction of bubble formation. Consequently, a cooling coil had to be used before adding the reducing reagent (L-cysteine) and passing the sample into the prereduction coil.

The prereduction PTFE coil was knotted and located in a water bath setup in the chimney of the microwave system, so that the bottom of the bath was in the interaction zone to keep the water boiling. At the exit of the reduction coil, another cooling system was used. The total time for acid digestion, prereduction, and hydride generation was 520 s and lead to a sample throughput of 7 sample/h. Dilute blood samples did not provide satisfactory As recovery results when using this procedure. The extent to which the organoarsenic compounds are attacked remains to be verified.

Mercury was determined (FI and cold-vapor generation AAS) in dilute blood sample by using a more simple setup of the reaction coil. A single reaction coil was used for digestion and located in the microwave interaction zone (40). The coil was made of perfluoroalkoxy polymer and had a length of 10 m and an id of 0.9 mm. The tube was knitted around a PTFE backbone tube, which was in turn folded to be fitted in the cavity. The knitted form was selected to reduce the axial dispersion and to obtain a high peak height sensitivity. A cooling coil was inserted between the reaction coil and the manifold. This technology was applied to human and bovine whole blood and lyophilized human reference whole blood.

Microwave-Assisted Production of Volatile Species

In contrast to microwave-assisted digestion, where it is usually crucial to avoid losses of volatile species, the focused-microwave technology can be used to enhance the production of volatile species such as halogens or cyanides for further determination. This enhancement is made possible by the use of an open vessel. Determination of total cyanides (i.e., alkaline cyanides and metal cyanides) in water, soil, and waste samples has been made possible by the production of gaseous HCN and further determination by spectrophotometry (41). Formation of HCN was obtained by addition of $CuSO_4$, $SnCl_2$, and H_2SO_4 into the flask. A 401 microwave system was used with either 50 mL of water or 5 g of solids in 50 mL of solution; 50% power was applied for 10 min. The total time was 15 min, including 5 min for cooling. HCN was taken up by a nitrogen flow and absorbed in a 0.25 N NaOH solution. Then, cyanide ions were transformed into cyanide chloride to be determined by spectrophotometry at 620 m. Limits of quantitation were 0.01 ng/L in water and 0.1 ppm in solids. Investigations are currently being conducted to apply these methods to the determination of halogens.

Microwave-Assisted Extraction

The extraction of organic compounds in natural matrices by using focused-microwave technology at atmospheric pressure is at an early stage of exploration (see Chapter 11 for closed systems). The first reports were made by Rocca et al. (42) on the analysis of essential oils in aromatized products such as candies and chewing-gums. For these samples, the quantitative results were in good agreement with results obtained by classical methods (Soxhlet extraction, simultaneous solvent extraction, or hydrodistillation). The extraction time (2–10 min) was dramatically reduced compared with the extraction time of conventional methods (3–4 h).

The focused-microwave, atmospheric-pressure extractions of organic compounds such as polycyclic aromatic compounds (PACs) and polychlorobiphenyls (PCBs) in environmental samples have recently attracted attention. The carcinogenicity of PAC is related to the structural arrangement of aromatic rings (i.e., benzo[a]pyrene is a carcinogenic compound, whereas benzo[b]pyrene is inactive). The toxicity of PCB is also related to the chlorination pattern of the phenyl moieties. The focused-microwave, atmospheric-pressure extractions of these stable organic contaminants also were investigated by Budzinski et al. (43–45) on marine sediments (standard reference material and environmental samples).

PACs were extracted from sediment samples under reflux with dichloromethane heated by microwave energy. This solvent was selected because it was currently used in conventional extraction methods (Soxhlet or reflux extraction methods). Various solvents (ethanol or acetone) with or without moisture were verified and did not exhibit different extraction recoveries.

The maximum recoveries of PACs (from three to six rings) in marine sediments were obtained for a power of 30 W and an extraction time of 10 min (43). Increases of the time and power did not increase the recovery efficiency and even induced a small decrease due to the degradation of the compounds.

Twelve PACs (from phenanthrene to benzo[g,h,i]perylene) were extracted from the marine sediment SRM 1941a by Soxhlet extractions and focused-microwave, atmospheric-pressure extractions (30 W, 10 min). When compared with the certified values, the Soxhlet method gave an average recovery of 96%, whereas the microwave extraction led to 83% (Figure 10). The standard deviations were in the same range both for Soxhlet and microwave extractions. The weakest recoveries were observed for low-molecular-weight compounds such as the anthracene and the phenanthrene (70% to 75%), which may be lost by evaporation during the microwave-assisted extraction. Superheating effects of extraction solvents may be responsible for this volatilization phenomenon. If the values of these two compounds are discarded, the recovery rate was in the range of 85%. These data were comparable with results previously published by Lopez-Avila and Young (46) on microwave extraction of aromatic compounds in closed vessels (recovery in the 65–85% range).

The focused-microwave, atmospheric-pressure extraction was also applied for the recoveries of 22 PCB (tetrachlorobiphenyl to decachlorobiphenyl) congeners in a marine sediment (44). The optimization of power and time that gave the best recovery efficiencies were 30 W and 10 min, respectively. The presence of water (10% in weight of the dry sediment) increased the recovery efficiency. Experiments exhibited a good agreement between values obtained for PCBs in a marine sediment by Soxhlet and microwave-assisted extraction (Figure 11).

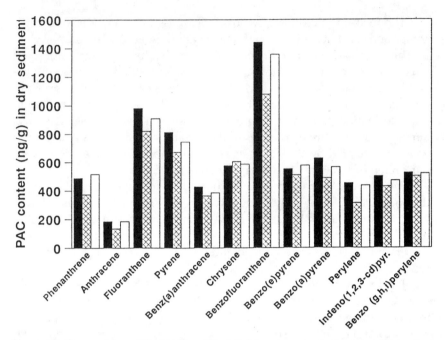

Figure 10. Comparison of the PAC certified (left bars) concentrations (ng/g) in dry reference marine sediment (SRM 1941a) with the focused-microwave, atmospheric-pressure assisted extraction (center bars) and the conventional Soxhlet extraction (right bars).

The focused-microwave-assisted, atmospheric-pressure extraction of organic compounds in natural matrices appears as an interesting alternative to conventional extraction techniques that use organic solvents heated by convection. The major advantages are:

- reduction of the extraction time by a factor of 50–150
- reduction of the solvent volume
- easiness and increased safety of sample handling in focused-microwave-assisted, atmospheric-pressure extraction when compared with microwave extraction in closed vessels under pressure.

The focused-microwave, atmospheric-pressure technology can also be used to assist leaching for the recovery of organometallic compounds in speciation analysis of complex matrices, such as sediments. The analysis time can be dramatically (20–100 times) reduced and the recovery efficiency from some species can be increased compared with conventional methods. Careful optimization of the extraction medium, power applied, and exposure time is required to preserve the stability of the organic compounds to be separated. This method has been applied to the separation of

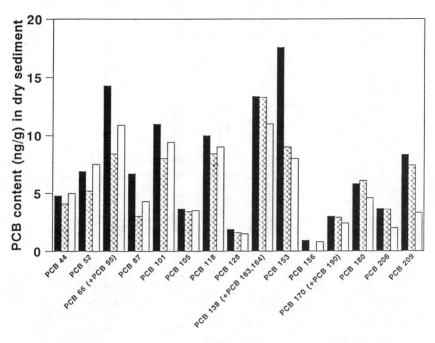

Figure 11. Comparison of certified (left bars) PCB concentrations (ng/g) in reference marine sediment (SRM 1941a) with the focused-microwave, atmospheric-pressure assisted extraction (center bars) and the conventional Soxhlet extraction (right bars).

organotin compounds. Compounds such as tributyltin (TBT) and triphenyltin (TPhT) are known to be toxic. Therefore, the separations of these products along with their products of degradation (mono- and dibutyltin and diphenyltin) are of particular importance. A 0.1–1-g sediment sample and 10 mL of 0.5 M acetic acid solution in methanol were placed in an extraction tube. The effect of the concentration of acetic acid on the recovery of organotin species is given in Figure 12. Preliminary investigations (*47*) indicated that the method proved to be efficient for the recovery of mono-, di-, and tributyltin from the certified PACS-1 and BCR 462 sediments. Phenyl and octyl tin derivatives were fairly stable in the microwave field. The extension of the method developed on biological matrices (algae, fish, oysters) in the near future is likely.

Conclusions

Focused-microwave, atmospheric-pressure technology presents some definitive advantages over the closed, pressurized systems, in particular in terms of safety and the possibility of adding reagents at any time during the microwave-assisted process. In addition to the digestion, which is of

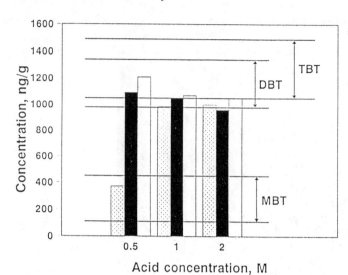

Figure 12. Effect of the concentration of acetic acid on the recovery of organotin from the PACS-1 sediment in methanol solution. Extraction procedure performed on 0.5 g of sediment in a 301 Prolabo system: power, 60 W; time, 3 min; volume of leaching agent, 10 mL; monobutyltin, left bars; dibutyltin, center bars; tributyltin, right bars; certified values, ——.

major concern, there is a significant potential for new applications such as extraction and leaching. However, there is still room for new developments. Evaluation of prototypes that use a feedback through the temperature instead of the power indicates that a more precise control of the digestion processes can be obtained.

Acknowledgments

I wish to thank D. Mathé from Prolabo for useful information and comments during the preparation of this manuscript and H. Budzinski, P. Garrigues, and O. Donard for providing their most recent results.

References

1.   Prolabo, 54 rue Roger Salengro, F-94126 Fontenay sous Bois Cedex.
2.   Commarmot, R.; Didenot, D.; Gardais, J. F. U.S. Patent 4,681,740, 1987.
3.   *Inductively Coupled Plasma Mass Spectrometry*; Jarvis, K. E.; Gray, A. L.; Houk, R. S., Eds.; Blackie: Glasgow, Scotland, 1992; pp 172–224.
4.   Feinberg, M. H. *Analusis* **1991,** *19,* 47.
5.   Schnitzer, G.; Pellerin, C.; Clouet, C. *Lab. Pract.* **1988,** *January*.
6.   Magaletta, R. L. AOAC Trace Analysis Seminar, New Brunswick, NJ, 1988.
7.   Grillo, A. C. *Spectroscopy* **1989,** *4,* 16.
8.   Schnitzer, G.; Pain, M.; Testu, C.; Chafey, C. *Analusis* **1990,** *18(10),* i20.

9.   De Kersabiec, A. M.; Vidot, F.; Boulègue, J.; Verhaeghe, I. *Analusis* **1990**, *18*, 214.
10. Santerre, A.; Mermet, J. M.; Villanueva, V. R. *Water, Air, Soil Pollut.* **1990**, *52*, 157.
11. Hocquellet, P.; Candillier, M. P. *Analyst* **1991**, *116*, 505.
12. Oles, P. J.; Graham, W. M. *J. Assoc. Off. Anal. Chem.* **1991**, *74*, 812.
13. Bermond, A.; Ducauze, C. *Analusis* **1991**, *19*, 64.
14. Carrelet, J. C.; Maillard, J. C.; Tournier, B. "Automatisation de la préparation d'échantillons en vue du dosage d'éléments chimiques radioactifs;" Internal report; Centre d'Etudes Atomiques de Valduc: Is sur Tille, France, 1991.
15. De Mauleon, N.; Berthoumieux, G.; Fedry, M. Presented at the Pittsburgh Conference, Chicago, IL, 1991; paper 231.
16. Gonon, L.; Mermet, J. M. *Analusis* **1992**, *20*, M26.
17. Duneman, L.; Meinerling, M. *Fresenius J. Anal. Chem.* **1992**, *342*, 714.
18. Krushevska, A.; Barnes, R. M.; Amarasiriwaradena, C. J.; Foner, H.; Martines, L. *J. Anal. At. Spectrom.* **1992**, *7*, 845.
19. Krushevska, A.; Barnes, R. M.; Amarasiriwaradena, C. J.; Foner, H.; Martines, L. *J. Anal. At. Spectrom.* **1992**, *7*, 851.
20. Pascal, B.; Rauline, A.; Longchamp, S. *Verre,* **1992**, *6*, 277.
21. Teissèdre, P. L.; Cabanis, M. T.; Cabanis, J. C. *Analusis* **1993**, *21*, 249.
22. Suard, C. L.; Feinberg, M. H.; Ireland-Ripert, J.; Mourel, R. M. *Analusis* **1993**, *21*, 287.
23. Feinberg, M. H.; Ireland-Ripert, J.; Mourel, R. M. *Anal. Chim. Acta* **1993**, *272*, 83.
24. Krushevska, A.; Barnes, R. M.; Amarasiriwaradena, C. *Analyst* **1993**, *118*, 1175.
25. Demesmay, C.; Ollé, M. *Spectra 2000* **1993**, *175*, 27.
26. Schnitzer G.; Soubelet, A.; Testu, C.; Vastel, C. Presented at the Pittsburgh Conference, Atlanta, GA, 1993; paper 1305.
27. Beaufils, D. "Development of New Applications of the A301 Microwave Digester Dissolution of Alloys in Metallurgy;" Internal Report; Institut de Soudure: Villepinte, France, 1993.
28. Thomas, B.; Quétel, C. R.; Donard, O. X. F.; Grousset, F.; Heussner, S. Presented at the Winter Conference on Plasma Spectrochemistry, Granada, Spain, 1993
29. De Kersabiec, A. M.; Vidot, F.; Boulègue, J.; Verhaegue, I. Presented at the Pittsburgh Conference, New Orleans, LA, 1992; paper 476.
30. Collins, L. W.; Lorentzen, E. M. L.; Kingston H. M. S.; Presented at the 1994 Pittsburgh Conference, Chicago, IL, 1994; paper 1145.
31. Theiller, G. "Minéralisation de produits végétaux certifiés par digesteur à micro-ondes Maxidigest Prolabo en vue du dosage d'éléments minéraux;" Internal Report; INRA: Villenave d'Ornon, France, 1990.
32. Thomas, P. "Comparaison entre les techniques de minéralisation par micro-ondes focalisées" Internal Report; Institut Pasteur: Lille, France, 1993.
33. Seres, 355 rue A. Einstein, BP 87000, F- 13793 Aix en Provence cedex.
34. Demesmay, C.; Ollé, M. Spéciation de l'arsenic par couplage HPLC et méthodes spectrales d'analyse élémentaire spécifique: application à des sols et sèdiments et des tissus biologiques;" Internal Report; Service Central d'Analyse du CNRS: Vernaison, France, 1993.

35. Lalere, B.; Donard, O. F. X. "Applications de la technique micro-ondes pour la spéciation des composés organo-stanniques dans l'environnement;" Internal Report; Université de Bordeaux: Bordeaux, France, 1993.

36. Tsalev, D. L.; Sperling, M.; Welz, B. *Analyst* **1992,** *117*, 1729.

37. Tsalev, D. L.; Sperling, M.; Welz, B. *Analyst* **1992,** *117*, 1735.

38. Welz, B.; Tsalev, D. L.; Sperling, M. *Anal. Chim. Acta* **1992,** *261*, 91.

39. Welz, B.; He, Y.; Sperling, M. *Talanta* **1993,** *12*, 1917.

40. Guo, T.; Baasner, J. *Talanta* **1993,** *40*, 1927.

41. Jeannot, R. "Dosage des cyanures totaux dans les eaux et les sols; Méthode de décomposition des cyanures par la technique micro-ondes focalisées;" Internal Report; BRGM: Orléans, France, 1994.

42. Rocca, B.; Arzouyan, C.; Estienne, J. *Ann. Fals. Exp. Chim.* **1992,** *911*, 347.

43. Budzinski, H.; Garrigues, P.; Mathé, D. Presented at the Pittsburgh Conference, New Orleans, LA, 1995; paper 1307.

44. Budzinski, H.; Pierard, C.; Garrigues, P.; Mathé, D. Presented at the Pittsburgh Conference, New Orleans, LA, 1995; paper 1301.

45. Budzinski, H.; Papineau, A.; Baumard, P.; Garrigues, P. C. R. *Acad. Sci.* Paris, série IIb, **1995,** *321*, 69.

46. Lopez-Avila, V.; Young, R. *Anal. Chem.* **1994,** *66*, 1097.

47. Donard, F.; Lalère, B.; Martin, F.; Lobinski, R. *Anal. Chem.* **1995,** *67*, 4250.

Chapter 6

Applying Flow Methodology to Microwave-Enhanced Reactions

K. E. Williams and Stephen J. Haswell*

Extending the application of microwave heating to include flow methodologies such as matrix digestion, analyte dissolution, chemical synthesis, and matrix–analyte modification forms the basis of this chapter. From the early development through the production of a commercial system, emphasis is placed on the inherent differences in both the hardware requirements and the chemical processes involved with this novel methodology. Selected examples have been drawn from published material to illustrate the practical aspects of coupling microwave systems with flow methodology which include numerous applications. The chapter demonstrates how flow methodology differs quite considerably from the more traditional batch closed vessel microwave heating system, in both approach and chemical capability. As a relative newcomer to the microwave field, flow methodology has considerable advantages to offer over the current techniques, and further developments in this field are expected to both extend and diversify the application of microwave-enhanced reactions in chemical, biological, and physical systems.

In the field of analytical chemistry rapid and efficient sample preparation procedures such as analyte extraction or matrix digestion (decomposition) are of paramount importance for reliable analytical methodology. The rapid development of automated instrumentation over the past 10 years has heralded a renaissance in analytical chemistry and offered more reliable and rapid forms of analyte detection. The limiting step or Achilles' heel of modern analytical methodology is clearly sample preparation, which in light of the financial and research investment that has gone into

*Corresponding author.

instrumental developments, still remains the poor relative of analytical chemistry. The introduction of microwaves to assist in matrix digestion or analyte extraction has without doubt gone a long way to improve the limitations of sample preparation (1). With only two exceptions all current systems commercially available for microwave-assisted sample preparation are based on closed-vessel devices that offer either a pressurised or nonpressurised mode.

Chapter 2 in this book describes in detail how such systems are used and the many advantages they offer. However, some of the major analytical disadvantages that still remain with such systems are generally brought about through the use of vessels and can be split into physical and chemical problems. The physical limitation comes from the fact that the discrete vessel systems require a high degree of manipulative handling, such as filling, closing, opening, and positioning the vessel in the microwave system used. Even though robotics can be used to assist in the automation of such procedural requirements, they remain cumbersome and time limiting and lack elegance. Coupled with the physical limitations are many potential chemical problems, including loss of volatile analyte during the digestion and opening of vessels, contamination, and dilution effects brought about through the need to operate with quantifiable weights and volumes. In addition to the robotics approach careful sample-handling protocols, the monitoring of reagents, and controlled atmospheres can negate many such problems.

The remaining limitation is that of time and the mere fact that most analytical measurement techniques offer a rapid throughput of samples, which often exceeds the process capability of the sample-preparation step. This shortfall can obviously be met by parallel sample processing, which whilst effective, does lack imagination.

In many areas of analytical chemistry the development of flow systems such as segment (autoanalyser) or flow-injection analysis (FIA) have gained great popularity. They offer contained and controllable chemical systems for which the chemistries involved can be easily adapted and modified. One area that until recently did not fully exploit flow systems was sample preparation and in particular the use of microwave energy in such a system. This chapter will review the developments of such methodology and considers some future trends.

First however, it is worth considering briefly the factors that underpin segmented flow and FIA systems. Historically, segmented flow systems pioneered by Skeggs (2) extended discrete batch detection methods (e.g., colorimetric methods) into an almost identical methodology, but one in which the batch vessels become slugs of liquid in a tube separated by air or an immiscible liquid. Therefore, chemical reactions akin to the batch or discrete vessel methodology could be performed in a dynamic mode whilst using close to thermodynamic reaction conditions. Thus, segmented flow converted traditional bench-based batch, essentially test tube methods, into automated tube flow technology. This development has led to a family of autoanalyser techniques that now find a large number of applications in analytical chemistry. At no time, however, were any real attempts made to adapt such systems to accommodate matrix decomposition or analytical extractions. Following segmented flow technology, FIA was pioneered by Ruzicka and Hansen (3). What emerged was a flow system in which a slug of liquid sample is manipulated through a series of tubes (the manifold), in which suitable chemistries are accommodated with subsequent passage to an on-line detector flow cell.

Unlike segmented flow the FIA approach did not attempt to create an encapsulated slug of analyte but rather aimed to produce a continuous stream of liquid carrier and analyte. FIA offers in general a more rapid, versatile, and precise system over that of segmented flow, in which chemical equilibrium is not a prerequisite for reproducible measurements. However, one of the consequences of adapting the FIA approach is that samples injected into the flowing stream will experience dispersion, which leads to a degeneration in the observed peak shape, notably an increase in peak tailing. In general FIA systems tend to be constructed from inert poly(tetrafluoroethylene) (PTFE) tubing of internal dimensions less than 1 mm. Flow rates are typically 0.5–6 mL/min, and tube lengths are relatively short (usually less than 5 metres). In essence one simply needs to have long enough residence time of an analyte in the manifold to produce a signal of sufficient sensitivity for the purpose. Once again, however, little attempt has been made across the wide application areas covered by FIA to include matrix preparation on-line. However, FIA has found many imaginative uses incorporating on-line processes such as stopped or reverse flow, on-line extraction, phase separation, solid-phase analyte removal and preconcentration, detector switching, and relatively complex chemistries.

The aim of this chapter is to give an overview of the development of microwave-enhanced flow systems in terms of relevant publications relating to current research by outlining the techniques developed and the theory and mechanisms involved for the flowing dynamic systems. The advantages of microwave-enhanced flow systems over batch processes will be considered, but these advantages basically include a significant reduction in sample preparation time, the ability to carry out reactions that would normally be too dangerous in a closed vessel because of sudden increases in temperature and pressure, and the capability to handle transient or readily decomposed samples or intermediates. Initial work in this field focused primarily on sample dissolution for elemental analysis. However, since these early publications, on-line microwave systems have been used for many different applications. These applications include drying, chemical synthesis, analyte derivatisation, oxidation, hydrolysis, on-line extraction, and other methodologies that usually involve either continuous or stopped-flow processes.

The rapid increase in the number of applications using microwave-enhanced flow systems led to the 1993 launch of the first commercially available microwave sample-preparation flow system (4). We anticipate that the number of applications of this methodology will grow considerably as more commercial systems become available over the next decade.

Applications of On-Line Microwave-Enhanced Processes

Since the first reports describing microwave-digestion flow systems the technology has been applied to a wide variety of processes ranging from the simplest drying techniques to organic synthesis. The following sections illustrate the effectiveness of microwave-enhanced chemistry in a range of applications selected to demonstrate the current technology available.

Sample Digestion, Dissolution, and Detection

The early research publications involving microwave flow systems described the combination of on-line sample dissolution in tandem with atomic spectroscopic techniques. The nebulizer systems of a typical flame atomic absorption spectroscopic (FAAS) instrument offered an ideal coupling tolerant to bubbles and a reasonably high dissolved solid content.

The earliest work reported in this field was by Burguera et al. (5), who described coupling microwave sample dissolution to FAAS by using flow injection analysis (FIA) methodology. The application was based on previous FIA studies for the analysis of metals in predigested blood serum, plasma, or whole blood. The methodology (Figure 1) involved the synchronous merging of reagent and sample followed by the mineralisation of blood in a Pyrex decomposition coil subjected to microwave irradiation. On-line analysis was performed by using FAAS.

The authors described the microwave as a source of intense energy, rapidly heating the samples. When combined with a suitable chemical reaction, sample mineralisation resulted.

The great advantage of this technique over the previous batch-based methods was that it minimized the problem of acid fumes generated during the digestion of biological samples common in the more traditional Teflon vessels used in such analyses. Gas evolution in the on-line method, however, did lead to some problems resulting in a pressure buildup and thus disturbance of carrier flow. Additional problems encountered were the inherent inhomogeneity of power distribution within the microwave cavity, and with only a small area taken up by the reaction coil, a high percentage of the microwave power within the oven cavity was not absorbed. Thus, the positioning and distribution of the reaction coil within the cavity was identified as an important parameter in achieving uniform heating of

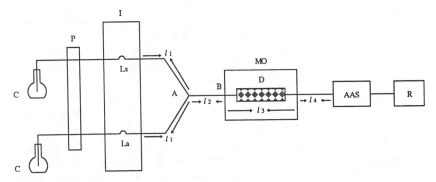

Flow diagram of the system used for the determination of copper, zinc, and iron in whole blood: (C) carrier solution; (P) peristaltic pump; (I) double injector; (D) decomposition tube; (MO) microwave oven; (AAS) atomic absorption spectrometer; and (R) recorder. The sample and reagent solutions fill loops Ls and La, which are the volumes introduced into the carrier streams; *l* 1–*l* 4 denote the lengths in metres of the various tubing.

Figure 1. First manifold developed for use in FI microwave-assisted sample dissolution as described by Burguera et al. (Reproduced with permission from reference 5. Copyright 1986 Elsevier.)

sample and reagents to ensure an efficiently controlled digestion. In addition to being inefficient, the large percentage of nonabsorbed power would probably result in a reduction in the lifetime of the microwave oven. Any resultant heating of the magnetron would cause a reduction in the microwave output.

The types of sample digested in the follow-up work by the same authors (6, 7) included liver and kidney tissues that, under conventional preparation methodology, required time-consuming digestion and extraction procedures and demanded stringent safety precautions to prevent contamination. Mineralisation of the biological samples was carried out in glass test tubes under the influence of microwave irradiation, during which time the acid fumes evolved were removed via a water aspirator and trap. This arrangement, to a certain extent, overcame the flow restrictions observed in the earlier work.

The procedure described (6) could be summarized as a batch dissolution method coupled to a flow-injection system, rather than a stopped-flow methodology. The procedure clearly offered advantages over conventional methods through a decrease in digestion time, and it considerably reduced the risk of sample contamination. The system overcame the earlier flow-disruption problems caused by excessive bubble formation, but the design of the method was complicated by increasing the number of steps during the operation procedure.

Good recoveries (>90%) were achieved by using the microwave technique described. The authors recognized the importance of the relationship between the amount of sample taken and the concentration volume of acid required and the parameters such as the microwave power and irradiation time. As will become apparent in the following examples, these variables are fundamental to the development of on-line microwave FIA methodology and highlight the important relationships between sample composition and mass on power absorption and subsequent digestion.

Carbonell et al. (8) focused on a slurry approach for the determination of metallic elements in solid samples. Slurried samples were taken up and digested on-line in a microwave oven, degassed, and passed through an ice chamber to allow condensation of vapor formed during the digestion. Following digestion, samples were passed via a rotary valve to the FAAS system for elemental determinations. Good recoveries of lead extracted from solid samples were achieved in only 5 min, representing a dramatic reduction in digestion time, giving the added advantage of reducing the loss of volatiles. The authors reported no disruption in the carrier flow but problems were encountered with the system blocking due to samples of a large particle size. This problem highlighted a limiting factor in such systems and identified that the particle size of the injected slurried sample should ideally be no larger than half the radius of the internal diameter of the tubing used. Also, the ability of a slurry to agglomerate and form larger particles can occur if the electrostatic forces of the material or the surface tension properties of the system used are inappropriate. Therefore, the surface tension of a system may need to be modified through the addition of surfactants. The agglomeration of slurry samples can also be a problem in the initial preparation of the sample prior to injection into the FIA system leading to poorer precision and even blockage of the injection port. Once again control of particle size and a small amount of surfactant can serve to alleviate the situation.

Work by Hinkamp and Schwedt (9) referred to an on-line microwave-digestion system and illustrated that applications are not just restricted to FAAS detection systems. The authors coupled the now familiar continuous-flow microwave-digestion technique with amperometric detection for the determination of phosphorus in waters. The chemistry involved was based on the methodology described by Fogg and Bsebsu (10) for the amperometric determination of orthophosphate by reduction of molybdophosphate at a glassy carbon electrode. Decomposition of sample was achieved in a crocheted PTFE coil similar to those used for postcolumn derivatisation reactions in high-pressure liquid chromatography (HPLC). The microwave irradiation encouraged hydrolysis or decomposition of the phosphorus contained in the sample. This result was achieved through the use of two reagents, potassium peroxodisulfate (for organic phosphorus) and dilute perchloric acid (for inorganic polyphosphates). The chemical processes involved a combination of both sample decomposition and hydrolysis of phosphorus species to orthophosphate. During their work the authors focused on the optimisation of the system used by recognising several important parameters involved not only in the detection system but also in the decomposition and the interaction between the two processes.

Optimised variables included residence time in the microwave cavity (governed by tube length and flow rate), pH of carrier stream, and reagent strengths. However, the effects of microwave power were not investigated in this particular study, but the power was held constant at 100% (650 W).

Problems encountered with this system included the production of gases (bubbles) during the decomposition, which caused disruptions in the detection system. Gas evolution also affected the flow pattern and the degree of reproducibility, a problem reported by other workers (5). A gas-permeable membrane was used postmicrowave to act as a degasser prior to mixing with the molybdate solution, and good recoveries and a rapid analysis time of 20 samples per hour were reported for organic phosphorus. However, only 60–70% recoveries were reported for the inorganic polyphosphates. These results concur with other phosphate investigations (11, 12), where adding an enzyme prior to decomposition was necessary to achieve maximum yields.

Karanassios et al. (13) described a stopped-flow microwave-digestion system for the dissolution of botanical reference materials. The authors focused on several operating parameters that primarily influence the efficiency of the dissolution technique: namely, microwave power, irradiation time, and the acid mixture.

The impetus behind the development of this method was to further reduce the labour-intensive process of batch microwave sample dissolution by using a flow system, thus increasing sample throughput and facilitating automation. However, this method did not involve the direct interface of the flow system with a detector, but merely collected the sample after dissolution. The design of the system was very similar to those described by previous authors and consisted of a peristaltic pump to carry the sample into the microwave cavity, a PTFE digestion coil, and a sample collecting vessel postmicrowave. The authors did not find it necessary to insert a cooling device or degasser; however, some risk may exist of volatile analyte loss with such a system when gaseous components emerge at the collection point. The technique has been described as a stopped-flow methodology because during digestion the sample slowly rotates inside the coiled tube, both clockwise and anticlockwise, because of pressure differences that developed on either side of the sample slug.

Results for the analysis of several botanical reference materials were reported to be close to 100% for the various elements selected. Problems were encountered, however, during short digestion times for the recovery of Al, but these problems were overcome by prolonged irradiation that increased the recovery to close to 100%. The system does have shortcomings, notably that the digestion coil became discoloured after prolonged use, particularly at valve joints and "T" pieces. Particulates from undigested sample appeared on the walls of the tubing and may have resulted in memory effects. The authors had to clean the tubing regularly and replace corroded joints. In their opinion it is not good safe practice to have joints and fittings within the microwave irradiation zone and this should at all costs be avoided.

The major feature of Karanassios's work was the clear indication that residence time in the microwave cavity is a function of flow rate and/or tube length, a point that will be returned to later in this chapter with reference to dispersion effects. In the work described by Karanassios (13) there was no carrier other than the slurry itself, and thus large volumes of acid were consumed. An additional limitation was the lack of an on-line cooling device, and delays were caused through the necessity of having to cool the digested samples prior to analysis by inductively coupled plasma–atomic emission spectroscopy (ICP–AES). Even though such problems may be considered to be trivial, they can be easily overcome at the system-design stage. The use of an aqueous carrier stream and the injection of acidic slurried sample via a rotary valve, for example, would offer a dramatic reduction in acid consumption and limit the corrosion of joints. An efficient cooling device on-line would also remove delays in collection. Indeed, applying both of these suggestions to the system facilitates the direct insertion of digested samples into the detection system.

This approach has proved successful for several authors in later publications involving, for example, on-line microwave sample dissolution coupled to FAAS (14–18) and ICP–AES–MS (18a) Similar methodologies were adopted consisting mainly of a carrier stream (usually aqueous), digestion on-line in PTFE tubing, cooling of the sample, and a direct interface with FAAS for elemental detection. The system described by Haswell and Barclay (14) is a good illustration of how the methodology has advanced in recent years. Many of the problems encountered by earlier workers have been negated, and the system consists of a more efficient cooling device as well as the addition of an on-line backflush filter and a pressure-regulated carrier stream (Figure 2). The work describes the optimisation of the digestion conditions: again, the important parameters were irradiation residence time, microwave power, and acid strength. After optimisation of the system, recoveries in the range of 94–107% were reported for Mg, Ca, Zn, and Fe in a number of organic-based reference materials.

One of the important developments of the Haswell system was the introduction of a pressure regulator. This development offered more control over the formation or release of bubbles arising from the boiling and gaseous release of components from the decomposition of a sample matrix. This control, in turn, improved flow properties. In earlier systems bubbles seriously disrupted the flow characteristics of the FIA manifold and led to significant dispersion of the analyte injection volume. One of the interesting aspects of the Haswell system is the importance it places on

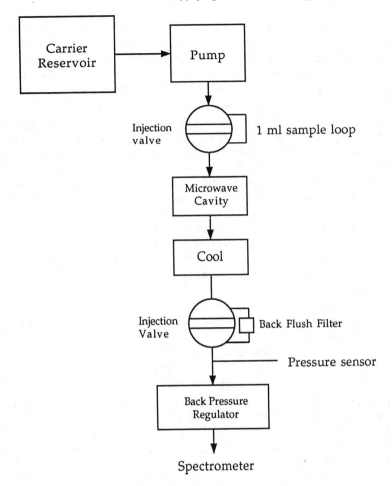

Figure 2. Schematic diagram of microwave-digestion FI system developed by Haswell and Barclay. (Reproduced with permission from reference 14. Copyright 1992 Royal Society of Chemistry.)

out-gassing and bubble control. The work demonstrated that for efficient digestion to take place a gas–liquid interface phase was required (a point noted in bomb systems also). This fact relates to the interfacial microwave effect described later in this chapter. In addition to the need for a gas phase in the digestion process, the work also illustrated that the formation of a segmented flow system minimised dispersion during the digestion process. Subsequent recondensing of the gaseous phase then resulted in recombination of the original injection slug with minimal peak broadening. The presence of bubbles thus aided digestion and allowed long tube lengths (30 metres) to be used. Therefore, residual time for a given flow rate (often dictated by the detection system) was increased without serious deterioration in the peak dispersion at the detector.

Similar reports were made by Carbonell et al. (15), who devised a more basic system than the proposed Haswell methodology. The method proved to be successful for the determination of Cu and Mn in solid samples, after the optimum irradiation time at maximum power had been determined. Again, as in previous studies, the methodology offers a rapid sample throughput, limited operator attention, and easier automation.

Tsalev et al. (16–18) used an on-line microwave system as a sample pretreatment step for hydride generation (HG) and cold vapour (CV) analysis of liquid samples by AAS. Here the chemistry involved was designed to liberate the analyte elements from an organic matrix. The analyte species then became present in a well-defined oxidation state suitable for vapour analysis by AAS (e.g., Hg(II), Se(IV), and Pb(V)). Previously, this type of sample pretreatment was carried out off-line and, as in many analytical systems, became the rate-determining step in the analysis. The paper illustrates the ability to interface the on-line microwave system with FI–CV–AAS and FI–HG–AAS. Much attention is given to monitoring of temperature postmicrowave and in the gas–liquid separator used.

Tsalev and co-workers found that temperatures between 50 °C and 90 °C could be achieved; however, problems were encountered because of peak broadening and loss of peak-height sensitivity due to dispersion. These problems were compensated for by the introduction of hot samples, thus improving peak-height sensitivity. However, the authors did report that overheating of samples (i.e., irradiating at higher power settings) did cause some problems with the aerosol formation that resulted in a corresponding reduction in precision. Also, limitations occurred because of the less-stable hydrides that decomposed at the higher temperatures. The best oxidising reagents were those containing bromate-bromide and peroxodisulfate. Careful optimisation of the digestion mixture was needed for each analyte, and the authors went on to successfully apply this technique to the analysis of mercury in urine and water samples. The basic concepts of on-line microwave heating employing FIA-type systems inevitably lead to extensions of the original methodology.

Balconi and co-workers (19) used an FIA microwave methodology for irradiation of well and river water samples to aid the oxidation step for the determination of chemical oxidation demand (COD) levels. Previous workers stressed that one of the main difficulties of using FIA for COD determination was the several hours of sample heating required. A batch method using microwaves as a heating source showed that an irradiation time of 7 min gave results comparable to conventional methods (18).

This dramatic reduction in heating time increased the possibility of applying this method to an FI system incorporating microwave heating. The manifold developed by Balconi used Teflon (PTFE and carbon) tubing as the reaction coil and a degassing unit to remove any bubbles produced during sample heating. The system describes a direct interface to a spectrophotometer for on-line detection. Optimisation was described in terms of microwave power settings.

Problems arose however in that the degassing unit of the system when the power setting of the microwave was above 216 W, resulting in interferences in the detection system. At lower power settings a longer irradiation time was required, and the system was not fully optimised. However, satisfactory results were achieved for the few samples analysed.

The problem of prolonged reaction times in FIA systems was brought to light again by Xu et al. (*21, 22*), who described the application of microwave irradiation for the enhancement of colour development of chromogenic reagents in the spectrophotometric determination of Pd and Rh. The habitually slow coordination reaction between noble metals and organic colour reagents can be accelerated by tens of minutes if conventional heating methods are applied. The application of such procedures in stopped-flow methodology, although effective, limits the sample throughput significantly. Thus, the authors displayed their interest in "coupling a microwave oven with a flow injection analyser for expanding the range in which flow injection analysis is used" (*21*). Results indicated improved reaction rates under microwave irradiation when compared with conventional heating methods. The authors described a simple experimental system and poignantly remarked that FIA is useful in studying the effects of microwave irradiation on chemical reactions. Xu and co-workers described such effects in terms of dispersion behaviour on the chromogenic reactions used in various solvents and under the influence of microwave irradiation in a flowing stream (*22*). They concluded that the dispersion coefficient of chromogenic reagents decreased and some chemical reactions were significantly accelerated. In the case of accelerated reactions, the influence of microwave irradiation either accelerated or had no adverse effect on reaction rate: that is, no fatalistic influences were observed.

One of the pioneering groups of the FI microwave process, Burguera and Burguera (*5*) expanded their study (*23*) on the determination of lead in biological materials. In this publication the rapid advancement of the methodology being developed for FIA microwave systems is apparent. The authors described a fully automated computer-monitored manifold (Figure 3) and discussed the design and operation of the automated on-line methodology with analysis by FIA–electrothermal atomization atomic absorption spectrometry (ETAAS). After some initial problems with the introduction of slurry into the FIA system, the authors described the optimisation of the manifold in three stages:

- the digestion and dissolution conditions
- FI parameters
- detector conditions

Although each step can be described as separate entities, the interaction among the three stages is also of great importance and must be considered during the optimisation process. For example, a more concentrated

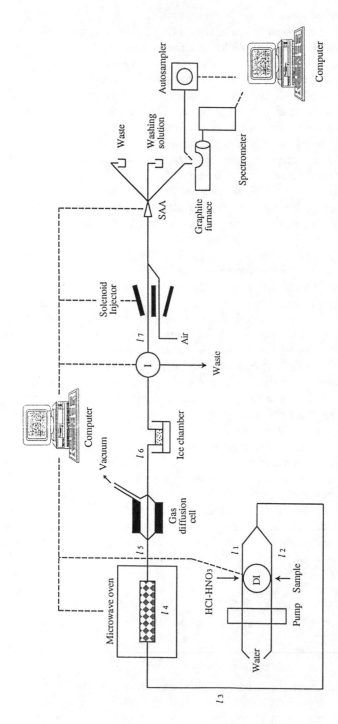

Flow injection- microwave oven-ETAAS manifold. (DI) six-way valve injector; (I) four-way valve injector; and (SAA) sampling arm assembly. l 1–l 7 indicate 1.0, 0.5, 1.0, 2.0, 0.5, 0.8, and 0.5 m tubing lengths, respectively.

Figure 3. Fully automated computer-monitored manifold developed by Burguera and Burguera (Reproduced with permission from reference 23. Copyright 1993 Royal Society of Chemistry.)

acid would encourage more rapid decomposition of organic matrices in a microwave oven. However, high concentrations of acid may cause inefficiency of the homemade degassing unit and also the possibility of oxidising the surface of the furnace, thus reducing its lifetime. Consequently, each of the three steps is described in detail and an in-depth profile of optimised experimental conditions is given. This paper is in strong contrast with the earlier publication (5) and highlights not only the improvement in technology but also the understanding of chemistries and mechanisms involved in the FIA microwave sample-dissolution process. The work has subsequently been extended to describe methods for the determination of iron and zinc in adipose tissue and cobalt in whole blood based on on-line microwave-assisted mineralization and ETAAS detection (24, 25).

The design of an on-line system using alternative heating sources is feasible and is indeed illustrated by Gluodenis and Tyson (26). The authors initially developed an on-line sample-dissolution methodology, with heating being supplied through resistive heating elements. However, the authors sought to improve this method by using microwave irradiation as a heating source, a glass reaction vessel, and a homemade pressure transducer (11). In this instance, the authors chose microwave heating primarily for the ability to control the energy flux into the sample. This trend illustrates an important contrast between conventional heating and microwave heating systems: namely, that the heating or energy transfer process can be made specific and directly focused to components in the digestion matrix with a high degree of control unlike a traditional passive, nonselective heating process. A general belief exists that this fact leads to the more efficient heating mechanism observed with microwave systems.

The manifold developed consisted of a small reactor vessel placed inside the microwave cavity into which slurried samples were introduced by an FIA sample handling system. The stopped-flow methodology was evaluated for the analysis of samples with high organic content. The system was pressurised, and the reaction column headspace was vented via a series of PTFE valves. After digestion and cooling, samples were flushed into calibrated flasks, diluted to volume, and analysed by FAAS and ICP–AES.

Results illustrate the advantages of this method over the previously described method (11): namely, a more rapid dissolution process and the ability to control the energy flux into the sample when using microwave irradiation. However, the Gluodenis and Tyson methodology has its limitations given that it was necessary to occasionally shut down the magnetron to avoid over-pressurisation. The manifold was not designed with any form of pressure regulation, and slight maintenance work was needed in replacing leaking valves every 2–3 months. In addition, erratic results were obtained for the determination of calcium when using the proposed methodology. This observation was thought to be due to interactions between the calcium in the sample and the borosilicate glass column. Residual silicates were also found in several samples because a hydrogen fluoride (HF) digestion could not be used due to the degradation of the glass reaction column. The use of PTFE tubing was not possible because the high pressures achieved would result in rupture. The authors also acknowledged that the design of a suitable interface for coupling to a detection system would be of great benefit.

The potential of using the on-line dissolution systems described in this chapter for process analysis is an attractive feature of the methodology. Several reports have already appeared describing the use of an FIA microwave manifold in water quality monitoring (12, 27–29). The importance of instrumentation for near real-time monitoring of water quality is well-emphasised by Hart et al. (12). Two examples of on-line analysis of water samples for total phosphate (27, 28) and urea (29) have been selected to illustrate the adopted methodology. In both cases, digestion of samples is carried out under continuous-flow conditions in thin-bore tubing (<1 mm id). Following digestion, samples are cooled on-line and then combined with further reagent streams for subsequent colorimetric detection.

In the work by Williams et al. (27) the phosphate detection is based on the reduction of heteromolybdophosphoric acid to molybdenum blue as an FIA peak at 690 nm. Results showed good comparison with conventional batch methods (Table I). The greatest advantages were a rapid total sample analysis time and facility for automation. Benson et al. (28) described an alternative phosphate system and preferred to use the molybdate–tin chloride reagent mixture rather than the molybdate–ascorbic acid reagents used by Williams et al. However, the authors did use the same digestion mixtures as those described by Hinkamp and Schwedt (9) earlier. In the paper by Schmitt et al. (29), urea is reduced on-line to ammonium ions, which on reaction with NaOH release gaseous ammonia. The gaseous ammonia passes through a gas-permeable membrane and causes a pH change in a bromothymol blue indicator stream, which is colorimetrically determined at 635 nm.

Although many papers have reported various applications of the on-line microwave systems, their primary usage is still considered to be for sample dissolution processes. A report by Kometa and Greenway (30) discussed on-line sample preparation for the determination of riboflavin mononucleotides in foodstuffs. Here, the authors sought a more rapid analysis time for water-soluble vitamins. The time-consuming process in this instance is separating the analyte from the complex food matrix. The microwave manifold described was similar to the system used by Williams et al. (27). In the method development, the effects of acid concentration, microwave power, and digestion coil length were investigated. Results showed that exposure to microwave irradiation provided an efficient on-line method for the liberation of flavins from complex food matrices. However, problems did occur when acids of higher concentrations

Table I. Analysis of Wastewater Samples by Batch and FI Microwave Methods for the Determination of Total Phosphate

Sample Number	Total Phosphate by Conventional Batch Method (ppm)	Total Phosphate by On-Line FI Method (ppm)	Precision (RSD) (n = 10) for FI Method (%)
1	10.7	10.5	1.4
2	14.9	14.4	1.9
3	17.0	16.0	4.8
4	11.1	11.8	1.4
5	2.2	2.4	1.6
6	4.2	4.4	1.8
7	5.8	6.3	2.9
8	20.9	21.1	2.3
9	14.3	15.6	3.1
10	6.4	6.0	2.6
11	31.7	30.5	1.4
Total analysis time	3 h	25 min	

NOTE: RSD is relative standard deviation.

SOURCE: Data are from reference 27.

were used because they caused the degradation of the riboflavin. The microwave method described (30) aided in reducing the full analysis time on ground food samples to approximately 20 min. In the field of food analysis the simple on-line saponification of water-soluble vitamins also is applicable to on-line microwave methodology (31). This research opens up new areas of application in the field of on-line analyte extraction such as pesticides from foods and environmental samples.

Organic Synthesis

Pioneering work in the field of on-line organic synthesis has been reported by works based at CSIRO, where the development of a continuous-flow organic reactor has been described (32). The MicroLab reports an operating time of many hours, flow rates of 20 mL/min at temperatures of up to 200 °C, and pressures to ca. 1200 kPa. These conditions are not easily achieved using conventional preparative equipment. Subsequent work has been reported involving microwave irradiation of organic reactions in flow systems, but only one recent publication uses this particular reactor (33).

Chen et al. (34) investigated the possibility of using a continuous-flow process for the organic synthesis of several compounds by using microwave irradiation. Previous studies reported that successful synthesis of organic compounds by microwave irradiation in closed and open vessels could be achieved, in many cases proving superior in both reaction rate and yield compared with conventional heating methods (35). The closed-vessel process tends to be limited by the high pressures and temperatures produced during irradiation, which may result in the need to limit yields to typically 1 g or less for fear that a rapid pressure buildup would lead to a violent explosion. Thus, the volume of the closed-vessel systems limits the maximum yield of organic product and does not make effective use of the efficient controllable heating process.

In the system described by Chen and co-workers (Figure 4), the reactants, held in a reservoir, were pumped into a Teflon reaction coil in a microwave oven for irradiation and then pumped out again to a product collection vessel. The resulting compounds were analysed by either HPLC or gas chromatographic (GC) methods. The expansion gases produced from the synthesis were not restricted because of the design of the manifold; thus, the risk of rapid pressure buildup was removed and as a result yields of greater than 20 g were reported for five different organic reactions.

These reactions were studied in detail and particular attention was paid to the effects of irradiation time on conversion of reactant to products. In each instance the continuous-flow method proved that the percentage conversion increased with increasing irradiation time. Also, the system was pressurised due to positive pressure from the pumps and an induced pressure increment from the irradiation process. This pressure effect was considered important in the enhanced rate of conversion observed. All reactions ran safely without explosion; however, for one of the methods the evaporation of methanol through the Teflon tubing resulted in solution spillage. The authors concluded that the on-line approach was feasible to large-scale organic synthesis for industrial use. Chen recognised the potential of such systems as a replacement for expensive or time-consuming conventional industrial processes. An example is use of the microwave process for the conversion of unwanted optical enantiomers into their racemic mixtures to recover the more useful enantiomers.

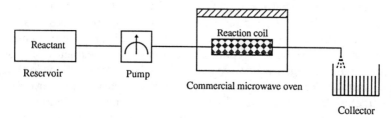

Schematic diagram of the continuous flow process for organic synthesis using microwave irradiation.

Figure 4. Schematic of microwave FI manifold for organic synthesis as described by Chen et al. (Reproduced with permission from reference 34. Copyright 1995 Royal Society of Chemistry.)

Work in the organic synthetic field represented a turning point in the development of the on-line microwave process, because it demonstrated that the general concept was not limited merely to a digestion or dissolution process but offered a more general philosophy to reaction methodology. The efficiency of the on-line microwave system for the synthesis of organic compounds has been attributed to several factors (*32, 34*). A well-documented fact since the early years of microwave research is that the heating of samples by microwave irradiation is achieved by means of two main processes: ionic conduction and dipole rotation. Chen attributes the enhancement of organic reactions observed in his research to both of these processes and, in addition, the *molecular agitation* or *stirring* of reactants. Further applications of microwave irradiation for the enhancement of organic reactions have been illustrated by Chen; several examples include the hydrolysis of proteins, racemisation of amino acids, and hydrolysis of sugars (*36*).

The high efficiency achieved in these reactions is reported by Chen to be not only associated with a temperature increase but also the perturbation in dipole moments induced by microwaves. Chen uses his molecular agitation theory to argue why in some cases, temperature elevation does not equate to observed rate enhancement; thus, the effect is nonthermal. The so-called nonthermal or non-Arrhenius effect is often argued to be nonexistent by many workers. The claim is made that many temperature measurements and experiments are carried out in the wrong environment and using the wrong equipment.

More recently, the so-called interfacial effect has been introduced by the authors to explain some phenomena observed in the microwave flow systems. Views on this subject support a non-Arrhenius effect, which appears to be more pronounced in thin-film FIA systems rather than the traditional bulk-batch methodology. No mechanism has yet been conclusively identified, but most of the workers in the field have expressed surprise at the apparent efficiency and products of the microwave reaction processes when using an FIA system. What is apparent from gas–solid, liquid–solid, and liquid–liquid interfacial reactions is that microwaves influence the surface reactions. In our opinion these surface effects are associated with interfacial charge induced by the microwaves, which seem to impart additional energy into the interfacial reaction. This interfacial effect is currently being studied by various research groups, but clearly colloidal-based systems are of interest together with membrane transfer and chromatographic, solid–gas and solid–liquid phase separations.

An example is the recent application for on-line preparation of fatty acid methyl esters (FAMES) for their subsequent GC analysis (*37*). This work has been carried out in our research laboratory and involves the esterification of fatty acids and other lipid extracts under the influence of microwave irradiation to their more-volatile FAMES and subsequent on-line membrane phase separation.

Other Applications of FI Microwave Systems

The simplest illustration of the compatibility of microwave usage in flow systems was the report of microwave irradiation used as an aid for a continuous-flow drying source to increase the drying rates of wheat grain, thus prolonging safe storage time (38). The field losses of wheat grain can be significantly reduced by harvesting at high-moisture content, but such practice requires drying of grain prior to storage. In this application microwaves were used successfully to preheat the grain at the inlet into a drier. This work emphasises the ease of adaptability of the microwave system being described and also the efficiency of microwave irradiation as a heating source.

Work by Canadian researchers has recently lead to the so-called microwave-assisted process (MAP), which is a novel extraction method based on microwave irradiation as a semiselective energy source. The process has been applied on both analytical-scale applications and industrial processing for commercial purposes (39).

The impetus behind this development is described by Paré (40) as being a high potential for commercial success, as well as the appeal of the "clean process technology" and applications associated with it. The first reported application of MAP was for the extraction of essential oils from plant products (40). Since this early success, other extraction processes have been applied to sample types such as water, soils, and animal and plant tissue, opening up a wide field of applications.

The mechanism behind the extraction process is thought to be based on the principle that the ability to absorb microwave energy varies from one sample to another, depending on chemical composition (in particular dipole characteristics) (39). This ability also varies with the frequency applied as well as with the temperature of the substance. Hence in theory, it is possible to effect a semiselective heating and thus selective extraction of different components in the system, depending on the dielectric properties. However, the experiments described to date were carried out at a fixed frequency of 2450 MHz. Two areas of extraction were discussed in detail: namely, liquid-phase and gas-phase extraction.

Alternative applications of the on-line microwave-enhanced systems will no doubt continue to be reported for device application areas. For example, Bond et al. (41) reported the application of on-line microwave heating in the field of heterogeneous catalysis. In this application, microwave irradiation was used not only to enhance catalytic reactions during catalytic reactions but also in catalyst preparation and characterisation. This work is discussed in detail elsewhere in this book (Chapter 10). However, catalytic processes stimulated by microwave irradiation will be of some importance in future industrial applications, and the rapid increase in publications using such processes illustrates one major field of future applications in microwave systems.

One of the most appealing aspects of on-line microwave technology is the ability to interface the sample treatment manifold with a variety of detectors depending on the analytical requirements. For example, Lopez et al. (42) used on-line microwave oxidation for the determination of organoarsenic compounds by HPLC–HG–AAS. Samples were introduced into the system and carried into the chromatographic module for separation. Oxidation of selected organoarsenic species was carried out in the microwave cavity in a PTFE reaction coil. The volatile arsine components were then carried into a gas–liquid separator and analysed by HG–AAS. The authors reported the successful determination of arsenite, arsenate, monomethylarsinate, dimethylarsinate, arsenocholine, and arsenobetaine in various water samples by using the proposed manifold. The microwave-oxidation system was a good alternative to photooxidation and thermooxidation. Using a similar approach but different chemistries, Pitts et al. (43) reduced selenium(VI) to selenium(IV) on-line in a hydrochloric acid medium by using microwave heating to accelerate the reduction step, thus enabling total seleniun to be determined. The authors described a method that with the microwaves off, selenium(IV) only was detected, but with the microwaves on total and hence selenium(VI) could be estimated.

Even though the concept of a coupled sample preparation detection system is attractive, problems with detector interfaces need to be addressed. In batch systems the spatial separation of the sample preparation procedure and the measurement method can account for difficult matrices that may cause interferences in the detector. A filtering or separation process is common. In principal with the on-line system everything injected into the carrier system is transported to the detector albeit after chemical processing. Many papers have considered the problems of matrix interference, and the use of phase separators is now being reported. These uses include gas-phase separation for urea determination as ammonia in sea water (29) as described earlier in this chapter. Even though the addition of phase separation further complicates the total system it offers more attractive methods for the development of a total analytical system.

Safety

One of the most important aspects to consider when developing new microwave-enhanced methodologies is safety, a subject approached by Cablewski et al. (33), who discussed organic synthesis in a microwave oven. Early reports of microwave-assisted organic synthesis illustrated the hazards and difficulties associated with such reactions. Lack of control of both temperature and pressure led to deformation of reaction vessels and in some cases explosion.

Clearly, a major problem to consider is a rapid increase in pressure during microwave-assisted reactions. Also, the engineering of equipment used must be adequate to safely handle the presence of organic and inorganic solvents. The authors also warn against the presence of joints and plastic or glass fittings in the microwave field because these represent a weak point where bursting can take place. In addition such components (notably poly(ether ether ketone) tubing) often become brittle. The best advice is to use a continuous length of PTFE tubing in the microwave and make all coupling to the tubing external to the oven or cavity system. If slurry work is being carried out then it is important to use particle sizes of less than half of the internal diameter of the carrier tubing. Blockage of the tubing by particulates can cause pressure buildup. More importantly, in a microwave field heating and charging of a solid sample can be rapid leading to potential hot spots and explosive situations. Practitioners are advised to avoid tight coils and bends in their reaction tubing which may trap particulates.

The concept of a continuous-flow microwave reactor for organic synthesis is an appealing one because many hazards associated with microwave batch reactions are more controllable. The closed flowing system does not need to hold large volumes of reagent in the microwave oven at any one time because samples are continuously moving and can be readily pressure controlled and temperature monitored. The microwave reactor has several advantages (including safe operation) over conventionally heated equipment. This fact is strongly supported by the work of Chen et al. (34, 36), which illustrates specific examples of successful syntheses carried out under such conditions.

Conclusions

This chapter has attempted to give a general review of the strategic development and current status of on-line microwave-enhanced chemical reactions and is in no way meant to give comprehensive coverage of every application reported to date. In addition some of the more important considerations in developing flow rather than batch microwave systems have been described. These considerations include careful selection of the chemistries employed, the use of pressure control, and the significance of microwave irradiation time and power. Perhaps one of the most significant factors of the flow systems developed is the indication that the nonthermal effects of microwave irradiation, in particular interfacial charging and molecular stirring, may be more pronounced in such systems. For those wishing to exploit the many advantages of an FIA system incorporating microwave-enhanced chemistry, the challenge is the ability to rationalise the introduction of a complete sample matrix into a closed system with relatively interference-free detection. The prize for developing such systems is considerable and current science and technology is now available. Let the challenge begin!

References

1. Abu-Samra, A.; Morris, J. S.; Koirtyohann, S. R. *Anal. Chem.* **1975**, *47*, 1475.
2. Skeggs, L. *Am. J. Clin. Path.* **1957**, *28*, 311.
3. Ruzicka, J.; Hansen, E. H. *Flow Injection Analysis;* John Wiley & Sons: New York, 1981.
4. Corporate Profile 43A, Society for Applied Spectroscopy: Frederick, MD, 1993.
5. Burguera, M.; Burguera, J. L.; Alarcón, O. M. *Anal. Chim. Acta* **1986**, *179*, 351.
6. Burguera, M.; Burguera, J. L.; Alarcón, O. M. *Anal. Chim. Acta* **1988**, *214*, 421.
7. De La Guardia, M.; Salvador, A.; Burguera, J. L.; Burguera, M. *J. Flow Injection Anal.* **1988**, *5(2)*, 121.
8. Carbonell, V.; De La Guardia, M.; Salvador, A.; Burguera, J. L.; Burguera, M. *Anal. Chim. Acta* **1990**, *238*, 417.
9. Hinkamp, S.; Schwedt, G. *Anal. Chim. Acta* **1990**, *236*, 345.
10. Fogg, A. G.; Bsebsu, N. K. *Analytst* **1982**, *107*, 566.
11. Gluodenis, T. J.; Tyson, J. F. *J. Anal. At. Spectrom.* **1992**, *7*, 301.
12. Hart, B. T.; McKelvie, I. D.; Benson, R. L. *Trends Anal. Chem.* **1993**, *12(10)*, 403,
13. Karanassios, V.; Li, F. H.; Liu, B.; Salin, E. D. *J. Anal. At. Spectrom.* **1991**, *6*, 457.
14. Haswell, S. J.; Barclay, D. A. *Analyst* **1992**, *117*, 117.
15. Carbonell, V.; Morales-Rubio, A.; Salvador, A.; De La Guardia, M.; Burguera, J. L.; Burguera, M. *J. Anal. At. Spectrom.* **1992**, *7*, 1085.
16. Tsalev, D. L.; Sperling, M.; Welz, B. *Analyst* **1992**, *117*, 1729.
17. Tsalev, D. L.; Sperling, M.; Welz, B. *Analyst* **1992**, *117*, 1735.
18. Welz, B.; Tsalev, D. L.; Sperling, M. *Anal. Chim. Acta* **1992**, *261*, 91. (a) Sturgeon, R. E.; Willie, S. N.; Methven, B. A.; Lam, J. W. H.; Matusiewicz, H. *J. Anal. At. Spectrom.* **1995**, *10*, 981.
19. Balconi, M. L.; Borgarello, M.; Ferraroli, R.; Realini, F. *Anal. Chim. Acta* **1992**, *261*, 295.
20. Jardin, W. F.; Rohweddwer, J. R. *Water Res.* **1989**, *23*, 1068.
21. Xu, Y.; Chen, X.; Hu, Z. *Anal. Chim. Acta* **1994**, *292*, 191.
22. Xu, Y.; Hu, Z. *Anal. Lett.* **1994**, *27(4)*, 793.
23. Burguera, J. L.; Burguera, M. *J. Anal. At. Spectrom.* **1993**, *8*, 235.
24. Burguera, J. L.; Burguera, M.; Carrero, P.; Rivas, C.; Gallignani, M.; Brunetto, M. R. *Anal. Chim. Acta* **1995**, *308*, 349.
25. Burguera, M.; Burguera, J. L.; Randon, C.; Rivas, C.; Carrero, P.; Gallignani, M.; Brunetto, M. R. *J. Anal. At. Spectrom.* **1995**, *10*, 343.
26. Gluodenis, T. J.; Tyson, J. F. *J. Anal. At. Spectrom.* **1993**, *8*, 697.
27. Williams, K. E.; Haswell, S. J.; Barclay, D. A.; Preston, G. *Analyst* **1993**, *118*, 245.
28. Benson, R. L.; McKelvie, I. D.; Hart, B. T.; Hamilton, I. C. *Anal. Chim. Acta* **1994**, *291*, 233.
29. Schmitt, A.; Buttle, L.; Uglow, R.; Williams, K. E.; Haswell, S. J. *Anal. Chim. Acta* **1993**, *284*, 249.
30. Kometa, N.; Greenway, G. M. *Analyst* **1994**, *119*, 929.
31. Luque-Perez, E.; Haswell, S. J. *Anal. Proc.* **1995**, *32*, 85.
32. Peterson, C. *New Sci.* **1989**, 44.
33. Cablewski, T.; Faux, A. F.; Strauss, C. R. *J. Org. Chem.* **1994**, *59*, 3408.

34. Chen, S. T.; Chiou, S. H.; Wang, K. T. *J. Chem. Soc. Chem. Commun.* **1990,** 807.

35. Caddick, S. *Tetrahedron* **1995,** *51,* 10403.

36. Chen, S. T.; Chiou, S. H.; Wang, K. T. *J. Chin. Chem. Soc.* **1991,** *38,* 85.

37. Williams, K. E. Ph.D. Thesis, University of Hull, 1995.

38. Radajewski, W.; Jolly, P.; Abawi, G. Y. *J. Agric. Eng. Res.* **1988,** *41,* 211.

39. Paré, J. R.; Bélanger, J. M. R.; Stafford, S. S. Environment Canada Ottawa, personal Communication, 1994.

40. Paré, J. R. *Bull. Can. Comm. Electrotechnol.* **1993,** *7(3).*

41. Bond, G.; Moyes, R. B.; Whan, D. A. *Catal. Today* **1993,** *17,* 427.

42. López-Gonzalvez, M. A.; Gómez, M. M.; Cámara, C.; Palacios, M. A. *J. Anal. At. Spectrom.* **1994,** *9,* 291.

43. Pitts, L.; Worsfold, P. J.; Hill, S. J. *Analyst* **1994,** *119,* 2785.

Chapter 7

Pressure-Controlled Microwave-Assisted Wet Digestion Systems

Günter Knapp, F. Panholzer, A. Schalk, and P. Kettisch

Two microwave-assisted decomposition systems—a cavity-type and a waveguide-type system—for pressure-controlled wet digestion of organic and inorganic materials are discussed. A special pressure-regulation system enables simultaneous pressure control in every digestion vessel used. The samples are digested in closed quartz or perfluoro alkoxy (PFA) vessels with pure nitric acid or acid mixtures at pressure levels up to 80 bar. By intensively cooling the vessels during decomposition, sample digestion can also be effected with sulfuric acid at a very high temperature. Both systems are able to decompose tough materials like crude oil, grease, coal, coke, and plastics fast and completely. Depending on the design of the apparatus up to six pressure-controlled digestion vessels are used simultaneously. Different vessel sizes are available for sample amounts up to 1.5 g. The digestion time depends on the sample matrix, the pressure, and the acids used and varies from 5–20 min. The cooling time is 10 min.

Remaining organic compounds are sources of systematic errors with some measuring techniques. Therefore, the amount of dissolved organic residue obtained on different digestion conditions will be indicated. Safety measures are very important for pressurized sample decomposition. Technical precautions and possibilities for the security check are depicted.

Pressurized microwave-assisted wet digestion is one of the most powerful technique for sample decomposition (*1–5*). The reasons for comparably good analytical results with this technique are

- enhancement of the oxidation potential of ashing reagents because temperatures above the boiling point of the reagent can be employed (ashing of tough organic materials with pure nitric acid)
- no volatilization of elements
- low consumption of high-purity reagents and therefore low blank values
- no contamination from external sources
- comparably small vessels made of inert materials causing low adsorption–desorption effects
- cutting decomposition time by microwave heating

Although the determination of elements in organic materials is one of the routine tasks that trace analysis has to deal with, a major problem still has to be overcome: namely, some of the elements that are to be determined occur at concentrations that are close to the detection limits of even highly efficient analytical methods. Analytical data, which are sufficiently accurate and reproducible, can often only be obtained if the samples are carefully prepared, which would above all include complete destruction of all organic matter. Even slight, unmineralized traces of organic sample constituents or artifacts of such substances can interfere with spectrochemical or electrochemical trace analyses. Often this interference means that it is impossible to obtain valid analytical statements.

To separate the excess organic matrix, which is a characteristic of all organic substances, a decomposition of the sample usually precedes the actual element determination process. Its objectives are as follows:

- conversion of the indefinitely bound trace elements to definite soluble compounds
- elimination of interference caused by the organic matrix, which is often a source of systematic errors
- isoformation of sample and standard

For this purpose, organic samples are usually treated with oxidizing acids. The carbon contained in the organic matrix is oxidized to form carbon dioxide, and thus it is separated from the element traces that are to be determined. The most important reagents are nitric acid and perchloric acid. They are either used on their own, mixed together, or in binary or ternary mixtures with sulfuric acid.

Of these reagents, perchloric acid is the strongest oxidizing agent for organic matter. In most cases, complete ashing was impossible without it. Working with perchloric acid, however, has a number of disadvantages. It is highly reactive to organic matter, which makes it an efficient ashing reagent, but the danger of an explosion is high (6). Therefore, it must never be used on its own, but only in mixtures together with other acids, primarily nitric acid. Even then, however, handling it can be dangerous, as many accidents have shown. Above all one should make sure that perchloric acid is not evaporated until all the organic matter has been ashed. To prevent perchloric acid fumes from escaping into the ventilation ducts, where it might form explosive mixtures with organic vapours, special exhaust systems have to be used.

Because of its explosive potential, perchloric acid should never be applied in closed systems, which means that losses of certain elements due to volatilization during the decomposition are impossible to prevent. For these reasons, one has to find less dangerous substitutes for perchloric acid.

Working with nitric acid is much safer. No danger exists that an explosive reaction might be effected by contact with organic substances. An additional bonus is that the available analytical-grade acid can easily be purified further by subboiling distillation (7). However, nitric acid is not a sufficiently strong oxidizing agent to completely mineralize organic materials in open systems. Depending on the sample material, 2–50% of the original carbon remains in the ashing residue (8). In addition carbon oxidation is partly unreproducible so that, depending on the degree of oxidation, the element determination will be interfered with to a greater or lesser degree by the organic decomposition products. In some instances the oxidation power of nitric acid is not even sufficient for the breakdown of biological substances into soluble components. Especially if the samples contain fats, therefore, large quantities of organic decomposition products remain undissolved. To fully exploit the advantage of nitric acid, pressure decomposition systems have to be used that permit the application of nitric acid at temperatures above the boiling point. As stated previously, perchloric acid should not be used in closed systems. Violent reactions could cause the detonation of the pressure vessel because the safety systems might be too slow to react to the pressure wave. The same concern applies to mixtures of nitric and sulfuric acids, which must not be used to mineralize fats if the process is effected in a closed system (9).

Even if nitric acid is used on its own, the amounts of organic samples that are added should be kept low so that the safety devices will continue to work properly even if violent reactions occur (10). The decomposition behaviour of a sample can be investigated by slowly increasing the sample weight.

Furthermore, an optimal ratio of sample weight to acid volume is important. The acid volumes (65–69% HNO_3) described in the literature as suitable for the decomposition of organic samples range between 0.4 and 2.0 mL per 200 mg of sample (11–13), depending on the sample matrix that is to be ashed. However, a simple and more generally applicable statement is possible: A volume of 2 mL HNO_3 is sufficient to mineralize an amount of organic sample that corresponds to a pure carbon content of 100 mg, no matter what type of substance is to be determined. Lower quantities of acid will suffice for biological substances with a low-fat content (8).

The sample weight is determined by the maximum pressure possible inside the decomposition vessel employed and the vessel size. The following examples refer to a decomposition system that withstands pressures of 20 bar and is fitted with a 35-mL poly(tetrafluoroethylene) (PTFE) vessel. A further assumption is that mineralization is effected at 180 °C. Under these conditions samples that correspond to a pure carbon content of 100 mg can be ashed. Regardless of the type of biological substance used, the pressure building up during the final phase of the decomposition process is around 20 bar. This pressure buildup is caused by the gases CO_2 (from the carbon contained in the sample) and NO/NO_2 (due to reaction with HNO_3) that form during the decomposition process and the vapour pressure of water and nitric acid (14). Thus, the appropriate sample weights of an organic substance are easy to calculate on the basis of its carbon content. Approximately 200 mg of freeze-dried tissue (ca. 50% C), 200–300 mg of freeze-dried plant substance (30–50% C), but only 120 mg of pure fat (ca. 78% C) can be used. If fresh substances are employed, the higher water content permits correspondingly greater sample weights.

Apart from temperature, time, and acid quantity, the proportion of sample weight to vessel volume is an important factor if oxidation of organic sample substances is to be as complete as possible. If, for instance, in a 35-mL crucible, the sample amount and the volume of nitric acid is reduced proportionally at constant temperature, the oxidation of carbon is less efficient in the case of samples that contain less than 50 mg of carbon. Increased residues of organic decomposition products result in the solution (15), even though the proportion sample weight:acid quantity has remained unaltered. The worst results are because NO_2 concentrations decrease in the liquid phase when the total pressure in the vessel drops because of the lower quantities used.

The NO_2, which is released when nitric acid is reduced, is thus largely implicated in the destruction of organic samples (15). For the practical application of decomposition under pressure this means that sample weights of less than 1.4 mg (related to C) per mL crucible volume are definitely not recommended if the digestion is carried out at the same temperature. No upper limit exists for the sample load, provided the decomposition system used can withstand the pressures building up during the process. However, higher pressures in the vessel do not give better oxidation results: At proportions above 1.4 mg C/mL vessel volume, the decomposition efficiency (i.e., the proportion of the organic substance oxidized during the process) is no longer affected by the pressure and thus independent of the sample weight. This result further implies that in the case of decomposition systems that have not been properly sealed so that NO_2 has been allowed to escape, carbon oxidation will deteriorate.

With pressure-controlled systems the situation is different. At constant pressure the temperature increases with decreasing sample weights resulting in a more efficient oxidation of the organic sample.

Neither increased reaction times nor increased volumes of acid will achieve complete mineralization of tough organic residues. Also, the maximum temperatures that are possible if PTFE or PFA vessels are used will not be sufficient for complete oxidation of organic compounds.

Organic residues usually do not affect atomic spectrometric methods as seriously as voltammetric techniques. The compounds are, however, a source of background signals in graphite-furnace atomic absorption spectroscopy (AAS) if they are not completely decomposed during the charring step of the temperature program. Finally, a source of error that affects all spectrometric methods with nebulization of sample solutions is interference due to changes in the viscosity and surface tension of the solution caused by organic substances. Appreciable differences of the nebulization rate of standard and sample solution cause wrong analytical results. The standard addition method can help to discover this type of error. Furthermore, organic residues may affect the hydride technique, the cold-vapour technique, and preconcentration methods. Under the conditions, which can be found in low-pressure microwave digestion, many organic substances can be dissolved to clear solutions. This result, however, gives no indication as to the remaining organic compounds. Sometimes one can see from the yellow colour that a large amount of organic substances is present in the solution.

With regards to trace analysis, an optimum solution is high-temperature decomposition with pure nitric acid in a closed system. The customary terms such as "pressure decomposition" or "microwave decomposition" are misleading because neither the pressure nor the microwave energy is responsible for the actual sample decomposition, but rather primarily decomposition temperature along with the choice of the oxidation agent are responsible. The necessary temperatures from 200–300 °C result in such high pressure in the digestion vessels due to the vapour pressure of the decomposition reagents that only high-pressure systems can be employed.

The temperature achieved during decomposition depends on the vessel material used. The optimum temperature range for PTFE or PFA vessels terminates at about 220 °C. At higher temperatures the lifetime of the decomposition vessels is reduced. The recommended maximum-use temperature for PTFE or PFA vessels is 250 °C. Vessels made of quartz can be used for decomposition at higher temperatures. Quartz glass is not only temperature-resistant but also has the best qualities for extreme element trace analysis (7).

For example, complete mineralization of organic compounds is essential in the case of trace analysis of metals by voltametric methods, and it is also highly recommended in all instances that require accurate, interference-free results. If perchloric acid is considered too hazardous and is therefore avoided, the only alternative left is decomposition with nitric acid at raised temperatures (above 200 °C). Decomposition with nitric acid requires a minimum temperature of 300 °C if the complete oxidation of organic samples is to be achieved (15–17). More than 99.9% of the original carbon content of the sample is oxidized when decomposition is performed at 300 °C. Because PTFE or PFA vessels at temperatures above 220 °C are subject to excessive softening, quartz vessels have to be used for this purpose.

From the presented results it follows that it is important to choose as high a decomposition temperature as possible. Temperature feedback control allows sample digestion at a certain oxidation potential of the ashing reagent. However, how pressure develops is not known exactly, so that work must be done far beneath the maximum system pressure to prevent blowing off the digestion vessel.

A similar problem exists with pressure-controlled systems if pressure is measured only in one vessel. Even if all vessels are equally filled, clear differences in pressure development in the individual vessels can occur. Therefore, work must be performed beneath the maximum system pressure in this case, too.

For practical use of decomposition systems in laboratories the best solution is a pressure control in all decomposition vessels because one can thus work at the maximum pressure possible for the system. This result means that the maximum temperature for each decomposition can be achieved and is generally advantageous with organic materials.

Safety Aspects for Wet Digestion in Closed Vessels

Even though few safety precautions were taken in case of explosion with the first-generation microwave decomposition devices, modern devices offer the user full safety in the event of a decomposition device exploding.

Parameters Influencing the Reaction Rate

The following parameters influence reaction speed and therefore must be observed:

- sample weight
- volume of the vessel
- temperature of the reaction mixture
- cooling intensity of the vessel
- heating rate
- kind and volume of decomposition reagent
- surface of the sample material

Influence of Sample Weight

Figure 1 shows the pressure curve of the decomposition of 100 mg and 150 mg of lubricating oil with 2 mL of nitric acid in a 35 mL quartz vessel at 80

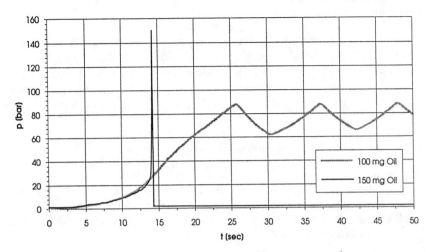

Figure 1. Pressure development during digestion of 0.1 g and 0.15 g lubrication oil with 2 mL nitric acid in a 35-mL quartz vessel.

bar. Because the experiment was performed with the pressure-controlled decomposition device (PMD) that will be described later, we see the pressure fluctuations caused by the control mechanism at 80 bar. With the 0.1-g sample the pressure rises evenly up to 80 bar where controlling begins. With the 0.15-g sample the reaction is so strongly exothermic that pressure rises to far in excess of 80 bar.

Volume of the Vessel

The larger the vessel volume compared with the sample amount used, the smaller the pressure peak is with a strongly exothermic reaction.

Influence of the Temperature

If one were to heat 0.2 g of finely ground charcoal with 2 mL nitric acid in a 35-mL quartz vessel in the microwave oven to about 180 °C, the sample is virtually indestructible. But if the same mixture is heated quickly to about 300 °C it may react explosively.

Influence of Cooling Intensity

With the "Super-Digest" described in a later section, the decomposition vessel can be cooled very intensively at high microwave power. If, for example, 1.5 g wheat flour are decomposed with 12 mL HNO_3 (65%) in an 80-mL quartz vessel with low vessel cooling, a strongly exothermic reaction ensues that carries on after switching off the microwave supply and leads to the destruction of the rupture disc. However, with intensive vessel cooling the reaction is stopped immediately after switching off the microwave.

Influence of Heating Rate

When aromatic compounds are decomposed with nitric acid, nitroaromatic compounds can be produced as explosive intermediates that are slowly decomposed further. If the production of nitroaromatic compounds ensues much faster by quickly heating the reaction mixture than does their decomposition, a dangerous accumulation of explosive substances can result.

Kind and Volume of Decomposition Reagents

The kind of oxidation reagent obviously has a great influence on reaction speed. The amount of the decomposition reagent does affect reaction speed in that a stoichiometric ratio between sample and reagent usually supplies the highest pressure peaks. If more oxidant is added, the pressure peaks drop off again.

Surface of Sample Material

Finely ground samples react considerably more quickly and produce higher pressure peaks than coarsely ground samples.

Sample Characteristics

Sample materials can be classified by their reactivity as follows.

Materials with Low Reaction Enthalpy

Typical representatives of this category are rock samples and inorganic samples with a low content of organic compounds. With regards to safety, these samples are easy to handle, which is why simple microwave-decomposition devices can also be used. However, a decomposition mixture that is in itself of low reactivity can produce explosive intermediates.

Materials with High Reaction Enthalpy

Typical examples are oils and fats, foods, metal powders, and plastics. Decomposing such samples requires decomposition devices with a high safety standard and skilled laboratory staff.

Safety Installations

The combination of different safety installations finally leads to safe microwave-decomposition equipment.

Time and Power Regulation

Practically all devices available today offer the possibility of preprogramming time and microwave power. Often a low power level is used at the beginning of a decomposition to allow samples with a high exothermic reaction to prereact slowly, and then samples are fully decomposed at high power. Without any device for pressure and temperature limitation, exact knowledge of the sample and of how it behaves in decomposition is absolutely essential. A linear (unpulsed) power regulation is desirable for the magnetron because even short power pulses can trigger strong reactions.

Time and power regulation without pressure and temperature feedback control is well-suited for samples that react very slowly or that release little energy. Exact knowledge of decomposition behaviour of the sample is important, or else method development should begin with very small sample amounts. This technique, however, cannot be recommended for organic sample material that reacts quickly.

Excess Pressure Protection

One very important element of safety for using decomposition bombs is the excess pressure protection. Decomposition bombs with no excess pressure protection are extremely dangerous and should not be used. Both excess pressure valves and rupture discs are usual. Excess pressure valves are still functional even after they have reacted, which means they open for a short time to let off the excess pressure and then reclose. One disadvantage of this method is that one cannot always tell if a valve has opened for a short time during decomposition or not. Unrecognized element losses can result. Furthermore, the excess pressure valve is no protection against explosion because of its small free cross-section.

Rupture discs with an accordingly large free cross-section, however, are very successful for explosion protection. Decomposition bombs with rupture discs should be provided with a pressure-control device, otherwise rupture discs blast very often.

Pressure or Temperature Control

The question is which systems can better meet the requirements with regards to safety and decomposition quality. From chemical reaction kinetics it follows that temperature not pressure is responsible for speed and completeness of decomposition. Raising the temperature by 10 °C, for example, leads to a doubling of reaction speed. As the oxidation effect of decomposition reagents increases with rising temperature, the concentration of resistant organic compounds in the decomposition solution decreases by the same degree. In standardized analytical methods, decomposition temperature is often prescribed.

In a closed system pressure increases with the temperature of the decomposition mixture. Because the decomposition vessels are designed for a certain maximum pressure, measuring decomposition pressure clearly takes priority over measuring temperature with regards to safety.

A comparison between pressure and temperature control shows that pressure control allows higher decomposition temperatures because the exact development of pressure during decomposition is not known. Thus, with temperature measuring samples must be decomposed at a pressure less than maximum to prevent blowing the rupture discs.

The speed of switching off the microwave supply is an equally important safety aspect. Because pressure transfer is far quicker than heat conduction, a pressure-controlled switch can react around 100 times faster. The typical reaction speed of a pressure-controlled switch is between 0.02 and 0.5 s, the speed of a temperature control is between 2 and 15 s. For this reason pressure control can often stop a fast exothermic reaction where temperature control would fail. The advantage of pressure control is clearly demonstrated when unknown samples must be decomposed.

Decomposition Bomb Jacket

The fluoropolymers used for decomposition vessels today have low mechanical strength at the necessary decomposition temperatures and must be supported by a jacket absorbing the mechanical strain. The bomb jacket must be microwave transparent, withstand high temperatures, have high mechanical strength, and absorb pressure blows. For quartz glass decomposition vessels the bomb jacket is not required for reasons of strength; it acts as a protection jacket.

Splinter Protection

Misuse (e.g., too great a sample weight) or unknown samples can be the cause of an explosion. With medium-pressure (35 bar) and high-pressure (>80 bar) systems, pressure peaks up to more than 10^4 bar can occur in the event of an explosion. Appropriate splinter protection (e.g., protective shield) on the microwave oven is a precaution for personal safety. The drastic effect of an explosion—test explosion with a 35-mL quartz vessel loaded with a detonator (Figure 2)—can be seen in Figure 3.

Reaction Vessel Cooling

In the decomposition of organic samples large amounts of heat possibly can be released. For example, in the oxidation of 0.2 g graphite with 2 mL nitric acid approximately 4 kJ are released after subtracting the energy for vaporization and heating. This energy is adequate to melt several grams of fluoropolymer if heat is not drawn off. Drawing heat off is usually effected via the side of the bomb jacket. Even fiberglass-reinforced high-performance plastics lose considerable strength at a higher temperature. With poly(ether ether ketone) reinforced with 30% glass fiber (PEEK–GF30), for example, the tensile strength sinks from 160 to 40 N/mm^2 when heated to 200 °C. This result means that with longer decomposition times (greater than 10 min) a clear reduction of protection capacity can ensue. To

Figure 2. Quartz vessel of PMD filled with a detonator (1 g pentaerythritol tetra-nitrate).

remedy this weakness the devices presented here are fitted with an interior cooling. That means the air flow is conducted between decomposition vessel and vessel jacket. In this way the high strength of the polymer is maintained even with a longer decomposition time.

Safety Regulations for Microwave-Assisted Decomposition Equipment

To guarantee effective protection for people and to avoid material damage the following points should be taken into account in developing microwave-assisted decomposition equipment:

1. Power control of the magnetron should be linear and not pulsed in order not to trigger any spontaneous reactions due to intensive microwave pulses.
2. Pressure should be monitored for each decomposition vessel and the microwave supply should be controlled; additional reaction temperature measuring is advantageous.
3. Rupture discs with high free cross-section are used as excess pressure protection for each pressure decomposition vessel.
4. Bomb jackets are used with high mechanical strength.
5. Splinter protection should be used in addition to the bomb jacket; additional shield in front of the oven door is advisable for user protection.

6.   Bomb jacket and reaction vessel cooling can avoid damage by over-
     heating.

Two microwave-assisted decomposition systems for high-temperature
digestion are described with which the previously mentioned safety criteria have been taken into account and which achieve maximum decomposition quality.

Description of Decomposition Systems

The basic idea of the pressure-controlled microwave-assisted decomposition systems presented is that the pressure in each decomposition vessel is
monitored and used to control the microwave supply. In this way, laboratory use with changing sample materials is much easier and convenient.
Venting of the vessels is avoided and beyond that the sample decomposition is always effected at the maximum possible temperature. The precondition for good analytical results is that the pressure sensor in the vessel
meets trace analytical requirements. No tubes may be used to transfer
pressure, and the sample may only come into contact with the vessel and
the smooth inert surface of the sensor. A material that does not cause any
blanks must be used for that part of the sensor that comes into contact
with the sample.

Figure 3. PMD after blowing up.

Two different pressure-controlled microwave-assisted decomposition systems, a cavity-type and a waveguide-type system, have been developed. With the cavity-type microwave system the entire pressure vessel is placed in a microwave oven. With the waveguide-type system only the lower part of the vessel in which the liquid is situated is placed in the center of a waveguide. In this way the pressure sensor is outside the microwave field.

With the cavity-type system there are either two stationary pressure vessels in a small microwave oven or a rotor with six pressure vessels in a big oven, according to the model. The vessel that reaches maximum pressure first turns off the magnetron. Equal filling of the decomposition vessels is a precondition for good decomposition results in all vessels. Venting of some vessels cannot occur, however, even if the vessels are not equally loaded.

With the waveguide-type system, each of the maximum of four pressure vessels has its own magnetron with waveguide. Therefore the vessels can be loaded with different sample materials and sample weights. Up to four microwave units can be combined and controlled by one single control unit.

A further important construction characteristic of both pressure-controlled systems is an intensive air-cooling device of the decomposition vessel. A stream of air is sucked through the gap between the decomposition vessel and the vessel jacket (*18*). Thus, decomposition with sulfuric acid can be effected in quartz vessels at a temperature of 350–400 °C. In PFA vessels decomposition can be effected at temperatures up to 250 °C. Following decomposition the pressure vessels are cooled to room temperature in 10 min by the stream of air. At the same time the pressure in the vessels drops greatly so that they can be removed from the decomposition device without any risk. This result is an important point for increasing the safety of microwave decomposition devices in the laboratory. Hot, pressurized decomposition vessels should never be removed from the oven.

Instrumentation

The decomposition equipment has been developed in cooperation with Anton Paar Corp., Austria and Prolabo, France.

Decomposition Vessels with Pressure-Control Device and Cooling System

In principle, the cavity-type system as well as the waveguide-type system use the same pressure-control mechanism. Only the six-vessel rotor system differs slightly.

Figure 4 shows the principle of the decomposition vessel unit with pressure-control mechanism and cooling system. The decomposition vessel unit is made of microwave-transparent plastic material. Quartz glass or PFA decomposition vessels can be placed within the vessel jacket. A pressure-control mechanism used to switch off the microwave energy at a defined pressure is built into the screw cap of the vessel unit. To follow the functional description look at the sectional drawing in Figure 4.

The vessel unit, filled with sample and acid, is placed in the microwave equipment. The connection for optical sensor [7] for pressure control is fitted to the screw cap [3]. The connector contains one transmitter and one receiver; the transmitter emits a red, luminous spot to the reflector pin [8], thus causing the receiver to activate the microwave supply. The microwave energy heats up the acid inside the decomposition vessel [1], a decomposition reaction with the sample will occur and a reaction pressure will be built up. The pressure pushes the stopper [4] and piston [5]

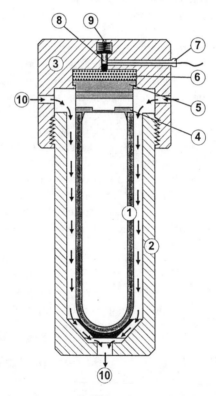

Figure 4. Principle of pressure-controlled and air-cooled decomposition vessel unit: 1, decomposition vessel; 2, vessel jacket; 3, screw cap; 4, stopper with rupture disc and sealing; 5, piston; 6, silicone oil; 7, connector for optical sensor; 8, reflector pin; 9, spring; 10, cooling air.

upward against the silicone oil [6]. This action causes the reflector pin [8] to move upward against the spring [9]. As soon as the reaction pressure of, for example, 80 bar is reached, the reflector pin [8] has moved upward so far that the reflector leaves the field of the optical sensor. The receiver detects this change of the optical signal and switches off the microwave supply. Both the temperature and the pressure inside the decomposition vessel fall, and the spring [9] again pushes the pin [8] downward. The reflector reenters the recognition field of the optical sensor, which activates the supply of microwave energy again. The hydraulic system consisting of piston [5], silicone oil [6], and reflector pin [8] causes the movement of stopper [4] and piston [5] for switching the microwave energy to be only about 0.1 mm. Therefore, the sealing of stopper [4] has a long lifetime. The sealing is of special importance for the bomb's functioning. Because of its particular shape, proper sealing always will be ensured. With increasing pressure, the lip will be increasingly pressed toward the vessel neck. This pressure-control mechanism has proved to be rigid and reliable. By using different springs, different cut-off pressures can be selected.

Overloading of the decomposition vessel may cause a persistent increase in pressure even after switching off the microwave supply. A built-in rupture disc in the stopper [4] will burst at a pressure of 135 bar, and excess pressure will escape through the bores of stopper [4] and screw cap [3]. Vessel jacket [2] serves as protection when an explosion should occur.

Figure 5 shows the principle of the rotor system. A rotor [1] made of microwave-transparent plastic can be loaded with up to 6 digestion vessel units. The digestion vessel units consist of the digestion vessel made of quartz or PFA [2], the vessel jacket [3], the screw cap [4], and the stopper with rupture disc and sealing [5]. In the upper part of the rotor a hydraulic system is installed. The pressure of each of the loaded decomposition vessels is transmitted via the piston [6] and the silicone oil [7] to the pressure transducer [8]. The pressure transducer [8] reflects a light beam from a laser diode. As soon as the selected reaction pressure in one of the decomposition vessels is reached, the pressure transducer [8] cuts off the reflected light beam. The receiver detects this change of the optical signal and switches off the microwave supply. After the pressure drops below the working pressure level the pressure transducer reflects the laser beam again and switches on the microwave supply. In the bottom part of the rotor there are channels [9] for a stream of cooling air to pass all the digestion vessels.

With this rotor system the particular vessel reaching the maximum pressure switches the microwave supply. During a digestion process different vessels switch the microwave corresponding to the progress in sample decomposition in the particular vessel. As a result the samples are

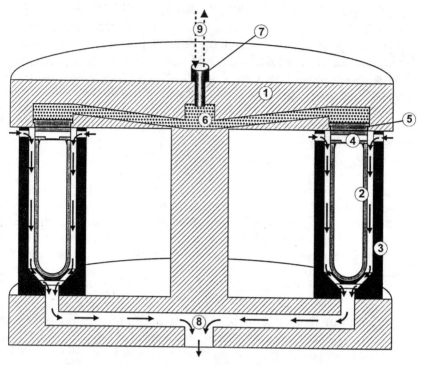

Figure 5. Scheme of the rotor system: 1, rotor; 2, decomposition vessel; 3, vessel jacket; 4, stopper with rupture disc and sealing; 5, piston; 6, silicone oil; 7, pressure transducer; 8, channel for cooling air; 9, light beam.

always digested at maximum possible pressure and at the resulting maximum temperature. With changing sample matrix, sample weight, and number of decomposition vessels no tedious method evaluation is necessary.

Cavity-Type Digestion Equipment

Figure 6 shows the principal of a cavity-type microwave digestion equipment with pressure-controlled digestion vessels and the cooling device. Two different versions of decomposition devices have been developed. A small microwave oven with two pressure-controlled vessels (pressurized microwave digestion [PMD]) and a big oven with a rotor system for six vessels (multiwave).

The microwave oven represents a reliable industrial product that is especially adapted to be used with the pressure-controlled decomposition vessels. Two decomposition vessels (PMD) or a rotor (multiwave) can be placed in the microwave oven. In addition to the microwave unit with programmable power and time, the instrument comes with a pressure-

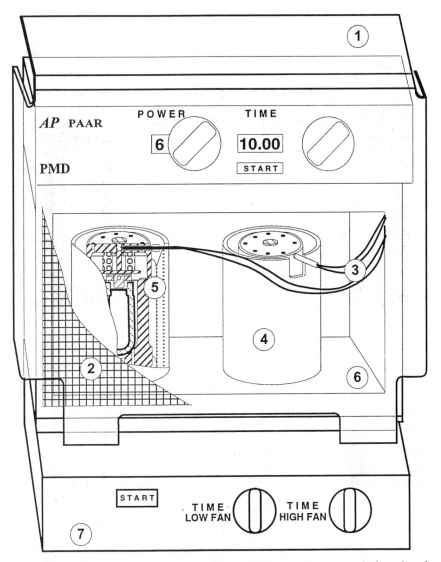

Figure 6. Scheme of cavity-type digestion equipment with pressure control and cooling device: 1, Plexiglas shield; 2, microwave oven; 3, fiber optics; 4, bomb assembly, total view; 5, bomb assembly, details; 6, ground plate with openings for exhaust adapter; 7, exhaust and cooling unit.

control mechanism to supervise the pressure inside the decomposition vessels by means of optical sensors and a cooling device.

The cooling device allows air to suck through the gap between the quartz or PFA vessel and the plastic jacket at low-stream or high-stream mode. Therefore, the sample inside a quartz vessel can be heated up to more than 300 °C, and the plastic parts of the decomposition bomb are kept at low temperature by means of low-stream cooling. After the decomposition procedure high-stream cooling is used to bring the vessel temperature down to room temperature within 10 min.

Waveguide-Type Digestion Equipment

The waveguide-type decomposition device (Super-Digest) consists of up to 4 independent magnetron units [1] with one pressure-controlled decomposition vessel [2] each, a control box [4], and a cooling unit [3] (Figure 7). Instead of a microwave oven, the Maxi-Digest by Prolabo is used. With this device the microwaves are directly focused on the decomposition solution and are not evenly distributed across the whole vessel. In this way up to 350 W microwave power can be transferred onto 5–10 mL sample solution. To be able to employ this high microwave power, the decomposition vessel is very intensively air cooled.

Design and operation of the pressure-controlled vessel unit are the same as described previously (Figure 4). With the Super-Digest 80-mL decomposition vessels made of quartz glass are used. The decomposition pressure is adjusted to 40 or 60 bar.

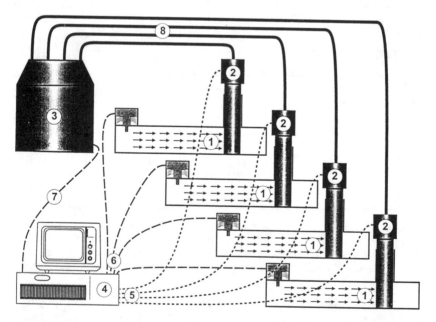

Figure 7. Scheme of a waveguide-type digestion equipment with pressure control and cooling device: 1, microwave-heating device; 2, pressure-controlled and air-cooled decomposition vessel; 3, cooling unit; 4, control box; 5, connectors for optical pressure control; 6, connectors for microwave power control; 7, connector for fan power control; 8, tubes for air cooling.

The cooling unit [3] is an important part of the Super-Digest. It serves to cool the pressure vessel [2] during decomposition and to quickly cool it off to room temperature after decomposition. At the same time acid vapours can be sucked off with it. The extreme suction power of the cooling unit [3] can be adjusted via the control device at 10 levels ranging from "cooling off" to "cooling maximum". Up to 4 microwave-decomposition vessels can be cooled simultaneously with a cooling unit.

The control box [4] controls up to four microwave heating devices [1] and one cooling unit [3]. A digestion program consists of up to nine steps. Each step contains separate values for microwave power, step duration, and cooling power. If more than one microwave heating device is attached the same program applies for each device. However, because the magnetron of each microwave heating device can be switched on and off independently by the appropriate decomposition vessel, different sample materials and sample amounts can be decomposed at the same time.

Decomposition Vessels

Depending on the decomposition problem varying digestion vessels are available, which differ in size, material, and maximum pressure level:

- for PMD: 35-mL quartz vessel at 90 bar; 50-mL PFA vessel at 34 bar; 120-mL PFA vessel at 15 bar
- for multiwave: 50-mL quartz vessel at 72 bar; 120-mL PFA vessel at 30 bar
- for Super-Digest: 80-mL quartz vessel at 60 bar

Reagents

All reagents were at least analytical grade (Merck, Darmstadt, Germany) or purified by subboiling distillation and were as follows: nitric acid (65%), nitric acid (fuming), hydrochloric acid (36%), sulfuric acid (conc.), hydrogen peroxide (30%), and hydrofluoric acid (40%).

Optimization of Parameters

The quality of a digestion procedure is described by the following parameters: recovery of elements, blank values, and completeness of the decomposition process. Often the digestion of organic samples is considered complete when no undissolved residue is left. Nevertheless such solutions can contain large amounts of dissolved organic compounds. For example the National Institute of Standards and Technology standard reference material (NIST SRM) 1577a, Bovine Liver, contains much protein. Particularly, phenylalanine is known to give nitroaromatic compounds during digestion with nitric acid, and these compounds are rather resistant to further attack by this acid (*19*). Therefore, NIST Bovine Liver has been chosen for systematic investigation and optimization of digestion parameters. The decomposition efficiency is expressed in DOC (dissolved organic carbon). The DOC content of the decomposition solution has been measured after dilution with distilled water. Decomposition period, amount of nitric acid (65%), sample load, and pressure level have been investigated employing a 35-mL pressure-controlled quartz vessel.

Decomposition Period

Decomposition efficiency as a function of time is shown in Figure 8. According to the investigations of Würfels (*15, 17, 20*) 196 mg Bovine Liver, corresponding to 100 mg pure carbon, was decomposed with 2 mL nitric acid (65%) in a 30-mL quartz vessel at a pressure level of 80 bar. The oxidation reaction was terminated after about 15 min. Decomposition efficiency is ~94% under the stated conditions and cannot be significantly increased even by lengthening the decomposition period. On the other hand, we see that 90% of the employed organic sample is oxidized after only 5 min.

Acid Volume

Decomposition efficiency as a function of volume of nitric acid (65%) is shown in Figure 9. A 196-mg sample of Bovine Liver was digested with

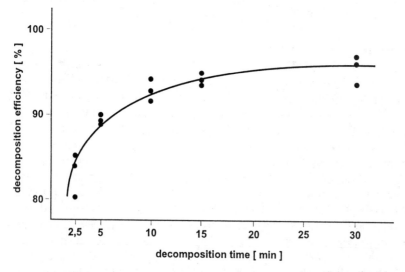

Figure 8. Decomposition efficiency as a function of time: 196 mg NIST 1577a Bovine Liver (100 mg C), 2 mL nitric acid (65%), 35-mL quartz vessel, 80-bar pressure.

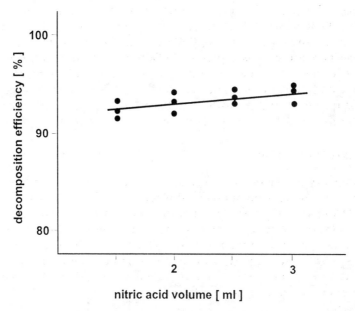

Figure 9. Decomposition efficiency as a function of acid volume: 196 mg NIST 1577a Bovine Liver (100 mg C), 10-min decomposition period, 35-mL quartz vessel, 80-bar pressure.

1.5–3.0 mL nitric acid (65%). Very little dependence of the decomposition efficiency on the volume of nitric acid exists within the range investigated. The reason for the remaining organic compounds is not a deficiency of oxidant but a low oxidation potential.

Pressure Level and Sample Load

Figure 10 shows the decomposition efficiency depending on decomposition pressure level and sample weight. The total pressure in the decomposition vessel consists of the sum of the single partial pressures. With increasing sample amount the partial pressure of CO_2 rises and causes a decrease in the partial pressure of HNO_3 and H_2O. Therefore, the boiling temperature and in this context the oxidation potential of the decomposition solution drop.

Quartz vessels of 35 mL were loaded with 100–180 mg of NIST Bovine Liver and 2 mL of nitric acid (65%). The decomposition pressure levels used were 30, 70, and 90 bar. As expected, Figure 10 shows that the decomposition efficiency decreases with increasing sample weight and decreasing pressure.

Possibilities To Improve Decomposition Efficiency

As can be seen from the previous investigations, the decomposition efficiency can only be insignificantly improved by increasing the amount of

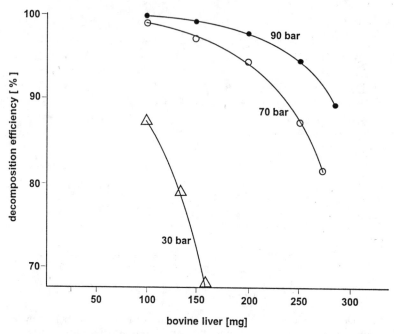

Figure 10. Decomposition efficiency as a function of pressure and temperature: Varying amounts of NIST 1577a Bovine Liver; 10-min decomposition period; 35-mL quartz vessel; 30-, 70-, and 90-bar pressure levels.

HNO_3 or by lengthening the ashing period. Clear reduction of the remaining carbon content can only be achieved by increasing the oxidation potential of the decomposition reagent. This result can be achieved by increasing the decomposition temperature or by using a different decomposition reagent with a greater oxidation potential.

If nitric acid (65%) is used for the decomposition of organic samples the reaction temperature for complete oxidation of certain compounds, for example nitroaromatic compounds, must be ~ 300 °C. As a result of the vapour pressure of the acid used, however, sample decomposition at such high temperatures can only be achieved in a high-pressure decomposition vessel with more than 80 bar.

Another way of achieving high temperatures in the decomposition mixture, without building up corresponding vapour pressure, is facilitated by the Super-Digest. Through the high microwave energy density in the decomposition solution (up to 350 W) many hot spots are apparently produced. By intensively cooling the decomposition vessel the heat is quickly taken away so that even 350-W microwave power can be applied for a long time at a pressure of up to 60 bar. Comparative measurements with the PMD and the Super-Digest clearly demonstrate the advantage of high-energy density in the waveguide. NIST Bovine Liver, 150 mg, was decomposed with 2 mL HNO_3 in a 35-mL quartz vessel in the PMD (10 min; power, 80%; pressure, 40 bar). In the Super-Digest 350 mg NIST Bovine Liver were decomposed in a 80-mL quartz vessel with 5 mL HNO_3 (4 min, 30% power and 10 min, 99% power; pressure, 40 bar). The decomposition efficiency is 80% after decomposition in the PMD and 95% in the Super-Digest.

The oxidation potential can also be improved by increasing the concentration of nitric acid. Fuming nitric acid (90–100%) is a powerful oxidation reagent for tough organic materials like coke or polyaromatic plastic materials. Another possibility to increase the oxidation power is the application of a mixture of sulfuric acid and nitric acid.

A decomposition efficiency of better than 99%, which is necessary for voltammetric element determination, can be obtained by means of the PMD. For this purpose 100 mg biological material can be completely decomposed in a 35-mL quartz vessel with 2 mL nitric acid (65%), 500 µL perchloric acid, and 50 µL sulfuric acid for 10 min at 80% power and 90 bar (21). Because the PMD was developed according to high safety standards, small amounts of perchloric acid can be used with no risk.

Recovery of Elements

Complete recovery of trace elements should be no problem with microwave-assisted wet digestion in closed quartz or PFA vessels as long as the vessels are tight. Venting of the vessels during the digestion procedure must be avoided. Many volatile element compounds, particularly halogen compounds, chromylchloride, etc., can be lost in such a case.

However, trace elements can be lost when precipitates are formed and adsorbed on the wall of the quartz vessel. PFA vessels hardly display this characteristic. Iron and tin are examples of problematic elements that precipitate as oxides on the surface of the quartz vessel during decomposition with nitric acid. To avoid this result halogens are added as complexing agents. An addition of 0.1–0.5 mL HCl to 2 mL HNO_3 is enough to keep iron in solution. With tin, 0.5–1.0 mL HCl are needed according to the sample matrix. Adding more than 0.5 mL HCl to 2 mL HNO_3 leads to a significant dilution of the oxidation agent HNO_3 and thus to an increase of organic residue.

Results and Discussion

Methods for the decomposition of organic materials by means of the PMD and the Super-Digest will be described and the consequence of different digestion parameters will be discussed. Methods for the decomposition of different biological materials with the PMD are depicted in Table I. The quality of the digestion process is expressed by visual appearance of the decomposition solution and by DOC. In the 35-mL vessel up to 0.3 g sample can be decomposed with 2 mL nitric acid to colorless solutions. The DOC values vary between <1% and 11%. To dissolve silica precipitate, 0.5

Table I. Degree of Mineralization of Biological Materials After Decomposition
with PMD Apparatus at Standard Conditions

Material	Sample Intake (g)	Decomposition Efficiency (%)	Remarks
Animal blood (dry)	0.19	89	
Animal muscle, SRM IAEA H-4	0.20	92	
Apple (wet)	1.0	95	
Beech leaves, SRM, BCR 100	0.21	>99	
Bovine liver, SRM, NIST 1577a	0.30	94	
Bovine muscle, SRM, BCR 184	0.20	96	
Brain tissue	0.70	95	
Charcoal, Merck 2186	0.10	>99	
Citrus leaves, SRM, NIST 1572	0.25	>99	little precipitate of silica
Cocoa	0.20	97	
L-Cystine	0.33	>99	
Glucose monohydrate	0.27	>99	
Hay powder, SRM, IAEA V-10	0.26	94	little precipitate of silica
Hop	0.30	98	
Lagorosiphon, SRM, BCR 60 API	0.29	>99	
Milk powder, SRM, BCR 150	0.30	97	
Mussel tissue, SRM, IAEA MA-M 2/TM	0.24	95	
Olea europea, SRM, BCR TP2	0.20	>99	
Olive leaves, SRM, BCR 62	0.30	99	
Ovomaltine	0.20	98	
Oyster tissue, SRM, NIST 1566	0.30	92	
Paper (cellulose)	0.30	>99	
Pig kidney, SRM, BCR 186	0.30	97	
Pine needles, SRM, NIST 1575	0.30	98	
Platihypnidium, SRM, BCR 61	0.29	>99	
L-Proline	0.19	98	
L-Sorbitol	0.25	99	
Spinach, SRM, NIST 1570	0.30	98	little precipitate of silica
Soylecithin	0.17	89	
Spruce needles, SRM, BCR 101	0.20	99	
Sunflower oil	0.15	95	
Wheat flour, SRM, IAEA V-5	0.25	98	
Wood (sawdust)	0.25	96	

NOTE: Conditions were as follows: 35-mL quartz vessel, 2-mL HNO_3 (65%), 0.2 mL HCl (36%) if iron has to be determined, 80-bar pressure, 80% power level, 10-min decomposition period, 10-min cooling time.

mL HF (40%) are added to the digestion solution and treated with ultrasonic waves for about 2 min.

For the determination of Pb, Cd, and Cu by means of differential pulse anodic stripping voltammetry, the DOC value of the digestion solution must be less then 1% (21). The quality of the determination of As in samples of marine origin with hydride-generation AAS (HAAS) is also strongly dependent on the digestion efficiency. Different As species are normally present in such samples, namely As-betaine, As-choline, MMA (monomethylarsenic), DMA (dimethylarsenic), As^{3+}, and As^{5+}. MMA, DMA, As^{3+} and As^{5+} can be measured by HAAS without any treatment. However, As-betaine and As-choline give no signal by HAAS. Complete destruction of these compounds by means of high temperature ashing is necessary for correct analytical results (22). Table II shows methods for complete digestion of biological materials (DOC, <<1%) by means of the PMD and using 35-mL quartz vessels (22).

For decomposition of resistant organic compounds with nitric acid, high temperatures of more than 250 °C are a precondition. Pressure-controlled high-pressure systems with cooled decomposition vessels are well-suited for this purpose. Table III shows the result of decomposition of tough organic materials by means of the PMD. For substances like calcined coke, glassy carbon, or graphite, fuming nitric acid or nitric acid–sulfuric acid mixtures have to be applied.

To be able to decompose larger amounts of organic samples the waveguide system Super-Digest with the 80-mL quartz vessel is suited. In Table IV the decomposition methods for various materials and the corre-

Table II. Complete Mineralization of Biological Materials for DPASV and HAAS by Means of PMD

Material	Sample Intake	Oxidant (mL)	Microwave Power (%)
Bovine muscle, SRM, BCR 184	0.10 g	2.0 HNO_3 0.5 $HClO_4$	80
Bovine liver, SRM, BCR 185	0.10 g	2.0 HNO_3 0.5 $HClO_4$	80
Cod muscle, SRM, BCR 422	0.10 g	2.0 HNO_3 0.5 $HClO_4$	80
Mussel tissue, SRM, BCR 278	0.10 g	2.0 HNO_3 0.5 $HClO_4$	80
Brown bread, SRM, BCR 191	0.10 g	2.0 HNO_3 0.5 $HClO_4$	80
Milk powder, SRM, BCR 63R	0.25 g	2.0 HNO_3 0.5 $HClO_4$	80
Lobster hepatopancreas, SRM, TORT-1	0.10 g	2.0 HNO_3 0.5 $HClO_4$	80
Spruce needles, SRM, BCR 101	0.10 g	2.0 HNO_3 0.5 $HClO_4$	80
Whole blood	1.0 mL	1.0 HNO_3 0.5 $HClO_4$	60
Urine	1.0 mL	1.0 HNO_3	80
Cooking oil	0.1 mL	2.0 HNO_3 0.2 H_2SO_4	60
Human breast milk	1.0 mL	1.0 HNO_3 0.5 $HClO_4$	60

NOTE: Conditions were as follows: 35-mL quartz vessel, 90-bar pressure, 10-min decomposition period, 10-min cooling time. DPASV is differential pulsed anodic stripping voltammetry and HAAS is hydride generation atomic absorption spectrometry.

Table III. Decomposition of Resistant Organic Materials by Means of PMD

Material	Sample Intake (mg)	Oxidant (mL)	Power (%)/ Time (min.)	Decomposition Efficiency (%)
Volatile organic solvents	50–100	2.0A+0.2C	I40/5 II80/10	
Polyethylene	100	1.0A+1.0D	80/20	98
PVC	100	1.0A+1.0D	80/15	99
Polystyrene	100	1.0A+1.0D	80/15	97
Polybutadiene rubber	100	2.0A+0.2C	80/15	96
Coal	100	2.0A	80/15	>99
Calcined coke	100	1.5A+1.5D	80/20	>99
Calcined coke	100	3.0B+0.2E	100/20	>99
Glassy carbon	100	3.0B+0.2E	100/30	>99
Pyrolytic graphite	50	3.0B+0.2E	100/20	>99
Crude oil	100	2.0A	100/10	90
			100/20	93
Tar	50	2.0A	100/15	96

NOTE: Conditions were as follows: 35-mL quartz vessel; pressure, 80 bar; digestion time and microwave power vary with sample material; cooling time, 10 min; oxidants, A, HNO_3 (65%); B, HNO_3 (100%) "fuming"; C, HCl (37%); D, H_2SO_4 conc.; E, HCl (37%) + HF (40%) (9+1, v/v).

Table IV. Decomposition of Organic materials with the Waveguide System "Super Digest"

Material	Sample Intake (g)	Oxidant	Program	Decomposition Efficiency (%)	Remarks
Milk powder	0,5		a	99	
SRM, BCR 150	1,0	2 mL C + 6 mL A	a	97	clear, yellowish
	1,0		c	98	clear, yellowish
Cocoa	0,25	2 mL C + 5 mL A	a	> 99	
	0,5			99	
Müsli	1,5	2 mL C + 8 mL A	a	93	
Pine needles SRM, NIST 1575	0,6	2 mL C + 6 mL A	a	96	
Saw dust	0,25	2 mL C + 5 mL A	a	> 99	
	0,5			> 99	
Bovine liver	0,3		a	90	
SRM, NIST 1577a	0,5	2 mL C + 6 mL A	a	98	
	1,0		b	95	
Pig kidney	0,8	2 mL C + 6 mL A	a	96	
	0,4		a	99	
Sunflower seed	0,8	2 mL C + 6 mL A	a	95	clear, yellowish
	0,8		c	96	clear, yellowish
	0,25		a	> 99	
	0,5		a	95	
Sunflower oil	0,5	2 mL C + 6 mL A	c	98	
	0,8		a	91	
	1,0		a	88	clear, yellowish
	0,15		a	> 99	
Lubrication oil	0,3	6 mL A	a	92	
	0,3		c	94	
Polystyrene	0,2	2 mL C + 5 mL A	d		clear, yellowish
	0,3	4 mL A + 2 mL D	b	> 99	
Polyethylene	0,5	2 mL C + 5 mL A	b	96	
	0,7	5 mL C + 5 mL A	c	96	
Polypropylene	0,5	2 mL C + 5 mL A	b	89	
			d	98	
PVC	0,3	6 mL A	b	98	
Latex	0,3	2 mL C + 5 mL A	a	> 99	
Polybutadiene	0,3	4 mL A + 2 mL D	a	97	
Coal	0,2	5 mL A	b	> 99	silica precipitate
Calcined coke	0,2	5 mL B	a	> 99	silica precipitate
		5 mL A + 2 mL D	d	> 99	silica precipitate

NOTES: Conditions were as follows: 80 mL quartz vessel, pressure 60 bar, oxidant and decomposition program vary, cooling time 6 min, oxidant: A: HNO_3 (65%), B: HNO_3 (100%) "fuming," C: H_2O_2 (30%), D: H_2SO_4 (conc.); decomposition programs (step/power level %/time min): a: 1/30/4, 2/99/10; b: 1/30/4, 2/99/20; c: 1/30/4, 2/99/30; d: 1/30/4, 2/99/60.

sponding results are given. With decomposition of large amounts of fine ground biological material, it is advantageous to wet the sample with H_2O_2 before adding nitric acid.

DOC values for a 60-bar system are comparably low. Here again, the effect of hot spots with the waveguide-type system is demonstrated. Some of these materials could not be decomposed in normal microwave oven equipment at a pressure of 60 bar.

To dissolve precipitated silica 5 mL diluted HF (1 mL HF + 4 mL H_2O) is added to the cool digested sample solution in the quartz vessel. After a short ultrasonic treatment a clear solution will be obtained. The quartz vessel is hardly attacked with this procedure.

A cavity-type digestion system with pressurized quartz vessels for 80 bar and a waveguide-type digestion system with quartz vessels for 60 bar are well-suited for powerful destruction of almost all organic materials. The precondition for a lower pressure level with the waveguide-type decomposition system is a high microwave density on the sample site and a high, efficient vessel cooling. A sophisticated pressure-control system for every vessel enables sample digestion at maximum possible temperature and a convenient handling.

References

1. Kingston, H. M.; Jassie, L. B. *Anal. Chem.* **1986,** *58,* 2534.
2. *Introduction to Microwave Sample Preparation: Theory and Practice;* Jassie, L. B.; Kingston, H. M., Eds.; American Chemical Society: Washington, D.C., 1988.
3. Matusiewicz, H.; Sturgeon, R. E. *Prog. Anal. Spectrosc.* **1989,** *12,* 21.
4. Dunemann, L.; Meinerling, M. *Fresenius' J. Anal. Chem.* **1992,** *342,* 714.
5. Knapp, G. *Mikrochim. Acta II* **1991,** 445.
6. Kahane, E. *Fresenius Z. Anal. Chem.* **1937,** *111,* 14.
7. Tschöpel, P.; Kotz, L.; Schulz, W.; Veber, M.; Tölg, G. *Fresenius Z. Anal. Chem.* **1980,** *302,* 1.
8. Würfels, M.; Jackwerth, E. *Fresenius Z. Anal. Chem.* **1985,** *322,* 354.
9. Tyler, L. J. *Chem. Eng. News* **1973,** *32,* 20.
10. Sah, R. N.; Miller, R. O. *Anal. Chem.* **1992,** *64,* 230.
11. Kotz, L.; Kaiser, G.; Tschöpel, P.; Tölg, G. *Fresenius Z. Anal. Chem.* **1972,** *260,* 207.
12. Kotz, L.; Henze, G.; Kaiser, G.; Pahlke, S.; Veber, M.; Tölg, G. *Talanta* **1979,** *26,* 681.
13. Stoeppler, M.; Müller, K. P.; Backhaus, F. *Fresenius Z. Anal. Chem.* **1979,** *297,* 107.
14. Würfels, M., personal communication.
15. Würfels, M. Ph.D. Thesis, Ruhr-Universität, Bochum, FRG, 1988.
16. Würfels, M.; Jackwerth, E.; Stoeppler, M. *Fresenius Z. Anal. Chem.* **1987,** *329,* 459.
17. Würfels, M.; Jackwerth, E.; Stoeppler, M. *Fresenius Z. Anal. Chem.* **1988,** *330,* 159.

18. Knapp, G.; Panholzer, F. U.S. Patent 5, 345, 066, 1994.
19. Pratt, K. W.; Kingston, H. M.; MacCrehan, W. A.; Koch, W. F. *Anal. Chem.* **1988,** *60,* 2024.
20. Würfels, M. *Mar. Chem.* **1989,** *28,* 259.
21. Schramel, P.; Hasse, S. *Fresenius' J. Anal. Chem.* **1993,** *346,* 794.
22. Schramel, P.; Xu, L-Q. *Fresenius' J. Anal. Chem.* **1991,** *340,* 41.

Chemistry Applications

Chapter 8

Microwave Heating in Organic Chemistry

An Update

George Majetich and Karen Wheless

This chapter serves to document the variety of applications of microwave heating in organic synthesis prior to the Spring of 1993.

In the seven years since the appearance of the first papers on organic synthesis using microwave heating (1, 2), the field has exploded and more than 100 papers have been published. This subject was first reviewed in 1989 by Giguere (3) and updated in 1991 by Abramovitch (4), but because of the remarkable growth of this field both these surveys are already dated.

This review serves to document the variety of applications of microwave heating in organic synthesis; its chronological development has been largely ignored, though the theory behind this technique has been described in a number of papers (5–12) and presentations (13). Although an attempt has been made to be comprehensive, the many uses of microwave heating in biochemical and biomedical applications (14), polymers (15), and inorganic chemistry (16)—each of which could be the subject of a separate treatise—have not been included. Because of the diversity of microwave systems used, the reader should consult the original references for descriptions of the instrumentation and settings employed. For clarity's sake, transformations carried out in commercial microwave systems capable of controlling the temperature and pressure of chemical reactions are marked with an asterisk (*). Finally, where possible, data from control experiments are provided to permit the comparison of microwave results with those obtained by using classical conditions. However, by optimizing

455

conventional reaction conditions, many reactions thought to require lengthy heating periods (and thus ideal candidates for microwave study) were either rapid transformations or did not require extensive heating.

Esterifications

One of the first organic reactions investigated in microwave chemistry was esterification. In 1985 Gedye, Smith, and Westaway (*17*) found that the Fisher esterification of benzoic acid in methanol could be carried out in 5 min with microwave heating, in 76% yield, by using a sealed tube in a kitchen-type microwave. At the same time Majetich and Casares independently carried out this reaction in comparable yield (81%) and reaction time (5 min), also using a commercial home microwave oven (unpublished work). In a control experiment this esterification required 80 min of refluxing to produce a 90% yield of methyl benzoate (*18*). Later, this same reaction was achieved by using a microwave system equipped with pressure control. At 50 psi and using an acid catalyst, this esterification occurs in 92% yield after only 1 min (*18*). *See* Scheme 1. Esterifications of benzoic acid were carried out by using *n*-propanol, *n*-butanol, and *n*-octanol catalyzed by *p*-toluenesulfonic acid (PTSA) as shown in Schemes 2–4 (*19*).

Loupy and co-workers have also investigated ester formation from carboxylate anions and primary alkyl halides. By using potassium benzoate, *n*-bromooctane, and Aliquat 336 (a phase-transfer reagent), the corresponding ester was obtained in 99% yield (Scheme 5). Esterification was also successful when these starting materials were adsorbed on alumina, but the yield was lower at 47% (12 min). These esterifications can be carried out in sealed-glass or Teflon vessels. Similar conditions were used for the esterification of potassium acetate with *n*-bromooctane (Scheme 6). The best results were obtained by using alumina as a solid support although silica gel was also effective (82% yield) (*20*). Comparable yields were obtained after 5 h of conventional heating at 100 °C on alumina or silica gel (93% or 70%, respectively). This technique was applicable for other esterifications by using hexadecyl bromide (75 s, 95%) or hexadecyl chloride (150 s, 90%) (*21, 22*). The same reactions with potassium acetate using catalytic amounts of Aliquat 336 gave identical yields.

Other esterifications are shown in Schemes 7–9. Cinnamic acid was esterified with 4-nitrobenzyl chloride in a pressurized Teflon vessel in 2 min (Scheme 7) (*17*). Both mesitoic acid and its potassium salt react with 1-octadecanol to give the corresponding ester in excellent yield (Scheme 8) (*19*). Alkyl tosylates also undergo esterification with carboxylate salts under microwave thermolysis, although these reagents give lower yields (Scheme 9) (*21*). Majetich and Hicks esterified the diacid shown subsequently by using methanol and sulfuric acid (Scheme 10). A control experiment for this Fisher esterification required 42 h for complete conversion.

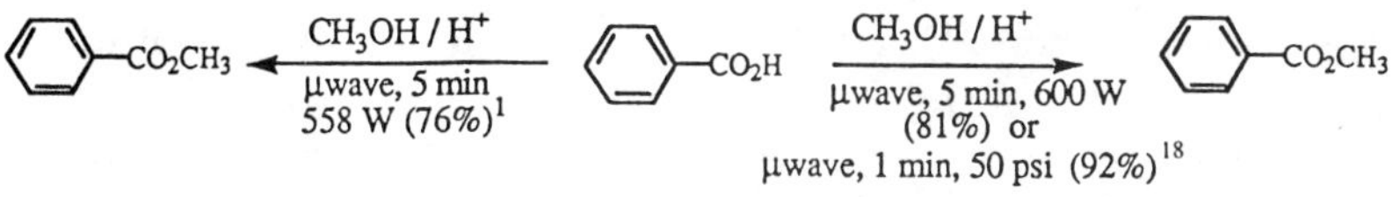

$$C_6H_5-CO_2CH_3 \xleftarrow[\substack{\mu wave, 5 \ min \\ 558 \ W \ (76\%)[1]}]{CH_3OH/H^+} C_6H_5-CO_2H \xrightarrow[\substack{\mu wave, 5 \ min, 600 \ W \\ (81\%) \ or \\ \mu wave, 1 \ min, 50 \ psi \ (92\%)[18]}]{CH_3OH/H^+} C_6H_5-CO_2CH_3$$

Scheme 1*.

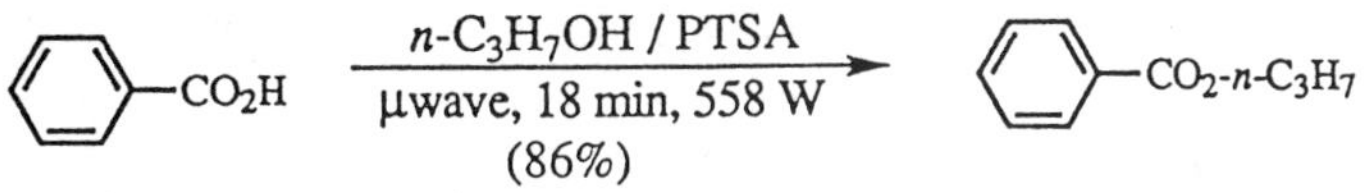

$$C_6H_5-CO_2H \xrightarrow[\substack{\mu wave, 18 \ min, 558 \ W \\ (86\%)}]{n\text{-}C_3H_7OH \ / \ PTSA} C_6H_5-CO_2\text{-}n\text{-}C_3H_7$$

Scheme 2.

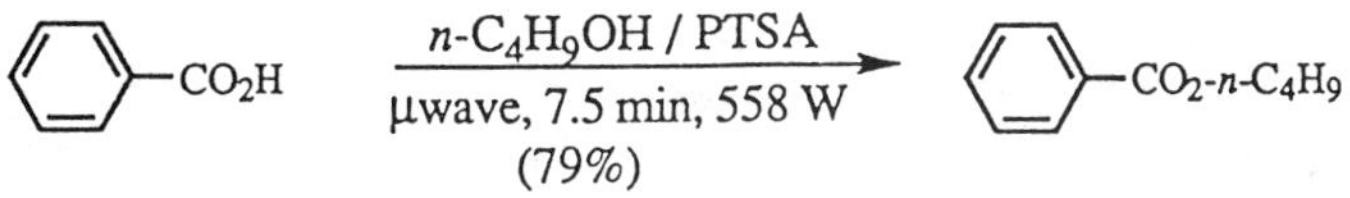

$$C_6H_5-CO_2H \xrightarrow[\substack{\mu wave, 7.5 \ min, 558 \ W \\ (79\%)}]{n\text{-}C_4H_9OH \ / \ PTSA} C_6H_5-CO_2\text{-}n\text{-}C_4H_9$$

Scheme 3.

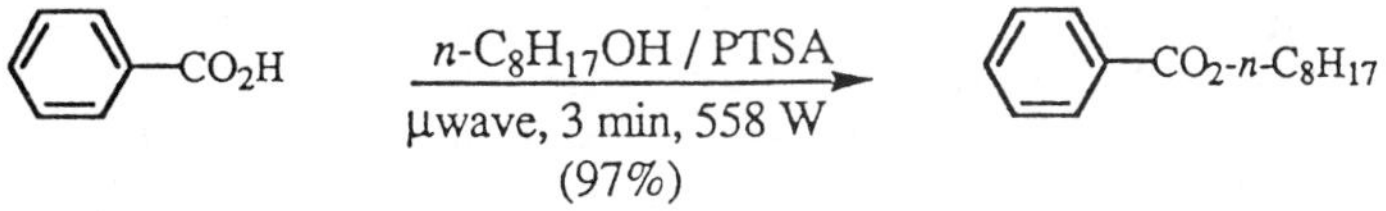

$$C_6H_5-CO_2H \xrightarrow[\substack{\mu wave, 3 \ min, 558 \ W \\ (97\%)}]{n\text{-}C_8H_{17}OH \ / \ PTSA} C_6H_5-CO_2\text{-}n\text{-}C_8H_{17}$$

Scheme 4.

$$C_6H_5-CO_2K \xrightarrow[\substack{\mu wave, 5 \ min, 600 \ W \\ (99\%)}]{n\text{-}C_8H_{17}Br \ / \ Aliquat \ 336} C_6H_5-CO_2\text{-}n\text{-}C_8H_{17}$$

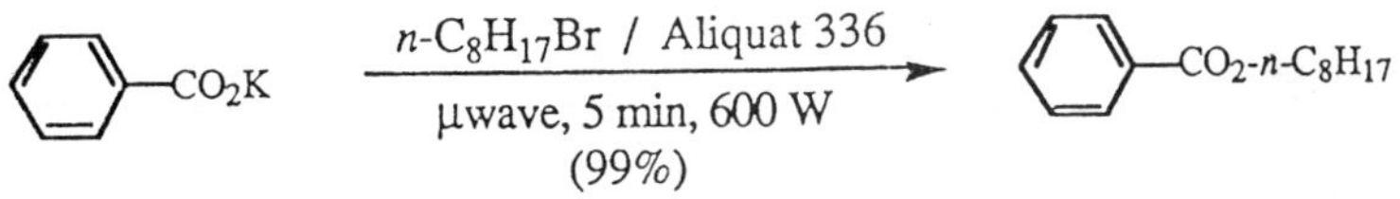

Scheme 5.

$$CH_3CO_2K \ + \ n\text{-}C_8H_{17}Br \xrightarrow[\substack{(91\%)}]{\mu wave, 10 \ min, 600 \ W} CH_3CO_2\text{-}n\text{-}C_8H_{17}$$

Scheme 6.

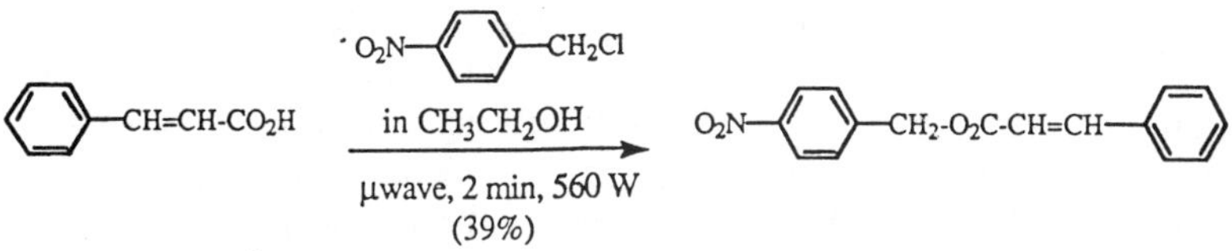

$$C_6H_5-CH=CH-CO_2H \xrightarrow[\substack{\mu wave, 2 \ min, 560 \ W \\ (39\%)}]{\substack{O_2N-C_6H_4-CH_2Cl \\ in \ CH_3CH_2OH}} O_2N-C_6H_4-CH_2\text{-}O_2C\text{-}CH=CH-C_6H_5$$

Scheme 7.

R = H or K

$$(CH_3)_2C_6H_2-CO_2R \xrightarrow[\substack{\mu wave, 45 \ sec, 600 \ W \\ (98\%)}]{\substack{C_{18}H_{37}OH \ / \\ Aliquat \ 336}} (CH_3)_2C_6H_2-CO_2C_{18}H_{37}$$

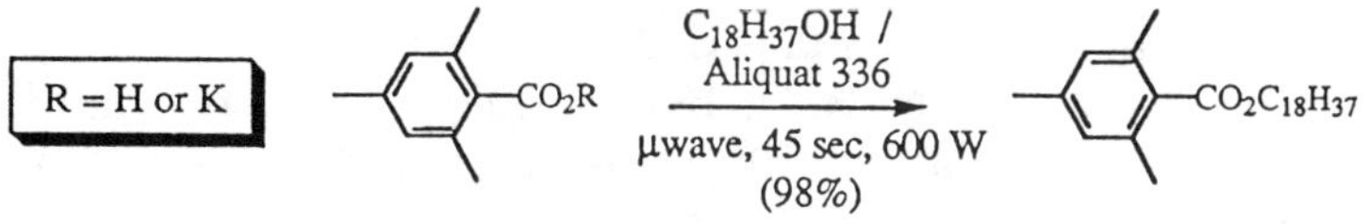

Scheme 8.

This lengthy reaction period was attributed to the low solubility of the starting diacid in methanol (*18*).

Wang and co-workers (*23*) developed a continuous-flow apparatus that pumps the reagents through the microwave cavity and allows the irradiation of volatile solvents without pressure buildup. The esterification of *p*-hydroxybenzoic acid with *n*-butanol by using this equipment was carried out in yield comparable with that obtained by using a closed reaction vessel (Scheme 11).

Transesterification has also been carried out by using microwave heating. Loupy and co-workers (*19*) treated methyl benzoate with *n*-octanol by using both acid and base catalysis (Scheme 12). The best acid catalyst for transesterification was PTSA, which gave a 97% yield of octyl benzoate in 2 min. The best basic conditions for transesterification used either K_2CO_3 or Aliquat 336. Sterically congested esters have also been prepared by using microwave heating. Octyl mesitoate was efficiently produced in only 10 min by using PTSA as a catalyst (Scheme 13). Although common esterifications have been emphasized, other esters have been prepared by using microwave technology. The acid-catalyzed reaction of 2-methylcyclopentane-1,3-dione with methanol gives enone **14a**, a vinylogous ester, in 86% yield (Scheme 14) (*18*). Enone **14a** typically requires overnight distillation with azeotropic removal of water to drive this reaction to completion.

Cycloaddition Reactions

Diels–Alder Reactions

Microwave-promoted Diels–Alder reactions occur cleanly and rapidly. Giguere, Majetich, and co-workers (*2*) first reported that the reaction between anthracene and maleic anhydride in *p*-xylene occurs in 3 min with microwave heating (Scheme 15). This reaction was initially carried out in a sealed tube with vermiculite packing employed as a safety precaution. However, the water of hydration of vermiculite absorbed the microwave irradiation and thereby heated the reaction vessel through convection. Bose and co-workers (*24*) achieved the same reaction in an open container by using the high boiling solvent diglyme to heat the reagents. In their hands, heating for only 1 min kept the solvent below its boiling point (162 °C) and afforded a 90% yield of the Diels–Alder adduct. The use of a microwave system capable of maintaining the pressure at 15 psi provided an 84% yield in 4 min by using dimethylformamide (DMF) as the solvent (*18*). These results are all comparable with those obtained by refluxing in *p*-xylene for a 10-min period (90% yield) (*2*).

The reaction of anthracene with dimethyl fumarate was one of the earliest triumphs of microwave-enhanced chemistry (Scheme 16). By using sealed-tube and oil-bath techniques, this transformation was reported to require 72 h of conventional heating for completion. In contrast, this same

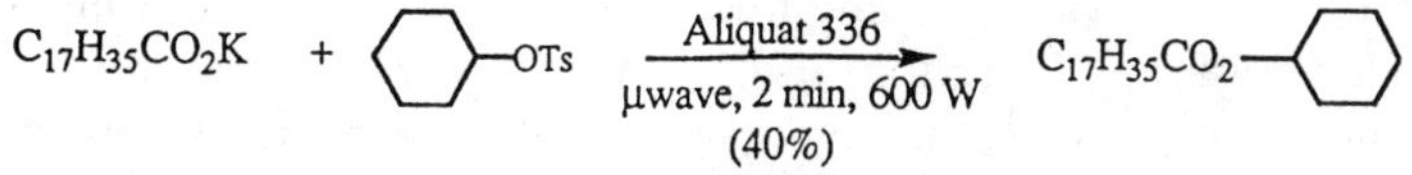

$C_{17}H_{35}CO_2K$ + [cyclohexyl]–OTs $\xrightarrow[\substack{\mu wave,\ 2\ min,\ 600\ W \\ (40\%)}]{\text{Aliquat 336}}$ $C_{17}H_{35}CO_2$–[cyclohexyl]

Scheme 9.

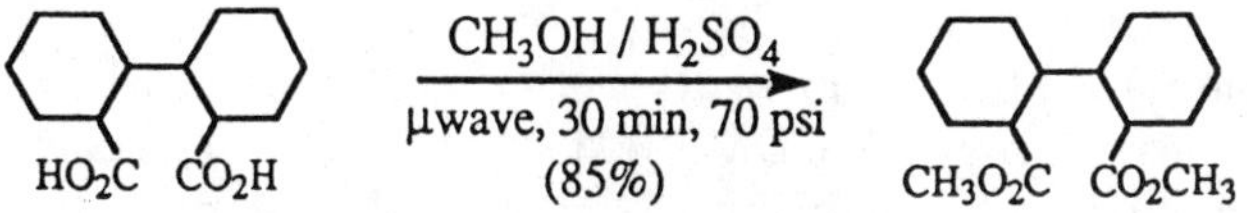

[bicyclohexyl with HO$_2$C CO$_2$H] $\xrightarrow[\substack{\mu wave,\ 30\ min,\ 70\ psi \\ (85\%)}]{CH_3OH\ /\ H_2SO_4}$ [bicyclohexyl with CH$_3$O$_2$C CO$_2$CH$_3$]

Scheme 10.

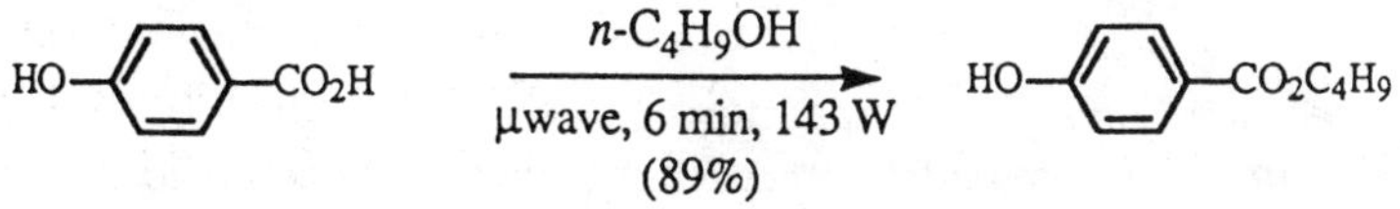

HO–[benzene]–CO_2H $\xrightarrow[\substack{\mu wave,\ 6\ min,\ 143\ W \\ (89\%)}]{n\text{-}C_4H_9OH}$ HO–[benzene]–$CO_2C_4H_9$

Scheme 11.

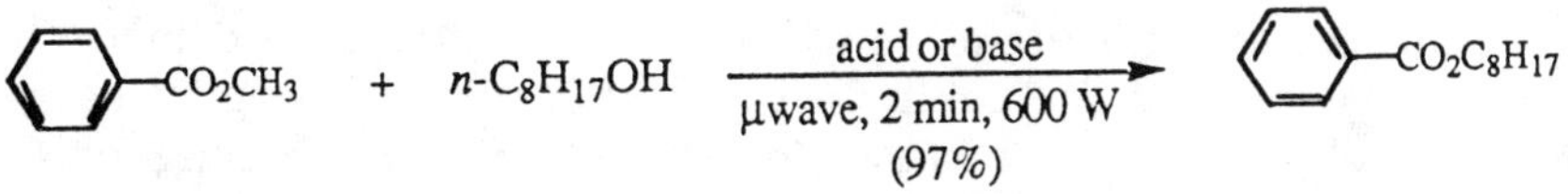

[benzene]–CO_2CH_3 + $n\text{-}C_8H_{17}OH$ $\xrightarrow[\substack{\mu wave,\ 2\ min,\ 600\ W \\ (97\%)}]{\text{acid or base}}$ [benzene]–$CO_2C_8H_{17}$

Scheme 12.

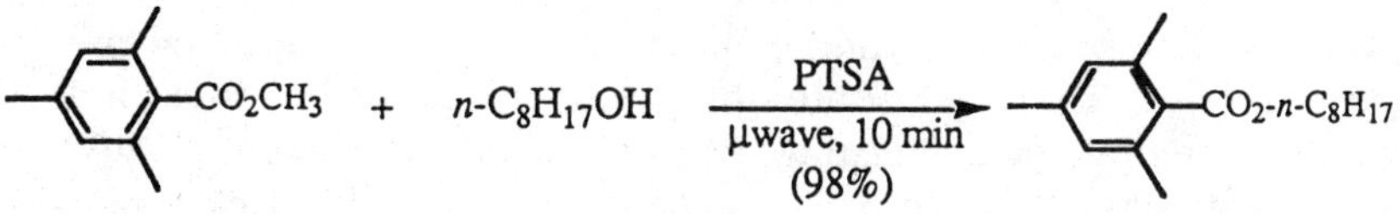

[mesityl]–CO_2CH_3 + $n\text{-}C_8H_{17}OH$ $\xrightarrow[\substack{\mu wave,\ 10\ min \\ (98\%)}]{PTSA}$ [mesityl]–$CO_2\text{-}n\text{-}C_8H_{17}$

Scheme 13.

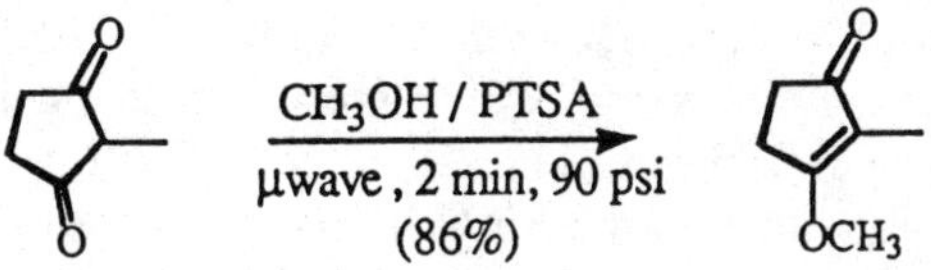

[2-methyl-1,3-cyclopentanedione] $\xrightarrow[\substack{\mu wave,\ 2\ min,\ 90\ psi \\ (86\%)}]{CH_3OH\ /\ PTSA}$ [3-methoxy-2-methylcyclopent-2-enone, OCH$_3$]

14a

Scheme 14*.

[anthracene] + [maleic anhydride] $\xrightarrow[\mu wave,\ 3\ min,\ 600\ W\ (92\%)]{\text{sealed tube, xylene}}$ [Diels-Alder adduct]

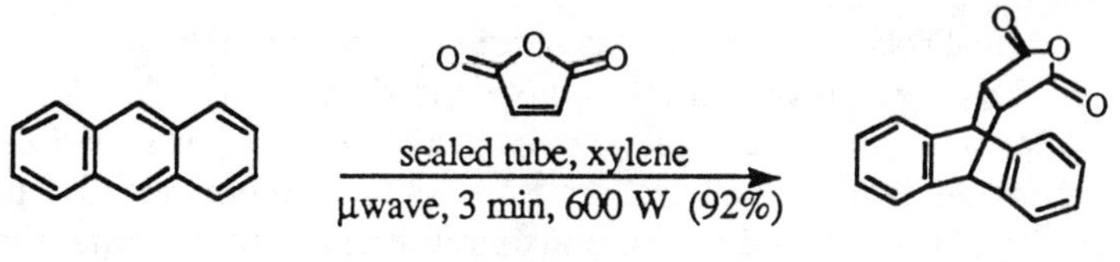

Scheme 15.

transformation was achieved by using microwave irradiation in 10 min in *p*-xylene (sealed tube) (*2*), in 12 min in DMF (at 20 psi) (*18*), and later in 5–10 min in an open vessel by using 1, 2, 4-trichlorobenzene as solvent (*24*).

Other examples of microwave-accelerated Diels–Alder reactions include the following reactions (*2, 18*):

- *E,E*,1,4-diphenyl-1,3-butadiene with diethyl acetylenedicarboxylate (sealed tube, neat, 12 min, 55%; DMF, 30 psi, 20 min, 58%) (Scheme 17)
- furan with diethyl acetylenedicarboxylate (sealed tube, neat, 10 min, 66%; DMF, 30 psi, 194–198 °C, 10 min, 86%) (Scheme 18)
- 2-cyclohexenone with 2,3-dimethyl-1,3-butadiene (sealed tube, neat, 15 min, 25%) (Scheme 19)

Intramolecular Diels–Alder reactions have been studied by using microwave heating. The cycloaddition reaction shown in Scheme 20 occurs in only 8 min with microwave heating, as opposed to 24 h by using oil-bath heating (*3*). Both processes give the same yield and the same ratio of diastereomers **20b**, **20c**, and **20d** (i.e., 63:21:16, respectively) (*25*).

Microwave heating is useful for multistep total synthesis, as demonstrated in 1990 by Fallis and Lei in their elegant synthesis of longifolene (**21a**). The Diels–Alder reaction shown in Scheme 21 required heating for 2.5 h but occurred in 97% yield (*26*). An improved synthesis of longifolene also used microwave heating to promote a different Diels–Alder reaction (*27*).

By using conventional heating methods the Diels–Alder reaction presented in Scheme 22 is complicated by polymerization of methyl vinyl ketone. By using microwave heating polymer formation was minimized. For this reaction Maat and co-workers (*28*) cut a hole in the back wall of the microwave oven through which a reflux condenser was fitted. By using this modified apparatus the reaction mixture was heated over a 24-h period at atmospheric pressure and using a cycle of 15 min of heating followed by 15 min without heating. The total yield of **22a** and **22b** was 32% in a 2.2:1 ratio, respectively. The same equipment was also used to achieve the cycloaddition shown in Scheme 23 (*29*).

Stambouli, Chastrette, and Soufiaoui (*30*) studied hetero-Diels–Alder reactions by using microwave irradiation. 2-Methyl-1,3-pentadiene reacted with methyl glyoxylate to give diastereomers **24a** and **24b** in 96% yield (Scheme 24), whereas traditional oil-bath heating at 140 °C required 6 h and gave only a 65% yield.

Other hetero-Diels–Alder reactions studied by using microwave heating include the reaction of 2-methyl-1,3-pentadiene with glyoxal 1,1-dimethylacetal without solvent (as shown in Scheme 25), or in benzene with zinc chloride as a catalyst (5 min, 82%) (*30*). Another reaction that was studied was that of 1-vinylcyclohexene with diethyl ketomalonate (Scheme 26) (*18*).

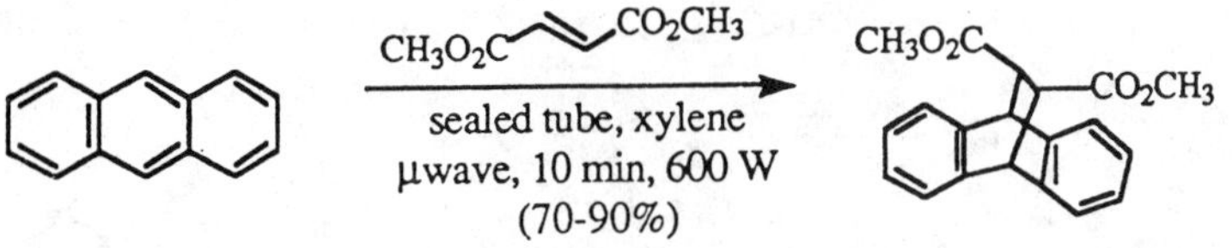

Scheme 16.

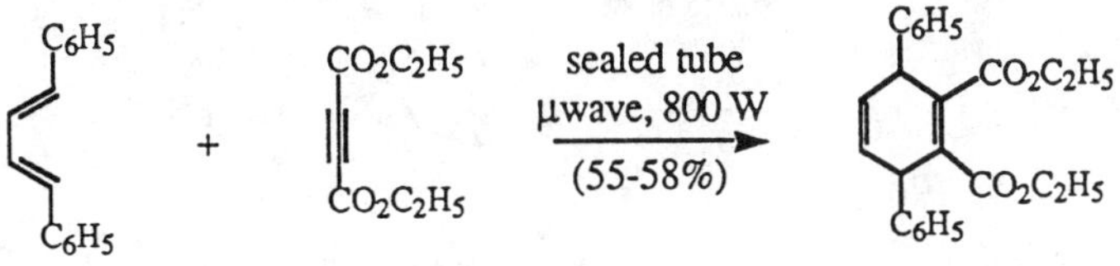

Scheme 17.

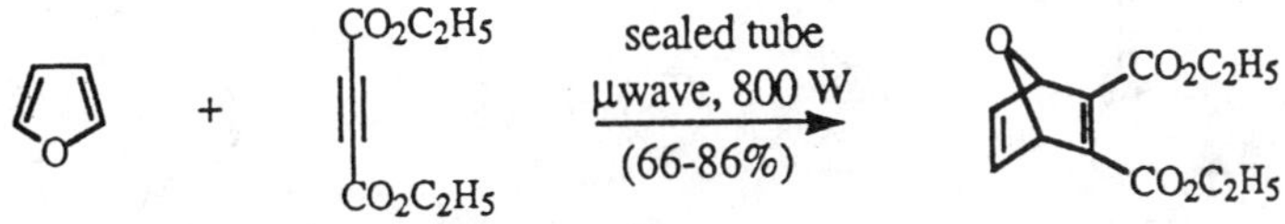

Scheme 18.

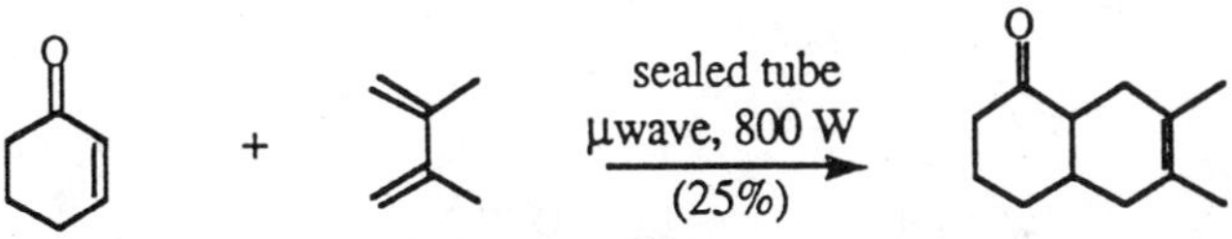

Scheme 19.

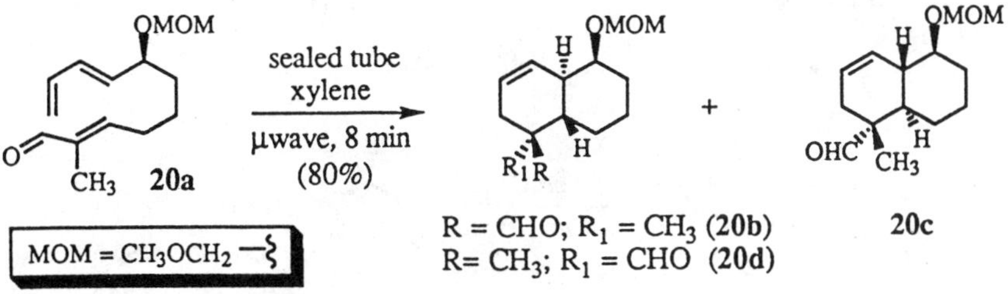

MOM = CH$_3$OCH$_2$—

R = CHO; R$_1$ = CH$_3$ (20b)
R= CH$_3$; R$_1$ = CHO (20d)

20c

Scheme 20.

Ene Reactions

The ene reaction is the reaction of an alkene with a reactive enophile, such as diethyl acetylenedicarboxylate or diethyl ketomalonate. In this reaction a new carbon–carbon single bond is formed, and the original double bond and a hydrogen are shifted through a cyclic transition state. An example of an intramolecular ene reaction is shown in Scheme 27. This reaction was carried out in 15 min under neat conditions (2) but required 8 h in DMF in

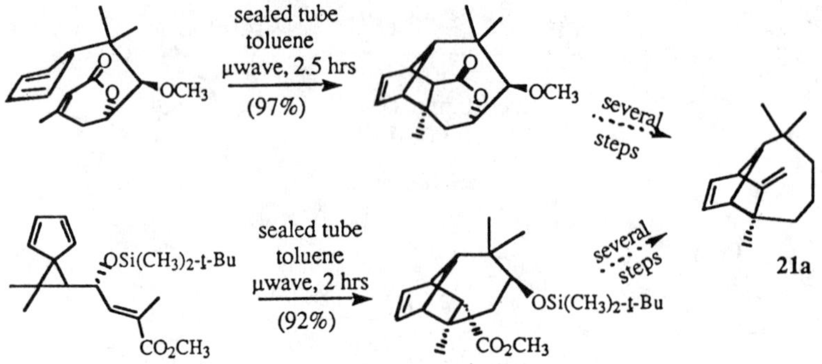

Scheme 21.

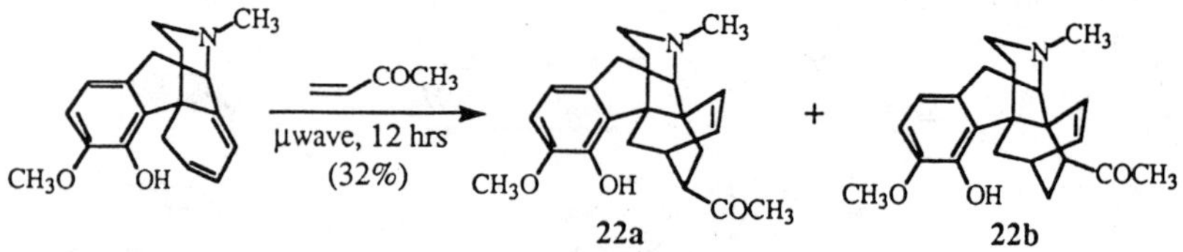

Scheme 22.

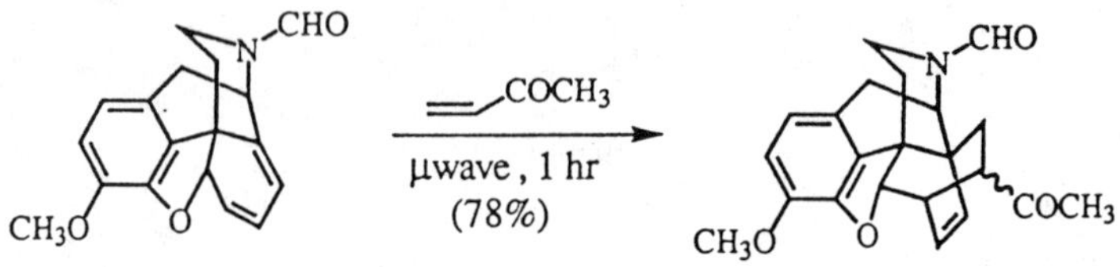

Scheme 23.

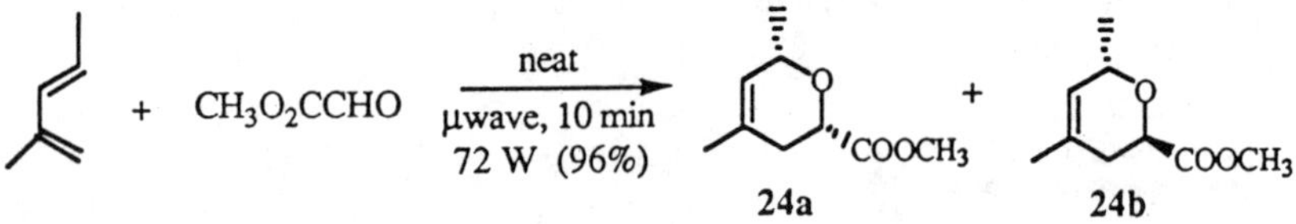

Scheme 24.

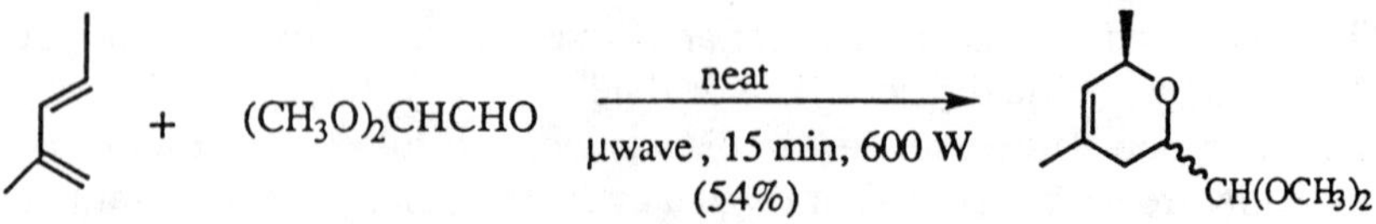

Scheme 25.

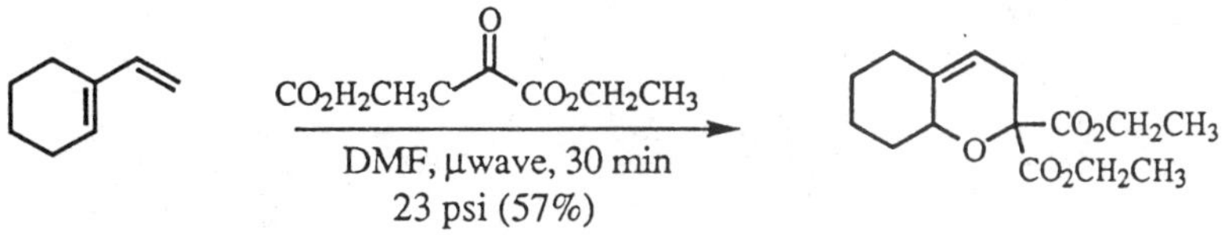

Scheme 26*.

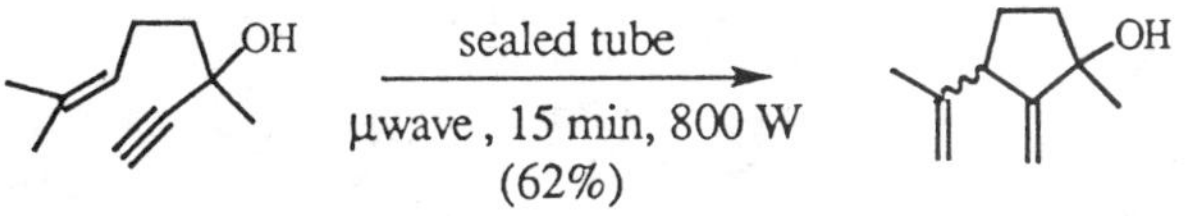

Scheme 27.

a Teflon vessel (73%) (*18*). This large reaction time difference occurs because the microwave-reaction pressure monitor prevents the reaction temperature from exceeding the temperature at which the Teflon reaction vessel deforms (~250 °C) (*31*). In contrast, sealed reaction vessels can reach reaction temperatures between 400–425 °C; hence, the shorter reaction times observed.

Enophiles can also contain heteroatoms, as illustrated in Scheme 28 using diethyl ketomalonate (*18*). In a final example of an intramolecular ene reaction, microwave heating of (+)-citronellal produces three ene products (Scheme 29) (*32*). Diene **29c**, albeit a minor product, is the result of the Zeolite-promoted dehydration of intermediate **29a** or **29b**.

Alder–Bong Reactions

The Alder–Bong reaction is the reaction of a 1,4-cyclohexadiene with an acetylenic enophile, usually diethyl acetylenedicarboxylate, to produce a highly functionalized polycyclic product. This reaction proceeds via a tandem ene and Diels–Alder pathway. When this reaction (Scheme 30) is performed in a sealed vessel without solvent, an 82% yield of diester **30a** is obtained after only 6 min of microwave heating (*33*); by using DMF as a solvent in an oven capable of controlling the pressure, the same reaction gives a yield of 49% after 20 min (Scheme 30) (*18*). These results are superior to those obtained by using conventional procedures (40 h, 150 °C, 14%). Five additional Alder–Bong examples are shown in Schemes 30–35. The use of 1,2-dimethyl-1,4-cyclohexadiene (**31a**) as the diene results in a mixture of two regioisomers in nearly quantitative yield (**31b** and **31c**, 4.6:1, respectively, Scheme 31). Yields of the Alder–Bong reaction are lower when a singly activated enophile is used (cf. Scheme 32). If the ene unit is contained in a bicyclic ring system, the product obtained is a tetracycle; therefore, this route is useful to polycyclic systems. When the 5,6-ring system in Scheme 33 reacts with dimethyl acetylenedicarboxylate (DMADC), an 80% yield of the Alder–Bong product is obtained in only 6 min (*33*).

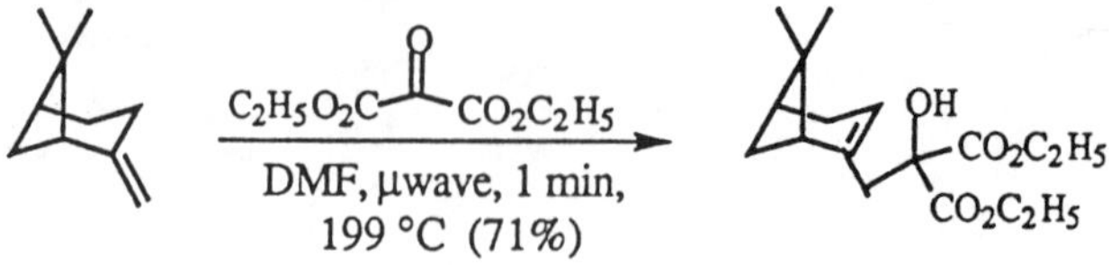
C2H5O2C–CO–CO2C2H5
DMF, µwave, 1 min,
199 °C (71%)
OH
CO2C2H5
CO2C2H5

Scheme 28*.

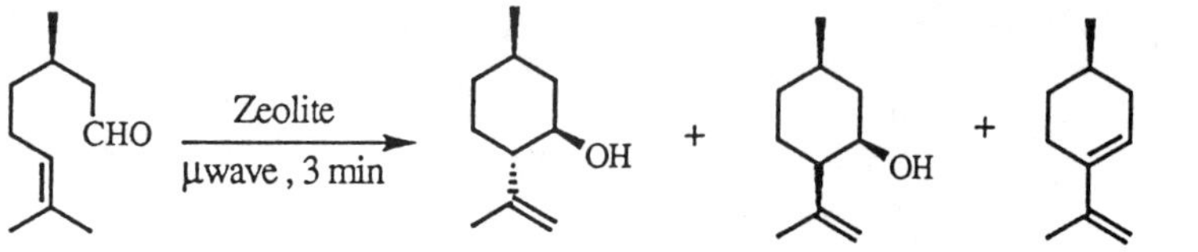
CHO
Zeolite
µwave, 3 min
OH
+
OH
+

29a (89%) **29b** (6%) **29c** (5%)

Scheme 29.

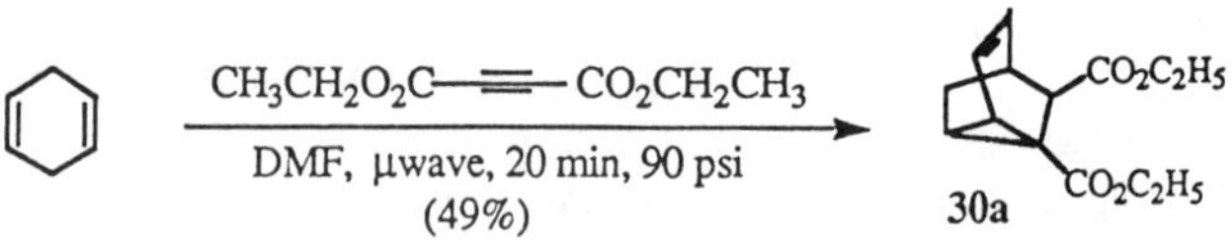
CH3CH2O2C≡CO2CH2CH3
DMF, µwave, 20 min, 90 psi
(49%)
CO2C2H5
CO2C2H5
30a

Scheme 30*.

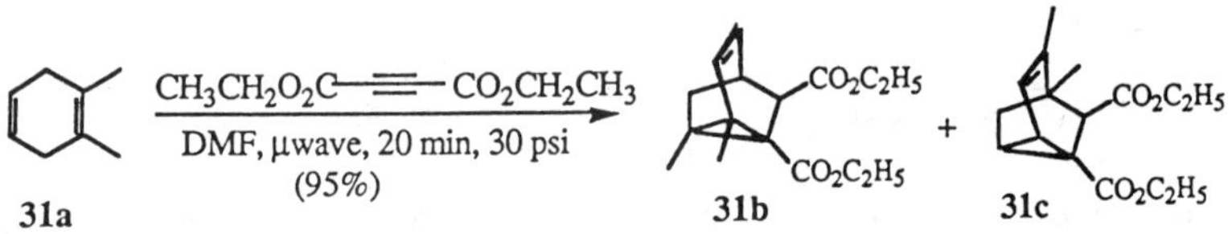
CH3CH2O2C≡CO2CH2CH3
DMF, µwave, 20 min, 30 psi
(95%)
31a
CO2C2H5
CO2C2H5
31b
+
CO2C2H5
CO2C2H5
31c

Scheme 31*.

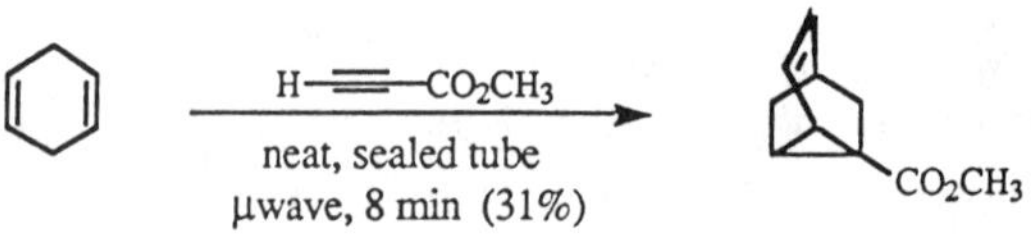
H≡CO2CH3
neat, sealed tube
µwave, 8 min (31%)
CO2CH3

Scheme 32.

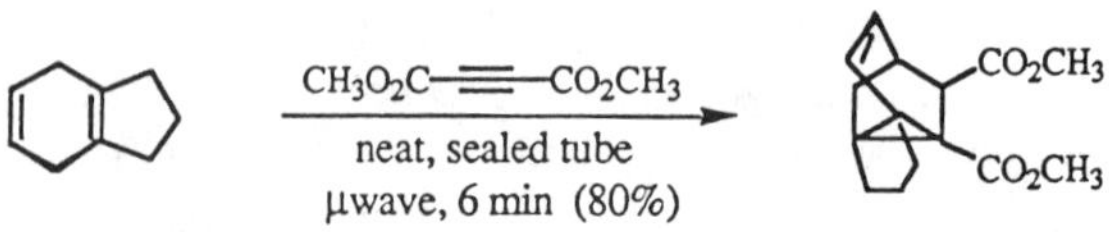
CH3O2C≡CO2CH3
neat, sealed tube
µwave, 6 min (80%)
CO2CH3
CO2CH3

Scheme 33.

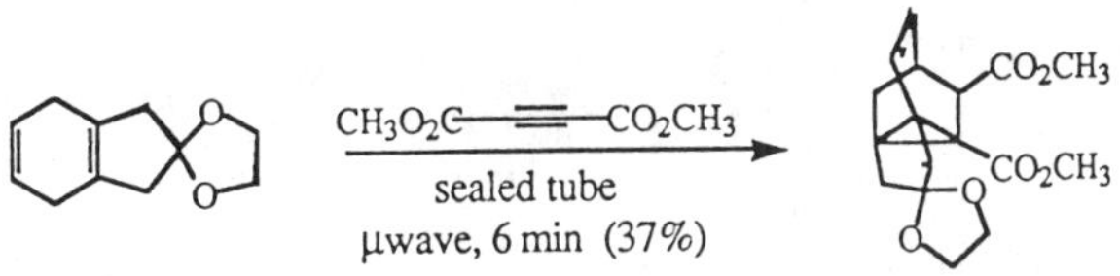

Scheme 34.

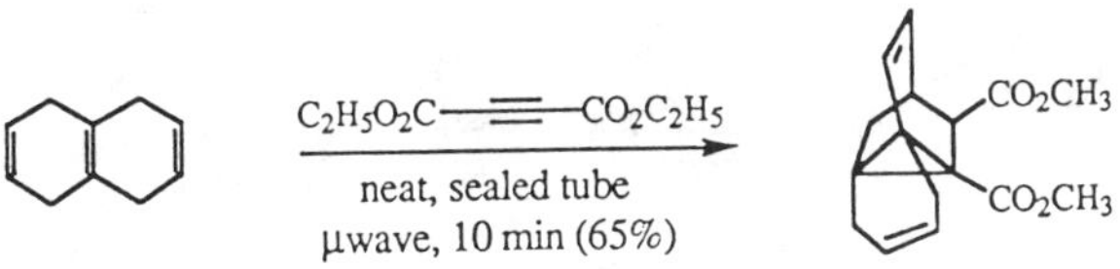

Scheme 35.

More functionalized tetracyclic systems can be made similarly, although the yields are lower. The ketal shown in Scheme 34 only gives a 37% yield on reaction with DMADC. The analogous 6,6-ring system gives a 65% yield (Scheme 35).

[3, 3]-Sigmatropic Rearrangements

The *ortho*-Claisen rearrangement is a [3, 3]-sigmatropic rearrangement of an allyl vinyl ether that occurs more rapidly in microwave ovens. The rearrangement of allyl phenyl ether has been studied under diverse conditions (Scheme 36). After microwave irradiation for 10 min in a sealed tube without solvent, the yield of 2-allylphenol (**36a**) was only 21% despite a reaction temperature between 325 °C and 361 °C. However, when the solvent DMF (which efficiently couples with microwave energy) (*11*) was added, the yield increased to 92% after 6 min of irradiation (*2*). When the starting ether was adsorbed on *y*-Zeolite, not only did the yield of the *ortho*-Claisen product decrease but by-products **36b** and **36c** were also formed (*32*). Rearrangements of other aryl allyl ethers have also been studied (Schemes 37–40). The methylenedioxy phenyl ether shown in Scheme 37 gave a 91% yield of the phenol product under pressure-controlled conditions (*18*). When both of the ortho positions of the allyl aryl ether are blocked, as illustrated in Scheme 38, the allyl substituent migrates to the para position (cf. phenol **38d**) as a result of a tandem *ortho*-Claisen–Cope rearrangement process [**38a** → **38b** → **38c** → **38d**]. As shown in Scheme 39, even when the ortho position is unsubstituted, some of the para product may be formed (cf. **39b**). Note that dihydrobenzofuran **39c** was also formed in 20% yield (*32*). A similar bicyclic by-product was formed when the para position of the starting allyl aryl ether was blocked (Scheme 40).

Another powerful [3, 3]-sigmatropic reaction is the orthoester Claisen rearrangement. This three-step process begins with a condensation to form

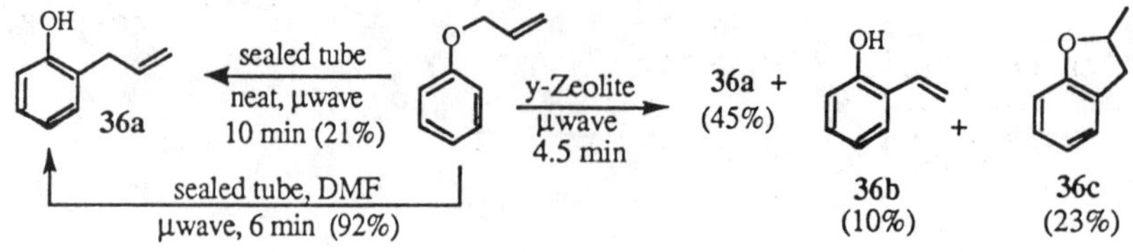

Scheme 36.

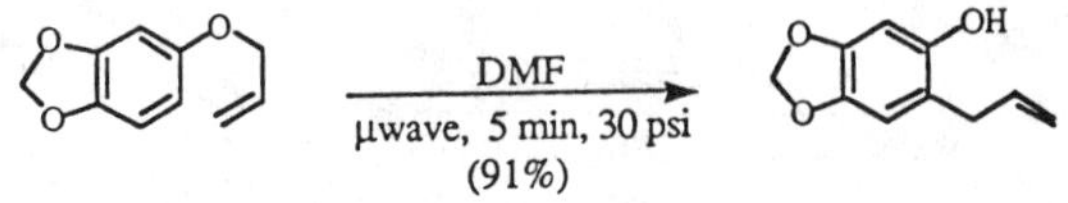

Scheme 37*.

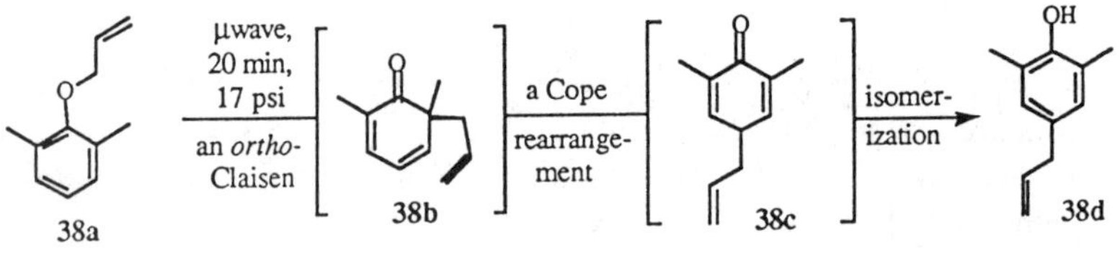

Scheme 38*.

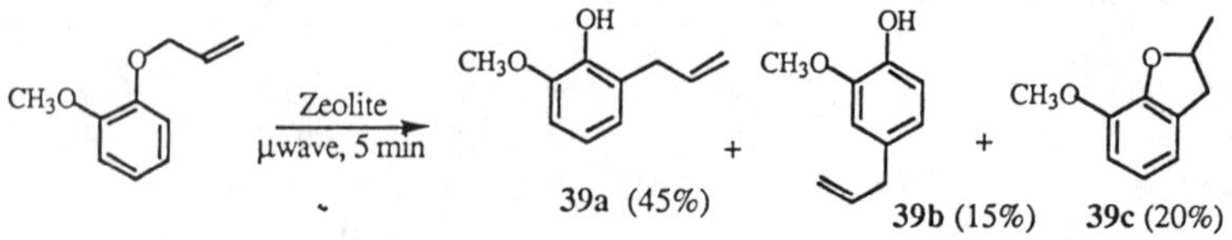

Scheme 39.

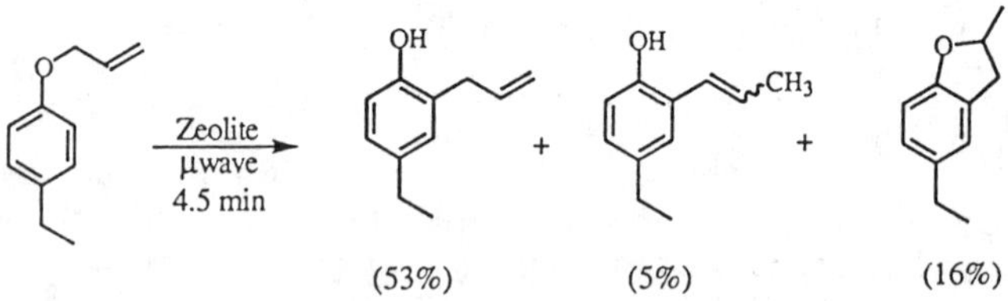

Scheme 40.

an orthoester (**41a**), followed by the elimination of a molecule of ethanol to give ketene acetal intermediate (i.e., **41b**) (R = CH_2CH_3). Rearrangement of this allyl vinyl ether culminates in the formation of the product, **41c**. Three additional orthoester Claisen examples are shown in Schemes 41–44. In the case of Scheme 43, Jones and Huber (34) found that thermolysis time was a function of catalyst and solvent used. Note that in the case of 2-butyn-1,4-diol, the [3, 3]-sigmatropic rearrangement of both propargylic alcohols results in the formation of a useful diene (Scheme 44) (35).

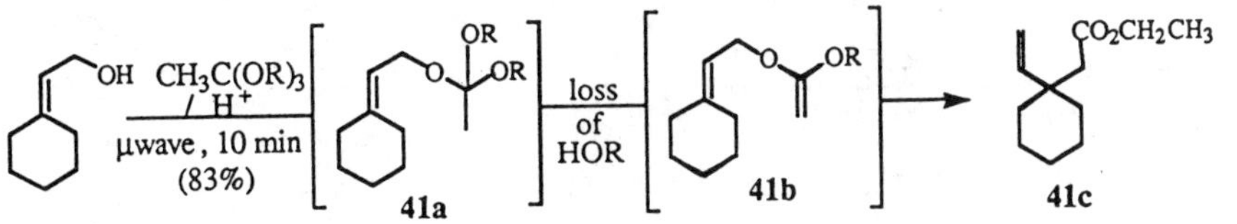

Scheme 41.

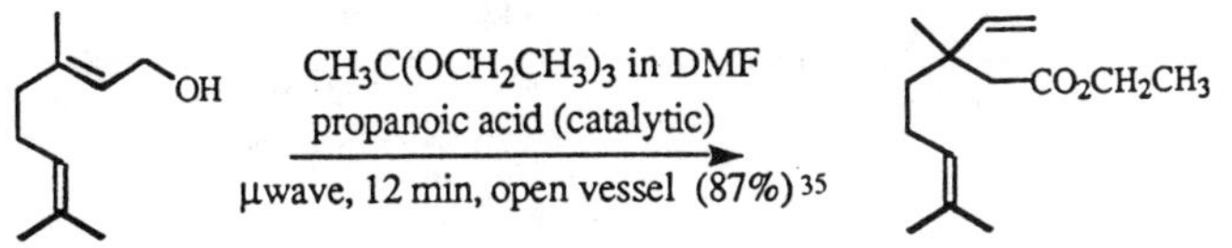

Scheme 42.

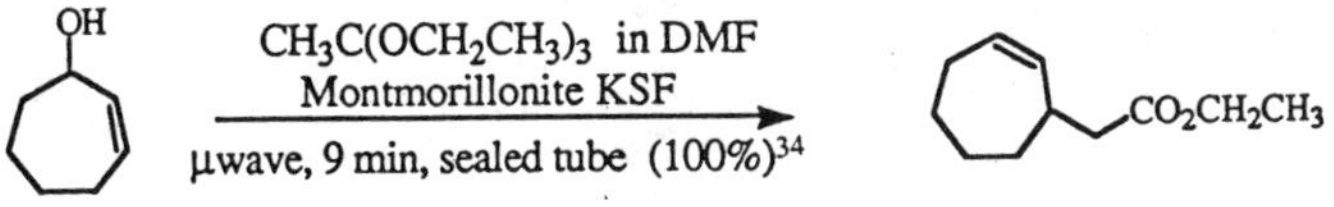

Scheme 43.

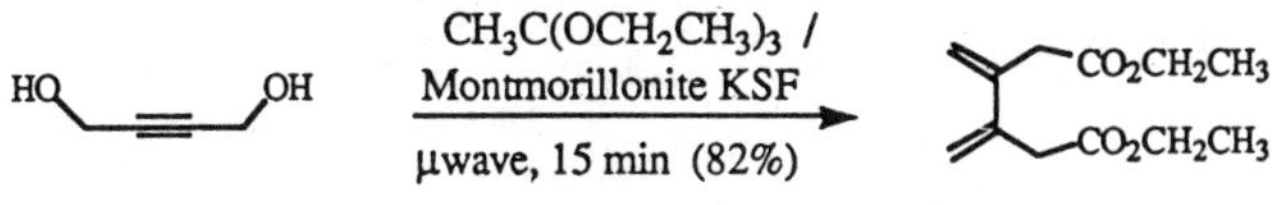

Scheme 44.

An example of a simple Claisen rearrangement using an allylic alcohol and ethyl vinyl ether gives three products in 95% total yield (Scheme 45) (*36*). Although aldehyde **45a** is the major product obtained by using microwave heating (75% yield), with conventional heating none of this product is observed.

Williamson Ether Synthesis

The Williamson ether synthesis, a versatile way to make unsymmetrical ethers, is highly accelerated by microwave heating. Six examples are summarized in Schemes 46–51. Under classical conditions, the reaction of benzyl chloride and *p*-cyanophenoxide ion takes 12 h and proceeds in 74% yield, but requires only 3 min by using microwave heating (Scheme 46) (*1*). By using a continuous-flow microwave unit, phenoxide ion reacted with benzyl chloride at lower yield (Scheme 47) (*24, 37*). The phenol can be used directly if a base catalyst is used. This reaction occurs in 96% yield (Scheme 48) (*18*). The aryloxyacetic acid derivative of 2-naphthol is prepared within 2 min in aqueous sodium hydroxide, albeit in only 14% yield

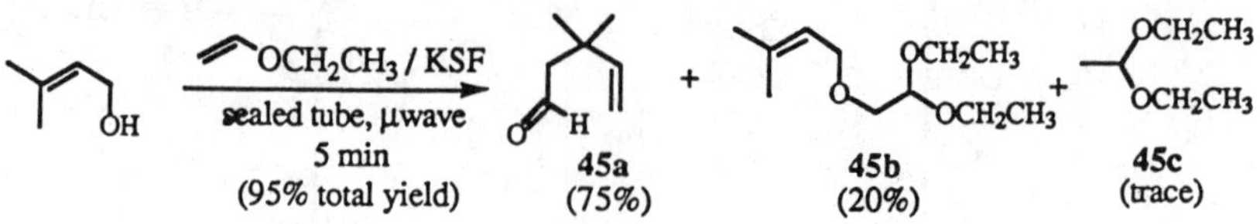

Scheme 45.

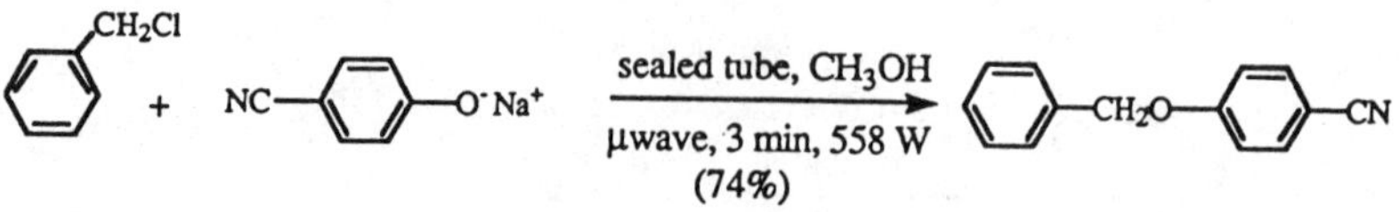

Scheme 46.

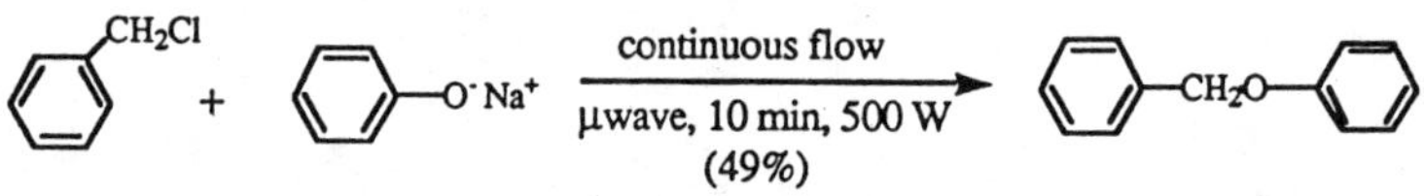

Scheme 47.

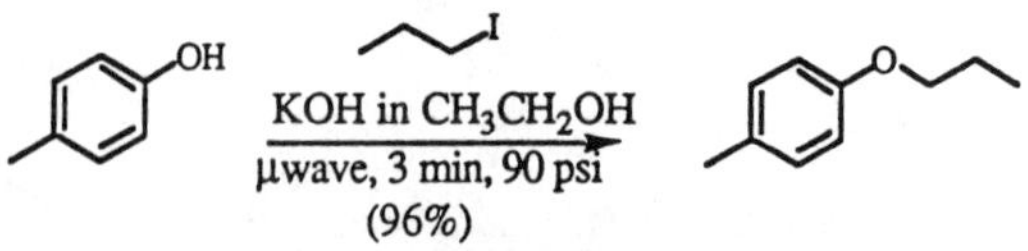

Scheme 48.

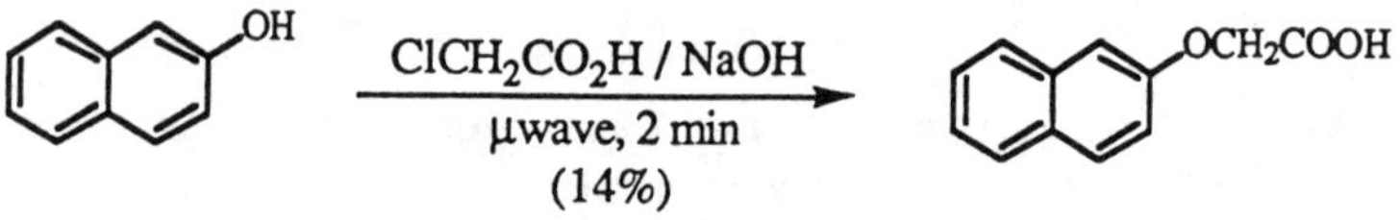

Scheme 49.

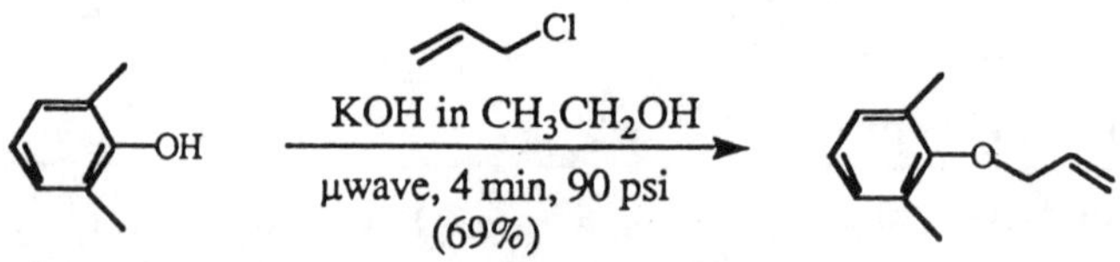

Scheme 50*.

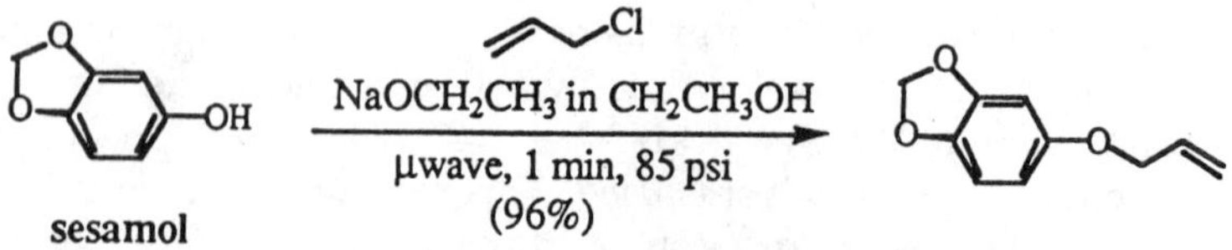

sesamol

Scheme 51*.

(*17*) (Scheme 49); conventional heating (60 min in refluxing ethanol) gives identical results. Other Williamson ether syntheses carried out in microwave ovens are the reactions of allyl chloride with 2,6-dimethylphenol and sesamol (Schemes 50 and 51, respectively) (*18*).

Substitutions

Simple S_N1 and S_N2 reactions are accelerated in microwave ovens. By using sodium bromide in aqueous acid, cyclohexanol is converted to cyclohexyl bromide within 10 min in 49% yield (Scheme 52). This rate is one-third of the time required for the reaction during conventional heating; moreover, the classical yield is only 33% (*18*). In a similar fashion diols such as 1,8-octanediol and 1,10-decanediol can be converted to the corresponding dibromides. During microwave heating, 1,8-octanediol yields 1,8-dibromooctane (Scheme 53), whereas 1,10-decanediol gives a mixture of mono- and dibromides **54a** and **54b** in the ratio of 1.5:1, respectively (*18*) (Scheme 54).

The preparation of an alkyl iodide from the corresponding bromide or chloride by treatment with sodium iodide in a ketone solvent, such as acetone or 2-butanone, is known as the Finkelstein halogen exchange reaction. A microwave-accelerated example is the reaction of *n*-hexyl bromide with sodium iodide, where *n*-hexyl iodide is obtained in 90% yield in only 4 min in 2-butanone at 20 psi (Scheme 55) (*3, 18*). *n*-Octyl bromide and cyclohexyl bromide give similar results under these conditions (Schemes 56 and 57, respectively) (*18*). Other examples of microwave-promoted substitution reactions are the conversions of *p*-(4-bromobutyl)anisole (Scheme 58) and 4-chlorobutyl acetate (Scheme 59) into the corresponding iodides in 8 and 15 min, respectively (*3*). Azides can be formed quickly from tosylates, as shown in the reaction of tosylate **60a** with sodium azide in DMF (Scheme 60) (*38*).

Phenyl sulfones can be synthesized from alkyl halides via simple substitution reactions (Schemes 61–67) (*39*). The following displacements were all carried out by adsorbing the alkyl halide and sodium phenylsulfinate onto alumina, followed by microwave heating in an open reaction vessel without solvent for 5 min. Most of the reactions gave high yields of the desired phenyl sulfone products; iodo- and diiodomethane gave the lowest yields at 52%, whereas the more electron-deficient reagents chloroacetonitrile and ethyl bromoacetate gave 73% and 99% yields, respectively. In Scheme 68 the sulfone formed under these conditions is the result of epoxide opening to generate a new epoxide, followed by a base-promoted elimination. Other useful substitution reactions include the formation of a substituted aniline from an aromatic chloride (Scheme 69) (*40*).

An important reaction of aromatic compounds is electrophilic substitution, wherein an electrophile reacts with an arene and substitutes for one

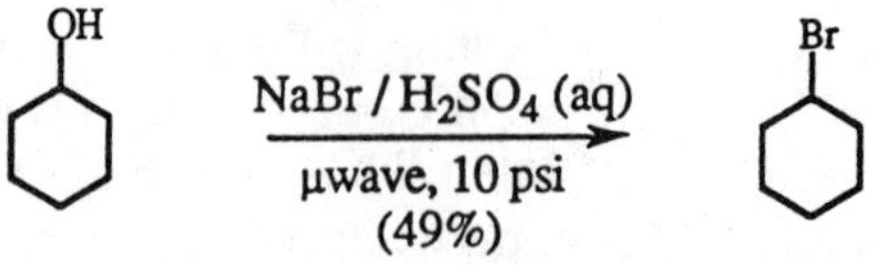
OH
$NaBr / H_2SO_4$ (aq)
µwave, 10 psi
(49%)
Br

Scheme 52*.

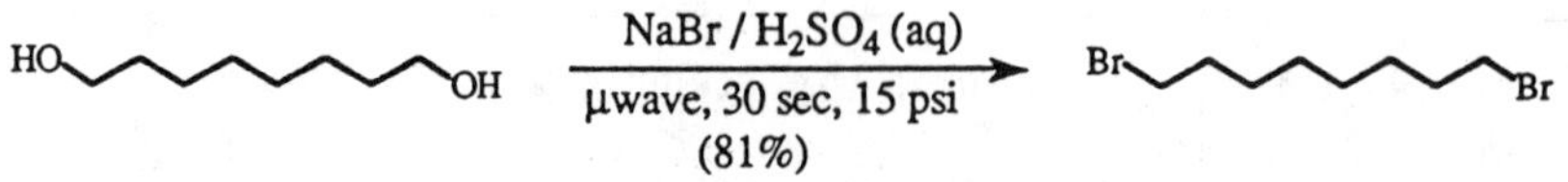
HO
OH
$NaBr / H_2SO_4$ (aq)
µwave, 30 sec, 15 psi
(81%)
Br
Br

Scheme 53*.

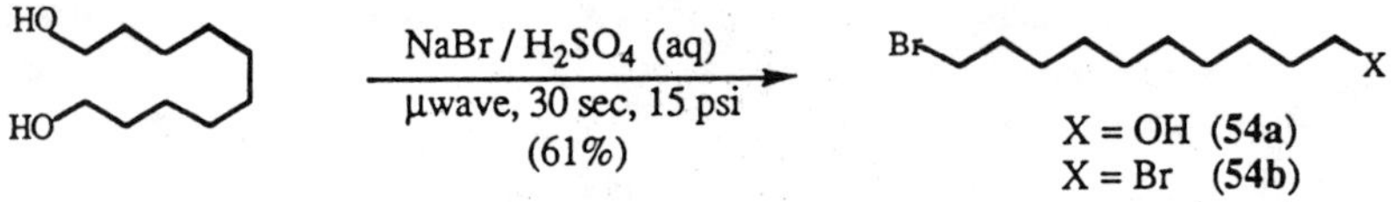
HO
HO
$NaBr / H_2SO_4$ (aq)
µwave, 30 sec, 15 psi
(61%)
Br
X
X = OH (54a)
X = Br (54b)

Scheme 54*.

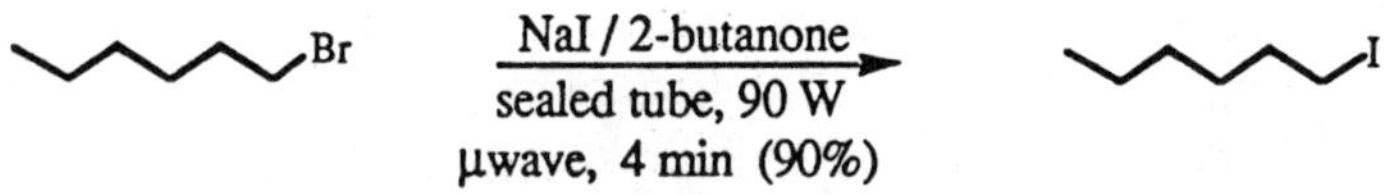
Br
NaI / 2-butanone
sealed tube, 90 W
µwave, 4 min (90%)
I

Scheme 55.

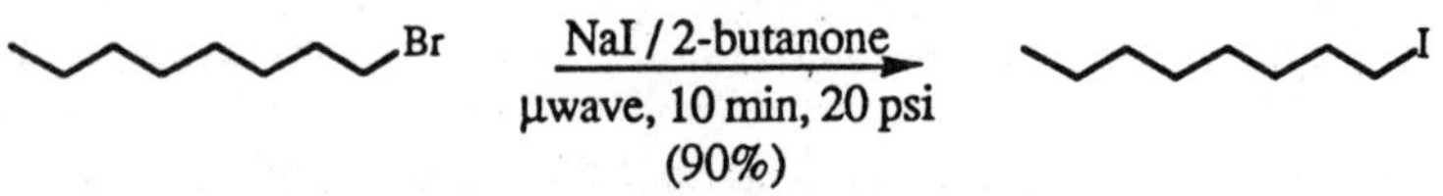
Br
NaI / 2-butanone
µwave, 10 min, 20 psi
(90%)
I

Scheme 56*.

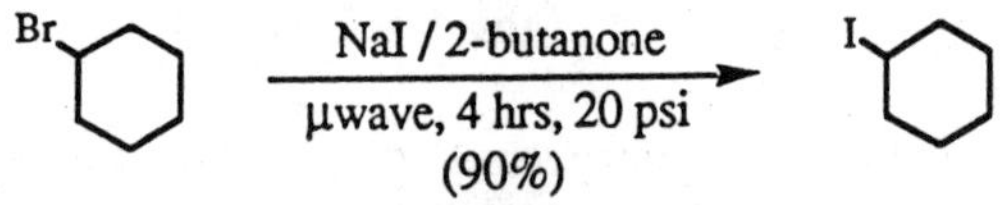
Br
NaI / 2-butanone
µwave, 4 hrs, 20 psi
(90%)
I

Scheme 57*.

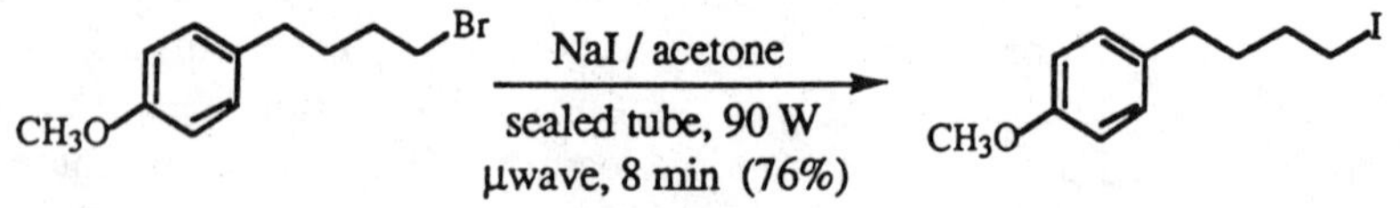
Br
CH_3O
NaI / acetone
sealed tube, 90 W
µwave, 8 min (76%)
I
CH_3O

Scheme 58.

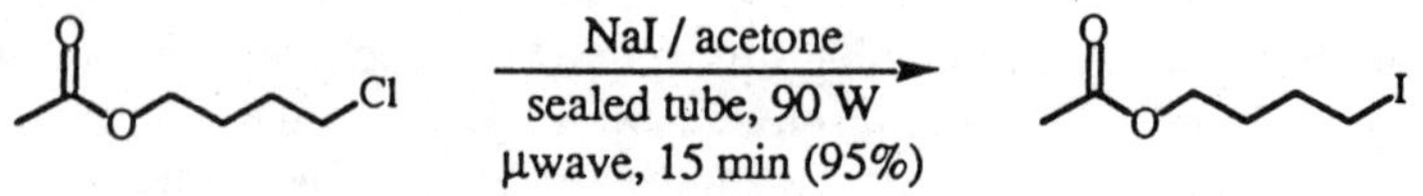
O
O
Cl
NaI / acetone
sealed tube, 90 W
µwave, 15 min (95%)
O
O
I

Scheme 59.

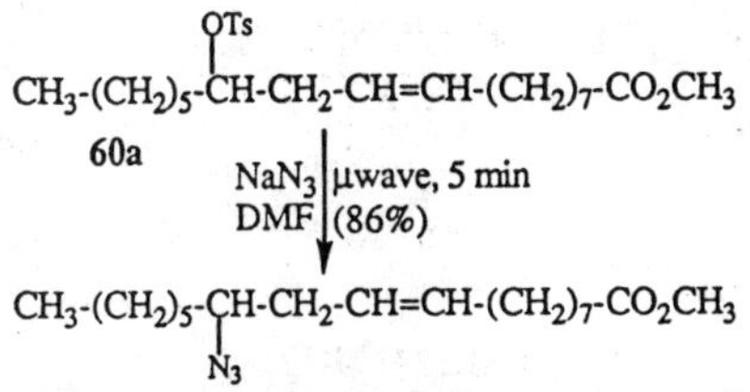

$$\underset{\text{OTs}}{\text{CH}_3\text{-(CH}_2)_5\text{-CH-CH}_2\text{-CH=CH-(CH}_2)_7\text{-CO}_2\text{CH}_3}$$

60a $\xrightarrow[\text{DMF}]{\text{NaN}_3 \mid \mu\text{wave, 5 min}}$ (86%)

$$\underset{\text{N}_3}{\text{CH}_3\text{-(CH}_2)_5\text{-CH-CH}_2\text{-CH=CH-(CH}_2)_7\text{-CO}_2\text{CH}_3}$$

Scheme 60.

$$\text{CH}_3\text{I} \xrightarrow[\substack{\mu\text{wave, 5 min, 160 W} \\ (52\%)}]{\text{C}_6\text{H}_5\text{SO}_2\text{Na on alumina}} \text{CH}_3\text{SO}_2\text{C}_6\text{H}_5$$

Scheme 61.

$$\text{CH}_2\text{I}_2 \xrightarrow[\substack{\mu\text{wave, 5 min, 160 W} \\ (52\%)}]{\text{C}_6\text{H}_5\text{SO}_2\text{Na on alumina}} (\text{C}_6\text{H}_5\text{SO}_2)_2\text{CH}_2$$

Scheme 62.

$$\text{NCCH}_2\text{Cl} \xrightarrow[\substack{\mu\text{wave, 5 min, 160 W} \\ (73\%)}]{\text{C}_6\text{H}_5\text{SO}_2\text{Na on alumina}} \text{NCCH}_2\text{SO}_2\text{C}_6\text{H}_5$$

Scheme 63.

$$\text{C}_2\text{H}_5\text{O}_2\text{CCH}_2\text{Br} \xrightarrow[\substack{\mu\text{wave, 5 min, 160 W} \\ (99\%)}]{\text{C}_6\text{H}_5\text{SO}_2\text{Na on alumina}} \text{C}_2\text{H}_5\text{O}_2\text{CCH}_2\text{SO}_2\text{C}_6\text{H}_5$$

Scheme 64.

$$\text{C}_6\text{H}_5\text{CH}_2\text{Cl} \xrightarrow[\substack{\mu\text{wave, 5 min, 160 W} \\ (99\%)}]{\text{C}_6\text{H}_5\text{SO}_2\text{Na on alumina}} \text{C}_6\text{H}_5\text{CH}_2\text{SO}_2\text{C}_6\text{H}_5$$

Scheme 65.

$$\text{C}_6\text{H}_5\text{COCH}_2\text{Br} \xrightarrow[\substack{\mu\text{wave, 5 min, 160 W} \\ (98\%)}]{\text{C}_6\text{H}_5\text{SO}_2\text{Na on alumina}} \text{C}_6\text{H}_5\text{COCH}_2\text{SO}_2\text{C}_6\text{H}_5$$

Scheme 66.

$$\text{CH}_2\text{=CH-CH}_2\text{Br} \xrightarrow[\substack{\mu\text{wave, 5 min, 160 W} \\ (99\%)}]{\text{C}_6\text{H}_5\text{SO}_2\text{Na on alumina}} \text{CH}_2\text{=CH-CH}_2\text{SO}_2\text{C}_6\text{H}_5$$

Scheme 67.

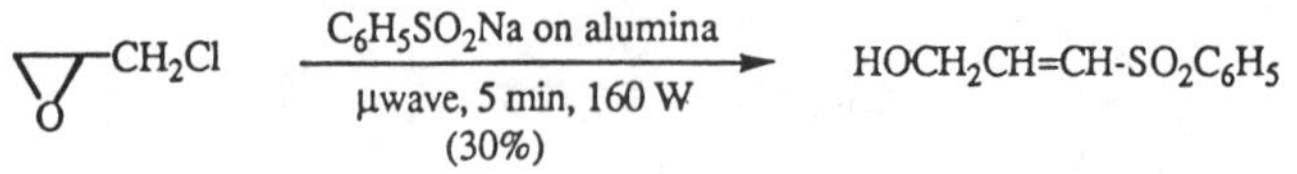

$$\text{\large$\triangle$}\!\!-CH_2Cl \xrightarrow[\substack{\mu\text{wave, 5 min, 160 W} \\ (30\%)}]{C_6H_5SO_2Na \text{ on alumina}} HOCH_2CH=CH\text{-}SO_2C_6H_5$$

Scheme 68.

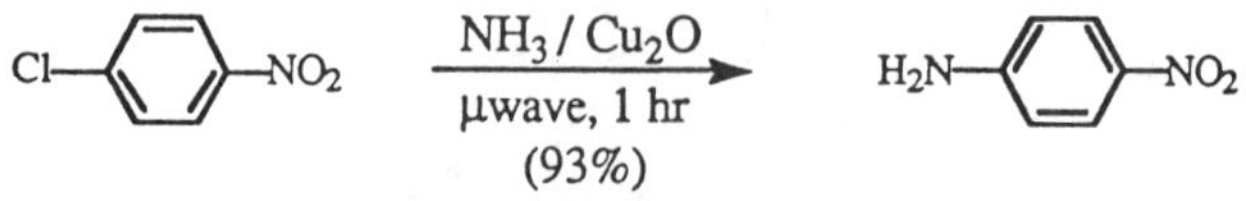

$$Cl\text{-}\langle\!\!\!\bigcirc\!\!\!\rangle\text{-}NO_2 \xrightarrow[\substack{\mu\text{wave, 1 hr} \\ (93\%)}]{NH_3 / Cu_2O} H_2N\text{-}\langle\!\!\!\bigcirc\!\!\!\rangle\text{-}NO_2$$

Scheme 69.

of the hydrogens. Perhaps the best known electrophilic aromatic substitution is the Friedel–Crafts alkylation reaction. Scheme 70 illustrates the Friedel–Crafts reaction of mesitylene and formaldehyde (*18*). This substitution is unusual in that the electrophilic species responsible for the formation of **70b** is the benzylic carbocation (**70a**), which results from addition of mesitylene to formaldehyde (an acylation) followed by loss of the aluminum-based leaving group.

Scheme 71 depicts an intramolecular Friedel–Crafts acylation to produce anthraquinone in 90% yield (*41*). Although this transformation is formally a substitution, it could also be regarded as a dehydration because a water molecule is lost during the acylation. The sulfonation of naphthalene is difficult to perform, even by using microwave techniques (Scheme 72) (*40*). However, with control of the temperature and pressure in the reaction vessel, this sulfonation takes place at 160 °C to provide sulfonic acids **72a** and **72b** in a ratio of 18.6:1. The preparation of substituted amines is often problematic. For example, 2-aminopyridine reacts with 2-bromopyridine in six min to give a tertiary amine in 20% yield (Scheme 73) by using microwave thermolysis (*42*). In contrast, conventional heating provided a 15% yield after refluxing for 16 h. Ligand exchange reactions can also be efficiently accelerated by using microwave irradiation, as shown in Scheme 74 (*43, 44*).

Microwave heating is especially useful for substitution reactions using radioactive atoms because even a small decrease in the reaction time can result in a substantial increase of the radiochemical yield. The exchange reaction of ^{131}I for iodine in Scheme 75 was achieved in a 90% yield when heated for a 5-min period or in a 78% yield when heated for only 1 min (*43, 44*). This result contrasts favorably with the percentage isotope exchange when using conventional heating (82%, 30 min at 135 °C). Similarly, in the examples shown in Schemes 76 and 77, ^{18}F can be substituted for the aryl nitro group.

One of the most difficult radioactive isotopes to incorporate into a molecule is ^{11}C, because of its short half-life ($t_{1/2}$ = 20 min). By using microwave heating, radioyields can be increased by 70–100%. In Stone-

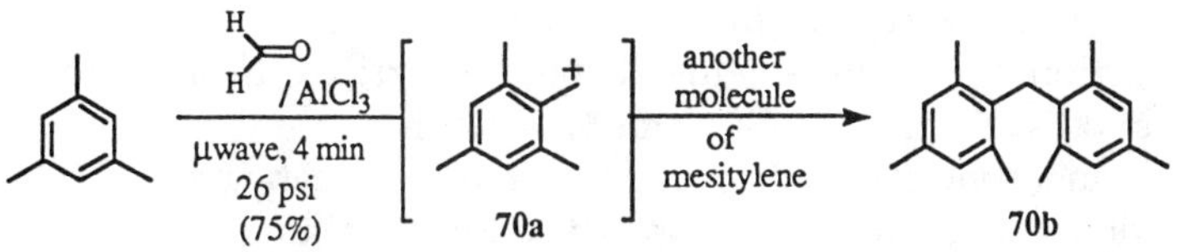
H
H
/ AlCl3
µwave, 4 min
26 psi
(75%)
70a
another
molecule
of
mesitylene
70b

Scheme 70*.

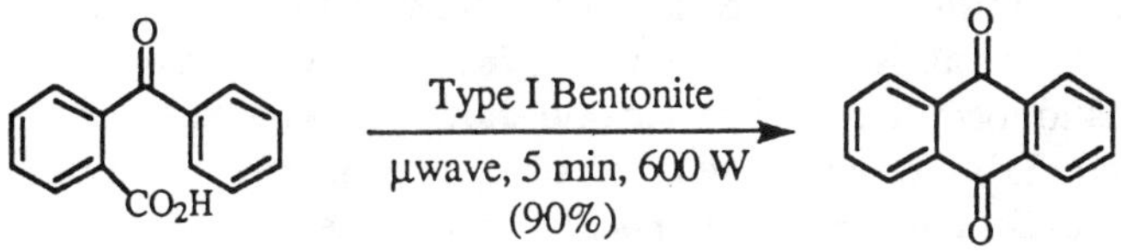
O
CO2H
Type I Bentonite
µwave, 5 min, 600 W
(90%)
O
O

Scheme 71.

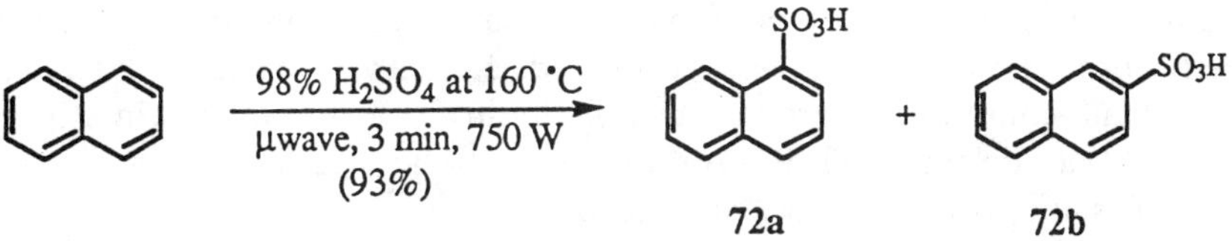
98% H2SO4 at 160 °C
µwave, 3 min, 750 W
(93%)
SO3H
+
SO3H
72a
72b

Scheme 72*.

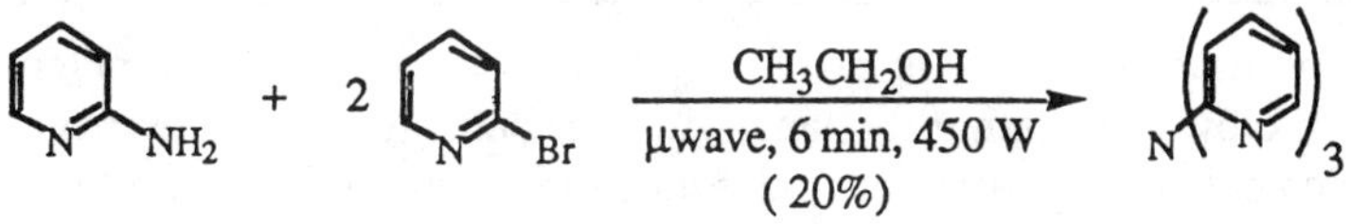
N
NH2
+ 2
N
Br
CH3CH2OH
µwave, 6 min, 450 W
(20%)
N
N
3

Scheme 73.

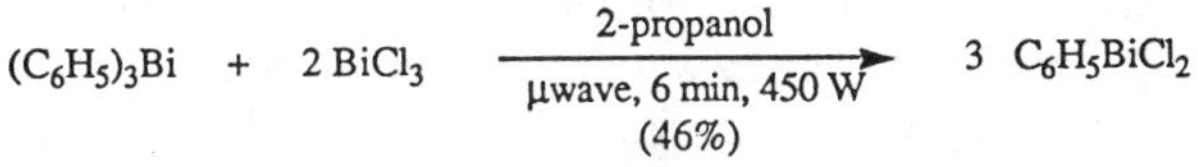
(C6H5)3Bi + 2 BiCl3
2-propanol
µwave, 6 min, 450 W
(46%)
3 C6H5BiCl2

Scheme 74.

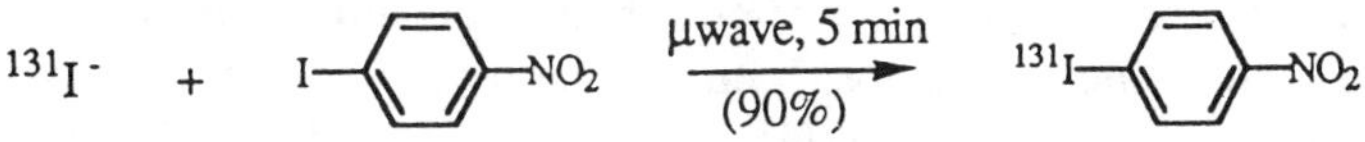
131I -
+
I
NO2
µwave, 5 min
(90%)
131I
NO2

Scheme 75.

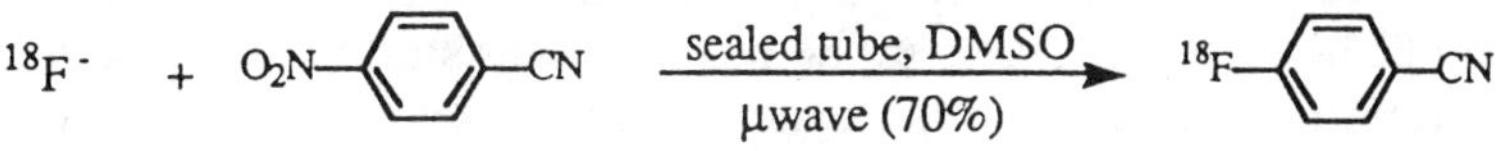
18F -
+ O2N
CN
sealed tube, DMSO
µwave (70%)
18F
CN

Scheme 76.

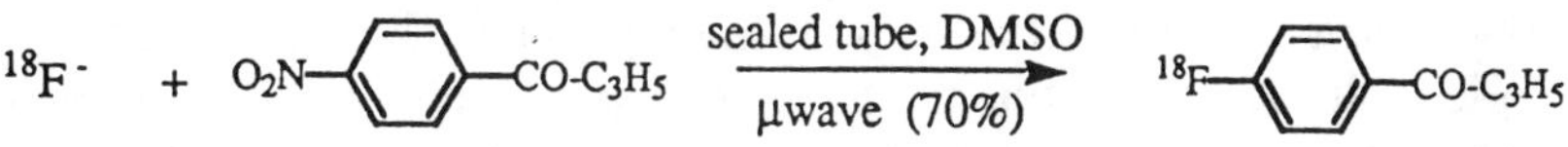
18F -
+ O2N
CO-C3H5
sealed tube, DMSO
µwave (70%)
18F
CO-C3H5

Scheme 77.

Elander and co-worker's (45) study, the microwave-accelerated substitution reaction shown in Scheme 78 can be carried out in methanol and water in 30 s and has a 60% yield. This result is comparable to the yield obtained in a control experiment, which requires heating at 90 °C by using conventional conditions for 7 min. The addition of unlabelled KCN to the reaction mixture has been reported to improve the radiochemical yield of nitrile **78a** to 74%. This result is curious because one would expect the radioactivity to decrease because of the competitive formation of nonradioactive material. Radioactive nitrile **78a** was hydrolyzed under acidic conditions to form the labelled lactone shown in Scheme 78.

Dimethyl dithioacetals are often used as protecting groups in organic synthesis. In Scheme 79, the two acidic methylene protons are substituted with sulfide groups to produce thioketal **79a**. This conversion was achieved in 94% yield by adsorbing the reagents on alumina and potassium fluoride, followed by brief microwave irradiation (46). By using copper(II) bromide supported on silica gel with the starting material dissolved in a small amount of chlorobenzene, the hydroxyl group of certain allylic alcohols can be replaced by bromine with allylic rearrangement. Three examples are provided in Schemes 80–82 (47).

Hydrolyses

Many organic compounds can be hydrolyzed quickly and efficiently by using microwave heating (Schemes 83–87). Benzamide was hydrolyzed to benzoic acid under aqueous acid conditions (Scheme 83). The hydrolysis of acetanilide gave good yields by using either acidic and basic catalysis (in aqueous methanol at 90 psi for 15 min, 91% yield, and for 45 min, 83% yield, respectively; Scheme 84) (18). The saponification of methyl benzoate required only 2.5 min of microwave irradiation (Scheme 85) (17). By using basic conditions, 2-cyanotoluene gave both the acid and amide products (**86a** and **86b** in a ratio of 1:19, respectively) (18). Finally, benzyl chloride is converted into benzyl alcohol as shown in Scheme 87.

Hydrolysis is particularly important for biochemical applications. For example, adenosine 5′-triphosphate can be hydrolyzed to adenosine 5′-diphosphate in the microwave (48, 49), and sucrose can be broken down into glucose and fructose (25, 37). Various types of starches can be hydrolyzed to produce 1,6-anhydro-b-D-glycopyranose (**88a**), albeit in low conversion (Scheme 88) (50). One of the first reactions studied by using microwave heating was the hydrolysis of amides by Gedye, Smith, and Westaway (Scheme 89). Ordinarily, this transformation requires more than 60 min at reflux to complete hydrolysis (90%) (13b). The amide linkage in peptide chains is easily cleaved by using aqueous HCl to give the hydrochloride salts of the constituent amino acids (18). Although three examples are presented in Schemes 90–92, other peptide and protein hydrolyses have been investigated more extensively (51–54).

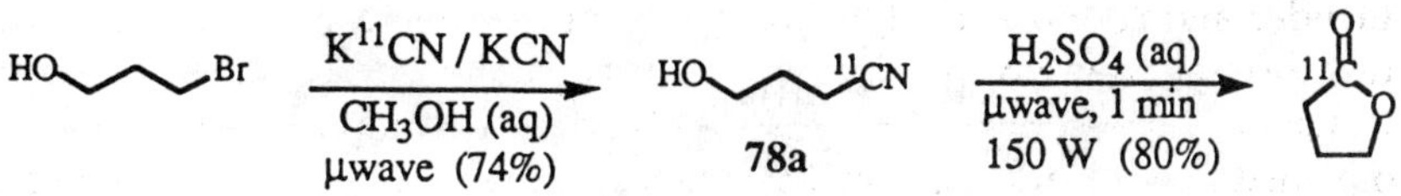

HO——Br $\xrightarrow[\substack{CH_3OH\ (aq) \\ \mu wave\ (74\%)}]{K^{11}CN\ /\ KCN}$ HO——^{11}CN $\xrightarrow[\substack{\mu wave,\ 1\ min \\ 150\ W\ (80\%)}]{H_2SO_4\ (aq)}$ (lactone with ^{11})

78a

Scheme 78.

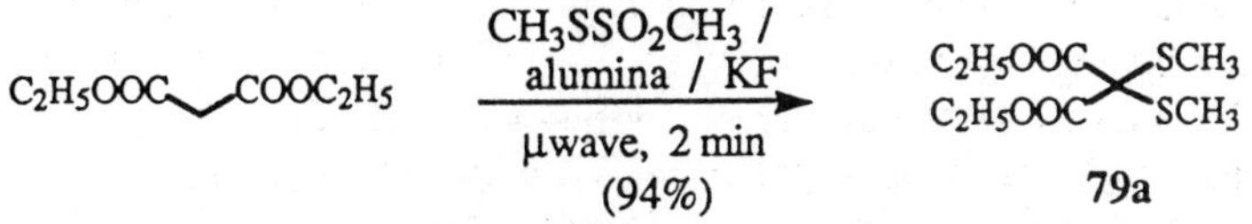

$C_2H_5OOC\diagdown\diagup COOC_2H_5$ $\xrightarrow[\substack{\mu wave,\ 2\ min \\ (94\%)}]{\substack{CH_3SSO_2CH_3\ / \\ alumina\ /\ KF}}$ $\substack{C_2H_5OOC \\ C_2H_5OOC}>\!\!<\substack{SCH_3 \\ SCH_3}$

79a

Scheme 79.

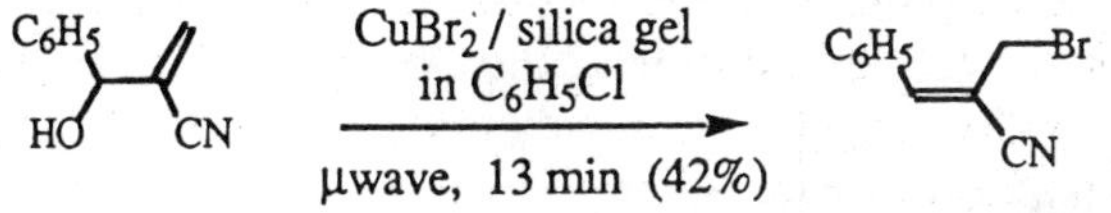

$\substack{C_6H_5 \\ \\ HO \diagdown CN}$ $\xrightarrow[\mu wave,\ 13\ min\ (42\%)]{\substack{CuBr_2\ /\ silica\ gel \\ in\ C_6H_5Cl}}$ C_6H_5—Br with CN

Scheme 80.

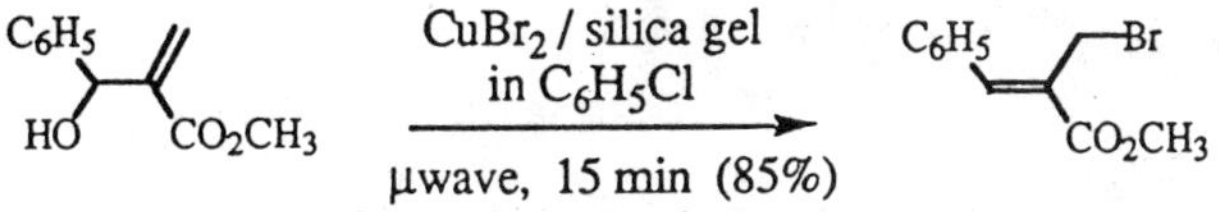

$\substack{C_6H_5 \\ \\ HO \diagdown CO_2CH_3}$ $\xrightarrow[\mu wave,\ 15\ min\ (85\%)]{\substack{CuBr_2\ /\ silica\ gel \\ in\ C_6H_5Cl}}$ C_6H_5—Br with CO_2CH_3

Scheme 81.

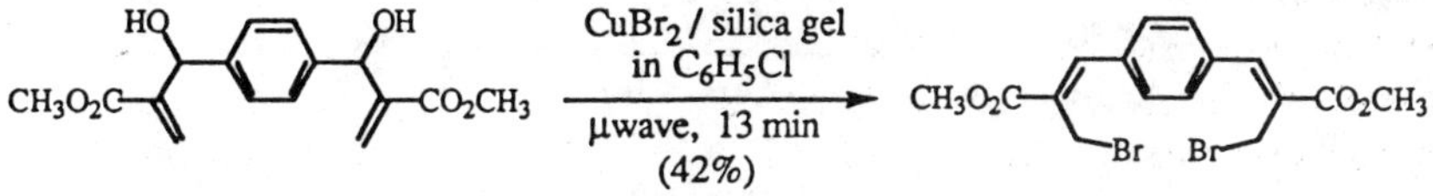

$\substack{HO \qquad\qquad OH \\ CH_3O_2C-\!\!\!\diagup\diagdown\!\!\!-C_6H_4-\!\!\!\diagup\diagdown\!\!\!-CO_2CH_3}$ $\xrightarrow[\substack{\mu wave,\ 13\ min \\ (42\%)}]{\substack{CuBr_2\ /\ silica\ gel \\ in\ C_6H_5Cl}}$ $CH_3O_2C\diagup\diagdown C_6H_4 \diagup\diagdown CO_2CH_3$ (Br Br)

Scheme 82.

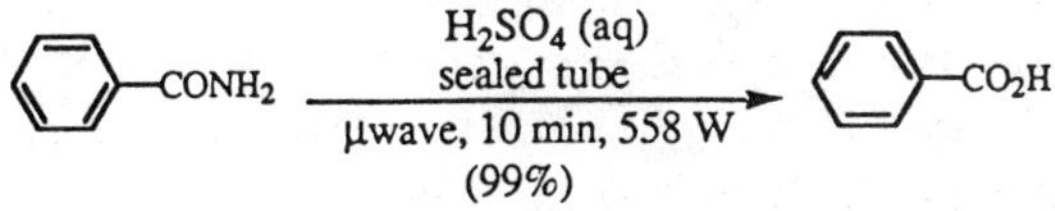

(C₆H₅)—$CONH_2$ $\xrightarrow[\substack{\mu wave,\ 10\ min,\ 558\ W \\ (99\%)}]{\substack{H_2SO_4\ (aq) \\ sealed\ tube}}$ (C₆H₅)—CO_2H

Scheme 83.

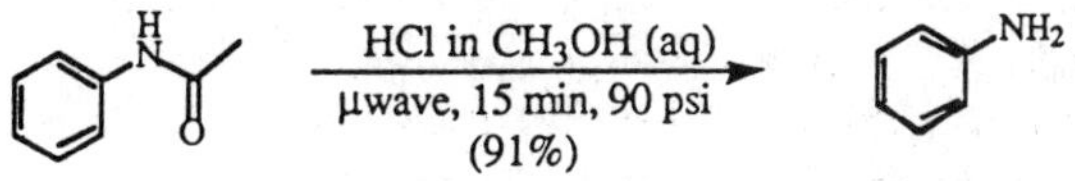

(C₆H₅)—$\overset{H}{N}\!\!-\!\!C(=O)CH_3$ $\xrightarrow[\substack{\mu wave,\ 15\ min,\ 90\ psi \\ (91\%)}]{HCl\ in\ CH_3OH\ (aq)}$ (C₆H₅)—NH_2

Scheme 84*.

(C₆H₅)—CO_2CH_3 $\xrightarrow[\substack{\mu wave,\ 2.5\ min \\ (84\%)}]{NaOH\ (aq)}$ (C₆H₅)—CO_2H

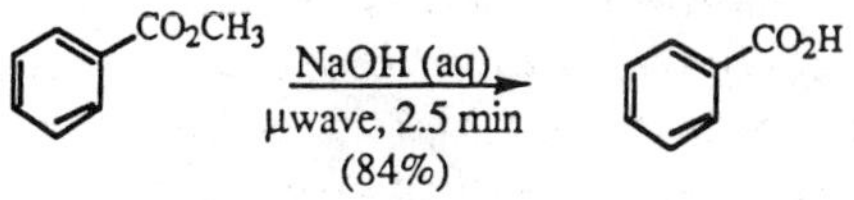

Scheme 85.

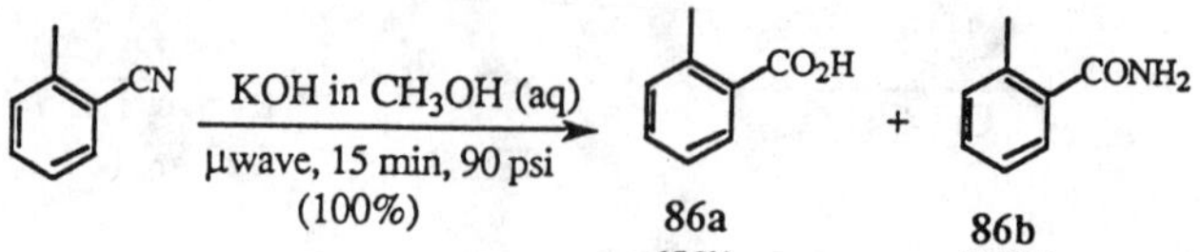

86a (5%) 86b (95%)

Scheme 86*.

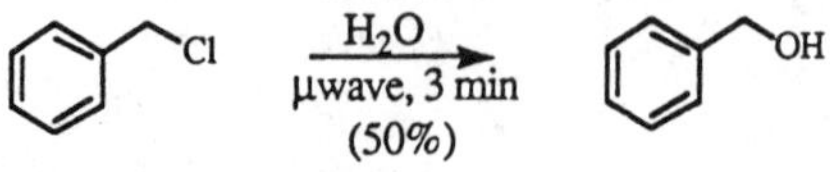

Scheme 87.

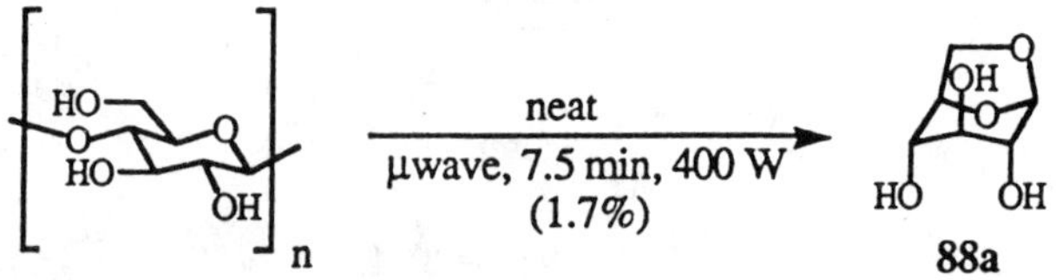

88a

Scheme 88.

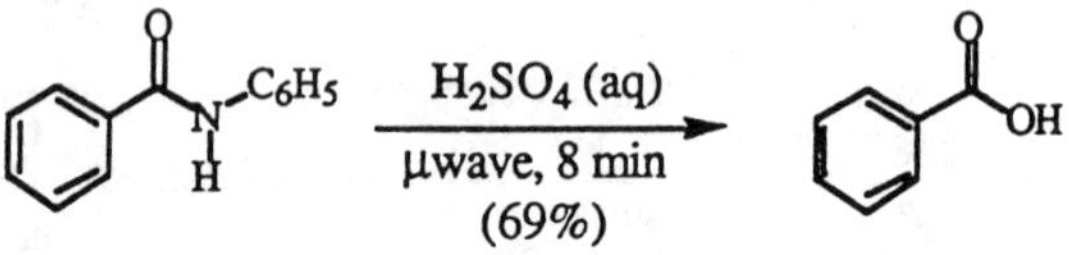

Scheme 89.

glyclglyclleucine —[HCl (aq); µwave, 15 min, 90 psi; (98%)]→ 2 glycine-HCl + leucine-HCl

Scheme 90*.

glyclglyclglycine —[HCl (aq); µwave, 3.5 hrs, 70 psi; (89%)]→ 3 glycine-HCl

Scheme 91*.

leucylleucine —[HCl (aq); µwave, 30 min, 70 psi; (98%)]→ 2 leucine-HCl

Scheme 92*.

Another important biochemical reaction that can be carried out in the microwave is the racemization of an amino acid chiral center. Chen, Wu, and Wang (*55*) racemized a number of amino acids in the microwave by heating them in the presence of acetic acid and benzaldehyde under nitrogen. All of the amino acids studied racemized completely. Indeed, L-alanine was racemized in 2 min with 100% yield (Scheme 93). These racemizations were also carried out by using Wang's continuous flow microwave apparatus (*23, 37*).

Other chiral centers in natural products have also been racemized by using microwave heating. For example, (–)-vincadifformine was converted to a racemic mixture in the microwave within 20 min (Scheme 94) (*56*). This racemization proceeds via a retro-Diels–Alder reaction (cf. **94a**), followed by an intramolecular Diels–Alder reaction.

Deprotections

The removal of various protecting groups on alcohols and carboxylic acids is often cleanly and easily accomplished in a microwave oven. The demethylations shown in Schemes 95 and 96 were both carried out in aqueous hydrobromic acid. The first reaction gave a 46% yield of phenol **95a**, whereas the second reaction gave a 96% yield of catechol **96a** (*18*). Both yields were about the same as those when demethylation was carried out conventionally; however, the reaction times were reduced from 52 h to 15 min for the deprotection shown in Scheme 95 and from 3 h to 5 min for the preparation of **96a** in Scheme 96.

Maat and co-workers (*28, 29*) used a microwave oven modified for reflux at atmospheric pressure to remove the protecting groups in their studies of opium alkaloids. The demethylation reaction in Scheme 97 was accomplished in 5 h by alternately heating the reaction for 15 min and then allowing the reaction to cool for an equal period of time. (The reason for this procedure is unclear.) The deprotection reaction shown in Scheme 98 is noteworthy because only the less sterically congested ether is demethylated (*57*). This reaction was carried out under basic conditions by using 1,2-ethanedithiol as both solvent and as the source of the nucleophilic species ($KSCH_2CH_2SH$). Most of the deacetylation reactions run in the microwave require the reactants to be adsorbed on alumina. This technique is especially appealing for deprotection reactions, because these reactions are usually clean and the resulting workup is minimal. For example, phenyl acetate was converted in 88% yield to phenol under these conditions (Scheme 99), whereas cyclohexyl acetate gave a 74% yield of cyclohexanol (Scheme 100) (*58*). Varma and co-workers (*59–61*) found that acetyl groups can be selectively removed from a compound by controlling the time of irradiation (Scheme 101). In the reaction after 30 s in the microwave only the aryl acetate group is lost; after 2.5 min of irradiation, however, both protecting groups are lost (in 92% yield) (*59*).

L-alanine —(acetic acid / benzaldyde; µwave, 2 min (100%))→ D,L-alanine

Scheme 93.

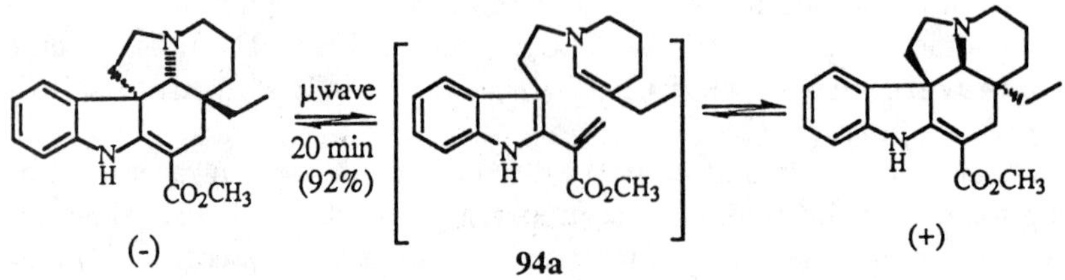

94a

Scheme 94.

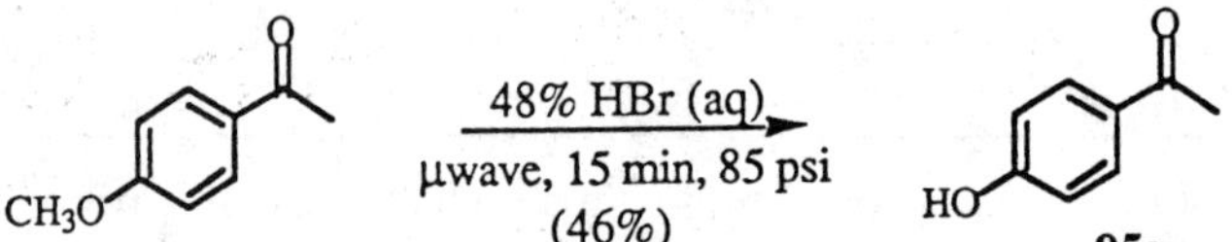

95a

Scheme 95*.

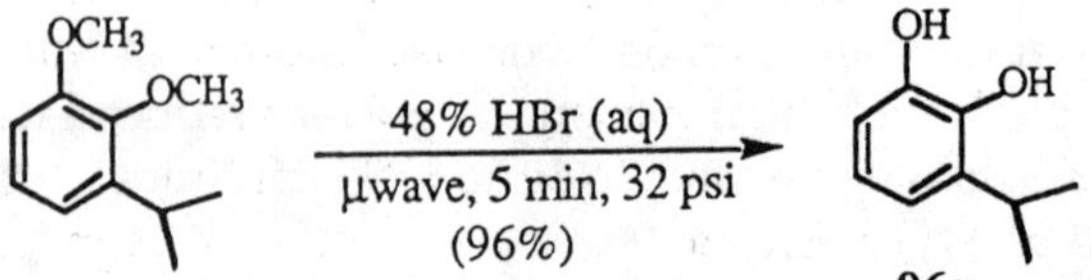

96a

Scheme 96*.

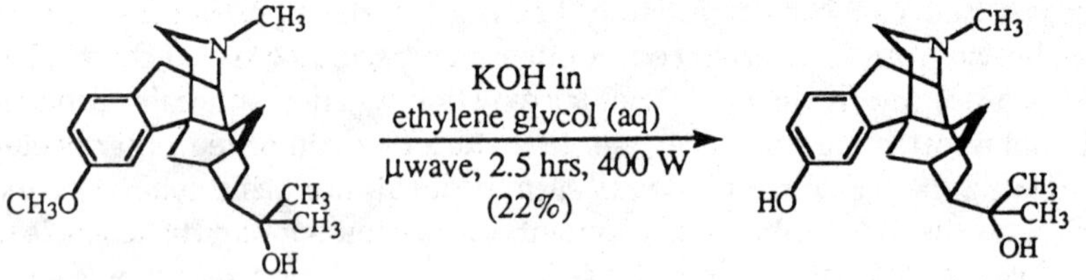

Scheme 97.

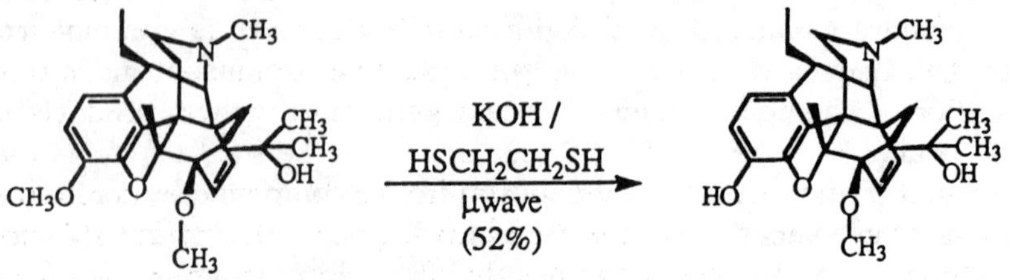

Scheme 98.

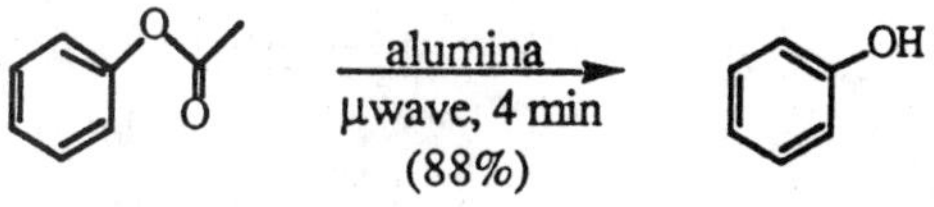

Scheme 99.

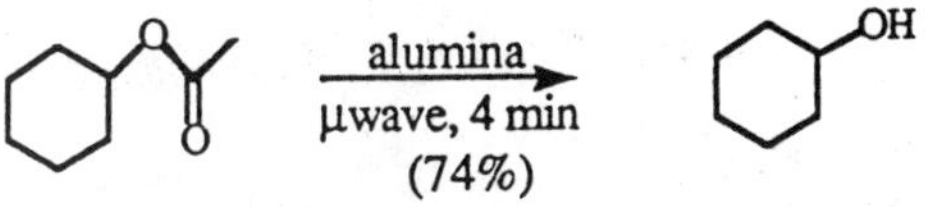

Scheme 100.

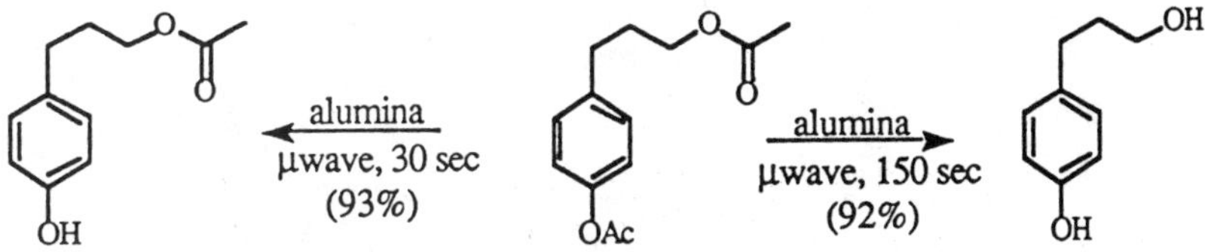

Scheme 101.

Pivaloyl esters have been removed in the same fashion as acetates (Schemes 102 and 103) (*58*). As with demethylation, when the starting material contains multiple protected hydroxyl groups, the most accessible ester is most easily removed. Benzaldehyde diacetates are also easily deprotected, the simplest example giving 98% yield in 40 s (Scheme 104) (*60*). Substituted arylaldehyde diacetates also give high yields of the unprotected aldehyde in less than 1 min. Note that an acetal unit is more easily removed in comparison with an acetyl group, as shown in Scheme 105. After 30 s of reaction, 4-acetoxybenzaldehyde (**105a**) is formed in 97% yield; in another 90 s, only 4-hydroxybenzaldehyde (**105b**) is present. Carboxylic acids are often protected as benzyl esters, which have been hydrolyzed by microwave heating by adsorbing the starting materials on acidic or neutral alumina prior to heating. The deprotection of benzyl benzoate and benzyl cinnamate occurs easily on acidic alumina (Schemes 106 and 107), whereas, deprotection of glycine benzyl ester occurs best when neutral alumina is used (Scheme 108) (*61*).

Benzyl protecting groups are routinely removed by hydrogenation. The lactam shown in Scheme 109 was added to palladium on carbon in ammonium formate and ethylene glycol (*62*). This mixture was then irradiated in an open vessel at a low power setting (to prevent boiling) for 2 min. Under these conditions, the deprotected alcohol was produced in 90% yield.

Another common protecting group that has been efficiently removed by using microwave heating is the *t*-butyldimethylsilyl group. This protecting group is most often removed by using fluoride ion or mild aqueous acid. In the microwave oven, adsorbing the starting ether on alumina,

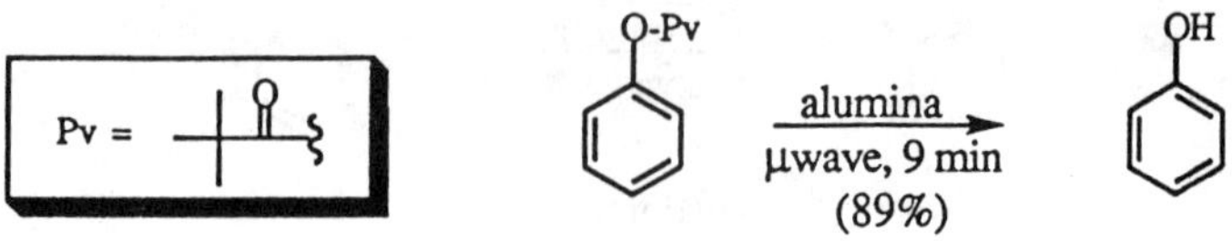

$$Pv = \text{(structure)}$$

Scheme 102.

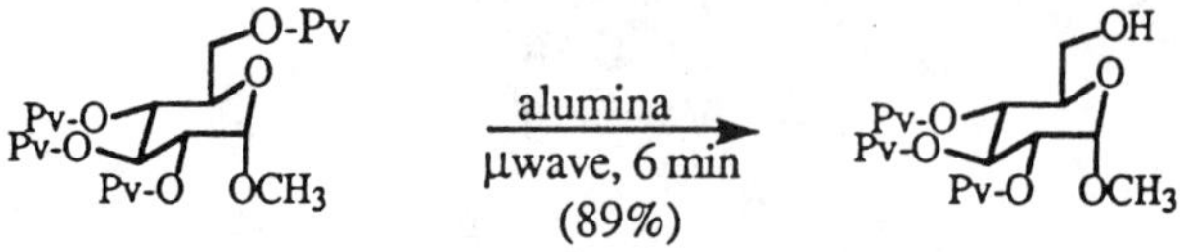

Scheme 103.

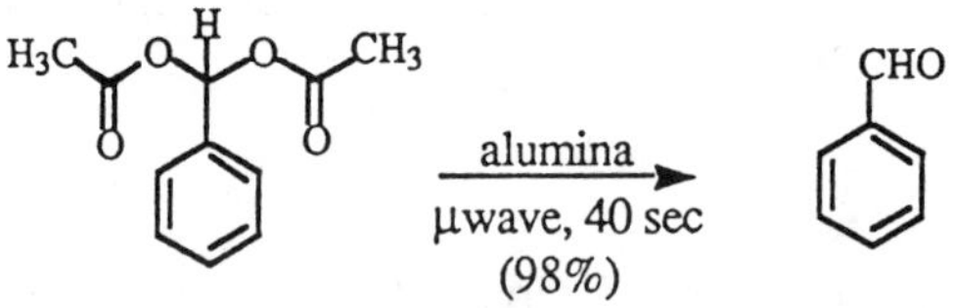

Scheme 104.

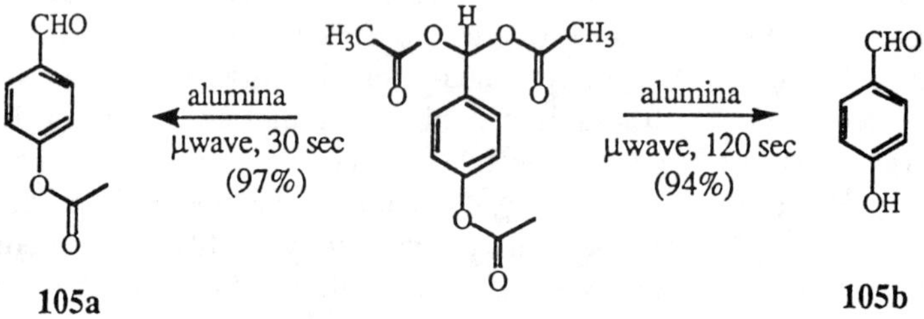

105a

105b

Scheme 105.

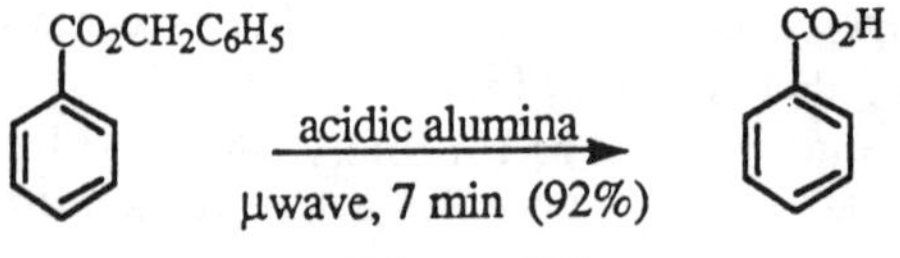

Scheme 106.

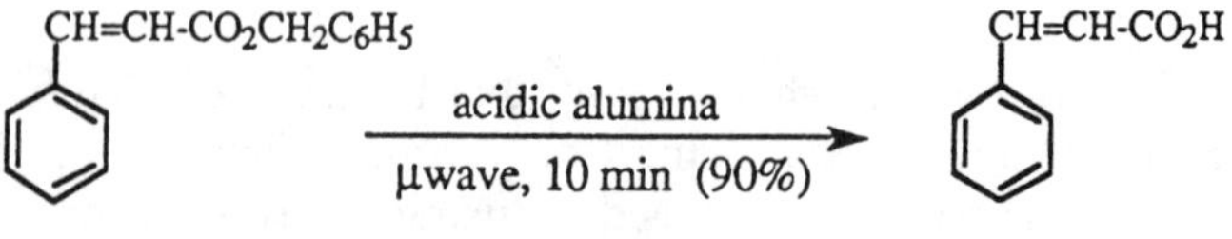

Scheme 107.

$$NH_2CH_2CO_2CH_2C_6H_5 \xrightarrow[\mu wave,\ 4\ min\ (95\%)]{\text{neutral alumina}} NH_2CH_2COOH$$

Scheme 108.

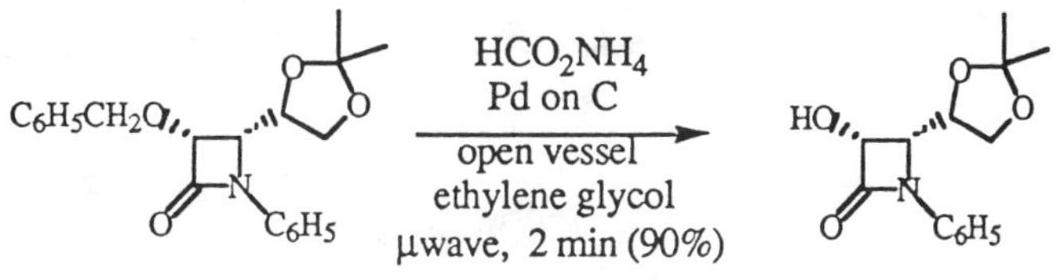

Scheme 109.

followed by irradiation, gives the alcohol moiety. For example, by using these conditions benzyl-*t*-butyldimethylsilyl ether was deprotected in 11 min in 82% yield (Scheme 110) (*63*). The bis-ether shown in Scheme 111 was also deprotected efficiently. (The singly deprotected species **111b** should be seen with shorter irradiation times, but was not reported.)

Dehydrations

In theory, microwave dehydration reactions should be difficult to drive to completion because the water cannot be easily removed as it is formed. Nevertheless, some dehydrations work well when carried out in the presence of molecular sieves and silica gel, which absorb the water as it is generated. Dehydration of the alcohol in Scheme 112 gives butenolide **112a** in 83% yield (*64*). In contrast, the microwave dehydration of 2-methylcyclohexanol adsorbed on molecular sieves occurs in low yield (Scheme 113) and gives a mixture of two of the three possible dehydration products (*32*). In the dehydration of tertiary alcohol **114a**, α-methylstyrene (**114b**) is formed in 98% yield after brief heating (Scheme 114). A longer irradiation time gives the dimerization product **114c** (*3*).

Aldehydes can be converted into nitriles by using hydroxylamine hydrochloride adsorbed on Mexican bentonite (*65*). The first step in this process is the acid-catalyzed formation of an oxime, which then undergoes catalyst-promoted dehydration. Yields for the reactions shown in Schemes 115 and 116 are 60% and 50%, respectively. However, if benzaldehyde is first adsorbed on alumina, and potassium fluoride and hydroxylamine hydrochloride are then added, heating the resultant mixture results in the formation of an oxime. This intermediate was treated in situ at room temperature with carbon disulfide to obtain the desired nitrile in high yield (*66*).

Cyclic anhydrides can be formed by means of microwave-promoted dehydration. The conversion of 1,2-cyclohexanedicarboxylic acid to its anhydride is swift when the diacid is adsorbed on montmorillonite KSF

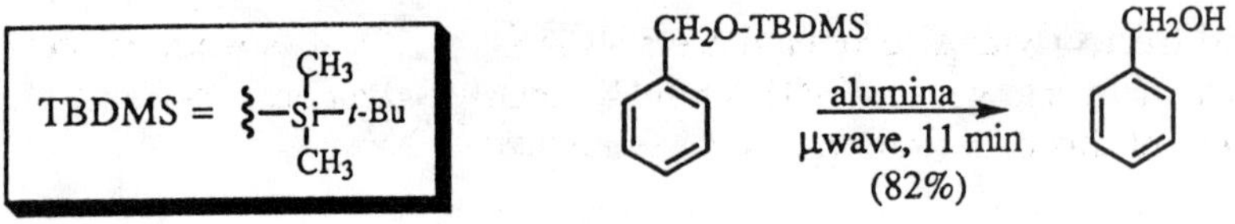

$$TBDMS = \{-\underset{\underset{CH_3}{|}}{\overset{\overset{CH_3}{|}}{Si}}-t\text{-Bu}$$

CH$_2$O-TBDMS

$\xrightarrow[\mu\text{wave, 11 min}]{\text{alumina}}$
(82%)

CH$_2$OH

Scheme 110.

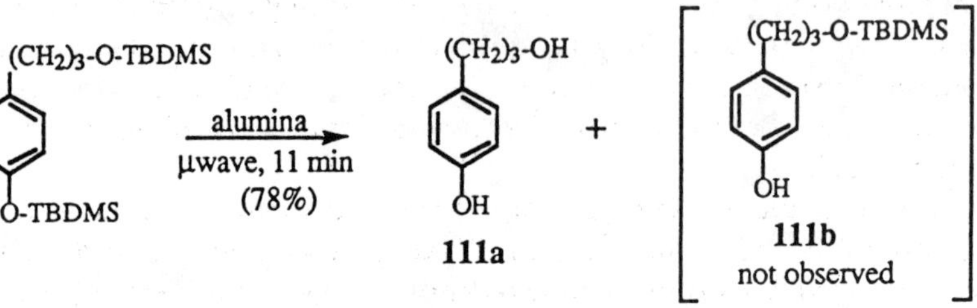

(CH$_2$)$_3$-O-TBDMS

O-TBDMS

$\xrightarrow[\mu\text{wave, 11 min}]{\text{alumina}}$
(78%)

(CH$_2$)$_3$-OH

OH

111a

+

$\left[\begin{array}{c}\text{(CH}_2\text{)}_3\text{-O-TBDMS}\\ \\ \text{OH}\\ \textbf{111b}\\ \text{not observed}\end{array}\right]$

Scheme 111.

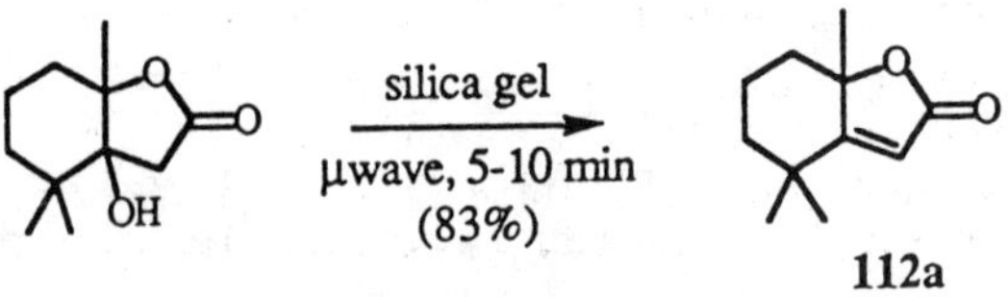

$\xrightarrow[\mu\text{wave, 5-10 min}]{\text{silica gel}}$
(83%)

112a

Scheme 112.

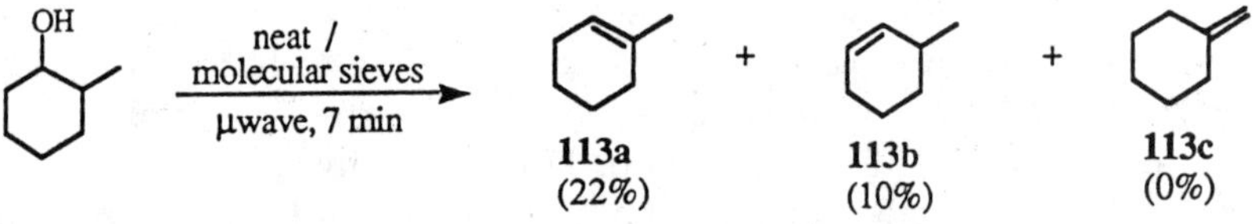

OH

$\xrightarrow[\mu\text{wave, 7 min}]{\text{neat / molecular sieves}}$

113a
(22%)

+

113b
(10%)

+

113c
(0%)

Scheme 113.

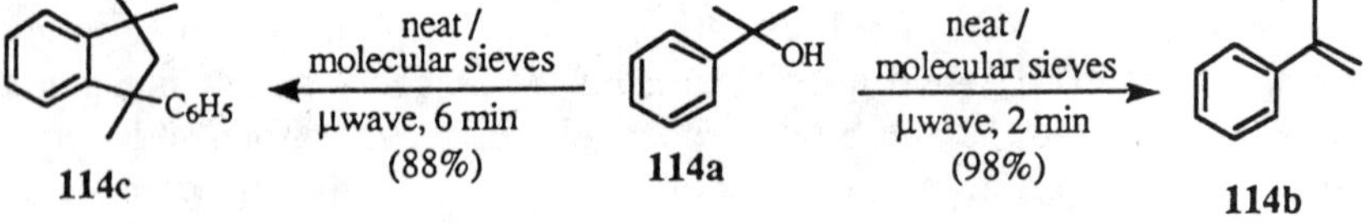

114c

C$_6$H$_5$

$\xleftarrow[\substack{\mu\text{wave, 6 min}\\(88\%)}]{\text{neat / molecular sieves}}$

OH

114a

$\xrightarrow[\substack{\mu\text{wave, 2 min}\\(98\%)}]{\text{neat / molecular sieves}}$

114b

Scheme 114.

CH=CH-CHO

$\xrightarrow[\mu\text{wave, 15 min (60\%)}]{\text{NH}_2\text{OH·HCl / Mexican bentonite}}$

CH=CH-CN

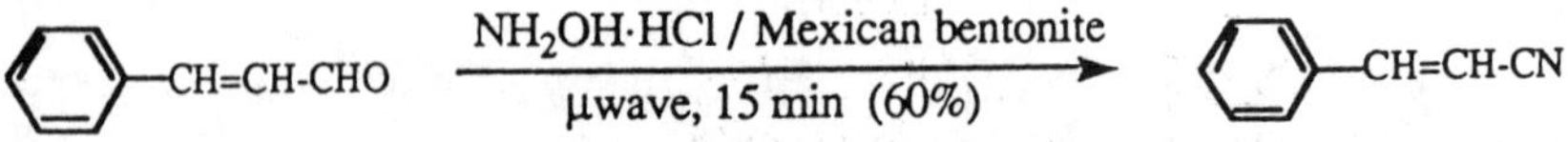

Scheme 115.

and treated with isopropenyl acetate (Scheme 117). This reaction occurs in 94% yield. However, if PTSA is used instead of isopropenyl acetate, the yield of this dehydration is lowered to 72% (*67*).

Oxidations and Reductions

Oxidations were among the first reactions studied in the microwave, and they often occur with yields comparable to or higher than their classical counterparts. The oxidation of toluene by using potassium permanganate (Scheme 118) produces benzoic acid in 40% yield, the same as that obtained under reflux conditions, but in 5 min of heating rather than 25 min (*1, 17*).

Most oxidations occur quite rapidly even at room temperature, and as such do not benefit from microwave heating. However, manganese dioxide is a selective oxidant whose reactions usually require long reaction times; therefore, these oxidations are good candidates for microwave heating. Several examples are shown subsequently (*18*). Diethyl ether, the solvent used in the oxidations, does not absorb microwaves; hence, a special ferrite–quartz vessel (Figure 1; 50-mL volume) was used that raised the solvent temperature by convection. Although the conversion of benzyl alcohol to benzaldehyde is trivial (Scheme 119), microwave heating shortens the reaction time from 8 h to 7 min. The oxidation of cinnamyl alcohol (Scheme 120) and the selective oxidation of the allylic alcohol in Scheme 121 are also expedited by using microwave irradiation. Alvarez and co-

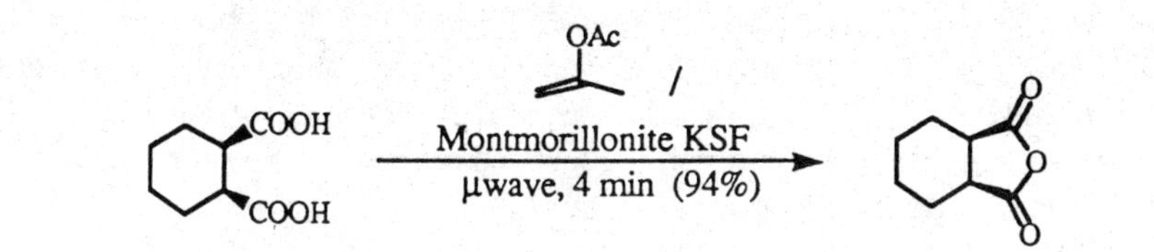

Scheme 116.

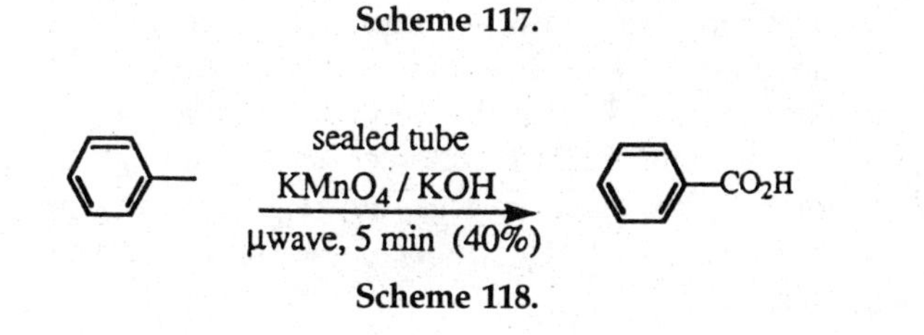

Scheme 117.

Scheme 118.

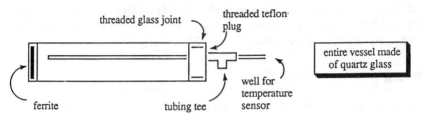

Figure 1. Ferrite-quartz vessel.

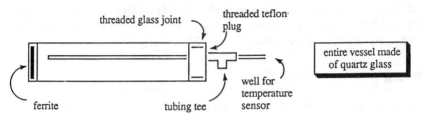

Scheme 119*.

Scheme 120*.

Scheme 121*.

workers (*68*) have also studied the oxidations shown in Schemes 119 and 120 by using manganese dioxide and bentonite in the absence of solvent (*68*). Oxidation of functionalized benzyl alcohols like *p*-methoxybenzyl alcohol and *m*-nitrobenzyl alcohol were achieved in nearly quantitative yield by using identical conditions.

In organic chemistry an oxidation is defined as a reaction that either decreases the hydrogen content of a molecule or increases its oxygen, nitrogen, or halogen content. The oxidation of Hantzsch dihydropyridines by using manganese dioxide absorbed on bentonite is an example of the first definition (cf. Scheme 122). In Scheme 123, oxidation not only generates the pyridine ring but also removes the isopropyl substituent (*69, 70*). However, in Scheme 124 both the 4-substituted and unsubstituted pyridines are formed.

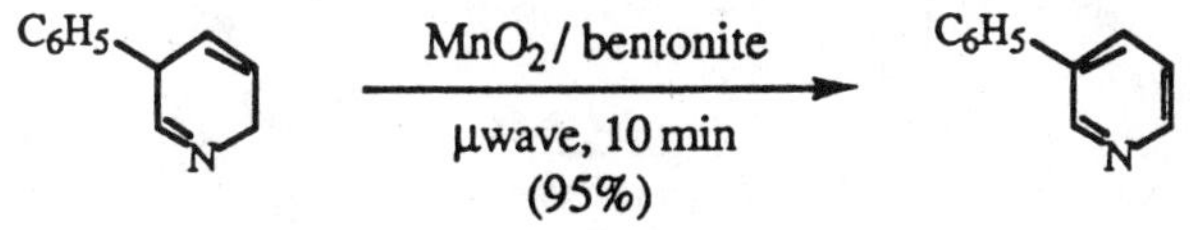

Scheme 122.

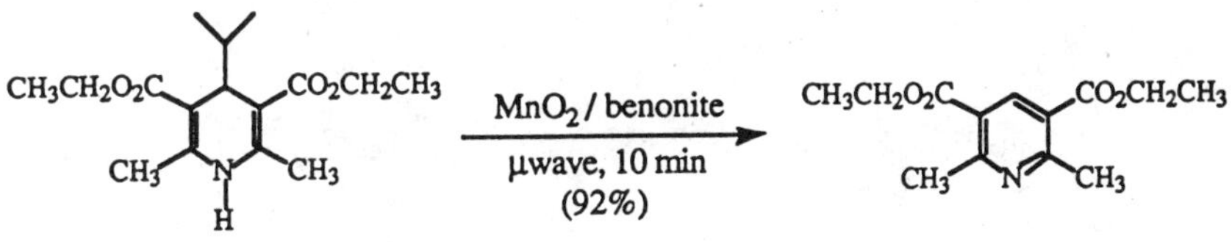

Scheme 123.

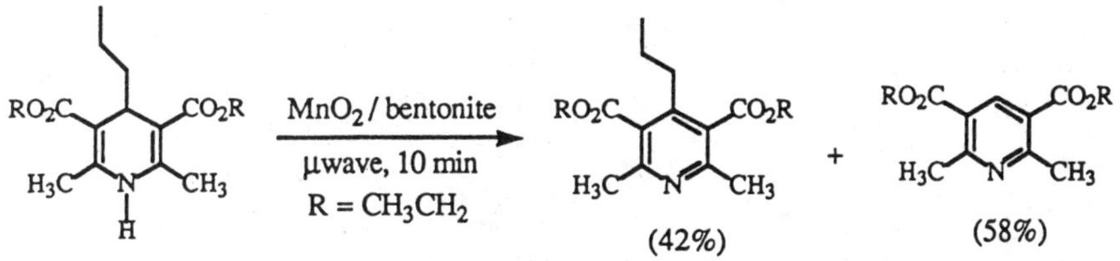

Scheme 124.

Scheme 125 details the oxidation of cyclohexadiene **125a** by using γ-Zeolite and copper oxide to produce *p*-cymene (*32*). The generality of this aromatization process remains to be determined. Conventional *meta*-chloroperbenzoic acid (*m*-CPBA) epoxidations take place at 0 °C but can often require long reaction times (*38*). By using microwave irradiation, however, epoxides are formed rapidly and in excellent yield (Scheme 126). The generality of this microwave-promoted procedure also needs to be determined more fully.

Hydrogenations can be performed in microwave ovens. Gordon and co-workers (*71*) observed that benzaldehyde is reduced to benzyl alcohol by using $Ru[PPh_3]_3(CO)–HCl$ as catalyst and formic acid, which serves as the hydrogen donor (Scheme 127). After heating for 16 min, the major product is formate **127b**, although a small amount of benzyl alcohol (**127a**) is also formed (4%). Esterification of the reduction product proceeds rapidly in the microwave, whereas conventional heating gives equal amounts of both **127a** and **127b**.

Microwave-accelerated hydrogenolysis was accomplished in 1991 by Bose and co-workers (*24*). Their procedure uses palladium on carbon as the reduction catalyst and ammonium formate as the reducing agent. Not only was the vinyl group in lactam **128a** reduced, but the benzylic C–N bond was also cleaved (Scheme 128). No yield was given although the "reaction was complete".

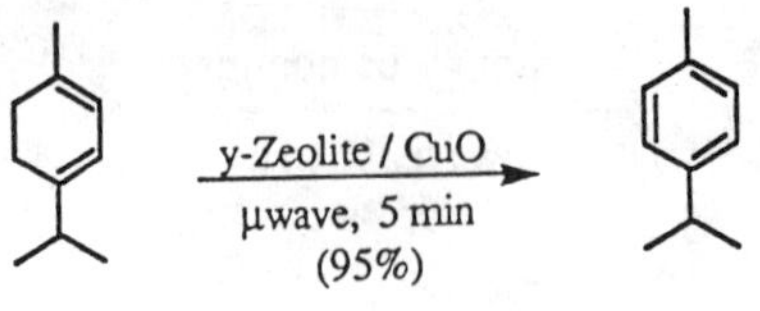
y-Zeolite / CuO
µwave, 5 min
(95%)

125a

Scheme 125.

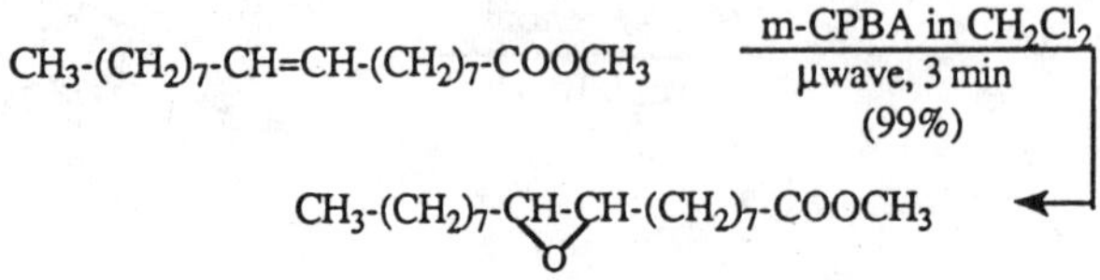
CH3-(CH2)7-CH=CH-(CH2)7-COOCH3
m-CPBA in CH2Cl2
µwave, 3 min
(99%)
CH3-(CH2)7-CH-CH-(CH2)7-COOCH3
O

Scheme 126.

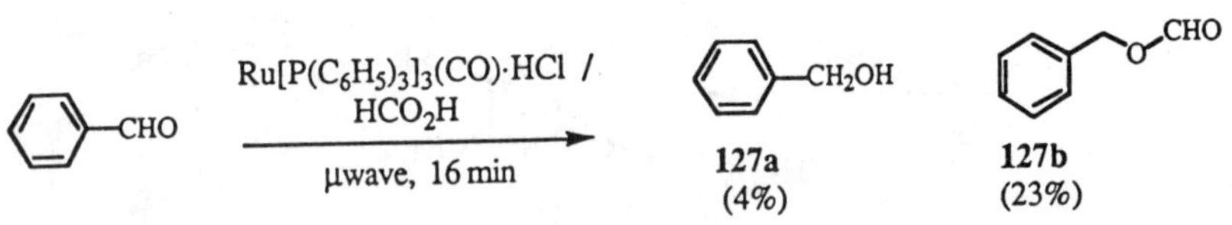
CHO
Ru[P(C6H5)3]3(CO)·HCl /
HCO2H
µwave, 16 min
CH2OH
127a
(4%)
O
CHO
127b
(23%)

Scheme 127.

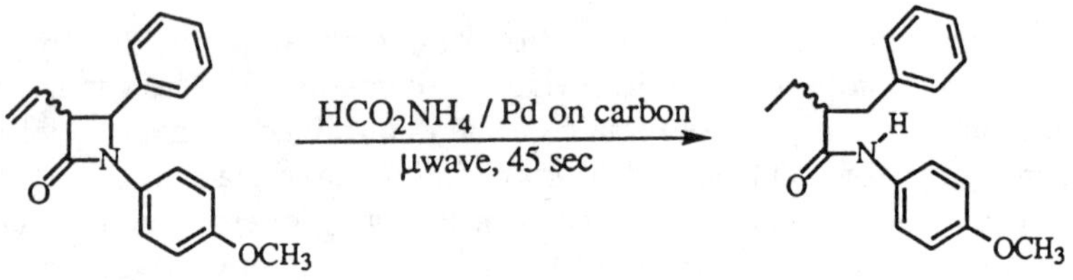
N
O
OCH3
HCO2NH4 / Pd on carbon
µwave, 45 sec
H
N
O
OCH3

128a

Scheme 128.

Condensation Reactions

Condensation reactions between aldehydes and activated methylenes were studied extensively by Villemin and Labiad (Schemes 129–137) (*72–74*). The starting materials are usually adsorbed on either alumina or Montmorillonite KSF clay to absorb the water produced in the course of the reaction, which forces these condensations toward completion. Many of the examples shown in Schemes 129–137 are "name reactions" or have interesting applications. For instance, the condensations shown in Schemes 132 and 133 are Knorr condensations, which produce pyrazolones **132a** and **133b**, precursors for antipyretic or analgesic drugs (*75*). Finally, Loupy and co-workers reported the condensation between benzaldehyde and pyrrole in their synthesis of *meso*-5,10,15,20-tetraphenyl-21*H*,23*H*-porphyrin

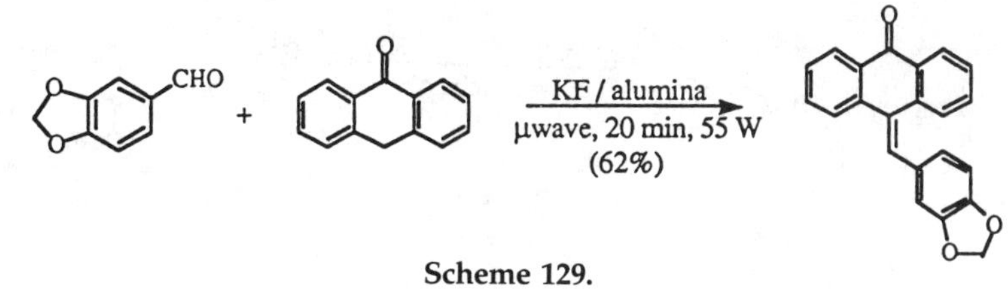

Scheme 129.

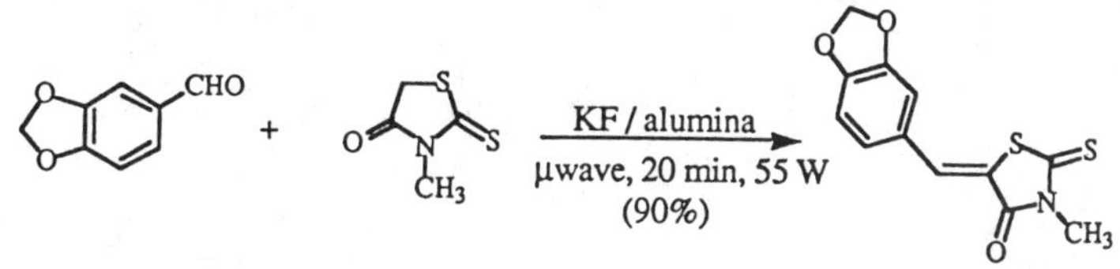

Scheme 130.

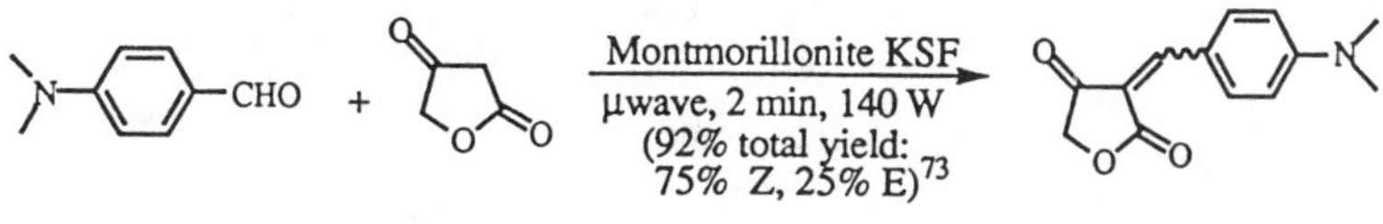

Scheme 131.

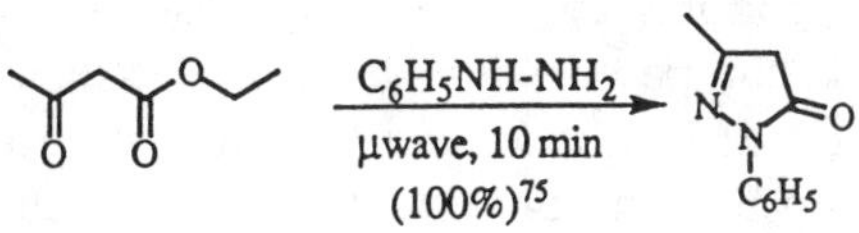

Scheme 132.

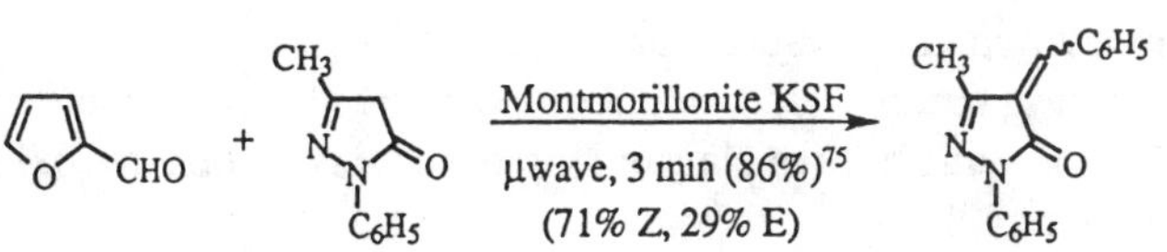

133a **133b**

Scheme 133.

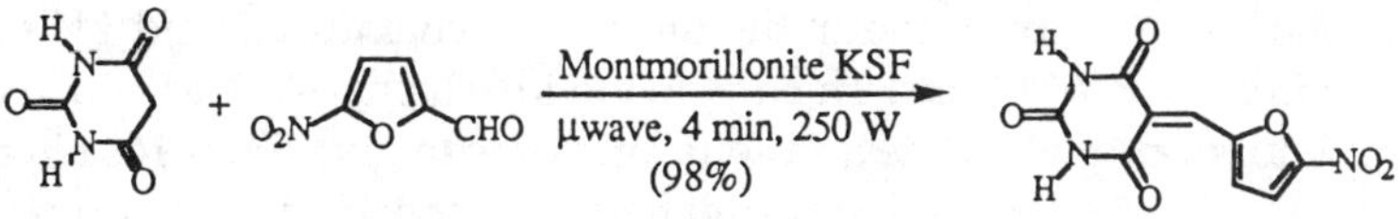

Scheme 134.

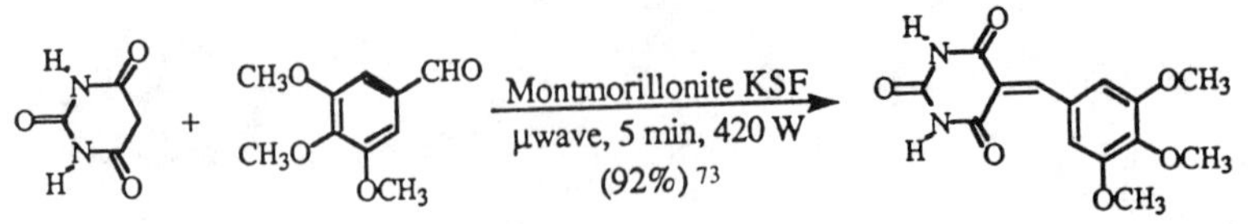

Scheme 135.

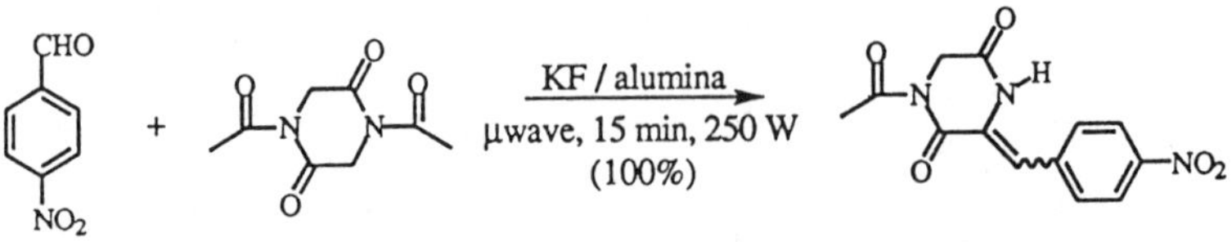

Scheme 136.

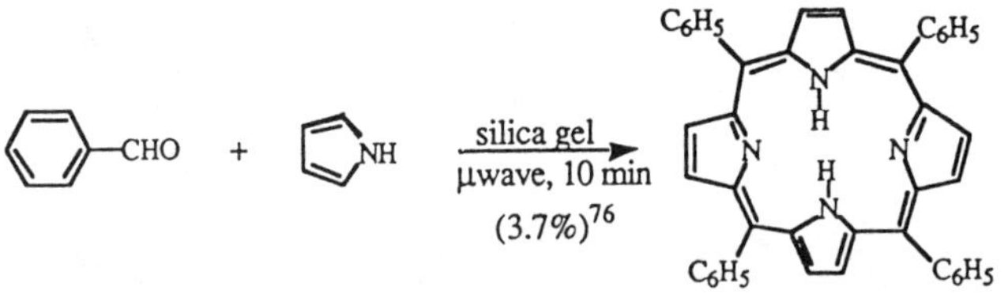

Scheme 137.

(Scheme 137), a sensitizer for the formation of singlet oxygen. Although the yields of porphyrins by using this technique are extremely low, the absence of other by-products and the ease of subsequent purification make this procedure viable compared with classical procedures (76). Knoevenagel condensations occur in good yield by using microwave irradiation (Schemes 138–140). In Scheme 140, the coumarin product is formed as a result of a lactonization of intermediate **140a**, the Knoevenagel product. Peptide bonds have been prepared in microwave ovens. Vazquez-Tato (79) found that amines couple with carboxylic acids in high yield in less than 5 min when catalyzed by acidic clay (Scheme 141). In the course of their pioneering studies, Gedye and co-workers (1) found that treatment of benzophenone with hydroxylamine gave the corresponding oxime in 71% yield (Scheme 142).

Heterocyclic Ring Formation

A number of examples of the formation of heterocyclic rings under microwave conditions can be found in the literature. Schemes 143 and 144 focus on the preparation of furans. The diepoxide in Scheme 143 was cyclized in the presence of propyl iodide and sodium iodide to give furan **143a** in 88% yield (42). (The role of the iodopropane is not specified or obvious.) By using classical heating conditions, this reaction takes 5 h and gives only a

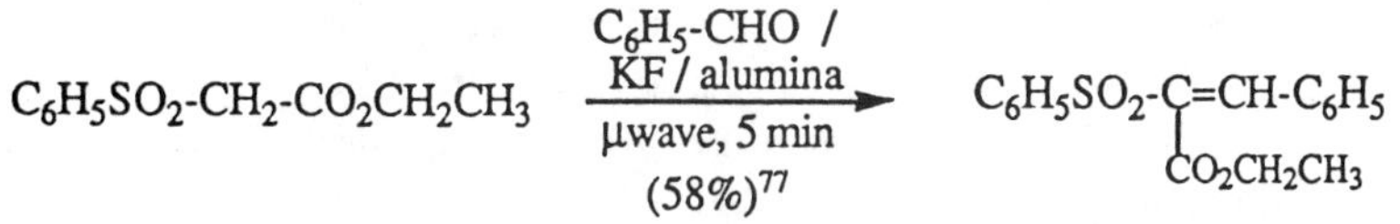

$$C_6H_5SO_2\text{-}CH_2\text{-}CO_2CH_2CH_3 \xrightarrow[\substack{\mu wave,\ 5\ min \\ (58\%)^{77}}]{\substack{C_6H_5\text{-}CHO\ / \\ KF\ /\ alumina}} C_6H_5SO_2\text{-}C{=}CH\text{-}C_6H_5$$

with $CO_2CH_2CH_3$ below the central carbon.

Scheme 138.

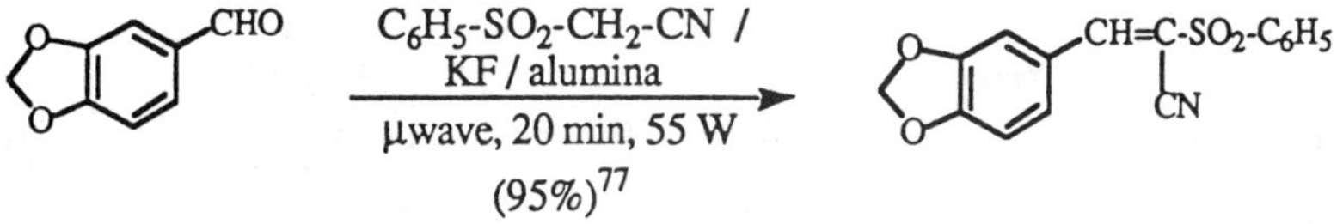

$$\xrightarrow[\substack{\mu wave,\ 20\ min,\ 55\ W \\ (95\%)^{77}}]{\substack{C_6H_5\text{-}SO_2\text{-}CH_2\text{-}CN\ / \\ KF\ /\ alumina}}$$

Scheme 139.

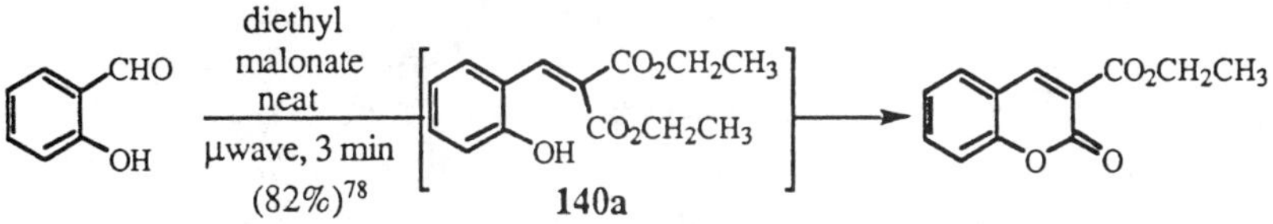

diethyl malonate
neat

$\xrightarrow[\substack{\mu wave,\ 3\ min \\ (82\%)^{78}}]{}$

140a

Scheme 140.

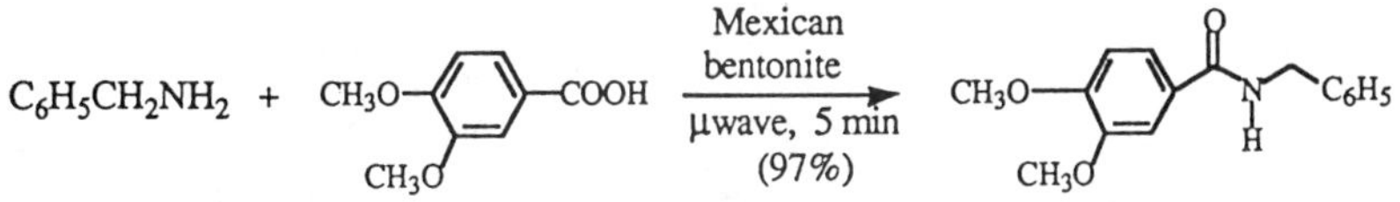

$$C_6H_5CH_2NH_2\ +\ CH_3O\!\!-\!\!\bigcirc\!\!-\!\!COOH \xrightarrow[\substack{\mu wave,\ 5\ min \\ (97\%)}]{\substack{Mexican \\ bentonite}}$$

with CH_3O substituents, product $CH_3O\!\!-\!\!\bigcirc\!\!-\!\!C(O)\text{-}N(H)\text{-}CH_2C_6H_5$

Scheme 141.

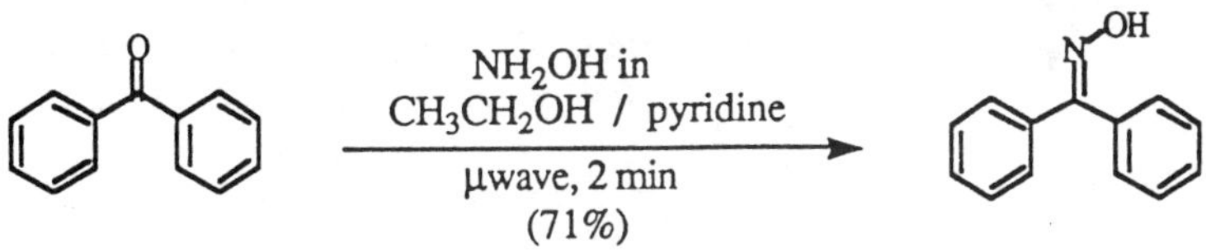

$$\xrightarrow[\substack{\mu wave,\ 2\ min \\ (71\%)}]{\substack{NH_2OH\ in \\ CH_3CH_2OH\ /\ pyridine}}$$

Scheme 142.

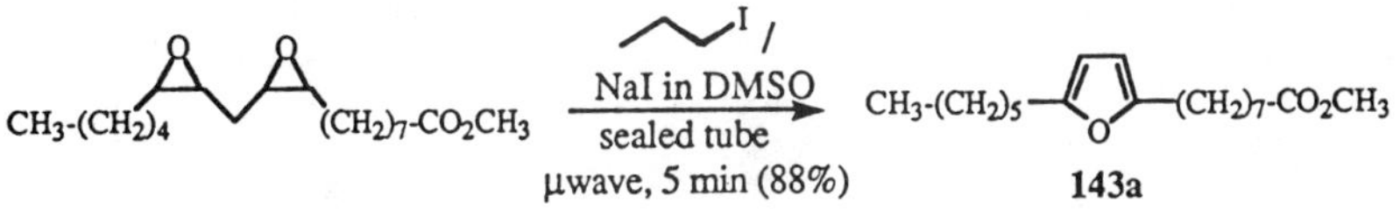

$CH_3\text{-}(CH_2)_4$ ... $(CH_2)_7\text{-}CO_2CH_3$

$$\xrightarrow[\substack{\mu wave,\ 5\ min\ (88\%)}]{\substack{NaI\ in\ DMSO \\ sealed\ tube}} CH_3\text{-}(CH_2)_5\!\!-\!\!\bigcirc\!\!-\!\!(CH_2)_7\text{-}CO_2CH_3$$

143a

Scheme 143.

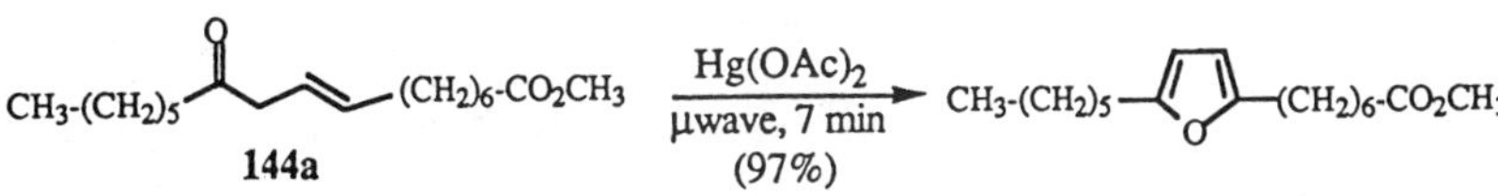

$CH_3\text{-}(CH_2)_5$ ... $(CH_2)_6\text{-}CO_2CH_3$

144a

$$\xrightarrow[\substack{\mu wave,\ 7\ min \\ (97\%)}]{Hg(OAc)_2} CH_3\text{-}(CH_2)_5\!\!-\!\!\bigcirc\!\!-\!\!(CH_2)_6\text{-}CO_2CH_3$$

Scheme 144.

43% yield. A furan can also be generated from conjugated enone **144a** in 97% yield after a brief exposure to mercuric acetate. Cyclic ethers such as tetrahydrofuran and 1,4-dioxane can be efficiently generated from the corresponding·diols by using a continuous-flow microwave setup (Schemes 145 and 146) (*23, 37*).

Schemes 147–156 contain a variety of microwave-promoted reactions that furnish nitrogen-containing heterocycles. The coupling of β-aminoalcohols or 1-ω-alkanediamines with ethyl levulinate over alumina give heterocycles **147a** and **148a**, respectively, in good yields after brief exposure to microwave heating. Conventional conditions typically involved several days at room temperature. Bose and co-workers developed efficient syntheses of a wide variety of heterocycles in open reaction vessels (cf. Schemes 149–154). Scheme 155 depicts a microwave-promoted Hantzsch pyridine synthesis (*77*), the one-pot condensation of an aldehyde with an alkyl acetoacetate and ammonia to afford substituted 1,4-dihydropyridines. Classical thermolysis conditions require long reflux periods (1– 12 h) to obtain poor yields (4–50%). Alvarez-Builla and co-workers have adopted Petrow's modified procedure (*78*), which involves the condensation of β-aminocrotonitrile with ketones, to produce these 1,4-dihydropyridine derivatives (Scheme 156) (*77*).

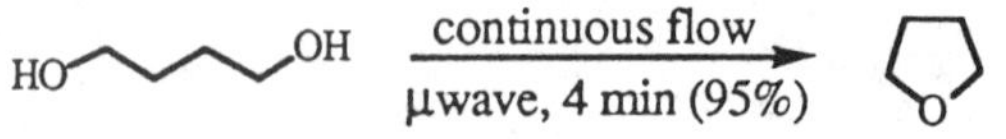

Scheme 145.

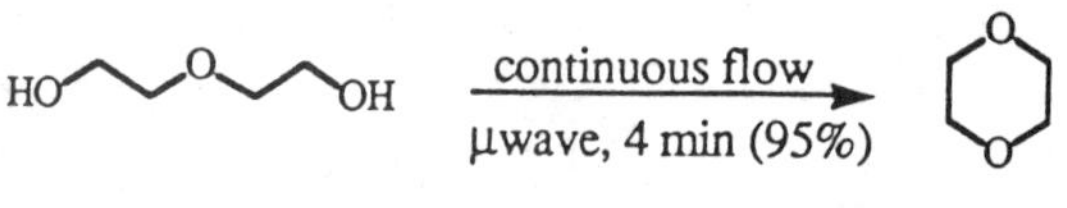

Scheme 146.

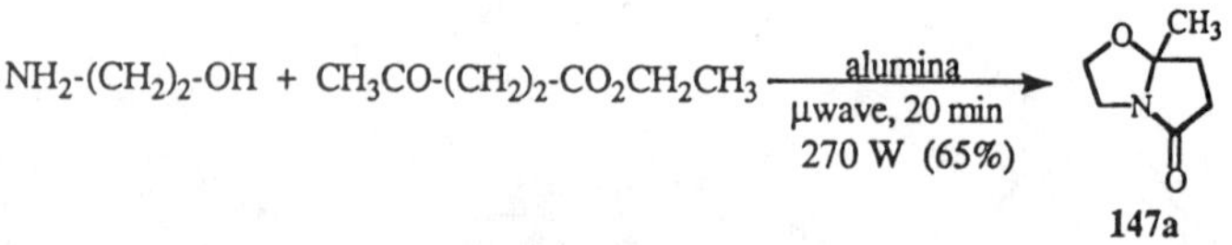

Scheme 147.

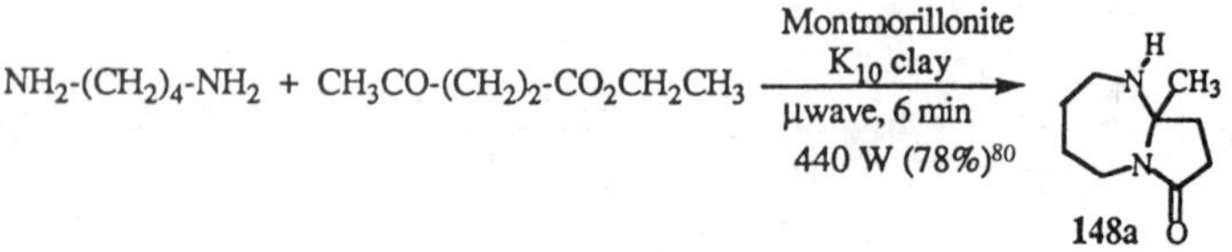

Scheme 148.

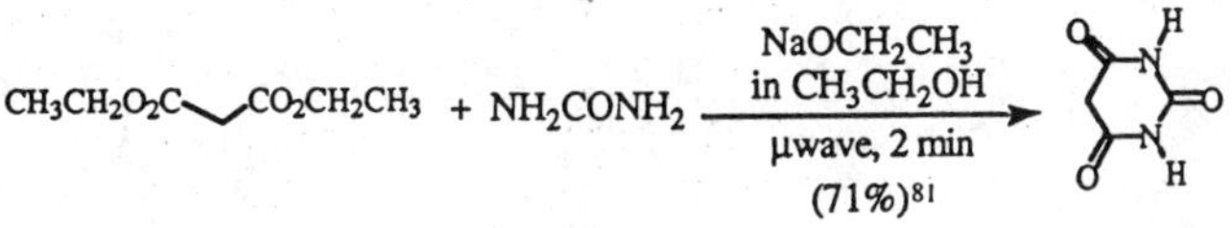

CH$_3$CH$_2$O$_2$C⁀⁀CO$_2$CH$_2$CH$_3$ + NH$_2$CONH$_2$ $\xrightarrow[\substack{\mu\text{wave, 2 min} \\ (71\%)^{81}}]{\substack{\text{NaOCH}_2\text{CH}_3 \\ \text{in CH}_3\text{CH}_2\text{OH}}}$

Scheme 149.

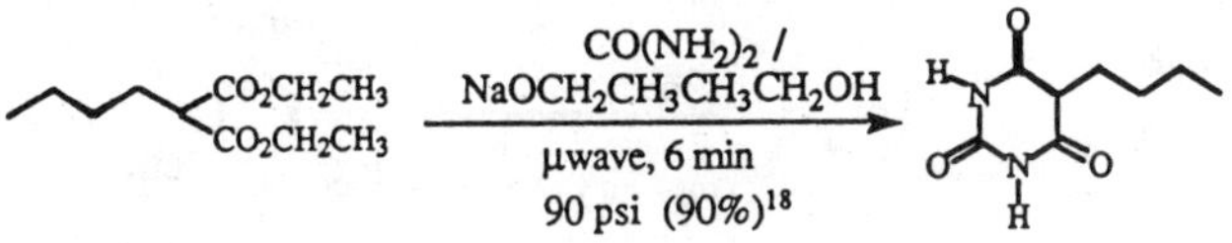

$\xrightarrow[\substack{\mu\text{wave, 6 min} \\ 90 \text{ psi } (90\%)^{18}}]{\substack{\text{CO(NH}_2)_2 \text{ /} \\ \text{NaOCH}_2\text{CH}_3\text{CH}_3\text{CH}_2\text{OH}}}$

Scheme 150.

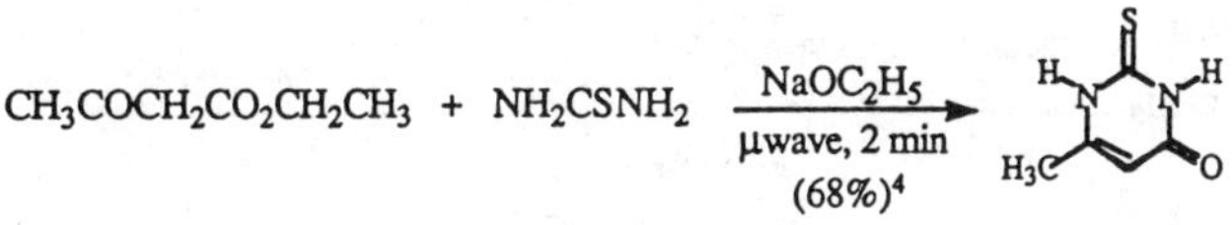

CH$_3$COCH$_2$CO$_2$CH$_2$CH$_3$ + NH$_2$CSNH$_2$ $\xrightarrow[\substack{\mu\text{wave, 2 min} \\ (68\%)^4}]{\text{NaOC}_2\text{H}_5}$

Scheme 151.

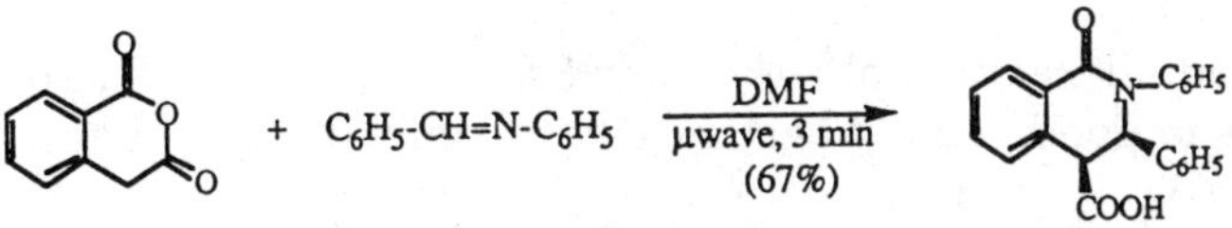

+ C$_6$H$_5$-CH=N-C$_6$H$_5$ $\xrightarrow[\substack{\mu\text{wave, 3 min} \\ (67\%)}]{\text{DMF}}$

Scheme 152.

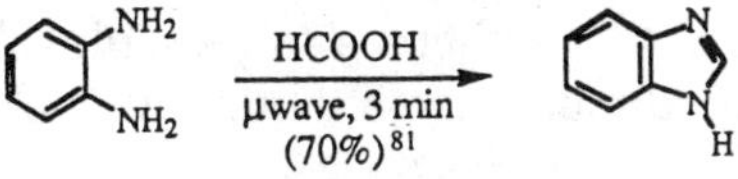

$\xrightarrow[\substack{\mu\text{wave, 3 min} \\ (70\%)^{81}}]{\text{HCOOH}}$

Scheme 153.

+ H$_2$NCH$_2$CO$_2$H $\xrightarrow[\substack{\mu\text{wave, 2 min} \\ (72\%)^{80}}]{\text{N(CH}_2\text{CH}_3)_3 \text{ in DMF}}$

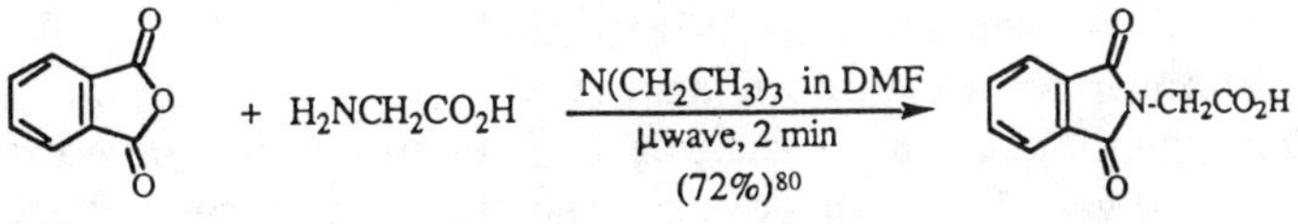

Scheme 154.

β-Lactams have been synthesized in microwave systems by using chemistry first developed by Ojima et al. (*82*). In the first synthesis shown, the uncyclized adduct **157a** in Scheme 157 is formed in 65% yield when the starting material is adsorbed onto K$_{10}$ Montmorillonite clay. When the reagents are adsorbed onto a mixture of 1,4,7,10,13,16-hexaoxacyclooctadecane (18 *crown* 6) and potassium fluoride, β-lactam **157b** is formed (*83*).

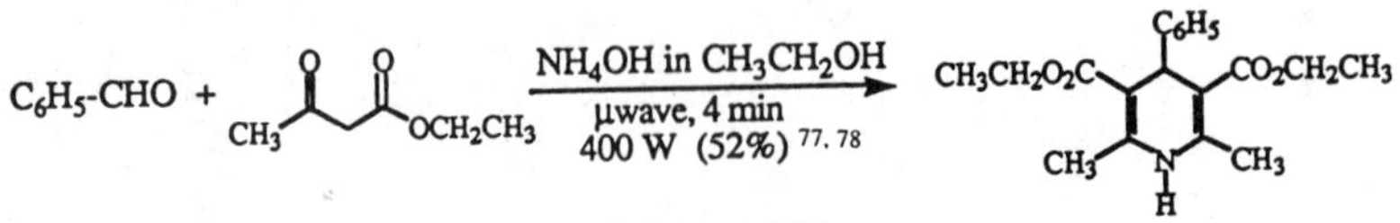

C_6H_5-CHO + [diketone structure with CH_3 and OCH_2CH_3] $\xrightarrow[\substack{\mu wave,\ 4\ min \\ 400\ W\ (52\%)\ ^{77,\ 78}}]{NH_4OH\ in\ CH_3CH_2OH}$ [dihydropyridine product with C_6H_5, $CH_3CH_2O_2C$, $CO_2CH_2CH_3$, CH_3, CH_3, N, H]

Scheme 155.

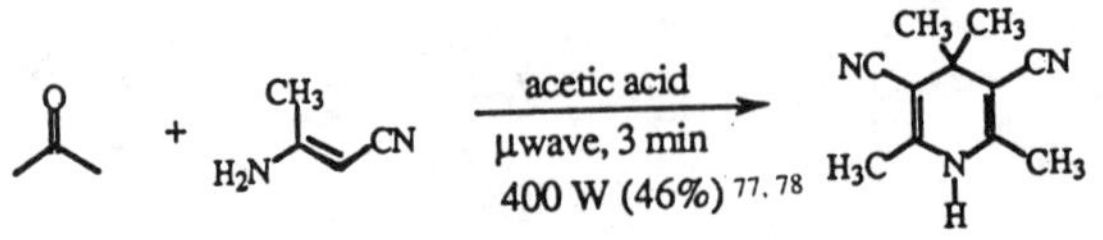

[acetone] + [H_2N structure with CH_3 and CN] $\xrightarrow[\substack{\mu wave,\ 3\ min \\ 400\ W\ (46\%)\ ^{77,\ 78}}]{acetic\ acid}$ [dihydropyridine product with CH_3 CH_3, NC, CN, H_3C, CH_3, N, H]

Scheme 156.

[structure 157a: C_6H_5, N-C_6H_5, CH_3, OCH_3, O] **157a** $\xleftarrow[\substack{\mu wave\ (65\%)}]{\substack{Montmorillonite \\ K_{10}\ clay}}$ [H_3C, $OSi(CH_3)_3$, OCH_3 structure] + [C_6H_5, N, C_6H_5 structure] $\xrightarrow[\substack{\mu wave\ (93\%)}]{18\text{-crown-}6\ /\ KF}$ [β-lactam: C_6H_5, C_6H_5, N, H_3C, O] **157b**

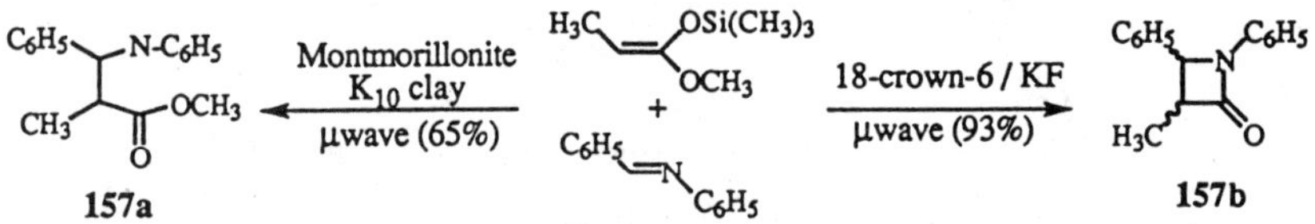

Scheme 157.

The two additional examples shown in Schemes 158 and 159 also produce β-lactams in good yield. β-Lactam **159a** is a mixture of syn and anti isomers (in a 35:65 ratio).

Bose and co-workers reported a microwave-promoted Bischler–Napieralski reaction. This condensation proved to be highly solvent dependent: in *p*-dichlorobenzene, the dihydroisoquinoline **160a** was isolated in 70–75% yield (*84*), whereas in *o*-dichlorobenzene an amidine (**160b**) was formed predominantly, although no yield was given (Scheme 160).

The Graebe–Ullman synthesis is extremely useful in building carazole subunits (Schemes 161–164). This two-step procedure involves an initial irradiation of triazole (such as **161a**) with 4-chloropyridine, followed by the addition of polyphosphoric acid (PPA) to the reaction mixture and additional heating, during which time nitrogen gas is evolved from the open reaction vessel (*85*). This two-step sequence produces carbazole **161b** in 71% yield. Alvarez-Builla and co-workers modified this chemistry so that it can be carried out in microwave ovens. The examples cited in Schemes 161–164 are just four of their many successful annulations. The Fischer indole synthesis, a powerful method for making indoles, has been independently investigated by several groups using varied conditions and microwave systems (Schemes 165–168). The annulation shown in Scheme 165 was carried out by Bose and co-workers (*84*) in a beaker by using KSF clay as the catalyst, whereas the reaction between cyclohexanone and phenylhydrazine (Scheme 166) was achieved by Majetich and Hicks (*18*) using pressure regulation. This reaction (Scheme 166) was also achieved by

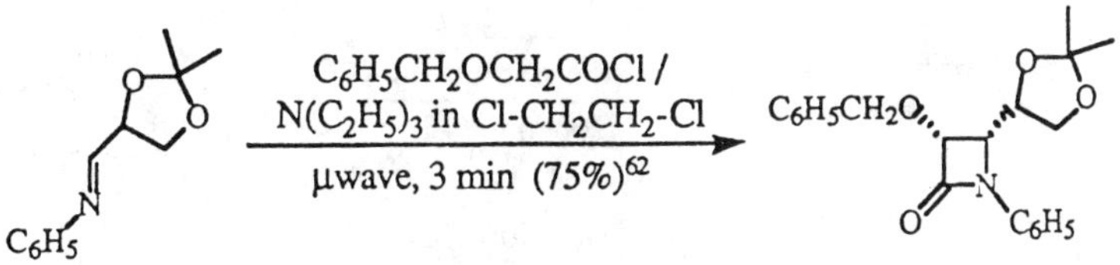

$$C_6H_5CH_2OCH_2COCl /$$
$$N(C_2H_5)_3 \text{ in } Cl\text{-}CH_2CH_2\text{-}Cl$$
μwave, 3 min (75%)[62]

Scheme 158.

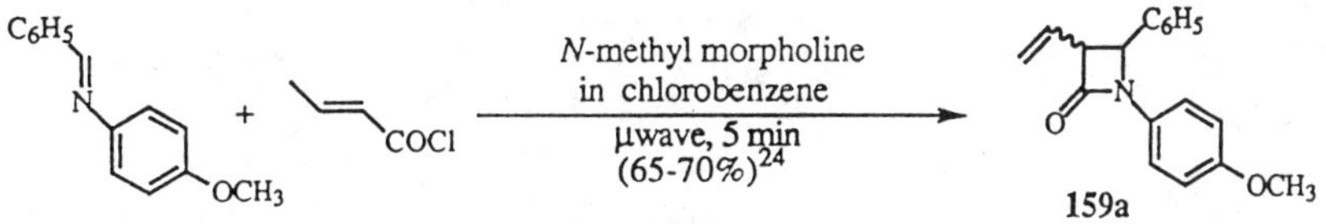

N-methyl morpholine
in chlorobenzene
μwave, 5 min
(65-70%)[24]

159a

Scheme 159.

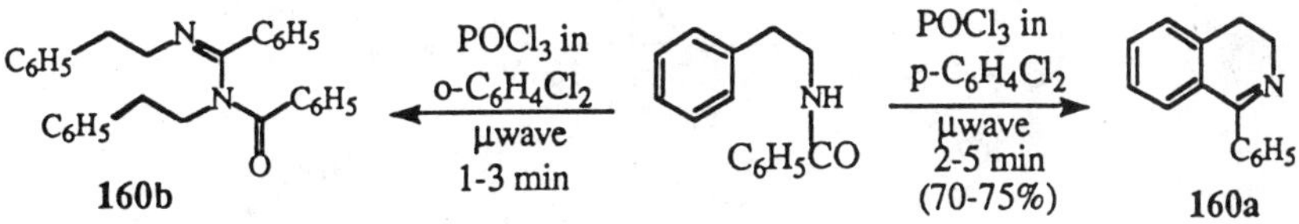

160b

POCl$_3$ in
o-C$_6$H$_4$Cl$_2$
μwave
1-3 min

POCl$_3$ in
p-C$_6$H$_4$Cl$_2$
μwave
2-5 min
(70-75%)

160a

Scheme 160.

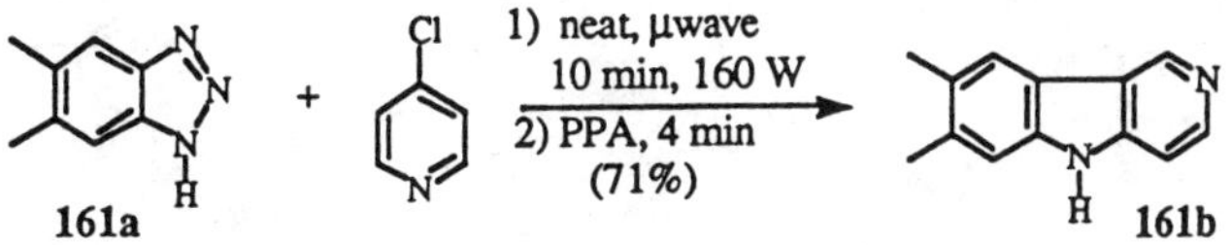

161a

1) neat, μwave
10 min, 160 W
2) PPA, 4 min
(71%)

161b

Scheme 161.

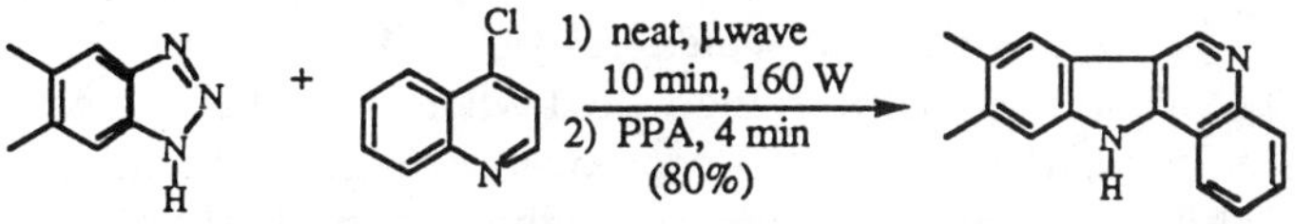

1) neat, μwave
10 min, 160 W
2) PPA, 4 min
(80%)

Scheme 162.

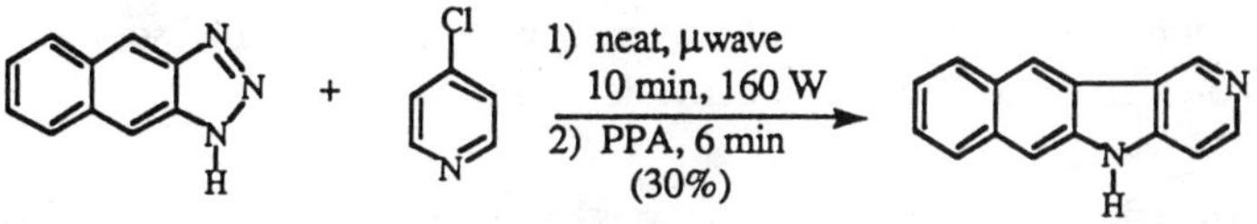

1) neat, μwave
10 min, 160 W
2) PPA, 6 min
(30%)

Scheme 163.

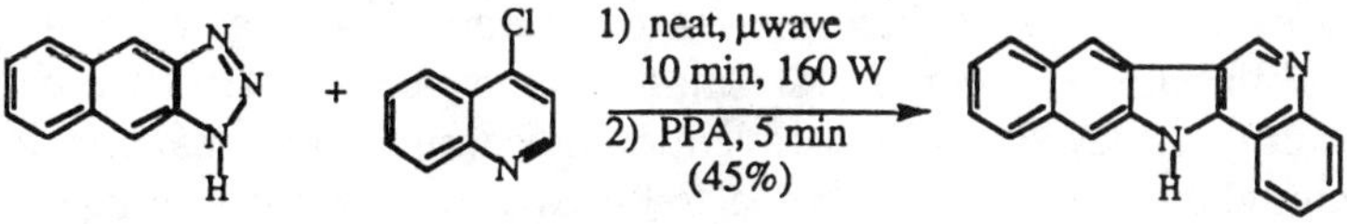

1) neat, μwave
10 min, 160 W
2) PPA, 5 min
(45%)

Scheme 164.

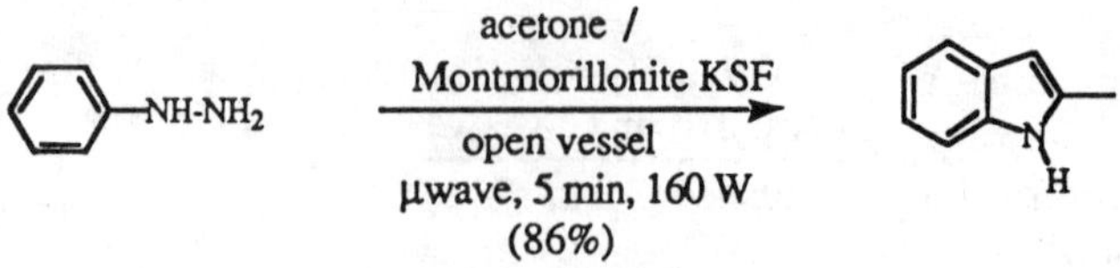

Scheme 165.

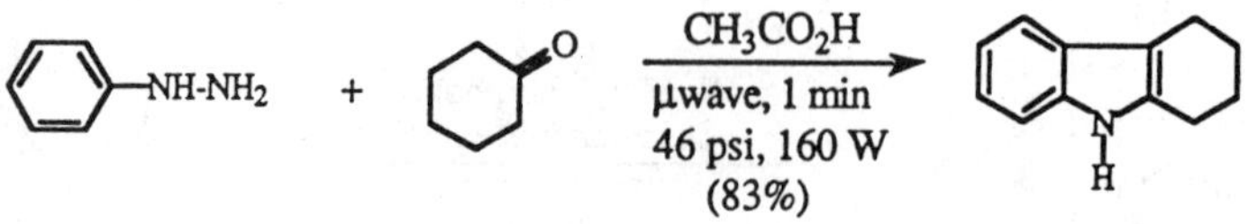

Scheme 166*.

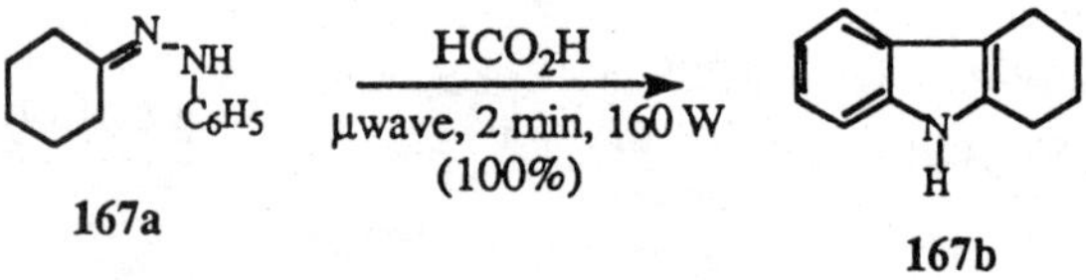

167a

167b

Scheme 167.

168a

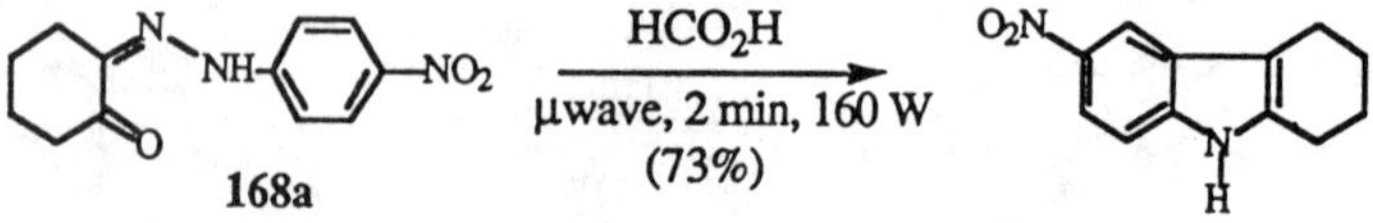

Scheme 168.

Villemin and co-workers in sealed vessels (86). The examples shown in Schemes 167 and 168 occur in high yield when carried out in concentrated formic acid in a Parr bomb (87). Moreover, Abramovitch and Bulman (87) observed that heating preformed phenylhydrazone **167a** with Montmorillonite KSF under identical conditions gave only trace amounts of the desired tetrahydrocarbazole (**167b**), whereas irradiation of preformed hydrazones (i.e., **167a** or **168a**)—in the presence of formic acid—facilitated their conversion to the Fischer indole products.

Miscellaneous Synthetic Transformations

Many common synthetic transformations are also microwaveable. Microwave irradiation can be used to remove esters or halides attached to the carbon alpha to a carbonyl unit. These dealkoxycarbonylations are usually difficult to achieve by using conventional heating conditions. The dealkoxycarbonylation shown in Scheme 169 uses a phase-transfer reagent (highest yields were obtained by using isosorbide) as well as lithium bromide

(88) and is an example of a Krapcho reaction. A second example of this reaction produces an alkylcyclohexanone (Scheme 170). Other β-keto esters have been decarboxylated by using this procedure (89).

The partial debromination of lactam **171a** in Scheme 171 under free-radical conditions after only 3 min gave a mixture of both the cis and trans isomers, although the cis product predominated (no yield was provided) (24). Many ring-opening reactions are accelerated by using microwave heating. In the example shown in Scheme 172, phenylthiocyclopropane **172a** opens on treatment with silver tetrafluoroborate to produce diene **172b** (90). This transformation is also a ring-expansion reaction, because the cyclohexane ring is enlarged. The furan ring in the fatty ester chain of Scheme 173 can be hydrolyzed to afford the corresponding 1,4-dione in 99% yield. Conventional procedures call for a 12-h reflux (84% yield) (38). Hydrolysis of epoxide **174a** in Scheme 174 by heating in dimethyl sulfoxide (DMSO) in a sealed tube for 5 min gives a mixture of methyl 9- and 10-oxosterates **174b** and **174c** in 72% yield (no ratio given).

A useful rearrangement involves the migration of a double bond into conjugation with another π-system to produce a more stable product, such

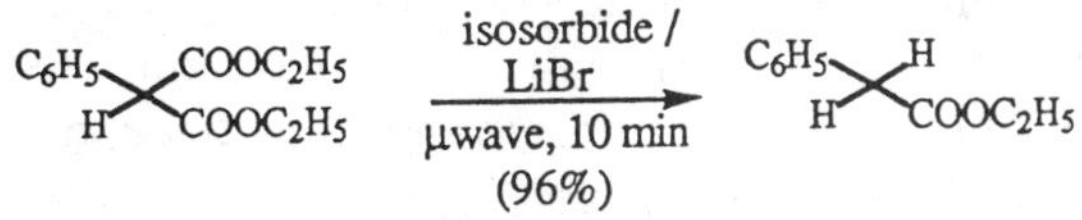

C_6H_5 —C(COOC_2H_5)(H)(COOC_2H_5) → isosorbide / LiBr, μwave, 10 min (96%) → C_6H_5 —C(H)(H)(COOC_2H_5)

Scheme 169.

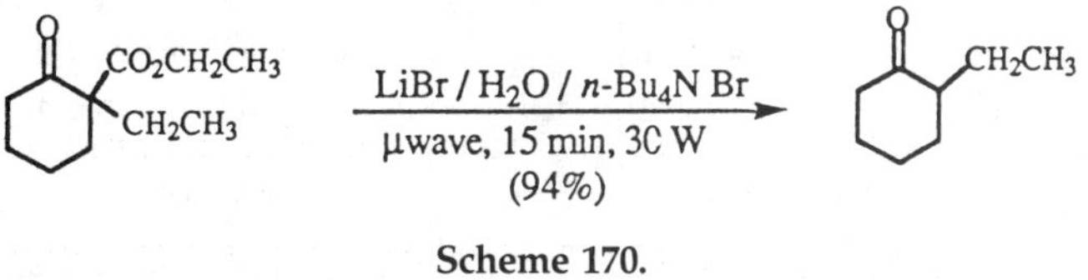

(cyclohexanone with $CO_2CH_2CH_3$ and CH_2CH_3) → LiBr / H_2O / n-Bu_4N Br, μwave, 15 min, 30 W (94%) → (cyclohexanone with CH_2CH_3)

Scheme 170.

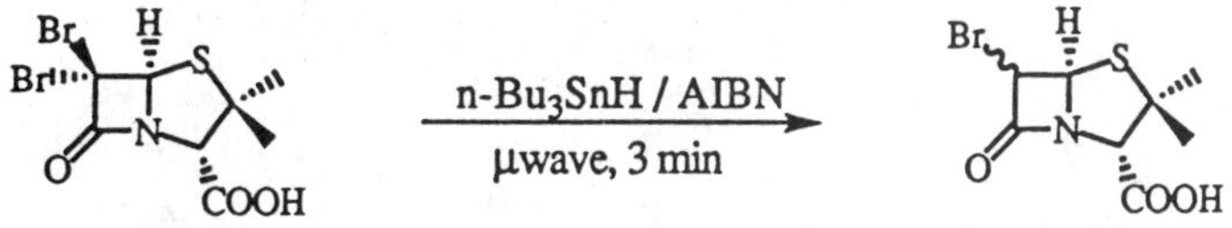

171a (β-lactam, Br, Br, H, S, N, O, COOH) → n-Bu_3SnH / AIBN, μwave, 3 min → (β-lactam, Br, H, S, N, O, COOH)

Scheme 171.

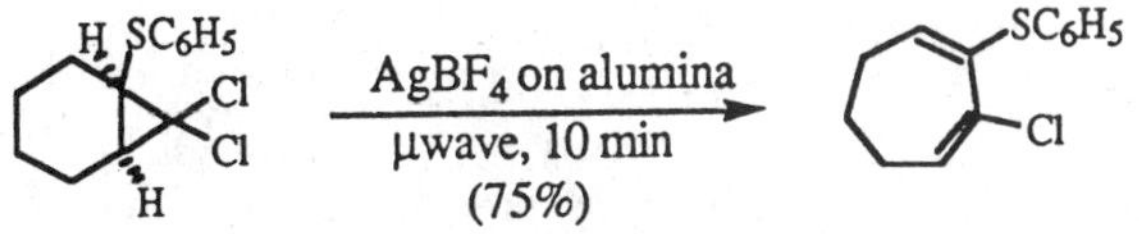

172a (H, SC_6H_5, Cl, Cl, H) → AgBF_4 on alumina, μwave, 10 min (75%) → **172b** (SC_6H_5, Cl)

Scheme 172.

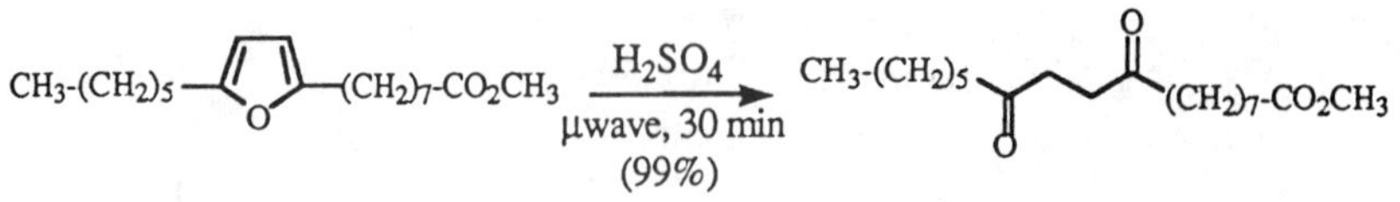

$CH_3-(CH_2)_5$ —[furan]— $(CH_2)_7-CO_2CH_3$ $\xrightarrow[\substack{\mu wave,\ 30\ min \\ (99\%)}]{H_2SO_4}$ $CH_3-(CH_2)_5$ —[diketone]— $(CH_2)_7-CO_2CH_3$

Scheme 173.

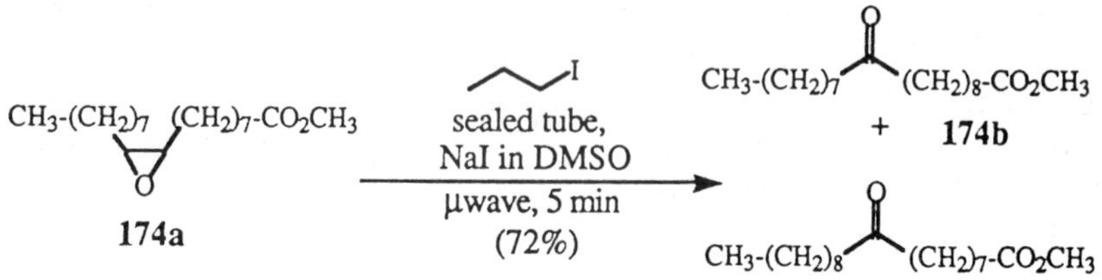

$CH_3-(CH_2)_7$ [epoxide] $(CH_2)_7-CO_2CH_3$

174a

$\xrightarrow[\substack{sealed\ tube, \\ NaI\ in\ DMSO \\ \mu wave,\ 5\ min \\ (72\%)}]{\text{[~~~~I]}}$

$CH_3-(CH_2)_7$ —[ketone]— $(CH_2)_8-CO_2CH_3$

$+$ **174b**

$CH_3-(CH_2)_8$ —[ketone]— $(CH_2)_7-CO_2CH_3$

174c

Scheme 174.

as the rearranged phenol shown in Scheme 175 (*18*). The double-bond isomerizations in Schemes 176 and 177 are additional examples of such rearrangements. In these two reactions the aryl ketone products result from enol-keto tautomerization of the rearranged intermediate. Scheme 178 depicts an example of an acid-catalyzed isomerization that ultimately results in the formation of a phenol (*9*). This transformation is an isomerization, not an oxidation, because the isopropenyl group ends up as an isopropyl group. The pinacol rearrangement, perhaps the best-known alkyl shift reaction, shown in Scheme 179 occurs in 99% yield when irradiated on aluminum Montmorillonite in an open container (15 min) (*20–22*). The second pinacol rearrangement shown (Scheme 180) also proceeds in excellent yield. Surprisingly, the pinacol rearrangements of cyclic diols have not yet been examined.

At first glance, the two transformations presented in Schemes 181 and 182 appear to be simple isomerizations. Instead these Montmorillonite-promoted reactions proceed via a solvolysis to generate benzylic carbocation **181a**, which is hydrolyzed via its allenic carbocationic resonance contributor (**181b**) to generate intermediate **181c**. Tautomerization of **181c** leads to the observed ketone product **181d**. Enone **182a** is produced by a similar mechanism. The efficient hydrosilylation of 2-vinylpyridine was reported by Abramovitch and Bulman using microwave irradiation as summarized in Scheme 183. Here, the reaction mixture was irradiated in a Parr microwave bomb for a 3-min period by using a series of 30-s bursts of heating and cooling. The yield of the isolated dichloromethylsilyl compound (**183a**) was 75% (*87*). When styrene is treated with carbon tetrachloride and copper iodide or copper diisopropylamine chloride, carbon tetra-

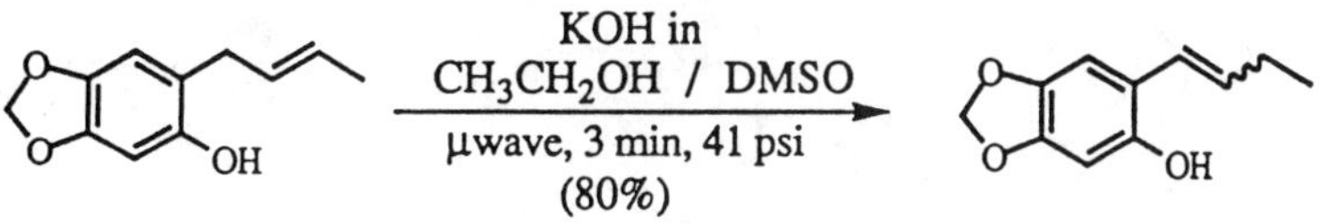
OH
KOH in
CH3CH2OH / DMSO
µwave, 3 min, 41 psi
(80%)
OH

Scheme 175*.

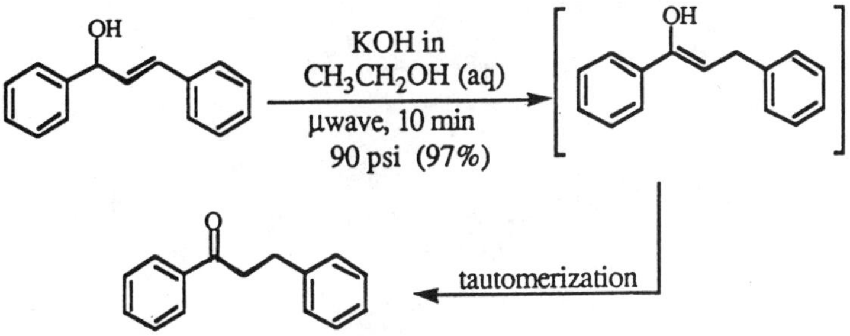
OH
KOH in
CH3CH2OH (aq)
µwave, 10 min
90 psi (97%)
OH
O
tautomerization

Scheme 176*.

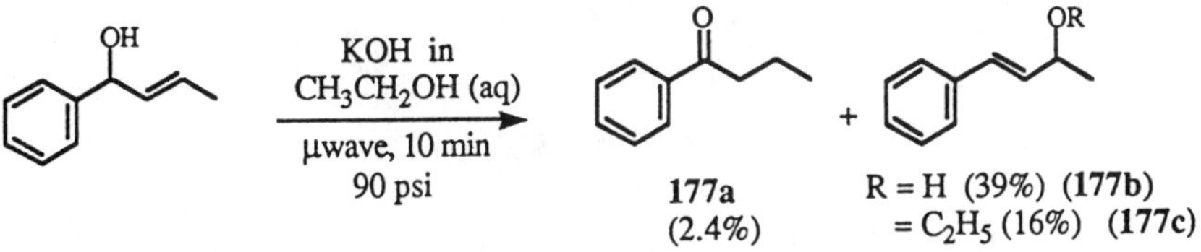
OH
KOH in
CH3CH2OH (aq)
µwave, 10 min
90 psi
O
OR
+
177a
(2.4%)
R = H (39%) (177b)
= C2H5 (16%) (177c)

Scheme 177*.

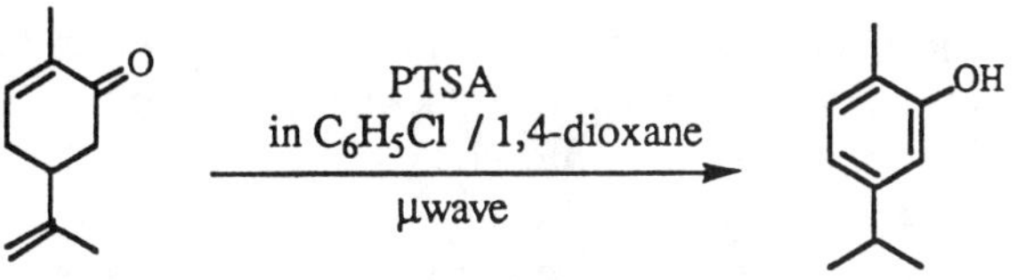
O
PTSA
in C6H5Cl / 1,4-dioxane
µwave
OH

Scheme 178.

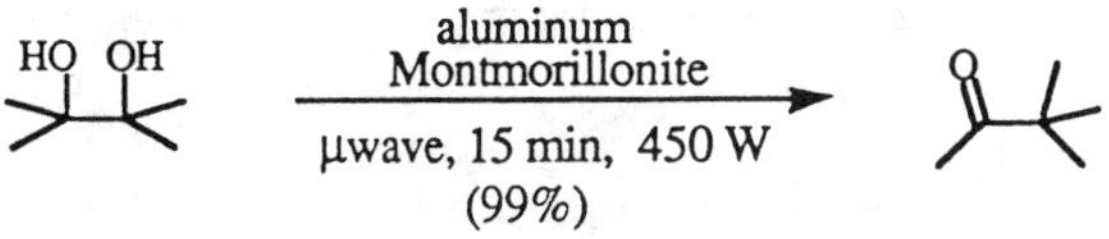
HO OH
aluminum
Montmorillonite
µwave, 15 min, 450 W
(99%)
O

Scheme 179.

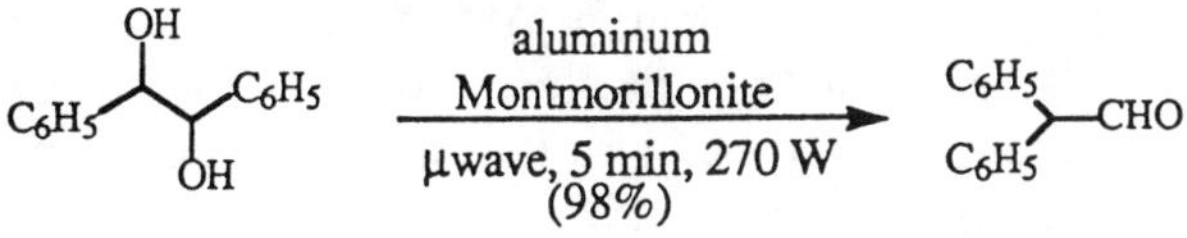
OH
C6H5
C6H5
OH
aluminum
Montmorillonite
µwave, 5 min, 270 W
(98%)
C6H5
CHO
C6H5

Scheme 180.

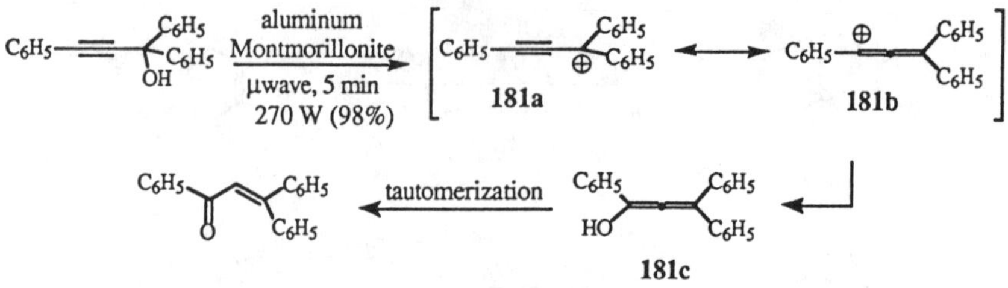

Scheme 181.

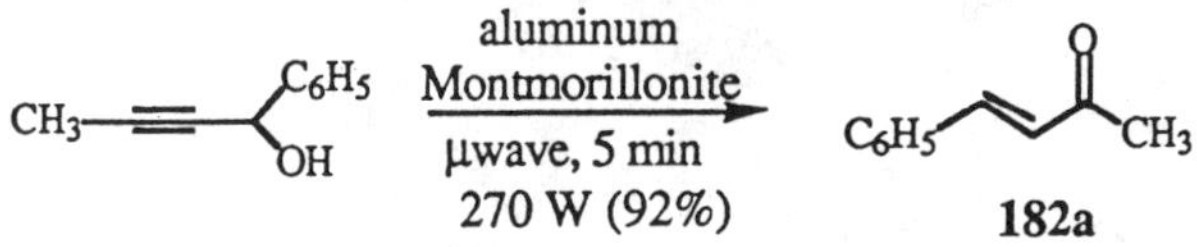

Scheme 182.

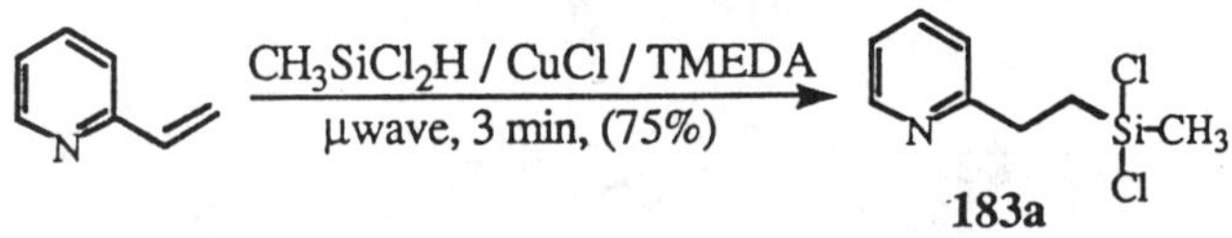

Scheme 183.

chloride adds to the double bond (Scheme 184) (91). This reaction occurs in 15–40 min by using acetonitrile as the solvent, because CCl_4 is a poor microwave absorber (11).

Microwave-initiated Beckmann rearrangements were investigated by Delgado and co-workers (65). These workers found that heating acetophenone and hydroxylamine in the presence of Mexican bentonite produced only a 16% yield of acetanilide (Scheme 185). In contrast, Majetich and Hicks (12) found that the pressure-regulated reaction of benzophenone, hydroxyamine, and triflic acid affords benzanilide in high yield (Scheme 186). Other examples are forthcoming (18). A Peterson olefination gave different products in the microwave, depending on the solid support used to adsorb the reagents (92). When cesium fluoride and K_{10} clay are used as the adsorbants, the expected olefinic product (187a in Scheme 187) is produced in 60% yield; however, using cesium fluoride and magnesium oxide as the adsorbants gives a mixture of 187a (14%) and alcohol 187b (86%, 3 min).

The olefination shown in Scheme 188 is more complex. By using potassium fluoride as the adsorbant, the silyl ether (188a) is produced in 62% yield. In contrast, when potassium fluoride and magnesium oxide are used, the three products shown are formed in a 6:25:69 ratio (8 min); using potassium fluoride with K_{10} clay gives only alkene 188a in 50% yield

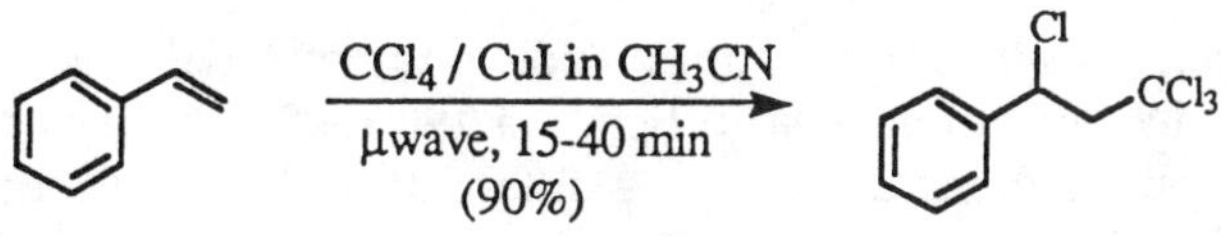

$$\text{CCl}_4 \text{ / CuI in CH}_3\text{CN}$$
μwave, 15-40 min
(90%)

Scheme 184.

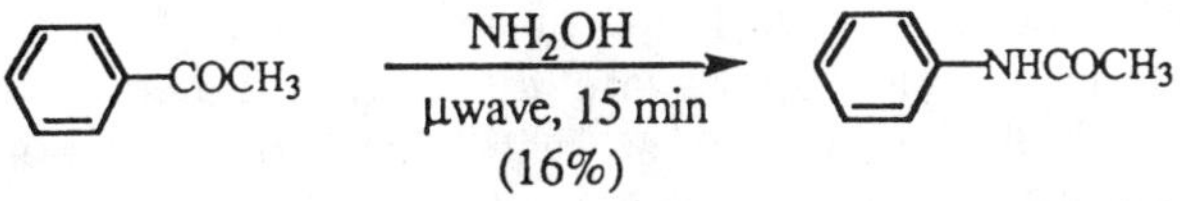

$$\text{NH}_2\text{OH}$$
μwave, 15 min
(16%)

Scheme 185.

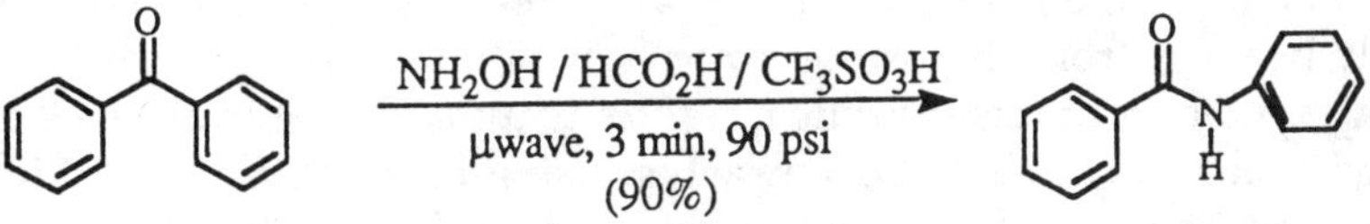

$$\text{NH}_2\text{OH / HCO}_2\text{H / CF}_3\text{SO}_3\text{H}$$
μwave, 3 min, 90 psi
(90%)

Scheme 186*.

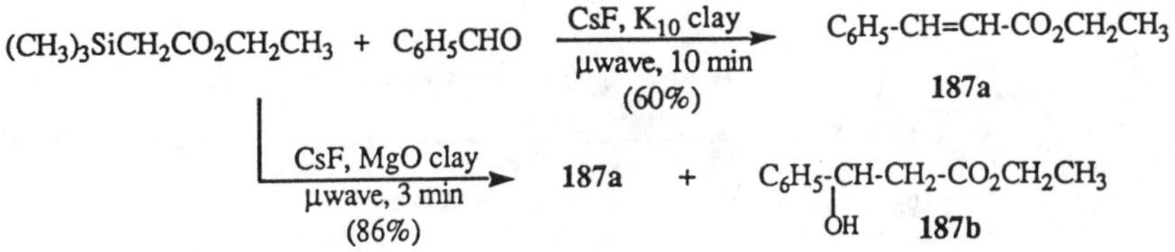

$(\text{CH}_3)_3\text{SiCH}_2\text{CO}_2\text{CH}_2\text{CH}_3$ + $\text{C}_6\text{H}_5\text{CHO}$ $\xrightarrow[\substack{\text{μwave, 10 min} \\ (60\%)}]{\text{CsF, K}_{10}\text{ clay}}$ $\text{C}_6\text{H}_5\text{-CH=CH-CO}_2\text{CH}_2\text{CH}_3$ **187a**

$\xrightarrow[\substack{\text{μwave, 3 min} \\ (86\%)}]{\text{CsF, MgO clay}}$ **187a**   + $\text{C}_6\text{H}_5\text{-CH-CH}_2\text{-CO}_2\text{CH}_2\text{CH}_3$
                                                        $\overset{|}{\text{OH}}$ **187b**

Scheme 187.

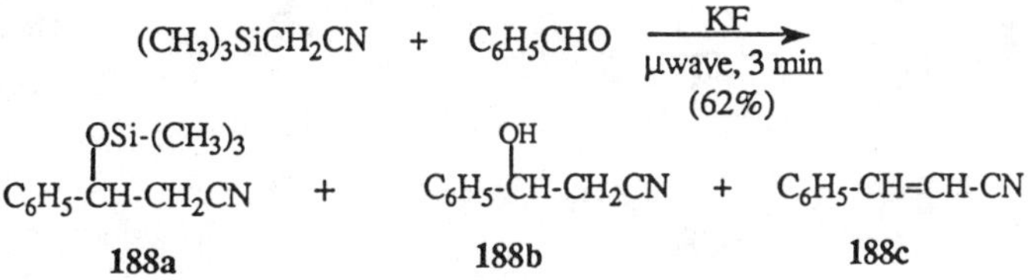

$(\text{CH}_3)_3\text{SiCH}_2\text{CN}$ + $\text{C}_6\text{H}_5\text{CHO}$ $\xrightarrow[\substack{\text{μwave, 3 min} \\ (62\%)}]{\text{KF}}$

$\overset{\text{OSi-(CH}_3)_3}{\underset{|}{\text{C}_6\text{H}_5\text{-CH-CH}_2\text{CN}}}$ + $\overset{\text{OH}}{\underset{|}{\text{C}_6\text{H}_5\text{-CH-CH}_2\text{CN}}}$ + $\text{C}_6\text{H}_5\text{-CH=CH-CN}$

188a   **188b**   **188c**

Scheme 188.

(3 min). These results suggest that the acidity of the adsorbant used may influence the product distribution.

The Büchner–Strecker reaction for preparing amino acids was used by Stone-Elander and co-workers (45) to incorporate ^{11}C into tyrosine (Scheme 189). The reaction was carried out in two steps: an initial irradiation of **189a** and labelled ammonium cyanide (30 s) followed by the addition of sodium hydroxide and further heating (30 s) to hydrolyze the nitrile group. The overall yield of labelled tyrosine in this two-step process was 60%. Base-catalyzed eliminations of secondary alkyl halides produced good yields of alkene products (70%, although no mention of the ratio of regioisomers was listed) (6). The balance of the material was the substitution product, an ether **190a** in Scheme 190 (30%). These microwave-promoted eliminations did not give different product distributions when compared with conventional heating.

Recently, Giguere and Herberich (93) devised an improved preparation of allyldiphenyl-phosphine oxide (**191a**) by using microwave heating. This one-step procedure, a variation of the Michaelis–Arbuzov reaction, produces **191a** in 86% yield after a brief thermolysis (Scheme 191). The best way to introduce fluorine into an aromatic ring is by means of the Schiemann (or Balz–Schiemann) reaction. Under standard thermal conditions, the fluoro-substituted arene shown in Scheme 192 is produced by heating diazonium fluoroborate **192a** in only 4% yield. This transformation obviously benefits from the use of a nonconventional heating source (18). Many syntheses of unsubstituted indoles require the removal of a 2-carboxyl group, a problematic step when using conventional conditions. Jones and Chapman (94) reported a procedure that employs microwave heating to effect decarboxylation of indoles in excellent yields (Scheme 193). Even though the use of quinoline as solvent gave the best results,

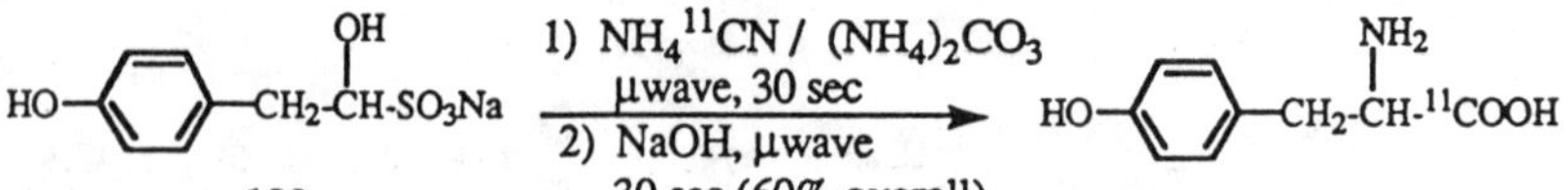

Scheme 189.

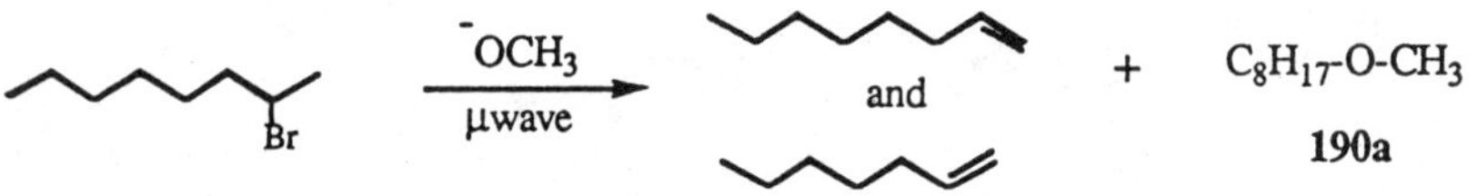

Scheme 190.

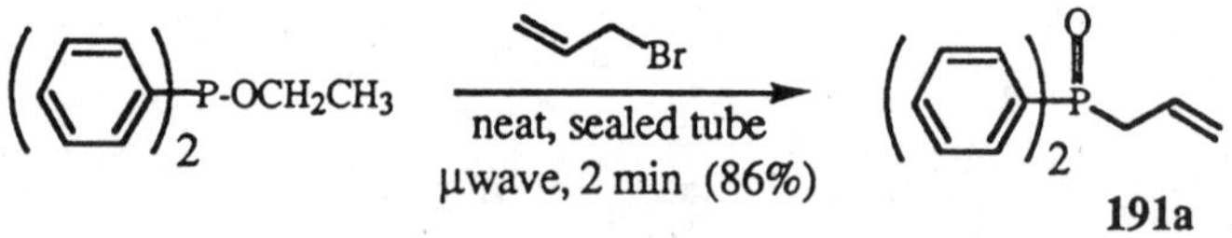

Scheme 191.

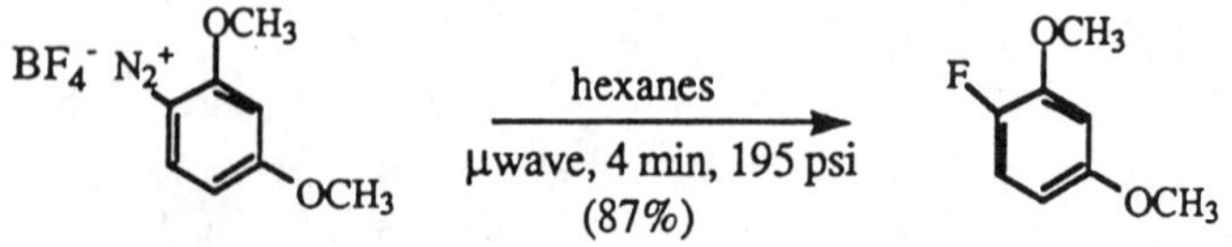

Scheme 192*.

high yields were also obtained in the absence of solvent or with copper catalysts and quinoline. Similar yields were obtained by using indoles with substituents on the aryl ring.

Hamelin and co-workers (95) reported an efficient synthesis of enaminoketones and acyl amines, as shown in Schemes 194–196. The 1,3-diketones and amines are first adsorbed onto K_{10} clay or silica gel in the absence of solvent and then irradiated, usually in an open vessel. When the reaction shown in Scheme 196 was carried out in a closed vessel, the expected enaminoketone was not produced. Instead, N-substituted acetamide **196b** and acetone are formed. Presumably, **196a** is initially generated and then water, which is not removed in a sealed vessel because it is in an open container, hydrolyzes the enaminoketone to give **196b**. In support of this conjecture, brief microwave heating of preformed **196a** in a closed vessel with dry silica gel gave only **196b**, albeit in low yield.

Conclusion

Although the use of microwave heating in organic synthesis is still in its infancy, this review documents that a large variety of fundamental organic

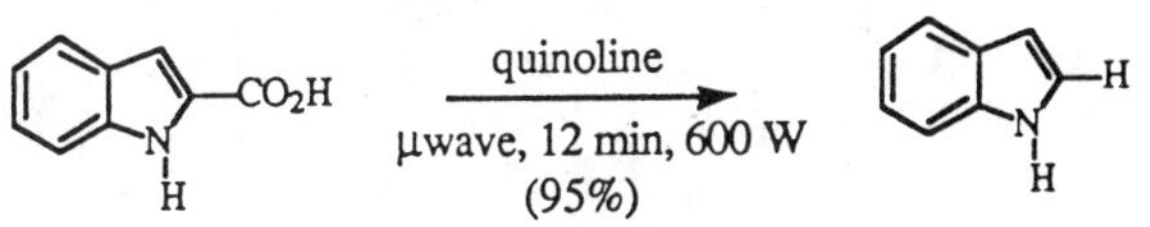

Scheme 193.

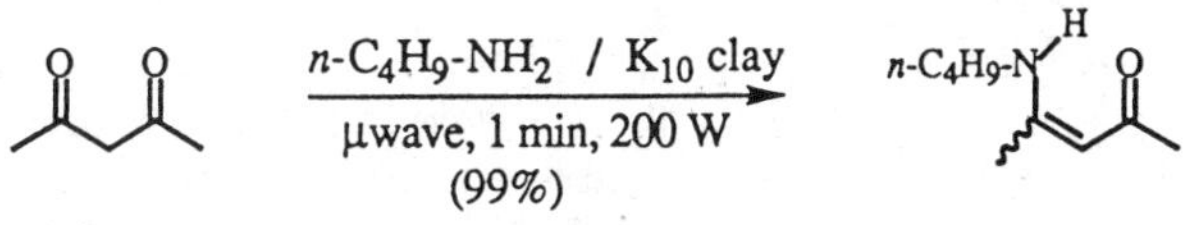

Scheme 194.

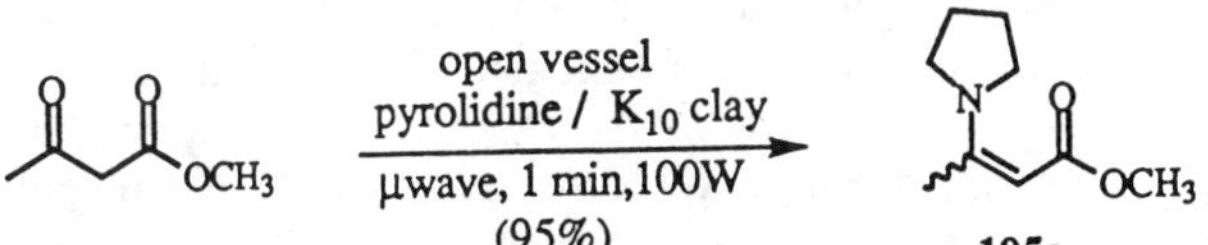

Scheme 195.

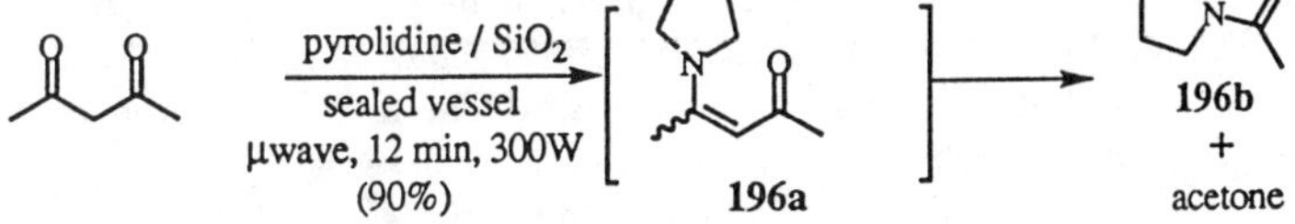

Scheme 196.

reactions can be achieved rapidly and in good yield by using microwave irradiation. The continued investigation of microwave-initiated organic reactions will be the focus of extensive activity. Moreover, we expect the use of continuous-flow instruments and improved reaction vessel design to fuel this growth.

Acknowledgments

Appreciation is extended to Edwin Neas and Michael Collins of CEM Corporation for the donation of equipment that we used to produce many of the results reported herein.

References

1. Gedye, R.; Smith, F.; Westaway, K.; Ali, H.; Baldisera, L.; Laberge, L.; Rousell, J. *Tetrahedron Lett.* **1986,** *27,* 279.
2. Giguere, R. J.; Bray, T. L.; Duncan, S. M.; Majetich, G. *Tetrahedron Lett.* **1986,** *27,* 4945.
3. Giguere, R. J. *Organic Synthesis: Theory and Application;* JAI Press: Greenwich, CT, 1989; p 103.
4. Abramovitch, R. A. *Org. Prep. Proced. Int.* **1991,** *23,* 683.
5. Mingos, D. M. P.; Baghurst, D. R. *Chem. Soc. Rev.* **1991,** *20,* 1.
6. Gedye, R.; Smith, F.; Westaway, K. *J. Microwave Power Electromagn. Energy* **1991,** *26,* 3.
7. Raner, K. D.; Strauss, C. R.; Vyskoc, F.; Mokbel, L. *J. Org. Chem.* **1993,** *58,* 950.
8. Laurent, R.; Laporterie, A.; Dubac, J.; Berlan, J.; Lefeuvre, S.; Audhuy, M. *J. Org. Chem.* **1992,** *57,* 7099.
9. Mingos, D. M. P. Chapter 1 in this volume.
10. Stuerga, D.; Gonon, K.; Lallemant, M. *Tetrahedron* **1993,** *228,* 6229.
11. Majetich, G.; Neas, E.; Hicks, R.; Hoopes, T. manuscript in preparation.
12. Majetich, G.; Hicks, R. *Res. Chem. Intermed.* **1994,** *20,* 61.
13. To date we have found that all microwave-accelerated organic reactions can be explained by simple thermal effects. These studies have presented at the following scientific conferences: (a) Hoopes, T.; Neas, E.; Majetich, G. Presented at The 42nd Southeastern Regional Meeting of the American Chemical Society, New Orleans, LA, December 7, 1990; paper ORGN 448. (b) Hoopes, T.; Neas, E.; Majetich, G. Presented at The 201st National Meeting of the American Chemical Society, Atlanta, GA, April 16, 1991; paper ORGN 231. (c) Majetich, G. Presented at 1st World Congress of Microwave Chemistry, Breukelen, Netherlands, October 3, 1992. (d) Majetich, G. Presented at The NSF/EPRI Workshop on Microwave Induced Reactions, Pacific Grove, CA, March 8, 1993. (e) Majetich, G. Presented at The Microwave Debate, sponsored by Nestle Ltd.: Lausanne, Switzerland, May 24, 1993. (f) Majetich, G. Presented at The Microwave Enhanced Chemistry Conference, Birmingham, England, September 29, 1993.
14. In addition to Abramovitch's review (reference 4) *see also:* Yu, H.-M.; Chen, S.-T.; Wang, K.-T. *J. Org. Chem.* **1992,** *57,* 4781.

15. In addition to Abramovitch's review (reference 4) *see also:* Lewis, D. A.; Summers, J. D.; Ward, T. C.; McGrath, J. E. *J. Polym. Sci., Part A: Polym. Chem.* **1992**, *30*, 1647.

16. In addition to Abramovitch's review (reference 4) *see also:* Mingos, D. M. P.; Baghurst, D. R. *Br. Ceram. Trans. J.* **1992**, *91*, 124.

17. Gedye, R. N.; Smith, F. E.; Westaway, K. C. *Can. J. Chem.* **1988**, *66*, 17.

18. Majetich, G.; Hicks, R. *J. Microwave Power Electromagn. Energy* **1995**, *30*, 27–45.

19. Loupy, A.; Petit, A.; Ramdani, M.; Yvanaeff, C.; Majdoub, M.; Labiad, B.; Villemin, D. *Can. J. Chem.* **1993**, *71*, 90.

20. Gutierrez, E.; Loupy, A.; Bram, G.; Ruiz-Hitzky, E. *Tetrahedron Lett.* **1989**, *40*, 945.

21. Bram, G.; Loupy, A.; Majdoub, M. *Synth. Commun.* **1990**, *20*, 125.

22. Bram, G.; Loupy, A.; Majdoub, M.; Gutierrez, E.; Ruiz-Hitzky, E. *Tetrahedron* **1990**, *46*, 5167.

23. Chen, S.-T.; Chiou, S.-H.; Wang, K.-T. *J. Chem. Soc. Chem. Commun.* **1990**, 807.

24. Bose, A. K.; Manhas, M. S.; Ghosh, M.; Shah, M.; Raju, V. S.; Bari, S. S.; Newaz, S. N.; Banik, B. K.; Chaudhary, A. G.; Barakat, K. J. *J. Org. Chem.* **1991**, *56*, 6968.

25. A sample of the trienal shown in eq 20 was provided to us by J. A. Marshall and Barry Shearer. For the synthesis and application of this trienal, *see* Marshall, J. A.; Shearer, B. G.; Crooks, S. L. *J. Org. Chem.* **1987**, *52*, 1236.

26. Lei, B.; Fallis, A. G. *J. Am. Chem. Soc.* **1990**, *112*, 4609.

27. Lei, B.; Fallis, A. G. *J. Org. Chem.* **1993**, *58*, 2186.

28. Linders, J. T. M.; Kokje, J. P.; Overhand, M.; Lie, T. S.; Maat, L. *Recl. Trav. Chim. Pays-Bas* **1988**, *107*, 449.

29. Linders, J. T. M.; Briel, P.; Fog, E.; Lie, T. S.; Maat, L. *Recl. Trav. Chim. Pays-Bas* **1989**, *108*, 268.

30. Stambouli, A.; Chastrette, M.; Soufiaoui, M. *Tetrahedron Lett.* **1991**, *32*, 1723.

31. These Teflon containers can be used at up to 14 atmospheres of pressure (200 psi) and at temperatures below 250 °C because these vessels deform at temperatures greater than 250 °C, often with vessel rupture.

32. Ipaktschi, J.; Bruck, M. *Chem. Ber.* **1990**, *123*, 1591.

33. Giguere, R. J.; Namen, A. M.; Lopez, B. O.; Arepally, A.; Ramos, D. E.; Majetich, G.; Defauw, J. *Tetrahedron Lett.* **1987**, *28*, 6553.

34. Huber, R. S.; Jones, G. B. *J. Org. Chem.* **1992**, *57*, 5778.

35. Srikrishna, A.; Nagaraju, S. *J. Chem. Soc., Perkin Trans. I* **1992**, 311.

36. Ben Alloum, A.; Labiad, B.; Villemin, D. *J. Chem. Soc., Chem. Commun.* **1989**, 386.

37. Chen, S.-T.; Chiou, S.-H.; Wang, K.-T. *J. Chin. Chem. Soc.* **1991**, *38*, 85.

38. Lie Ken Jie, M. S. F.; Yan-Kit, C. *Lipids* **1988**, *23*, 367.

39. Villemin, D.; Ben Alloum, A. *Synth. Commun.* **1990**, *20*, 925.

40. Abramovitch, R. A.; Abramovitch, D. A.; Iyanar, K.; Tamareselvy, K. *Tetrahedron Lett.* **1991**, *32*, 5251.

41. Bram, G.; Loupy, A.; Majdoub, M.; Petit, A. *Chem. Ind.* **1991**, 396.

42. Ali, M.; Bond, S. P.; Mbogo, S. A.; McWhinnie, W. R.; Watts, P. M. *J. Organomet. Chem.* **1989**, *371*, 11.

43. Hwang, D. R.; Moerlein, S. M.; Lang, L.; Welch, M. J. *J. Chem. Soc., Chem. Commun.* **1987**, 1799.

44. Hwang, D.; Dence, C. S.; Gong, J.; Welch, M. J. *Appl. Radiat. Isot.* **1991,** *42,* 1043.

45. Thorell, J.; Stone-Elander, S.; Elander, N. *J. Labelled Compd. Radiopharm.* **1992,** *31,* 207.

46. Villemin, D.; Ben Alloum, A.; Thibault-Starzyk, F. *Synth. Commun.* **1992,** *22,* 1359.

47. Gruic, A.; Foucaud, A.; Moinet, C. *New J. Chem.* **1991,** *15,* 943.

48. Sun, W. C.; Guy, P. M.; Jahngen, J. H.; Rossamondo, E. F.; Jahngen, E. G. E. *J. Org. Chem.* **1988,** *53,* 4414.

49. Jahngen, E. G. E.; Lentz, R. R.; Pesheck, P. S.; Sackett, P. H. *J. Org. Chem.* **1990,** *55,* 3406.

50. Straathof, A. J. J.; van Bekkum, H.; Kieboom, A. P. G. *Recl. Trav. Chim. Pays-Bas* **1988,** *107,* 647.

51. Margolis, S. A.; Jassie, L.; Kingston, H. M. *J. Autom. Chem.* **1991,** *13,* 93.

52. Chan, S. T.; Chio, S. H.; Wang, T. K. *Int. J. Pept. Protein Res.* **1987,** *30,* 572.

53. Jassie, L. Chapter 10 in this volume.

54. Engelhart, G. Chapter 12 in this volume.

55. Chen, S.-T.; Wu, S.-H.; Wang, K.-T. *Int. J. Pept. Protein Res.* **1989,** *33,* 73.

56. Takano, S.; Kijima, A.; Sugihara, T.; Satoh, S.; Ogasawara, K. *Chem. Lett.* **1989,** 87.

57. Woudenberg, R. H.; Oosterhoff, B. E.; Lie, T. S.; Maat, L. *Recl. Trav. Chim. Pays-Bas* **1992,** *111,* 119.

58. Ley, S. V.; Mynett, D. M. *Synlett* **1993,** 793.

59. Varma, R. S.; Varma, M.; Chatterjee, A. K. *J. Chem. Soc., Perkin Trans. I* **1993,** 999.

60. Varma, R. S.; Chatterjee, A. K.; Varma, M. *Tetrahedron Lett.* **1993,** *34,* 3207.

61. Varma, R. S.; Chatterjee, A. K.; Varma, M. *Tetrahedron Lett.* **1993,** *34,* 4603.

62. Banik, B. K.; Manhas, M. S.; Kaluza, Z.; Barakat, K. J.; Bose, A. K. *Tetrahedron Lett.* **1992,** *33,* 3603.

63. Varma, R. S.; Lamture, J. B.; Varma, M. *Tetrahedron Lett.* **1993,** *34,* 3029.

64. Subbaraju, G. V.; Manhas, M. S.; Bose, A. K. *Tetrahedron Lett.* **1991,** *32,* 4871.

65. Delgado, F.; Cano. A. C.; Garcia, O.; Alvarado, J.; Velasco, L.; Alvarez, C.; Rudler, H. *Synth. Commun.* **1992,** *22,* 2125.

66. Villemin, D.; Lalaoui, M.; Ben Alloum, A. *Chem. Ind.* **1991,** 176.

67. Villemin, D.; Labiad, B.; Loupy, A. *Synth. Commun.* **1993,** *23,* 419.

68. Martinez, L.; Garcia, O.; Delgado, F.; Alvarez, C.; Patino, R. *Tetrahedron Lett.* **1993,** *34,* 5293.

69. Alvarez, C.; Delgado, F.; Garcia, O.; Medina, S.; Marquez, C. *Synth. Commun.* **1991,** *21,* 619.

70. Delgado, F.; Alvarez, C.; Garcia, O.; Penieres, G.; Marquez, C. *Synth. Commun.* **1991,** *21,* 2137.

71. Gordon, E. M.; Gaba, D. C.; Jebber, K. A.; Zacharias, D. M. *Organometallics* **1993,** *12,* 5020.

72. Villemin, D.; Labiad, B. *Synth. Commun.* **1990,** *20,* 3207, 3333.

73. Villemin, D.; Ben Alloum, A. *Synth. Commun.* **1990,** *20,* 3325, 3333.

74. Villemin, D.; Labiad, B. *Synth. Commun.* **1990,** *20,* 3213.

75. Petit, A.; Loupy, A.; Maillard, P.; Momenteau, M. *Synth. Commun.* **1992,** *22,* 1137.

76. Villemin, D.; Ben Alloum, A. *Synth. Commun.* **1991,** *21,* 63.

77. Alajarin, R.; Vaquero, J. V.; Garcia Navio, J. L.; Alvarez-Builla, J. *Synlett* **1992,** 297.

78. Petrow, J. A. *J. Chem. Soc.* **1946,** 884.

79. Vazquez-Tato, M. P. *Synlett* **1993,** 506.

80. Pilard, J. F.; Klein, B.; Texier-Boulet, F.; Hamelin, J. *Synlett* **1992,** 219.

81. Bose, A. K.; Manhas, M. S.; Ghosh, M.; Raju, V. S.; Tabei, K.; Urbanczyk-Lipkowska, Z. *Heterocycles* **1990,** *30,* 741.

82. Ojima, I.; Inana. S.; Yoshida, K. *Tetrahedron Lett.* **1977,** 3643.

83. Texier-Boullet, F.; Latouche, R.; Hamelin, J. *Tetrahedron Lett.* **1993,** *34,* 2123.

84. Manhas, M. S.; Bari, S. S.; Raju, V. S.; Ghosh, M.; Bose, A. K. *Abstracts of Papers,* Fourth Chemical Congress of North America, New York, American Chemical Society: Washington, DC, 1991; ORGN-195.

85. Molina, A.; Vaquero, J. J.; Garcia-Navio, J. L.; Alvarez-Builla, J. *Tetrahedron Lett.* **1993,** *34,* 2673.

86. Villemin, D.; Labiad, B.; Ouhilal, Y. *Chem. Ind.* **1989,** 607.

87. Abramovitch, R. A.; Bulman, A. *Synlett* **1992,** 795.

88. Loupy, A.; Pigeon, P.; Ramdani, M.; Jacquault, P. *J. Chem. Res.* **1993,** 36.

89. Barnier, J. P.; Loupy, A.; Pigeon, P.; Ramdani, M.; Jacqualt, P. *J. Chem. Soc., Perkin Trans, I.* **1993,** 397.

90. Villemin, D.; Labiad, B. *Synth. Commun.* **1992,** *22,* 2043.

91. Adamek, F.; Hajek, M. *Tetrahedron Lett.* **1992,** *33,* 2039.

92. Latouche, R.; Texier-Boullet, F.; Hamelin, J. *Tetrahedron Lett.* **1991,** *32,* 1179.

93. Giguere, R. J.; Herberich, B. *Synth. Commun.* **1991,** *21,* 2197.

94. Jones, G. B.; Chapman, B. J. *J. Org. Chem.* **1993,** *58,* 5558.

95. Rechsteiner, B.; Texier-Boullet, F.; Hamelin, J. *Tetrahedron Lett.* **1993,** *34,* 5071.

Chapter 9

Use of Microwave Heating in Undergraduate Organic Laboratories

George Majetich, Rodgers Hicks, and Edwin Neas

In this chapter, prepared in 1993, we present the following results: A comparison of a Williamson ether synthesis performed by a class of nonscience majors using conventional heating versus microwave heating; the adaptation of a dozen common organic reactions to this new technology; the modification of a known four-step synthesis of the diterpene carpanone; and an analysis of the advantage of incorporating microwave heating into undergraduate teaching programs.

The use of microwaves for heating chemical reactions has increased dramatically over the past seven years (*1*). Our contributions to this active field focus on the use of microwave heating to reduce the time necessary to achieve esterifications, hydrolyses, ether preparations, Diels–Alder reactions, Claisen rearrangements, oxidations, ene reactions, and halide exchange reactions (*2, 3*). This new technology has been enthusiastically embraced by industry and by those involved in the preparation of radioactively labelled medicinal agents with short half-lives (*4*).

Many organic reactions performed in instructional laboratories require heating for extended periods of time. We reasoned that microwave technology could easily be incorporated into undergraduate organic laboratories and would thereby increase the amount of time available for other learning experiences, such as further characterizations, purifications, or additional synthetic transformations. Indeed, several common undergraduate organic laboratory experiments were studied using microwave heating. In this review we present the following results:

- a comparison of a Williamson ether synthesis performed by a class of nonscience majors using conventional heating versus microwave heating

- the adaptation of a dozen common organic reactions to this new technology

- the modification of a known four-step synthesis of the diterpene carpanone, which undergraduate students can complete in four laboratory periods by using microwave heating, including the requisite purifications and characterizations

- an analysis of the advantages of incorporating microwave heating into undergraduate teaching programs

Equipment

Our early work was safely carried out in sealed thick-walled (3-mm) glass tubes by using domestic microwave units (2a). Unfortunately, neither the pressure nor the temperature can be controlled by using these ovens, so they are unsuitable for instructional or research purposes.

The microwave oven used for this study is equipped with a pressure-monitoring device and is marketed for the acid digestion of samples (Figure 1) (5). The magnetron supplies 630 ± 50 W of power, and effective power levels of 0–100% of the maximum value are achieved with 1-s cycle times. This instrument is designed so that the heating stops when a set pressure is reached; thus, the pressure monitor functions as a baristat and controls the reaction temperature indirectly. By using the carousel shown, as many as 12 reaction vessels may be heated simultaneously; however, the pressure and temperature of only one vessel can be monitored (6). Comparison of individual yields over a dozen fully loaded carousels indicates that pressure control allows reaction completion in all vessels with adequate elapsed time.

Organic chemists are accustomed to regulating the temperature of a reaction rather than pressure. Surprisingly, monitoring the temperature of a microwaved reaction in situ presented a dilemma. Mercury or alcohol thermometers could not be used because both mercury and alcohol absorb microwave energy and give erroneous readings. Standard thermocouples have a metal sheathing that heats up in a microwave field and gives biased temperatures. The solution to this problem was to monitor the temperature with a fiber-optic temperature probe (7). This probe operates by sending light of a known wavelength down a glass fiber cable until it reaches a sensing element, which reflects it back. The sensor is made of a material with a very high refractive index. As the temperature of the probe environment changes, the sensor refractive index changes, altering the color of the reflected light. Through internal computation, the reflected light is then correlated to a temperature that is digitally displayed. There-

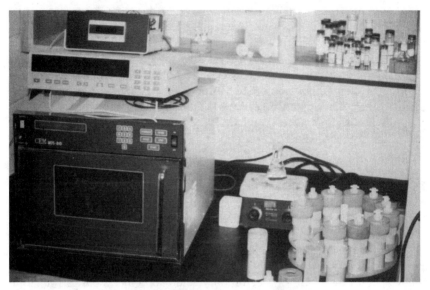

Figure 1. The microwave system.

fore, because there are no metallic components, the probe is transparent to the microwave field and gives accurate temperature measurements.

All experiments were performed in sealed Teflon [poly(tetrafluoroethylene)] acid digestion vessels. These vessels can be used at up to 18 atm of pressure (252 psi) and at temperatures below 200 °C. The vessels are inert to all of the chemicals used in this work. However, these Teflon reaction vessels deform at temperatures greater than 250 °C, often with vessel rupture, and therefore they require the previously mentioned pressure-control mechanism.

Figure 1 also illustrates how these reaction vessels appear both before and after assembly. The four basic components of these vessels are

- a polyetherimide outer shell
- a Teflon insert into which the reagents and solvents are placed (the volume is 250 mL)
- a seal cap with a pressure-release mechanism
- a threaded closure collar.

The seal cap fits on top of the Teflon insert to seal the reactants in and to provide a complete Teflon exposure to the contents. The pressure relief mechanism on the seal cap consists of a pressure sensitive rupture disc. If the internal pressure exceeds 250 psi the disc will rupture and the contents will safely vent into the oven exhaust system and presumably out into a fume hood.

A reaction vessel was modified to monitor the temperature and pressure. The seal cap on this vessel contains both the rupture release mechanism and a quartz sheath, sealed at one end, which serves to isolate the sensitive fiber optics cable from the reaction medium. Unlike Pyrex or soft glass, quartz does not absorb microwave radiation. The connector used for the quartz sheath also connects a Teflon tube to a pressure monitor that controls the microwave power input. The ability to control the reaction pressure and temperature, coupled with a knowledge of the limitations of the Teflon reaction vessels, allows one to use microwave systems safely for research and teaching.

Comparison of Heating Methods

To confirm the usefulness of microwave heating in undergraduate laboratories, we arranged for a class of students registered for Chemistry 261: Organic Chemistry for Non-Science Majors to be divided into two groups. Both groups of students carried out the preparation of 4-methylphenoxyacetic acid (1) by means of a common Williamson ether synthesis (8). One group used the conventional procedure, heating mantles and reflux condensers, whereas the second group charged the microwave reaction vessels with the necessary reagents and used the microwave oven as a heat source. Typical results are given in Table I. Moreover, all students in the microwave group finished at least 30 min earlier than the students in the control group.

Chart I presents a comparison of reactions carried out by using microwave heating versus conventional heating for 12 typical organic reactions. The experimental procedures investigated include a Friedel–Craft alkylation, two Fisher esterifications, two Diels–Alder reactions, two condensa-

Table I. Comparison of Results of Williamson Ether Synthesis

Actual Yield (g)	Yield (%)	Actual Yield (g)	Yield (%)
Student results by using microwave heating		Student results by using standard reflux	
2.8	73	1.96	51
2.79	73	3.22	84
3.28	86	0.19	5
2.51	66	2.43	64
3.95	103	3.00	78
3.50	92	3.07	80
2.31	60	2.99	78
2.94	77	1.57	41
3.33	87	1.88	49

tion reactions, a Beckmann rearrangement, a Williamson ether synthesis, the conversion of an alcohol into an alkyl halide, a Fischer indole synthesis, and an amide hydrolysis. All of these reactions are common reactions studied in introductory laboratory courses in organic chemistry (*8–11*). Note that the times required for reactions using microwave heating are significantly less than those using conventional procedures. Moreover, only four of these experiments required modification. For example, in the condensation reaction of benzaldehyde and fluorene to prepare 9-benzalfluorene (**7**), the solvent called for in the literature procedure is benzyl alcohol. Unfortunately, benzyl alcohol is not an efficient absorber of microwaves. Simply changing the solvent to ethanol overcomes this difficulty without complicating the reaction.

Likewise, the synthesis (*12*) of barbituric acid derivative **11** and the Diels–Alder reaction of anthracene with maleic anhydride required changing the solvent from xylene to dimethylformamine (DMF), which strongly couples with microwave energy. DMF has the additional advantage of being miscible with water so that on dilution of the reaction mixture with water the products precipitate and are easily isolated by filtration. Finally, the original procedure for the preparation of propyl *p*-tolyl ether (**9**) calls for using 25% aqueous NaOH and a phase transfer catalyst. We observed that microwave heating does not provide the agitation present in a refluxing solution and consequently the phase transfer agent is ineffective. Instead, the use of sodium hydroxide in ethanol provides a homogeneous system that overcomes this problem. In general, we have found that a wide variety of common undergraduate organic experiments are readily amenable to microwave heating in Teflon reaction vessels provided the reaction temperature does not exceed 250 °C.

Microwave-Based Synthesis of Carpanone

A short, elegant synthesis of carpanone (**14**, Scheme 1), a tropical tree product containing 20 carbon atoms and five contiguous asymmetric centers, was achieved by Chapman and co-workers in the early 1970s (*13*). This synthesis already has been presented as an undergraduate exercise in an early edition of Fieser and Williamson's laboratory manual (*9a*).

The synthetic sequence and the experimental conditions Chapman used are shown in Scheme 1. The first reaction, addition of allyl chloride to sesamol (**15**), is another example of a Williamson ether synthesis. Phenyl allyl ether **16** undergoes an *ortho*-Claisen rearrangement on thermolysis to yield 2-allylsesamol (**17**), which, in turn, is isomerized by strong base to 2-propenylsesamol (**18**). 2-Propenylsesamol is oxidatively coupled by using cupric ion to generate in situ enone **20**, which undergoes a hetero-Diels–Alder reaction to directly produce carpanone. In our experience more than four laboratory periods are required to complete this synthesis, with two additional laboratory periods needed to carry out purifications or spectroscopic analysis.

Chart I. The Experiments Examined

A = reaction temperature and pressure; B = reaction time; C = reaction solvent; D = reaction yield

Transformations and Reagents	Microwave Heating	Conventional Heating

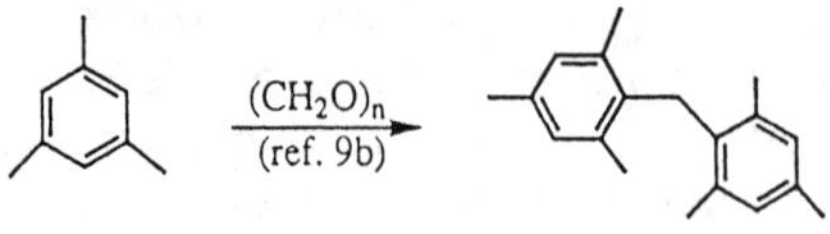

mesitylene → dimesitylmethane (2)
$(CH_2O)_n$ (ref. 9b)

A: 135 °C / 42 psi
B: 4 minutes
C: formic acid
D: 75%

reflux (101 °C)
2 hours
formic acid
95% (crude)

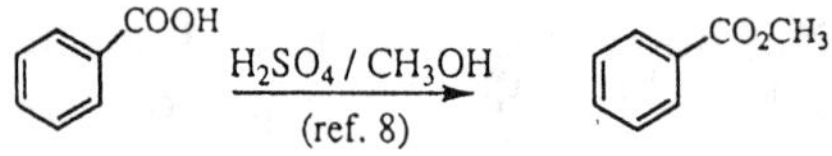

benzoic acid → methyl benzoate (3)
H_2SO_4 / CH_3OH (ref. 8)

A: 120 °C / 50 psi
B: 2 minutes
C: methanol
D: 90%

reflux (65 °C)
1 hour
methanol
62%

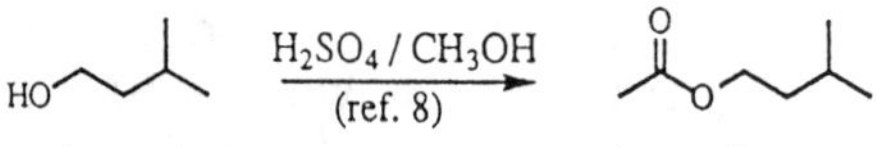

isoamyl alcohol → isoamyl acetate (4)
H_2SO_4 / CH_3OH (ref. 8)

A: 140 °C / 70 psi
B: 2 minutes
C: acetic acid
D: 100%

reflux (118 °C)
1 hour
acetic acid
80-90%

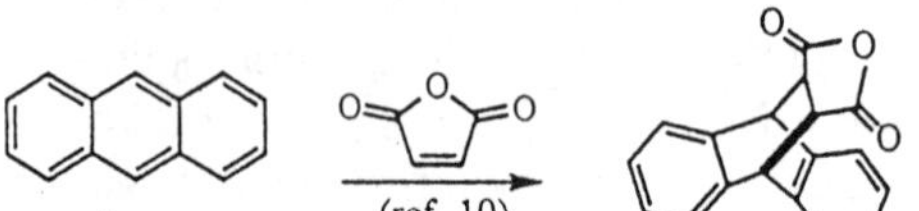

anthracene (ref. 10) → (5)

A: 190 °C / 15 psi
B: 4 minutes
C: DMF
D: 85%

reflux (144 °C)
1 hour
xylene
92%

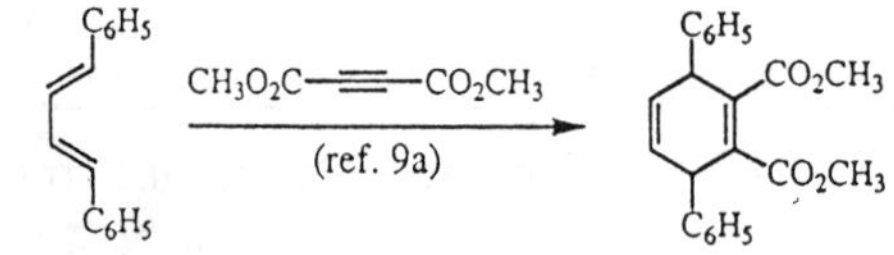

1,4-diphenyl-butadiene
CH_3O_2C—≡—CO_2CH_3 (ref. 9a) → (6)

A: 198 °C / 30 psi
B: 20 minutes
C: DMF
D: 58%

reflux (216 °C)
30 minutes
triglyme
95-100% (crude)

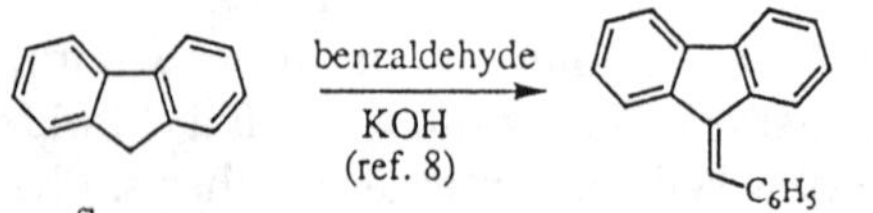

fluorene
benzaldehyde KOH (ref. 8) → 9-benzalfluorene (7)

A: 135 °C / 80 psi
B: 2 minutes
C: ethanol
D: 60%

100 °C
90 minutes
benzyl alcohol
not listed

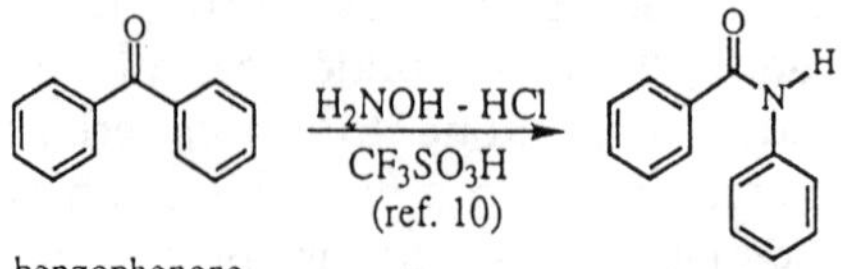

benzophenone
H_2NOH - HCl CF_3SO_3H (ref. 10) → benzanilide (8)

A: 170 °C / 90 psi
B: 2 minutes
C: formic acid
D: 99%

reflux (125 °C)
90 minutes
formic acid
70%

Chart I. Continued

A = reaction temperature and pressure; B = reaction time; C = reaction solvent; D = reaction yield

Transformations and Reagents	*Microwave Heating*	*Conventional Heating*

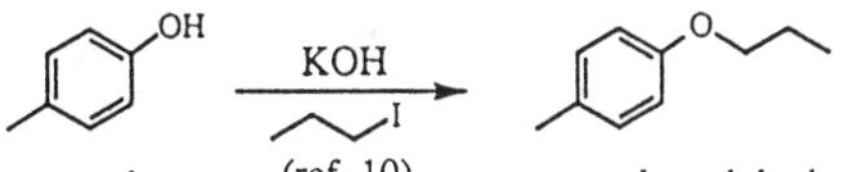

p-cresol →(KOH, I, ref. 10)→ propyl p-tolyl ether **(9)**

A: 131 °C / 90 psi
B: 4 minutes
C: ethanol
D: 96%

reflux (115 °C)
45-60 minutes
H$_2$O & catalyst
89% yield

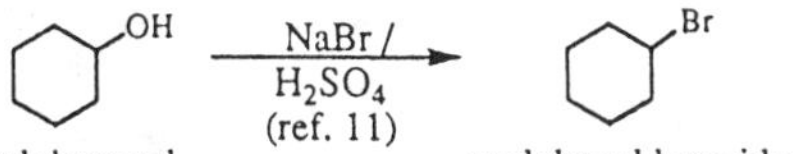

cyclohexanol →(NaBr / H$_2$SO$_4$, ref. 11)→ cyclohexyl bromide **(10)**

A: 140 °C / 10 psi
B: 10 minutes
C: water
D: 49%

reflux (118 °C)
30 minutes
water
33%

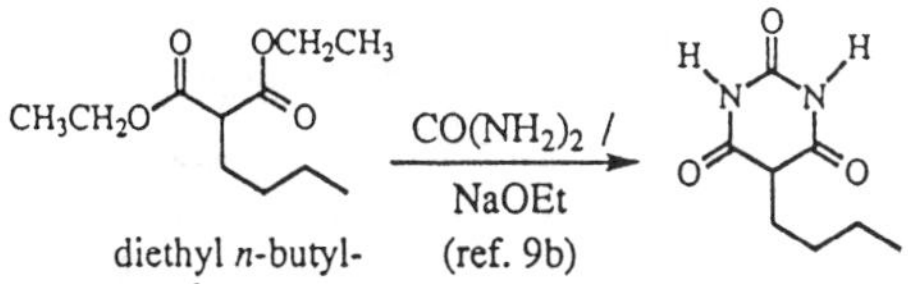

diethyl n-butyl-malonate →(CO(NH$_2$)$_2$ / NaOEt, ref. 9b)→ **(11)**

A: 142 °C / 90 psi
B: 6 minutes
C: ethanol
D: 80%

reflux (78 °C)
2 hours
ethanol
63%

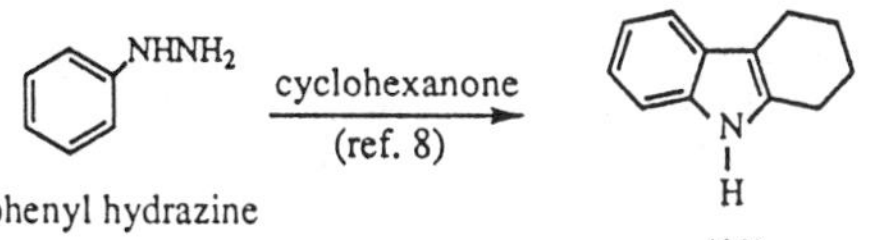

phenyl hydrazine →(cyclohexanone, ref. 8)→ **(12)**

A: 175 °C / 46 psi
B: 1 minute
C: acetic acid
D: 99%

reflux (78 °C)
15 minutes
acetic acid
100%%

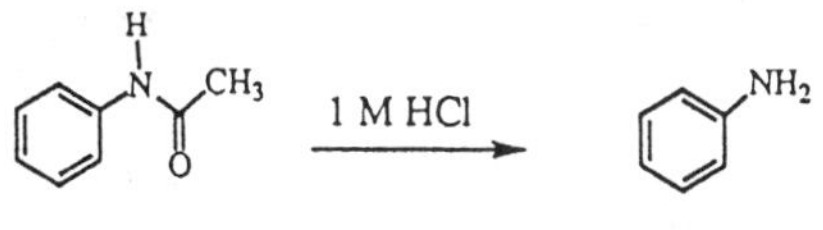

acetaniline →(1 M HCl)→ aniline **(13)**

A: 141-117 °C / 90 psi
B: 15 minutes
C: methanol / H$_2$O
D: 91%

reflux (79 °C)
4 hours
methanol / H$_2$O
98%

Using microwave heating, we were able to minimize the heating periods required so that the entire synthesis, complete with purifications and characterizations, can be carried out within four 4-h laboratory periods (Scheme 2). Conditions were developed so that the Williamson ether synthesis requires less than 3 min of heating, whereas the *ortho*-Claisen rearrangement is complete after 5 min of heating. At 195 °C, the base-catalyzed isomerization of **17** is finished in only 5 min by using potassium hydroxide instead of potassium *t*-butoxide. Here, ethanol is used as a cosolvent because dimethyl sulfoxide alone does not have a high enough vapor pressure at 170 °C to operate the oven pressure controller. The final step did not require modification.

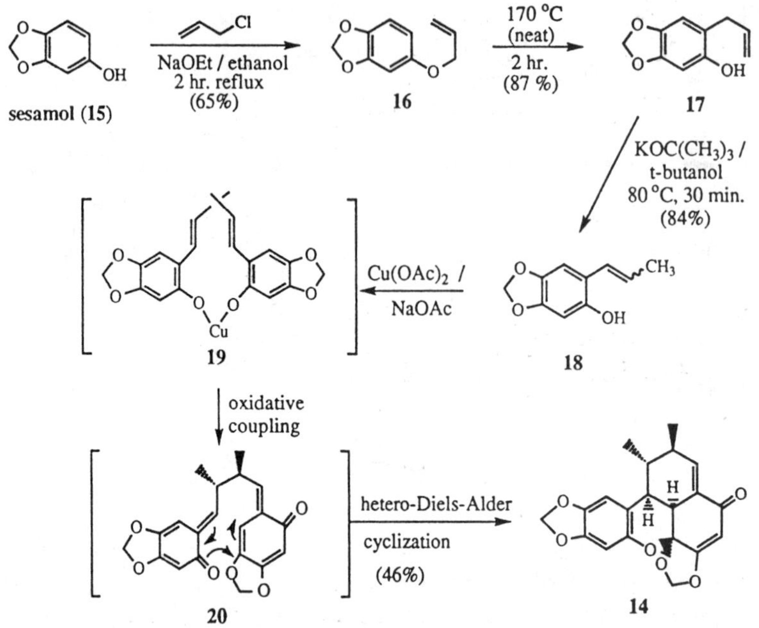

Scheme 1.

Because of the significant lab time saved, the students were able to purify their products by using flash chromatography (*14*) and to characterize each of the compounds they prepared by ^{1}H NMR and ^{13}C NMR spectroscopies, including a distortionless enhancement by polarization transfer (DEPT) experiment to establish the multiplicities of the carbon spectrum. We are presently examining other multistep syntheses where this technology should be applicable.

Advantages of Microwave Heating

Use of this technology has the potential to benefit students and instructors as well as the educational institution. The use of microwave technology can result in a more productive and less frustrating laboratory experience for the student. The reaction yields tend to be higher because thermal decomposition is minimized during the relatively short heating periods employed. Should students need to repeat an experiment, they have sufficient time to do so—*in the same lab period*—thereby avoiding frustration when things do not work as they are supposed to. Cleaning the Teflon reaction vessels is extremely simple, because even the worst residue does not adhere to the Teflon and is easily removed by using acetone. Thus, students using the microwave reaction vessels have fewer items to clean up at the end of the period. The benefit of these advantages to student morale is obvious.

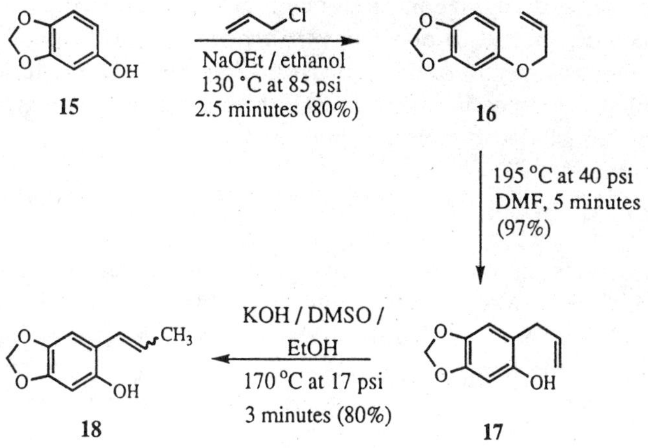

Scheme 2.

Ordinarily the first organic lab experiments are rarely related to the material being presented in lecture. As the course develops, the laboratory experiments often have no relationship to the material being covered in class. Reducing the time spent heating reactions offers instructors the opportunity to incorporate more examples of what they are covering in class into the lab itself.

Because the reactions are run in Teflon reaction vessels, the opportunity for glassware related injuries are lessened. Because heating is provided by the absorption of microwaves rather than an open flame, solvents can be used that are normally avoided because of the fear of fire.

Although the purchase of the initial equipment is presently expensive ($10,000), this investment is partially offset because stirring plates, heating mantles, glass round-bottomed flasks, and condensers may no longer be needed. This purchase can also result in minimizing the costs related to outfitting and maintaining the labs. In addition to a safer source of heating, instructors do not have to spend time checking that heating mantles and rheostats are operational. Furthermore, many undergraduate labs are not equipped with adequate hood space. Ordinarily this shortfall means that only a few students can safely carry out experiments at a given time. Because the microwave unit itself is designed with its own exhaust system, should any fumes be generated during heating they can be vented by means of a duct system, connected from the back of the oven into an existing hood or exhausted through either an open window or a table canopy hood.

Conclusions

Microwave heating has several advantages over conventional methods in terms of time, safety, reducing student frustration, and potential savings in equipment costs and ventilation requirements. More importantly, students may use the time saved to perform purifications or additional experiments, or to record spectra of their products. Alternatively, the time saved by using microwave heating may be used to pursue multistep syntheses that are often precluded because of time constraints.

Experimental Procedures

All of the reactions reported here were carried out in digestion vessels and a Model MDC-81D oven equipped with a pressure-monitoring system sold by CEM Corporation (5, 15). A MetriCorp fiber-optic system was used to monitor the temperature (7). Although we operated the microwave oven within a fume hood, additional safety measures were not necessary.

Unless otherwise indicated, a power setting of 35% is sufficient for a single reaction vessel. Full power should only be used for batches containing 12 reaction vessels. The pressure and temperature of only one of the carousel reaction vessels need to be monitored.

As a safety precaution, after heating is complete the reaction vessel is permitted to cool in the oven for at least 5 min before the carousel is removed. Further cooling may then be completed by immersing the carousel in an ice bath.

All solvents and reagents were distilled prior to use. References to ether imply diethyl ether, whereas all chromatography was performed over silica gel. Rotary evaporators operating at water aspirator pressure were used to concentrate organic phases. Finally, control experiments were run by using conventional heating for those reactions that required modifications to facilitate the use of microwave heating. These studies are presented where relevant.

Preparation of 4-Methylphenoxyacetic Acid (1)

A microwave reaction vessel is charged with 5 g of potassium hydroxide, 10 mL of water, 2.5 g of *p*-cresol, and 7 mL of 50% aqueous 2-chloroacetic acid. Microwave heating at 140 °C and 30 psi for 2 min is sufficient for complete conversion. After cooling and venting, the reaction mixture is acidified with 5 mL of concentrated HCl. The resulting solution is extracted with two 25-mL portions of diethyl ether and the combined organic layers are washed with 25 mL of water. The 4-methylphenoxyacetic acid is extracted into two 25-mL portions of 5% sodium carbonate leaving any unreacted *p*-cresol in the organic phase. The combined aqueous extracts are acidified with concentrated HCl. Suction filtration and air drying provide 4-methylphenoxyacetic acid as a white solid in 70% yield. The product is sufficiently pure that recrystallization is not necessary.

Preparation of Dimesitylmethane (2)

A microwave reaction vessel is charged with 3.65 mL of mesitylene, 0.37 g of paraformaldehyde, and 2.3 mL of formic acid (90%). The resulting mixture is heated to 135 °C at 42 psi for 4 min. After heating, the vessel is permitted to cool to below 60 °C so that the vessel may be vented safely. Crystallization of the product is slow and may require seeding. The crude dimesitylmethane is collected by suction filtration and purified by washing with ice cold toluene–methanol (14:1), followed by saturated aqueous sodium carbonate and recrystallization by using toluene–methanol (14:1). A typical yield is 75%.

Preparation of Methyl Benzoate (3)

A microwave reaction vessel is charged with 0.5 g of benzoic acid, 3 mL of anhydrous methanol, and 0.2 mL of concentrated sulfuric acid (95–98% ACS-grade reagent). The vessel is sealed and heated to 120 °C at 50 psi for 2 min. After heating, the vessel is permitted to cool and then is carefully vented. Workup consists of diluting the cold reaction mixture with 15 mL of diethyl ether and washing with 5-mL portions of water, saturated aqueous sodium bicarbonate, and brine. The organic layer is dried over anhydrous sodium sulfate, filtered, and concentrated to obtain pure methyl benzoate (3) in greater than 90% yield.

Preparation of Isoamyl Acetate (4)

A microwave reaction vessel is charged with 2.5 mL of acetic acid, 2.0 mL of isoamyl alcohol, and 0.5 mL of concentrated sulfuric acid (95–98% ACS-grade reagent). The reaction mixture is heated to 140 °C at 70 psi for 2 min, cooled, and vented. Workup consists of diluting the cold reaction mixture with 15 mL of diethyl ether and washing with 5 mL portions of water, saturated aqueous sodium bicarbonate, and brine. The organic layer is dried over anhydrous sodium sulfate, filtered, and concentrated to obtain pure isoamyl acetate (4) in quantitative yield.

Diels–Alder Reaction Between Anthracene and Maleic Anhydride

Anthracene (2.0 g), maleic anhydride (1.0 g), and 25 mL of dimethylformamide are added to a microwave reaction vessel and heated to 190 °C for 4 min at 15 psi. After cooling and venting, 25 mL of water are added to the reaction mixture to precipitate the Diels–Alder adduct. Unfortunately, the crude adduct, isolated by using suction filtration, also contains unreacted anthracene. This contaminant can be removed by recrystallization from xylene. A typical yield of pure 5 is 85%.

Diels–Alder Reaction Between 1,4-Diphenylbutadiene and Diethylacetylenedicarboxylate

1,4-Diphenylbutadiene (1.00 g) and diethylacetylenedicarboxylate (1.23 g) in 15 mL of DMF are heated at 70% power for 20 min, during which time the reaction temperature declines from 198 °C to 194 °C while at 30 psi. The reaction mixture is permitted to cool and is diluted with water and extracted four times with ether. The combined ether extracts are washed with saturated brine, dried over anhydrous magnesium sulfate, and concentrated at reduced pressure. The residue is chromatographed on silica gel (hexanes:ether, 10:1) to isolate adduct **6** as a pale yellow solid (1.05 g, 58%).

In the control experiment using conventional heating, an identical reaction mixture was refluxed until thin-layer chromatographic (TLC) analysis showed complete consumption of the limiting reagent (6 h). Identical workup and chromatography gave 1.22 g (67%) of adduct **6**.

Preparation of 9-Benzalfluorene (7)

Potassium hydroxide (1.25 g) dissolved in 12 mL of absolute ethanol is placed in a microwave reaction vessel that is then charged with 2.5 mL of benzaldehyde and 2.5 g of fluorene. The reaction mixture is heated to 135 °C for 1 min at 80 psi and then cooled to below 70 °C before venting. Addition of water produces a brown solid that may be recrystallized from hexane to give pure 9-benzalfluorene (**7**) (mp 75–76 °C). A typical yield of pure **7** is 60%.

Preparation of Benzanilide (8)

A microwave reaction vessel charged with 1.0 g of benzophenone, 0.5 g of hydroxylamine hydrochloride, and 5.0 mL of triflic acid–formic acid solution is heated at 170 °C for 1 min at 90 psi. (The triflic acid–formic acid solution is prepared by adding 2 drops concentrated triflic acid per 5 mL 90% formic acid. Because of the hazardous nature of triflic acid the instructor should prepare this solution in advance.) After cooling, the reaction mixture is extracted with one 30-mL portion of methylene chloride, followed by two additional 10-mL portions. The combined organic extracts are dried over anhydrous sodium sulfate, filtered, and evaporated. The crude solid may be recrystallized from 95% ethanol to give pure benzanilide (**8**) (mp 62–64 °C) in approximately 80% yield.

Williamson Synthesis of Propyl p-Tolyl Ether (9)

A microwave reaction vessel is charged with 8.0 mL of p-cresol, 7.5 mL of propyl iodide, and 20 mL of a 10% potassium hydroxide in absolute ethanol solution. The resulting mixture is heated to 131 °C for 4 min at 90 psi. After cooling and venting, the reaction mixture is diluted with an equal volume of water and extracted with 50 mL of diethyl ether. Unreacted p-cresol may be removed by washing with 20 mL of aqueous NaOH followed by 20 mL of water. Crude propyl p-tolyl ether is obtained by washing the ethereal phase with brine, drying over anhydrous sodium sulfate, filtering, and concentrating at reduced pressure. Pure **9** is isolated in approximately 65% yield by vacuum distillation (bp 210 °C at 1 atm).

Preparation of Cyclohexyl Bromide (10)

Cyclohexanol (2.00 g) is added to a solution of 3.1 g of NaBr in 7 mL of concentrated sulfuric acid (95–98% ACS-grade reagent) and 7 mL of water. The mixture is heated at 70% power to 10 psi. This pressure is maintained for 10 min during which time the temperature rises from 137 °C to 140 °C. After cooling the mixture it is diluted with water and extracted with hexanes. The organic phase is washed with water containing a trace of NaHSO$_3$, washed with brine, and then extracted twice with concentrated sulfuric (95–98% ACS-grade reagent). The resulting mixture is washed again with water, dried over anhydrous magnesium sulfate, and filtered. Subsequent concentration at reduced pressure gave 1.60 g (49%) of cyclohexyl bromide (**10**) as an amber colored oil.

Preparation of n-Butyl Barbituric Acid (11)

Commercially available sodium ethoxide (1.4 g) is dissolved in 10 mL of absolute ethanol, and the resulting ethoxide solution is added to 4.3 g of diethyl *n*-butylmalonate in a microwave reaction vessel. A solution of 1.2 g of urea in 12 mL of ethanol is then added to the reaction vessel, and the reaction mixture is heated for 6 min at 142 °C at 90 psi. After cooling, the mixture is acidified with 10% aqueous HCl. After the volume is reduced to one-half by evaporation at reduced pressure, crude *n*-butylbarbituric acid crystallizes in approximately 80% yield. (A typical crude melting range is 195–202 °C.) Further purification may be effected by recrystallization from boiling water. The melting point of pure **11** is 209–210 °C.

Preparation of 1,2,3,4-Tetrahydrocarbazole (12)

Freshly distilled phenylhydrazine (5.0 mL), cyclohexanone (5.5 mL), and glacial acetic acid (40 mL) are combined in a microwave reaction vessel and heated to 175 °C for 1 min at 46 psi. Cooling the reaction vessel to below 100 °C requires approximately 7 min. Addition of 70 mL of water, followed by cooling (0 °C), results in the formation of crystalline 1,2,3,4-tetrahydrocarbazole **12** in nearly quantitative yield. If required, this product may be recrystallized from methanol (mp 117–119 °C).

Hydrolysis of Acetanilide

Acetanilide (1.00 g) is added to 20 mL of 1 N HCl in methanol–water (1:1). The mixture is heated to 90 psi for 15 min during which time the temperature declines from 141 °C to 117 °C. The cooled mixture is diluted with saturated aqueous $NaHCO_3$ until no longer acidic. Following extraction with ether (4×20 mL), the organic phases are dried over anhydrous sodium sulfate, filtered, and concentrated to give 571 mg of aniline (**13**) in 83% yield.

In a control experiment, an identical mixture was refluxed until TLC analysis indicated complete consumption of acetanilide (4 h). Identical workup gave 679 mg of **13** (98%).

Preparation of Sesamol Allyl Ether (16)

Sodium (0.5 g) is dissolved in 13 mL of absolute ethanol and placed under an atmosphere of dry nitrogen. The resulting solution is transferred to the Teflon-sleeved reaction vessel that is then capped. A solution of 2.5 g of sesamol (**15**) in 5 mL of absolute ethanol and 2.1 g of allyl chloride are added to the reaction vessel by syringe through one of the tubing fittings. The mixture is heated at 35% power for 1 min at 85 psi during which time the temperature remains at 130 °C. The cooled mixture is diluted and extracted three times with ether. The combined organic extracts are washed with 5% aqueous NaOH and water, dried over anhydrous magnesium sulfate, filtered, and concentrated to give ether **16** (3.11 g, 96%) as a light brown oil.

As a control experiment, an identical mixture was refluxed for 1 h until TLC analysis indicated complete consumption of the limiting reagent (i.e., sesamol). Identical workup provided 2.9 g (89%) of **16**.

Preparation of 2-Allylsesamol (17)

Sesamol allyl ether (5.12 g) in 10 mL of DMF is heated at 50% power and 40 psi to 195 °C. This temperature is maintained for 5 min. After cooling, the mixture is diluted with water and extracted three times with ether. The combined organic layers are extracted twice with 5% aqueous NaOH, and the resulting aqueous solution is acidified (pH < 2) by 10% aqueous HCl addition and then extracted with ether. The combined ethereal extracts are dried over anhydrous sodium sulfate and concentrated to give 4.96 g of allylsesamol (97%) as amber crystals.

In the control experiment using conventional heating, an identical mixture was refluxed for 72 h. The mixture was subjected to identical workup to give 2.08 g (72%) of **17**.

Preparation of 2-Propenylsesamol (18)

2-Allylsesamol (200 mg) dissolved in 10 mL of dimethyl sulfoxide is added to a solution of KOH (500 mg) in 5 mL of ethanol. The mixture is heated at 35% power to 17 psi. This pressure is maintained for 3 min during which time the temperature remains at 170 °C. The cooled mixture is acidified (pH < 2) by addition of 10% aqueous HCl, diluted with water, and extracted three times with 30-mL portions of ether. The organic phases are washed with water, dried over anhydrous magnesium sulfate, and concentrated. The residue is chromatographed on a silica gel column (eluted with hexanes–ether, 1:1) to give 160 mg (80%) of **18** as light brown crystals.

In the control experiment using conventional heating, an identical mixture was refluxed 4 h and subjected to identical workup and chromatography to give 179 mg of **18** (90%). In both cases a trace of unreacted 2-allylsesamol remained, which was chromatographically inseparable from the product.

Preparation of Carpanone (14)

2-Propenylsesamol (1.0 g) prepared in the previous step is dissolved in a solution of 5 mL of methanol and 10 mL of water containing 3 g of sodium acetate. To this mixture, a solution of 1 g of cupric acetate in 5 mL of water is added dropwise with stirring. The mixture is stirred for 5 additional minutes and the resulting suspension is filtered. The filtrate is diluted with twice its volume of water and extracted with three 35-mL portions of ether. The combined ether extracts are washed with two 50-mL portions of brine and dried over anhydrous magnesium sulfate. The drying agent is removed by filtration and the ether is evaporated. The resulting crude residue is taken up in 2 mL of warm carbon tetrachloride and set aside to crystallize. If crystals do not form overnight, seed crystals are added and the resulting solution is allowed to stand to promote crystallization. On the basis of X-ray crystallographic studies, each molecule of carpanone cocrystallizes with a molecule of carbon tetrachloride (mp 185 °C). Thus, the theoretical yield of this reaction is 1.4 g. Typical yields ranged from 30% to 40%.

Acknowledgments

Appreciation is extended to CEM Corporation for the donation of equipment to carry out this investigation. Gratitude is also extended to John Barbaro, the supervisor for the undergraduate organic laboratories at The University of Georgia (UGA), for his assistance in carrying out the Williamson ether synthesis comparison. Special thanks are also extended to Brenda Smith, Scott Wagner, James White, and Ning Zhang—four UGA CHM 342 students—for volunteering to carry out the carpanone synthesis in place of their regularly scheduled experiments.

References

1.   For reviews of microwave heating in organic synthesis, *see:* (a) Majetich, G; Wheless, K. Chapter 9 in this volume. (b) Abramovitch, R. A. *Org. Prep.*

Proced. Int. **1991**, *23(6)*, 683-711. (c) Mingos, D. M. P.; Baghurst, D. R. *Chem. Soc. Rev.* **1991**, *20*, 1. (d) Giguere, R. J. *Organic Synthesis: Theory and Applications;* JAI Press: Greenwich, CT, 1989; Vol. 1, pp 103–172.

2.   (a) Giguere, R. J.; Bray, T. L.; Duncan, S. M.; Majetich, G. *Tetrahedron Lett.* **1986**, *27*, 4945. (b) Giguere, R. J.; Namen, A.; Lopez, B.; Arepally, A.; Ramos, D.; Majetich, G.; Defauw, J. *Tetrahedron Lett.* **1987**, *28*, 6553.

3.   *See* also: (a) Gedye, R. N.; Smith, F. E.; Westaway, K. C.; Ali, H.; Baldisera, L.; Lasberge, L. Rousell, J. *Tetrahedron Lett.* **1986**, *27*, 279. (b) Gedye, R.; Smith, F. E.; Westaway, K. C. *Can. J. Chem.* **1988**, *66*, 17. (c) Bose, A. K.; Manhas, M. S.; Ghosh, M.; Raju, V. S.; Tabei, K.; Urbanczyk-Lipkowska, Z. *Heterocycles* **1990**, *30*, 741. (d) Bose, A. K.; Manhas, M. S.; Ghosh, M.; Shah, M.; Raju, M.; Bari, S. S.; Newaz, S. N.; Banik, B. K.; Chaudhary, A. G.; Barakat, K. J. *J. Org. Chem.* **1991**, *56*, 6968.

4.   (a) Hwang, D. R.; Moerleing, S. M.; Lang, L.; Welch, M. J. *J. Chem. Soc., Chem. Commun.* **1987**, 1799. (b) Stone-Elander, S. A.; Elanderr, N. *Appl. Radiat. Isot.* **1991**, *42*, 885. (c) Thorell, J. -O.; Stone-Elander, S.; Elander, N. *J. Label. Cmpd. Radiopharm.* **1992**, *31*, 207.

5.   For additional information regarding the MDS-81D microwave unit used, contact: CEM Corporation. Phone (800) 726–3331. Since this work was carried out, CEM has producted a new unit, Model MDS-2000, having a built-in thermooptic temperature contol device in addition to the built-in pressure control device.

6.   The turntable in this unit changes the direction of rotation every 360° so that the monitoring cables do not become entangled.

7.   A MetriCor (Model 1400) fiberoptic multisenor system was used. For additional informations, contact: MetriCor, 18800 142nd Ave NE, Woodinville, WA 98072 Phone: (206) 483–557.

8.   Durst, H. D.; Gokel, G. W. *Experimental Organic Chemistry*, 2nd Ed.; McGraw-Hill: New York, 1987.

9.   (a) Fieser, L. F.; Williamson, K. L. *Organic Experiments;* 3rd Ed.; Heath: Lexington, MA, 1975. (b) Fieser, L. F.; Williamson, K. L. *Organic Experiments,* 6th Ed.; Heath: Lexington, MA, 1987.

10.   Mayo, D. W.; Pike, R. M.; Butcher, S. S. *Microscale Organic Laboratory*, 2nd Ed.; Wiley: New York, 1989.

11.   Ault, A. *Techniques and Experiments for Organic Chemistry*, 2nd Ed.; Holbrook: Boston, MA, 1976.

12.   Barbituric acid itself and derivatives bearing one alkyl substituent are physiologically inactive. The wide variety of commercially important barbiturate drugs are disubstituted derivatives.

13.   Chapman, O. L.; Engel, M. R.; Springer, J. P.; Clardy, J. C. *J. Am. Chem. Soc.* **1971**, *93*, 6667.

14.   Still, C.; Kahn, M.; Mitra, A. *J. Org. Chem.* **1978**, *43*, 2923.

15.   The experiments reported herein have also been duplicated in other commercial microwave systems having pressure and temperature monitoring capabilities.

Chapter 10

Microwave-Assisted Inorganic Reactions

David R. Baghurst and D. Michael P. Mingos[*]

In contrast to the high level of interest shown by organic chemists in microwave processing techniques, inorganic chemists have been more conservative. Only in the past two years has a significant growth in the number of publications occurred in this area. The aim of this chapter is to review the applications of microwave-heating techniques in inorganic chemistry, to discuss how new inorganic microwave applications may evolve, and to emphasise the specific advantages of microwave heating over conventional heating techniques.

The number of applications of microwave-heating techniques in chemistry is growing very rapidly. Application areas include chemical analysis, polymer chemistry, organic synthesis, and ceramic processing (1–4).

Mechanism of Microwave Heating

The fundamental mechanisms of microwave absorption (1) by the majority of liquids and solids is well-established. For example, polar liquids interact with the electric-field component of microwave electromagnetic radiation through their dipoles. Therefore, the mechanism of microwave heating in common solvents such as ethanol, methanol, and water is dipolar relaxation.

Conduction due to mobile electrons, holes, or ions is the most important absorption mechanism in those solid materials that absorb microwave energy efficiently (CuO, NiO, V_2O_5, etc.). Even though the conduction contribution at room temperature may be small, many solids become good electrical conductors at high temperatures. This result is particularly true of room-temperature insulators such as pure alumina and semiconductors such as NiO.

*Corresponding author.

In ionic solutions the dipolar contribution to microwave interaction may be supplemented and even dominated by a conduction contribution. Mobile ions in solution can migrate under the influence of the oscillating electric field associated with microwaves. As the frequency of electromagnetic radiation is lowered the time scale for stimulating ionic motion increases. As a result the ionic contribution would be expected to be larger at lower frequencies. From the simple model the size, mass, number, and mobility of the ions as well as the polarity and viscosity of the solution will all be important factors in determining the extent of interaction between the microwave radiation and the solution in question. Illustrative data for various concentrations of NMe_4Cl in methanol is given in Figure 1. At high concentrations the dielectric loss, a measure of the interaction between a material and an oscillating electric field, in the microwave region is dominated by conduction losses. However, as the concentration is reduced a dielectric relaxation is revealed and the dielectric loss peaking is close to the domestic-microwave-oven operating frequency of 2.45 GHz.

A dipolar contribution can also arise in solid materials. Polar molecular entities within a solid such as water may be only partially bound to the solid structure, allowing them to interact with a microwave field. This interaction is responsible for the rapid melting of ice (the small fraction of free water absorbs microwaves strongly, whereas the solid ice itself is transparent to microwaves). Dipolar contributions may also arise because of the distribution of vacancies, interstitial atoms, defects, and impurities

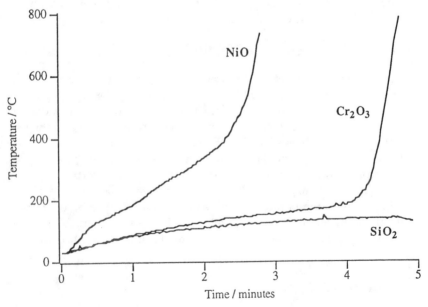

Figure 1. Dielectric loss of ionic solutions of NMe_4Cl in MeOH.

across grain boundaries within the solid that can give rise to fluctuating dipoles under the influence of microwave radiation.

Predicting Rate of Microwave Heating

For cases where no significant interaction occurs between the microwave magnetic field component and the materials under investigation, the complex permittivity, ϵ^*, alone is necessary in quantifying the interaction between the material and the radiation. The complex permittivity can be expressed in terms of a real part ϵ', the dielectric constant, and an imaginary part ϵ'', the dielectric loss. Unfortunately the relationship between the dielectric properties of a material and its heating rate in a microwave field is an extremely complex one. One expression that is often quoted is (1)

$$\text{heating rate} = \frac{\text{constant} \times \text{dielectric loss} \times \text{frequency} \times \text{field intensity}}{\text{specific heat capacity} \times \text{density}} \quad (1)$$

The derivation of this relationship involves assumptions that are invalid for the majority of microwave-heating applications, namely that the material absorbs all of the energy that penetrates the material and the intensity of the radiation is constant throughout the material. The expression serves to establish the complexity of the problem. The field intensity in an empty multimode microwave cavity will vary in a complex fashion from point to point within the cavity. Even in a single-mode cavity where the empty field distribution can be calculated relatively easily, the introduction of dielectric material in the form of a sample will distort the field distribution out of all recognition. Not only will the field distribution vary with position inside the cavity but it will also vary inside the material under test. As the dielectric properties alter, perhaps as a consequence of heating or a chemical reaction, the field distribution changes.

The electric field distribution, dielectric loss, sample density, and specific heat capacity can all be expected to vary with the temperature of the material. Indeed even the frequency of the magnetron would be expected to drift within its operating range in cases where the load and magnetron are poorly isolated. Given that the physical properties of the material are well-known over the range of temperature, quantitative predictions can be made concerning the rate of microwave heating for specific geometries.

The ability to predict temperature evolution has been the focus of a number of investigations. In particular, the thermal runaway phenomenon by which the temperature of a material increases very rapidly beyond some threshold temperature has been studied (5, 6). In Figure 2, 55-mm high *tall-form* fireclay crucibles were filled with samples of Cr_2O_3, SiO_2, and NiO and were heated in a 650-W domestic microwave oven. Each inorganic oxide behaves in a unique fashion reflecting its microwave

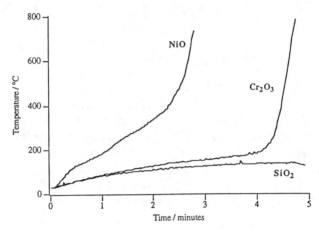

Figure 2. Temperature profiles for microwave heating of a range of inorganic oxides.

dielectric properties. Nickel oxide, a room temperature semiconductor, absorbs microwave energy strongly and rapidly achieves temperatures in excess of 700 K. At the opposite extreme, silica, a poor microwave absorber, achieved only 400 K even after 5 min heating. However, for chromium oxide a rapid acceleration in the heating rate occurs once the material achieved a temperature of about 440 K.

Some of the best modelling results have come from studies by Davis and co-workers (7–10), where the electric field and thermal conduction equations were solved simultaneously by using Maxwell's equation and a finite element model in both one and two dimensions. Temperature-dependent dielectric and thermal properties for a range of materials have been accounted for by using fifth-order polynomials. In this work the thermal runaway observed in Nylon 66 strands was modelled. No theoretical model is complete without experimental verification, and the group observed the irregular variation of heating rate with sample geometry for cylindrical samples irradiated by a plane wave (Figure 3). Rather than finding a smooth variation in power absorption and heating rate with increasing cylinder diameter, certain sample sizes seemed to be favoured and led to high absorption rates. The effect falls off with increase in cylinder diameter and is the result of a kind of 'resonant' absorption of microwave energy for sample sizes of radii 0.5, 0.82, and 1.18 cm.

Interpreting Observed Rate Accelerations

Reactions in Solution at Atmospheric Pressure

Much early enthusiasm for microwave chemistry arose from reports of dramatic rate enhancements that were observed for a wide range of reaction classes. The existence of a specific microwave effect has, and contin-

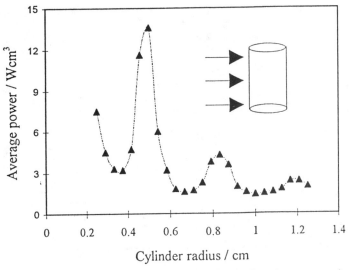

Figure 3. Average power absorbed in cylindrical samples of water as a function of sample radii (electric field is oriented along the long axis of the cylinder).

ues to be, a source of some controversy and has even been the sole subject of at least one meeting (*11*). With the recent development of more reliable ways of monitoring the reaction temperature and pressure in a microwave field many of the early results in support of a microwave effect can be satisfactorily explained on purely thermal grounds.

In early work on the applications of microwave heating to the synthesis of organometallic and coordination compounds in a microwave reflux system, we found rate enhancements on the order of 5–20-fold in comparison to our experience of conventional reflux syntheses. These reactions, as will be discussed in detail subsequently, were performed in organic solvents, mostly alcohols. However, experiments initially by Majetich, Neas, and co-worker (*12*), and later by ourselves (*13*) and Whan and co-workers (*14*) have shown that solvents at reflux in microwave cavities can boil at temperatures considerably higher than their accepted conventional boiling points. In Table I, a selection of microwave boiling points is given.

We proposed (*13*) a model that can be used to explain the boiling point increase seen for some solvents. Conventional nucleate boiling relies on the existence of cavities, pits, and scratches on the surface of the vessel in contact with the liquid in which bubbles can form. Bubble inception further relies on the existence of a vapour embryo trapped inside the crevice by the bulk liquid. When the temperature of the surrounding liquid is at least equal to the saturation temperature corresponding to the pressure in the vapour embryo, bubble growth occurs. In the simplest model of bubble growth it is proposed that evaporation occurs all around the bubble–liquid interface. The energy for evaporation is supplied by a superheated

Table I. Effect of Microwaves on Observed Boiling Points of a Range of Solvents

Solvent	Normal bp (K)	Microwave bp (K)	Change (K)	Heating Rate (K/s)	Initial Power Absorption (%)
Water	373	377	+4	1.01	>95
Ethanol	352	376	+24	2.06	90
Methanol	338	357	+19	2.11	>95
Dichloromethane	313	328	+15	2.16	75
Tetrahydrofuran	339	354	+15	2.04	—
Acetonitrile	354	380	+26	2.36	>95
2-Propanol	355	373	+18	2.11	>95
Acetone	329	354	+25	2.23	90
1-Butanol	391	405	+14	1.87	80
1,2-Dimethoxyethane	358	379	+21	2.54	—
Diglyme	435	448	+13	2.17	—
Ethyl acetate	351	368	+17	1.78	70
Acetic anhydride	413	428	+15	1.97	—
Isoamyl alcohol	403	422	+19	1.92	85
2-Butanone	353	370	+17	2.57	—
Chlorobenzene	405	423	+18	2.63	90
Trichloroethylene	360	381	+21	1.54	45
Dimethylformamide	426	443	+17	2.18	>95
1-Chlorobutane	351	373	+22	2.59	>95
Isopropyl ether	342	358	+16	1.90	50

NOTES: Power values are % power absorption to nearest 5% assuming full power, 650 W. Dashes indicate data not available.

liquid layer that surrounds the bubble. Once the forces holding the bubble in place are overcome the bubble is released from the cavity and boiling occurs. As the volume of the superheated liquid layer is increased the number of active embryo sites increases as the layer temperature approaches the saturation temperature of more and more sites.

This model emphasises the importance of available crevices with dimensions of the order of micrometers and a superheated layer adjacent to these crevices. In conventional heating thermal energy is supplied to the vessel walls themselves by a heating mantle, oil bath, etc. The walls of the vessel in turn heat the liquid in the vessel. In some regions superheated layers will exist adjacent to embryo sites. Thus, boiling occurs from the surface of the containment vessel at suitable nucleation sites. The inversion of the temperature profile observed when the vessel is heated by microwaves accounts for the observation of high boiling points. In microwave heating the outer walls of the flask are continuously cooled by convective air flow. The layer of solvent immediately adjacent to the walls may be somewhat below the temperature of the bulk liquid. When the bulk temperature rises to the boiling point, heating of the layer of liquid adjacent to the walls of the vessel approaches the saturation temperature of embryo sites.

As a consequence of this model, the differences in behaviour observed for water and the common organic solvents may be explained. The number of vapour embryos trapped in the surface of the vessel is determined in part by the ability of the solvent to wet the surface. Most organic solvents wet the surface of glassware well and diminish the vapour trapping and retention capabilities. As a result, the number of nucleation sites is small and a high temperature is required before boiling can occur. On the other hand water wets surfaces poorly and large numbers of active sites will be available.

That different groups report different tables of microwave boiling points can probably be attributed to the fact that the microwave-heating equipment used for each study was unique. In the light of the described model factors that would be expected to influence the degree of microwave superheating include the nature of the solvent itself, the size of sample studied, the amount of microwave power used on the experiment, and the nature of the surface of the containment vessel. Although the boiling point elevations observed by Majetich and Neas are rather larger than those we report (by 5–10 K), they used a more powerful microwave source and larger solvent volumes.

Reactions in Solution at Elevated Pressure

In studies on the acceleration of chemical syntheses under pressure during microwave heating, very dramatic rate enhancements have been observed (10^2 or greater). Reactions under pressure occur at higher temperatures no matter what source of heating is chosen. Typical graphs of reaction temperature versus reaction pressure are shown in Figure 4 for methanol and ethanol in equipment that will be described later. Note that while an increase in reaction pressure from 1 to 2 atm (1 atm = 101325 Pa) is large in terms of the temperature gained, this advantage begins to fall off at higher pressures. From this type of data and a simple Arrhenius-type analysis (*see* Table II), the order of rate enhancements achieved when reactions are performed under pressure can be predicted.

We studied the temperature versus pressure profiles achieved in a glass reaction vessel for a range of solvents under both microwave and conventional thermal heating conditions (*15*). Within the error of our experiments (±2 K), no difference existed between the temperatures under the two heating techniques. This result is in agreement with our crude model of microwave boiling point elevation because liquids in sealed containers do not boil. Instead an equilibrium is achieved between molecules in the liquid and gas phase. The properties of the vessel surface are not involved.

Rate of Solid-State Processes Under Microwave Heating

The situation in the solid state is far from resolved (*16, 17*). Sintering experiments on the mullitization and densification of [$3Al_2O_3 + 2SiO_2$]

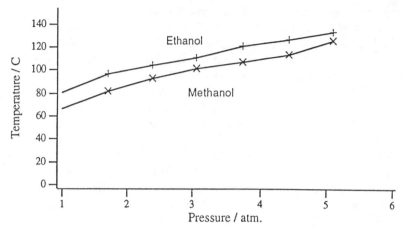

Figure 4. Temperature versus pressure for methanol and ethanol.

powder compacts have failed to confirm the presence of a specific micro-wave effect (*18*) while in sintering studies on partially stabilised zirconia microwave enhanced sintering is clearly demonstrated with densification occurring at a lower temperature when a high frequency electric field is applied (*19*).

The overall rate of any chemical reaction is governed by the speed of the slowest step or steps. For many solid-state reactions the slowest step is the diffusion of the reactants. Any process that can increase the bulk diffu-sion coefficient could lead to a dramatic enhancement in the observed rate of reaction. Many authors continue to claim that some solid-state processes occur either at lower temperatures or much faster under microwave heat-ing than they do by conventional heating. Specifically, much of the micro-wave work on the processing of ceramic materials and glasses suggests that enhanced transport properties occur when dielectric heating effects are applied.

Studies on the sintering of Al_2O_3–TiC composites have shown that it is possible to achieve a density of 93% at 2023 K where conventionally a tem-perature of around 2153 K would be required (*20*). Comparative data for microwave and conventional sintering of alumina indicates more rapid densification in a microwave field. The apparent activation energy for microwave sintering of alumina (160 kJ/mol) is less than one third that observed for conventional sintering (575 kJ/mol). Normal grain growth in dense fine grained Al_2O_3–0.1% MgO has been studied under both conven-tional furnace and 28 GHz microwave furnace annealing conditions (*21*). The microstructural changes that occurred were the same for both sets of samples: bubble microstructures were observed and the aspect ratios and shape factors did not change during the annealing cycles. The kinetics of grain growth were greatly increased by the 28 GHz microwave anneal, for

Table II. Rate Enhancements Expected on the Basis of Arrhenius-Type Behaviour

Temperature (K)	Rate Constant, k (1/s)	Time to 90% Complete
300	1.55×10^{-7}	4126 h
350	4.76×10^{-5}	13.4 h
400	3.49×10^{-3}	11.4 min
450	9.86×10^{-2}	23.4 s
500	1.43	1.61 s

NOTE: Values are based on a first-order reaction with an activation energy, E_a of 100-kJ/mol and a preexponential factor of $A = 4 \times 10^{10}$ mol^{-1} s^{-1}.

example, the grain growth rate at 1773 K in the microwave furnace was the same as the rate at 1973 K in the conventional furnace. The activation energy for grain growth was reduced by the microwave anneal from 590 to 480 kJ/mol.

Cation diffusion in solid right circular cylinders of Pyrex has been studies by back scattered electrometer microscopy profiling (22). Each cylinder was placed in a 50 cm^3 Pyrex crucible and was packed with either CoO (conventional case) or a concentrated solution of $Co(NO_3)_2 \cdot 6H_2O$ (microwave case). Diffusion occurred to a depth of 40 mm in a sample heated conventionally to 1198 K for 60 min. However using microwave heating for just 15 min at a maximum temperature of 1023 K the diffusion depth was found to be 70 mm. Additionally several cobalt rich phases had developed in the microwave-heated sample that were absent in the conventional sample. Microwave processing at 1308 K for just 10.5 min gave a diffusion depth of some 200 mm.

Although there are many supporters of a specific microwave effect in the solid state others believe that our knowledge of the temperature inside a solid-state reaction during microwave heating may be in error. Some claim that the temperature cannot be measured with any level of certainty by using the available equipment. Even though substantial problems with temperature measurement have existed in the past the situation is improving. Others suggest that because microwave heating is to some degree selective some components of a reaction mixture may absorb a much larger fraction of the available microwave energy. Consider, for example, microwave absorption by metal or graphite fibers inside a block of poorly absorbing polymer. Not only would the microwaves be absorbed selectively through the graphite or carbon fibers, but the heating effect would be expected to be concentrated in the surface of the fibers because microwaves cannot penetrate very far inside strongly absorbing materials. In this model the average temperature recorded by a thermocouple might be a rather poor estimate of the true temperature of the boundary between the fibers and polymer matrix. Arguments such as these have been used to explain the improvements achieved in carbon fiber–epoxy bonding in composites processed by using microwave-heating techniques (23). The authors of this particular paper believe that there results are consistent with an interface temperature some 75 K higher than the bulk temperature. In similar work a finite element model has been developed to predict the temperatures differences that can be generated when using short pulses of microwave energy to heat mixtures of pyrite and calcite (24).

Recently a number of papers have appeared that aim to explain theoretically why diffusion processes would seem to be accelerated under microwave conditions. A stochastic analysis has been applied to the mobility of ions in ceramic materials. New barrier values are found to be in better agreement with diffusion experiments (25). Another group developed a model for nonthermal effects on ionic motion by proposing that the microwave radiation couples into low (microwave) frequency elastic lattice oscillations, generating a nonthermal phonon distribution that enhances ion mobility and thus diffusion rates (26). The effect of the microwaves is not to reduce activation energy but rather to make the use of a Boltzmann thermal model inappropriate. The group developed a simplified linear oscillator model to qualitatively explore coupling from microwave photons to lattice oscillations. The linear mechanism possibilities include resonant coupling to weak bond surface and point defect modes and nonresonant coupling to zero-frequency displacement modes. Nonlinear mechanisms such as inverse Brillouin scattering are suggested for resonant coupling of electromagnetic and elastic travelling waves in crystalline solids. The models suggest that nonthermal effects should be more pronounced in polycrystalline forms and at elevated bulk temperatures.

Summary

Many inorganic reactions in solution at atmospheric pressure will be accelerated by a factor of up to 10 if the reactions are performed by using microwave heating rather than conventional heating techniques. Where long reflux times are usual, the microwave-heating technique provides an invaluable tool for reducing the time scale. In the teaching laboratory a wider range of inorganic reactions can be studied within the same time frame.

These reactions are accelerated compared with reactions at atmospheric pressure regardless of the heating technique. No specific advantages exist in using microwave-heating techniques if the reactants remain in solution because the same temperature is achieved in conventional heating. However, the availability of equipment originally designed for analytical work should see a growth in the use of microwaves for reactions under pressure because of the greater convenience.

It is not possible at this stage to state unequivocally whether a specific microwave effect exists in some solid-state processes. Some of the reactions for which a specific microwave effect has been proposed probably have been performed with poor temperature control and the effects may be purely thermal.

Microwave Equipment for Laboratory Chemistry

Until recently workers interested in developing new chemical applications of microwave heating have had to rely solely on their own ingenuity in designing suitable apparatus. Much early work was performed in domestic microwave ovens with appropriate modifications for the introduction of inlet and outlet ports; reflux and pressure systems; and monitoring systems such as thermocouples, gas thermometers, and pressure transducers. Although established equipment is available from a number of suppliers for the analytical applications of microwave heating, only in the past few years has equipment been marketed for synthetic applications. Unfortunately, the price of this equipment is high. For those workers wishing to try microwave chemistry for the first time, the best route remains a home-built modified domestic microwave oven. At the other end of the scale workers with established microwave know-how may wish to upgrade their systems for single-mode work, tuned multimode cavity, or larger scale applications.

Modifying a Domestic Microwave Oven

Ideally a domestic microwave oven for laboratory work should have a switched mode power supply based on the latest compact inverter technology. These ovens are capable of delivering a range of true powers rather than average powers based on a duty cycle. In other words, 50% power should be just that rather than an average produced by 15 s at full power and 15 s off. For our own work we favour ovens with as few luxuries as possible because they are in general easier to modify and at lower cost. As far as power is concerned the requirement obviously varies with the application. For work with polar solvents the power required can be calculated if the heat capacity, volume of solvent, and required heating rate are known in advance. Microwave heating of most organic solvents is greater than 85% efficient on volumes above 50 cm^3. A sturdy construction is also an advantage, especially where work under pressure is likely!

Ports on microwave cavities (sometimes called chokes) are necessary to prevent microwave leakage. Most chokes are based on conducting cylinders such as copper or brass tubes that are well-earthed to the walls of the microwave cavity, perhaps by bolting down a flanged mount. Microwave ports can be regarded as cylindrical waveguides. Given the dimensions of a port and the frequency of the radiation (2.45 GHz) it is possible to calculate the attenuation achieved with a particularly port geometry with equations available from any electromagnetic engineering text. A point to note, though, is that if the port is designed to take a reflux condenser, account should be made for the rather different dielectric properties of solvents compared with air. If the reflux is vigorous a condenser can become an efficient means of transmitting microwave energy from the cavity because the microwaves are guided by the cylindrical condenser waveguide. A microwave oven modified for the study of organometallic reactions under microwave heating has been described (27).

The methods available for measuring temperature inside a material under microwave heating are still far from ideal. Well-grounded thermocouples can be used to measure the temperature inside a solid or powder sample (Figure 5). Grounding the sheath of the thermocouple prevents the wires of the thermocouple itself from coming into contact with the microwave field. In the example shown the sheath of the thermocouple is mounted on a base plate that can be bolted to the floor of the microwave cavity. For applications where aqueous and acidic solutions are in use grounded thermocouples have been successfully used. However, for work with organic solvents the use of thermocouples is not advised because of the risk of sparking if the surface of the thermocouple sheath should become damaged or the grounding contact deteriorate.

For solutions and solid applications where thermocouples are inappropriate, fluoroptic thermometry, infrared pyrometry, and a simple gas thermometer are all options. The limitations of fluoroptic thermometry (28) are the high cost involved, the delicate nature of the probes, and the limited temperature range for each type of probe. Infrared pyrometry is limited in that only the surface temperature can be measured, and for accurate measurements material emmisivity must be known over the entire range of study. A simple gas thermometer of the type described by Whan et al. (29) is probably the cheapest solution for solution work. However, we have found it necessary to recalibrate the thermometers between each experiment to obtain repeatable measurements.

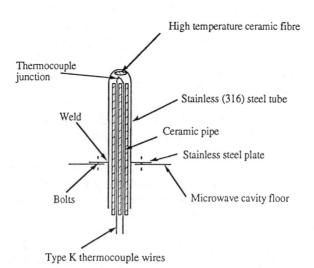

Figure 5. Thermocouple arrangements for studying reactions in a microwave field.

Reaction pressure is a more readily monitored and controlled parameter in a microwave experiment because the pressure transducer itself can be mounted outside the microwave oven. Typically the pressure can then be transmitted from the reaction vessel to the transducer via a glass or Teflon tube. Commercial temperature and pressure monitor–controllers can then be used to control the reaction environment by, for example, interrupting the microwave oven circuitry with a relay. Personal computers with appropriate expansion cards can also be employed to control experiments and log data. A vessel suitable for inorganic work has been described in the literature (30). The advantages of this system over those based on Teflon autoclaves are reduced cost, increased lifetime, higher available temperature range to >573 K, and the transparency of the Pyrex vessel, which allows the reaction progress to be monitored.

For reaction mixtures that are viscous and where stirring is necessary, a pressure system incorporating stirring was described by Strauss and co-workers (31). A domestic microwave oven modified for use in heating mineral acids in open vessels has been described by Pougnet (32). An extraction unit was built into an existing microwave cavity to prevent vapours reaching the oven electronics where a spark could lead to an explosion. The cavity walls have also been lined to prevent corrosion.

When modifying microwave ovens for laboratory use a reliable hand-held microwave monitor is mandatory. Currently, according to legislation the maximum allowed exposure to microwave radiation is 10 mW/cm^2 for up to 2 min in any 1-h period. Short-term exposure is permissible up to 25 mW/cm^2. The most sensitive parts of the human body to prolonged microwave exposure are the eyes and the male sex organs. Coagulation of protein in the lens of the eye leads to visible white flecks and grey cataracts. Damage to the spermatic ducts may lead to temporary or permanent sterility.

Advanced Microwave-Heating Equipment

The electromagnetic field distribution used for microwave heating can be defined more accurately by using a single-mode cavity as the applicator. Such cavities come in a number of shapes and geometries, although rectangular cavities based on sections of commonly available rectangular waveguide and cylindrical cavities based on metal tubes are the most common. A sample rectangular TE_{10N} ($N \geq 1$) mode cavity is illustrated in Figure 6. The important features of the cavity are the coupling iris or loop through which the microwaves enter the cavity and the dimensions of the cavity are determined through the position of the tuning plunger that controls the resonant frequency (33). By adjusting both the iris and cavity dimensions it is possible to selectively excite specific electromagnetic modes.

The value of moving to single-mode cavities is that focused heating, particularly applicable to work on ceramic bonding, and high power densities for processing relatively poorly absorbing materials then become

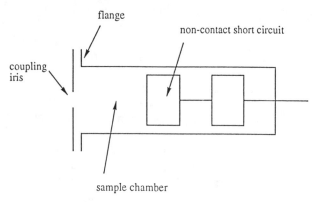

Figure 6. A TE_{10N}-mode rectangular cavity.

possible. As the properties of the materials processed in such equipment changes, the cavity length and coupling characteristics must be continuously monitored. Poor coupling not only results in low efficiency but can also lead to magnetron damage if circulators are not used to isolate reverse power. Automatic methods have been described previously, but the problem is far from trivial particularly in processing materials that exhibit sharp changes in dielectric properties (34, 35).

An improvement in the efficiency of microwave heating in multimode cavities can also be made by similar tuning techniques (36). Other authors have tackled the problem of multimode oven-field inhomogeneity by moving to high microwave frequencies of 28 GHz. At higher microwave frequencies the wavelength is somewhat shorter resulting in a smaller distance between peaks and troughs in the electric field. The result is that the thermal conductivity of the material can smooth out temperature differences between hot and cold areas on this smaller scale. Unfortunately, sources of microwave radiation at frequencies other than the domestic microwave oven frequency of 2.45 GHz are too expensive at this time to be commercially viable.

Other methods for smoothing the field distribution include mixing a very large number of microwave single modes in a controlled fashion (37). Alternatively, by moving away from the use of cavity magnetrons to travelling wave tubes it is even possible to obtain a smoother field by continuously varying the frequency of the applied electromagnetic field (38).

An unusual idea has also been described where microwave and ultrasound sources are combined in the same reactor. The apparatus has been used to study two classes of organic reactions (pyrolysis and esterification), and rate enhancements over conventional reactions were reported (39).

Applications of Microwave in Inorganic Chemistry

For clarity the applications of microwave-heating techniques to inorganic chemistry have been divided according to the nature of the materials formed. Thus, classical coordination compounds such as Vaska's complex $[Ir(CO)Cl(PPh_3)_2]$ will be discussed separately from the hydrothermal synthesis of zeolite A. In terms of the apparatus used to perform the reactions (sealed Teflon vessels in domestic microwave ovens), the reactions are very similar. Additionally, in both cases the mechanism by which the reaction mixtures are heated is the strong microwave absorption by the reaction solvent: dimethylformamide in the case of Vaska's complex and water in the case of zeolite A. In both cases the initial reaction solutions contain high concentrations of ionic species and as a result there will be both dipolar and conductivity contributions to the interaction with microwave radiation.

Metal Oxides

Aspects of ceramic processing such as sintering and joining have been described in several excellent reviews and are beyond the scope of this chapter (*40–43*). We focus on chemical transformations. Calcination, carbothermal, hydrothermal, and pyrolysis reactions have all been studied by using microwave-heating techniques. Calcination and carbothermal routes generally involve forming intimate mixtures and sometimes pressed compacts of the reactants. One or more of the reactants will be a strongly microwave-absorbing semiconducting or conducting solid. Hydrothermal routes are generally reliant on the microwave absorption by ionic aqueous solutions in sealed pressure vessels.

Calcination

Microwave heating of solid-state reaction mixtures is the most established application in inorganic chemistry. In early work it was shown that a range of inorganic oxides could be rapidly synthesised in domestic microwave ovens, with minor modifications for controlling the reaction atmosphere and protecting the cavity from the large temperatures generated inside the samples (*44*). Samples of superconducting $YBa_2Cu_3O_{7-\varsigma}$ were synthesised with high onset temperatures (96.5 K) from stoichiometric mixtures of Y_2O_3, CuO and $Ba(NO_3)_2$ (*45*). In further work on these important materials other workers have compared the morphology of samples produced by both conventional and microwave techniques (*46, 47*).

Single-mode techniques have been used to process other ceramic oxides. The specific advantage of microwave processing by using these techniques is that extremely fast heating rates can be achieved that are difficult to generate by any other method. As a result samples often have unique morphologies. Morphology is important in influencing bulk physical properties such as dielectric properties and hardness. Samples of $Ba_{0.65}Sr_{0.35}TiO_3$ were synthesised under fast firing conditions. Even though conventional calcination required process temperatures of 1423 K for 5 h, comparable samples were produced in 45 min by microwave heating to 1373 K. The microwave products developed a fairly uniform and fine microstructure with promising dielectric properties (48). Relatively low powers of 300–400 W were used to synthesise $MgAl_2O_4$ and ZrO_2 dispersed matrix composites in a cylindrical TM_{012} mode cavity (49). Again, fast firing conditions that can be routinely achieved through microwave-heating techniques led to fine-grain structures.

Other chemical transformations have been reported in single-mode microwave-heating experiments by Gasgnier et al. (50). Thus, V_2O_3 has been converted into a mixture of higher oxides including V_6O_{13} and V_2O_5, anatase (TiO_2) has been converted into rutile, and Cu_2O and Pb_3O_4 have been reduced to Cu and PbO, respectively. All of these reactions were completed within 10 min.

Carbothermal Reactions

The industrial potential of microwaves as an alternative form of heating for carbothermal processes has been investigated. Reduced oxides of titanium and vanadium have been synthesised by microwave heating of TiO_2 and V_2O_5, respectively, with amorphous carbon (1). Reduction of Fe_2O_3 to iron metal itself has also been studied (51). Here, the conventional carbothermic process has not been successful in practice because of the inability of conventional heating processes to supply heat to the interior of composite pellet samples. Reaction cold centres result in low gas concentrations and poor reaction rates. The volumetric nature of microwave heating could lead to a more successful process. The applicability of the technique in extraction of iron from hematite and magnetic ores is likely to depend strongly on the type of ore in question. In preliminary work final products ranging from simple white cast iron to complex super-hard alloy irons have been produced (52). The technique has been applied to a wider range of oxide minerals (53).

Very rapid, selective, single-step deoxygenation of layer and chain containing oxides MoO_3, CrO_3, V_2O_5, α-$VOPO_4$ $2H_2O$, and $Ag_6Mo_{10}O_{33}$ has been achieved by carbothermal reduction using graphitic carbon (54). The products are MoO_2, Cr_2O_3, VO_2, VPO_4, and a mixture of Ag and MoO_2, respectively. Although conventional methods for preparing these materials are tedious, the microwave method is simple, fast, and yields homogeneous products of good crystallinity.

Hydrothermal Reactions

Hydrothermal microwave techniques were first developed for the synthesis of zeolites, and these applications will be discussed shortly. Subsequently hydrothermal work has been directed toward the synthesis of a range of ceramic oxide powders (55) and even samples of superconducting $YBa_2Cu_3O_{7-\varsigma}$ (56). In the former paper the syntheses of crystalline uranary oxides such as TiO_2, ZrO_2, and Fe_2O_3 and binary oxides such as $KNbO_3$ and $BaTiO_3$ are described. Additionally, the workers report the synthesis of a new layered Al_2O_3 phase that they successfully intercalated with ethylene glycol. The effect of varying experimental parameters such as chemical species, pH, time, and reaction temperature have been studied through measurements of crystal size, morphology, and level of agglomeration. Microwave hydrothermal processing may prove to be a valuable process in the low-temperature production of fine ceramic powders.

Curiously, the products of conventional hydrothermal reactions are not always the same as the microwave reactions. In the synthesis of titania powders from 0.5 M $TiCl_4$, Komarneni et al. (55) reported that the microwave reaction (2 h at 437 K) gave a pure rutile phase. The conventional preparation with the same experimental parameters gave anatase with only a small amount of rutile. Much longer reaction times were necessary to produce a predominantly rutile product, and even after 72 h some anatase was present along with amorphous material. In this specific case, the intermediate anatase, which is itself a partial absorber of microwaves, likely may be heated directly. As a result the temperature recorded in the aqueous phase may not be a good indication of the temperature of the titania precipitate.

The same workers also reported the synthesis of electroceramic powders (57), $BiFeO_3$ (58), and hematite (59). In the synthesis of hematite from 1 M $Fe(NO_3)_3$, whereas microwave hydrothermal processing led to the crystallisation of hematite, the conventional reaction under the same conditions led to the crystallisation of a mixture of goethite and hematite.

Pyrolysis Reactions

The precursors for pyrolysis reactions are combustible metal-organics such as metal alcoholates, esters, B-diketonates, amides, imides, or mixed esters and amides. During the early stages of heating reaction mixtures based on these materials, a rapid endothermic reaction occurs that generates sufficient heat to drive the reaction to completion very rapidly. Because the reactions can be completed very rapidly, unique and unusually morphologies and chemistries can be achieved. Additionally, because the precursor mixtures are generally formed by sol–gel techniques, complete intimate mixing of the reactants can be achieved. This result leads, in principle, to more homogeneous products than can be achieved by reacting granular materials.

Conventional pyrolysis reactions can be an unreliable means of performing a reaction suffering from poor repeatability. The rationale for this fact is that because conventional furnaces heat materials from the outside-in, the outer layers of material in a sample may react completely before the inner layers have ignited. Because the atmospheres associated with pyrolysis reactions are generally reducing (high levels of CO), the inner layers of a sample may be oxygen deficient compared with the outer layers, and a gradient-type ceramic material will result. Microwave pyrolysis provides a solution to this dilemma by providing a means to simultaneously ignite the inside as well as the outside of a sample. For those special cases where gradient-type materials are desired, the microwave characteristics of the sample can be tuned to provide the correct gradient.

A large range of ceramic materials was produced in this fashion by Willert-Porada and co-workers (60) including titania-, alumina-, and zirconia-based ceramic oxides. A review of the work is available (60). Other authors have adapted the techniques to the synthesis of the Tl-based 2212 and 2223 high-temperature superconductors (61).

Metal Borides and Carbides

A range of metal borides (CrB, Fe_2B, ZrB_2) was synthesised by direct combination of the elements in an inert atmosphere in a modified domestic oven (62). Metallic implements can lead to dramatic and destructive arcing when placed in a microwave field. For this reason, reactions involving finely divided metal powders have only recently begun to attract attention. By dispersing fine metal powders within less strongly absorbing materials, such as boron, it is possible to use microwave heating of the metal powder to drive synthetic reactions. The syntheses of metal borides are particularly challenging reactions for the microwave chemist because conventionally very high reaction temperatures of the order 1900–2500 K are required for syntheses by the direct reaction of the elements. Additionally, the metal borides often show metallic conductivity that rivals that of the metals from which they were prepared, and to increasing difficulties in the microwave syntheses are increased as the reaction proceeds because of arcing.

Microwave-assisted carbothermal reduction of the corresponding oxides has been used for the synthesis of TiC and TaC (63). For both reactions, an increase seems to occur in the reaction rate compared with the conventional heating process. In the case of TaC the increase corresponds to a 3-fold increase in the preexponential factor.

Metal Chalcogenides

The potential for coupling microwave radiation selectively to some phases has been the motivation for work on the desulfurisation of coal (64–66). Mined coal contains large quantities of microwave absorbing FeS_2, which can be removed in a two-stage process. Insoluble FeS_2 is decomposed to FeS by heating to >573 K. FeS is removed by washing in dilute HCl. If the FeS_2 can be selectively heated to >573 K the organic coal matrix will be preserved.

The production rather than the removal of metal chalcogenides was the focus of other work (Table III) (*67*). In a typical reaction the chalcogenides S, Se, or Te (1–5 g) were mixed with metal powders Cr, Mn, Sn, Fe, and Ta (1–5 g). The mixtures were sealed in 10-cm silica tubes under vacuum. When the silica tube was placed on a firebrick in a multimode cavity and exposed to 650-W microwave power, the temperature rose rapidly to 1100–1300 K and led to pure samples of metal chalcogenides in less than 10 min.

The total reaction time is not controlled by the rate of heating of the metal powder but by the volatility of the chalcogenide. Volatile sulfur tends to sublime to cooler parts of the tube away from the metal particles. For these reactions it is necessary to cool the sample following 30-s heating at full power and then to shake the tube to redistribute the reactants. This process is repeated several times leading to a total process time of up to 30 min. The fact that the metal particles heat so rapidly enables the reaction to occur before significant sublimation of the chalcogenide from the reaction site. In conventional heating low heating rates are necessary to prevent tube explosions resulting from the pressure build-up associated with the unreacted chalcogenides. Interestingly, the presence of unreacted sulphur is indicated by a blue plasma in the silica tube. The loss of this plasma provides an excellent guide to reaction completion.

The scope of these syntheses was extended to include several phases of chalcopyrite semiconductors (*68*). Polycrystalline samples of $CuInS_2$, $CuInSe_2$, and CuInSSe were prepared from stoichiometric mixtures of the pure elements by microwave radiation. The reactions were performed in sealed quartz tubes in as few as 3 min.

Table III. Metal Chalcogenides Synthesised by Microwave Techniques

Compound	Metal (g)	Chalcogen (g)	Time (min)	Anneal (h)
CrS	2.08	1.28	8.5	17
Cr_7S_8	2.08	1.50	8.5	10
Cr_2S_3	2.08	1.94	8.5 + 20	1
CrSe	2.60	3.95	5	5.5
Cr_3Se_4	2.60	5.27	5 + 5	8
Cr_2Se_3	2.08	4.78	5	3
Cr_2Te_3	1.734	6.35	<1	14
α-MnS	2.74	1.60	1	10 min
Fe_7S_8	2.79	1.60	1 + 10	1
TaS_2	4.52	1.60	20	12 + 12
SnS_2	2.37	1.60	3	10 min

Metal Halides

Microwave-heating techniques can also be employed in reactions between metal powders and gases (69). A range of metal chlorides, oxychlorides, and bromides was synthesised by using either a metal sample suspended inside a vessel containing the gaseous reactant or a fluidized bed-type reactor. The technique has considerable potential where small samples of high purity specimens are required. By combining metal powders with high-oxidation-state metal chloride it is also possible to synthesise solid-state cluster compounds by reactions of the type shown in Reaction 1 (67).

Intercalation Compounds

The intercalation of organic and organometallic compounds into layered oxide and sulfide structures has attracted considerable interest in recent years. The kinetics of the intercalation processes are often slow, and consequently even after refluxing for several days these reactions do not always proceed to completion. Ultrasound has been shown to increase the rates of intercalation reactions but can result in loss of crystallinity of the samples as with the thermal methods. This result makes structural characterisation of the intercalated compounds by using X-ray powder techniques problematical.

By using a sealed Teflon vessel several pyridine intercalates of the layered mixed oxide α-VO(PO$_4$) were prepared (70). In a typical experiment α-VO(PO$_4$) and the pyridine (or an equivalent volume of pyridine in xylene) were placed in the Teflon autoclave and exposed to 650 W of 2.45-GHz microwave radiation for several minutes. The length of each microwave exposure was limited to about 5 min by the pressure limitations of the autoclave. The maximum temperature reached was 473 K. Dramatic rate enhancements of the order 10^2–10^3 were generally observed (Table IV).

An additional advantage of the microwave technique is the quality of the X-ray powder data obtained from the more crystalline microwave products. Blank experiments established that whereas α-VO(PO$_4$) • 2H$_2$O itself does not strongly absorb microwaves, the intercalated products do. Therefore, in the initial stages of the reaction the high-temperature rise associated with the solution is primarily responsible for the rate enhancement. These efficient heating effects lead to pressures of about 50 atm and temperatures of 473 K within 1 min. In the latter stages of the reactions autocatalytic effects associated with the absorption of microwaves by the product may contribute to the completion of the intercalation reactions. These autocatalytic effects would have no analogs in conventional conductive heating.

[119]Sn Mössbauer spectroscopy studies have shown that attempts to intercalate three aryl tin compounds [Ph$_3$SnCl, (Ph$_3$Sn)$_2$O, and Ph$_2$SnCl$_2$]

$$MoCl_6 + Mo \longrightarrow [Mo_6Cl_8]Cl_4 \qquad \{equivalent\ to\ MoCl_2\}$$

Reaction 1

Table IV. Comparison of Syntheses of $VO(PO_4) \cdot 2H_2O$ Intercalates by Using Microwave and Conventional Thermal Techniques

	Microwave		Conventional		Expansions[b]	
Conditions: Guest[a]	t (min)	Stoichiometry	t (h)	Stoichiometry	c (Å)	Δc (Å)
Pyridine	5	0.84	36	0.35	9.55	5.44
4-Methylpyridine	3	0.86	12	0.60	10.56	6.45
4-Phenylpyridine	2 × 5	0.85	64	0.51	12.23	8.12[c]

NOTE: Δc represents the difference in the length of the c axis for the intercalation compound and that of anhydrous $VOPO_4$.

[a] Host is $VO(PO_4) \cdot 2H_2O$.
[b] Coordinated water is lost during the reaction of $VOPO_4 \cdot 2H_2O$ ($c = 7.41$ Å).
[c] Another phase is also observed with a layer expansion of 9.89 Å.

into the synthetic smectite clay laponite under ambient conditions result in the formation of tin(IV) oxide pillared clays (*71, 72*). The Mössbauer data indicate that the effectiveness of conversion to tin oxide pillars is in the order $Ph_3SnCl > (Ph_3Sn)_2O \sim Ph_2SnCl_2$. The organic product of the pillaring reaction is benzene. The new materials are novel because the pillaring was achieved by means of neutral precursors rather than by sacrificial reaction of the exchanged cation. The intercalation–pillaring reactions are much more rapid when carried out in Teflon containers in a domestic microwave oven. In a typical experiment 1.0 g laponite and 0.3 g $(Ph_3Sn)_2O$ were mixed with ethanol (10 cm³). The temperature of the oven space was set to 413 K, and the mixture was subjected to five 1-min bursts of microwave radiation.

Combined ^{119}Sn Mössbauer and X-ray photoelectron spectroscopic studies suggest that, in the microwave experiments, Ph_3SnCl has a higher initial affinity for the clay surface than $(Ph_3Sn)_2O$ but that the chloride undergoes hydrolysis on the surface once sorbed. Addition evidence exists for the considerable mobility of Mg^{2+} within the laponite lattice during microwave heating.

Microwave hydrothermal processing in combination with the polyol process has been used to prepare Ag, Pt, and Pd intercalated montmorillonite as a potential supported catalyst (*73*). The metal-intercalated composites contain subnanometer metal particles introduced into the layers with some larger 5–100-nm crystals on the surfaces of the montmorillonite grains. The metal-intercalated clays showed higher BET (Brunouer–Emmett–Teller isotherm) water surface areas than N_2 surface areas as a result of the inaccessibility of N_2 to the interlayers.

Zeolites and Related Compounds

In early hydrothermal work in sealed Teflon vessels the syntheses of a range of zeolite materials were successfully studied, including zeolite A

(1). The possibility of using microwave-heating techniques for cation exchange in zeolites was also reported. More recently, this work has been extended by Arafat and co-workers (74) who used Teflon autoclaves and microwave heating for the synthesis of zeolite Y and ZSM-5. They found that microwave-heating techniques allowed a longer range of Si:Al ratios to be prepared for zeolite Y. When zeolite synthesis mixtures containing quaternary ammonium templates are heated in the microwave field, both gel dissolution and the degradation of the template are accelerated, and the fast formation of gel spheres as well as of zeolite result.

Crystalline, thermostable molecular sieve MCM-41 with hexagonal channel porosity was synthesised in a temperature-controlled microwave from aged precursor gels in about 1 h (75). A rapid synthesis of solid coagular Na aluminate was described (76). Hydroxyapatite (calcium phosphate) materials also were synthesised (77). Porous hydroxyapatite ceramics with porosity up to 73% were successfully fabricated within a few minutes under optimum conditions.

Coordination Compounds

As discussed previously a range of inorganic reactions can be studied conveniently by microwave reflux techniques. Reactions in ethanol, methanol, acetonitrile, and some other organic solvents would be expected to be accelerated by using microwave heating compared with conventional heating techniques in cases where efforts are made to reduce the number of available nucleation sites for boiling. In other words, antibumping granules, porous Teflon stirrer bars, and fast streams of gas into the reaction mixture are undesirable in achieving the best results. Reactions in water would not be expected to be greatly accelerated because for water the elevation of the reflux temperature is small—of the order of a few degrees.

In studies on the synthesis of ruthenium-based catalyst precursors from $RuCl_3$ and PPh_3 in MeOH, comparable yields of $[RuCl_2(PPh_3)_3]$ were obtained in 30 min by microwave reflux and 3 h of conventional reflux (27). The rate enhancement achieved by microwave processing is on the order expected on the basis of superheating of between 20 and 30 K.

To obtain more dramatic accelerations, reactions in minutes rather than days, it is necessary to work in sealed vessels. Consider the syntheses involving substitutionally inert transition metal ions and particularly d^3, d^8, and low-spin d^6, which conventionally require long and tedious synthetic procedures (see Table V) (78, 79).

By using a modified glass reaction vessel several compounds have been made by improved routes, in high yields and in accelerated times (see Table VI) (30). For example, the synthesis of $[Ru(9S3)_2]^{2+}$ (9S3 = 1,4,7-trithiacyclononane) was previously possible only by the prior conversion of inert $RuCl_3$ into the labile intermediates $[Ru(SO_3CF_3)_3]$ or $[Ru(Me_2SO)_6]^{2+}$. The

Table V. Coordination Compounds Synthesised in Sealed Teflon Vessels

| | | Microwave | | Thermal | |
| | | Yield | Time | Yield | Time |
Reactants	Products	(%)	(s)	(%)	(h)
CrCl$_3$ 6H$_2$O, urea, aq. EtOH, dipivalolylmethane	Cr(DPM)$_3$	71	40	65	24
IrCl$_3$ xH$_2$O, PPh$_3$, DMF^a	Ir(CO)Cl(PPh$_3$)$_2$	70	45	85–90	12
IrCl$_3$ xH$_2$O, 9S3, MeOH	IrCl$_3$(9S3)	98	16	—	2
K$_2$PtCl$_4$, tpy, H$_2$O	[PtCl(tpy)Cl] 3H$_2$O	47	2 × 30	—	24–100
HAuCl$_4$, tpy, H$_2$O	[AuCl(tpy)Cl] 3H$_2$O	37	2 × 30	—	24
RuCl$_3$ xH$_2$O, DMF	[RuCl(CO)(bpy)$_2$]Cl	70	3 × 20	—	168
RuCl$_3$ xH$_2$O, 9S3, MeOH	[Ru(9S3)]$^{2+}$	49	6 × 25	—	—

NOTE: 9S3 is 1,4,7-trithianonane; tpy is 2,2′,2″-terpyridine; bpy is 2,2′-dipyridyl; DPM is 2,2′,6,6′-tetramethyl-3,5-heptadionato; DMF is 1,1′-dimethylformamide.

^{a}The reaction product was at times contaminated with Ir(CO)Cl$_3$(PPh$_3$)$_2$, which was reduced to Ir(CO)Cl(PPh$_3$)$_2$ with zinc in DMF.

Table VI. Coordination Compounds Synthesised in a Sealed Glass Vessel at up to 10 Atmospheres

Product	Solvent	Time (min)	Yield (%)	Temperature (K)
Mo$_2$(acac)$_4$	CH$_3$CO$_2$H	30	65	468
[Ru(9S3)$_3$](PF$_6$)$_2$	MeOH	70	96	390
[Ru(bpy)$_3$](PF$_6$)$_2$	MeOH	10	87	406
[Ru(C$_7$H$_8$)Cl$_2$]$_2$	EtOH	9	66	408
[Rh(cod)Cl]$_2$	EtOH–H$_2$O (4:1)	0.5	84	413
W(CO)$_4$(CH$_3$CN)$_2$	CH$_3$CN	19	70	414
(Mo$_6$Cl$_8$)ac$_2$Cl$_2$^{a}	CH$_3$CO$_2$H	8	79	428

NOTE: cod is cyclooctadiene; 9S3 is 1,4,7-trithiacyclononane; bpy is 2,2′-dipyridyl.

added advantage of this glass vessel is that it is possible to observe the progress of reactions through colour changes.

Organometallic Compounds

From a range of synthetic studies on organometallic compounds similar results have been obtained to those discussed for coordination compounds. For reflux procedures, rate enhancements leading to microwave reaction in minutes versus conventional reaction in hours have been found for reactions in solvents exhibiting strong microwave superheating (Table VII) (27). For reactions is sealed vessels, rate enhancements on the order of microwave reaction in seconds versus conventional reflux reaction in days have been found for reactions in polar solvents (Table VIII) (80). Whereas in the sealed-

Table VII. Organometallic Compounds Synthesised by Microwave Reflux

Reactants	Products	Microwave Yield (%)	Time (min)	Thermal Yield (%)	Time (h)
RhCl$_3$ xH$_2$O, C$_8$H$_{12}$, EtOH–H$_2$O (5:1)	[Rh(cod)Cl]$_2$[a]	87	25	94	18
RuCl$_3$ xH$_2$O, EtOH, 1,3-cyclohexadiene	[Ru(C$_6$H$_6$)Cl$_2$]$_2$	89	35	95	4
RuCl$_3$ xH$_2$O, EtOH, α-phellandrene	[Ru(p-cymene)Cl$_2$]$_2$	67	10	65	4

NOTE: cod is 1,5-cyclooctadiene.

Table VIII. Organometallic Syntheses in Sealed Vessels

Reactants	Products	Microwave Yield (%)	Time (s)	Thermal Yield (%)	Time (h)
RhCl$_3$ xH$_2$O, C$_8$H$_{12}$, EtOH/H$_2$O (5:1)	[Rh(cod)Cl]$_2$	91	50	94	18
RhCl$_3$ xH$_2$O, EtOH/H$_2$O (5:1), norbornadiene	[Rh(C$_7$H$_8$)Cl]$_2$	68	35	—	—
IrCl$_3$ xH$_2$O, C$_8$H$_{12}$, EtOH/H$_2$O (5:1)	[Ir(C$_8$H$_{12}$)Cl]$_2$	72	45	72	24
RhCl$_3$ xH$_2$O, MeOH, C$_5$H$_6$[a]	[Rh(C$_5$H$_5$)$_2$]$^+$	62	30	—	—
RuCl$_3$ xH$_2$O, aq. EtOH, 1,3-cyclohexadiene	[Ru(C$_6$H$_6$)Cl$_2$]$_2$	89	35	95	4

NOTE: C$_8$H$_{12}$ is cyclooctadiene; C$_5$H$_6$ is cyclopentadiene.

[a]Methanolic NH$_4$PF$_6$ added to the reaction mixture to obtain the yellow salt.

vessel case the accelerations can be rationalised on purely thermal grounds, the syntheses are much more convenient by the microwave route.

By using a different type of Teflon autoclave vessel, various posttransition metal organometallic compounds were synthesised (81). Organometallic compounds have also been synthesised in nonpolar solvents by using microwave radiation (82). The reaction mixtures include microwave-absorbing metal powders to provide the heat for the reactions to take place. In Reaction 2, a typical reaction, an 88% yield of the adduct was obtained in only 3 min of microwave heating. Microwave techniques also were used for the synthesis of ferrocenyl-substituted heterocycles (83).

Carboranes

Reaction 3 has been studied by both conventional reflux techniques and by microwave heating in a sealed-glass vessel (84). Whereas no isolable quantities of material were recovered in conventional reflux techniques, the sealed-glass heating produced excellent yields of the two novel isomeric platina-carboranes. The reaction was completed in 30 min at a pressure of 10 atm and had a final temperature recorded of around 408 K. These new compounds provide an important insight into the mechanisms of thermally induced rearrangements of metallocarboranes.

$$C_6H_6 + Fe(C_5H_5)_2 \xrightarrow[\text{(ii) HPF}_6]{\text{(i) AlCl}_3/\text{Al}} [Fe(\eta\text{-}C_5H_5)(\eta\text{-}C_6H_6)]PF$$

Reaction 2

$$cis\text{-}[Pt(PMe_2Ph)_2Cl_2] + [7\text{-}Ph\text{-}7,8\text{-}nido\text{-}C_2B_9H_{11}]^-$$

$$1\text{-}Ph\text{-}3,3\text{-}(PMe_2Ph)_2\text{-}3,1,11\text{-}PtC_2B_9H_{10} + 11\text{-}Ph\text{-}3,3\text{-}(PMe_2Ph)_2\text{-}3,1,11\text{-}PtC_2B_9H_{10}$$

Reaction 3

Summary

Microwave-heating techniques were applied to a wide range of inorganic reaction types. In solution-phase reactions convenient techniques have been developed to allow the study of reflux reactions and reactions at elevated pressures. Microwave reflux reactions in a range of organic solvents occur at a rather higher temperature than those obtained conventionally, a result leading to rate enhancements on the order of 5–10 times. Further work will exploit the virtues of microwave over conventional heating: namely, the volumetric nature of microwave heating, selective coupling of microwave energy by components of a reaction mixture, and improved energy efficiency. Observed enhancements in the rate of solid-state reactions remain a topic of some controversy.

References

1. Mingos, D. M. P.; Baghurst, D. R. *Chem. Rev.* **1991**, *20*, 1.
2. Abramovitch, R. *Org. Prep. Proced. Int.* **1991**, *23*, 685.
3. Gedye, R.; Smith, F.; Westaway, K. *J. Microwave Power Electromagn. Energy* **1991**, *26*, 3.
4. Gedye, R.; Rank, W.; Westaway, K. *Can. J. Chem.* **1991**, *69*, 706.
5. Roussy, G.; Mercier, A.; Thiebaut, J.-M.; Vanbourg, J.-P. *J. Microwave Power* **1985**, *20*, 47.
6. Kenkre, V. M.; Skala, L.; Weiser, W. M.; Katz, J. D. *J. Mater. Sci.* **1991**, *26*, 2483.
7. Ayappa, K. G.; Davis, H. T.; Davis, E. A.; Gordon, J. *AICHE J.* **1991**, *37*, 313.
8. Ayappa, K. G.; Davis, H. T.; Crapiste, G.; Davis, E. A.; Gordon, J. *Chem. Eng. Sci.* **1991**, *46*, 1005.
9. Ayappa, K. G.; Davis, H. T.; Davis, E. A.; Gordon, J. *AICHE J.* **1992**, *38*, 1577.
10. Barringer, S. A.; Davis, E. A.; Gordon, J.; Ayappa, K. G.; Davis, H. T. *AICHE J.* **1994**, *40*, 1433.

11. *Proceedings: Microwave Induced Reactions Workshop;* Burka, M.; Weaver, R. D.; Higgins, J., Eds.; Electric Power Institute: Palo Alto, CA, 1993.
12. Hoopes, T.; Neas, E.; Majetich, G. *Abstr. Pap. Am. Chem. Soc.* **1991,** *201,* 231.
13. Baghurst D. R.; Mingos, D. M. P. *J. Chem. Soc., Chem Commun.* **1992,** 674.
14. Bond, G.; Moyes, R.; Pollington, S.; Whan, D. *Chem. Ind.* **1991,** 686.
15. Baghurst D. R.; O'Sullivan, K., Imperial College of Science, Technology and Medicine, unpublished results.
16. Johnson, D. L. *J. Am. Ceram. Soc.* **1991,** *74,* 849.
17. Meek, T. T. *J. Mater. Sci. Lett.* **1987,** *6,* 638.
18. Piluso, P.; Gaillard, L.; Lequeux, N.; Boch, P. *J. Eur. Ceram. Soc.* **1996,** *16,* 121.
19. Wroe, R.; Rowley, A. T. *J. Mater. Sci.* **1996,** *31,* 2019.
20. Tian, Y. L.; Johnson, D. L.; Brodwin, M. E. *Proceedings of the First International Conference on Powder Processing;* Orlando, FL, November 1987.
21. Janney, M. A.; Kimrey, H. D.; Schmidt, M. A.; Kiggans, J. O. *J. Am. Ceram. Soc.* **1991,** *74,* 1337.
22. Meek, T. T.; Blake, R. D.; Katz, J. D.; Bradbury, J. R.; Brooks, M. H. *J. Mater. Sci. Lett.* **1988,** *7,* 928.
23. Hook, K. J.; Agrawal, R. K.; Drzal, L. T. *J. Adhes.* **1990,** *32,* 157.
24. Salsman, J. B.; Williamson, R. L.; Tolley, W. K.; Rice, D. A. *Miner. Eng.* **1996,** *9,* 43.
25. Kenkre, V. M.; Kus, M.; Katz, J. D. *Phys. Rev. B: Condens. Matter* **1992,** *46,* 13825.
26. Booske, J. H.; Cooper, R. F.; Dobson, I. *J. Mater. Res.* **1992,** *7,* 495.
27. Baghurst, D. R.; Mingos, D, M. P. *J. Organometal. Chem.* **1990,** *384,* C57.
28. Bowman, R. R. *IEEE Trans. Microwave Theory Tech.* **1976,** *24,* 43.
29. Bond, G.; Moyes, R.; Pollington, S.; Whan, D. *Meas. Sci. Technol.* **1991,** *2,* 571.
30. Baghurst, D. R.; Mingos, D. M. P. *J. Chem. Soc., Dalton Trans.* **1992, 1151.**
31. Constable, D.; Raner, K.; Somlo, P.; Strauss, C. *J. Microwave Power Electromagn. Energy* **1992,** *27,* 195.
32. Pougnet, M. A. B. *Rev. Sci. Instrum.* **1993,** *64,* 529.
33. Metaxas, A. C. *J. Microwave Power Electromagn. Energy* **1990,** *25,* 16.
34. Bernard, P.; Marzat, C.; Miane, C. *J. Microwave Power Electromagn. Energy* **1988,** *23,* 218.
35. VanKoughnett, A. L.; Wyslouzil, W. *J. Microwave Power* **1971,** *6,* 25.
36. Catrysse, J., Katholieke Industriele Hogeschool West, Oostende, personal communication, 1991.
37. Chang, J.; Brodwin, M. *J. Microwave Power Electromagn. Energy* **1993,** *28,* 32.
38. Garard, R. S.; Fathi, Z.; Wei, J. B. *Ceram. Trans.* **1995,** *59,* 117.
39. Chemat, F.; Poux, M.; Dimartino, J. L.; Berlan, J. *J. Microwave Power Electromagn. Energy* **1996,** *31,* 19.
40. Metaxas, A. C.; Binner, J. G. P. In *Advanced Ceramic Processing and Technology;* Binner, J. G. P., Ed.; Nottingham, 1990; Vol. 1, p 1.
41. Sutton, W. H. *Am. Ceram. Soc. Bull.* **1989,** *68,* 376.
42. Sheppard, L. M. *Am. Ceram. Soc. Bull.* **1988,** *67,* 1656.
43. *MRS Symposium Proceedings;* Sutton, W. H.; Brooks, M. H.; Chabinsky, I. J., Eds.; Material Research Society: Pittsburgh, PA, 1988; Vol. 124.
44. Baghurst, D. R.; Mingos, D. M. P. *J. Chem. Soc., Chem. Commun.* **1988,** 829.
45. Baghurst, D. R.; Chippindale, A. M.; Mingos, D. M. P. *Nature (London)* **1988,** *332,* 311.

46. Bosi, S.; Beard, G.; Moon, A.; Belcher, W. *J. Microwave Power Electromagn. Energy* **1992**, *27*, 75.

47. Warrier, K. G. K.; Varma, H. K.; Mani, T. V.; Damodoran, A. D. *J. Am. Ceram. Soc.* **1992**, *75*, 1990.

48. Selmi, F.; Guerin, F.; Yu, X. D.; Varadan, V. K.; Varadan, V. V.; Komenarni, S. *Mater. Lett.* **1992**, *12*, 424.

49. Patil, D.; Mutsuddy, B.; Gerard, B. *J. Microwave Power Electromagn. Energy* **1992**, *27*, 49.

50. Gasgnier, M.; Loupy, A.; Petit, A.; Jullien, H. *J. Alloys Compd.* **1994**, *204*, 165.

51. Standish, N.; Worner, H. *J. Microwave Power Electromagn. Energy* **1990**, *25*, 177.

52. Barnsley, B. P.; Reilly, L.; Jones, J.; Eshman, J. *1st Australian Symposium on Microwave Applications;* Australian Power Institute: Sydney, Australia, 1989; p 49.

53. Worner, H. K.; Bradhurst, D. H. *Fuel* **1993**, *72*, 685.

54. Vaidhyanathan, B.; Ganguli, M.; Rao, K. J. *J. Mater. Chem.* **1996**, *6*, 391.

55. Komarneni, S.; Roy, R.; Li, Q. H. *Mater. Res. Bull.* **1992**, *27*, 1393.

56. Ounaies, Z.; Selmi, F.; Varadan, V. K.; Varadan, V. V.; Megherhi, M. *Mater. Lett.* **1993**, *17*, 13.

57. Komarneni, S.; Li, Q.; Stefansson, K. M.; Roy, R. *J. Mater. Res.* **1993**, *8*, 3176.

58. Komarneni, S.; Menon, V. C.; Li, Q. H.; Roy, R.; Ainger, E. *J. Am. Ceram. Soc.* **1996**, *79*, 1409.

59. Komarneni, S.; Menon, V. C.; Li, Q. H. *Ceram. Trans.* **1996**, *62*, 37.

60. Willert-Porada, M. *MRS Bull.* **1993**, *18*, 51.

61. Bayya, S. S.; Snyder, R. L. *Physica C (Amsterdam)* **1994**, *235–40*, 543.

62. Mingos, D. M. P.; Baghurst, D. R. *Br. Ceram. Trans. J.* **1992**, *91*, 124.

63. Binner, J. G. P.; Hassine, N. A.; Cross, T. E. *Ceram. Trans.* **1995**, *59*, 335.

64. Weng, S. H. *J. Appl. Phys.* **1993**, *73*, 4680.

65. Weng, S. H.; Wang, J. *Chin. Sci. Bull.* **1992**, *37*, 1603.

66. Weng, S. H.; Wang, J. *Hyperfine Interact.* **1992**, *71*, 1395.

67. Whittaker, A. G.; Mingos, D. M. P. *J. Chem. Soc., Dalton Trans.* **1992**, *2751*, 1995, 2073.

68. Landry, C. C.; Barron, A. R. *Science (Washington, D.C.)* **1993**, *260*, 1653.

69. Whittaker, A. G.; Mingos, D. M. P. *J. Chem. Soc., Dalton Trans.* **1993, 2541.**

70. Chatakondu, K.; Green, M. L. H.; Mingos, D. M. P.; Reynolds, S. M. *J. Chem. Soc., Chem. Commun.* **1989**, 1515.

71. Berry, F. J.; Ashcroft, R. C.; Beevers, M. S.; Bond, S. P.; Gelders, A.; Lawrence, M. A. M.; McWhinnie, W. R. *Hyperfine Interact.* **1991**, *68*, 261.

72. Ashcroft, R. C.; Bond, S. P.; Beevers, M. S.; Lawrence, M. A. M.; Gelder, A.; McWhinnie, W. R.; Berry, F. J. *Polyhedron* **1992**, *11*, 1001.

73. Komarneni, S.; Hussein, M. Z.; Liu, C.; Breval, E.; Malla, P. B. *Eur. J. Solid State Inorg. Chem.* **1995**, *32*, 837.

74. Arafat, A.; Jansen, J. C.; Ebaid, A. R.; van Bekkum, H. *Zeolites* **1993**, *13*, 162.

75. Wu, C. G.; Bein, T. *J. Chem. Soc., Chem. Commun.* **1996**, 925.

76. Wang, J.; Song, H.; Hong, P. *Huaxue Shijie* **1995**, *35*, 180.

77. Fang, Y.; Agrawal, D. K.; Roy, D. M.; Roy, R. *J. Mater. Res.* **1992**, *7*, 490.

78. Baghurst, D. R.; Cooper, S. R.; Greene, D. L.; Mingos, D. M. P.; Reynolds, S. M. *Polyhedron* **1990**, *9*, 893.

79. Greene, D. L.; Mingos, D. M. P. *Tran. Met. Chem.* **1991**, *16*, 71.

80.   Baghurst, D. R.; Mongos, D. M. P.; Watson, M. J. *J. Organomet. Chem.* **1989,** *368,* C43.

81.   Ali, M.; Bond, S. P.; Mbogo, S. A.; McWhinnie, W. R.; Watts, P. M. *J. Organomet. Chem.* **1989,** *371,* 11.

82.   Dabirmaneh, Q.; Roberts, R. M. G. *J. Organomet. Chem.* **1993,** *460,* C28.

83.   Puciova, M.; Ertl, P.; Toma, S. *Coll. Czech. Chem. Commun.* **1994,** *59,* 175.

84.   Baghurst, D. R.; Copley, R. C. B.; Fleischer, H.; Mingos, D. M. P.; Kyd, G. O.; Yellowlees, L. J.; Welch, A. J.; Spalding, T. R.; O'Connell, D. *J. Organomet. Chem.* **1993,** *447,* C14.

Chapter 11

Applications of Microwaves in Catalytic Chemistry

Gary Bond and Richard B. Moyes

Heterogeneous catalysts find widespread use throughout the chemicals industry. The interaction of these complex materials with microwave fields provides interesting and markedly different results compared to conventional heating methods. This chapter reviews work carried out using microwaves to provide improvements in catalyst preparation, catalyst characterisation, and the stimulation of catalytic reactions.

Catalysts are well-known in industry for their ability to speed up reactions, but they not only speed up reactions, they provide pathways by which the selectivity of the reaction may be improved. This increased selectivity is the most important factor in catalysis, because a required product is obtained, rather than a mixture with side products of little value. This result clearly has advantages from the economic and the environmental points of view. Thus, a major objective in catalytic research is the improvement of selectivity. Catalytic reactors usually contain a small mass of solid catalyst, relative to the amount of reaction that they can stimulate, over which reactant gases or liquids can pass. Microwave heating has apparent benefits in catalytic systems where gases react on solid surfaces. By using single-mode devices, the heating can be concentrated on the catalyst mass to provide relatively cool surroundings that can enhance selectivity.

For microwave heating to be applied to catalytic systems, the catalyst must itself absorb the microwave energy or be mixed with an absorber. Many active phases of known catalytic materials absorb microwave energy readily to different, sometimes large, extents, whereas the supports in common use, silica and alumina, do not. Wan et al. (1) in Canada demonstrated that a range of catalysed processes are accelerated when exposed to microwave energy. These reactions include the catalytic conversion of cyclohexene to benzene, catalytic cracking, decomposition of organic halides, decomposition of methane to ethene and hydrogen, and steam reforming. Most of these reactions require relatively high temperatures compared with those reported for the microwave-heated systems.

A patent (2) exists on the microwave catalytic stripping of cracking catalysts, and a number of detailed papers on the platinum-catalysed cracking of 2-methylpentane (3, 4) suggest that the catalyst is at a higher temperature than that expected from the bulk measurements. Roussy et al. (5) published a detailed account of the drying of zeolites with microwaves that suggests a mechanism for the accelerated preparation of such materials. In the field of organic chemistry there has been an extensive development of *dry* organic reactions, where the reactants are adsorbed at hydroxylated surfaces and react on irradiation with microwave energy (6, 7).

Moyes, Bond, and Whan (8) demonstrated that the oxidative coupling of methane to form ethene and ethane can be achieved at bulk temperatures of 300–400 K less than in the conventionally heated systems and across a range of coupling catalysts. This result is supported by more recent work in the selective conversion of isopropanol to acetone or to propene (9).

The selectivity to one or other product depends on the acidity or basicity of the surface and both processes are shown to be accelerated by microwave heating when compared with conventional heating. This result has led the authors to suggest that specific sites on the catalysts are preferentially heated by microwave energy, whereas the bulk of the solid is at a lower temperature (9). Temperature measurement has difficulties, because most probes for temperature measurement themselves interact with the electromagnetic field and are at best unreliable. Remote sensing has its own difficulties, in that IR sensors detect the surface temperature rather than that of the bulk. Nevertheless, reasonably reliable systems are available that indicate that an error cannot exist in the measurement that explains the difference in apparent reaction temperature.

In general evidence exists that catalysed reactions are accelerated by microwave heating when compared with conventional heating. A number of explanations of this effect exist and further research could yield substantial dividends for these industrially important systems.

Single-Mode Microwave Apparatus

Most people are now familiar with the domestic microwave oven that operates as a multimode cavity (*see* Chapters 2 and 5) in which microwaves are generated by a magnetron and pass to the cavity (food compartment) by way of a short section of waveguide. The cavity may contain a mode stirrer to aid reflection of the microwaves in a random manner and to produce a uniform microwave field within the cavity. Equipment based on this basic design is now widely available for use in the laboratory and can be highly sophisticated. It can monitor and regulate pressure within poly(tetrafluoroethylene) (PTFE) vessels and can monitor sample temperature by means of fiber-optic devices. However, traditionally this type of device has suffered from a fairly crude means of power control that consists of switching the microwave supply on and off. The time scale of this switching is of the order of several seconds, thus making this form of equipment unsuitable for continuous reactions in flowing systems.

With the exception of the catalyst drying studies, the majority of the work described in this review has been carried out in single-mode microwave cavities. Because some readers will not be familiar with the design of a single-mode cavity, a brief description is included here. Slight differences will occur in the design of single-mode cavities in different laboratories, but these differences will be minor. The diagrammatic representation of a single-mode cavity and associated control equipment is shown in Figure 1. This equipment can be considered to consist of five main units:

- a control unit that is capable of varying the microwave power
- a microwave generator
- a section of waveguide into which the reactor is placed
- a means of monitoring the temperature of the sample while it is being irradiated
- a personal computer with suitable analogue to digital capabilities (optional)

The waveguide itself can be considered as five separate sections (Figure 2). Microwaves produced by the magnetron enter the launch section

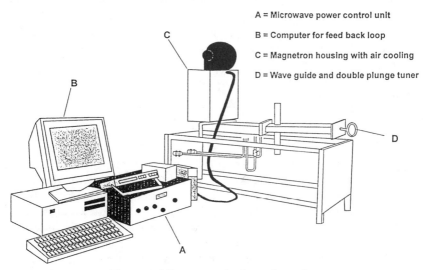

A = Microwave power control unit

B = Computer for feed back loop

C = Magnetron housing with air cooling

D = Wave guide and double plunge tuner

Figure 1. Microwave single-mode cavity.

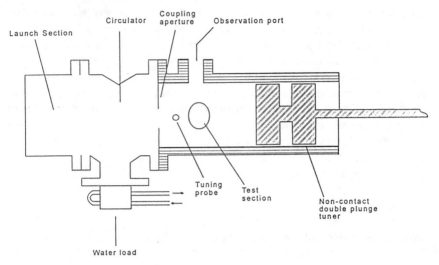

Figure 2. Waveguide section of the microwave single-mode cavity.

and then the test section through the circulator. The function of the circulator is to reflect the microwaves through 120°. This reflection causes microwaves exiting the launch section to be directed into the test section, and microwaves that are not absorbed by the sample are then turned through a further 120° into a water load. This prevents reflected microwaves from reentering the launch section where they would cause overheating and eventual failure of the magnetron. The function of the contactless double-plunge tuner is to reflect the microwaves efficiently, thus establishing a standing wave with the maximum in the E-field at the point at which the sample is mounted in the test section. Tuning of the cavity is achieved by varying the position of the double-plunge tuner, by way of a screw-thread mechanism, so that a maximum in signal is obtained from a tuning probe.

Catalyst Preparation

Supported Metal Catalysts

Supported metal catalysts are among the most common type of catalyst used in industry. The catalyst consists of the metal, which is the active component, distributed on the surface of the support. The support usually takes the form of a stable, inert oxide: silica or alumina are two of the most commonly used supports. The function of the support is to help maximize the distribution of the metal, hence increasing the metal surface area per unit mass of metal. The support must also have good mechanical properties because fragmentation of the support can cause problems in industrial reactors, processes, and products. Industrial catalysts are usually in the form of pellets. Generally, two methods of preparing the supported metal catalyst exist. The first method is to impregnate a preformed pellet of support material with a solution of metal in the form of a soluble salt. The pellet is then dried, after which the pellet is heated more strongly to convert the metal salt to the oxide (calcining). Once in place in the reactor the metal oxide is reduced to give the active catalyst. The second method is to coprecipitate the metal and the support material. The precipitate is then dried and calcined prior to pelletization.

Impregnation of Preformed Supports

Once the support has been impregnated with the aqueous metal salt (i.e., nickel), the next stage is to dry the sample. This stage is where problems arise, because when the pellets are heated by conventional methods moisture is initially removed from the external surface of the pellet that produces the required moisture gradient for outward flow (10) (Figure 3A). This moisture gradient causes the redistribution of nickel ions as water is drawn from the wet interior to the dry exterior. Therefore, nickel tends to concentrate on the external surface of the pellet. The difference in moisture content between the surface and the interior results in the dry surface being placed under tension, which causes the pellets to be weakened.

Microwave drying of ceramics has shown that when a wet body is exposed to microwave radiation the section that has the highest moisture content absorbs the microwaves most strongly and therefore becomes the hottest part of the body. Hence, the rate of evaporation is the greatest from the wettest region. This effect results in a very small moisture gradient within the body (1) (Figure 3B). Bond et al. (11) have applied this effect to the preparation of supported nickel catalysts. A study comparing conventionally dried and microwave dried alumina pellets (in the form of rings) which had been impregnated with nickel nitrate in order to produce a catalyst with 10% loading of nickel. The dried pellets were sectioned and analysed using a scanning electron microscope. The energy dispersive X-ray analysis facility was used to produce dot maps displaying the distribution of nickel within the pellets. The results from this study are shown in Figure 4 and clearly show that a more homogeneous distribution of the nickel is achieved by using microwave drying. Measurements of the crushing strength indicated that the microwave dried pellets were stronger than those produced by conventional drying, probably because of the moisture levelling achieved in microwave drying.

An alternative approach to the preparation of supported metal catalysts has been adopted by Zerger et al. (12), who used a microwave-induced argon plasma to decompose metal carbonyls that then impregnated zeolite supports. The authors found that the microwave discharge method allowed them to produce much more highly dispersed metal atom

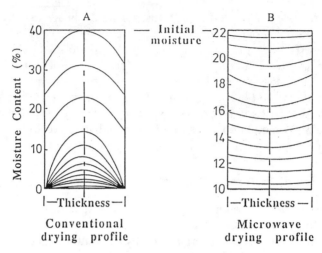

Figure 3. Moisture profiles for both microwave and conventional drying of ceramic bodies.

clusters than was possible by other methods. The technique had several other advantages: for example, the methodology was generally applicable so several supports and types of metals could and indeed have been used.

The size of the metal particles can be controlled by optimising

- time of microwave discharge
- amount of metal
- location of the metal in the preparation apparatus
- type of metal species
- type of support

Preparation of Coprecipitated Catalysts

To date no studies have been reported in the literature concerning the use of microwave radiation in the production of coprecipitated catalysts; however, the potential of such an approach is enormous. Industry at present typically uses dated technology that uses separate drying and calcining equipment. Klanding and Horn (13) have shown that microwave drying of aqueous solutions sprayed into a microwave cavity can result in the formation of a solid that consists of uniformly small particles. This process could readily be applied to the preparation of catalytic material.

Zeolite Catalysts

Zeolite catalysts have become increasingly important industrial catalysts. These crystalline aluminosilicates have the general formula $M_y(AlO_2)_x(SiO_2)_y \bullet zH_2O$, where M is a monopositive cation (e.g., sodium

Microwave dried   Conventionally dried

Figure 4. Nickel dot maps showing the distribution of nickel in catalyst particles dried by both microwave and conventional methods.

or ammonium) and y equals x. Arafat et al. (*14*) used microwave radiation in the preparation of the zeolites ZSM-5 and zeolite Y. Using conventional heating these syntheses take from 10 to 50 h, but with the aid of microwave radiation the materials have been prepared in 30 min. The microwave syntheses were performed in a PTFE autoclave, and the temperature within the autoclave was 140 °C.

Several attempts have been made to synthesize zeolite Y with a low aluminium content. However, it has been shown that a high Si:Al ratio and good crystallinity were incompatible. The use of microwave radiation has enabled the preparation of uniformly sized zeolite Y crystals with Si:Al ratios of up to 5.

Catalyst–Materials Characterisation

Some of the most widely used techniques employed to characterise catalysts and other materials involve heating the sample using a linear temperature program. Chemical and structural changes can be determined by either monitoring the gas evolved or observing temperature changes that occur relative to an inert standard, both of which can be associated with endothermic or exothermic processes. These techniques can suffer from the problem of inefficient heat transfer from the furnace wall to the sample. This problem results in temperature gradients being established within the sample, the outcome of which is a reduction in the resolution of the analytical technique.

Microwave energy provides a means of heating samples more homogeneously than by conventional means. As a result the resolution of the technique should increase. Thermal analysis using microwave radiation was investigated by two groups of workers. Karmazsin et al. (*15, 16*) attempted to perform differential thermal analysis using microwave radiation. The experiments were performed in a single-mode microwave cavity. The sample, in this case calcium hydrogen phosphate dihydrate ($CaHPO_4$ $2H_2O$), and the reference, alumina, were mounted inside the section of waveguide. The temperature of the sample and reference was monitored using thermocouples in such a way that they were said not to interact with the microwave field (17). The dehydration of the sample was then studied using microwave radiation. The authors observed what appeared to be a two-step exothermic loss of water associated with the dehydration. We consider this interpretation somewhat dubious because the loss of water is an endothermic process. The reason for an exotherm is probably the result of the water of crystallization being removed and interacting strongly with the microwave field, the result of which would be an increase in the temperature of the sample relative to that of the reference.

Bond et al. (18) also used microwave radiation to perform thermal analysis. These authors adopted a different approach in that they linearly increase the temperature of the sample within the cavity. The temperature is continuously monitored using either a fiber-optic probe or an IR pyrometer, the output from which is linked to a personal computer. The computer in turn sends a signal to the microwave generator to regulate the microwave power as is necessary to maintain a linear heating rate. Two methods of monitoring phase changes were employed, either by connecting the sample in series with a thermal conductivity detector and monitoring the gas evolved or by suspending the sample from a balance and monitoring weight changes. The authors also found that a third method of determining phase changes could be employed by monitoring the microwave power required to maintain a linear heating rate. Figure 5 shows the decomposition of a mixed Cu–Zn–Al hydroxycarbonate, which is a precursor to a methanol synthesis catalyst. This figure demonstrates that monitoring of microwave power can provide a highly sensitive method of determining this decomposition without the added complexity of evolved gas analysis. Figure 6 provides a second example of the use of monitoring microwave power, this time for the dehydration of $CaHPO_4 \cdot 2H_2O$. The sample was suspended from an electronic balance and the weight change was monitored. The two-stage decomposition is reflected in the plot of microwave power by peaks A and B. Clearly further work is required to fully determine the potential of microwave radiation in the characterisation of catalysts and other material; however, early results appear to be promising.

Catalytic Reactions

Reactions of Hydrocarbons Under Microwave Irradiation

The reactions of a wide variety of hydrocarbons have been studied under microwave irradiation. However, the most studied system is the reaction of methane to form higher hydrocarbons or oxygenated products. This fact reflects the large effort being made by the catalytic community to find a single-step process for the use of the world's large resources of natural gas.

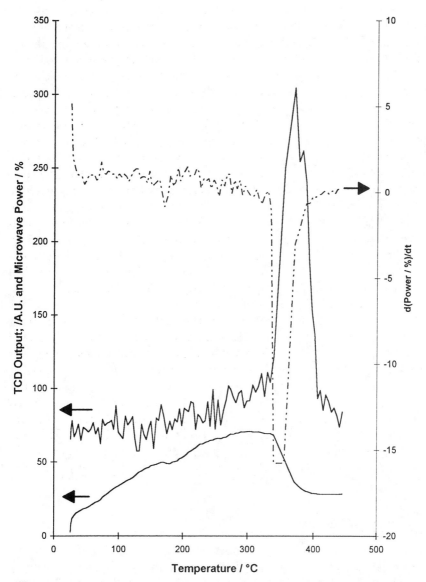

Figure 5. Microwave decomposition of a Cu–Zn–Al hydroxycarbonate using a heating rate of 15 °C/min.

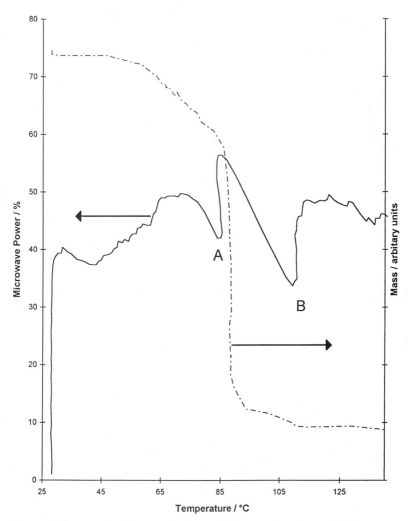

Figure 6. Dehydration of calcium hydrogen phosphate dihydrate.

Reactions of Methane

The existence of vast world-wide reserves of natural gas, of which methane is the major constituent, has focused attention on the possibility of converting it directly to other forms of fuel and more valuable chemicals. The need for such a direct conversion arises at least in part from the remoteness of the reserves and the relatively high costs involved in the transportation of gaseous fuels over long distances. In many cases this results in waste of methane as "flare off" where total oxidation results in the formation of carbon dioxide, which is highly environmentally undesirable.

One form of conversion of methane to more valuable products is achieved by steam-reforming the methane to produce a mixture of carbon monoxide and hydrogen known as Syngas. Syngas can then be converted to a range of useful products: for example, by reaction over a copper catalyst to produce oxygenates (methanol and formaldehyde) or by using an iron catalyst to produce higher hydrocarbons. The drawback of this technology is that it is multistep, and steam-reforming is highly energy-intensive.

Clearly, a single-step conversion of methane to more valuable products would be of immense industrial significance and economic advantage. Currently, two pathways are being extensively investigated by research groups throughout the world. The first pathway is the oxidative coupling of methane to yield C2 and higher hydrocarbons, whereas the second pathway is the direct partial oxidation of methane to produce methanol and other oxygenates. Despite extensive research programs no commercially viable process has been arrived at for either process. The main problem is that methane is a particularly stable molecule, and consequently high temperatures are required to initially activate the methane. Once activated the problem that then arises is that most products are then more reactive than methane, a result causing difficulties in the prevention of total oxidation of the methane.

Formation of Higher Hydrocarbons

A number of studies comparing conventional and microwave radiation to stimulate the oxidative coupling on methane have been carried out by Bond et al. (*8*). A number of basic oxides have been examined; the results from all the catalysts tested confirm that microwave stimulation of the reaction results in C2 formation being attained at lower temperatures and with increased selectivity. The results from reaction over a 1% Sr 15% La–MgO catalyst are contained in Figures 7A and 7B.

This significant reduction in reaction temperature exceeds all attempts that have so far been made to improve catalyst activity when conventional heating is employed. Additionally microwave irradiation results in an increased ratio of ethene:ethane, which is desirable. The explanation for the observed result is that irradiation of the catalyst results in the formation of *hot spots* that have a temperature very much in excess of the bulk catalyst. The result is that activation of the methane takes place at these hot spots; the highly reactive intermediates that are formed are less likely to undergo total oxidation because of the relative low temperature of the surroundings. Roussy et al. (*19*) have made a similar study using $SmLiO_2$ and supported $SmLiO_2$ as the catalyst. They confirmed that the use of microwave radiation to stimulate the catalyst is beneficial.

Wan et al. (*1*) used pulsed-microwave radiation to study the reaction of methane in the absence of oxygen. The reaction has been performed

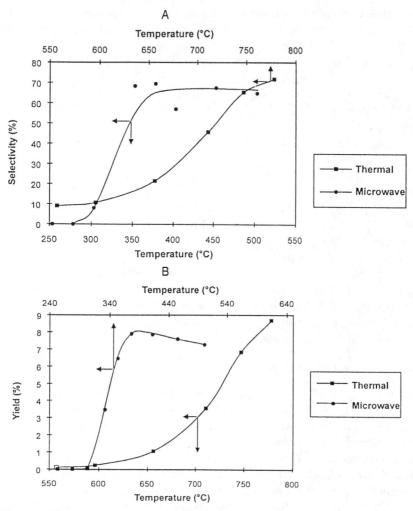

Figure 7. Selectivity and yield to C2 hydrocarbons for the oxidative coupling of methane.

over a series of nickel catalysts, and the products produced appear to be a function of both the catalyst used and the power and frequency of the microwave pulses. A Ni–SiO$_2$ catalyst has been reported to produce 93% ethyne, whereas a Ni (1 μ)/Ni (100 mesh) catalyst under the same conditions of irradiation produced no ethyne but 83% ethene and 8.5% ethane. By altering the microwave pulse cycle a complex mixture of hydrocarbons up to and including C6 and C7 aromatics were produced.

Microwaves have been used to generate plasmas in methane by using pressures in the 5–50 Torr range. The radicals produced in such a system were then allowed to react over a nickel catalyst resulting in the production of ethane, ethene, and ethyne (20).

Formation of Oxygenates from Methane

The formation of oxygenates from the reaction of methane and oxygen under conditions of conventional heating has received considerably less attention than the coupling reaction, despite the potential products being more desirable. The reason for this is that initial results have been far from promising. The problems encountered are again poor selectivity and the formation of large amounts of CO_x.

The use of microwave radiation in the formation of oxygenates from methane has shown some encouraging results. Zerger et al. (*21*) have used microwaves to generate a plasma in an atmosphere containing methane and oxygen. The plasma produced was then passed over a metal or metal oxide catalyst and resulted in the formation of C2 hydrocarbons and some oxygenates. Wan et al. (*22*) used water vapor as the oxidant in preference to oxygen. Their results show that acetone, isopropyl alcohol, methanol, and dimethyl ether can be produced. Reaction of propane yields C3 oxygenates.

Formation of Hydrogen Cyanide

Hydrogen cyanide (HCN) is a critical component in many industrial processes: for example, in the manufacture of pharmaceuticals and polymers. Many of these processes require only a small amount of HCN, which at present is supplied by shipping liquid HCN in cylinders. The commercial processes for producing HCN are clearly dangerous, and because of the potential hazards in transporting HCN manufacturers are becoming less happy about shipping HCN in cylinders.

Other methods of small-scale production of HCN require significant process equipment and require disposal of residual material. Wan and Koch (*23*) developed a method for the production of HCN on a small scale from the reaction of methane and ammonia. The reaction was carried out over a series of Pt–Al_2O_3, Ru–Al_2O_3, and carbon-supported catalysts under the presence of pulsed-microwave radiation. Conversions in excess of 90% have been reported.

More Reactions of Hydrocarbons Under Microwave Irradiation

A number of studies have been concerned with the use of microwave radiation to stimulate catalytic reactions of hydrocarbons. The topics covered are wide ranging and tend to be little more than exploratory studies. However, a continuing underlying trend is that the use of microwave radiation can cause significant changes in catalyst activity to that observed when conventional methods of heating are employed.

The isomerisation of 2-methylpentane has been studied over a 0.2% Pt–Al_2O_3 catalyst by Thiebaut et al. (*3*). The catalysts exhibit an activity as if the platinum, the active phase, is 25 °C above the temperature of the alumina support. However, the product selectivity under microwave conditions shows a marked change that cannot be accounted for in terms of temperature alone. In a later publication these authors account for these differences in terms of the microwave irradiation altering the platinum particle size and morphology (4). The reaction of cyclohexene over a catalyst consisting of a mixture of nickel powder and a CaNi alloy has been studied using two microwave systems (24). The first system used was an industrial microwave oven with variable power (80–1000 W). Microwave irradiation was carried out for up to 70 s in the form of 3–10-s pulses. A second experimental arrangement used an antenna to deliver microwaves directly to the reactor. This system operated using a maximum power of 100 W. Irradiation was carried out in 5-min cycles for a total of between 10 and 15 min.

The bulk temperature of the catalyst bed was monitored by inserting a thermocouple once the microwaves had been turned off. By monitoring the cooling of the sample, together with measurements of the thermal conductivity of the catalyst, the bulk temperature of the catalyst could be calculated at the point of switching off the radiation. The results from the study conclude that hydrogen was the better carrier gas for the reaction. The main products were cyclohexane and benzene, and small amounts of cracking products, mainly ethene, were also found. Longer irradiation times appear to favor the production of benzene. Comparison with results from conventionally heated experiments confirms the suggestion that the surface temperature of the metal was significantly greater than the recorded bulk temperature.

The hydrogenation of alkenes was studied by Wan et al. (25) using $CaNi_5H_5$ as the catalyst. The reaction can be performed in the absence of gas-phase hydrogen using very short irradiation times. However, the metal hydride does become depleted of hydrogen and requires regeneration. In the same paper Wan described microwave experiments on the water–gas shift reaction, in which carbon monoxide and water react to produce hydrogen and carbon dioxide. This reaction is performed on a commercial scale where high temperatures are required to facilitate the reaction. In the presence of a microwave field and using a nickel catalyst the authors reported that the reaction occurred while the outside of the reactor was raised to 35–40 °C.

The reactions of alcohols have also been studied under conditions of microwave irradiation (9). Alcohols can react in one of two ways, either by dehydration to produce an alkene or by dehydrogenation to produce either an aldehyde or a ketone. The first process occurs at acidic sites on a catalyst whereas the second occurs at basic sites. The reaction of a range of alcohols has been studied over a series of alkali-metal-doped catalysts. The supports used include silicalite, coconut charcoal, and graphite. Because the catalysts were doped with alkali metal, they exhibited the characteristics of basic catalysts when heated conventionally. However, if the reactions are carried out under conditions of microwave irradiation then two effects were observed. First, the temperature required for a given reaction to occur was reduced by approximately 150 °C, and this change was independent of the alkali metal or the support used. Secondly, the selectivity to the production of alkene increased as the concentration of surface hydroxyls on the support increased. This result was not observed when conventional heating was employed; in the case of conventional heating the nature of the alkali metal is of primary importance. The explanation that has been postulated to describe these observations is that the microwaves are efficient at dehydrating the surface hydroxyls of the support. Therefore, the reaction will then proceed in such a way as to generate water to rehydrate the support, and hence the increased selectivity to the alkene is observed.

Environmental Catalysis

Concern is growing about the effect chemical wastes are having on our environment. Acidic gases such as nitrogen oxides and sulfur dioxide result in the formation of acid rain. Other gaseous pollutants include carbon dioxide, emissions of which are adding to the green-house effect, and halogenated hydrocarbons, which destroy the ozone layer. As a result ever-tighter legislation is being brought to bear on industry to reduce emissions. This legislation is stimulating the development of new technologies for the removal and disposal of waste materials.

The decomposition of SO_2 and NO_2 by using microwave radiation has been studied over a range of nickel and copper catalysts (26). Experiments involving NO_2 produced gaseous oxygen and nitrogen as the major products. Experiments involving SO_2 resulted in the formation of oxygen; the sulfur remained as a solid although the nature of this was not determined.

Wan et al. (27) have shown that carbon dioxide can be reacted over supported metal catalyst in the presence of water vapor to yield alcohols and other oxygenated products when the catalyst is irradiated with microwaves. The bulk temperature of the catalysts are reported to be in the 220–350 °C range. Consideration of the thermodynamics of the reactions involved indicates that it is not possible for these reactions to occur at such low temperatures. The only plausible explanation is that the surface temperature of the metal is several hundred degrees higher than the bulk temperature of the catalyst. The decomposition of organic halides has been studied by a number of groups. Wan et al. (28), for example, studied the decomposition of chloromethane over a metal catalyst that was irradiated with pulsed-microwave radiation. The major product of the reaction was methane; however, if water was present quantities of methanol and dimethyl ether were also formed. In addition the halogen present combined with the metal catalyst to form a metal halide. These experiments were carried out by reacting gases over the metal catalyst. A different approach has been adopted by Bond et al. (29), in which they added a metal catalyst to a flask containing chlorocyclohexane. The flask was mounted inside a microwave oven and connected to an externally mounted reflux condenser. The flask was then irradiated by using sufficient power to maintain the chlorocyclohexane at its boiling point. The main products for this reaction were cyclohexene and hydrochloric acid; however, no evidence existed for the formation of a metal chloride.

Conclusion

The number of laboratories around the world investigating the potential of using microwave radiation to stimulate chemical reactions is increasing. The results contained in this chapter are limited to the use of microwaves to stimulate catalytic reactions. These results appear to have one underlying feature: namely, that irradiation appears to result in catalysts behaving as if they are hotter than the recorded temperature. The reason for this observation can only be postulated because further fundamental work is required to determine how microwaves interact with complex materials such as heterogeneous catalysts. One credible theory is that the microwaves selectively heat different parts of the catalyst. For example, in the case of a supported metal catalyst it is plausible that the microwave energy is selectively absorbed by the metal crystallites in preference to the support material. The result will be hot metal particles on a relatively cool support. Attempts to determine the spatial temperature profile of such systems will tend to give an average value for the metal and support and hence will be significantly lower than that at the active center.

Microwaves may also play an important role in preventing catalyst deactivation. Two of the main reasons for catalyst deactivation are sintering of the metal particles in supported metal catalyst and also excessive carbon deposition on a variety of catalysts. A catalyst operated under conditions of microwave irradiation, as discussed previously, will have hot metal particles and relatively cool support. This relatively cool support may act as a barrier to the migration of metal particles and hence reduce sintering and so prolong catalyst activity. Carbon is a very lossy material (i.e., it absorbs microwaves very efficiently); hence, any carbon that is deposited on the surface of the catalyst will be strongly heated. If the reaction medium contains air or hydrogen then the carbon will be removed in the form of carbon dioxide or methane, respectively. Evidence exists for this reaction with zeolite catalysts (30), where the catalyst activity has been shown to decay as carbon is deposited. Once the carbon deposition has reached a certain level it appears to absorb microwave energy strongly and is subsequently removed resulting in an increase in the activity of the catalyst.

A number of problems must be overcome if the potential of microwave radiation is to be realized on a commercial scale. First, one must be able to scaleup the highly successful laboratory experiments to a size suitable for use in industry. Interestingly, microwave engineers have overcome these difficulties in the fields of food and polymer processing; therefore, the potential exists for scaleup for chemical reactors. The second problem concerns suitable materials. Catalytic reactors in the laboratory tend to be constructed from silica, which is an ideal material because it is virtually transparent to microwave radiation and has a melting point in excess of 1000 °C. Unfortunately, glass reactors are not suitable for use on an industrial scale because they are too fragile and cannot withstand the pressures involved in many catalytic processes. PTFE, which is commonly used in pressure vessels designed for use within microwave cavities, has a limited range of operating temperatures that makes it unsuitable for many catalytic reactions. One possibility to overcome these problems may be to mount the catalyst directly into a section of waveguide. The third problem is that of cost; microwave processes must be able to compete with their conventionally heated counterparts, and this may not be possible in terms of energy input. However, if microwave processes are sufficiently more selective than current methods, then they will become increasingly attractive. As we approach the 21st century, pressure on the chemical industry is growing to develop 100% selective processes as legislation concerning the treatment and disposal of chemical waste becomes more stringent. The future of microwaves may lie in the area of speciality chemicals where the energy costs are insignificant compared with the value of the product.

References

1. Wan, J. K. S.; Husby, T. H.; Depew, M. J. *Microwave Power Electromagn. Energy* **1990,** *25,* 32
2. U.S. Patent 4,968,403, 1988.
3. Thiebaut, J. M.; Roussy, G.; Medjram, M.; Seyfield, L.; Garin, F.; Maire, J. *Chim. Phys.* **1992,** *89,* 1427.
4. Thiebaut, J. M.; Roussy, G.; Medjram, M.; Seyfield, L.; Garin, F.; Maire, J. *Catal. Lett.* **1993,** *21,* 133.
5. Roussy, G.; Zoulalian, A.; Charreyre, M.; Thiebaut, J. M. *J. Phys. Chem.* **1984,** *88,* 5702.
6. Villemin, D.; Labiad, B.; Ouhilal, Y. *Chem. Ind.* **1989,** 60.
7. Villemin, D.; Labiad, B. *Synth. Commun.* **1990,** *20,* 3333.
8. Bond, G.; Moyes, R. B.; Whan, D. A. *Catal. Today* **1993,** *17,* 427.
9. Bond, G.; Moyes, R. B.; Theaker, I.; Whan, D. A. *Top. Catal.* **1994,** *1,* 177.
10. Metaxas, A. C.; Binner, J. G. P. *Advanced Ceramic Processing and Technology;* Binner, J. G. P., Ed.; Noyes Publications: Park Ridge, NJ, 1990; Vol. 1.
11. Bond, G.; Moyes, R. B.; Pollington, S. D.; Whan, D. A. *Proceedings of the 10th International Congress on Catalysis;* Guczi, L.; Solymosi, F.; Tétényi, P., Eds.; Akadémiai Kiadò: Budapest, Hungary, 1993.

12. Zerger, R. P.; McMahon, K. C.; Seltzer, M. D.; Suib, S. L. *J. Catal.* **1986,** *99,* 498.

13. Klanding, W. F.; Horn, J. E. *Ceram. Int.* **1990,** *16,* 99.

14. Arafat, A.; Jansen, J. C.; Ebaid, A. R.; van Bekkum, H. *Zeolites* **1993,** *13,* 162.

15. Karmazsin, E.; Barhoumi, R.; Satre, P.; Gaillard, F. *J. Therm. Anal.* **1985,** *30,* 43.

16. Karmazsin, E.; Barhoumi, R.; Satre, P. *J. Therm. Anal.* **1984,** *29,* 1269.

17. Karmazsin, E.; Barhoumi, R.; Satre, P.; Gaillard, F. *C. R. A. F. C. A. T Bruxelles* **1984,** *15,* 1.

18. Bond, G.; Moyes, R. B.; Whan, D. A. *Catalysis and Surface Characterisation;* Dines, T. J.; Rochester, C. H.; Thomson, J., Eds.; The Royal Society of Chemistry: London, 1992.

19. Roussy, G.; Thiebaut, J.; Souri, M.; Kiennemann, A.; Maire, G. *Quality Enhancements Using Microwaves;* 28th Microwave Symposium Proceedings Pub. IMPI; Industrial Microwave Power Institute: Clifton, VA, 1993. pp 79–82.

20. Suib, S. L.; Zerger, R. P. *J. Catal.* **1993,** *139,* 383.

21. Zerger R. P.; Suib, S. L.; Zhang, Z. *Abs. Am. Chem. Soc.* **1992,** *203,* 344.

22. Tse, M. Y.; Depew, C.; Wan, J. K. S. *Res. Chem. Intermed.* **1990,** *13,* 221.

23. Wan, J. K. S.; Koch, T. A. *Res. Chem. Intermed.* **1994,** *20,* 29.

24. Wolf, K.; Choi, H. K. J.; Wan, J. K. S. *AOSTRA J. Res.* **1986,** *3,* 53.

25. Wan, J. K. S.; Wolf, K.; Heyding, R. D. *Catalysis on the Energy Scene;* Kaliaguine, S.; Mahay, A., Eds.; Elsevier Science Publishers: Amsterdam, Netherlands, 1984.

26. Tse, M. Y.; Depew, M. C.; Wan, J. K. S. *Res. Chem. Intermed.* **1990,** *13,* 221.

27. Wan, J. K. S.; Bamwenda, G.; Depew, M. C. *Res. Chem. Intermed.* **1991,** *16,* 241.

28. Dinesen, T. R. J.; Tse, M. Y.; Depew, M. C.; Wan, J. K. S. *Res. Chem. Intermed.* **1991,** *15,* 113.

29. Bond, G.; Moyes, R. B.; Wan, D. A.; Wilkes, T., University of Hull, United Kingdom, unpublished material.

30. Bond, G.; Moyes, R. B.; Pollington, S. D.; Wan, D. A., University of Hull, United Kindgom, unpublished material.

Chapter 12

Microwave-Assisted Solvent Extraction

L. Jassie, R. Revesz, T. Kierstead, E. Hasty, and S. Matz

The topic of microwave-assisted solvent extraction is introduced with a review of classical extraction methods followed by a section on modern instrumental techniques for solid–liquid separations. A discussion of microwave theory of heating and solvent compatibility follows, with emphasis on those features unique to microwave extractions and how they differ from traditional Soxhlet, sonication, reflux, and shaking extraction methods. Modern microwave-extraction instrumentation is surveyed with special emphasis on safety considerations. Microwave-assisted extraction applications are discussed as they apply to natural products, plastics and polymers, and environmental pollutants in soils and sediments. The chapter concludes with a brief look at possible future directions for this technology.

Separating one substance from another has preoccupied the chemist since ancient times. Prying precious metals from rocks or the essence of myrrh and frankincense from natural products like tree bark, ancient civilizations had few techniques to help accomplish these extractions. Even today, separating components of mixtures is still a formidable and time-consuming task. Separation science deals variously with substances in solution or homogeneous liquids that partition themselves according to physical principles of size, charge, similarity, and dissimilarity, among others. Although we freely borrow vocabulary and operating concepts from all of the separation sciences, we concern ourselves here only with solid–liquid separations and focus on the extraction of a substance or similar groups of substances, the solute, from solid materials, the matrix, by means of dissolution in a compatible solvent.

Classical Methods of Liquid Extraction of Solids

Classical solid–liquid extraction techniques have common features that permit comparison across techniques. This section briefly describes fundamental chemical and physical interactions that are important to these techniques and serves to focus attention on those parameters in each approach that can be optimized to improve extraction efficiency.

Classical solvent extraction may be thought of as a phase transfer of solute from one phase into another, such as from an aqueous phase into organic solvent as in a liquid–liquid extraction or a phase transfer from solids to liquid solutions. Desorption is the transfer of a substance from the solid phase to solution. Adsorption of an analyte like polycyclic aromatic hydrocarbons (PAHs) onto soil particles from dilute aqueous solutions, for example, is dependent on their distribution between the solid and liquid phases (1):

$$K_d = C_s/C_w \tag{1}$$

where K_d is the distribution coefficient, C_s is the concentration of the species such as PAH in the solid, and C_w is the species concentration in the liquid and assumes a linear sorption isotherm. Changing the liquid-phase concentration, C_w, requires a new K_d value for the target analyte. A traditional partition (distribution) coefficient, K_p or K_d, based on octanol and water affinity in liquid solutions may offer guidance to the solvation power of a solvent for that analyte. That is, a large K_p would be a more powerful solvent and more likely to accumulate the target analyte.

Soxhlet Extraction

Soxhlet extractions typically employ solid–liquid ratios that range from 1:10 to 1:50. Such large solute:solvent ratios allow even minimally soluble analytes to be dissolved. The difficulty is that even under the most favorable solution conditions, a target analyte might not be desorbed no matter how well matched chemically the solvent and solute are. Physical problems such as compacting, channeling, and variable particle size as well as inaccessibility and the inability of even the best solvent to compete with tightly bound solute (2) limit solvent extraction efficiency.

Good Soxhlet solvents tend to be low-boiling liquids that are easy to evaporate during analyte recovery steps. Because the Soxhlet system is at atmospheric pressure, the thermal energy of the extracting solution is often below the solvent boiling point. At these levels important rate advantages due to temperature are not seen. As a result, these open-vessel atmospheric extractions may take up to 16–20 h to reach an acceptable level of solute recovery. It is not often appreciated that the temperature of pure solvent that leaches the matrix during distillation is slightly below the boiling point because it is cooled by a cold-water condenser. Nevertheless, the solute or target analyte is exposed to pure, clean solvent on each pass. Soxhlet extractions benefit from nearly unattended operation although lengthy extractions may require topping off the solvent from time-to-time.

An automated rapid-Soxhlet apparatus (3) permits extractions to be reduced to 1–2 h. The Soxtec device features a thimble with sample that is immersed in boiling solvent for approximately half the extraction time and then is raised and extracted similarly to the traditional Soxhlet technique for an additional 30–60 min. Extraction times are reduced by up to 90%. Both sample size and solvent quantity are smaller, although matrix-to-solvent ratios are similar to ordinary Soxhlet ratios.

Tumbling and Shaking

Mixing, shaking, or tumbling combinations offer a simple, effective, but time-consuming and imprecise method of extraction. Sample-to-solvent ratios are similar to Soxhlet ratios, and extractions are normally run overnight, usually at room temperature. Occasionally, shaking may be accomplished with heated platforms. Temperatures are barely above room temperature because little pressure buildup occurs. Although sample manipulation is reduced, this technique requires nearly as much time as the Soxhlet method.

Sonication

Sonication is an extraction approach that uses ultrasonic frequencies to disrupt or detach the target analyte from the matrix. Horn type sonic probes operate at pulsed powers of 400–600 W in the sample solvent container. The same solvent container also could be immersed in warmed water baths for transmission of the wave, but these extractions are less efficient. Sonication is rapid and effective for certain situations because cavitation raises the temperature at the particulate surface creating a localized superheating (4) even though bulk heating is minimal. Temperature effects coupled with vibrational and torquing forces permit extractions to occur in minutes versus hours. Because only one sample at a time is processed, sample throughput is low despite the rapidity of extraction; precision tends to be poor. Sample sizes may average 30 g and total solvent volumes range from 150–300 mL. Depending on the level of contamination, soil samples as small as 2 g in 10–30 mL of solvent can be used (5), especially for screening.

Reflux Extraction

Reflux extractions are popular in polymer applications and share the sample immersion in hot solvent common to the microwave technique. Under reflux conditions solvents are at their atmospheric boiling point, which is frequently <100 °C. These atmospheric pressure and temperature methods are lengthy and labor intensive. For all classical extraction methods, solvent selectivity, in general, is low: that is, solvents with high capacity tend to have low selectivity.

Modern Instrumental Methods

Supercritical Fluid Extraction

Supercritical fluids have been used as solvents for the extraction of plant materials, environmental samples, polymers and foods (6–8). Most supercritical fluid extraction (SFE) instrumentation incorporates supercritical carbon dioxide with and without organic solvent modifiers. SFE is analyte-selective, and the extraction power can be fine-tuned by adjusting the supercritical fluid density and by adjusting temperature and pressure. Solvating power of supercritical fluids can be adjusted by adding more polar solvent modifiers, such as acetone or methylene chloride. The technique is highly matrix and analyte dependent and must be optimized for each type of material and analyte. It is a relatively fast technique and has extraction times of less than 1 h. A number of SFE manufacturers offer a wide variety of automated and manual extraction instruments. Samples are placed into a high-pressure high-temperature cartridge, and the supercritical fluid is passed through the sample and then depressurized in an analyte-soluble solvent or collected onto an absorbant trap for recovery.

Accelerated Solvent Extraction (ASE)

Accelerated solvent extraction is a liquid–solid extraction process performed at elevated temperatures, usually between 50–200 °C and pressures of 1500–2000 psi (9). Any type of solvent or solvent mixture may be used, and solvent volumes of less than 15 mL are required for a 10-g sample. Samples are placed in a cartridge and the hot pressurized liquid solvent is passed through the sample, cooled, and then collected. Currently, only one type of accelerated solvent extraction system is available in the market place and it is an automated instrument. The sequential technique is applicable to polymers, plant and animal tissue, foods, and environmental samples. The U.S. Environmental Protection Agency's (EPA) Office of Solid Waste has proposed a method for extracting semi-volatile organic compounds using this technology for the third update to the second edition of SW-846 (10).

Microwave-Assisted Extraction

Microwave-assisted extraction (MAE) is the process of using microwave energy to heat solvents that are in contact with solid samples and to partition compounds of interest from the sample into the solvent. Extractions are performed in closed or open microwave-transparent vessels in which the extracting solvent and sample are combined and then uniformly exposed to microwave energy. Microwave (dielectric) heating in solution occurs by 3 mechanisms:

- a single solvent or mixture of solvents that have high dielectric loss coefficients
- solvent mixtures of high and low dielectric loss
- a susceptible sample that has a high dielectric loss in a low dielectric loss solvent

Partitioning may occur by any one of these heating mechanisms or as a combination of 2 or 3 mechanisms.

Dielectrically Heated High Loss-Coefficient Solvents and Mixtures

In a homogeneous polar solvent, dielectric heating occurs through dipole rotation (*11*) and may result in partitioning at a relatively low temperature, such as 50 °C or up to 200 °C depending on the solvent susceptibility to microwave energy. Partitioning of compounds in extraction (solute into the solvent) may consist of more than one process: desorption at the matrix–solvent interface followed by diffusion of the analyte into the solvent (*2*). A microwave-compatible solvent or reagent such as isopropanol normally couples with the electromagnetic (EM) field, and through dielectric relaxation mechanisms transfers heat to the solvent medium. Such highly polar microwave-susceptible solvents are characterized by large dielectric-loss tangents and loss coefficients. This property results in the elevated bulk temperatures associated with microwave heating. In MAE, the matrix, such as soil, biological, botanical, and many minerals, does not generally absorb microwave energy. High-temperature extractions accomplished in closed vessels may result in pressures up to 200 psi (~14 bar) within the containers. Both temperature and pressure influence the extraction rate, and for safety reasons it is important to monitor the extraction-solvent temperature. This monitoring may be accomplished by a combination of temperature measurement and feedback control of the microwave source. MAE is usually performed with solvent volumes of less than 50 mL and extraction times of less than 30 min. Multiple simultaneous extractions are often performed. The technique is applicable to most types of samples and can be accomplished in laboratory microwave systems now available from several manufacturers.

Microwave Heating with Energy-Transparent Solvents

One example of microwave extraction using energy-transparent solvents is exemplified by the microwave-assisted process (MAP). The process features a partitioning mechanism in which the sample, a biological material, is a good dielectric in the presence of a low-dielectric, poorly heating solvent. It can best be understood by comparing and contrasting it to traditional microwave-assisted (solvent) extraction where the often dry matrix does not generally absorb microwave energy. Intrinsic moisture is an essential component in MAP because the water superheats and eventually causes the cell membranes to rupture and propel the cell contents into surrounding nonabsorbing cooler solvent where they dissolve. The dispersed moisture content for this process was originally specified at 40–90% (12).

Another example was the first microwave-extraction experiments where botanical materials with both polar and nonpolar solvents were treated with microwave energy at repeated 30-s exposures (13, 14). Hexane, a nonmicrowave-absorbing solvent, was used to extract biologicals, food crops, and prepared foods (13). In this early work, the author did not specify the water content and it is presumed to cover processes outside the range of MAP by using the described mechanism. In the patent describing this process (12) microwave heating is applied to similar biological and botanical matrices that contain substantial intrinsic moisture. Although 40–90% is the specified range, materials containing as little as 20% moisture are amenable to the process (15). These materials are suspended in a specifically nonabsorbing microwave solvent such as hexane, benzene, or isooctane. The weakly or nonsusceptible microwave solvent acts as a solubilizing medium for substances expressed from the matrix by the action of microwave heating.

In general, the phenomenon is hypothesized to involve localized heating of the water present, which is a good absorber of microwave energy, so that the water superheats. Extraction is often less than 2 min. Once at or above its boiling point, the (super)heated water causes cell-membrane rupture, and as steam makes its way through the interstitial spaces of the solid, transport of the target analyte is effected. The MAP effect can be produced in dry samples by incorporating water into the sample (12, 13, 15). One of the principle features of the process is the lower temperatures observed in the microwave-extracted materials in contrast to volumetric heating usually experienced in traditional solvent procedures. These lower temperatures and the migration of the analyte into the cooler surrounding solvent where heat is dissipated may account for the fact that little degradation of analytes is observed (12, 13, 16).

Microwave-Energy-Mediated Gas-Phase Extractions

Another type of MAP microwave extraction where energy is selectively applied to the sample matrix rather than the solvent environment can be found in the gas-phase extractions of benzene from a water solution in a closed container having some headspace (16, 17). In this system, energy absorbed by water is transformed into heat that is conducted to the benzene, which in turn is warmed sufficiently to volatilize into the headspace because its partial vapor pressure and heat of vaporization as well as its heat capacity are smaller than water, which favors its volatility. Gases absorb microwave energy to a far lesser degree than liquids; hence, the liquid-phase heating is generally observed without affecting the gas phase. Sampling of the headspace by gas chromatography (GC) can identify the compounds present. If a soil containing a mixture of pollutants, such as PAHs and phenols, is irradiated by microwaves, the more polar analytes selectively absorb energy and are volatilized preferentially from the solid matrix. Conversely, a wet semisolid that is irradiated with microwaves would heat sufficiently to volatilize the pollutants with the highest vapor pressures first. A headspace sampling device was described (16, 17) that features a moveable semipermeable membrane fitted inside a headspace sampling device that can concentrate gases into a given volume prior to GC analysis. Microwave extraction in both liquid–solid and liquid–vapor formats is a clean process technology using smaller quantities of less toxic organic solvents than traditional extraction methodologies, thereby reducing waste disposal and handling.

Solvent Polarity and Dielectric Compatibility

Proper selection of the organic solvent is the key to successful microwave-assisted extraction. Solvent choice is dictated by the solubility of the analytes of interest in the solvent, the interaction of the solvent and the matrix, and the microwave absorbing properties of the solvent. The major advantage of MAE is the rapid delivery of energy to the total volume of the solvent and its subsequent rapid heating. Important physical parameters as they relate to microwave heating are presented in this section to aid in understanding the microwave absorptivity characteristics of organic solvents and the temperature capabilities of such solvents when heated with microwave energy.

Polarity

The magnitude of the solvent dipole moment is the main factor that correlates with the microwave-heating characteristics of the organic solvent. The larger the dipole moment the more vigorously the solvent molecules will oscillate in the microwave field. Polar solvents such as alcohols, ketones, and esters strongly couple (absorb) microwave energy. Benzene, xylene, and straight-chain aliphatic hydrocarbons are nonpolar. They do not interact with the microwave field and do not heat. Inspection of dipole moment values in Table I suggests that acetone with a dipole moment of 2.69 or acetonitrile with a dipole moment of 3.44 will rotate easily when exposed to an alternating electric field of microwave energy. This oscillation produces collisions with surrounding molecules and energy is transferred with subsequent heating. For microwave solvent extraction to be

Table I. Selected Physical and Dielectric Constants of Organic Solvents

Solvent	bp (°C)	vp	(kPa)	ε'	Dipole Moment	tan ∂ × 10⁻⁴
Methylene chloride	40	436	58.2	8.93	1.14	
Acetone	56	184	24.6	20.7	2.69	
Methanol	65	125	16.7	32.7	2.87	6400
Tetrahydrofuran	66	142	19.0	7.58	1.75	
Hexane	69	120	16.0	1.88	<0.1	
Ethyl acetate	77	73	9.74	6.02	1.88	
Ethanol	78			24.3	1.69	2500
Methyl ethyl ketone	80	91	12.1	18.51	2.76	
Acetonitrile	82	89	11.9	37.5	3.44	
2-Propanol	82	32	4.27	19.92	1.66	6700
1-Propanol	97	14	1.87	20.33	3.09	~2400ᵃ
Isooctane	99	49	6.54	1.94	0	
Water	100	760	101.4	78.3	1.87	1570
Methyl isobutyl ketone	116	20	2.67	13.11	—	
Dimethyl formamide	153	2.7	0.36	36.71	3.86	
Dimethyl acetamide	166	1.3	0.17	37.78	3.72	
Dimethyl sulfoxide	189	0.6	0.08	46.68	3.1	
Ethylene glycol	198			41.0	2.3	10,000
N-Methyl pyrrolidinone	202	4.0	0.53	32.0	4.09	

NOTES: Temperatures °C were determined at 101.4 kPa (69); vapor pressures were determined at 25 °C; dielectric constants were determined at 20 °C; dipole moments were determined at 25 °C. Tan ∂ values (ε" and ε') are from reference 19.

ᵃValue was determined at 10 °C (20).

effective, the solutions or the sample must heat when exposed to microwave energy. Therefore, certain criteria must be taken into consideration before a solvent or combination of solvents is chosen for microwave extraction.

Dielectric Compatibility

Dissipation or dielectric loss coefficient, ϵ'', is the physical parameter that describes the ability of a material to heat when placed in the microwave field. The larger the loss factor or coefficient, the more optimal the heating. Dielectric loss coefficient is a measure of the ability of the material to transform the EM energy to heat through internal mechanical motion and is wavelength dependent. Short wavelengths heat intensely and at surfaces, whereas longer wave lengths heat less intensely over longer distances. The dielectric constant, ϵ', is the ability of a material to slow the velocity of EM radiation. Loss tangent, or tangent delta (tan δ), is a ratio of the dielectric loss coefficient, ϵ'', and the dielectric constant, ϵ'. This ratio is a more accurate measure of a material's thermal performance in microwave fields than the dielectric constant alone.

   Although difficult to measure, dielectric loss coefficients are the physical parameters that most influence the tan δ value of a material, because large loss coefficients produce large loss tangents. Optimal heating occurs at or near the resonance frequency of the material (18); because ϵ' decreases as resonance approaches, it is not the critical dielectric parameter. Loss tangents are frequency dependent; thus, a material that heats well at microwave frequencies may demonstrate negligible heating in other portions of the electromagnetic spectrum. Another way to understand the role of the loss coefficient as the important physical constant that governs dielectric heating is to look at the equation that describes the power dissipated as heat in a unit volume of material (18):

$$P/V = CE^2 f \epsilon'' \tag{2}$$

where C is a constant; the power per unit volume, P/V, is directly proportional to the frequency of the wave, f, the square of the electrical field strength in the material, E^2 (V^2/m^2), and the dielectric loss factor, ϵ''. The physical and chemical parameters, including dielectric constants and loss tangents for some commonly used solvents (19), are listed in Table I.

When MAE is conducted in closed vessels, the temperatures achieved will be greater than the atmospheric boiling points of the solvents. Table II lists the atmospheric boiling points of common organic solvents and their mixtures, and the temperatures that are achieved at 175 psig (1 psig = 6.895 kPa). Nonpolar aliphatic solvents such as hexane and cyclohexane indeed do not heat and therefore would not be considered suitable high-temperature microwave-extraction solvents. For most other solvents, the temperatures achieved inside the closed container are two to three times

Table II. Solvent Boiling Point and Microwave-Heated Closed-Vessel Temperature Comparison

Solvent	bp (°C)	Closed-Vessel Temperature (°C)
Dichloromethane	39.8[a]	140
Acetone	56.2[a]	164
Methanol	64.7[a]	151
Hexane	68.7	
Ethanol	78.3[a]	164
Cyclohexane	80.7[a]	
Acetonitrile	81.6[a]	194
2-Propanol	82.4[a]	145
Petroleum ether	35–52[a]	
Acetone:hexane (1:1 v/v)	52[b]	156
Acetone:cyclohexane (70:30 v/v)	52[b]	160
Acetone:petroleum ether (1:1 v/v)	39[b]	147

NOTE: Closed-vessel temperature values were determined at 175 psig.

[a]Values are from reference 61.
[b]Values were determined experimentally.

higher than atmospheric boiling points. For example, dichloromethane (DCM), which normally boils at 39.8 °C, can be heated to 140 °C. Miscible mixtures of polar and nonpolar solvents will heat in the microwave because the polar solvent heats the nonpolar solvent by conduction. A mixture of acetone:hexane (1:1) heats quickly, but the heating rate and the final temperature of the solvent mixture are less than the heating rates and the final temperature of the same volume of acetone alone. Conversely, polar solvents such as acetone and DCM are heated in excess of 100 °C above their normal boiling points. Not only are higher than atmospheric bp temperatures attainable, but as seen in Figure 1, they are reached in a matter of minutes. Similar temperatures are achieved for mixtures of acetone and cyclohexane and acetone and petroleum ether. Elevated solvent temperatures increase extraction efficiency and reduce extraction time.

The elevated temperature of the solvent increases the solubility of the analytes of interest in the extraction solvent and also increases the desorption kinetics of the analytes from the matrix being extracted. All mass-transport phenomenon are sped up at elevated temperatures and therefore influence the rates of microwave-heated extractions. The major benefit of microwave heating is the speed and efficiency of the delivery of energy to the organic solvent. The ability to work in a closed container at elevated pressures and temperatures is also advantageous because volatile analytes are retained.

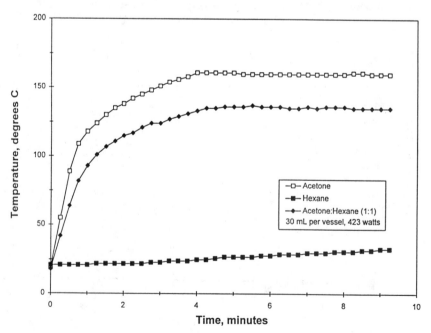

Figure 1. Simultaneous heating rates for acetone, hexane, and acetone:hexane (1:1).

In microwave solvent extraction, the solvent is in constant contact with the solute and the matrix surface. Without agitation or stirring however, there is little opportunity for the solvent to percolate through the matrix as in a classic Soxhlet extraction or in cartridge SFE. In Soxhlet extractions, matrix and solute are at a lower temperature than the boiling solvent and are extracted with clean, condensed solvent. In microwave extractions the matrix is constantly bathed by hot liquid. When solvent boiling occurs, either inside cells (*12*) or at superheated sites (*20*), the sample–solvent mixture may be constantly agitated.

For microwave extraction of organic pollutants, various solvent combinations that are more or less polar depending on the target analyte have been proposed (*19, 21, 22*). Although hexane may be the solvent of choice for most petroleum hydrocarbons and PAHs, it will not readily solubilize polar organophosphorous compounds. Cosolvents such as methanol or acetone are commonly added to hexane and toluene to improve extraction of more polar materials. Polar matrices, or those that contain substantial polar components, may be extracted with nonpolar extracting solvents at microwave frequencies. Polar materials such as water and elemental-carbon-containing polymers, self heat when exposed to microwave energy. Heat lost from the sample is transferred by conduction to the surrounding solvent, which may be transparent. Figure 2 compares the heating rates of hexane, dry soil in hexane, ashed dry soil in hexane, and soils with <0.1%

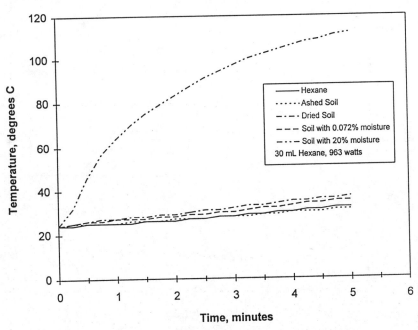

Figure 2. Heating rate for hexane and hexane with soil.

and with 20% moisture irradiated at 2450 MHz with 963 W of output power for 5 min. Negligible heating is observed for all combinations except where large amounts of water are present. Similarly, Figure 3 shows that a 3-g sample of neoprene rubber in hexane heats well at similar applied powers and times even though hexane itself is relatively unaffected by microwave frequencies. Solvent choices are thus unlimited.

Microwave-Instrumentation Considerations in Solvent-Extraction Safety

Because of the flammable nature of many organic solvents the potential for fire and explosion when such solvents are heated in the microwave field must be given high priority. When polar organic solvents or mixtures of polar and nonpolar solvents are heated in a closed vessel to as much as 100 °C above their normal boiling point, pressures in excess of 100 psi are common, and the potential for accidents is greatly increased.

Every microwave solvent-extraction instrument should have redundant safety features, each acting as a backup for the other to prevent any possible fire or explosion from occurring in the microwave cavity. Instruments must be designed to eliminate ignition sources within the cavity, to

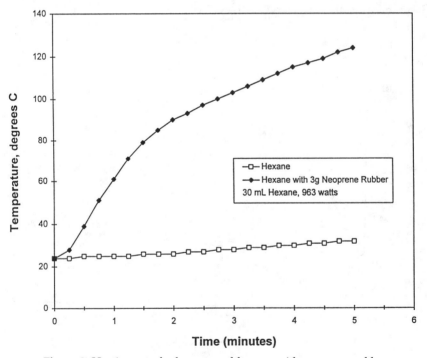

Figure 3. Heating rate for hexane and hexane with neoprene rubber.

contain solvent, and to detect and remove possible leaking solvent. Figure 4 is an illustration of the interior of a CEM MES-1000 microwave-extraction system and shows a single vessel in the 12-position turntable. Safety aspects of the instrument illustrated include

1. an exhaust fan to evacuate the cavity air volume approximately once every second. If the exhaust fan fails or air flow is blocked in the exhaust duct, then the air-flow switch located in the system exhaust interrupts electrical power to the magnetron.

2. a solvent detector that monitors the possible presence of solvent within the instrument cavity. This detector has the capability of shutting off the microwave source immediately at concentrations below the lower explosive limit (LEL) or limit of flammability (LOF) for that particular solvent.

3. a Teflon ceiling inside the cavity held in place with polypropylene dart clips to minimize the possibility of high-energy electrical discharges from the mode stirrer in the top of the cavity. The potential for electrical arcing can be eliminated by using nonconducting materials and removing bare metal edges and points.

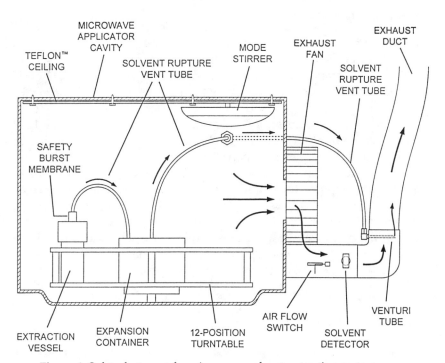

Figure 4. Safety features of a microwave solvent-extraction system.

4. the integral safety burst pressure membrane of the vessel that is connected to the central expansion container through the vent tubing. In the event of a membrane rupture due to excess internal pressure in the vessel, hot solvent gases are removed from the microwave cavity through the rupture vent tube.

5. a venturi tube in the cavity exhaust duct establishes negative pressure in the collection container drawing vapors into the exhaust system. Microwave-extraction systems must be capable of containing the liquid solvent and solvent vapors so that they cannot escape into the cavity while the magnetron is on.

An additional safety feature, not shown in the diagram, is a terminal isolator located in the instrument waveguide. This device diverts reflected microwave energy to a dummy load that eliminates damage to the magnetron from overheating and reduces electrical arcing potential within the cavity. A mechanical safety shield has also been added to the unit. It is lowered into place after the door is closed and is designed to prevent the door from opening and expelling vessels or hot solvent in the event of an accident or catastrophic system failure.

Vessel Design and Materials

Closed vessels for microwave solvent extraction should be constructed of materials that are transparent to the EM radiation and inert to the solvents. If solvent-incompatible materials must be used, then the vessel design should be such that solvent contact with these materials is minimized. Figure 5 illustrates a CEM standard lined extraction vessel and a control extraction vessel comprised of Teflon perfluoroalkoxy (PFA) inner liners, seal covers, vent fittings, ferrules, nuts, and vent tubes. Vessel bodies and caps are made from Ultem, a polyetherimide. Cap and cover for the control vessel are modified to allow connection of a pressure-sensing tube and a temperature probe for monitoring both the internal pressure and temperature of the vessel. These containers must be thermally compatible with the microwave instrument to insure controlled, even heating of all vessels within the microwave cavity. This control is usually accomplished for multiple vessels by evenly spacing vessels in the carousel and rotation through the pattern of a 360° oscillating turntable. Figure 6 shows a turntable system containing 12 extraction vessels.

Temperature and Pressure Control

Finally, the instrument must be capable of monitoring and controlling temperature and pressure within the extraction vessels to guard against

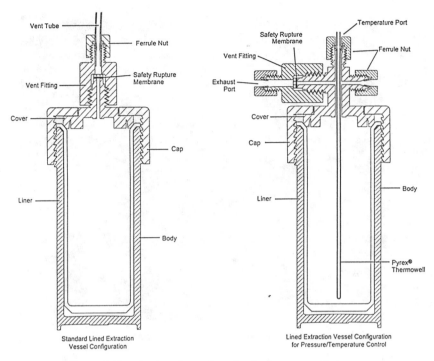

Figure 5. Standard and monitored microwave extraction vessels.

overheating and over-pressurization of the vessels. This protection is usually accomplished by properly matching vessel-compatible temperature and pressure detectors (sensors) with the instrumentation software. Temperature- and pressure-controlled extraction systems allow for precise heating of solvents, which in turn permits reproducible extraction conditions and the protection of thermally labile analytes. Such microwave systems are commercially available.

Applications

Although MAE has probably been applied to most situations in which traditional Soxhlet extractions are performed, three applications are presented in depth. They are the most broadly used and are discussed in historical context beginning with natural products, followed by plastics and polymers and lastly environmental pollutants. Unique applications are also included.

Natural Products

In 1986, Ganzler and co-workers published the first of three papers (*13, 23, 24*) exploring the use of microwave energy to extract various compounds

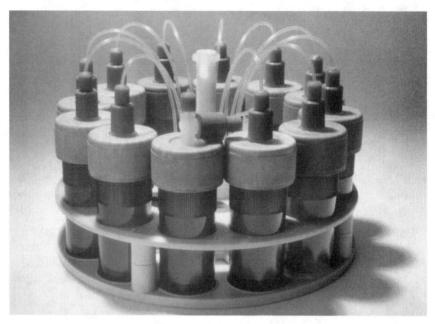

Figure 6. Fully assembled carousel with lined microwave-extraction vessels.

from soils, biologicals, and botanicals prior to chromatographic analysis. Vicine and convicine antinutritives were extracted from fava beans with a 1:1 methanol:water mixture irradiated with microwave energy for multiple 30-s intervals. Extracts yielded approximately 20% more than Soxhlet-extracted samples without loss of the analyte. Gossypol, another antinutritive, was extracted from cottonseeds into a 1:1 methanol:water mixture with similar efficiency. Crude fat was extracted from yeast, lupine, maize, cottonseed, and meat flour with the same solvent system and with hexane from Robaby, an infant formula.

   Lupin alkaloid and drug metabolites were also extracted by Ganzler et al. (*14*) from lupin seeds and rat feces, respectively, into solutions of methanol:acetic acid and methanol:ammonia with about 20% greater efficiency than by Soxhlet. These microwave extractions of antinutritives, crude fat, pesticides, and drug metabolites were performed in open containers at temperatures below the solvent atmospheric boiling points. The authors concluded that high microwave-extraction efficiency was correlated with both good microwave heating of the solvent and its ability to solubulize the analyte.

   In using microwave energy to extract components from biological materials, like fat from processed food (*13*) or peppermint (*Mentha piperita*) oil from mint leaves (*12*), the liquid–solid extraction depends on the presence of a solvent compatible with the target analyte. The original extractions of essential oils from botanicals, pesticides and oils from animal tissue, and organics from filters (*12, 13*) were all performed in open vessels with selected solvents that preferably do not absorb microwave energy so as not to compete with any natural or added moisture present in the matrix. Such extractions produced high-quality product with less tissue damage in far less time than classical Soxhlet or steam distillation methods. Scanning electron micrographs of mint leaves showed less destruction of the cell than for similar Soxhlet extractions (*16*). High-quality essences are efficiently extracted from solids because only the target analyte and intercellular and intracellular water are susceptible to microwave radiation. This process resembles steam distillation, which Dauerman and Windgasse (*25*) suggested is the operative process in microwave-irradiated wet-soil remediation for toxic wastes. The technique has been shown to extract a wide variety of substances from soil, animal tissues, and processed foods as well as from solutions (*13, 15*).

Bichi et al. (*26*) extracted pyrrolizidine alkaloids from dried plants of *Senecio palvadosos* and *Senecio cordatus* into methanol by using microwave closed vessels at temperatures between 65–100 °C. These extractions are comparable to the Soxhlet and are accomplished in less time. Extraction of pyrrolizidine alkaloids from dried plants of the *Senecio* family are accomplished in 20 and 30 min microwave extractions in 25 and 50 mL of methanol, respectively. Temperature-feedback capability permits highly reproducible extractions. By extracting at the same temperature every time, the same extract composition is produced and the purity of the extracts is similar. The qualitative and quantitative aspects of the GC analyses of these extracts are nearly identical from run to run, and microwave extracts are qualitatively and quantitatively identical with chromatographs from the Soxhlet extracted material (*26*).

Chen and Spiro (*27*) studied the heating characteristics of microwave-assisted extraction of rosemary, *Rosmarinus officinalis,* and peppermint, *Mentha piperita piperata,* leaf materials. MAP using transparent hexane as the solvent, as well as a more traditional microwave-extraction technique with an absorbing solvent such as ethanol and mixtures of the two were used. Temperature rises were dependent on the solvent and the leaves; more effective extractions in nonpolar solvents were accomplished when the weight of leaves was increased, whereas more microwave power was needed to elevate the temperature of the polar solvent being evaluated.

The fungal metabolite ergosterol and fatty acids were successfully isolated from fungal hyphae, spores, mushrooms, and other contaminated natural products (*28*). In only 5 s of microwave irradiation, 70% of the ergosterol was extracted whereas only 30–40 s was required to achieve maximum recoveries. In the presence of methanol and aqueous sodium hydroxide, extraction and saponification were accomplished simultaneously and rapidly under microwave conditions. Compared with classical solvent and SFE processes, the author felt the MAE technique was simple, fast, reliable, and gave more consistent results (relative standard deviation (RSD) <4%).

Unique Natural Product Extraction

A recent application of microwave extraction for the analysis of drugs in biologicals has been reported by Franke et al. (29). Lidocane, methadone, diazepams, and propoxyphenes were extracted from human serums into a toluene–isoamyl alcohol–n-heptane mixture in an atmospheric microwave system. Extraction efficiencies and precision were largely comparable to traditional liquid–liquid extractions but accomplished more rapidly, with less solvent and with less analyst exposure to the solvent load.

Additives in Polyethylene and Other Polymers

Additives such as antioxidants and UV stabilizers are compounded into polyethylene (PE) and other polymers to protect them during processing and end-use applications. Slip agents, antistatic agents, antiblock agents, flame retardants, and pigments are also incorporated to impart particular properties to the polymer (30, 31). A fast and reliable method for determining the additive concentration is essential to maintain consistent quality during production. Typically, extraction is accomplished by refluxing the polymer in an appropriate solvent for 1–48 h (32, 33). Following extraction, the sample is cooled, exchanged if necessary, filtered, and placed in an autosampler vial for high-pressure liquid chromatographic (HPLC) analysis. In some cases, ultrasonic exposure reduces the extraction time (34, 35). Recent studies show however, that 30–60 min sonication is required to quantitatively extract common antioxidants from polyethylene but this method is not as efficient as 60-min reflux heating with the American Society for Testing and Materials extraction method (32, 36).

After extraction, these antioxidants, UV stabilizers, and slip agents are analyzed by reversed-phase HPLC because it gives accurate, reproducible results and is capable of analyzing a large number of compounds simultaneously (37–40). Technological advances allow HPLC instruments to run unattended or with minimal operator intervention. The current rate-limiting step in additive analysis is the sample extraction. Suitability of microwave solvent extraction for this application was first shown by Freitag and John (41) who obtained excellent recoveries of Irgonox from polypropylene (PP) and PE in less than 8- and 5-min exposures, respectively, to heated solvent.

To determine additives and antioxidants that have been incorporated during formulation, it is helpful to know the solubility parameter, delta, to choose an appropriate solvent to dissolve the polymer (42). Solvent parameter delta is roughly correlated to hydrogen-bonding strength and polymer dissolution is achieved by matching the delta value of the polymer to that of the solvent (43). In many cases, a complete dissolution is not achieved. Rather, the solvent swells the polymer by diffusion into the crystalline lattice and this swelling allows the physically blended additives to dissolve. However, swelling is finite; the better the solvent the more swelling occurs. Polymers in the liquid state require true solution (42), which occurs predominantly when polymer-chain lengths are small, on the order of 5000–10,000 Da.

Criteria for choosing a solvent for microwave extraction of additives from a polymer for HPLC analysis are similar to those for extracting pollutants from soil, that is the additives must be soluble in the solvent. The polymer must swell to release trapped additives *and* the solvent must be compatible with the HPLC mobile phase so that solvent exchange is not required before analysis. Common extraction solvents such as alcohols, especially 2-propanol, are good reflux extraction solvents especially for additives in low-density polyethylenes (LDPE). These solvents also heat rapidly in the microwave. In contrast, solvents typically used to extract additives from high-density polyethylene (HDPE) are less polar, such as hexane, heptane, cyclohexane, toluene, and isooctane and heat poorly in the microwave. Thus, for HDPE, a binary solvent mixture consisting of a nonpolar solvent to swell the resin and a polar solvent that will heat rapidly in the microwave are required.

Common antioxidants like BHT (2,6-di-*tert*-butyl-4-methylphenol), Irganox 1016 (octadecyl-3,5-di-*tert*-butyl-4-hydroxy hydrocinnamate), and Irganox 1010 [tetrakis(methylene-3,5-di-*tert*-butyl-4-hydroxyhydrocinnamate)methane] are efficiently extracted in 20 min by using two different solvent combinations and heating programs. BHT is a small, volatile phenolic antioxidant; Irganox 1016 is a medium-size antioxidant; and Irganox 1010 is a large, difficult-to-extract antioxidant. Together they represent a range of molecular sizes and stabilities. A 50:50 mixture of cyclohexane:2-propanol irradiated at 325 W or a 98:2 mixture of methylene chloride:2-propanol irradiated at 168 W are both fast and give reproducible recoveries (Table III) (34). To achieve 90% extraction efficiency, however, stirring was required every 5 min. Antioxidants like Irganox 1010, Irgafos 168

Table III. Microwave Extraction Recoveries of Additives from High Density Polyethylene

Additive	Cyclohexane:IPA (1:1)	CH_2Cl_2:IPA (98:2)
BHT	451 (90%)	455 (91%)
1010	454 (91%)	459 (92%)
1076	480 (96%)	474 (95%)

NOTE: Values are concentrations in μg/g.

SOURCE: Reprinted with permission from reference 33. Copyright 1991.

[tris(2,4-di-*tert*-butylphenyl)phosphite] and Chimassorb 81 are extracted with >90% efficiency in 6 min with a 50:50 acetone:*n*-heptane mixture (*41*). If the extracts contain small polar antioxidants and UV stabilizers and are analyzed by reversed-phase HPLC, then solvent exchange is required because heptane is not miscible with the acetonitrile:water gradient typically used to separate antioxidants. Thus, common additives such as BHT, BHEB (2,6-di-*tert*-butyl-4-ethylphenol), Irganox MD1024, and Tinuvin P [2-(2'-hydroxy-5'-methylphenyl)benzotriazole] normally cannot be analyzed without a solvent-exchange step.

Problems were found in comparing microwave extractions with reflux extractions because 5 g of 20-mesh polyethylene swells slightly in 50 mL when heated for 60 min at reflux conditions. By using the same solvents in microwave heating, such as, isopropanol and isobutanol for LDPE and cyclohexane for HDPE, the polymer melts at the elevated pressure and temperature conditions and then resolidifies on cooling with significant swelling. A solvent-to-polymer ratio of 25 mL:2.5 g avoids this problem. Equivalent extraction efficiencies are obtained in 10 min for both 20- and 30-mesh grinds (Table IV). The optimized parameters are 2.5 g of 20-mesh polymer in 25 mL of isopropanol heated at 25% power (168 W) for 10 min. Table V shows no loss under these conditions. Ten-minute extractions are sufficiently rugged to give reproducible results (Figure 7). Small and mid-size additives take 5 min and longer heating times do not increase the recoveries. Thus, 10-min heating is sufficient to ensure that all additives have been extracted, but it is brief enough that smaller, more volatile additives are not lost. Extraction results with microwave heating in isobutanol or isopropanol compared with a 60-min reflux extraction also show good reproducibility (<3% RSD for all additives analyzed). Recovery and precision data for several additives are given in Table V.

Table IV. Comparison of Reflux and Microwave Extraction Solvent Efficiencies of Additives in Different Mesh Size Low Density Polyethylene

| | Microwave | | |
Additive	30 mesh	20 mesh	Reflux/20 mesh
BHEB	468 (1.50)	450 (1.42)	452
BHT	443 (2.00)	416 (2.64)	421
Erucamide	463 (3.87)	480 (1.84)	491
Irganox 1076	498 (1.72)	494 (2.38)	513
Irganox 1035	439 (2.05)	459 (0.97)	463
Irganox 1010	399 (2.57)	404 (3.65)	399
Isonox 129	488 (2.19)	499 (1.29)	505
TNPP	481 (0.67)	482 (0.69)	496
No. of extractions (*n*)	3	6	1

NOTE: Values are concentrations ($\mu g/g$); parenthetical values are relative standard deviations (RSD, %) for *n* extractions.

Table V. Comparison of Reflux and Microwave Extraction Solvent Efficiencies for Additives in Linear Low Density Polyethylene

| Additive | Microwave (mg/L) | | Reflux (mg/L) Isobutanol |
	Isopropanol	Isobutanol	
BHT	382 (2.28)	372 (1.59)	394
Erucamide	501 (4.92)	435 (4.56)	423
Isonox 129	198 (1.25)	196 (0.92)	188
Cyasorb UV531	421 (2.73)	420 (0.56)	426
Irganox 1010	405 (2.27)	407 (1.21)	397
Irganox 1076	443 (1.73)	422 (0.46)	414
Irgafos 168	458 (2.58)	449 (0.50)	454
n	6	6	2

NOTE: Parenthetical values are %RSD for n extractions.

Significant time savings are achieved by using multivessel extractions. Quantitative extractions for 2.0 g of (LD)PE are achieved in ~21 min with 25 mL of 2-propanol in 12 vessels simultaneously. For additives like BHEB and BHT, Erucamide, Irganox 1076 (octadecyl-3,5-di-*t*-butyl-4-hydroxyhydrocinnamate) and DLTDP (dilaurylthiodipropionate) at 475–500 ppm (confirmed by mass spectrometry and Fourier transform IR spectroscopy), the RSD ranges from 0.5–3.53% (Table VI). For Irganox 1010,

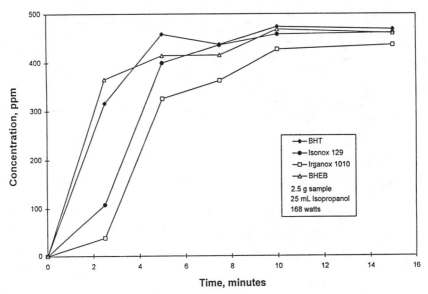

Figure 7. Influence of time on microwave-extraction efficiencies for additives in low-density polyethylene.

Table VI. Multivessel Microwave Extraction Recoveries of Additives in Low Density Polyethylene by Using 6 and 12 Vessels

Additive	6 Vessels	RSD (%)	12 Vessels	RSD (%)
BHEB	482[a]	0.50	477	1.05
BHT	465	1.83	457	2.88
Erucamide	460	1.77	445	2.86
Irganox 1010	274	5.15	274	6.23
Irganox 1076	482	1.07	472	1.35
DLTDP	476	3.53	505	2.16

[a]Value is average concentration (µg/g).

confirmed at 275–300 ppm, the RSD is 5.15%. Imprecision increases, however, for most additives on going from 6 to 12 vessels, but it is still <6.5%. By using experimental-design techniques, the optimal conditions that give extraction efficiencies of >90% of all additives with 7:3 acetone:cyclohexane are accomplished in 10 min at 375 W. Figure 8 shows that at these conditions all 5 additives targeted at 500 ppm are nearly quantitatively extracted within 9 min. From Figure 8, this result corresponds to having optimum conditions in the vessel at the maximum pressure and temperature for about 4 min. Table VII shows that 11 additives dissolved in the 7:3 acetone:cyclohexane mixture and heated at 375 W for 10 min are neither

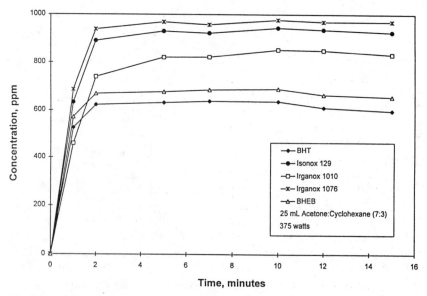

Figure 8. Influence of time on microwave-extraction efficiencies for additives in high-density polyethylene.

Table VII. Effect of Microwave Heating on Polymer Additives in Acetone–Cyclohexane (7:3)

Additive	Conc. Before Heating (mg/L)	Conc. After Heating (mg/L)
Tinuvin P	310	310
BHT	11.8	11.9
BHEB	9.50	9.58
Cyasorb UV531	11.7	11.7
Topanol CA	6.15	6.21
Irganox 1035	14.2	14.0
Isonox 129	10.4	10.5
Irganox 3114	11.0	11.1
Ultranox 626	9.58	9.23
Irganox 1010	9.10	9.36
Irganox 1076	9.42	9.68

NOTE: Heating was performed at 375 W for 10 min.

lost nor degraded during the extraction procedure. Similar results are found for the same additives extracted with isopropanol. This procedure is rugged enough to extract HDPE resins of differing crystallinity and melt indices, which ordinarily could make it difficult to extract additives reproducibly. Recovery and precision data for HDPE additives presented in Table VIII show the microwave method has an RSD of <4%.

The three solvents used for extracting additives from PE behave differently during microwave heating. At extraction temperatures near 140 °C, the acetone:cyclohexane mixture reaches pressures in excess of 100 psig, as the result of the high vapor pressures of the solvent. At the same temperature, 2-propanol and isobutanol, in contrast, reach only 78 psig and ~35 psig, respectively. Because cyclohexane dilutes the acetone and does not absorb microwave energy, substantially more microwave power is required (~540 W (80%) versus ~270 W (40%)) to heat 6 vessels of mixed solvent compared to 6 vessels of only 2-propanol to 140 °C.

Environmental Pollutants

Pollutants and Pesticides

Soil and sediment samples comprise a large fraction of environmental materials that must be analyzed for hazardous substances. In the laboratory, analysts seek to determine if these hazardous materials, such as pesticides, can be leached from the solids. The underlying assumption is that extractable materials are those that are labile in the environment and, if hazardous, constitute a threat to life forms. Hydrocarbon contaminants, such as total petroleum hydrocarbons (TPHs), PAHs, polychlorinated biphenyls (PCBs), chlorinated pesticides, and nitrogen- or phosphorous-containing compounds are among the most frequently determined pollut-

Table VIII. Comparison of Reflux and Microwave Extraction
Efficiencies for Additives in High Density Polyethylene

Additive	Target	Reflux	Microwave	RSD (%)
BHT	150	140	157	2.45
Isonox 129	200	200	197	1.26
Irganox 1010	200	154	199	2.33
Irganox 1076	200	187	199	1.60
Irgafos 168	200	195	208	1.39
BHT	650	607	627	1.22
Isonox 129	800	746	793	2.78
Irganox 1010	800	702	817	2.72
Irganox 1076	800	814	806	0.88
Irgafos 168	650	564	570	3.24

NOTE: Values are concentrations ($\mu g/g$).

ants in soils and sediments. By nature, soils are highly variable in composition and although no two are exactly alike, some common analytical problems exist (1).

For example, recoveries of alkanes from wet clays by SFE are lower than from dry clay, perhaps because the clay matrix is swollen by the moisture that inhibits the mass transfer of the analyte out of the matrix (2, 44). Analyte may also be physically trapped within channels by water present in clay particles. Water may also reduce the surface area available for penetration of the extractant. Microwave extractions of PAHs, in contrast, are improved when soil is wet, probably because the water absorbs microwave energy and may cause local superheating of the soil, which facilitates desorption of the analyte. In other soil-test systems with both microwave extraction and SFE, pollutant spikes are more easily removed than native material (21, 44). The majority of MAE papers have been devoted to the analysis of pollutants extracted from soils and sediments.

PAHs and Other Semivolatile Compounds

Successful microwave extraction of PAHs from soils has been accomplished in several solvent systems. Early experiments at the normal boiling point of methylene chloride showed recoveries of >85% in just 15 min, and complete recovery was achieved in 4 h at 40 °C, a savings of 12 h (45). The influence of extraction time on recovery at a fixed temperature (Figure 9) suggests that 10 min is not adequate at 100 °C; however, at greater than 15 min, compounds may be lost. Fluoranthene is as efficiently extracted in 10 min as at 20 or 30 min without loss. Higher molecular weight compounds are recovered with >90% efficiency at 15 min, and recoveries improve slightly when extractions are extended to 20 min (Figure 10). Pyrene and chrysene are recovered quantitatively in 15 min but may deteriorate slightly when held at 100 °C for more than 15 min.

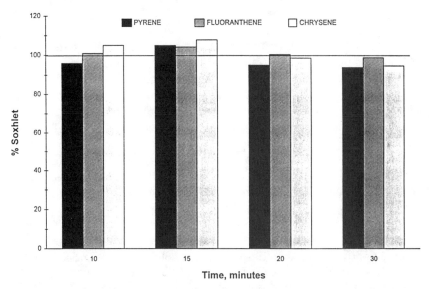

Figure 9. Influence of time on microwave-extraction recovery of PAHs in Marine Sediment SRM1941a with methylene chloride at 100 °C.

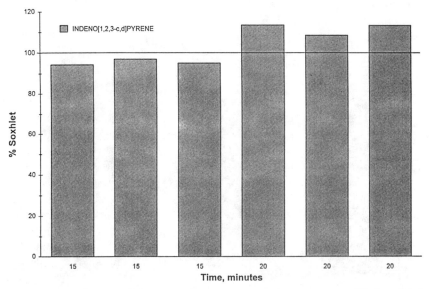

Figure 10. Influence of time on microwave-extraction recovery of Indeno[1,2,3-c,d]pyrene in Marine Sediment SRM1941a with methylene chloride at 100 °C.

Increasing the temperature to 85 °C and 100 °C for a 15-min extraction generally improves recovery for 3-, 4-, and 5-ring compound (Figure 11). Chrysene is extracted well at all temperatures in contrast to pyrene, which requires higher temperatures. Cooling to room temperature in a closed container is important to retaining volatile analytes. Naphthalene recovery, normally problematic in traditional extractions because of its easy volatility, is improved with higher temperatures (Figure 12).

Optimization of extraction parameters clearly shows that for the majority of the 16 PAHs of interest to the EPA (46), maximum recovery is achieved at 15 min (Figure 9). Poor recovery was originally thought to be due to analyte degradation as the result of prolonged exposure to heat or to volatility at the higher temperatures. In fact, some compounds are poorly extracted because of solvent incompatibility and matrix effects (21). For larger molecular weight compounds like indenopyrene, the recovery increases with longer heating. Although results for larger multiring compounds are not as consistent as for fewer-ring compounds, they still agree within 10% (45). If larger compounds take longer to extract because they are more tightly bound, then desorption, not solubility, may indeed be the rate-limiting factor. This result is supported by recent results (21, 44, 47) showing that PAHs spiked onto soil are recovered more quickly than native pollutants are from clays and soils. Rapid at first, recovery becomes asymptotic and reaches the limiting value of the spike long before the native pollutant (44).

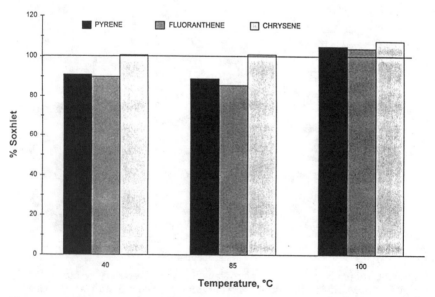

Figure 11. Influence of temperature on microwave-extraction recovery of PAHs in Marine Sediment SRM1941a with methylene chloride for 15 min.

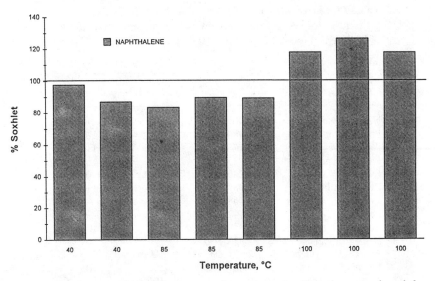

Figure 12. Influence of temperature on microwave-extraction recovery of naphthalene in Marine Sediment SRM1941a with methylene chloride for 15-min.

Extraction results from ~5 g of sediment with 30 mL of methylene chloride are shown in Table IX. Average reproducibility is similar for both small and large sample sizes, but small improvements in accuracy and reproducibility are possible with larger samples because the concentrations are not near the detection limit. Well-mixed samples as small as 1 g can give reproducible results; therefore, mapping a site with numerous smaller samples may be more informative than to select larger samples that need to be homogenized to accurately estimate contamination levels.

Because many environmental (solid) samples are wet and may contain as much as 50% moisture, it is important to understand the influence of water on samples extracted with microwave heating. Water is a good absorber of microwave energy and its presence increases the heating rate from room temperature to 100 °C in 130 s to ~75 s at 604 W (45). Crucial extraction temperatures are reached sooner in wet samples and the recovery of all analytes is improved in the presence of water (Table X). In contrast, Lopez-Avila et al. (21) found that the presence of additional water more than doubled the time needed for dry soils in solvent to reach an appropriate temperature. Average recoveries of PAHs for both wet and dry matrices were around 85% with 10% RSD. Recoveries of nearly all pollutants extracted from both wet and dry commercial-reference-material soils were greater with MAE than with traditional EPA methods (22). By using SFE, the recoveries of functional amines such as atrazine, cyanazine, and diuron from wet soil samples extracted into chloroform improved (2). Evidence shows that heating wet soils with microwave energy induces a

Table IX. Microwave-Heated Methylene Chloride Extraction Efficiencies for PAHs from Different Size Samples (1 and 4.5 g) of Marine Sediment (SRM 1941a) after 15 min. at 100 °C

		Soxhlet Recovery (%)	
PAH	µg/g (± 1 SD) Soxhlet[a]	1 g	4.5 g
Naphthalene	906 ± 65	121	—
Phenanthrene	480 ± 11	108	103
Anthracene	170 ± 5	118	118
Fluoranthene	925 ± 40	104	102
Pyrene	839 ± 32	105	100
Benz[a]anthracene	387 ± 17	108	112
Chrysene	468 ± 19	108	—
Perylene	442 ± 22	—	97
Benzo[k]fluoranthene	379 ± 20	100	103
Benzo[a]pyrene	622 ± 30	100	105
Benzo[ghi]perylene	487 ± 31	100	—
Ideno[1,2,3-cd]pyrene	621 ± 36	96	105
Mean ± RSD		106 ± 8	105 ± 6

NOTE: Values are mean of three dry samples of spiked (deuterated) sediment in 30 mL of solvent: 1 spiked solvent blank and 2 solvent blanks in 120-mL Teflon PFA vessels. Heating was done at 632 W.

[a]These values are from acetone–hexane extractions.

steam distillation effect in closed containers (25), a factor that may account for the improved analyte recovery.

Studies conducted at room temperature, 80 °C, 115 °C, and 145 °C show that optimum analyte recoveries are achieved at elevated temperatures and no significant difference exists between 5-, 10-, and 20-min extractions (21). Most importantly, recoveries appear to be matrix dependent, a common occurrence as well for metals in soils (48). Lopez-Avila and co-workers (21) found that the presence of a matrix, whether soil or sediment, has little effect on the heating rate of (1:1) hexane:acetone when extracting soils containing PAHs, phenols, basic or neutral compounds, and organochlorine pesticides. At a constant mass-to-volume ratio, however, 14 min are required for 10 mL of the hexane:acetone to reach 151 °C, nearly 7 times as long as it takes for 30 mL with 6 g of soil to reach a similar temperature. Ultimately, recovery of PAHs and selected basic or neutral compounds is adversely affected when the solvent volume and sample masses are both increased. Reduced recoveries of 15–30% for PAHs in the presence of a soil matrix are attributed to volatility during concentration steps and to measurement errors (21). PAHs were also extracted from a spiked topsoil into methylene chloride and acetone:hexane (1:1) in an open microwave system (49). Overall recoveries are comparable to the closed-vessel microwave values, and little difference exists in the efficiency between the two solvent systems.

Table X. Microwave-Heated Methylene Chloride Extraction Efficiencies for PAHs from Wet and Dry Samples of Marine Sediment (SRM 1941a) after 10 min at 100 °C

		Soxhlet % Recovery		
PAH	Soxhlet[a]	Dry	Wet	Wet/ Na$_2$SO$_4$
Napthalene	906 ± 65	94	—	96
Phenanthrene	480 ± 11	102	108	98
Anthracene	170 ± 5	109	138	118
Fluoranthene	925 ± 40	100	104	96
Pyrene	839 ± 32	97	105	96
Benz[a]anthracene	387 ± 17	114	118	111
Chrysene	468 ± 19	100	—	—
Perylene	442 ± 22	68	—	87
Benzo[k]fluoranthene	379 ± 20	66	91	75
Benzo[a]pyrene	622 ± 30	77	109	104
Benzo[ghi]perylene	⟩487 ± 31	—	—	—
Indeno[1,2,3-cd]pyrene	621 ± 36	—	113	111
Mean ± RSD		93 ± 17	111 ± 14	99 ± 13

NOTE: Values are mean of three 1-g (dry) samples of spiked (deuterated) sediment in 30 mL of solvent: 1 spiked solvent blank and 2 solvent blanks in 120-mL Teflon PFA vessels. Wet samples are ~50% moisture; wet/Na$_2$SO$_4$ contains 1:1 sodium sulfate to water. Heating was done at 632 W.

[a]These values are from acetone–hexane extractions, μg/g.

Method performance was a function of matrix but the dependence on concentration was uncertain (*21, 50*). Reextracting a portion of already extracted material, Lopez-Avila et al. (*50*) were able to show that whatever could be extracted in 10 min had been extracted. For other organic compounds such as dibenzofuran, carbazole, and other nitrogen-containing compounds extraction recoveries were ~85%. Some volatile compounds like 1,2-dichlorobenzene and 2-methylphenol gave expected low recoveries, but in most cases recoveries were higher than those achieved with EPA-approved methodology. There was indication for a majority of the 20 compounds examined that recoveries from a wet matrix were marginally higher by MAE than by conventional extractions.

Ninety-five compounds listed in U.S. EPA SW-846 *Test Methods for Evaluating Solid Waste* Method 8250 (*51*) were examined for efficacy and stability to MAE extraction (*52*). Pollutants were always more easily recovered from freshly spiked soil samples (water was added to insure good mixing and more closely resemble a real-world sample) when compared with aged samples, and recovery decreased with age with more spread. Here too, 5-g samples in 30 mL of the acetone:hexane (1:1) mixture were extracted at 115 °C for 10 min at 100% power. For semivolatile compounds in the Method 8250 list, like PAHs and nitroso-amines, recoveries of 79 of the 95 compounds extracted from freshly spiked top soil were 80–120%. When comparing freshly spiked to 24-h aged samples, a downward trend is observed and higher recoveries are seen for compounds like acridine, 2,4-dinitrophenol, and 2-picoline in wet, aged matrices. Basic compounds from the list like aromatic amines freshly spiked onto an ERA reference soil were extracted with ~70% efficiency by using acetonitrile as the microwave extractant (53).

Barnabas, Dean, and co-workers (22) studied two different solvent systems in their evaluation of MAE for its efficiency in the extraction of PAHs from contaminated land soils and a consensus soil. Five grams of soil were extracted with 30 mL of DCM and compared with extractions with variable composition acetone–hexane mixtures. During the initial development of methodology, arcing was observed inside the microwave vessels. This result was alleviated by reducing the sample size to 2 g and increasing the solvent volume to 40 mL. Experimental-design parameters included temperatures from 40–120 °C and times from 5–20 min in volumes between 30 and 46 mL. Microwave and Soxhlet recoveries were comparable but RSDs for MAE were less than one-third those of Soxhlet. Moreover, the mixed-solvent experiment showed that for PAHs, the best recovery was obtained in 100% acetone. These results were, on average, higher than those extracted with DCM. Repeatability is excellent (2.4%).

On the basis of the results of their Central Composite Design (CCD) study, the authors concluded that "for a particular soil, simple microwave operating parameters may have little influence on the amount of analyte recovered. Whereas, the nature of the matrix in which the analytes are bound can have a profound effect on the recovery of the compounds" (22). This observation points to the generally robust nature of microwave solvent extraction: MAE can be applied successfully for 5–20 min, at any one of several elevated temperatures in solvent volumes of 30–40 mL to give results comparable to classical techniques. They also speculated that higher recoveries would be obtained from analyte spiked soils than from real-world (aged) soils as Lopez-Avila et al. (52) found and as has been reported for SFE (44). These authors suggest the effect of stronger binding of analyte to the matrix in native samples as the reason, which is a standard tenet of soil chemistry (1, 54). When a completely different soil matrix was extracted with both acetone and DCM under the same experimental conditions, there was no difference in the amount of PAHs recovered by each solvent. This result contrasts with their earlier results on the different soil. Such variations in recovery for different soils under the same operating conditions confirms that recoveries by microwave extraction may indeed be largely due to soil matrix effects.

Organophosphorous Pesticides

When 47 organophosphorous compounds from the EPA Method 8141A list were extracted from freshly spiked and variously aged soil, 35 were recovered at 80–120%. A few, such as, dichlorovos and fensulfathion, had 50–80% recoveries. Naled and HMPA were among those poorly recovered at < 20%. Spiked samples aged 14 days fared better; 37 were recovered at >80%, a few at 50%, and some at >120%. Of samples aged 21 days, 35 compounds had acceptable recoveries between 80–120%. Naled and phosmet, however, were recovered below 50%. Recovery of poorly extracted compounds from freshly spiked soils always improved in briefly aged topsoils. That group includes monochrotophos, naled, TEPP, phosphamidon, trichlorfon, and HMPA. This result was attributed to the presence of water in the matrix. Continued aging reversed the trend overall because recoveries of these same analytes decreased after 21 days (*52*). This anomalous behavior was attributed to microbial degradation.

With just 2–3 mL of ethanol, Ganzler (*13*) was able to extract bromophos and parathion from 0.5–1.0-g soil samples heated with microwaves. Analyte recoveries were equivalent to or higher than comparable Soxhlet extractions, and preparation times were greatly reduced.

Aqueous extractions of organophosphorous herbicides with microwave heating have shown some success because of their increased solubility in water compared to PAHs or organochlorines. Steinheimer (*55*) described a microwave-assisted extraction of atrazine and its principle metabolites from agricultural soils with 0.35 N HCl solution.

Imidazolinones, a new class of low-use-rate herbicides that pose a reduced environmental risk, are finding increased use for protection of a wide variety of agricultural crops. Monitoring the persistence of these compounds in the environment, mainly soils, has been a challenge because of the lengthy extraction and cleanup required. Using a radionuclidic imazethapyr as a surrogate for the class, Stout et al. (*56*) developed a 1-h extraction procedure using 20 mL of 0.1 M ammonium acetate–ammonium hydroxide reagent at pH 10 for 20 g of silt, sandy, and clay loams. A cleaner extract afforded by MAE coupled with a more selective eluent with SPE cartridge allowed them to combine gas chromatography with electron-capture negative chemical ionization mass spectrometry to improve the sensitivity of the system. These improvements led to a reduction in the amount of soil that needed processing. The authors were able to validate the method on spiked control (aged field treated) samples with a linear response over the 0.5–5 ppb of analyte, and recoveries averaging 92 ± 13%. The ability to process 12 samples simultaneously yielded an overall reduction in time per sample from ~2 h to ~10 min.

Organochlorine Pesticides, PCBs, and Phenols

Despite the ban on DDT and related chlorinated pesticides, aldrin, endrin, dieldrin, lindane, alpha-chlordane, heptachlor, and DDT breakdown products are generally resistant to further weathering and have persisted in the environment. For this reason, their presence in the environment must still be monitored and MAE offers an efficient and reproducible alternative technique to classical methods (57). Following EPA SW-846 Method 8081 protocol for spiking and analysis of chlorinated pesticides, Fish and Revesz (57) achieved 95% recoveries for a matrix spike in 5 g of clean soil extracted for 15 min with 30 mL of acetone:hexane (1:1). They also found increasing recoveries on going from 80 °C to 120 °C. When 3 g of a commercial reference material was extracted in the same solvent system at 120 °C for 15 min, the results were all in the advisory range. However, chlordane and heptachlor, were lower than expected. By changing to a 3:2 acetone:hexane solvent ratio, more like the azeotropic vapor in a Soxhlet, the recoveries improved. Increasing the solvent volume to 50 mL and the extraction time to 20 min were additional modifications made to the procedure to improve recoveries. The microwave method was sensitive enough to easily and completely recover as little as 1 ppb of pollutant and at the same time to insure that with proper cleaning and good laboratory practice, no detectable carryover was observed even by 100 ppm of another pollutant.

Onuska and Terry (58) also successfully extracted aldrin, dieldrin, and DDT from soils and sediments in a 1:1 mixture of isooctane:acetonitrile irradiated for five 30-s intervals. Maximum recovery increased with extraction time: seven 30-s extractions are optimum. Except for p,p'-DDD, as Table XI shows, extraction efficiencies of chlorinated pesticides from moist contaminated soils are >80% with isooctane and are achieved in multiple 2.5-min microwave exposures (58). Recoveries were more efficient than customarily obtained with Soxhlet extractions. The best recoveries of microwave-extracted pesticides endrin and aldrin from wet soil are obtained with isooctane:acetonitrile (1:1) mixtures. Recoveries of the pesticides increase as matrix moisture content increases to a maximum of 15% water and then tapers off; recovery in the dry state is poor (14). Here too, the presence of moisture improves extraction efficiency. When the same soils are extracted by Soxhlet, the recoveries of dry and wet soils are both ~90%.

Organochlorine pesticide compounds listed in EPA Method 8081 were spiked onto 5 g of topsoil, clay soil, sand, and organic compost matrices with standard solutions (59). These included BHC (hexachlorocyclohexane) isomers, DDT and DDE analogs, as well as endrin and dieldrin. Sand was the cleanest matrix; that is, it gave the highest recoveries and the best precisions. Interfering coextractable materials from the topsoil and composted materials are believed to interfere with the analysis (59) and lead to high recoveries of endrin and other compounds. When 45 organochlorine pesticides and arochlors listed in Method 8081 were freshly spiked onto a series of soils and variously aged, 38 of the 45 compound recoveries from the freshly spiked soil were between 80 and 120%. Most of the compounds were recovered at similar levels after aging for 24 h, although captafol,

Table XI. Microwave-Extraction Recoveries of Chlorinated
Pesticides from Sediments

Pesticide	Recovery (%) (n = 5)
Hexachlorobenzene	91.7 ± 4.8
α-BHC	83.8 ± 5.0
Lindane	86.7 ± 4.9
Aldrin	93.8 ± 4.7
trans-Chlordane	89.3 ± 4.8
cis-Chlordane	85.8 ± 4.5
Dieldrin	88.8 ± 4.1
p,p´-DDE	83.8 ± 4.9
Endrin	91.5 ± 4.9
p,p´-DDD	74.0 ± 3.0
o,p-DDT	81.7 ± 4.3
p,p´-DDT	81.6 ± 4.2
Methoxychlor	95.3 ± 3.9
Mirex	81.2 ± 4.0
Decachlorobiphenyl	84.0 ± 4.9

Source: Reprinted with permission from reference 58. Copyright 1993.

captan, and dichlone were poorly recovered. Recoveries of these compounds also decreased with increased aging time, a phenomenon attributed in part to microbial degradation. Nevertheless, they are higher than comparably extracted samples by Soxhlet or sonication techniques by about 7% (50). In comparison with recoveries from pure solvents, analytes recovered from soils suggest that deterioration is not a major problem for either the organochlorine pesticides or phenols.

With consistent recoveries in the 80–120% range for many chlorinated compounds extracted from both fresh and 24-h aged spiked samples by using microwave heating, the authors concluded that certain compounds are not stable under thermal conditions of extraction. Poor recoveries of similar compounds by all classical methods as well support this suggestion (52, 59). For the organochlorine pesticides, the presence of matrix soil still gives acceptable yields: 83–117% with an attrition for χ-chlordane and other BHC isomers (21).

Because organochlorine pesticides persist in the environment and are lipophilic they accumulate along the food chain. Monitoring programs tracking this activity have recently reported the application of the MAP technique to improve sample-preparation activities required in connection with pesticide analysis and assessment of accumulations. Recoveries in excess of 90% were obtained for 6 samples in <40 min from samples of gray seal blubber and spiked pork fat (60).

Successful extraction of PCB aroclor types 1248, 1254, and 1260 followed by enzyme-linked immunosorbent assay (ELISA) was demonstrated by Lopez-Avila et al. (50) to compare favorably with traditional extraction techniques. Moreover, no degradation of PCBs occurs during heating solvent–soil suspensions with microwave energy (*50, 59*). In fact, freshly spiked chlorinated pesticides are almost quantitatively extracted in 5 min at room temperature (*21*). More importantly, 10-min extractions of up to 12 samples simultaneously with MAE combine favorably with the speed and simplicity of ELISA techniques, thus making them suitable for consideration as field screening and monitoring methods (*50, 53*).

Phenolic compounds did not degrade when recovered from solvent only, and in the presence of a reference soil recoveries were reduced although still acceptable (80–111% vs. ~70%). Dinitrophenol recoveries were never more than 17%, however. In a wide-ranging study (*21*), the recovery of microwave-extracted phenols such as pentachlorophenol and butyl benzylphthalate plasticizer is compared with extractions by EPA-approved methods. Microwave methods are superior in nearly every case. Although increased pentachlorophenol recovery was seen, the presence of moisture does not substantially affect recovery of many polar pollutants and shows only marginally improved recoveries of higher molecular weight PAHs. Time to temperature is again more rapid in the presence of higher moisture content. Phenol recovery improves in many cases at the higher temperatures: 80 °C, 115 °C, and 145 °C. When comparing the recovery of spiked solvent with that of solvent and soil combinations, the presence of soil negatively biases the recovery of most phenols, much like PAHs reported earlier. Degradation is thought to be the cause of poor performance for substituted dinitrophenols compared with volatility and measurement error that may account for lower molecular weight PAH losses.

Standard solutions of phenolic compounds spiked onto 5 g of topsoil, clay soil, sand, and organic compost matrices behave much like the organochlorine pesticides (*21, 59*). In addition, 10 g of a certified reference soil containing 4-chloro-3-methylphenol, 2,4-dichlorophenol, 2-methylphenol, 3-methylphenol, pentachlorophenol, and 2,4,6-trichlorophenol were extracted in 30 mL of acetone:hexane (1:1) for 10 min at 115 °C. More than 70% of the determinations gave >70% recovery. 2,4-Dinitrophenol and 4,6-dinitro-2-methylphenol were degraded in all soils except sand, which again demonstrates the profound influence of matrix on extraction. In all cases, the MAE results were slightly higher than classical extraction technique results and had significantly better precision: 3% on average compared with 15% for Soxhlet and 20% for sonication.

Total Petroleum Hydrocarbons

Total petroleum hydrocarbon (TPH) extraction from environmental samples such as soils is a relatively easy task by closed-vessel microwave techniques. As the data in Table XII show, complete microwave extraction of TPHs from 5 g of soil with (1:1) acetone:hexane can be accomplished in 15

Table XII. Comparison of Microwave-Extraction Recoveries with Classical Methods for Total Petroleum Hydrocarbons from Soil

Parameter	Microwave Closed-Vessel Extraction[a]	7-h Soxhlet Extraction[b]	24-h Soxhlet Extraction[b]	Sonication Extraction[c]
TPH (%)	1.32 ± 0.01	1.33 ± 0.02	1.34 ± 0.01	0.96, 0.92
n	11	6	4	2

NOTES: Values are from gravimetric determination of 5-g samples. Sample was service station contaminated soil.
[a]Extraction conditions were as follows: 30 mL acetone-hexane (1:1, v/v), 150 °C, 15 min.
[b]Extraction conditions were as follows: 100 mL acetone-hexane (1:1, v/v).
[c]Extraction conditions were as follows: 400 mL Freon-113, 20-g sample, modified SW846/EPA Method 9071.

min at 150 °C. The recovery data are comparable to 7- and 24-h Soxhlet-extracted TPH values (*61*). Diesel-range organics also can be quantitatively extracted from moist soil in 1:1 acetone:hexane in 15 min at 150 °C. Microwave extracts in acetone:hexane (1:1) compare favorably with sonication extracts for the same solvent pair. Analysis by gas chromatography with flame ionization detection (GC/FID) at 220 mg/kg shows that microwave-extraction recoveries are 100–113% compared with 100% by sonication. For other dry reference soils containing 990 mg/kg No. 2 diesel fuel extracted by both microwave and sonication into acetone:methylene chloride (1:1), replicate microwave recoveries averaged 95% compared with 78% by sonication. Sonications require 30–500 mL of solvent per sample; microwave extractions are accomplished in 30 mL of solvent. Despite ample solvent volume available for dissolution, sonication extraction efficiency is inferior to the microwave extraction in the more polar solvent combination. This result demonstrates the advantage of polar solvent interactions in microwave fields.

Gravimetric analyses of TPHs extracted from 5 g of reference soil material containing 2140 mg/kg gave recovery of 2230 mg/kg in 30 min compared with 2171 mg/kg by Soxhlet for 7 h. Marginally improved recovery is observed when the 5-g samples of hydrocarbon-contaminated soil are extracted with 30 mL of solvent (acetone:hexane 1:1 preferred) for 15 min at increasing temperatures between 125 °C and 160 °C. Recovery is slightly reduced at the highest temperature, and 150 °C appears to be optimum. When extraction times at 150 °C are varied from 5, 10, 15, and 30 min, recovery is virtually the same. In this case, extraction time is not a critical parameter. When the extraction volume is varied from 25 to 50 mL, the smallest volume gives the poorest recovery. However, no significant difference exists between 30 and 50 mL. Choice of solvent is important to reproducible recovery. The acetone:hexane combination is the most efficient, probably as the result of greater compatibility of the nonpolar hexane solvent with the nonpolar TPH analytes. DCM is the least efficient of the individual solvents. Because PAHs are quantitatively extracted from soils by using acetone:hexane, simultaneous extractions of TPHs and PAHs probably could be performed on a single soil sample.

Organometallics

Synthetic compounds combining the special properties of metals with a hydrocarbon moiety as a vehicle for their introduction into chemical formulations such arsones and selenones, arsines, organotins, and organmercurics have become widespread in the environment since the early 1960s. For example, roxarsone, a di-substituted phenylarsonic acid, has been used as a growth promoter in animal food production. Triphenyl tin (TPT) and tributyl tin (TBT) are added to marine antifouling paints as growth inhibitors. Analysis of matrices for such compounds is an important environmental monitoring function, and recently Croteau et al. (62) developed a rapid MAP procedure to extract 3-nitro-4-hydroxyphenylarsonic acid, a feed additive, from animal tissue. Recoveries by the MAP technique were consistently higher than sonication and homogenization for liver, kidney, or muscle tissue. These results were accomplished in a 9-s exposure by using ethanol and acetic acid as the solvent system.

In contrast, aquatic environments are the normal repository for organotin compounds, and they accumulate both in sediments and local marine life. Donard and co-workers (63) found that the highly substituted TPTs and TBTs exposed to microwave heating degrade in purely aqueous (fresh and saline) environments mainly to inorganic tin, although they persist in isooctane and methanol and could be extracted rapidly and efficiently from wet sediments in methanol acidified with acetic acid. This polar solvent enhanced the extraction of organotins from sediment matrix although persistence of the TBT and DBT was inversely dependent on acid concentration with monobutyl tin (MBT), the longest lived at high concentration. All species recoveries were highly matrix dependent, as seen with other pollutants.

Unique Environmental Applications

Although microwave-heated solvent extractions have been applied to traditional Soxhlet-type extraction for nearly 10 years, microwave energy has been applied to a broad range of specialized activities since the 1970s. For example, Bosisio et al. (64) used microwave power to thermally extract 86% of total bitumen, a crude oil, from Athabasca tar sand. Solvents were not used. Instead, direct heating of the petroleum compounds was effected and the crude distillate was collected in a cold trap.

A sequential extraction scheme to assess the binding characteristics of metals such as lead in soil fractions was described by Mahan et al. (65). Metals partitioned into fractions by the type of solution and the pH used in leaching. As many as six fractions were typified by their vastly different metal-binding character. When microwave heating was used, nearly all of the fractions were leached more quickly and more efficiently and the results were comparable to conventional methods. Tessier's extractions of calcium, copper, iron, and manganese from a lagoon sediment gave similar results for concentrations of metals in respective fractions. Extractions were faster with microwave heating than conventional heating, especially when ammonium acetate was substituted for magnesium chloride (66).

In 1988, Dauerman and his group (25) began treating hazardous wastes with microwave energy. Organic pollutants could be removed from soils irradiated with microwave energy because microwave treatment induced a steam distillation effect due to the ubiquitous presence of moisture in soils. The steam passed through an organic solvent trap or scrubber where the pollutant was removed.

Conclusions and Future Trends in Microwave-Extraction Chemistry

Published data on microwave-assisted extractions of pollutants from environmental matrices, of additives from plastics, and of essential ingredients from natural products are comparable or often superior to previous data obtained from classical sonication and Soxhlet extraction techniques. MAE methods are highly versatile, broadly applicable, always faster, and use less solvent than traditional methods, usually with better precision. Because microwave heating depends on the presence of polar molecules or ionic species, successful applications of MAE depend on the dielectric susceptibility of both solvent and matrix. And, as in many chemical reactions, temperature is the most important extraction parameter governing successful recoveries. Moist samples often give better recoveries or yields because the water is a good absorber of microwave energy. Such materials heat more quickly and possibly to higher temperatures than dry materials in single-solvent systems.

Because solvent polarity and dielectric compatibility are mutually inclusive, tailoring microwave extractions to other existing liquid–solid interactions is suggested as an approach to new methodology. For example, aprotic, universal solvents such as dimethyl sulfoxide and acetonitrile have large dipole moments (3.1 and 3.44, respectively). Such solvents are good microwave absorbers. Protic solvents, on the other hand, are generally acidic (e.g., acetic acid) or amphiprotic (like water and alcohol) and have protons and electron pairs available for bonding. Large dipole moments are the rule here, too. If the target analyte is soluble in these reagents, then such solvents should be evaluated as extractants because they are likely to couple well with microwave energy and afford good extractions.

If the dielectric parameters are considered structural attributes of the material and thought of in terms of molecular-relaxation mechanisms, then a search for liquids having dielectric relaxation constants at or near the wavelength of interest may more-accurately predict microwave compatibility and heating. For example, 1-pentanol has a dielectric relaxation (τ) of 6.4×10^{-11} (20). Maximum absorption and maximum heating occur when the frequency (f) (18)

$$f = \frac{1}{2\pi\tau} \qquad (3)$$

The frequency of maximum heating for 1-pentanol is 2.49×10^9, very close to 2.45×10^9 MHz. At 2450 MHz, tangent delta (tan δ) for this solvent is 0.504, compared with tan δ = 0.102 for water (20). Such liquids may enhance extractions through more efficient heating. Thus, dielectric parameters should be inspected for clues to promising candidates for extractions applications. Quite possibly, analytes such as phenols and amines distributed in nonpolar matrices such as hydrocarbons may be extracted into marginally polar solvents that are immiscible with the hydrocarbon if the relaxation constant is matched to the heating frequency. Aqueous extractions can benefit from microwave heating in the same way that organic-solvent-extracted solids has benefited. Trace organics such as PAHs distributed in water should dissolve in an appropriate solvent such as methylene chloride because microwave-heated water warms the solute at the same time that the solvent is warmed and increases its dissolving power. Onuska and Terry (67) successfully extracted PCB congeners into isooctane from water in their development of a model system for microwave liquid–liquid extractions. The excellent compatibility of water with microwave energy and the ability to attain 200 °C with pressures of 200–1200 psi in today's microwave equipment makes it an excellent candidate for solvent extractions of other solid matrices. At low temperatures and pressures the polarity of water is high and diminish with increasing temperature and pressure—a property that makes it a highly versatile extractant. Water is the ultimate environmentally friendly solvent: it is safe, nontoxic, and easily disposed.

Acknowledgments

We appreciate the assistance of Melinda Hays in the HPLC determinations of PAHs and of Stephen Wise for providing the soils for the PAH study and for helpful discussions. S. Matz appreciates the team effort and support provided by Cynthia Collis and John Kahn who carried out much of the laboratory work for the study.

References

1. Manahan, S. E. *Environmental Chemistry,* 5th ed.; Lewis Publishers: Boca Raton, FL, 1991; Chapters 15–19.
2. Hawthorne, S. B.; Galy, A. B.; Schnitt, V. O.; Miller, D. J. *Anal. Chem.,* **1995,** *67,* 2723–2732.
3. Lopez-Avila, V.; Bauer, K.; Milanes, J.; Beckert, W. J. *AOAC Int.* **1993,** *76,* 864–880.
4. Margulis, M. A. *Principle of Sonochemistry: Chemical Reactions in Acoustic Fields;* M. Vyssh. Shk.: Moscow, Russia, 1984.
5. *Test Methods for Evaluating Solid Waste, SW–846 Method 3550A (Sonication),* 3rd ed.; U.S. Environmental Protection Agency: Washington, DC, 1994.
6. Williams, D. F. *Chem. Eng. Sci.* **1981,** *36,* 1769–1788.
7. McNally, M. E. P. *Anal. Chem.* **1995,** *67,* 308A–315A.
8. Majors, R. E. *LC–GC* **1996,** *14(2),* 88–96.
9. Richter, B. E.; Jones, B. A.; Ezzell, J. L.; Porter, N. L.; Avdalovic, N.; Pohl, C. *Anal. Chem.* **1996,** *68,* 1033–1039.
10. *Hazardous Waste Management System: Testing and Monitoring Activities;* EPA 40 CFR Part 260, 264, 265; U.S. Environmental Protection Agency, Washington, DC, July 25, 1995; Vol 60, p 142.
11. Neas, E. D.; Collins, M. J. In *Introduction to Microwave Sample Preparation: Theory and Practice;* Kingston, H. M.; Jassie, L. B., Eds.; American Chemical Society: Washington, DC, 1988; Chapter 2.
12. Pare, J. R. J. U. S. Patent 5, 002, 784, March 26, 1991.
13. Ganzler, K. ; Salgo, A.; Valko, K. *J. Chromatogr.* **1986,** *371,* 299–306.
14. Ganzler, K.; Szinai, I. ; Salgo, A. *J. Chromatogr.* **1990,** *520,* 257–262.
15. Pare, J. R. J. U.S. Patent 5, 458, 897, October 17, 1995.
16. Pare, J. R. J; Belanger, J. M. R.; Stafford, S. S. *Trends Anal. Chem.* **1994,** *13(4),* 176–184.
17. Pare, J. R. J. U.S. Patent 5, 377, 426, January 3, 1995.
18. Thuery, J. *Microwaves: Industrial, Scientific and Medical Applications;* Artech House: Boston, MA, 1992; Chapter I.3.
19. Jassie, L.; Hays, M.; Wise, S. Presented at Pittsburgh Conference and Exposition, Chicago, IL, 1994; paper 1146.
20. Neas, E. *Proceedings of the First World Congress on Microwave Chemistry;* Industrial, Medical, and Instrumentation Section, International Microwave Power Institute: Clifton, VA, 1992.
21. Lopez–Avila, V.; Young, R.; Beckert, W. F. *Anal. Chem.* **1994,** *66,* 1097–1106.
22. Barnabas, I.; Dean, R.; Fowlis, I.; Owen, S. *Analyst* **1995,** *120,* 1897–1904.
23. Ganzler, K.; Bati, J.; Valko, K. *International Eastern European–American Symposium of Chromatography Published Proceedings;* Kalasz, H.; Ettre, L. S., Eds.; Akedemiai Kiado: Budapest, Hungary, 1986; pp 435–442.
24. Ganzler, K.; Salgo, A. *Lebensm. Unters. Forsch.* **1987,** 274–276.
25. Dauerman, L.; Windgasse, G. *J. Microwave Power & E² **1992,** *27(1),* 23–32.
26. Bichi, C.; Beliarab, F. F.; Rubiolo, P. *Lab. 2000* **1992,** *6,* 36–38.
27. Chen, S.; Spiro, M. *J. Microwave Power Electromagn. Energy* **1994,** *29,* 4.
28. Young, J. *J. Agri. Food Chem.* **1995,** *43(11),* 2904–2910.
29. Franke, M.; Winck, C. L.; Kingston, H. M. *Forensic Sci. Int.* **1996,** *81,* 51–59.

30. Greek, B. *Chem. Eng. News* **1988**, *June 13*, 35.

31. Stevens, M. J. *Chem. Ed. 70*, 535.

32. "Test Method for Determination of Phenolic Antioxidants and Erucamide Slip Additives in LDPE Using Liquid Chromatography," *Annual Book Standard;* Dolmage, D. L., Ed.; ASTM STP D1996, Vol 08. 01 Plastics; American Society for Testing and Materials: Philadelphia, PA, 1993.

33. Haney, M. A.; Dark, W. A. *J. Chromatogr. Sci.* **1980**, *18*, 655–659.

34. Nielson, R. C. *J. Liq. Chromatogr.* **1991**, *14*, 503–519.

35. Brandt, H. J. *Anal Chem.* **1961**, *33*, 1390.

36. DeMenna, G. J.; Edison, W. J. *Novel Sample Preparation Techniques for Chemical Analysis–Microwave and Pressure, Dissolution, Chemical Analysis of Metals;* Coyle, F. T., Ed.; ASTM STP 994; American Society for Testing and Materials: Philadelphia, PA, 1987; p 45.

37. Schabron, J. F.; Fenska, L. E. *Anal. Chem.* **1980**, *52*, 1411–1415.

38. Schabron, J. F.; Smith, U. J. *J. Liq. Chromatogr.* **1982**, *5*, 613–624.

39. Schabron, J. F. *J. Liq. Chromatogr.* **1982**, *5*, 1269–1276.

40. Majors, R. E.; Johnson, E. L. *J. Chromatogr.* **1978**, *167*, 17–30.

41. Freitag, W.; John, O. *Angew. Makromol. Chem.* **1990**, *175*, 181–185.

42. Shinoda, K. *Principles of Solution and Solubility;* Marcel Dekker: NY, 1978; p 147.

43. Dack, M. R. J. *Solutions and Solubilities;* Interscience: New York; Vol. 2, p 438.

44. Langenfeld, J. J.; Hawthorne, S. B.; Miller, D. J.; Pawliszyn, J. *Anal. Chem.* **1995**, *67*, 1727.

45. Jassie, L.; Hays, M.; Wise, S. Presented at the Pittsburgh Conference and Exposition, Atlanta, GA, 1993; paper 171.

46. *Test Methods for Evaluating Solid Waste*, 3rd ed.; SW–846 Method 8100, proposed Update II; U.S. Environmental Protection Agency: Washington, DC, 1986.

47. Emery, A. P.; Chesler, S. N.; MacCrehan, W. A. *J. Chromatogr.* **1992**, *606*, 221–228.

48. Milacic, R.; Stupar, J.; Kozuh, N.; Korosin, J. *Analyst,* **1992**, *117*, 125–130.

49. Lopez–Avila, V.; Benedicto, J. *Trends Anal. Chem.* **1996**, *15(8)*, 334–341.

50. Lopez–Avila, V.; Young, R.; Benedicto, J.; Charan, C. *Environ. Sci. Technol.* **1995**, *29(10)*, 2709–2712.

51. *Test Methods for Evaluating Solid Waste*, 3rd ed.; proposed Update II, Methods 8250, 8081 and 8141A; U.S. Environmental Protection Agency: Washington, DC, 1986.

52. Lopez–Avila, V.; Young, R.; Benedicto, J.; Ho, P.; Kim, R. *Anal. Chem.* **1995**, *67(13)*, 2096–2102.

53. Lopez–Avila, V.; Young, R.; Teplitsky, N. *J. AOAC Int.* **1996**, *79(1)*, 142–156.

54. Barrer, R. M. *Zeolites and Clay Minerals as Sorbants and Molecular Sieves;* Academic: London, 1978; pp 407–466.

55. Steinheimer, T. R. *J. Agric. Food Chem.* **1993**, *41*, 588–595.

56. Stout, S. J.; daCunha, A. D.; Allardice, D. G. *Anal. Chem.* **1996**, *68(4)*, 653–658.

57. Fish, J.; Revesz, R. *LC–GC,* **1996**, *14(3)*, 231–234.

58. Onuska, F. I.; Terry, K. A. *Chromatographia* **1993**, *36*, 191–194.

59. Lopez–Avila, V.; Young, R.; Kim, R.; Beckert, W. *J. Chromatogr. Sci.* **1995**, *33*, 481–484.

60. Hummert, K.; Vetter, W.; Luckas, B. *Chromatographia* **1996**, *42(5/6)*, 300–304.

61. Hasty, E.; Revesz, B. *Am. Lab.* **1995,** *February,* 66–74.

62. Croteau, L.; Akhtar, M.; Belanger, J.; Pare, J. *J. Liq. Chromatogr.* **1994,** *17(13),* 2971–2981.

63. Donard, O.; Lalere, B.; Martin, F.; Lobinski, R. *Anal. Chem.* **1995,** *67,* 4250–4254.

64. Bosisio, R. G.; Cambon; Chavarie, C.; Klvana, D. *J. Microwave Power* **1977,** 301–307.

65. Mahan, K. I.; Foderaro, T. A.; Garza, T. L.; Martinez, R. M.; Maroney, G. A.; Trivisonno, M. R.;   Willging, E. M. *Anal. Chem.* **1987,** *59,* 938–945.

66. Gulmini, M.; Ostacoli, G.; Zelano, V.; Torazzo, A. *Analyst* **1994,** *119,* 2075–2080.

67. Onuska, F. I.; Terry, K. A. *J. High Resol. Chromatogr.* **1995,** *18,* 417–422.

Biochemistry Applications

Chapter 13

Microwave Hydrolysis of Proteins and Peptides for Amino Acid Analysis

W. Gary Engelhart

Leucine was found by Proust and then Braconnot and Braconnot found Glycine in 1819–1820. It was many years after 1820 before it became clear that hydrolysis in boiling acid was the best way to break a protein down into its constituents.

John T. Edsall

This chapter focuses on the use of microwave heating for rapid acid hydrolysis of proteins and peptides prior to amino acid analysis. Initial research in the field and subsequent development of the microwave instrumentation, labware, and techniques necessary for high-sensitivity amino acid analysis are discussed.

In 1953 after nearly 10 years of work Frederick Sanger and colleagues at Cambridge University determined the complete amino acid sequence and structure of a protein for the first time. This accomplishment is a milestone in biochemistry and won Sanger a Nobel Prize in 1958. Today the amino acid sequences of more than 2000 proteins have been determined (1).

The first step in protein structure determination is the isolation and purification of the protein to be studied. The second step is identification of the protein's constituent amino acids. This second step involves performing an acid-hydrolysis procedure to break peptide bonds and quantitatively release the constituent amino acids prior to analysis by liquid chromatography.

Sanger studied the structure of insulin, a polypeptide hormone that is easily extracted and recovered in pure form from the pancreases of animals. Analysis of the 51 amino acid residues of bovine insulin was not a simple matter in Sanger's time. Because of the limitations of chromatographic equipment and techniques available in the late 1940s complete analysis of a protein hydrolysate required approximately two weeks. In the 1950s ion-exchange chromatography reduced amino acid analysis time to one week (2).

Preparation of protein hydrolysates has become the rate-limiting step in amino acid analysis. The widely used conventional protocol developed by Hirs, Stein, and Moore in the 1950s involves heating samples in liquid 6 N HCl at 110 °C for periods of 24 h or more (3). Therefore, the amount of time required for sample preparation was consistent with the analysis that followed.

Advances in high-performance liquid chromatography (HPLC) instrumentation now allow accurate amino acid analysis on minute amounts (<100 picomoles) of sample in less than 1 h. Unfortunately, the improved sensitivity and separation times of state-of-the-art amino acid analyzers are offset by the potential for contamination during sample transfer steps, run-to-run variability, and time associated with many conventional hydrolysis procedures.

Protein Structure and Acid-Hydrolysis Techniques

A major portion of the complexity and difficulty involved in amino acid analysis is attributable to the preparation of protein hydrolysates. The sequence of amino acids that composes a protein, as well as the degree of protein folding and its overall macromolecular structure, all affect the rate of peptide-bond hydrolysis and resulting yield of amino acids for a given hydrolyzing agent. Proteins exhibit several levels of structural organization (primary, secondary, tertiary, and in some cases quaternary structure) (Figure 1). Primary structure refers to the sequence of amino acids present. A chain of amino acids is referred to as a polypeptide that forms helices, sheets, or other shapes representing the secondary structure of the protein. Interactions among various sections of the polypeptide lead to further folding and coiling into a complex three-dimensional structure representing the protein tertiary structure. A final quaternary level of protein organization may arise when several separate polypeptide structures bind together into a single large protein molecule (4).

Protein composition and structure can impede the interaction of hydrolyzing agents with peptide bonds, slowing or preventing cleavage in two ways. First, the bulky side chains of aliphatic amino acids—valine (Val), leucine (Leu), isoleucine (Ile), and alanine (Ala)—may cause steric hindrances that slow hydrolysis of certain peptide linkages. The structural formula of an α-amino acid is illustrated in Figure 2a. The functional side chains and three letter abbreviations for the 20 naturally occurring amino acids are shown in Figure 2b. Amino acid sequences containing Ile–Ile, Ile–Leu, and Val–Val bonds are particularly resistant to hydrolysis. Second, the three-dimensional macromolecular structure of the protein may limit unfolding of the protein and exposure of peptide bonds to the hydrolyzing agent. For these reasons a hydrolyzing agent may work slowly on one portion of a protein and rapidly on another (5). Amino acids that are more readily cleaved and liberated are subject to oxidative degradation while

Primary Protein Structure

gly pro thr gly thr gly glu ser lys cys pro leu met val lys val leu asp

ala val arg gly ser pro ala ile asn val ala val his val phe arg lys ala

ala asp asp thr trp glu pro phe ala ser gly lys thr ser glu ser gly glu

leu his gly leu thr thr glu glu gln phe val glu gly ile tyr lys val glu ile

Secondary Protein Structure

Tertiary Protein Structure

Quaternary Protein Structure

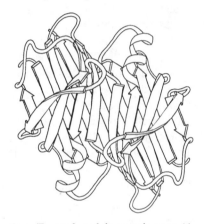

Figure 1. Levels of protein structural organization. (Reproduced from reference 4.)

hydrolysis continues in other sections of the protein. The amino acids serine (Ser) and threonine (Thr) are acid labile and are progressively destroyed over time.

Method development is required to arrive at a set of hydrolysis conditions (time and temperature) that yields acceptable recoveries of as many different amino acids as possible from the sample protein. Method development typically entails conducting a time-course study for periods of 12, 24, 48, 72, and 96 h at a constant temperature of 110 °C. The hydrolysate from the 24-h hydrolysis will exhibit the best recoveries of sensitive amino acids and the lowest recoveries of aliphatic amino acids. Conversely, the protein hydrolysate from the 96-h run will exhibit complete (or the greatest degree of) cleavage of the bonds between aliphatic amino acids and the lowest recovery of sensitive amino acids. The kinetics of protein hydrolysis and amino acid degradation reactions force analysts into adopting a

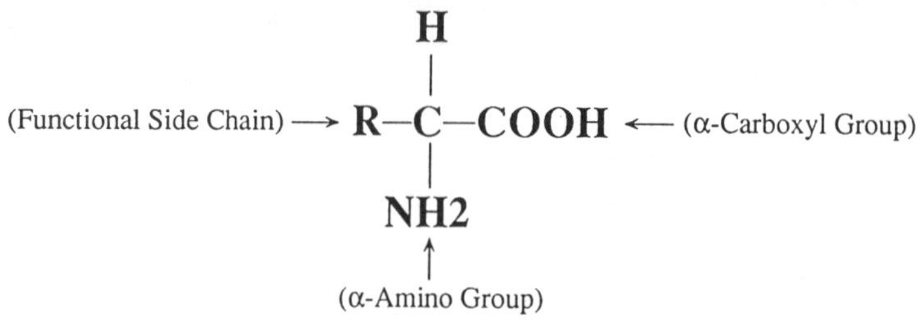

$$H$$
$$|$$
(Functional Side Chain) $\longrightarrow$ R—C—COOH $\longleftarrow$ (α-Carboxyl Group)
$$|$$
$$NH2$$
$$\uparrow$$
(α-Amino Group)

Figure 2a. Structural formula of an α-amino acid.

method that reflects a compromise in time and temperature conditions rather than an optimized method. In some cases the analyst may employ two hydrolysis runs of different duration to achieve acceptable recoveries of labile and hydrophobic amino acids in a sample protein (5).

Many combinations of hydrolyzing agents, hydrolysis conditions, and protective agents have been devised to overcome problems such as incomplete hydrolysis, destruction of certain specific amino acids, and oxidative degradation of liberated amino acids (5–7). Hydrochloric acid is the most widely used agent for hydrolyzing proteins. One of two basic methods is employed, liquid-phase or vapor-phase hydrolysis. In liquid-phase hydrolysis a protein sample is dissolved directly in a 6 N HCl solution and heated. The liquid-phase method works well when a large amount of protein is available or when complex matrices such as grain, tissue samples, or food must be analyzed. In the vapor-phase method a dry protein sample in an open vial is placed inside a vacuum dessicator along with a separate reservoir of 6 N HCl. The dessicator is evacuated of air and then heated in a laboratory oven, and the protein sample is contacted and hydrolyzed by acid vapor. The vapor-phase method results in significantly lower levels of background contamination and is the preferred approach for high-sensitivity analysis, because amino acid contaminants present in the acid are not volatilized and do not come in contact with the sample.

Acid hydrolysis by either method destroys certain amino acids, (cysteine, cystine, tryptophan, asparagine, and glutamine), and partially destroys other amino acids (serine, threonine, methionine and tyrosine). Asparagine and glutamine are converted into aspartic acid and glutamic acid, respectively, during acid hydrolysis and can be derivitized and quantitated in these forms. Elimination of oxygen from the hydrolysis container is an important consideration to prevent oxidation of liberated amino acids, particularly methionine, tyrosine, cysteine, serine, and threonine. Protective agents may also be added to 6 N HCl to prevent destruction or conversion of some amino acids. Phenol is added to prevent halogenation of tyrosine, and 2-mercaptoethanol prevents the oxidation of methionine. A prehydrolysis reaction step is required to convert cysteine and cystine to acid-stable forms for accurate quantitation. Tryptophan is destroyed dur-

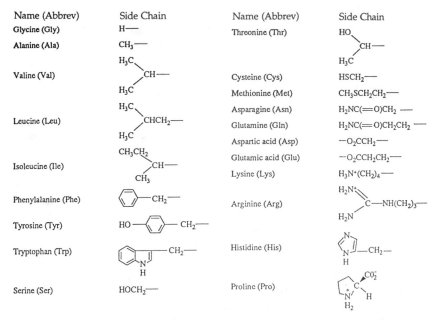

Figure 2b. Functional side chains and abbreviations of the 20 naturally occurring amino acids.

ing hydrolysis in 6 N HCl, and alternative hydrolyzing agents such as bases or sulfonic acids are required to recover this amino acid.

Rationale for Microwave-Assisted Hydrolysis of Proteins

Several factors provided the impetus for research into microwave-assisted acid hydrolysis of proteins

- the importance of amino acid analysis in biomedical research
- the potential for time savings and improved laboratory productivity
- the possibility of simplifying and reducing the sources of error and variability in existing methods
- scientific acceptance of microwave acid-digestion techniques for the preparation of atomic absorption and emission spectroscopy samples

Amino acid analysis is a cornerstone of protein chemistry as evidenced by award of the 1972 Nobel Prize in Chemistry to William Stein and Stanford Moore for pioneering the development and application of the technique. The importance of high-sensitivity amino acid analysis has steadily increased in biotechnology and pharmaceutical research to keep abreast of advances in peptide synthesis, protein purification, and recombinant DNA technology. Amino acid analysis serves multiple purposes in the protein chemistry field. The technique is routinely used to determine the amount and composition of purified proteins and synthetic peptides and to confirm the composition of peptides produced by enzymatic and chemical cleavage methods for microsequencing via Edman degradation or mass spectrometry (8, 9).

Modern HPLC instrumentation permits amino acid analysis of a protein hydrolysate in less than an hour. Preparation of protein hydrolysates has become the rate-limiting step and bottleneck in amino acid analysis. Characterization of synthetic peptides by amino acid analysis is an essential quality-control check. Generally, amino acid analysis is too slow to permit monitoring of solid-phase peptide synthesis (SPPS) while it is progressing (10). For this reason amino acid analysis is typically performed on the complete peptide chain following cleavage from the synthesis resin. In SPPS the growing chain of amino acids is subjected to multiple treatments and washes with chemical reagents. This process sometimes leads to incomplete amino acid coupling reactions and accumulation of undesireable side-reaction products. When this result occurs the purity and bioactivity of the peptide is compromised for research or therapeutic purposes. Lengthy conventional hydrolysis procedures preclude in-process monitoring of SPPS to detect incomplete or erroneous coupling reactions. The possibility of using microwave hydrolysis to monitor the progress of SPPS has been theorised because the technique would permit hydrolysis and analysis to be accomplished in less time than a single amino acid residue coupling reaction (10, 11).

In addition to being the rate-limiting step in amino acid analysis, hydrolysis is also the source of the greatest errors and contamination problems. The high degree of automation of HPLC instrumentation contrasts sharply with the manually operated hydrolysis apparatuses that analysts must rely on. Operator skill and experience in maintaining stringent cleanliness of samples, vials, reagents, and the lab environment ultimately determines the sensitivity and working range of analysis. The analyst's difficulties of handling and transferring samples without losses or contamination are compounded by the limitations of hydrolysis apparatuses. Although an apparatus such as a vacuum dessicator may permit initial evacuation of air from the hydrolysis chamber, measurement of vacuum conditions throughout the run is typically not possible. A leak occurring during a lengthy hydrolysis procedure and the resulting oxidative degradation of amino acids is only detected during chromatographic analysis.

The widely used Pico-Tag workstation from Waters and Reacti-Therm heating block system from Pierce have this limitation and require hydrolysis and analysis of a well-characterized protein or peptide standard to detect if a hydrolysis run has been compromised by a leak (12). The heating method of a hydrolysis apparatus and the means of temperature measurement and control are other potential sources of variability in hydrolysis. Sample vials placed in different locations in an electrical-resistance heating block may not experience equivalent temperatures. None of the commercially available hydrolysis devices directly measures acid temperature. Temperature of the metal heating block, oil bath, or convection oven is measured and used to control hydrolysis conditions. Indirect measurement of temperature is less accurate, and the resulting heating characteristics and gradients can cause incomplete hydrolysis and variable amino acid recoveries.

The use of microwave heating to accelerate acid decomposition of samples for atomic spectroscopy was first described in 1975 (13). Microwave digestion was an accepted sample-preparation technique when researchers began conducting studies on microwave hydrolysis of proteins in the late 1980s. Indeed, this body of knowledge on microwave heating of acid mixtures was directly applicable to microwave hydrolysis and facilitated research efforts and subsequent instrument development.

Scientific Basis of Microwave Hydrolysis Techniques

The first published study on the use of microwave heating for acid hydrolysis of proteins and peptides appeared in 1987 (14). In this work, researchers at Taiwan National University, Institute of Biochemical Sciences, performed liquid-phase hydrolysis on 0.2–0.5-mg samples of ribonuclease A and insulin B chain in 6 N HCl and a 1:1 proprionic acid: 12 N HCl mixture. The acid and samples were placed inside custom-made Teflon vials, with silicon septa and Teflon caps. Each vial was flushed with nitrogen gas by using two small needles for gas inlet and outlet prior to capping and placement into a larger Teflon container. Samples were heated for periods of 1–7 min by using a Whirlpool domestic microwave oven, Model MW3500 XM, at 80% power. After microwave heating the hydrolysates were pipetted into acid-cleaned autosampler vials and evaporated to dryness by using a vacuum centrifuge prior to analysis.

The amino acid recoveries obtained following microwave hydrolysis of insulin B chain for periods of 3, 4, and 5 min were in good agreement with both the theoretical 30-residue composition and values obtained by using the conventional hydrolysis protocol (6 N HCl at 110 °C for 24 h). Amino acid recoveries for the larger 124-residue ribonuclease A protein standard also corresponded well with the known composition and values from conventional hydrolysis. The researchers lacked a means of directly measuring acid temperature inside the hydrolysis vials during microwave heating and estimated the temperature of the 6 N HCl to be 174–176 °C based on microwave-heating experiments conducted with compounds of known melting points. Similarly, a means of measuring acid vapor pressure inside the hydrolysis vials was also unavailable. In fact, the researchers indicated that their initial attempts to use sealed Pyrex tubes for microwave hydrolysis were unsuccessful because the tubes would burst because of internal vapor pressure.

Once the feasibility of microwave-assisted acid hydrolysis of proteins had been demonstrated a multitude of fundamental questions remained to be answered before the technique could gain acceptance in the protein chemistry field. Does microwave energy destroy amino acids liberated during hydrolysis? Does microwave heating lead to a higher degree of amino acid racemization than conventional heating methods? Can the same hydrolyzing and protective agents be used in microwave hydrolysis? Was vapor-phase microwave hydrolysis of proteins possible for minute amounts of sample?

Microwave Irradiation and Amino Acid Stability

Microwaves are a nonionizing form of electromagnetic radiation of insufficient quantum energy to disrupt chemical bonds and therefore should not destroy amino acids that have been liberated during acid hydrolysis. To confirm this fact, the research group in Taiwan conducted microwave heating experiments in which a mixture of 17 individual amino acid standards was heated in 6 M HCl for periods of 2, 4, and 8 min and then analyzed to determine the extent of degradation (*15*). Amino acids were shown to be as stable during microwave heating conditions as in acids heated by conventional means.

Microwave Hydrolysis and Racemization of Amino Acids

The characteristic structure of amino acids is a tetrahedral arrangement of functional groups around an α-carbon atom. Amino acids therefore can be found in two mirror-image stereoisomer forms referred to as D and L isomers. Only L-amino acids are present in natural proteins. The bioactivity of a synthetic peptide may be altered or lost because of epimerization of amino acids during synthesis. The purity and homogeneity of peptides is therefore an important consideration in research applications and for therapeutic usages. Analytical techniques such as HPLC and capillary electrophoresis can be used to separate and identify peptide diastereoisomers. Racemization of amino acids can occur during peptide synthesis or during acid hydrolysis of the peptide. Determining the extent of racemization and when it occurred in the synthesis and analysis processes is essential quality-control information. If microwave hydrolysis induced greater racemization of amino acids than conventional hydrolysis methods it would be unsuitable for certain applications. Researchers at Hoffman La-Roche in Basel, Switzerland, first examined this aspect of microwave hydrolysis (*16*). In this work, conventional hydrolysis of peptides containing racemization-prone L-phenylglycyl-glycine in 5.7 M HCl at 110 °C resulted in a 19.5% conversion of this amino acid to D-phenylglycine. By contrast, microwave hydrolysis for 3 min resulted in formation of only 0.4% of this D-isomer.

A deuterium labelling technique exists for distinguishing D-amino acid isomers formed during acid hydrolysis from those formed during peptide synthesis (*17*). By using deuterium chloride as a hydrolyzing agent any racemization of amino acids leads to incorporation of deuterium at the α-carbon atom of the affected amino acids. This change, indicative of hydrolysis-induced racemization, can be detected by ion monitoring using gas chromatography–mass spectrometry (GC–MS). In 1993 a research group from Vrije Universiteit Brussel used this technique to study racemization occurring during microwave hydrolysis of peptides (*18*). Three peptides prepared by SPPS and checked for purity by HPLC and NMR analysis were used in this study. Microwave hydrolysis was performed by the liquid-phase method. Sample peptides (0.5–2.0 mg), 300 μL of 6 N deuterium chloride, and phenol were added to argon-purged Teflon vials. A Philips-Whirlpool Model 764, 850-W domestic microwave oven was used to heat the vials. Conventional liquid-phase hydrolysis was performed with the same amounts of peptide and hydrolyzing–protective agents in 1-mL Pierce Reacti-Vials and heated for 24 h in a Reacti-Therm heating module maintained at 110 °C.

Peptide-1 contained the sensitive amino acids Thr, Cys, Tyr, and Trp. A microwave-hydrolysis time-course study was conducted for periods of 5, 10, 15, 30, and 60 min at 450 W power to ascertain the reaction time that would give an amino acid yield profile comparable with conventional hydrolysis. Microwave hydrolysis for 30 min was determined to give an amino acid yield comparable with 24-h hydrolysis. Racemization increased with increasing microwave-hydrolysis times but was lower than that observed by using the conventional hydrolysis method. Peptide-2, [Tic3]deltorphin B, contained a difficult-to-hydrolyze Val–Val bond that required 60 min at 850 W for complete hydrolysis.

On the basis of results with the 3 subject peptide samples and 10 other peptides analyzed in their lab, the researchers concluded that microwave hydrolysis for 30 min at 450 W yields acceptable amino acid recoveries with minimal racemization. Racemization observed with microwave hydrolysis was consistently lower than that observed with conventional acid hydrolysis.

Alternative Hydrolysis Agents, Protective Additives, and Derivatization Chemistries

A number of studies have been conducted using alkaline solutions, sulfonic acids, and protective agents to determine if these reagents that permit quantitation of amino acids destroyed during acid hydrolysis with 6 N HCl are equally effective in microwave hydrolysis (*15, 19–25*). Generally speaking, the same acids, protective agents, and derivatization chemistries can be employed in microwave-hydrolysis techniques.

Microwave heating and hydrolysis deviate from conventional heating and hydrolysis methods in several subtle ways. First, a reservoir of liquid acid must be maintained in the bottom of the hydrolysis vessel to ensure efficient microwave heating. If too little acid is used, a nonequilibrium heating condition will develop inside the vessel in which the acid is vaporized into the headspace and heating is interrupted until the vapor cools and recondenses back into the liquid state allowing microwave heating to resume. When this phenomena occurs the target hydrolysis temperature may be impossible to attain. For this reason slightly more acid may need to be used for microwave hydrolysis than is used for conventional heating and hydrolysis techniques. The volume of the microwave-hydrolysis vessel determines the minimum amount of acid required to maintain efficient, uninterrupted microwave heating. (A 45–60-mL microwave-hydrolysis vessel for multiple samples will operate efficiently with as little as 5 mL of acid.)

Second, microwave heating of alkaline solutions inside a closed vessel concentrates these solutions as water vapor partitions into the vessel headspace. This process causes precipitation of salts and formation of crystal deposits on vessel walls at the liquid level line. These crystal deposits will absorb microwave energy, causing localized heating that can char and damage a fluoropolymer vessel and lead to possible failure. This problem is most likely to be encountered during hydrolysis of milligram to multigram food samples in large volume fluoropolymer vessels using small amounts of alkaline solution. An interesting approach to microwave heating and alkaline hydrolysis of food samples for tryptophan determination was described in 1993 (25). Lyophilized samples of beef, trout, powdered milk, and chicken and veal baby foods were weighed into glass autosampler vials, 50 mg of sample in each vial, along with 100 mg of lithium hydroxide and 1 mL of water. The autosampler vials were assembled with a screw cap and perforated Teflon septum. The assembled vials were then placed inside a Teflon perfluoroalkoxy (PFA) microwave vessel and heated for 18 min at 175 °C. Tryptophan recoveries obtained following microwave hydrolysis with lithium hydroxide were comparable with those obtained by conventional hydrolysis with barium hydroxide for 12 h at 110 °C.

The third difference between microwave and conventional heating methods involves the use of argon or nitrogen gas to blanket samples with an inert, anaerobic atmosphere following a vacuum-evacuation procedure to minimize oxidative degradation of sensitive amino acids during hydrolysis. If a high vacuum (<100 millibar) is pulled on a microwave-hydrolysis vessel and that vessel is then purged with a noble gas such as argon, a plasma discharge condition may result during microwave heating (indicated by a blue glow inside the vessel) (26, 27). Nitrogen is recommended as the purge gas in microwave-hydrolysis procedures.

Vapor-Phase Microwave Hydrolysis

Commercial amino acid analyzers are highly sensitive, capable of detecting amino acids in the low picomole range (<10 picomoles) with UV detection, and into the femtomole range by using fluorescence detection (*12, 28, 29*). Even though the ability to detect amino acids at these levels is beneficial for research purposes, elimination of background contamination from free amino acids poses a major challenge. Amino acids are common contaminants in acids, organic solvents, airborne particles, and water. Fingerprints on glassware, pipette tips, etc., can introduce high levels of background contamination (*5, 6*). A single fingerprint on a dry glass surface contains 17 different amino acids (*30*). Vapor-phase hydrolysis is the preferred approach when the amount of sample available for analysis is limited (<1 nanomole), because the sample is contacted and hydrolyzed by pure acid vapor and condensate minimizing contamination.

The first reports documenting vapor-phase microwave hydrolysis on minute amounts of purified peptides and proteins were presented by two independent research groups at the Third Symposium of the Protein Society in Seattle, Washington, in July of 1989 (*19, 31*). Researchers from Hoffman-LaRoche conducted vapor-phase hydrolysis on three protein samples: melittin, bovine insulin B-chain, and recombinant human interferon α-2a. Melittin and insulin B-chain contain difficult-to-hydrolyze Val–Leu, Leu–Ile, and Leu–Val bonds. Samples between 10 pmol and 10 nmol were pipetted into 300-μL autosampler vials and vacuum dried. These open vials were placed inside a Teflon–tetrafluoromethaxil (TFM) microwave vessel specially modified for vapor-phase hydrolysis (Figure 3). The vials rested on a perforated fluoropolymer tray above a small cup containing 5 mL of 6 N HCl. A condensation dome was installed above the vials to prevent liquid acid condensing on the underside of the vessel cover from dropping into the vials. The vessel was heated inside a 1200-W microwave digestion system (Milestone MLS, Leutkirch, Germany) for 10 min at 900 W and 20 min at 300 W. Conventional liquid-phase hydrolyses in 6 N HCl at 110 °C for 24 h and liquid-phase microwave hydrolysis for 4 min at 900 W were conducted to assess the comparative hydrolytic efficiency of vapor-phase microwave hydrolysis. Amino acid recoveries indicated that vapor-phase microwave hydrolysis completely hydrolyzed the samples, including the hydrophobic peptide linkages. The amino acid profiles were generally consistent with the theoretical composition of each protein and the results obtained by conventional and liquid-phase microwave-hydrolysis methods. Serine, threonine, and tyrosine recoveries obtained with the vapor-phase hydrolysis methods were lower than by the conventional liquid-phase method.

A collaborative research effort between Hewlett Packard and CEM Corporation on vapor-phase microwave hydrolysis was contemporaneous with the previously described research of Hoffman-LaRoche (*31*). In this

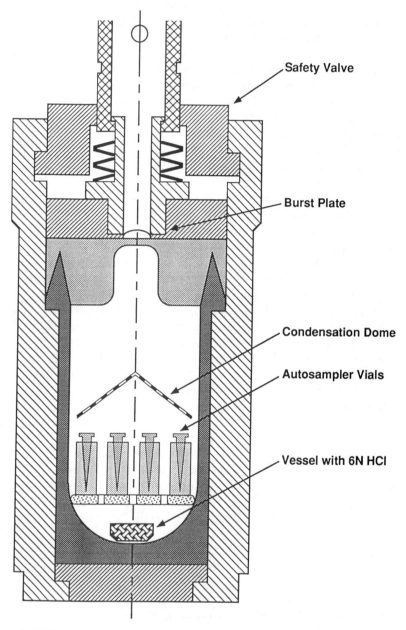

Figure 3. PTFE vessel used for vapor-phase microwave-hydrolysis experiments.

study 20-, 40-, and 80-pmol samples of methionyl human growth hormone, m-HGH (Pro-tropin, Genentech) were hydrolyzed to determine if vapor-phase microwave hydrolysis was suitable for high sensitivity amino acid analysis employed in protein chemistry research. Aliquots of a 20-pmol/µL m-HGH solution in 0.1 N HCl were pipetted into 100-µL glass autosampler vials and vacuum dried. All vials were pyrolyzed at 465 °C to eradicate amino acid contaminants before use. Vials containing the dried m-HGH samples and vacuum dried 0.1 N HCl blanks were introduced into a special tray positioned above liquid acid inside a Teflon PFA microwave vessel. The vessels were placed in liquid nitrogen to freeze the acid and prevent "bumping" of the liquid acid during the subsequent vacuum evacuation and nitrogen purge steps. The vessels were heated inside an MDS-81D microwave digestion system (CEM, Matthews, NC) for periods of 2, 4, 6, 8, 10, and 12 min. Microwave-hydolysis runs were repeated with protective agents added to the 6 N HCl (5% v/v β-mercaptoproprionic acid and 0.5% v/v phenol) to prevent oxidation of labile amino acids.

Vapor-phase hydrolyses were conducted on 80-, 160-, and 320-pmol samples of m-HGH at 105 °C for 24 h by using 6 N HCl, with and without protective agents, in a vacuum dessicator heated inside a laboratory convection oven for comparison. Amino acid analysis showed no significant differences between results obtained with 20-, 40-, or 80-pmol samples, results indicating the microwave-hydrolysis technique was suitable for high sensitivity analysis of minute amounts of protein. After 2 min of microwave hydrolysis samples were only 5–10% hydrolyzed. The frozen state of the acid at commencement of microwave heating apparently delayed release of acid vapor and hydrolysis. Hydrolysis of m-HGH was complete after 8–10 min of microwave heating. The molecular weight of m-HGH is 22,061 Da and it contains only one Leu–Val hydrophobic peptide bond that is resistant to hydrolysis. The researchers recognized that larger protein molecules containing more hydrophobic linkages might require longer periods of microwave heating for complete hydrolysis and examined this issue in subsequent work (32).

This work is also significant because it represents the first time acid temperature and vapor pressure were directly measured during microwave hydrolysis. In-situ temperature measurements of liquid 6 N HCl were made during microwave heating by using a Luxtron Model 750 fiber-optic thermometry system (Luxtron Corp., Santa Clara, CA). After 5.5 min of microwave heating 10 mL of 6 N HCl inside a 120-mL vessel, a temperature of 178 °C was attained and generated a vapor pressure of 132 psig. All microwave hydrolyses of m-HGH were conducted at these conditions. An acid temperature of 178–180 °C was maintained by feedback control of vapor pressure inside the vessel at 140 psig by using an external module containing a pressure transducer. When acid vapor pressure reached the 140 psig control set point, power to the magnetron was cycled on and off to maintain the target pressure within ±1 psig.

Development of Microwave Hydrolysis Systems

Temperature Control

Much of the early research on microwave-assisted acid hydrolysis of proteins was conducted by using domestic microwave ovens intended for home cooking (*11, 14, 15, 18, 21, 22*). A means of directly measuring and controlling acid temperature and vapor pressure inside sealed hydrolysis vessels located in a microwave field was unavailable to these researchers. Acid temperatures were estimated by microwave heating and observation of compounds with known melting points or measuring the temperatures of liquids in open containers immediately after a period of microwave heating (*11, 14, 15, 18*).

Acceptance of microwave hydrolysis in scientific and commercial applications would depend on the development of microwave systems with the capability to measure, control, and document acid temperature for two major reasons. First, differences in the molecular weight, amino acid composition, and structure of proteins necessitate method development to establish optimum hydrolysis conditions for a given protein or peptide. Acid temperature is a critical parameter in method development because it directly influences hydrolytic efficiency and reaction kinetics. Optimization and validation of microwave-hydrolysis methods is impossible without a means of accurately measuring and controlling acid temperature.

Second, documentation of acid hydrolysis conditions is important information for quality control and regulatory purposes. Temperature data confirm the proper hydrolysis procedure was followed. If a variation in the expected amino acid profile of a protein sample is observed, temperature data is useful in determining if the hydrolysis step is the source of the problem, or if the problem is attributable to other causes (reagents, instrumentation, protein purity, etc.). The U.S. Food and Drug Administration (FDA) requires amino acid analysis of all synthetic peptides and recombinant protein products intended for human therapeutic use (i.e., insulin, human growth hormone, etc.). Good manufacturing practice (GMP) regulations govern both the production and quality control of these therapeutic agents. Analytical methods used to ensure product quality and safety must be validated through testing and documented as written standard operating procedures (SOPs). Acid temperatures (usually heating-device control set points) are defined in the SOPs of conventional acid hydrolysis methods. A laboratory contemplating adoption of a microwave-hydrolysis method for preparation of quality-control samples would need to document acid temperature to validate the method or establish equivalency to an existing SOP. Once a microwave-hydrolysis SOP was validated, temperature data from every hydrolysis run performed on production batches of the protein product must be recorded, cross-referenced to amino acid analysis data, traceable to the batch, and available for review by FDA inspectors.

Direct measurement and control of acid temperatures inside sealed sample vessels located in a microwave field was problematic when experiments on microwave hydrolysis were first conducted. An indirect method (pressure control) represented the state-of-the-art technique available for controlling acid temperatures in microwave digestion systems during this time period. The first commercial system for microwave hydrolysis was introduced in 1990 (33). The system consisted of an MDS-81D microwave digestion system (CEM Corp.), external valve panel for vacuum evacuation and nitrogen purging of sample vessels, and a pressure control module. Acid temperature and vapor pressure data, derived in separate microwave-heating experiments at the factory, were provided to enable the user to select a pressure-control set point corresponding to a desired hydrolysis temperature. The inability to directly measure and confirm acid-hydrolysis temperatures constrained adoption of microwave hydrolysis in many applications. The results of a study comparing amino acid recoveries from vapor-phase microwave hydrolysis of lysozyme using indirect pressure control of acid temperature to the recoveries obtained with a Waters Pico-Tag system underscore the critical importance of direct temperature measurement and control for method development and validation (34). Amino acid recoveries from hydrolysis at 0.621 MPa (90 psig) for 20 and 40 min and 0.689 MPa (100 psig) after 10, 20, and 40 min varied from 63 to 111% and relative standard deviation values were <8.0%.

In 1991 microwave digestion and hydrolysis systems with onboard temperature-control capability became available. Temperature measurement is performed by one of two alternative methods in these laboratory systems: fiber-optic thermometry or thermoelectrically using a special grounded–shielded thermocouple probe.

A simplified schematic of a microwave-hydrolysis system employing the fiber-optic thermometry technique appears in Figure 4. A fiber-optic probe inserted into the thermowell of a hydrolysis vessel exits the microwave cavity via a bulkhead connector, leading to a control board mounted on the main central processing unit board of the microwave system. The control board transmits a pulse of light down the fiber-optic probe to a phosphor sensor at the tip. After a light pulse excites the phosphor, it emits a fluorescent signal. The decay rate of this fluorescent signal is temperature dependent and allows determination of temperature via a calibration table stored in the computer memory. Temperature measured inside the hydrolysis vessels is used as a feedback control signal to cycle microwave power on and off and maintain a user-selected hydrolysis temperature.

A microwave-hydrolysis system equipped with a special acid-resistant thermocouple is pictured in Figure 5. The cover of the hydrolysis vessel contains three ports to accommodate a thermocouple and two fluoropolymer tubes for preliminary vacuum evacuation and nitrogen purging. The tubes exit the microwave cavity and connect to a valve manifold connected to an acid-resistant vacuum pump and nitrogen supply source. A

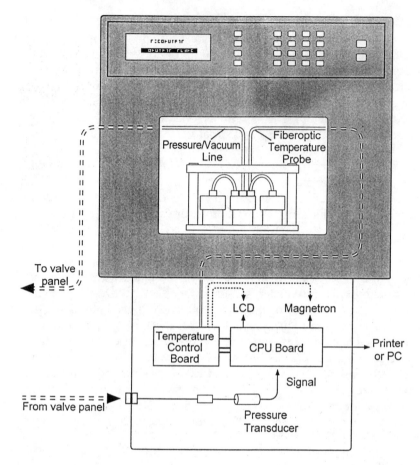

Figure 4. Schematic of microwave-hydrolysis system with fiber-optic thermometry system.

temperature curve from controlled microwave heating of 6 N HCl at 160 °C is reproduced in Figure 6. Two separate stages of microwave heating were employed and are represented at the top of the graph. Continuous, unpulsed microwave power (250 W) was used in the initial stage to minimize hysteresis in temperature control after the transition to the second heating stage by using 500 W of pulsed power.

Labware for Microwave Hydrolysis

The labware used in microwave hydrolysis must be constructed of materials possessing a unique combination of properties. The dielectric, mechanical, and thermal properties of a material must be considered along with acid resistance and absence of oxygen-trapping microscopic voids.

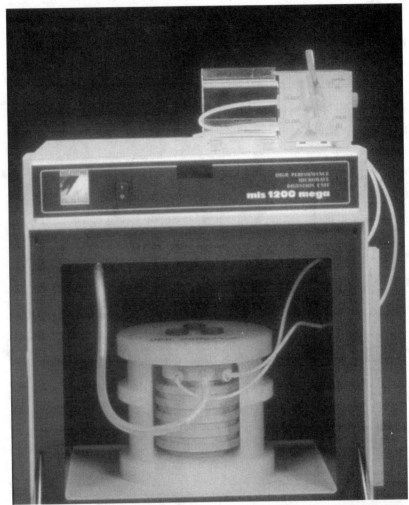

Figure 5. Microwave-hydrolysis system with acid-resistant thermocouple.

The elevated vapor pressures generated during acid hydrolysis can rupture sealed, glass, hydrolysis tubes. This safety concern is a common one with conventional hydrolysis devices despite heat-transfer mechanisms that result in a gradual rise in acid temperature and vapor pressure. The rapid temperature and vapor-pressure increases resulting from direct microwave heating of acids make this safety concern more acute and preclude the use of sealed glass tubes in microwave hydrolysis. Vessels for microwave hydrolysis are constructed of fluoropolymers capable of withstanding internal pressures greater than 100 psig.

T [°C]   P [Bar]

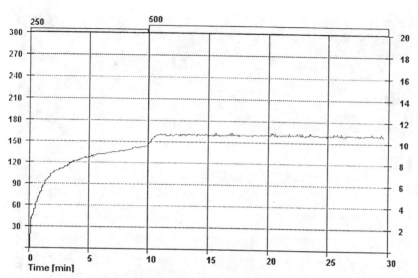

Figure 6. Temperature profile from two-stage (250–500 W) microwave heating of 6 N HCl at 160 °C. (Milestone MLS.)

Fluoropolymers such as Teflon PFA or tetrafluoromethaxil are preferred materials for their mechanical strength and low porosity. Injection molding and isostatic pressing fabrication techniques eliminate microscopic voids that can trap and subsequently release oxygen during hydrolysis, destroying liberated amino acids. Internal vessel components used for holding autosampler vials during vapor-phase hydrolysis are also constructed of these materials.

Vessels for liquid-phase microwave hydrolysis are the same basic design used in microwave digestion. Large samples (0.1–2.0 g) of grain, meat, or biological tissue are immersed in acid that directly contacts the fluoropolymer vessel walls. Vessels may also have tubing connections for vacuum evacuation, nitrogen purging, and temperature monitoring and control. A set of liquid-phase microwave-hydrolysis vessels is shown in Figure 7.

Vessels for vapor-phase microwave hydrolysis contain internal trays for holding glass autosampler vials and a covering piece to prevent liquid acid from condensing and dropping into the open vials. Examples of vessels for vapor-phase microwave hydrolysis are shown in Figures 8 and 9. Vial trays with numbered positions are desirable to identify samples and blanks (e.g., Figure 10), because wax and ink markings may be lost in contact with acid vapor and introduce contamination. If a tray lacks numbered positions, a diamond stylus can be used to mark vials.

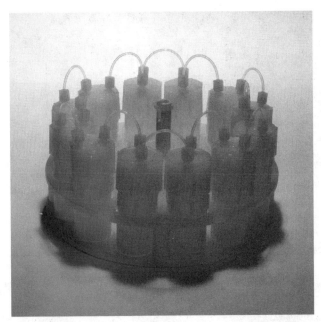

Figure 7. Microwave vessels for liquid-phase hydrolysis interconnected for vacuum evacuation and nitrogen purging.

Interchangeable vial trays that are compatible with the autosampler vials of all HPLC amino acid analyzers are available from microwave-hydrolysis system manufacturers. In vapor-phase microwave hydrolysis, samples are hydrolyzed and analyzed in the same vials. This approach eliminates the possibility of losses or contaminant addition during transfer steps, which may occur in devices that use flame-sealed hydrolysis tubes or different hydrolysis and analysis vials.

Polypropylene vials are not suitable for microwave hydrolysis because of thermal limitations of this material. Microwave-hydrolysis procedures are generally conducted at temperatures between 130 °C and 175 °C. The melting point of polypropylene is 168–171 °C, and the maximum service temperature is 100–105 °C.

Poly(vinylidene difluoride) electroblotting membranes are used to recover peptides and protein fragments that have been separated by sodium dodecyl sulfate–polyacrylamide gel electrophoresis. While some microwave experiments have been conducted by using these membranes, the service limitations of poly(vinylidene difluoride) have not been adequately defined.

Safety Considerations in Microwave Hydrolysis

Microwave hydrolysis involves sample vessels containing heated, pressurized acids. Many of the same safety considerations and practices detailed

Figure 8. TFM vessel, vial tray, and condensation cover for vapor-phase microwave hydrolysis.

in Chapter 16 of this book regarding microwave acid digestion therefore are directly relevant. Analysts considering adopting microwave-assisted hydrolysis techniques should review this safety information before proceeding.

Researchers conducting early microwave-hydrolysis experiments attempted to use flame-sealed glass ampoules and vials. The rapid rise in acid temperature and vapor pressure from microwave heating caused these vials to rupture violently (14). The primary safety consideration in microwave hydrolysis is selection and use of a sample vessel specifically designed for microwave heating of acids. Accordingly, the vessel must be microwave transparent, capable of withstanding elevated acid temperature and vapor pressure conditions, and equipped with an overpressurization relief device.

The use of domestic microwave cooking ovens for microwave-assisted acid hydrolysis is also problematic and should be avoided. Domestic microwave ovens are not constructed of acid-resistant materials and lack fume exhaust systems. Consequently, acids and reagents released into the cavity may chemically attack the microwave cavity door seal and safety interlocks that protect against microwave emission. Many of the acids and reagents employed in protein hydrolysis procedures are toxic and hazardous (e.g., HF

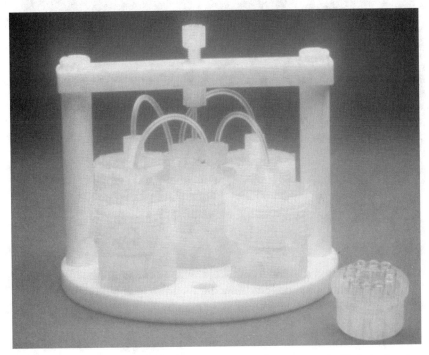

Figure 9. Teflon PFA vessels and vial tray for microwave hydrolysis.

and sulfonic acids). The risk of operator exposure to such hazardous substances is unnecessarily increased by the use of a domestic microwave oven.

Applications of Microwave Hydrolysis

The field of amino acid analysis is quite broad and encompasses diverse scientific and commercial applications. The worldwide market for amino acid analysis instrumentation and consumables is in excess of $30 million per year (35). A preliminary acid-hydrolysis sample-preparation step is not required in every application, such as the determination of abnormal concentrations of free amino acids in plasma and physiological fluids for diagnosis of diseases. Microwave hydrolysis is potentially applicable in every instance where hydrolyzates are analyzed. Approximately 100 microwave-hydrolysis systems are currently in use in research and quality-control laboratories worldwide (36).

The largest concentration of applications are found in the pharmaceutical and biotechnology industries. Samples include synthetic peptides, recombinant proteins, protein-derived implant materials, and special dietary formu-

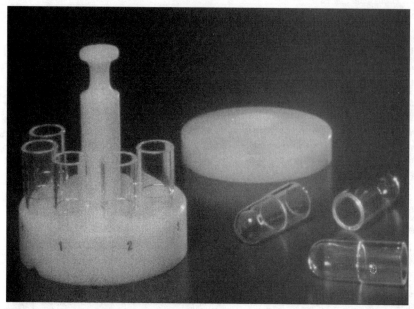

Figure 10. Vial tray for vapor-phase microwave hydrolysis with numbered positions for sample identification.

lations for infants and medical patients. Microwave hydrolysis is being employed to support research and development of new therapeutic agents and for quality control of commercial products following validation studies. The sweetener aspartame (L-aspartyl-L-phenylalanine methyl ester), insulin, and human and animal growth factors are a few examples.

An important area of application for microwave hydrolysis is in molecular biology and protein chemistry research. High-sensitivity amino acid analysis used in these fields is not a simple or routine task. Operator skill, knowledge, and experience are important considerations in achieving satisfactory results. For this reason many universities and research institutes operate core laboratories to perform the analytical procedures that support the molecular biology and protein chemistry research of colleagues. This setup also eliminates costly, inefficient duplication of expertise and instrumentation. The Association of Biomolecular Resource Facilities (ABRF), to which core laboratories belong, periodically conducts performance studies to assess the ability of the member labs to successfully perform analytical tasks associated with protein characterization. The results of one such ABRF study on the accuracy and precision of amino acid analyses conducted at core laboratories were reported at the third symposium of the Protein Society (37). Core laboratories at several universities have microwave-hydrolysis systems to support the large number of samples they handle (35).

Another application of microwave hydrolysis is the rapid preparation of samples for collagen analysis. Collagen is an extracellular protein found in connective tissues and constitutes approximately 25–30% of the total protein in animals. Collagen analysis is performed for medical research purposes and to assess the quality of protein in meat products.

Degradation of collagen in connective tissues is a factor in a number of medical conditions such as rheumatoid arthritis, osteoarthrosis, and scleroderma. The progressive formation of cross-links between collagen fibers is the suspected cause of the reduced muscle flexibility, joint stiffening, and increasing brittleness of bones associated with aging. Medical research institutes in the United States and Italy have microwave-hydrolysis systems to hydrolyze tissue samples.

Several European countries have established limits on the amount of connective tissue that may be present in meat products. Collagen, the principal constituent in connective tissue, is a low quality protein of little nutritional value. Collagen contains a high concentration of the amino acid hydroxyproline, whereas other muscle proteins contain virtually no hydroxyproline. By analyzing meat products for hydroxyproline content the amount of connective tissue present can be determined. This test involves hydrolysis of a 1–2-g sample of meat in hydrochloric or sulfuric acid before measuring the hydroxyproline content by HPLC analysis or a photometric technique (e.g., International Organization for Standardization Method 3496.2). Comparative studies conducted by research institutes in Sweden and Germany have shown no significant differences among the results obtained by using a 30-min liquid-phase microwave-hydrolysis procedure and standard AOAC International and NMKL (Nordic Methods Committee) methods employing 8–16 h hydrolysis procedures (*38, 39*). Several universities and meat product companies have microwave systems for this purpose.

Nutritional research studies examine the use of protein in humans, domesticated animals, and wildlife by amino acid analysis of the diet, biological tissues, physiological fluids, and waste products. Several microwave-hydrolysis systems are in use at universities for this purpose. One example of this application is a nutritional study to optimize the production of mohair in goats (*40*). The use of microwave hydrolysis for amino acid determination in biomass and proteins isolated from fermentation biomass recently was reported (*41*).

Amino acid analysis and atomic spectroscopy are among the techniques employed by art institutes to determine the source of artists' materials, to facilitate preservation and restoration of paintings, and for better understanding of historical methods and materials. Microwave systems are uniquely suited for preparing samples for both these analytical techniques and are installed at several art institutes including the Belgian Institute of Art in Brussels.

Several microwave sample preparation procedures have been described for closely related protein- and biochemical-analysis applications. A standard test for determining the concentration of micro amounts of protein in solution is the BCA (bicinchoninic acid) assay. This assay involves reaction of Coomassie blue reagent with protein in a BCA–alcohol medium to form a protein–dye complex that can be measured at 595 nm to derive the protein concentration. Formation of the protein–dye complex normally requires 30–120 min. Microwave heating has been used successfully to accelerate this reaction and reduce assay time to 20 s (42).

A microwave-hydrolysis technique for rapidly hydrolyzing bile acid methyl esters and isolating bile acids has been described (43). Bile acids are steroid carboxylic acids bound in peptide linkages to the amino acids taurine or glycine and are recovered by alkaline hydrolysis. Free bile acids serve as starting materials for the synthesis of therapeutically important steroid hormones (44).

Summary

Acid hydrolysis of protein samples has been the rate-limiting step in amino acid analysis since commercial HPLC amino acid analyzers became available beginning in the late 1960s. Microwave hydrolysis is an alternative to conventional hydrolysis techniques that fundamentally changes this situation. Protein hydrolysates can now be prepared in less time than a single chromatographic run without compromising accuracy and precision. The lengthy multiday time-course studies required to develop hydrolysis methods using conventional techniques can be streamlined with microwave hydrolysis. The technique is particularly advantageous in cases where two or more hydrolyses are performed to achieve optimum recoveries of labile and refractory amino acids. For instance, microwave-hydrolysis runs of 20- and 40-min duration can be performed in place of one- and two-day conventional hydrolyses.

Microwave-hydrolysis systems with reliable means of temperature control have been developed, and the technique is suitable for quality control of protein products and documenting compliance with relevant GMP, GLP, and International Organization for Standardization guidelines. In fact, microwave hydrolysis is the only technique in which acid temperature and vapor pressure are measured directly, in situ. Other techniques measure the temperature of a heating device such as a metal heating block or oven surrounding a hydrolysis chamber.

The evaluation and adoption of microwave-hydrolysis techniques by recognized national and international standard organizations will be an important factor in gaining general acceptance of these techniques. An example is a microwave-hydrolysis study conducted at National Institute of Standards and Technology (NIST) on standard reference material (SRM

926, Bovine Serum Albumin) (*45*). Research at NIST laid the scientific foundation for the acceptance of microwave acid-digestion techniques and NIST spearheaded development of U.S. Environmental Protection Agency microwave methods. Similarly, NIST documented research and microwave-hydrolysis methods for protein standard reference materials would increase usage of microwave hydrolysis on FDA regulated protein products and issuance of AOAC International methods.

Microwave hydrolysis accelerates the rate of reaction without altering the chemistry of amino acid analysis. The same acids, protective agents, and derivatization chemistries can be used in microwave hydrolysis permitting direct replacement of an existing technique following a very brief time-course study to establish equivalency.

Whereas vapor-phase microwave hydrolysis is well-suited for high-sensitivity amino acid analysis in biotechnology applications, the technique can also be expected to play an increasing role in other areas of protein chemistry. The improved laboratory productivity and analytical turnaround times made possible by microwave hydrolysis are as beneficial to those analyzing food and grain products as to those analyzing high-value therapeutic peptides and proteins.

References

1. Lubert, S. *Molecular Design of Life;* W. H. Freeman and Company: New York, 1989; pp. 22–23.
2. Moore, S.; Stein, W. H. *Science (Washington, D.C.)* **1973,** *180*, 458–464.
3. Hirs, C. H. W.; Stein, W. H.; Moore, S. *J. Biol. Chem.* **1954,** *211*, 941–950.
4. Pines, M. *The Structures of Life—Discovering the Molecular Shapes That Determine Health or Disease;* NIH Publication No. 88–2778; Office of Research Reports: Washington, DC, 1988; pp 21–33.
5. Blackburn, S. In *Amino Acid Determination Methods and Techniques,* 2nd ed.; Blackburn, S., Ed.; Marcel Dekker, New York, 1978; pp 7–36.
6. Ozols, J. In *Methods in Enzymology;* Academic: Orlando, FL, 1990; Vol. 182, pp 587–601.
7. *Sequencing of Proteins and Peptides;* Allen, G., Ed.; Laboratory Techniques in Biochemistry and Molecular Biology Vol. 9; Elsevier: Amsterdam, Netherlands, 1989.
8. Smith, J. A. In *Techniques in Protein Chemistry;* Hugli, T. E.; Academic: Orlando, FL, 1989, pp 251–254.
9. Shively, J. E.; Paxton, R. J.; Lee, T. D. *Trends Biochem. Sci.* **1989,** July, 246–252.
10. Atherton, E.; Sheppard, R. C. *Solid Phase Peptide Synthesis—A Practical Approach;* The Practical Approach Series; IRL Press: Oxford, England, 1989; pp 107–110.
11. Yu, H. M.; Chen, S. T.; Chiou, S. H.; Wang, K. T. *J. Chromatogr.* **1988,** *456,* pp 357–362.
12. Chang, J. Y.; Knecht, R.; Jenoe, P.; Vekemans, S. In *Techniques in Protein Chemistry;* Hugli, T. E., Ed.; Academic: Orlando, FL, 1989; pp 305–313.

13. Abu–Samra, A.; Morris, J. S.; Koirtyohann, S. R. *Anal. Chem.* **1975**, *47*, 1475–1477.

14. Chen, S. T.; Chiou, S. H.; Chu, Y. H.; Wang, K. T. *Int. J. Pept. Protein Res.* **1987**, *30*, 572–576.

15. Chiou, S. H.; Wang, K. T. *J. Chromatogr.* **1989**, *491*, 424–431.

16. Lahm, H. W.; Lergier, W.; Manneberg, M.; Knorr, R. *J. Protein Chem.* **1988**, *7*, 258–259.

17. Liardon, R.; Lederman, S. *Anal. Chem. Symp. Ser.* **1984**, *21*, 7–17.

18. Peter, A.; Laus, G.; Tourwe, D.; Gerlo, E.; Van Binst, G. *Pept. Res.* **1993**, *6*, 48–52.

19. Manneberg, M.; Lahm, H. W. *Advances in Amino Acid Analysis*; Third Symposium, The Protein Society: Seattle, WA, July 1989; Poster paper M–208.

20. Woodward, C.; Engelhart, W. G. Presented at the Fourteenth International Symposium on Column Liquid Chromatography, Boston, MA, 1990; Poster paper P455.

21. Pecavar, A.; Prosek, M.; Fercej–Temeljotov, D.; Marcel, J. *Chromatographia* **1990**, *30*, 159–162.

22. Engelhardt, H.; Kramer, M.; Waldhoff, H. *Chromatographia* **1990**, *30*, 523–526.

23. Miller, C.; Woodward, C. Presented at the American Peptide Symposium, Cambridge, MA, 1991; Poster paper P–417.

24. Grimm, R. *Amino Acid Analysis of Protein and Peptide Hydrolyzates—Evaluation of Accuracy*; Hewlett Packard Application Note, Publication No. 12–5091–4585E; Hewlett Packard: Waldbronn, Germany, 1992.

25. Carisano, A. *Ind. Aliment.* **1993**, *32*, 346–348.

26. Woodward, C. Hewlett Packard Company, private communication, 1989.

27. Lautenschlager, W. MLS–Mikrowellen Labor Systeme GmbH, private communication, 1994.

28. Bidlingmeyer, B. A.; Cohen, S. A.; Tarvin, T. L. *J. Chromatogr.* **1984**, *336*, 93–104.

29. *Guaranteed Amino Acid Analysis*; Publication No. 12–5962–3514E; Hewlett Packard: Waldbronn, Germany, 1991.

30. Spackman, D. H. In *Methods in Enzymology*; Academic: Orlando, FL, 1990; Vol. 11, p 3.

31. Gilman, L. B.; Woodward, C. In *Current Research in Protein Chemistry: Techniques, Structure, and Function*; Villafranca, J. J., Ed.; Academic: Orlando, FL, 1990; pp 23–36.

32. Woodward, C.; Gilman, L. B.; Engelhart, W. G. *Int. Lab.* **1990**, *20*, 40–46.

33. Engelhart, W. G. *Am. Biotech. Lab.* **1990**, *8*, 30–34.

34. Tatar, E.; Khalifa, M.; Zaray, G.; Molnar-Perl, I. *J. Chromatogr. A*, **1994**, *672*, 109–115.

35. Wilkinson, G.; Wilkinson, B. *Anal. Instr. Ind. Rep.* **1993**, *10*, 3.

36. Engelhart, W. G. Milestone MLS, unpublished data, 1993.

37. Crabb, J. W.; Ericsson, L.; Atherton, D.; Smith, A. J.; Kutny, R. In *Current Research in Protein Chemistry: Techniques, Structure, and Function*; Villafranca, J. J., Ed.; Academic: Orlando, FL, 1990; pp 49–61.

38. Bauer, F. *Microwave Digestion of Proteins for Rapid Determination of Hydroxyproline in Meat Products*; Proceedings of the 6th European Conference on Food Chemistry; Gesellschaft Deutscher Chemiker: Frankfurt a.M., Germany, 1991; Vol. 2, pp 552–556.

39.   Kolar, K.; Berg, H. *Microwave Hydrolysis for Rapid Determination of Hydroxyproline in Meat and Meat Products;* Swedish Meat Research Institute Publication # 0924a; Swedish Meat Research Institute: Kävlinge, Sweden, 1990.

40.   Hart, S. Langston University, private communication, 1992.

41.   Joergensen, L.; Thestrup, H. N. *J. Chromatogr. A* **1995,** *706,* 421–428.

42.   Akins, R. E.; Tuan, R. S. *BioTechniques* **1992,** *12,* 496–499.

43.   Dayal, B.; Salen, G.; Dayal, V. *Chem. Phys. Lipids* **1991,** *59,* 97–103.

44.   *Concise Encyclopedia of Biochemistry;* Scott, T.; Brewer, M., Eds.; Walter de Gruyter: Berlin, Germany, 1983.

45.   Margolis, S. A.; Jassie, L.; Kingston, H. M. *J. Auto. Chem.* **1991,** *13,* 93–95.

Chapter 14

Microwave Methods
for Sample Preparation
in Pathology

L. P. Kok and Mathilde E. Boon

A new scientific truth does not triumph by convincing its opponents and making them see the light, but rather because its opponents eventually die, and a new generation grows up that is familiar with it.

Max Planck

Although the first report on the use of microwaves in pathology dates back to 1970, it took 15 years before the theoretical basis of the physics applied to the diagnostic samples was explored. The insight obtained allowed us to exploit microwave techniques in all the steps of sample preparation in pathology. This result led not only to often dramatically shortened processing times and less use of reagents, but also to improved quality of microscopical images. Of recent date are applications involving the use of vacuum, and antigen retrieval. In this chapter we present an overview and discuss these two microwave techniques in more detail.

Health care relies on good diagnoses. Many diagnoses are made in a laboratory environment, with the help of microscopy. Much of our research effort in the past decade has been devoted to the use of microwaves in the steps preceding the visual inspection of specimens by the microscopist, who diagnoses on the basis of an image. The image is created by the chemical process of staining and optics. In cytologic samples, one deals with cells and cell clusters removed from the tissue. In histology, one deals with tissue. The scientific basis for the full procedure of sample preparation lies in cytochemistry and histochemistry, respectively.

Both in cytology and histology the first step toward microscopy is fixation of the sample. In histology, the fixation enables the tissue to withstand the hazards of histoprocessing. In cytology, the fixation serves likewise to conserve the architecture of the cell and its components. Fixation can be done with or without chemicals. In both cases the use of microwaves can be beneficial. Many chemical fixatives are based on cross-linking of proteins. Using microwaves combined with a chemical fixative, one may enhance diffusion of chemicals into the biological material. Moreover, one may increase the reaction rates of the binding of fixatives and of cross-linking. One also can use coagulation as a means of fixation. A common method of coagulation of biological proteins is microwave-induced heating.

For microscopy one needs to prepare thin objects. In cytology a monolayer of cells is perfect for light microscopy: The height of cells on a microscopical slide is of the order of magnitude of ten wavelengths of the light. Accordingly, it is not necessary to reduce the thickness of the cytological sample. In histology, to have a similar thickness, one must cut tissue pieces on a microtome. In fact, a variety of techniques has been developed to obtain thin slices of material. For example, frozen sections (cryosections) are cut on a microtome from a piece of frozen tissue. Unfortunately, the quality of these sections in general is not as good as paraffin or resin sections. Therefore, the mainstream of preparatory techniques involves embedding of tissue material in a cuttable medium like paraffin or plastic. The first step of these histoprocessing methods is dehydration. Before the paraffin can be brought to the places where water was present before dehydration, some intermediate step or combination of steps is necessary. The paraffin blocks containing the sample are cut and the slices (sections) are placed on microscopical glass slides to allow staining. Unstained slides have little contrast. Therefore, cytological and histological slides must be stained to allow microscopy.

Immunochemistry is a branch of chemistry that deals with the chemical aspects of immunology. Immunological principles are exploited to stain specific proteins (antigens) in tissue. Only since the 1960s has immunostaining been a part of the pathologist's armamentarium, whereas many of the classical staining procedures can be traced back to as early as the 19th century. We shall review the use of microwaves in both classical staining and immunostaining. Many of the principles of the preparatory steps for light microscopy can be followed for electron microscopy and immunoelectron microscopy.

Since 1985 we developed (and followed the development of) microwave techniques in each of the aforementioned areas of pathology. Clearly, temperature (and therefore control and measurement of temperature) plays an important role in these techniques. Recently, we have begun to explore the effects of variation of pressure (vacuum and high pressure) to improve existing procedures, and we have had great success. Therefore, in the following sections we shall discuss in detail a vacuum–microwave technique for histoprocessing. In addition, in this chapter we focus on the second new development of microwaves in pathology, that is, on microwave antigen retrieval. Other applications (freeze techniques, decalcification, all analog procedures in electron microscopy, resin embedding, DNA in-situ hybridization, enzyme-linked immunosorbent assay, DNA amplification, embryology, botany, and entomology) are discussed in our book

(*1*). We shall pursue the footsteps of the pathologist and follow the corresponding order of events. We do not discuss the hazards of the use of microwaves. For that, we refer to the literature (*see*, for example, reference 2 and sources quoted there).

Microwaves and Fixation

The first step of the pathologist is fixation of the sample. This step should be done directly following the surgical procedure in which the sample is taken from the (living) patient. An exception here is the autopsy. For establishing the cause of death and extent of the disease (for instance, cancer) tissue samples for microscopy are taken from the corpse.

Stabilization of Unfixed Tissue by Microwave Treatment

Microwave treatment of unfixed tissue in saline solution is often called *fixation*. However, Marani et al. (*3*) introduced the term *stabilization* when no chemical fixatives were involved in the microwave treatment of tissue to prepare it for the hazards of embedding. In this chapter we will use the term *fixation* exclusively when fixatives are used, and *stabilization* if only the physical effects of microwaves are applied.

The microwave exposure elevates the temperature throughout the tissue (if the tissue block is not thicker than 5 cm), and this change results in coagulation of the proteins, including the soluble ones. Small antigens may diffuse into the solute but to a lesser degree than when immersion fixation is applied.

When one stabilizes tissue in the microwave unit, one has to be careful not to increase the temperature too much: it is wise to keep the temperature between 45 °C and 55 °C. The tissue should be placed in a solution (0.9% NaCl, phosphate-buffered saline, pH 7.4, or tris(hydroxymethyl)aminomethane buffer, 0.05 M HCl, pH 7.4), and the temperature probe should be inserted into the solution. Exposure time depends on the type of tissue and its size.

Hsu et al. (*4*) reported that microwave stabilization of tissue improved the quality of the DNA recovered for use in nucleic acid analysis in paraffin blocks. Microwave stabilization also preserved the epitope of the hepatitis B surface antigen considerably better than formalin. We expect that many more related papers concerning the advantage of omitting formalin and replacing it by microwave stabilization to preserve labile viral antigens will appear in the scientific literature.

Microwave-Stimulated Fixation with Fixatives

Microwave exposure can be used to shorten diffusion of fixation reagents into the tissue and to accelerate the chemical process involved in the fixation process. By using microwave exposure properly, the fixative can be optimally exploited.

Formaldehyde Fixation and Microwave Exposure

Formalin is a 37% (wt/wt) solution of formaldehyde in water. Formalin fixation varies from laboratory to laboratory with respect to concentration, pH, buffers of the fixative, and temperature and length of the fixation process. All these factors strongly influence outcome. Maximum binding of formaldehyde to tissue occurs at alkaline pH. Secondary cross-linking reactions involving the formation of methylene bridges are optimal at an acidic pH. However, there is a paradox in the working mechanism of formaldehyde: it penetrates the tissue rapidly, but it fixes very slowly. For biopsies, optimal fixation time is 24 h. When the tissue remains in the formalin, the process of cross-linking continues and results in excessive cross-linking.

Formaldehyde-fixative solutions contain little formaldehyde (HCHO): instead, they contain mainly methylene glycol formed by the reaction between formaldehyde and water (*see* Reaction 1). The active fixing component is formaldehyde, but it is mainly methylene glycol that penetrates, and at a fast rate. The little formaldehyde in the tissue is mainly formed from the methylene glycol. The formaldehyde binds very slowly with the tissue, and hence it disappears at a slow rate. To maintain chemical equilibrium, more formaldehyde is formed by the dehydration reaction (*see* Reaction 2).

$$HCHO + H_2O \rightarrow CH_2(OH)_2$$
Reaction 1

$$CH_2(OH)_2 \rightarrow HCHO + H_2O$$
Reaction 2

The three steps of formalin fixation can be influenced by microwave exposure and the concomitant, almost instantaneous, and homogeneous increase in temperature: first, diffusion of methylene glycol into the tissue; second, formation of formaldehyde by dehydration of methylene glycol in the tissue; and third, binding of formaldehyde to the proteins by cross-linking. Hot formaldehyde fixation was used already by Ehrlich in the 19th century (5). When the fixative is heated, Steps 2 and 3 are completed where methylene glycol is present: that is, the periphery of the tissue block. Further diffusion into the center is then hindered by the thus-created dense protein network. The same is true for working at microwave-induced elevated temperatures; also, the dense protein network interferes with diffusion. Hopwood et al. (6) have shown in a model system that microwave exposure in the presence of aldehydes resulted in an intramacromolecular cross-linking.

We concluded (7) that for formaldehyde fixation the effect of microwave exposure can only be optimal when the methylene glycol is present throughout the tissue when temperature increase starts. In practice, this implies that the tissue samples of a maximum thickness of 5 mm must be immersed in the formaldehyde solution for at least 4 h prior to the microwave step. The soaking time of thicker pieces of tissue should be adapted. The microwave step of formalin-soaked tissue takes 2–5 min. Finally, we should keep in mind that simultaneously some microwave stabilization is achieved. This result might explain why immunostaining is better when compared with conventional formalin fixation (8).

Kryofix Fixation and Microwave Exposure

Kryofix (E. Merck, Darmstadt, Germany, Prod. Nr. 5211) is a coagulant fixative containing poly(ethylene glycol) (PEG). When ethyl alcohol is used on its own, without PEG, excessive shrinkage of the tissue results. For biopsies, optimal fixation time is 6 h. After this period, the tissue does not undergo further changes (in contradistinction to the formalin case). In the Leiden Cytology and Pathology Laboratory, all histologic specimen are fixed in Kryofix, making it a completely formalin-free laboratory, which is a boon for both clinicians and laboratory personnel. Both ethyl alcohol and PEG diffuse rapidly into the tissue block. Also for Kryofix, it is better to soak the tissue block prior to microwaving. Microwave exposure of 2–5 min of soaked tissue is sufficient to complete fixation throughout the tissue block. Kryofix is well-suited for immunocytochemistry.

Histoprocessing

The second step of the pathologist is histoprocessing of the fixed sample. Until the advent of microwaves, histoprocessing remained virtually unchanged for almost a century. It was a time-consuming process (16–72 h) involving dehydration through baths of graded alcohols, graded solutions containing a clearing agent (often xylene), and finally a number of paraffin baths. Clearly, diffusion of reagents into and out of the tissue is the key factor: the viscosity of liquids decreases with temperature, and microwaving is a very effective means to heat liquids throughout their entire volume (internal heating).

Since 1984, all histoprocessing in our laboratory has been done using microwave techniques, in which fast processing became possible because microwaves were used to stimulate diffusion of the heated reagents. We have described our experiences in a number of papers (9–13) and in the three editions of our *Microwave Cookbook*, appearing in the years 1987 through 1992 (1). This series of publications reflects the changing level of microwave technology and instrumentation and our increased experience.

In the microwave procedures, it was no longer necessary to use graded solutions, thus the samples could be placed directly in the alcohol bath (100%), then into the intermedium (100%), and then into the molten paraffin

(100%). This way the number of steps involved could be reduced from 12–20 to a mere three (alcohol 100%, intermedium 100%, molten paraffin).

Microwaves act strongly on polar (dielectric) media. In fact, the new microwave technology could be used optimally by selecting reagents on the basis of their molarity. For instance, isopropanol, found as an intermedium in books written in a period when time was an unimportant factor, was chosen as clearing agent because, in contrast to the presently more commonly used xylene, it is highly polar. We note that xylene is on the list of chemicals to be banned from pathology laboratories.

However, the isopropanol was not completely removed in the paraffin step. Therefore, for our routine samples we introduced a vacuum step immediately after the microwave paraffin step (Sec. 11.7 of reference 1). The samples in the molten paraffin were placed in a sealable container in which vacuum could be drawn. Indeed, when applying suction, the remaining traces of isopropanol formed bubbles in the paraffin escaping from the samples. In practice, this setup was not completely perfect, because the evaporation required so much heat that the paraffin could not always be kept above melting temperature.

We have used the previously described three-step microwave histoprocessing for almost 10 years; but clearly, it would have been ideal to combine microwaving and drawing vacuum. For this process, we needed special equipment. This chapter describes a novel two-step vacuum–microwave embedding technique in which this process has been achieved, thanks to a specially designed microwave unit. We have exploited extensively the fact that the boiling temperature of liquids featuring in histoprocessing (water and isopropanol) decreases when the liquid is placed in vacuum. Accordingly, we use vacuum (i.e., low pressure) to pull out the molecules. Moreover, we can do this at relatively low temperatures, so that the tissue is protected against the negative effects of overheating. For 72 biopsies (thickness, 2–4 mm), histoprocessing with the vacuum–microwave method takes 40 min. (The conventional histoprocessing methods are typically carried out overnight in a highly automated cycle.) Larger specimens ($4 \times 6 \times 1$ cm^3) require merely 4 h vacuum microwaving. (For the larger specimens conventional histoprocessing requires days.)

By optimally exploiting the possibilities of microwaving, histoprocessing can finally be reduced to a mere two steps: isopropanol and paraffin. Thus, there is no ethyl alcohol step. Moreover, shrinkage of tissue, which hampered other methods employing an ethyl alcohol step, could be totally avoided. Extensive bubbling out of reagents, especially in the paraffin step, is all-important. It is achieved thanks to stepwise decrease of pressure to the final low value of 100 hPa: that is, merely one-tenth of atmospheric pressure. The vapor bubbles are extremely small and cause no damage to the tissue that is visible macroscopically or microscopically. The resulting optimal paraffin impregnation gives blocks that are easy to cut, even for tissues that are notoriously difficult, for example, tissues containing fat, such as breast. Moreover, large blocks can be cut comfortably on a routine paraffin microtome.

Two-Step Procedure

The samples must be well-fixed before they can be histoprocessed. The tissue blocks cut (thickness, 2–4 mm) from the tissue samples are placed in microwave-transparent plastic cassettes. The procedure is given schematically in Figure 1.

The cassettes are placed directly from the fixative into isopropanol. In this procedure, the 100% ethyl alcohol step is skipped. Simultaneous application of microwaves–vacuum is chosen in such a way that temperature does not rise too high (i.e., no detrimental effects for the final morphology), but at the same time clearing is as complete as possible. In this step, the sample is also partially dehydrated because isopropanol is a weak dehydrator. Thus, temperature-controlled microwave exposure is used. A temperature of 55 °C is programmed. The pressure has been selected to go down to 400 hPa in a short time, and the total clearing step has been determined at 20 min. After the isopropanol step, the samples are placed directly in molten paraffin. The subsequent paraffin step has been determined to take 18 min in total. The first 13-min microwave exposure has been selected at the power level of 400 W and temperature control of 60 °C. The pressure is brought down gently during the last 5 min according to the graph given in Figure 2, to a final low value of 100 hPa.

In the first step (isopropanol) the temperature of 55 °C is below, but close to, the boiling points of the principal reagents water and isopropanol. If the tissue is fixed in the ethyl alcohol containing Kryofix, the ethyl alcohol is removed from the tissue under these circumstances. In the second step, the temperature of the paraffin is 60 °C, thus at 100 hPa the temperature is well above the boiling point of water and isopropanol (Figure 3). In these final 5 min, the isopropanol and the traces of water left behind in the tissue evaporate from the tissue and escape through the molten paraffin, which in its turn replaces the isopropanol (and water) in the tissue.

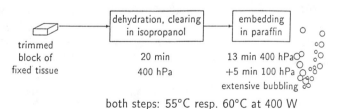

Figure 1. Schematic processing sequence for paraffin embedding of biopsies. For large blocks, see text.

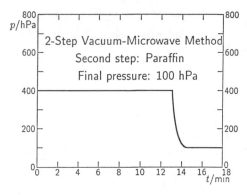

Figure 2. Pressure and power during Step 2 of embedding.

Giant Blocks

Sometimes the pathologist wants to embed large pieces of tissue, measuring as much as $4 \times 6 \times 1$ cm^3, for instance to establish the extent of carcinoma in breast. The procedure for these large specimens is completely analogous to that for small samples described in the previous section. The giant paraffin blocks that are thus obtained can be cut on a routine microtome (Microm, Heidelberg, HM340). The times of the two steps must be adapted to the size of the specimen. Keeping in mind the basic law of diffusion, that $\langle x^2 \rangle$ t (average square distance of the diffusing reagents is proportional to time t), one can use as a rule of thumb that if a block is 3× as thick, one needs to use times that are 9× as long. The basic idea is that all diffusing reagents should reach their sites throughout the tissue. Clearly, the constant of proportionality in the mathematical equation is essentially the diffusion constant, which is enormously temperature dependent (and hence can be manipulated by microwaves) (1). In fact, the temperature dependence is of the well-known Arrhenius type. Obviously, the smallest size of height, width, and depth is the key factor to determine whether diffusion will have progressed sufficiently.

Practice and Results

We have used the two-step method routinely in our laboratory with great satisfaction. The microscopical quality of the sections (cut at 2 μm and 4 μm) is comparable with, or slightly better than conventionally processed tissue fixed under the same conditions. The method can also be used for the formalin-fixed tissue that can be directly placed in the isopropanol bath. If formalin fixation is less than 24 h, the tissue samples should first undergo microwave-stimulated fixation.

For fatty tissue, such as breast, the two steps may have to be taken a bit longer, because fat is notorious for its adherence to water. In case the

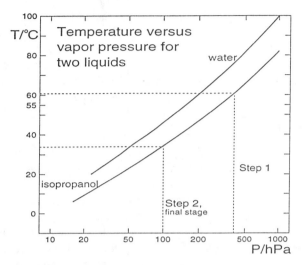

Figure 3. The relationship between temperature (°C) and vapor pressure (Pa) for various liquids. At a given pressure (horizontal) one finds the boiling temperature at the vertical scale. Note that in the isopropanol step (Step 1) temperature is *below* but very close to boiling temperature. Observe, however, that in the last stage (cf. Figure 2) of the paraffin step (Step 2) the temperature is *above* boiling temperatures of water *and* isopropanol. All full curves are accurate within line thickness and taken from the standard tables of the physics and chemistry literature.

technician is not completely satisfied with the resulting paraffin embedding, which will be noticed when the blocks are cut (if there is incomplete impregnation of the tissue it is difficult to cut) the block is placed back in molten paraffin and the paraffin step is simply repeated, including the vacuum of 100 hPa. Now, the traces of water left in the tissue leave it, and allow complete paraffin impregnation. With this treatment, the blocks have become easy to cut and perfect sections with optimal morphology are produced.

   The complete absence of shrinkage during histoprocessing of the giant blocks is noteworthy (we could establish this in the giant blocks with great precision (*13*)). This result is highly important for quantitative pathology where objective measures are required: for example, for depth of invasion or volume of tumor. In earlier experiments, in which the temperature of ethyl alcohol was 60 °C, we did see considerable shrinkage of these large blocks (up to 30%). Note that in the two-step method no ethyl alcohol is used! Equally important is that on completion of the procedure the paraffin has invaded everywhere in the tissue. As a result the paraffin blocks are easy to cut. As remarked earlier, fatty tissue tightly binds water; therefore, it is difficult to dehydrate the samples. In the conventional histoprocessing methods, often special measures are required such as including an acetone or chloroform step; both steps influence in a negative way the final microscopical image. This process is not needed at all in our vacuum–microwave method. For tissues containing a lot of fat, for instance samples from breast, the cutting is easier than for conventionally processed tissue. Even the large tissue blocks can be cut on a routine paraffin microtome at a thickness of 4 μm (often, for giant blocks processed with the conventional methods, one needs to use microtomes of special design).

Microwave Unit

The two-step vacuum–microwave procedure can be carried out in the laboratory microwave unit, the LAVIS 1000 microMED, available from Milestone [24010 Sorisole (Bergamo), Italy] (Figures 4–6). The magnitude of the inflow of air at A (in Figure 4) determines the level of vacuum in the vacuum chamber C, which is placed entirely inside the microwave cavity B and drawn by the vacuum pump D. For the best obtainable vacuum we use an electric valve E to seal inflow completely. With this electric valve at position *closed*, one can easily draw a minimum pressure of 100 hPa. Fine tuning of the pressure is possible by manually manipulating the flow meter inlet A.

For our two vacuum–microwave steps, a specially designed vacuum chamber is used. It consists of microwave-transparent materials (partly polypropylene and partly Teflon) that have the required mechanical strength properties for low vacuum. The lid of the vacuum chamber contains an inlet for air (or, if desired, inert gas) and an outlet for air plus vapors. Tubes through holes in the walls of the unit chamber connect these valves with the atmospheric air (or inert-gas bottle), and the vacuum pump, respectively. Air (or inert gas) can enter the vacuum chamber through a tube that can be opened and closed by the operator or by an electric valve, which happens only in the last phase of the paraffin step. During the vacuum–microwave steps, the valve of the air flow is opened such that the vapors in the vacuum chamber (i.e., isopropanol) are rinsed out and greatly increases the efficiency of the procedure.

The instrument is equipped with software-regulating power output, temperature, and time (the LAVIS microMED). A temperature probe or an

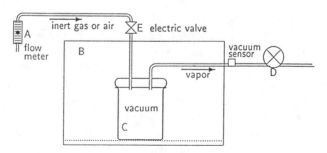

Figure 4. Schematic view of microwave-vacuum unit.

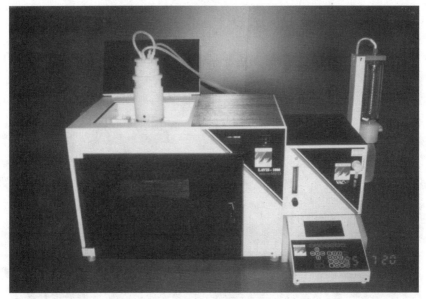

Figure 5. General view of the microwave unit microMED, closed.

IR sensor can be used during microwaving to control power level and hence temperature. Even distribution of the microwave energy over the specimen is achieved by using a Twist-Set turntable on which the vacuum chamber is fitted, moving the specimen during microwaving. Microwaving the specimen under vacuum conditions is achieved by using a vacuum unit, the VAC-60 Standard System, consisting of a pump outside the oven and connected to it the vacuum chamber closed by the Teflon lid (*see* Figures 4 and 6). The specimens in plastic cassettes are put in a specially designed holder. Each cassette contains one tissue block. The holder used can contain 30 cassettes, or 72 cassettes, depending on the chosen model (Figures 7 and 8). The loaded holder is put in a glass container fitting in the vacuum chamber. In each step, 400 mL of the reagent are brought into the glass container. The vacuum chamber is then closed and placed on the Twist-Set turntable in the center of the microwave unit. The glass container absorbs some microwave energy, and thus warms up particularly in the paraffin step (because paraffin itself absorbs little microwave energy).

Physics of Vacuum Microwaving

The use of vacuum in conventional histoprocessing technique seems confined to merely the paraffin step. Bancroft and Stevens (*14*) mentioned this option and advised a nonimmediate drop in pressure to a final value not lower than 400–500 hPa. By contrast, we do use vacuum in both histoprocessing steps, and we do go down further in pressure, to 100 hPa in the

Figure 6. General view of the microwave unit microMED, open, with vessel inside cavity showing thermocouple (temperature probe).

last part of the final step. The final pressure in our method is at least six times lower than the minimum pressure in commercial conventional microwave processors. Only at such low pressures evaporation of reagents is achieved.

Liquids like ethyl alcohol and isopropanol can be brought to boiling at relatively low temperatures provided pressure is sufficiently low (Figure 4). Therefore, we employ a vacuum chamber with pressures as low as one-tenth of atmospheric pressure. Evaporation requires energy. Therefore, to boil liquid, one has to supply heat. Only when sufficiently heated does the liquid start to form bubbles throughout. We use microwaves to heat the liquid under vacuum. In fact, microwaves can travel freely through vacuum, and have the additional advantage that they warm the relatively small objects under consideration in a uniform manner: that is, they warm the objects internally. This result explains why microwaves are preferable over IR waves for heating under vacuum, because with this type of radiation only the outer layer of the container will be heated.

The tissue samples that we want to dehydrate contain water and ethyl alcohol if it is used in the fixative. At normal pressure water boils at 100 °C, and ethyl alcohol boils at 78 °C (Figure 4). At sufficiently low pressure, the boiling temperature of water can be at or below physiological level. (This result can be read from Figure 4, and from Table I, where in addition we relate pressure to atmospheric height.) We used this fact in

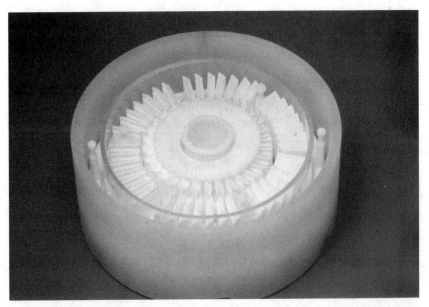

Figure 7. Vacuum container with glass holder. The cassettes containing the samples are placed in the glass holder.

previous work to dehydrate tissue by purely physical (i.e., nonchemical) means: no ethyl alcohol was used (*12*), and the physically dehydrated tissue could be placed directly into the plastic embedding medium. This method is of interest for research, but for routine diagnostic practice it is impractical because the samples must be precisely of the same size and consistency to achieve predictable results. Therefore, for our routine biopsies, initially we maintained dehydration with 100% ethyl alcohol (*13*). The size requirements in this case are less critical. From subsequent work and experience we learned that it is possible to delete the 100% ethyl alcohol step altogether, and to start immediately with the isopropanol bath, which is both a (weak) dehydrator and a clearing agent. Accordingly, vacuum–microwave histoprocessing could be brought back to the two steps described in this section.

When the fixed samples are placed directly in the isopropanol bath, they contain a mixture of water and ethyl alcohol or water and formalin. In this step we want to select temperature and pressure such that one obtains good diffusion of water in the two alcohols and vice versa, and enhanced dehydration. Therefore, we have chosen a combination of temperature and pressure so that we are close to the boiling point of this mixture, and we remain at such a low temperature that the tissue still can withstand the hazards of the treatment without adverse effects. We control the temperature at 55 °C by using the temperature probe and control the pressure at 400 hPa.

Figure 8. Detail of components (polypropylene container, glass container, and tetrafluoromethoxil (TFM) racks).

In the second step we use microwave exposure for 18 min at a 400-W maximum (Figure 1) to ensure a temperature of 60 °C (the melting temperature of paraffin is 56–58 °C). The pressure starts at 400 hPa and is brought down during the final phase to the low value of 100 hPa. During the last stage of the paraffin step, we have extensive formation of small gas bubbles of isopropanol (and the remnants of ethyl alcohol and water) in the fluid paraffin.

Temperature control relies on a well-established control technique. A sample temperature curve and report is shown in Figure 9.

In all steps we use the principle of flushing the vacuum chamber with air (which is microwave transparent). The air acts as an effective transport medium for the molecules we wish to be removed (water, alcohol, and isopropanol). The required pressure of 400 hPa is controlled quantitatively by setting the flow meter (A in Figure 4) at the appropriate level.

Classical Staining

In classical staining, cell and tissue components are bound to dyes. As a result, a component can be recognized by its color, for instance DNA by the blue color of hematoxylin (H) and protein by the red color of eosin (E). Accordingly, in routinely H and E stained sections a cell nucleus is blue, and the rest of the cell is imaged in red.

Table I. Boiling Temperature of Water at Various Pressures and Corresponding Atmospheric Heights

Temperature (C)	Pressure (hPa)	Height (Km)
140	3612	
120	1984	
100	1013	0
70	313	8.88
46	100	16.2

Staining of tissue is based on two factors: diffusion of the dye into the cells and binding of the dye to the substrate. In both, microwave exposure can be beneficial in speeding up the staining process. Diffusion is a physical process. This process can be enormously accelerated by microwave exposure. Binding of stains to cell substrates is a physical–chemical process, and the role of microwave exposure depends on various factors.

In the staining process, affinity is of paramount importance. Affinity is the tendency of the stain to move from the staining solution into the tissue. Any factor in the staining system, stain, and solvent cosolutes, which favors transfer of the stain into the cell, is a contribution to stain–cell affinity.

The simplest form of stain–cell attractive forces are the Coulombic forces that play a role in the binding of ionic stains to cellular substrates of opposite electric charge. An example is the interaction of azure B (in the Giemsa method) to DNA. These reactions are temperature dependent: at higher temperatures, the binding is accomplished in a shorter period of time. The amount of dye ion able to enter a given tissue substrate will depend not only on the sign of the charges of dye and tissue but also on their magnitude and, further, on the amount of nondye electrolyte present in the dye bath and on the ability of the tissue substrate to swell or shrink.

In the dye solution and tissue, the temperature rises rapidly when the microwave exposure is started. This rise results in shorter staining times. Charged molecules are excited by microwaves. So, polar and charged dye molecules (heavy metal ions!) react extremely well to microwave exposure. In fact, in the field of metallic staining the very first success of microwave technology was reported by Brinn (15). He reduced the time-consuming traditional periodic acid–methenamine silver stain according to Jones from 180 min to 20 min. Avid *microwavers* as we are, we further reduced staining time to 10 min by also performing the periodic-acid step in the microwave unit. More applications of metallic-staining methods were reported in the literature and in our cookbooks (1). In general, we observed that for metal-staining methods the microwaved slides had better contrasts, more intense staining, and less nonspecific staining compared with the conventional methods. Also for the nonmetallic stainings, we found superior results for the microwave methods. Particularly for the thin plastic sections, the increased signal-to-noise ratio is highly beneficial.

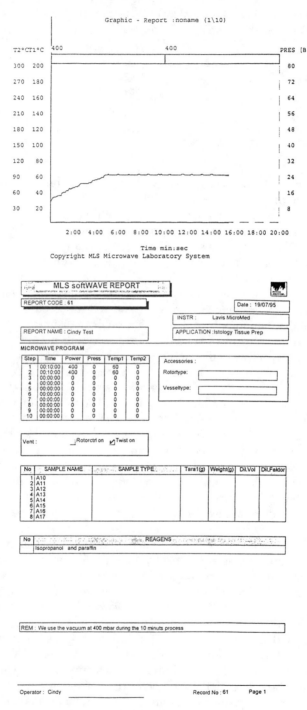

Figure 9. Temperature curve and report.

Papers on microwave-stimulated staining methods often do not mention the temperatures of the staining solutions. This deficiency is a pity, because we know that temperature is very important. For most staining procedures, there is a temperature optimum. Suurmeijer et al. (16) found in their laborious trial-and-error approach that for most nonmetallic stains the optimum is between 55 °C and 60 °C, whereas the optimal temperature for metallic stains is much higher (between 75 °C and 95 °C). For a third group (the Schiff reaction in the Feulgen staining method), it is wise to keep the temperature around 30°. For practical reasons, we mainly use a temperature setting of 60 °C, but for the Schiff reaction we have to use a low temperature setting of 30 °C. Where possible, we stain in the microwave unit with the temperature probe in the staining jar to define the temperature exactly.

In the literature one often encounters the expression *standing time*: that is, a period of time after exposure during which the slides are left in the staining solution. In our opinion there is nothing magical about standing time, it is merely some time of exposure to the heated solution. In some procedures this method is practical. This standing time adds to the staining time. In these cases, staining time is warming-up time plus keeping-warm time plus standing time.

Antigen Retrieval for Immunostaining

In immunostaining, the epitope of the protein in the sample (antigen) must bind to the (primary) antibody. Similar antigen–antibody reactions take place in life cells and tissues in immunological, physiological, and pathological processes; hence, the term *immunostaining*. Commercial firms bring monoclonal (very specific) and polyclonal (less specific) antibodies on the market for the immunostaining methods. In current diagnostic practice, immunostaining is widely used. Initially, immunostaining with monoclonal antibodies was only possible on cryostat sections that are not cross-linked by formalin. This result is no problem in research, but for diagnostic pathology immunostaining is only attractive if it can be performed on routinely formalin-fixed paraffin sections. Many immunostainings need special treatments of paraffin sections, such as enzyme digestion. The enzymes digest proteins or networks of proteins (produced by formalin fixation) that cover the epitope, and the thus-freed epitope is allowed to bind with the antibody. For this process the (in our eyes slightly incorrect) term *antigen retrieval* is commonly used.

Difficulties in standardization of enzyme treatments provided a powerful incentive for Shi et al. (17) to develop a new technique with the requirement that it should be more powerful, more widely applicable, and easier to use than enzyme digestion. These authors have reached these goals by developing the microwave antigen-retrieval method. Because they and many other authors have been so successful, microwave antigen retrieval has revolutionized diagnostic pathology. In 1995, in The Netherlands there is hardly a pathology laboratory left where microwave antigen retrieval is not practiced.

Shi et al. (17, 18) reported that their microwave antigen-retrieval techniques were triggered by biochemical studies indicating that the chemical reactions that occur between protein and formalin may be reversed, at least in part, by high-temperature heating or strong alkaline hydrolysis. Accordingly, they worked in alkaline solutions. Interestingly enough, later authors showed that for many antigens the method even works better if low pH values are employed. Since 1991, when the paper of Shi et al. came into print, more than 100 articles on microwave antigen retrieval have been written from all over the world and demonstrate its value for routine immunohistochemistry. More than 200 antibodies have been tested, and excellent immunohistochemical staining results have been obtained. For only a few antibodies, the use of the antigen-retrieval method showed no improvement. Microwave antigen retrieval can be extremely cost saving, because when applied, (expensive) antibodies may be further diluted. Shi et al. (17, 18) reported that some antibodies may be diluted to more than 1:1000! A breakthrough in diagnostic pathology is realized through microwave antigen retrieval allowing pathologists to *see* oncogenes, proliferation proteins, and hormone receptors in their routine paraffin sections.

Formalin fixation causes modification of antigens due to chemical alteration of their primary, secondary, or tertiary structure, such that the modified antigens can no longer bind to the antibody. Moreover, because of the induced changes in the protein containing the epitope, or cross-linking the neighboring proteins, the epitopes may be masked. Not all antigens react the same on formalin fixation. Particularly, the staining of intermediate filament proteins and leucocyte subsets is reduced or lost by formaldehyde fixation. Even in the group of keratins there are differences: basic, high-molecular-weight keratins (AE3 epitopes) are more susceptible than acidic, low-molecular-weight keratins (AE1 epitopes) to the masking effects of formalin fixation. In addition, the masking effects depend on length of fixation. Short formaldehyde fixation leads predominantly to loose binding (e.g., as Schiff's bases), whereas longer fixation results in tight links (e.g., as methylene bridges). By simple washing in buffer overnight, some epitopes in tissue fixed for a short time are already retrieved, probably because the loosely bound formaldehyde was removed by this relatively gentle treatment.

The retrieval solution is not the only important parameter. The temperature reached in the retrieval solution is also important. For most antigens, the solution should be brought to boiling, but for some the temperature recommended can be as low as 60 °C.

For each antigen, the best combination of antigen-retrieval solution (including its pH and molarity), heating time, and number of heating cycles must be established. Thus, no universal rules can be applied to *all* antigens. For some antibodies, prolonging the heating time over 20 min results in higher background staining. We experienced that it is of paramount importance not to cool the slides too fast after the microwave steps.

Retrieval Solutions

Lead and Zinc Salt Solutions

Shi et al. (*17*) placed their formalin-fixed paraffin sections in a metal-salt solution (lead thiocyanate or zinc sulfate) for microwave boiling. Among 52 antibodies tested with their method, 39 antibodies demonstrated increased immunostaining, 9 antibodies showed no change, and 4 antibodies showed reduced staining of the treated sections. These authors noted that

- microwave heating was more effective than conventional heating
- the best results were obtained when saturated lead thiocyanate solution or 1% zinc sulfate was used, although boiling in distilled water also had some effect
- their antigen-retrieval method appeared to be superior to enzyme digestion of tissue sections, in particular for the recovery of keratins
- in some cases background staining was seen with zinc sulfate but not with lead thiocyanate

Using BioGenex's antigen-retrieval system (lead thiocyanate), we found staining occurred with keratin antibodies (e.g., AE3), which otherwise performed poorly with routinely processed tissue even after trypsin digestion. Thus, for these antibodies we could substantiate the statement from Shi and his colleagues that for some antigens microwave lead antigen retrieval is superior. However, for other keratin antibodies, like AE1 and 5 D3, it was not superior to trypsinization.

Aluminum Salt Solution

Testing other metal salt solutions for antigen retrieval, we obtained very promising results with a 4% aluminum chloride solution. In formalin-fixed and paraffin-embedded tissue, the immunostaining results obtained with a 4% aluminum chloride solution were comparable to those achieved with the commercially available lead thiocyanate solution (BioGenex), at least as far as keratins and Vimentin are concerned. Recently we found that the immunoreactivity for CA 125 (detected with monoclonal antibody OC 125) is also strongly enhanced with the microwave–aluminum chloride method. In addition, we were successful in recovering the nuclear proliferation-associated proteins detected by MiB-1 and PCNA. Evers and Uylings (*19*) found that for the 5 antibodies they examined the aluminum salt solution gave the best results, followed by distilled water.

Citrate Buffer Solution

Cattoretti et al. (20) used citrate buffer for their method (0.01 M citrate buffer, pH 6, adjusted with 2 N NaOH). These investigators compared microwave boiling with enzyme digestion on formalin-fixed paraffin sections for more than 150 antibodies. Enhanced immunostaining was found for 43 antibodies. In addition, immunostaining of some very useful antigens, like a pan B-cell marker (HM 57) and the proliferation marker MiB-1, was now possible on paraffin sections. We tested citrate buffer pH 6.0 for the retrieval of MiB-1 in formalin and Kryofix-treated tissue and obtained excellent results in all cases. Without microwave antigen retrieval, MiB-1 gives negative staining even in tissue fixed in the ethanol-based fixative Kryofix. In this case, breaking of (formalin-induced) cross-links cannot be held responsible for the retrieval of MiB-1, because Kryofix is a noncross-linking fixative. Shi et al. (17, 18) reported that for MiB-1, the pH of the citrate buffer should be either above 6.0 or under 3.0. A maximal antigen-retrieval–immunohistochemistry staining of MiB-1 could be obtained by using a low pH (pH 1.0–2.0).

For several antigens such as L26, PCNA, AE1, EMA, and NSE, the pH of the citrate buffer solution is unimportant (18). Other antigens (e.g., MiB-1) show no staining at all between pH values 3.0 and 6.0, but strong staining under and above this range. Still other antigens (MT1 and HMB45) showed negative or very weak focally positive immunostaining with a low pH (1.0–2.0), but excellent results in the high pH range. In short, there are several essentially different pH patterns.

Formic Acid Solution

The first authors using an *acid* antigen-retrieval solution were Hashimoto and his co-workers (21), who used formic acid. With this method, they succeeded in staining kuru plaques at a 1:10,000 antibody dilution.

Distilled Water

Distilled water can also be used as an antigen-retrieval fluid. It is used in many experiments, but often proved slightly less effective than when special antigen-retrieval solutions are used (17).

Microwave and Autoclave Heating

Shin et al. (22) were the first to report that autoclave pretreatment of paraffin slides enhances immunoreactivity of formalin-fixed tissue. A household pressure cooker can also be used for antigen retrieval. Recently, we have used a pressure vessel in the microwave unit to bring the temperature to 120 °C in 30 s (*see also* Table I, where the pressure for this temperature is given). With this procedure, microwave antigen-retrieval time could be brought down from 20 min to a mere 2 min.

Microwave-Stimulated Enzyme Treatment

As stated earlier, immunoreactivity of certain antigens can be restored with enzyme digestion. A variety of enzymes including pepsin, trypsin, chymotrypsin pronase, and protease type XIV can be used for this type of antigen retrieval. Length of exposure and concentration of the enzyme solution should be adapted to the degree of tissue cross-linking, thus on length of formalin fixation.

The action of enzyme digestion may include removal of macromolecules hindering access to the epitope, and conversion of nonimmunoreactive precursors to immunoreactive forms. In principle, enzyme treatment can be enhanced by microwaving. Here, it is wise to keep the temperature at physiologic levels, although some enzyme incubations can be performed at the unphysiologically high temperature of 50 °C. The sections should be placed in a cuvette containing the enzyme solution, the solution brought to the chosen temperature in the microwave unit, and kept at that temperature for 2 to 5 min.

Immunostaining

In immunostaining various methods of binding are involved.

1. Antibody–antigen, which is a key and keyhole bonding. The bonds can be very strong in some reactants, but weak in others.
2. Chemical conjugation for the attachment of labels, such as peroxidase and colloidal gold or FITC in the fluorescence methods, for the formation of complexes such as the streptavidin–peroxidase complex.
3. Binding of protein A to the antibody.
4. Binding of the streptavidin to the biotin of the biotinylated antibody.

To obtain a reaction product, the reagents must diffuse into the tissue. To prevent background staining the unbound reagent must be washed out after incubation (also by diffusion). In diffusion, the large size of the molecules is a decisive factor. In particular the linking of the antibody to the antigens (incubation) requires often a very long time. In conventional immunostaining, incubations are often done overnight.

To the avid microwaver it is clear that immunostaining is a promising field, in which microwave exposure might produce a significant shortening of the procedure. All incubation, washing, and blocking steps seem to be good candidates for microwaving. It is even conceivable that by shortening the incubation times, the background staining can be diminished.

Leong and Milios (23) were the first to publish their findings. They used microwave exposure to accelerate the incubation of the primary antibodies against T and B lymphocytes in the ABC procedure. (A is avidin, B is biotin, and C is complex; avidin is linked chemically to the primary antibody, and biotin in its turn is chemically linked with the avidin–peroxidate.) Incubation of the performed avidin–biotin complex was done outside the unit. Chiu and Chan (24) accelerated three steps of the PAP (peroxidase antiperoxidase, *see* reference 1) method in the microwave unit.

1. incubation with the primary antibody
2. incubation with swine–antirabbit serum
3. incubation with the PAP complex

They achieved a minimal nonspecific staining and could work with higher dilutions. They warned that the sera should not boil and dry.

Notwithstanding these positive reports, we experienced that the application of microwave exposure is far from simple. The linking of antibody to antigen is often a very delicate process and many complexes prove to be unstable when the temperature becomes too high and when the microwave exposure too intense. Therefore, to perform immunoreactions in the microwave unit, certain precautions must be taken to keep the system in hand, and some steps (with instable complexes) must be performed outside the microwave unit.

The use of a water load in the microwave unit during incubation allows better control of temperature rise as a function of time. The greater the volume of water, the better the control. This control is needed when droplets are used for incubation. The quality of staining achieved in microwave incubations depends on the primary antibody. Here, the number of available antigens and their sensitivity to microwave exposure are important. Our experiences have taught us that primary antibodies can be divided into poor performers, which should not be used for microwave procedures (i.e., MiB-1), and good performers (i.e., Vimentin), which allow immunostaining to be performed within minutes instead of hours or days (25).

References

1. Kok, L. P.; Boon, M. E. *Microwave Cookbook for Microscopists—Art and Science of Visualization;* Coulomb Press: Leiden, Netherlands, 1992.
2. Marani, E.; Boon, M. E.; Horobin, R. W. *J. Neurosci. Meth.* **1994,** *55,* 111–117.
3. Marani, E.; Boon, M. E.; Adriolo, P. J. M.; Rietveld, W. J.; Kok, L. P. *J. Neurosci. Meth.* **1987,** *22,* 97–101.
4. Hsu, H. C.; Peng, S. Y.; Shun, C. T. *J. Virol. Meth.* **1991,** *31,* 251–261.
5. Ehrlich, P.; Lazarus, A. *Die Anämie I.;* Abt Holder: Wien, Austria, 1898.
6. Hopwood, D.; Yeaman, G.; Milne, G. *Histochem. J.* **1988,** *20,* 341–346.
7. Boon, M. E.; Gerrits, P. O.; Moorlag, H. E.; Nieuwenhuis, P.; Kok, L. P. *Histochem. J.* **1988,** *20,* 313–322.
8. Login, G. R. *Am. J. Med. Technol.* **1978,** *44,* 435–437.
9. Boon, M. E.; Kok, L. P.; Ouwerkerk-Noordam, E. *Histopathology* **1986,** *10,* 303–309.

10. Kok, L. P.; Boon, M. E. *Rapid Method for Cell-Block Preparation.* U.S. Patent 4 656 047, 1986.

11. Kok, L. P.; Visser, P. E.; Boon, M. E. *Histochem. J.* **1988,** *20,* 323–328.

12. Kok, L. P.; Boon, M. E. *Eur. J. Morphol.* **1994,** *32,* 86–94.

13. Kok, L. P.; Boon, M. E. *Histochem. J.* **1995,** *27,* 411–419.

14. Bancroft, J. D.; Stevens, A. *Theory and Practice of Histological Techniques,* 3rd ed.; Churchill Livingstone: Edinburgh, Scotland, 1990.

15. Brinn, N. T. *J. Histotechnol.* **1983,** *6,* 125–129.

16. Suurmeijer, A. J. H.; Boon, M. E.; Kok, L. P. *Histochem. J.* **1990,** *22,* 341–346.

17. Shi, S. R.; Key, M. E.; Kalra, K. L. *J. Histochem. Cytochem.* **1991,** *39,* 741–748.

18. Shi, S. R.; Gu, J.; Kalra, K. L.; Chen, T.; Cote, R. J.; Taylor, C. R. *Cell Vision* **1995,** *2,* 6–22.

19. Evers, P.; Uylings, H. B. M. *J. Neurosci. Meth.* **1994,** *55,* 163–72.

20. Cattoretti, G.; Dominoni, F.; Fusilli, F.; Zanaboni, O. *Histochem. J.* **1992,** *24,* 594.

21. Hashimoto, K.; Mannen, T.; Nukina, N. *Acta Neuropathol.* **1992,** *83,* 613–617.

22. Shin, R. W.; Iwaki, T.; Kikamoto, T.; Taeishi, J. *Lab. Invest.* **1991,** *64,* 693–702.

23. Leong, A. S.-Y.; Milios, J. *J. Pathol.* **1986,** *148,* 183–187.

24. Chiu, K. Y.; Chan, K. W. *J. Clin. Pathol.* **1987,** *40,* 689–692.

25. Boon, M. E.; Kok, L. P. *Micron* **1994,** *25,* 151–170.

Accessing Sample Preparation Information on the WWW

Chapter 15

SamplePrep Web

Analytical Sample-Preparation and Microwave-Chemistry Resource Center

Stuart J. Chalk, H. M. (Skip) Kingston,
Peter J. Walter, Kristen McQuillin, and Jason Brown

This chapter has several objectives. It introduces the Internet as a medium for communications on, and a resource for, analytical chemistry, in particular sample preparation and microwave-enhanced chemistry. It also covers some general information about accessing and using the Internet and the World Wide Web, as well as places of interest to the analytical chemistry community. A description of the SamplePrep Web site and associated Internet resources as well as the philosophy of its creation are also included. The discussion of the SamplePrep Web site is only meant to serve as a general guide because the site will evolve with participation of the users.

Just as life seems to be getting faster and distances seem to have shrunk, so to has the need for timely information become more compelling. With the growth in global communication, in particular the Internet, the need is being met and yet is still underused.

The fields of sample preparation and microwave chemistry are very interdependent. Traditional sample-preparation tools have remained virtually unchanged for over a century. They were considered state of the art until the application of microwave technology. Microwave-sample preparation, and now microwave-enhanced chemistry, is emerging as the new standard tool for effectively converting real samples into solutions that analytical instruments can accurately analyze. The use of microwaves is driving the field of sample preparation with new technology and improving the quality of many traditional sample-preparation digestions and extractions.

Sample preparation is becoming recognized as a legitimate area of specialization in analytical chemistry. It is a technology with new and rapidly developing equipment just as was formerly witnessed in the areas of chromatography and spectroscopy. Where chromatography and spectroscopy have matured and are more established, stable, and are recognized as subdisciplines in analytical chemistry, sample preparation is still being defined and is expanding and growing at an incredible rate. This specialized field has been created by the progress, development, and reliance on instrumental analysis.

In the second half of the twentieth century, a shift occurred in the fundamental nature of what we now call analytical chemistry. Previously single-analyte wet-chemical analysis procedures relied on hot-plate sample preparation. These procedures are now being replaced by sophisticated instrumental methods that require equally sophisticated sample preparation. The close of this century brings us both new sample-preparation and information tools meeting the needs of analytical chemistry.

As an ongoing experiment, we have created SamplePrep Web, a central resource site on the Internet, to take advantage of the immediacy and global nature of the information superhighway. To help the rapid developments in this area, it is important that we provide support to the field while it is growing and not to wait until it has matured and stabilized. Most scientists wish to contribute to their field of expertise and interest. SamplePrep Web was developed and put together with these ideas in mind. This concept guarantees that the site will change as it matures and as scientists from all over the world contribute to its knowledge base.

Included on the World Wide Web (WWW) site, and related Internet components of SamplePrep Web, is support for sample preparation from a variety of aspects such as sample dissolution, sample extraction, reaction chemistry, applications, synthesis, literature, standards, and clean chemistry. Eventually the site will be linked to an electronic journal on sample preparation that will help focus effort on this important area. This support is an electronic *library* put on-line to answer the perceived need for rapid communication of ever-changing ideas about sample preparation.

Sample-preparation technologies will hopefully mature more rapidly and with more stability if support is provided early, and this new global information resource is the perfect vehicle to accomplish this. It is time to experiment—the thing scientists do best. To that end, this chapter has been prepared to provide a guide to this growing resource in which you are invited to contribute and participate. It is truly an experiment and not a rigid institution. It will change and we welcome innovation and interaction. It will change as it meets the needs that now exist and it will develop as the fields of sample preparation and microwave chemistry develop. Our ideas of some of the areas that will be supported and developed are shown in Figure 1.

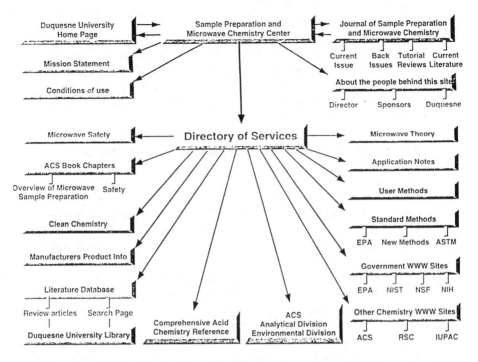

Figure 1. Outline of the SamplePrep Web WWW site at Duquesne University.

The following text has been written to introduce you to the Internet and get you *on-line*. It is not meant to be a thorough treatise on the Internet, but rather an introductory guided tour. The vastness of the Internet is covered much more comprehensively elsewhere (1, 2). As the technologies that this site supports mature, and as the technology it uses advances, the section related specifically to the SamplePrep Web site will become somewhat outdated, even obsolete by the time this book is published. As we write this, we ourselves cannot predict the end of this experiment, but can only provide the purpose for its inception. We invite all scientists to contribute as opportunity and time permit, and hope everyone enjoys and uses the site.

The Information Superhighway

What Is the Internet?

"What is the Internet?" is a question many people ask. More recently though, few people ask, "Where did it come from?" The Internet, or more simply Net, as it is called by those who frequent it, has so invaded people's everyday lives that many of those who use it cannot remember a time when it did not exist. A common misconception is that the Internet is a place. Literally, the Internet is a conglomeration of interconnected networks, hence the term *Internet*.

The Internet, as it exists today, represents a cooperative effort around the globe to establish a means by which information and ideas are freely shared. Ironically, the concept for wide area networking was born out of a military effort to distribute U.S. computing and scientific resources to make them less vulnerable to a single attack.

In 1957, the United States formed the Advanced Research Projects Agency (ARPA) as an arm of the Department of Defense (DoD) to develop technology applicable to the military. In 1969, the DoD commissioned ARPA to investigate networking. The first node on the ARPA network (ARPANET) was at UCLA with the Stanford Research Institute soon following.

The fundamental premise behind the ARPANET network architecture was "The physical layer of the network is unreliable". That may seem an odd basis for a networking design, but the original implementors were faced with the prospect of designing a communications system that could withstand outages at various points along the way (such as bomb attacks) and still function. Today these outages are more likely to be caused by construction workers digging up a communications cable than a bomb strike.

The concept of internetworking remote sites so appealed to researchers and scientists that the ARPANET began to expand to nonmilitary sites, mainly universities, and the Internet networking standard began to evolve. To that end, growth on the Internet has been explosive in the past 14 years. In 1981, there were 213 hosts connected via the Internet. At the beginning of 1995, that number had climbed to just over 4 million hosts.

This type of exponential growth fostered the development of all of the modern Internet utilities in use today including: FTP, Telnet, Gopher, WAIS, Usenet News, and finally, the World Wide Web.

Why Use the Internet?

Access to Information

The Internet is the largest computer network in the world. You can find information ranging from university catalogues to how to purchase groceries to how to write a program in C++ or html. You will find information relating to your profession, no matter what your discipline is.

You can read newspapers on-line, check the card catalog of the local library, read the latest White House press release, check the weather forecast, read a book or movie reviews, order flowers, visit an electronic bookstore, or get information on the latest software. A brief list of chemistry-related resources on the Internet is provided at the end of this chapter. Information and *links* to other sites that specialize in many areas of chemistry will be provided on the SamplePrep Web site so that this information is constantly up to date.

Global Communications

Over half of the world's nations have sites on the Internet. Even more have access to e-mail gateways to exchange electronic mail with Internet sites. You can communicate with friends and colleagues in Russia, India, Mexico, Europe... even Antarctica! Collaborating with researchers around the world could not be easier.

In addition to being able to send e-mail all over the globe, you can communicate through Netnews (discussion forums) or *live* via Internet Relay Chat (IRC) or Talk.

Research

With every major university in the United States and Europe connected to the Internet, you can be sure to find information that will help you with your research. Hundreds of libraries provide access to their card catalogs. Also, many free and commercial databases are accessible via the Internet.

Education

Using the Internet is a great learning experience. Not only will you learn things from the content of what you are reading, but you will become more proficient at navigating computer tools and forming searches for information. Your writing and typing skills may improve, too!

Business

The Internet can be an effective medium for business communications. Services such as Gopher and the World Wide Web are excellent ways to distribute information about products or services. E-mail is a great tool for communicating with business contacts and customers.

However, beware of violating the established *Netiquette*. Sending unsolicited e-mail (junk mail) to people, or posting your advertisement to every newsgroup will earn you a very bad reputation among the Internet community.

Personal Development

If you take advantage of all the Internet has to offer, you will broaden your horizons. You will have the opportunity to learn about different cultures from the people who live in them, meet people from different walks of life, and expose yourself to new ideas and opinions.

Entertainment

For some downtime you can enjoy the less academic aspects of the Internet. Enjoy talking with people? Try Internet Relay Chat! Do you like to play games? Connect to a Multi-User Dungeon, a chess match, backgammon, or fantasy game. Read the humor newsgroups! Surf the World Wide Web in search of graphics, sound, and movies.

How To Connect to the Internet

Of course, none of these tools is available to you without some sort of access to the Internet. Fortunately, access is becoming easier and easier to obtain.

If you are affiliated with a college or university, chances are good that you have access to a computer account with a dedicated Internet connection. Many businesses are beginning to provide Internet access to their employees; check with your computer systems department to find out what is available.

If you do not have Internet access at school or work, worry not. Hundreds of Internet service providers (ISPs) around the country specialize in offering affordable access to individuals and small businesses. Major on-line services, such as America On-Line, Prodigy, and others, also provide access to the Internet. *See* Tables I and II for a brief list of ISP contacts.

No matter who provides your access, you will need to determine what sort of access you have. There are several types.

Terminal Dial-In

The most basic service type, terminal dial-in, gives you text-based access to Internet tools. At home or work, with a computer and modem, you can connect to a machine that provides access to electronic mail, Telnet, file transfer, Netnews, Gopher, and the World Wide Web. The only requirements for terminal dial-in are a computer, modem, and software to dial the modem. Your ISP may be able to provide you with special graphical user interface (GUI) software to make navigating the Internet a little bit easier (Figure 2).

SLIP and PPP

A step up from terminal dial-in, SLIP (Serial Line Internet Protocol) and PPP (Point-to-Point Protocol) give your computer a direct connection to the Internet while you are dialed into your service provider. You will need a computer, modem, and a variety of software installed on your machine. Popular packages (at the time of this writing) include Internet-in-a-Box for Windows, and MacPPP (plus related client programs) for the Macintosh. When you use SLIP or PPP, Internet tools are installed directly on your

Table I. Internet Service Providers (ISPs) in the United States

National Service Providers	Access Number	Contact E-mail Address
AGIS (Apex Global Information Services)	313-730-1130	info@agis.net
ANS	703-758-7700	info@ans.net
Concentric Research Corporation	800-745-2747	info@cris.com
CRL Network Services	415-837-5300	sales@crl.com
Delphi Internet Services Corporation	800-695-4005	info@delphi.com
Global Connect, Inc.	804-229-4484	info@gc.net
Information Access Technologies	510-704-0160	info@holonet.net
Institute for Global Communications	415-442-0220	igc-info@igc.apc.org
Liberty Information Network	800-218-5157	info@liberty.com
MIDnet	800-682-5550	info@mid.net
Moran Communications	716-639-1254	info@moran.com
NETCOM On-Line Communications	408-554-8649	info@netcom.com
Netrex, Inc	800-3-NETREX	info@netrex.com
Network 99, Inc.	800-NET-99IP	net99@cluster.mcs.net
Performance Systems International	800-827-7482	all-info@psi.com
Portal Information Network	408-973-9111	info@portal.com
SprintLink–Nationwide 56K–45M access	800 817 7755	info@sprint.net
The ThoughtPort Authority Inc.	800-ISP-6870	info@thoughtport.com
WareNet	714-348-3295	info@ware.net
Zocalo Engineering	510-540-8000	info@zocalo.net

NOTE: For a more complete list of providers in different U.S. area codes, *see* reference 3.

SOURCE: Susan Estrad's *"Connecting to the Internet"*, published by O'Reilly and Associates, or visit **http:/romney.mtjeff.com/colossus/list/htm** from the Providers of Commercial Internet Access directory (POCIA). Copyright 1995 by Celestin Company, Inc. All rights reserved. To retrieve the complete list, send a blank e-mail message to info@celestin.com with the subject "SEND POCIA.TXT". For more information about Celestin Company and its products, send a blank message with no subject to info@celestin.com.

machine, giving you the advantage of pictures, sounds, and menus/ buttons that you are familiar with in Windows, Macintosh, or other graphical environment (Figure 3).

Dedicated Connections

Dedicated connections are most typically used by medium to large organizations to provide access for a number of people simultaneously. Dedicated connections require a server computer, special equipment to accept and route the Internet data traffic, and usually a staff person or two for administration (Figure 4).

Overview of Internet Tools and Definitions

Tools for Information Gathering

Telnet. Telnet is one of the three building blocks of the Internet. Telnet allows you to log into a machine on the Internet where you have an account, or one that provides a public service.

Table II. International Service Providers

Country	Service Providers	Access Number	Contact E-mail Address
Australia	AusNet Services Pty Ltd	+61 2 241 5888	sales@world.net
Austria	Net4You	+43 4242 257367	office@net4you.co.at
Belgium	Infoboard Telematics	+32 2 475 22 99	info@infoboard.be
Bulgaria	EUnet Bulgaria	+359 52 259135	info@bulgaria.eu.net
Denmark	DKnet/EUnet Denmark	+45 3917 9900	info@dknet.dk
Finland	Clinet Ltd	+358 0 437 5209	clinet@clinet.fi
France	French Data Network	+33 1 4797 5873	info@fdn.org
Germany	Point of Presence GmbH	+49 40 2519 2025	info@pop.de
Ghana	Chonia Informatica	+233 21 66 94 20	info@ghana.net
Greece	Foundation of Research	+30 81 221171	forthnet-pr@forthnet.gr
Hong Kong	Asia On-Line Limited	+852 2866 6018	info@asiaonline.net
Hungary	iSYS Hungary	+36 1 266 6090	info@isys.hu
Iceland	SURIS/ISnet	+354 1 694747	isnet-info@isnet.is
Ireland	Cork Internet Services	+353 21 277124	info@cis.ie
Israel	NetVision LTD.	+972 550330	info@netvision.net.il
Italy	ITnet S.p.A.	+39 10 6563324	info@it.net
Japan	Internet Initiative Japan	+81 3 3580 3781	info@iij.ad.jp
Kuwait	Gulfnet Kuwait	+965 242 6728	info@kw.us.com

Luxemburg	EUnet Luxemburg	+352 47 02 61 361	info@luxemburg.eu.net
Mexico	Internet de Mexico S.A.	+52 5 3602931	info@mail.internet.com.mx
Netherlands	Hobbynet	+31 365361683	henk@hgatenl.hobby.nl
New Zealand	Actrix Networks Limited	+64 4 389 6356	john@actrix.gen.nz
Norway	Oslonett A/S	+47 22 46 10 99	oslonett@oslonett.no
Poland	PDi Ltd. - Public Internet	+48 42 30 21 94	info@pdi.lodz.pl
Romania	EUnet Romania SRL	+40 1 312 6886	info@romania.eu.net
Russia	GlasNet	+7 95 262 7079	support@glas.apc.org
Singapore	Singapore Telecom Limited	+65 7308079	admin@singnet.com.sg
Slovakia	EUnet Slovakia	+42 7 725 306	info@slovakia.eu.net
South Africa	Aztec	+27 21 419 2690	info@aztec.co.za
Spain	Servicom	+34 93 580 9396	info@servicom.es
Sweden	NetGuide	+46 31 28 03 73	info@netg.se
Switzerland	Internet ProLink SA	+41 22 788 8555	info@iprolink.ch
Ukraine	Crimea Communication Centre	+380 0652 257214	sem@snail.crimea.ua
United Kingdom	Pavilion Internet plc	+44 1273 606072	info@pavilion.co.uk
Venezuela	Internet Comunicaciones c.a.	+58 2 959 9550	info@ccs.internet.ve

Source: Reproduced from the Providers of Commercial Internet Access directory (POCIA). Copyright 1995 by Celestin Company, Inc. All rights reserved. To retrieve the complete list, send a blank e-mail message to info@celestin.com with the subject "SEND POCIA.TXT". For more information about Celestin Company and its products, send a blank message with no subject to info@celestin.com.

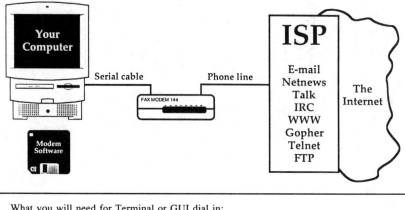

What you will need for Terminal or GUI dial in:	
Computer	Macintosh™, PC™ (DOS™ or Windows™), Amiga™, etc.
Modem	2400 bits per second (bps) minimum
Software to dial modem	ISP's GUI software, Procomm, Microphone, Telix, etc.
Account with an ISP	See Table 15.1

Figure 2. Schematic of terminal or GUI dial-in to the Internet.

FTP. File-transfer protocol (FTP) is the second of the three building blocks of the Internet. FTP allows you to move files from one Internet machine to another.

Gopher. A menu-based Internet information tool that was developed at the University of Minnesota. It combines the functions of Telnet and FTP into an easy to use tool that novices find more friendly than the somewhat cryptic building blocks. In addition to providing Telnet and FTP links, Gopher also includes a search program called Veronica (Very Easy Rodent-Oriented Network Index to Computerized Archives) that does keyword searches of the items in *gopherspace*, the collection of information servers that make up the whole of Gopher.

World Wide Web. The World Wide Web (WWW), as will be discussed later, is the most recent addition to the Internet information retrieval tools. Its popularity, ease of use, and functionality have made it the medium of choice for researchers, companies, and individuals who wish to "stake their claim" on the Internet. Because literally hundreds of new WWW documents appear daily on the Internet, several efforts have been put forth by different groups to organize the information on the WWW into some type of searchable form. Some of the more popular WWW indexes can be found at the following locations:

- *Yahoo*: http://www.yahoo.com/ A subject-oriented listing of Web resources
- *EINet's Galaxy*: http://www.einet.net/ Another subject-oriented list of Web servers

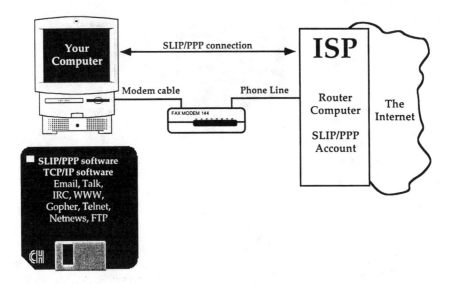

What you will need for SLIP/PPP dial-in:

Computer	Macintosh™, PC™ (with Windows™), Amiga™, etc.
Modem	14,400 bits per second (bps) minimum
Internet software	Internet-In-a Box, FreePPP, Chamelon, Trumpet Winsock, etc.
SLIP/PPP Account with an ISP	See Table 15.1

Figure 3. Schematic of SLIP/PPP dial-in to the Internet.

- *Alta Vista*: http://www.altavista.digital.com/ Web and Usenet information, 11 billion words, 22 million pages
- *GNN's Whole Internet Catalog*: http://www.gnn.com/ Yet another subject-oriented listing of information resources
- *WebCrawler*: http://www.webcrawler.com/ America Online's Web search engine
- *Lycos, Catalog of the Internet*: http://www.lycos.com/ Carnegie Mellon's searchable database, 34.1 million uniform resource locators (URLs)
- *DejaNews*: http://search.dejanews.com/ Large Usenet news search engine

The WWW offers access to documents, sounds, graphics, movies, and animation. Many graphical browsers exist for the World Wide Web; some popular ones include *Netscape*, *Mosaic*, and *Microsoft Explorer*. On the Web, everything is laid out in a HyperText format. Instead of selecting items from a menu, as you do with Gopher, you choose highlighted words (hotlinks) to connect to what you want. With graphical interfaces, you can also

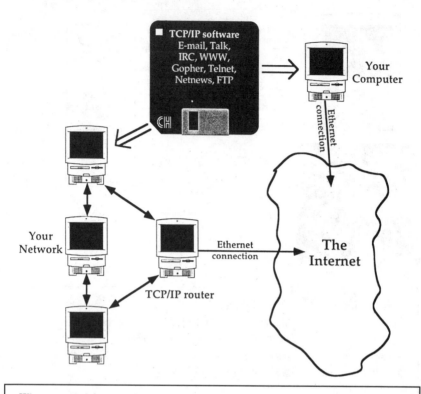

Figure 4. Schematic of dedicated Internet connection.

click on pictures to establish links. Information in the Web is specified by using addresses called URLs. A URL is a uniform resource locator and can point your Web browser to information from Telnet, FTP, gopher, and other Internet sites. For a more extensive introduction to the WWW, *see* the section entitled *Navigating the World Wide Web*.

As we move into the twenty-first century, information is fast becoming a very valuable commodity. The Internet's ability to deliver that information in a content-rich environment has been the primary reason for its growth and popularity. It is revolutionizing publishing and will change communications the same way the printing press did.

Tools for Communication

Electronic Mail. E-mail is the final of the three building blocks of the Internet. It gives you the ability to send messages, memos, and other text to individuals and groups on the Internet and other networks. Mailing lists are a special e-mail service that let a group of people with a similar interest correspond with one another as a group.

Netnews. A generic term for a collection of public discussion forums. Each forum or newsgroup focuses on a single topic. Millions of people worldwide have access to Netnews and participation occurs from the novice to the expert levels. A newsgroup is much like a mailing list, except that it is public and you use a tool called a newsreader to access the messages, rather than your e-mail program.

Talk. This program allows you to converse directly with an individual on the Internet. You start the talk program and give it an e-mail address. If the person whose e-mail address you enter is logged into his or her account, the talk program on their machine will alert them that you would like to converse. They can then connect to you and you both share a live, typed conversation.

Internet Relay Chat. Internet Relay Chat, known more commonly as IRC, is a live group communication tool, much like a citizens band radio for typists. When you use IRC, you can talk to anyone who is on the same channel as you—from one person to hundreds of people. Channels can be made private or invite-only, allowing you to use IRC as a virtual classroom or discussion session.

Navigating the World Wide Web

The WWW is a vast expanse of interconnected vessels of information residing on the Internet. Therefore, the utility and usefulness of the WWW are often measured against how easily one can move from site to site in a manner that best suits an individual's train of thought.

To that end, a plethora of different WWW browsers has appeared spanning nearly every computer system and operating system available today. Two of the most popular WWW browsers currently in use are *Netscape* and *Internet Explorer*. The semantics of all WWW browsers are common to all, so this section will focus on *Netscape*, the most common.

Both *Netscape* and *Internet Explorer* are available for the Apple Macintosh, PC-compatible computers running Microsoft Windows, and selected UNIX systems (e.g., Sun and Silicon Graphics). *Mosaic* is also available for the Virtual Memory System VMS operating system using the X-windows graphical user interface.

Using a WWW Browser

To use a WWW browser, you must have the following essential elements:

- computer and operating system that support one of the current WWW browsers
- Internet connection that can either be a true ethernet network connection or a dial-up SLIP (serial line internet protocol), or PPP (point-to-point protocol) connection
- WWW browser installed on your computer

For Macintosh or Microsoft *Windows* users, the typical method of launching *Netscape* is to locate the icon that represents the application, point to it with the mouse, and double click. *See* Figure 5 for an example of how *Netscape*'s application icon typically shows up in Microsoft *Windows* or on an Apple Macintosh.

Netscape may be obtained either directly from the developers, Netscape Communications Inc. or via the Internet by using the FTP protocol at the site ftp.netscape.com. Fully functional copies of the software are available free of charge for educational institutions. Commercial and individual users are charged a small fee.

The WWW is organized by *sites*. Sites store information for you to browse, retrieve, or print. The information stored at the sites is organized into pages and hereafter they will be referred to as *Web pages*. Each site's opening or welcome page is generally referred to as its *home page*.

When *Netscape* is initially launched, it creates a window and displays a starting Web page. This page is usually the home page for Netscape Communications Inc. (Figure 6).

We will now take a minute to get familiar with a few of the most commonly used controls in *Netscape* and then explore the *Analytical Sample Preparation and Microwave Chemistry Center*, a.k.a. *SamplePrep Web*, at Duquesne University.

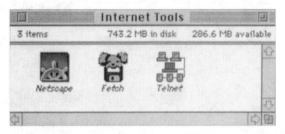

Figure 5. Netscape application icons for Windows and Macintosh.

To successfully use a WWW browser, there are only a few concepts to become familiar with. As was mentioned earlier, the WWW is broken up into discrete WWW servers scattered about the Internet. Each of these servers contains a number of pages which are interconnected in a Hyper-Text format. Before you can begin exploring the information at a WWW site, you must know its address and understand how to convey that address to a WWW browser.

Uniform Resource Locators

WWW sites are located on the Internet by using the URL address format as mentioned previously. URLs consist of a *type* of service, a *hostname* where that service is available, and the *location* of a specific file at the site. For example, the URL for the SamplePrep Web server at Duquesne University is composed as follows:

http://www.sampleprep.duq.edu/sampleprep

This format is easily understood by looking at the individual components present in this URL.

http://

This portion is the type of service. HTTP stands for HyperText transport protocol. HTTP is the protocol WWW servers use to exchange information with WWW browsers. Other types of services that you might encounter are ftp://, for file-transfer-protocol, gopher://, for the Gopher protocol, and news://, for the Usenet-news protocol.

Figure 6. Netscape home page

www.sampleprep.duq.edu

This portion is the host that is serving the WWW information. Hostnames in URLs adhere to the normal rules for Internet hostnames.

/sampleprep

This portion is a pathname that locates the home.html Web page on the www.sampleprep.duq.edu Web server.

Now, let us take a look at the *Netscape* controls and find out where to enter the URL (Figure 6). The *Netscape* interface provides a field-labeled location (Figure 7) where you may enter a valid URL. Alternatively, you can select the *Open Location* option located under the *File* menu item in the *Netscape* menu bar or click the *Open* button located in the first row of buttons beneath the menu bar. *See* Figure 8 for an example of the SamplePrep Web home page.

Bookmarks

Unless you frequently use only a few URLs, they can be difficult to remember. To counteract this problem, *Netscape,* as well as other WWW browsers, allows you to create a list of *bookmarks* for Web pages that you think you will revisit. To create a bookmark with *Netscape,* you select the menu option *Add Bookmark* under the *Bookmarks* menu item in the menu

Netscape: SamplePrep Web(TM) Home Page
Netsite: http://www.sampleprep.duq.edu/sampleprep/

Figure 7. Netscape controls.

Figure 8. The SamplePrep Web home page.

bar. This action will store the URL of the page that you are currently viewing in your list of bookmarks. To view your bookmarks, select the menu option *View Bookmarks* under the *Bookmarks* menu item (Figure 9).

The bookmarks window stores the title of each bookmarked page along with its URL. Once you have created a bookmark, you may return to that page by using a couple of different methods. You may double-click the title of the page, or single-click the title to select it, and then click the *Go To* button at the top of the bookmarks window.

HyperText Links

Web pages are created by writing files using the HyperText markup language (HTML). As such, Web pages generally contain regions of text or images that serve as links to other locations within the current document; to other documents in the current WWW server; to documents on other WWW servers; or to files that may be retrieved such as graphics, sounds,

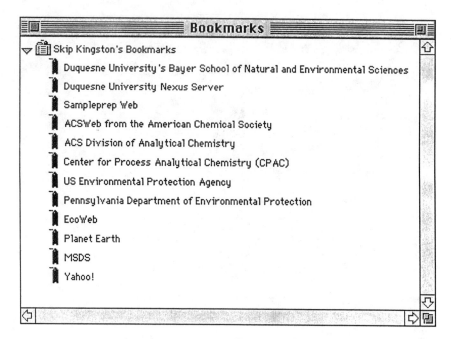

Figure 9. Netscape bookmarks window.

animations, application software, or other data. Web pages themselves may consist of any combination of text, graphics, and HyperText links.

The SamplePrep Web home page (Figure 8) contains a banner graphic, some introductory text, and links near the bottom to access the main directory, and take you to other pages on the SamplePrep Web server. HyperText links usually stand out visually from the rest of the surrounding text by using a different text color or underlining the text. To follow one of the links, place the mouse over the text that represents the link and single-click. Clicking on the *SamplePrep Web Directory of Services* link at the bottom of the SamplePrep Web home page retrieves the *Directory of Services* page (Figure 10).

Once you begin to follow HyperText links within Web pages, it becomes necessary to be able to backtrack through pages you have seen so that you can take different branches on a WWW server. Good Web page designs often will have a link at the bottom of each page that gives you several navigation options. Those options are typically, *previous page, home page,* or *next page.* At the bottom of the directory page on the SamplePrep Web WWW server you will find a link to the *SamplePrep Web home page.* Clicking this link will return you to the SamplePrep Web page you accessed first. In the absence of these types of links on a page, *Netscape* provides you with controls for performing the forward and backward navigation of Web pages.

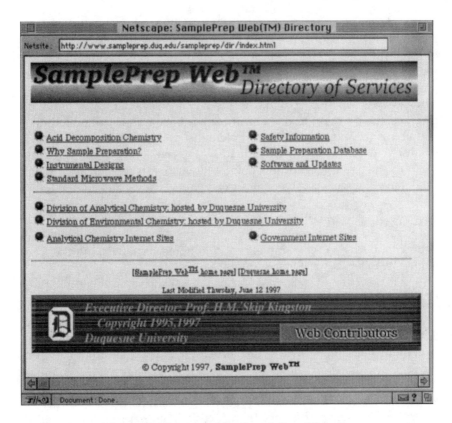

Figure 10. Directory of the SamplePrep Web site.

At the top of the *Netscape* window, just below the menu bar, is a row of buttons that contains buttons labeled *Back, Forward,* and *Home.* The *Back* button takes you to the last page you were viewing. Once you have begun to move backward through pages that you have seen, the *Forward* button will take you forward again. The *Home* button returns you to the starting page that *Netscape* displayed when the application was initially launched.

How To Use the SamplePrep Web Site

Accessing SamplePrep Web

Access to the WWW site is only available by using a WWW browser such as *Netscape, Mosaic,* or *Internet Explorer* as described in the previous section. On your first visit to SamplePrep Web (http://www.sampleprep.duq. edu/sampleprep) you will see our home page. You can look at the directory of services, background, organizers, mission statement for SamplePrep Web as well as accessing Duquesne University's home page.

In addition to the Web site, you can go to ftp.sampleprep.duq.edu and go into the /pub/sampleprep/epa subdirectory where copies of a number of standard U.S. Environmental Protection Agency (EPA) methods are located. If you need help with a particular question regarding sample preparation or microwave chemistry, you can send us an e-mail message at info@www.sampleprep.duq.edu. We will answer your query either by directing the question to one of our technical experts or suggesting that you access the Web site for a complete discussion of the topic you are interested in. Please bear in mind that this information is written based on the Web site as it stands at this point and when you read this text it will likely have changed as it continues to grow.

Entering the Main WWW Site

After connecting to the SamplePrep Web home page, click on the *SamplePrep Web Directory of Services* link.

Directory of Services

Navigating through the SamplePrep Web site is very easy. Clicking highlighted text in a document takes you to other places in the site and in each document you can return to the directory. A brief summary of our expected sections of the site is listed next.

- *Acid Decomposition Chemistry.* An extensive compilation of acid chemistry that is necessary for acid decomposition and element stabilization in sample preparation, discussed in detail later (3).
- *Why Sample Preparation?* Discussion on the principles of sample preparation.
- *Instrumental Designs.* Discussion and examples of the fundamental instrument and vessel designs and commercial instrumentation in microwave sample preparation.
- *Standard Microwave Methods.* Compilation and summary of standard and proposed standard microwave methods worldwide including AOAC, ASTM, EPA, and other national and international standard methods. When available, the method can be downloaded in several different computer text formats.
- *Safety Information.* Safe use of microwave digestion vessels. Vessel design, manufacturers, materials, and specifications. Case studies of accidents involving microwave equipment. Reagent and reagent mixture safety information.
- *Sample-Preparation Database.* Listings of review articles, courses, and books on aspects of sample preparation and microwave chemistry.
- *Software and Updates.* Complementary microwave software aids. Updates to existing microwave instrumentation software.

- *Division of Analytical Chemistry.* The American Chemical Society (ACS) Division of Analytical Chemistry *hosted by Duquesne University.* Hotlinks to analytical chemistry related sites.
- *Division of Environmental Chemistry.* The American Chemical Society (ACS) Division of Environmental Chemistry *hosted by Duquesne University.* Hotlinks to environmental chemistry related sites.

Following is a list of potential sites in the future.

- *Clean Chemistry for Sample Preparation.* Practical design of clean apparatus. Production of clean acids and reagents. Appropriate storage containers and levels of extractable ions. Analytical procedures for working in a clean environment.
- *Literature.* Access to full text copies of articles of interest. Searchable bibliographic database of sample preparation and microwave chemistry.
- *SamplePrep* Electronica. Access to a completely electronic journal. Electronic reprints from the on-line microwave journal. Sample-preparation standardization software. Microwave calibration software. Standard method software
- *Questions and Answers.* Frequently asked questions (FAQ) list. Answers to specific user questions. E-mail link to info@www.sampleprep.duq. edu for sending specific questions to experts around the world (e.g., safety).

Acid Decomposition Chemistry

The acid decomposition chemistry site is an electronic and continually expanding version of the dissolution reagents section of Chapter 2 (3). The information is a compilation of acid chemistry information from the literature. Although it was in part derived from literature on classical acid decomposition literature, it is critical for the development of both open and closed vessel microwave and classical acid digestion method development. This information has utility for all types of environmental, industrial, medical, and analytical applications. This site has the highest information density currently available and is reminiscent of data compiled during the Manhattan Project.

The site is constructed to be continually expanded and improved. Recommendations and contributions to the database from scholarly acid chemistry literature will be integrated and acknowledged. Please submit scholarly and preferably peer reviewed data and references for inclusion. If the reference is in a language other than English, please send a copy of the paper and if possible an English translation.

The site was originally compiled by Peter J. Walter as part of his doctoral dissertation. The goal is to provide chemists throughout the world with a key reference source. The vision is shared by H. M. (Skip) Kingston and the data's sponsor, Milestone Inc.

When *Acid Decomposition Chemistry* is selected from the SamplePrep Web's *Directory of Services,* a general introduction and description of the site appears. Page down and select *Go To Acid Decomposition* to continue to the main screen of acid decomposition chemistry, *see* Figure 11. From this page the acid decomposition chemistry can be viewed in three distinct forms. By using a mouse, the pulldown menu, in Figure 11 labeled *Periodic Table of Elements,* can be activated and one of the reactivity properties of the acids can be selected. For example, selecting *Volatile from HF* brings up Figure 12. A total of 21 figures that cover various reactivity properties of the entire periodic table with either nitric, hydrochloric, hydrofluoric, sulfuric, and perchloric acids or hydrogen peroxide are selectable. At the bottom of Figures 11 and 12, there are icons for the left and right sides of a periodic table. Selecting one of these icons will bring up a window with that side of the periodic table with all of the reactivity parameters for the specific acid. These tables are electronic versions of Figures 10–15 in Chapter 2. The third view is activated by selecting an individual element, which brings up a window as in Figure 13. This time the entire acid chemistry information is displayed for all reagents for the selected element.

The majority of this information is text and graphics that you browse through. However, some components (still in development) require input from the user as described below.

Software, Demos, and Updates

Listings of current versions of software, demos, and software updates will be hotlinked to the files on the SamplePrep Web FTP site. Clicking on the name of the software will bring up a brief description of the file, the operating system it works on, and its size in kilobytes. Click the download link to transfer the software to your machine (this link is available on nearly every Web browser). When downloading shareware we encourage you to register the software and support the evolution of inexpensive, useful software. (Note: Software will be checked for computer viruses before it is added to the software library.)

Database Searching Option

Several options will be available to the user in the future. You will either decide to read the review articles on sample preparation and microwave chemistry or search the database of sample-preparation bibliographic information. Searches are performed by entering selected keywords in the keyword field or combining them with *AND, OR,* and *NOT* (simple Boolean logic). Parentheses can be used to nest Boolean searches.

Search Example:

Find references containing EPA Method 3050 or 3052 in *Talanta* for all years except 1994

Search string: ((3050 OR 3052) AND EPA AND Talanta)

NOT 1994)

Information available to be searched will include the journal title, article title, authors, year, volume, issue, page numbers, and keywords. No abstract information is currently available.

Successful matches to the search string will be displayed in a new document with the title of each article highlighted as a hotlink. You can scroll through the list and decide which references you need to look at and download a copy of this list to your local machine (most Web browsers support this feature). In the future, it is planned that clicking on a title of interest will bring up a brief summary or bibliographic information and keyword information. Potentially, a FULL TEXT button will take you to the full text copy of the article if you have set up an account to obtain this

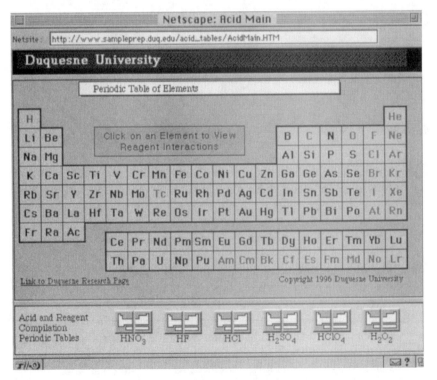

Figure 11. The Acid Decomposition Chemistry home page.

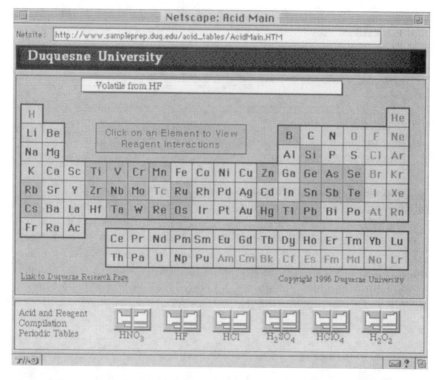

Figure 12. Compilation of elements reported in the literature to be volatile from hydrofluoric acid.

service (checking the box on the registration form). Information on using the full text service will be on-line when these become available.

Questions and Answers

If you have new information to contribute, wish to report a safety problem, or if the question you have cannot be answered by reading either the FAQ or user questions and answers, you can send an e-mail message to our information hotline info@www.sampleprep.duq.edu. At the bottom of any document, click on the graphic, enter your e-mail address, the subject, and a brief message, then click SEND. If the question requires an expert in a specific area, one of our panel of experts will be forwarded your inquiry and will respond to your question directly as soon as he or she is able. If the question is pertinent enough, it and the reply may be added to the user's question list in the FAQ section. (Note: Check to see if your Web browser supports sending e-mail messages directly. If not, you can always send us an e-mail message at the above address by using a separate e-mail application.)

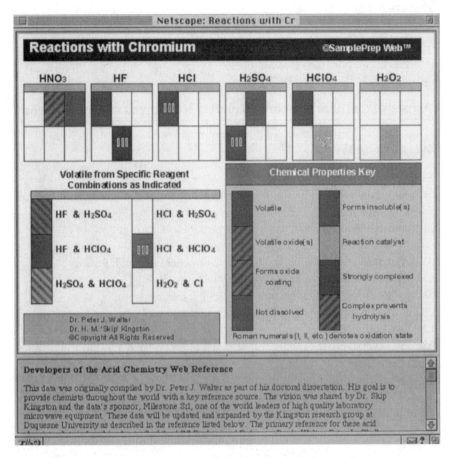

Figure 13. Reactivity characteristics of chromium with six common dissolution reagents.

SamplePrep *Electronica*

The Web version of a new journal is planned to be published and will be affiliated with the SamplePrep Web site. It may also be published in hard copy form, but will not contain the additional electronic components. You will be able to download copies of the articles by using your Web browser or print them out if you wish. You will only be able to get to this area if you have paid for a subscription with the collaborating publisher.

Again, please be aware that the above information is based on how we envision the site to be at the time of writing. It will undoubtedly be somewhat different in its actual production and implementation. Visitors are advised to read any updates to this chapter that are provided through the site. Please participate in this experiment.

References

1. The internet Society. http://www.isoc.org/ (accessed May 1996).
2. Zen and the Art of the Internet. http://www.cs.indiana.edu/docproject/zen/zen-1.0_toc.html (accessed May 1996).
3. Scott, F. J., Jr. *Am. Lab.* **1997**, *April 20.*

Appendix. Chemistry and Related Resources on the Internet

World Wide Web

SamplePrep Web	http://www.sampleprep.duq.edu/sampleprep
WWW Virtual Library	http://www.chem.ucla.edu/chempointers.html
Clearinghouse for Chemical	http://www.indiana.edu/~cheminfo/
Instructional Materials	ACS http://www.acs.org/
	http://pubs.acs.org/
ACS Polymer Division	http://www.chem.umr.edu/~poly/
Chemical Abstracts Service	http://www.cas.org/
Society for Applied Spectroscopy	http://esther.la.asu.edu/sas/
OCLC	http://www.oclc.org/
Chemistry Abroad	http://www.ch.cam.ac.uk/ChemSitesIndex.html
MSDS (Material Safety Data Sheets)	http://www.enviro-net.com/technical/msds/
	http://www.chem.uky.edu/resources/msds.html
	http://www.msc.CORNELL.edu/helpful-data/msds.html
	http://www.its.ilstu.edu/chemsafety/msds.html
	http://haz1.siri.org/msds/

Gopher

ACS	gopher://acsinfo.acs.org/
U.S. EPA	gopher://gopher.epa.gov/
National Institute for Standards and Technology Publications	gopher://gopher-server.nist.gov/
Science and Technology Information Center: National Science Foundation	gopher://stis.nsf.gov/
National Institutes of Health	gopher://gopher.nih.gov/

Electronic Mail

*   SamplePrep Web: Information and questions to info@www.sampleprep.duq.edu
*   CHEME-L, the Chemical Engineering Mailing List: Send e-mail to listserv@psuvm.psu.edu. In the body of the message, type: subscribe cheme-l firstname lastname.

- CCL, Computational Chemistry Mailing List: Send e-mail to oscpost@oscsunb.osc.edu. In the body of the message, type: send help from chemistry.
- Process-L, Sampling and Analysis Process Mailing List: Send e-mail to maiser@fs4.in.umist.ac.uk. In the body of the message, type: subscribe process-l.

Usenet Newsgroups

alt.drugs.chemistry	Pharmaceutical chemistry
sci.chem	Chemistry
sci.chem.analytical	Analytical chemistry
sci.chem.electrochem	Electrochemistry
sci.chem.organomet	Organometallic chemistry
sci.techniques.spectroscopy	Spectroscopy
sci.engr.chem	Chemical engineering

FTP

SamplePrep Web	ftp://ftp.sampleprep.duq.edu/
Computational Chemistry Archive	file://infomeister.osc.edu/pub/chemistry
Diane Kovacs AcadList files	ftp://ftp.cac.psu.edu/pub/internexus/
Sheffield Chemistry Software Archive	ftp://ftp.shef.ac.uk/pub/uni/academic/A-C/chem

Note: The above sites/URLs were checked as of May 31, 1997, but may have changed at the time you read this text.

Laboratory Microwave Safety

Chapter 16

Laboratory Microwave Safety

H. M. (Skip) Kingston, Peter J. Walter, W. Gary Engelhart, and Patrick J. Parsons

The unsafe conditions and practices that may occur in the chemical laboratory related to operation of microwave systems are evaluated. A detailed discussion of relevant equipment standards, safety-code requirements, and general safety guidelines relating to laboratory microwave systems is presented. Laboratory safety considerations, including chemical hazards that have been identified in the literature, are extensively discussed. Case studies are presented for specific types of accidents that have previously occurred to assist in preventing repetition of similar incidents. A mechanism for periodic updating of safety information, including the capability for microwave system users to input their own contributions and for others to access this information, is provided.

The majority of unsafe conditions or practices that can arise during use of laboratory microwave systems are avoidable. As described throughout this text, many diverse chemical procedures are now performed in either atmospheric-pressure or closed pressurized-vessel microwave systems. Microwave techniques introduce unique safety considerations that are not encountered by the analyst in other nonmicrowave methods. Differences in operating conditions between traditional laboratory practices and microwave-implemented methods should be examined before applying microwave energy to heat reagents or samples.

Accepted procedures and good laboratory practices pertaining to microwaves are scattered throughout the literature, as are safety suggestions for using reagents to decompose samples, synthesize compounds, and extract analytes. Complete coverage of this literature and the range of possible sample types is beyond the scope of this chapter; however, general guidelines and some specific examples of unsafe conditions previously reported are provided. Microwave-enhanced chemistry is not

697

exempt from traditional safety considerations; sources on chemical laboratory safety should be consulted for documentation of explosive mixtures, toxic chemical behavior, and reagent-handling precautions. Examples of literature sources that can be consulted are as follows: for specific mineral acids decomposition reactions, references 1–4 and Chapter 2 of this book; and for flash points, autoignition points, and flammability hazards of solvents used in synthesis and extraction, references 5 and 6. These hazards are temperature-dependent and specific conditions are documented to prevent these situations. Lists of incompatible chemical combinations are available (7–9), and general chemical and laboratory hazard lists have been compiled to assist in safety evaluations (10–12).

Specific standards that apply to microwave laboratory equipment are evaluated extensively as they relate to the apparatuses used in microwave-enhanced chemistry. Previous instances of serious, extreme, and explosive reactions that have occurred by using laboratory microwave equipment are included to prevent repetition of these known hazards. Other safety concerns unique to microwave systems such as the direct effects of microwave energy and special design and performance characteristics are presented to prevent inappropriate equipment configurations and usage.

Microwave interactions mechanisms producing heat were described in detail in Chapter 1. Microwave energy is absorbed not only by polar solutions (e.g., mineral acids, organic solvents, reactants, and aqueous mixtures) producing heat and accelerating chemical reactions, it is also absorbed by some sample molecules, container materials, and surfaces of an apparatus that may not be intended to heat during a reaction. Microwave energy at 2450 MHz may also be absorbed by mammalian tissue. All laboratory microwave equipment is designed to shield the analyst and prevent such exposure.

Safety Standards and Regulations

Microwave Exposure

Standards, limits, and ranges of tolerance have been established for microwave radiation exposure in most of the industrial world. The United States, Russian Republics, Germany, Belgium, Denmark, France, Italy, United Kingdom, Poland, the former Czechoslovakia, Canada, Australia, Sweden, European Economic Community, military and governmental organizations in these countries, as well as international organizations have all established safety standards (13, 14). An underlying reason for the large number of exposure standards is the manner in which they are defined (e.g., by electromagnetic energy frequency, duration of exposure, body mass, and time or periodicity of exposure).

Studies into the biological effects of microwave radiation exposure have been extensively detailed (~1000 references) in several reviews dealing with scientific, industrial, and medical applications (*13–15*). Overall, the effects on human tissue are thermal in nature and relate to overheating of exposed tissue. The underlying protective principle of several standards is derived from data on the amount of energy necessary to raise human skin and tissue temperatures to biologically significant levels. Exposure to energy such as sunlight is basically a surface phenomenon; however, microwave energy penetrates the skin into subcutaneous tissue and therefore also raises the temperature level of tissue and blood (*13–15*).

Table I provides an indication of microwave half-power penetration depth in a dielectric material such as tissue and the energy associated with particular frequencies. Energy variations among frequencies is the major reason why there is not a single standard for exposure to microwave energy (different frequencies of microwave energy penetrate to different depths and result in different amounts of energy being absorbed).

A diagram (Figure 1) has been prepared to illustrate a single unified set of exposure criteria based on the American National Standards Institute (ANSI), American Conference of Governmental Industrial Hygienists (ACGIH), and the International Radiation Protection Association (IRPA) standards (*16*). As can be seen, the standard is frequency- (wavelength-) specific, including the three most commonly used laboratory (and commercial) microwave frequencies of 2450, 915, and 27 MHz.

Like ionizing-radiation regulations, the standards for microwave exposure have developed over a period of time with a series of criteria, each taking into account additional limiting factors. A comprehensive history of the exposure limits for scientific, industrial, and medical applications was compiled in 1992 (*13*) and provides insight into development of standards in many countries.

The potential of microwave energy to cause ionizing radiation effects can be evaluated by examining its energy in comparison with other forms of electromagnetic radiation and common chemical bond energies. Compare a list of nominal energies in electron volts (eV) for the major forms of electromagnetic energy, Table II, and selected common organic bond energies in Table III. Microwave radiation does not have sufficient energy to be

Table I. Microwave Energy Depth of Penetration for Human Tissues

Frequency (GHz)	Penetration Depth (cm)	Energy (μJ/cm)
0.915	3.03	17.3
2.450	2.05	20.6
3.0	1.97	20.9
30.0	0.078	143.3
100.0	0.032	376.4
300.0	0.023	579.1

SOURCE: Data are from reference 13, p 482.

mW cm⁻²

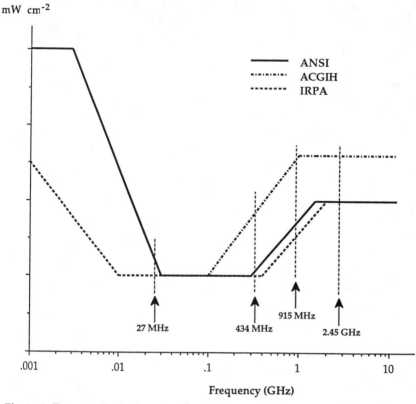

Figure 1. Exposure limits for microwave radiation from ANSI, ACGIH, and IRPA standards (*16*).

classified as ionizing radiation. Microwave energy is two orders of magnitude below the energy necessary to disrupt bonds of common organic molecules. This is not to say that there are not other biological effects, or significant interactions of electromagnetic radiation that have been and that are still under investigation, but it does dispel the notion of classical bond ionization (*13*).

Currently, in the United States, microwave energy exposure from laboratory and domestic equipment at 2450 MHz is limited to 5 mW/cm² at a distance of 5 cm from any surface of a product, or from an insulated wire inserted through any hole into an energy-containing space. These performance standards are incorporated in the Radiation Control for Health and Safety Act, a Federal Law enacted in 1968. These regulations were promulgated in 1970, became effective in 1971, confirmed in 1974, and are administered by the Center for Devices and Radiological Health (CDRH) of the U.S. Food and Drug Administration (FDA). The regulations are contained in Title 21 of the *Code of Federal Regulations* (CFR), Part 1030.10. Although no microwave leakage limit exists for other microwave heating products,

Table II. Microwave Energy in Comparison with Other Electromagnetic Energy

Radiation Type	Typical Frequency (MHz)	Quantum Energy (eV)
Gamma ray	3.0×10^{14}	1.24×10^6
X-ray	3.0×10^{13}	1.24×10^5
UV	1.0×10^9	4.1
Visible	6.0×10^8	2.5
Infrared	3.0×10^6	0.012
Microwave	2450	0.0016
Radio	1	4×10^{-9}

such as laboratory and scientific units, manufacturers of these products are subject to other FDA regulations (21 CFR 1002-1004), such as reports of design and quality control, including radiation safety measures; reports of accidental radiation exposures to users, service, or production personnel; and recall of any product that is found defective (i.e., presents a risk of radiation injury to any person). Such reports should be addressed to FDA (FDA, Center for Devices and Radiological Health, 5600 Fishers Lane (HFZ-312), Rockville, MD 20857).

Manufacturers subject to the FDA regulations include original manufacturers, importers, and persons who remanufacture products for distribution to others. Remanufacturing includes adapting a product for a new intended use, such as converting household cooking ovens for laboratory use, and reselling them, but it does not include user modification of a product once purchased. Twenty-nine states have their own regulations and some have adopted the latest ANSI guideline of 5 mW/cm^2 maximum exposure at a frequency of 2450 MHz. Reviewers of these exposure standards seem to be satisfied that the ANSI C95 committee recommendations on standards are adequate and consistent with current understanding of the biological effects of microwave energy fields (13).

Additionally, in the United States, scientific, industrial, and medical products are covered under regulations by the Occupational Safety and Health Administration (OSHA). They require that the maximum exposure to radio frequency (RF) energy for an operator in a safe work place is <10 mW/cm^2 averaged over a 6-min period (established in 1970 in 29 CFR 1910.97). OSHA regulations also require that, if microwave equipment is

Table III. Chemical Bonds with Related Energies

Chemical Bond Type	Chemical Bond Energy (eV)
H–OH	5.2
H–CH$_3$	4.5
H–NHCH$_3$	4.0
H$_3$C–CH$_3$	3.8
PhCH$_2$–COOH	2.4
Hydrogen bond (water)	0.21

modified or the integrity of a safety device is violated, the product must be demonstrated to be safe by measuring the microwave radiation exposure potential. Many references to other international standards are available and should be consulted if your geographical location places the use of microwave under other jurisdictions (13).

Most laboratory microwave equipment surpasses (lower than) the protective requirements in emission standards. As long as damage, wear, or misuse have not lessened the effectiveness of the instrument, all the exposure limits of various national/international standards are met or exceeded.

Function of Microwave-Digestion Systems

Microwave-digestion systems are not analytical instruments. Functionally, they are chemical reaction systems for acid decomposition of samples, producing solutions suitable for introduction into common analytical instrumentation.

Most chemical reaction systems are built for conducting a specific, well-defined, reaction, or for the study of a group of reactions, such as catalytic hydrogenations. Reaction conditions and operating parameters are specified in detail so that appropriate engineering decisions can be made regarding system design and safety.

Microwave-digestion systems differ from other types of chemical reaction systems in two important ways:

1. Microwave reaction and digestion systems are general-purpose systems. Reactants and reaction conditions are not specified and are unknown to the manufacturer in many cases.
2. Heat transfer in conventional chemical-reaction systems with external heating jackets is indirect via conduction and convection. The temperature of reactants rises slowly and internal cooling coils can be used to remove heat from exothermic reactions and to moderate reaction rates. Heat transfer in microwave-digestion systems is via direct absorption of microwave energy by reactants inside the pressure vessel. Energy transfer is instantaneous and the temperature of reactants rises rapidly. Threshold activation temperatures for exothermic reactions are attained more quickly in microwave-heated systems. A temperature control device may stop further microwave heating when a set point value is reached, but there is no effective means of cooling and removing heat from exothermic reactions inside microwave-digestion vessels.

Microwave equipment specified by manufacturers and the literature rarely list pressures in the International System of Units (SI) units of pascals (Pa). Instead pressures are commonly listed in atm, bar, or psi. The following pressure unit conversions may be helpful (5): 1 atm = 1.01325 × 10^5 Pa, 1 atm = 1.01325 bar, 1 bar = 1 × 10^5 Pa, 1 atm = 14.69595 psi, 1 psi = 6894.76 Pa. Pressures reported throughout this chapter are in the units specified by the equipment, standard, or case study.

Codes and Standards Relevant to Microwave-Digestion Systems

A number of existing safety codes and standards are relevant to micro-wave-digestion systems and vessels (17–19). The National Fire Protection Association (NFPA) has issued a standard, NFPA 45 *Fire Protection for Laboratories Using Chemicals*, which contains criteria for classifying laboratory hazards. Equipment design, construction, sizing, and safety measures required to deal with each class of hazard are defined in this standard.

If any of the five conditions defined in Section 2.3 "Laboratory Work Area and Laboratory Unit Explosion Hazard Classification" exist, the laboratory is considered to contain an explosion hazard. For instance, laboratories operating closed-vessel microwave-digestion systems may be considered to have an explosion hazard because two of the five classification criteria may be present: exothermic oxidation reactions and high-pressure reactions (19).

A large percentage of the acid digestion procedures performed in microwave systems involve the decomposition of organic compounds or matrices having endothermic heats of formation with oxidizing acids (the precise definition of an exothermic oxidation reaction) (20). Guidelines for distinguishing between high- and low-pressure reactions are defined in NFPA 45 based on a reaction vessel operating pressure and volume. These two parameters are plotted as a curve and reproduced here in Figure 2. Paragraph C-4.5.2 of NFPA 45 states: "Reactions that produce pressures above the curve in Figure C-4.5 should be classified as high pressure reactions". According to this criterion, a single microwave-digestion vessel with a capacity of 0.1 L, rated for operation above 400 psi (27 atm) would fall in the high-pressure category.

Pressure conditions generated during microwave acid digestion are the result of two factors; microwave heating, which raises acid temperature and vapor pressure, and accumulation of gaseous decomposition products (CO_2, NO_x, SiF_4) of the reaction inside the vessel.

The general reaction mechanisms responsible for formation of noncondensable gaseous decomposition products during microwave acid digestion are:

$$(CH_2)_x + 2HNO_3 \rightarrow CO_2(g) + 2NO(g) + 2H_2O \tag{1}$$

$$SiO_2 + 4HF \rightarrow SiF_4(g) + 2H_2O \tag{2}$$

Increasing sample weight produces more gaseous decomposition products within the fixed volume of the vessel and results in higher pressure conditions. An understanding of these reaction mechanisms is impor-

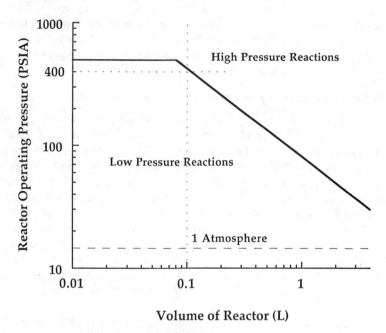

Figure 2. Parameters for distinguishing between high- and low-pressure reactions. (Reproduced with permission from reference 19. Copyright 1991 National Fire Protection Association.)

tant for selection of microwave-digestion vessels with appropriate pressure ratings for the type and amount of sample being decomposed, as the following example illustrates.

Pressure Calculation for Microwave Digestion of Glucose

$$C_6H_{12}O_6 + 8HNO_3 \xrightarrow{\Delta} 6CO_2(g) + 8NO(g) + 10H_2O$$

Assuming:
Vessel: 100 mL volume, 10 mL of nitric acid, 90 mL vapor phase, and 30 atm capacity
Liquid and vapor temperatures of 180 °C and 150 °C, respectively
Neglecting other chemical reactions, vapor pressure of water, and assuming ideal gas laws
Then:
Vapor pressure of nitric acid at 150 °C is 20.1 atm (21)
Residual pressure capacity for digestion products is 9.9 atm

$$\frac{(1 \text{ atm})(x \text{ L})}{(298 \text{ K})} = \frac{(9.9 \text{ atm})(0.090 \text{ L})}{(423 \text{ K})}$$

ΔΔ

Yields 0.63 L of gas

$$(X \text{ g glucose}/180 \text{ g mol}^{-1})(14 \text{ mol } CO_2 \text{ and NO})(22.4 \text{ L mol}^{-1}) = 0.63 \text{ L}$$

Therefore:

0.36 g of Glucose will cause this vessel to vent

This calculation experimentally corresponds well with Figure 27 in
Chapter 3 in which 0.306 g of motor oil produced a pressure of ~24 atm.

According to NFPA 45 hazard-analysis criteria, laboratories operating
microwave-digestion systems may be considered to contain an explosion
hazard. An explosion is defined in NFPA 45 as follows: "(1) a violent
bursting, as of a pressurized vessel or (2) an extremely rapid chemical
reaction with the associated production of noise, heat, and violent expansion of gases". Container failure is also defined in NFPA 45, "When a container is pressurized beyond its burst strength, it may violently tear asun-
der (explode)".

NFPA 45 contains three recommendations for protection against
explosion hazards arising from reactions conducted above atmospheric
pressures that are relevant to microwave-digestion systems. The first of
these recommendations is, "High-pressure experimental reactions should
be conducted behind a substantial fixed barricade that is capable of withstanding the expected lateral forces".

In the case of microwave-digestion systems, the cavity, door, and
structural frame of the system must serve as the primary protective barricade for operators in the case of vessel explosions. The force exerted on
these system elements from a microwave-digestion vessel venting or failing can be calculated and used by engineers to design the system to withstand such an event. The volume of the microwave cavity, digestion vessel
volume, relief-device venting pressure, and surface area of the structural
component that the resulting force acts on are the principle calculation factors. An example of the force exerted on the door of a domestic microwave
oven resulting from the venting of a single 30 atm (440 psi) vessel follows.

Calculation of Door Excess Pressure on Vessel Venting

A single 100 mL vessel (at 30 atm) vents into a microwave cavity

Assuming:

Microwave Cavity Volume: 30 L

Surface Area of Microwave Door: 900 cm^2

Vessel Volume: 100 mL

Relief Device Venting Pressure: 30 atm = 30 kg/cm^2

Then:

$$(30 \text{ kg cm}^{-2})(0.1 \text{ L}) = (x \text{ kg cm}^2)(30 \text{ L})$$

yields a force of 0.1 kg/cm^2 onto the walls and door of the cavity.

Acting on a cavity door surface area of 900 cm^2

$$(0.1 \text{ kg cm}^{-2}) (900 \text{ cm}^2) = (90 \text{ kg})$$

The force of 90 kg is exerted on the door!
(refer to Case #12 below for the result of a vessel bursting at 200 psi).

The manufacturers of domestic microwave ovens have not designed the door of their products to be capable of withstanding this magnitude of force. The force generated by the venting or failure of a microwave-digestion vessel is capable of removing the door of such a system and turning it into a secondary missile. Case #12, presented later in this chapter, describes such an event when a domestic microwave door was used by a laboratory microwave company.

The functions of explosion-resistant shields and barriers defined in NFPA 45 are "(a) withstand the effects of an explosion; (b) vent over-pressures, injurious substances, flames, and heat to a safe location; (c) contain missiles and fragments; and (d) prevent the formation of secondary missiles caused by failure of hood or shield components" (19).

For a microwave-digestion system to be an effective protective barrier for laboratory personnel it must be designed to withstand the force resulting from venting or failure of one or more vessels used inside. Systems that cannot confine venting or failing vessels and have cavity doors that can become detached, secondary missiles do not meet the first NFPA 45 recommendation for protective measures against explosion hazards.

The second recommendation in NFPA 45 for explosion protection is, "Reaction vessels should be built of suitable materials of construction and should have an adequate safety factor". A third recommendation states, "All reaction vessels should be provided with a pressure-relief, valve, or rupture disc" (19).

Pressure Vessel Design and Safety Codes

At this time, no design or safety code has been specifically developed for microwave-transparent vessels heated inside multimode or single-mode microwave cavities. In the absence of such a code, manufacturers are responsible for establishing prudent design practices for their products.

The American Society of Mechanical Engineers (ASME) has developed a comprehensive body of American National Standards for industrial process equipment. ASME *Boiler and Pressure Vessel Code* Section VIII, Division 1, defines engineering principles, design calculations, and safety factors for metal pressure vessels. Although the ASME Code does not specifically encompass microwave-transparent plastic pressure vessels, some of the design and safety principles it contains represent prudent engineering practices, and are applicable to such vessels.

The ASME Code requires that all pressure vessels, irrespective of size or operating pressure, be equipped with a pressure-relief device that prevents pressure from rising more than 10% above the maximum allowable working pressure (MAWP) of the vessel. Two basic types of pressure-relief devices are recognized in the code; reclosing- and nonreclosing-type safety devices. By ASME Code definition: "A pressure-relief valve is a pressure-relief device designed to reclose and prevent further flow of fluid after normal conditions have been restored. A nonreclosing pressure-relief device is a pressure-relief device designed to remain open after operation". Rupture disks are nonreclosing-type pressure-relief devices. Both types of pressure-relief devices are employed on microwave-digestion vessels (22–27).

Pressure-Relief Devices

Pressure-relief setting and venting capacity are the critical performance parameters of a pressure-relief device. Formulas and test procedures for calculating and validating these performance parameters are explicitly defined in the ASME Code. The requirements stipulate that, when a vessel is equipped with a single pressure-relief device, it must be set to allow operation up to the vessel's maximum allowable working pressure.

The first microwave-digestion vessels were equipped with external valves to protect them from over-pressurization. Figure 3 demonstrates the design of this vessel, the pressure-relief valve, and the measurement system. The pressure tube exits the microwave cavity through a wavelength attenuator cut-off to prevent loss of microwave radiation. The valve relief pressure can be set to correspond to the vessel pressure range. Pressure inside the vessel is transmitted to both the external spring-loaded relief

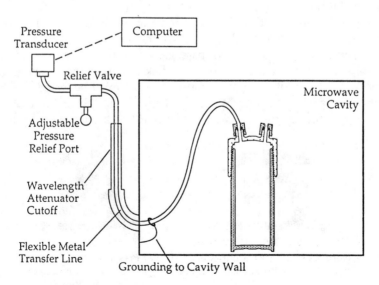

Figure 3. Microwave vessel with pressure-relief adjustable valve external to the vessel and system.

valve for safety and the transducer for measurement and recording (28). In the case of over-pressurization, the vessel permits a preset release of pressure above the valve limit and reseals, protecting the integrity of the vessel. The gas is expelled external to the unit; however, the content of the vessel may, or may not, be compromised depending on the amount of escaping gas and whether any liquid or analyte vapor is expelled with the gas. This type of mechanism is still used for prototype and specialty vessels, but it has been largely replaced with integrated vessel relief devices.

A microwave-digestion vessel equipped with a nonreclosing-type pressure-relief device is depicted in Figure 4. In this design, the vessel is protected from over-pressurization by a fluoropolymer rupture membrane (22). The membrane functions as a rupture disk that bursts to prevent vessel over-pressurization and failure. Once the membrane has burst, the vessel depressurizes to atmospheric pressure and remains open. The use of nonreclosing-type pressure-relief devices has certain disadvantages in microwave-digestion systems. First, the vessel contents are forcibly expelled and lost for analysis. Second, the cavity exhaust system evacuates noncondensable vapors; however, certain vessel contents may condense or deposit on surrounding vessels and other components in the cavity. If the microwave run is not aborted, condensed liquids and deposited solid materials will absorb microwave energy, heat, and possibly damage other vessels and system components. Acid, or other absorbing reagents, remaining in the open vessel will also continue to heat and gradually outgas until the vessel is dry. When using rupture disks it is advisable to terminate a run on the basis of the nature of the reagents (e.g., corrosiveness or flammability).

The membrane burst pressure (relief setting) is a function of material properties of the membrane, dimensions of the membrane's effective venting area, and geometry of the vent passage behind the membrane. Because the properties and thickness of rupture disk materials (both metals and plastics) vary from lot to lot, each lot of material must be tested to establish its burst pressure rating. Section UG-127 of the ASME Code details requirements for testing and rating rupture disks, accounts for lot-to-lot

Normal Operating Conditions   Over-Pressurization Condition
Internal Pressure < Burst Disk Strength   Burst Disk Breaks and Excess
                                          Pressure Vents

Figure 4. Microwave vessel with nonreclosing pressure-relief mechanism.

variability, stating, "Every rupture disk shall have a stamped bursting pressure within a manufacturing design range at a specified temperature, shall be marked with a lot number, and shall be guaranteed to burst within 5% of its stamped bursting pressure at coincident disk temperature".

A microwave-digestion vessel equipped with a reclosing pressure-relief device is depicted in Figure 5. In this vessel design, a torque wrench is used to apply sealing force equivalent to the maximum allowable working pressure. The applied torque compresses a spring in a thrust plate on top of the vessel's fluoropolymer cover. This spring-loading arrangement functions as a reclosing-type relief-valve mechanism. When pressure inside the vessel exceeds the sealing force applied by the torque wrench, the cover will lift to vent excess pressure and reclose, resealing at the maximum allowable working pressure initially set by the torque wrench. Non-condensable vapors creating the over-pressure condition, (e.g., CO_2, NO_x, etc.), are released into the cavity and evacuated by the exhaust system. This type of mechanism frequently reseals after venting without compromising the sample integrity. This reclosing-type relief-valve design was developed and patented specifically for microwave vessels (*29*). The pressure-relief device setting is mathematically calculated to be equivalent to the MAWP and is applied by the use of a calibrated torque wrench. This design provides rapid and minimal torque strain of the vessel during over-pressurization events.

The venting capacity of relief devices on microwave-digestion vessels is especially important. The instantaneous heating of reactants and rapid attainment of threshold temperatures for activation of exothermic reactions precludes cooling and heat removal for moderating reactions. Accordingly, the rate of pressure rise from acid vapor and accumulation of decomposition gases is significantly faster in microwave systems than in conventional acid digestion bombs. For proper vessel over-pressurization

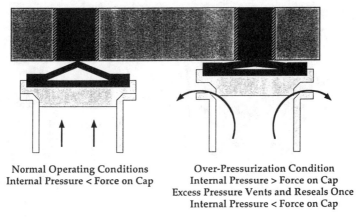

Normal Operating Conditions
Internal Pressure < Force on Cap

Over-Pressurization Condition
Internal Pressure > Force on Cap
Excess Pressure Vents and Reseals Once
Internal Pressure < Force on Cap

Figure 5. Diagram of MDR vessel in normal and venting modes.

protection, the rate of vapor and gas removal must be equal to, or greater than, the rate of vapor and gas generation (*30*). Section UG-131 of the ASME Code provides theoretical formula to calculate the venting capacity of nonreclosing relief devices for various media. A safety factor coefficient of 0.62 is then applied to the calculated value to ensure that the device is sized conservatively.

The majority of laboratory microwave-digestion systems in use today has multimode cavities with doors for introduction of microwave-transparent plastic vessels. The vessels contain pressure and the microwave system functions as a heating and fume evacuation device and protective barrier for laboratory personnel. The extent to which existing safety codes and standards apply to these systems has been documented to assist the chemist in their safe and reliable use.

Recently, a new type of microwave system (Figure 6) has been developed that combines microwave heating and high-pressure vessel technology with increased safety (*31*). This microwave-heated autoclave is designed for conducting chemical reactions at pressures and temperatures up to 200 bar (2900 psi) and 350 °C. The 4.2 L pressure vessel is constructed of forged stainless steel and has been hydrostatically proof tested according to the German Technische Überwachung Verein (TÜV). The vessel meets ASME Code design requirements and has a safety factor >4× the maximum allowable working pressure. All safety relief devices employed on the vessel system comply with TÜV and ASME requirements. The vessel interior is protected by a titanium nitride coating for acid and chemical resistance. Continuous (i.e., nonpulsed) microwave

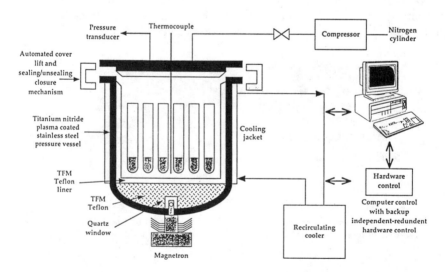

Figure 6. The ultraCLAVE microwave autoclave for high-pressure chemical reactions.

energy, at operator selectable settings of 0–1000 W, is delivered into the vessel through a microwave-transparent port. Samples are not processed in closed vessels; rather the unit applies pressure to the samples in open vessels. This eliminates unsafe conditions where vessels are at unknown temperatures and pressures.

A unique class of microwave pressure digestion systems is flow-through or stopped-flow reactors. In these systems, pressure is typically built up within a Teflon tube (the reactor) contained in a microwave cavity. Without warning, these tubes, fittings, and other parts are prone to blow-outs. Safe operation advises making the entire tubing inside the microwave out of a single piece (without connections) to minimize connector-related failures (*see* Chapter 6). Frequently these Teflon tubes are reinforced against stretching and bulging, that weaken the tubing, with braiding constructed form of polyetherimide. Even though ruptures still occur in these armored tubes, they are less frequent.

Recently, a new approach to flow-through microwave digestion that uses an external gas pressurization outside the Teflon tubing was developed (*32*). The externally applied pressure is computer-controlled to balance the pressure inside the digestion tubing. Balancing the pressure minimizes the stress on the Teflon tubing and effectively minimizes the primary safety hazard of flow-through microwave-digestion systems.

Safe Usage of Laboratory Microwave Instrumentation

General Safety Issues

Proper usage of laboratory microwave equipment is the responsibility of laboratory personnel. It is possible to render the safety devices of many instruments and vessels ineffective by carelessness or misuse. It is the responsibility of the analyst to follow good laboratory practices and the manufacturers' instructions when assembling, using, and maintaining the equipment. For example, by placing a microwave system inside a fume hood, where exhausted acid fumes may get circulated around the unit, the designed physical isolation of the electronics from the cavity is defeated. Accelerated corrosion of the electronics, including the safety interlock mechanisms and control circuits, can result. Chemical vapors should always be transported away from the unit, or the cavity air swept away to an exhaust hose, fume extraction or neutralization system or hood. Deterioration of the waveguide, door seals, or cavity walls can provide leak paths for the escape of microwave energy as well as degradation of the equipment.

The hazards associated with inappropriate use of microwave equipment cannot be entirely prevented by interlocking devices and other safeguards. However, the risk can be minimized if the analyst continually inspects the system to ensure that the equipment is maintained in safe working order. If any portion of the microwave unit such as a door seal or vessel casing becomes damaged by a catastrophic event such as an acid spill, prolonged wear, or impact, the safety of the equipment should be reevaluated before it is returned to service.

In addition to compliance with microwave energy leakage standards, safety interlock devices are required to prevent accidental exposure on all commercial and consumer microwave equipment (33). These interlocks protect against initiating or continuing the emission of microwave energy into the cavity if the microwave system door is open or misaligned. Safety devices should never be removed or defeated on any microwave equipment, but especially on laboratory systems. Other components important for safe operation, such as wavelength attenuators in atmospheric-pressure systems, door seals, or waveguides (if the unit's cover is removed) should be inspected and tested for microwave leakage if corrosion is noted or if a vessel vents and reagents have prolonged contact with nonresistant parts of equipment.

Figures 7 and 8 show a vessel that has not failed, but is about to do so and should be permanently removed from use. Even though the crack and nitric acid degradation are obvious in this example, many other stresses, or chemical interactions, are less obvious, and diligence is required on the part of the analyst to maintain a safe working environment. We are asked frequently how long a vessel can be used. The answer depends on how much you are stressing it each time it is used, on how it has been maintained, and on how much residual pressure capability has been designed into that particular model by the manufacturer. Some manufacturers have vessel designs that are very rugged and last for hundreds of uses; others have designs that are not as robust and do not last if taken to the upper limit of their specifications for more than a dozen uses. Frequently, degradation is obvious but occasionally there are no obvious warning signs of an impending vessel failure. In these cases, secondary safety systems outside the vessel such as doors, exhausts, cavity structure, preventive measurement devices, and active cutoff switches are required to handle catastrophic vessel failures.

Good Laboratory Practices and Common Sense

A seemingly innocuous event that has been observed by one of the authors on several occasions is the dropping of a vessel casing from a laboratory bench. In each instance, the analyst retrieved the casing, placed it back among the others on the bench, and failed to inspect it for damage such as cracks or chips. Polyetherimide (in most cases) vessel casements are brittle because of their high mechanical strength; therefore, they must be inspected for damage before every usage. This example is clearly the responsibility of the analyst using the equipment.

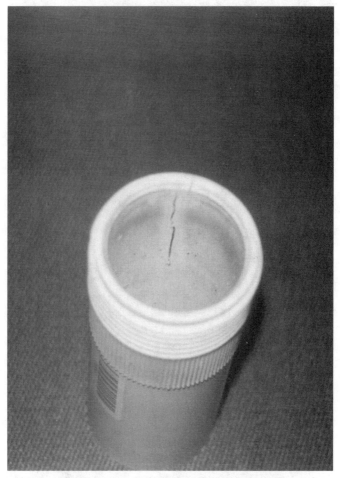

Figure 7. Vessel with internal physical wear.

Another incident involved the deliberate removal of the mesh screen from the inside of a microwave door to allow the technician to see the reaction in all-Teflon perfluoroalkoxy (PFA) vessels more clearly. The perforated metal screen functions as a microwave barrier that prevents passage of microwave radiation at 2450 MHz. Once removed, the plastic window is transparent to microwave radiation, which is transmitted into the laboratory. Whereas this example is obvious to some, to others who do not understand the proper use of this component it is not. Education and consultation are always prudent before using any scientific instrument, but especially laboratory microwave systems.

Figure 8. Vessel with external physical wear.

Open-Vessel Microwave Safety

The use of a single-mode microwave system operating at atmospheric pressure alleviates some safety concerns. However, safety cannot be taken lightly when dealing with such systems. The presence of hot reagents, open to the laboratory, still needs great attention.

The use of a fume scrubber is a requirement for any laboratory that is performing microwave digestions or extractions (leaching) at close to the boiling point of the acids used. Not only does a fume scrubber stop aggressive fumes or volatile toxic species from entering the laboratory environment; but it also mitigates the effects of vigorous reactions that may occur on reagent additions to hot solutions. With the microwave waveguide open to the environment this also minimizes the possibility of acid attack on the waveguide and possible feedback of microwave energy to the magnetron or microwave leakage to the environment. Careful cleaning of the scrubber transfer line is essential after use, because gases can condense in the line and can cause problems in subsequent runs, especially if perchloric acid or other condensable reagents are used.

Problems with the waveguide and magnetron assemblies can also be caused if any artifacts drop down into the waveguide so as to disturb the microwave field pattern. Leakage of microwave radiation is a serious potential problem if a microwave-conductive material is introduced into the microwave vessel. Thus, under no circumstances should mercury thermometers, thermocouples, thermistors, or metal stirrers be inserted into the waveguide. These devices conduct microwave energy from the cavity on their surface and radiate it into the laboratory.

The analyst should make sure that all safety interlocks are always operational and should be very careful when removing a digestion vessel from the cavity, because the vessel itself can be as hot as the solution it contains.

Muffle Furnace Safety

Microwave muffle furnace systems are unique combinations of new and old technology. These systems are usually constructed of a strong microwave-absorbing material with an insulator to create a small muffle furnace that can be quickly heated and cooled within a dedicated uncoated microwave oven cavity. The sample or matrix is not directly heated by microwave energy but through convection and conduction heating from the muffle furnace. The microwave muffle furnace has not been fundamentally altered from its original concept described previously (34, 35).

The sample is fused or ashed in traditional quartz, porcelain, nickel, or platinum crucibles that are shielded from microwave radiation by the strong microwave-absorbing furnace. The safety considerations are usually similar to classical muffle furnace systems, with a few exceptions. Although the furnace heats rapidly it also cools very rapidly, thereby minimizing the hazards of exposing the analyst to high-temperature devices. Specific reaction temperature protocols can be developed for individual samples for reaction optimization and safety because of the rapid heating method.

Sulfuric acid is frequently used as an ashing aid especially in the pharmaceutical industry. When performed in a microwave muffle furnace the sulfuric acid fumes have no direct exit from the furnace so they attack and damage the muffle furnace. In these cases, the muffle furnace typically must be replaced in less than 1 year. A manufacturer designed a muffle furnace with a vacuum port from a quartz ceiling of the furnace that connects to an acid-neutralization system (36). As the sulfuric acid fumes are generated, they are evacuated outside the entire microwave system into a condenser where the sulfuric acid is neutralized in a collection flask. Additionally, the sulfuric acid muffle furnace was designed with a chemically resistant thermocouple measurement system.

Intermixing Equipment of Different Manufacturers

In general, the intermingling of different manufacturers' equipment is inadvisable. Under certain circumstances it may be possible if the analyst has an excellent understanding of the limitations of each piece of equipment and consults each manufacturer prior to use. The divergent pressure conditions produced in vessels with different thermal insulation characteristics are described in Chapter 3 and clearly indicate why using vessels of dissimilar design for the same digestion protocol can (and has) produced disastrous results (see Case #17). Interestingly, the intention of the analyst was to be as safe as possible by minimizing the pressure developed inside the higher-pressure vessels. Paradoxically, the nonintuitive nature of microwave/thermal interactions caused an uncontrolled situation to occur that was catastrophic.

Modification of Microwave Equipment

Sometimes microwave equipment must be modified to produce special research configurations that permit the direct addition of reagents or the measurement of temperature and pressure inside a microwave cavity. Modifications should be performed only by those trained in high-voltage circuitry and who understand the shielding requirements of microwave equipment. It is highly advisable to consult the manufacturer about proposed modifications, especially those involving the waveguide, mode stirrer, metrology devices, vessels, and electronics.

After any modification of microwave equipment, it is important to ensure that no path for radiation leakage from the microwave cavity has been created. Measurements to detect microwave radiation must always be performed in and around the modified portion of the equipment. A microwave survey meter with good sensitivity is the recommended method of evaluating such alterations for leakage (see Appendix 1). The unit should be run at full power with no load. If an appliance-grade microwave oven is being evaluated, a small load (~100 mL of water) should be in the unit at the time of testing to prevent possible damage to the magnetron.

A common modification to microwave equipment is the attachment of a wavelength attenuator port to permit safe introduction of tubing, thermocouple wires, and fiber optics into the microwave cavity. These devices have been used in many configurations. Two types of attenuators tested by us are rigid stainless-steel tubing (28) and flexible, tin-plated copper braid (37). The common construction feature of both attenuators is metallic tubing constructed of conducting metal with the smallest diameter possible to allow a tube, wire, or other device to pass through its inside diameter. The effectiveness of small-diameter holes as a barrier to microwave energy is the basis of the perforated metallic screens located at air intakes and grid on the inside of an otherwise transparent microwave door.

A cut-off device is effective only if the diameter is small and the length exceeds the 12.25 cm wavelength of microwave radiation at 2450 MHz. Dimensions of the devices used by us were approximately 0.7 cm inside diameter and between 35 and 70 cm in length. However, many various dimensions of cut-off devices are possible. The attenuator must be grounded by making electrical contact with the microwave cavity wall around the hole and the unit grounding circuitry. Metallic devices passing through the attenuator must also be grounded at the microwave cavity wall to prevent them from acting as an antenna for transport of the electromagnetic energy out of the cavity, totally defeating the wavelength attenuator cut-off (28). Experiments in which thermocouple wires were passed through the attenuator without grounding were conducted and energy fields >10 mW/cm^2 were measured at the end of the attenuator. Six separate attenuator designs were tested and found to prevent microwave radiation from escaping (<0.01 mW/cm^2, the limits of the detection equipment).

Chemical and Acid Safety

Acid temperatures that are achieved after 5–30 min of heating by using traditional methods are attainable within seconds in a microwave cavity, greatly reducing reaction times. The coupling of electromagnetic energy with the molecular dipoles of a sample may aid the digestion, extraction, and synthesis and further accelerate the specific reaction. With rapid heating, the evolution of reaction products may occur too quickly and in too large a quantity to be vented or contained. Thus, caution should be used not to speed up a reaction to such an extent that safety of the individual or structural limits of the equipment will be compromised.

A comprehensive review of all unsafe chemical reactions that may occur in microwave chemistry is beyond the scope of this chapter. These uncontrollable reactions are a function of the reagent or combination of reagents reacting with the sample. Reference books are available identifying classes of incompatible chemicals, toxicity of substances, and general laboratory safety guidelines (7, 8). Sources are also available that discuss the reactivity of the decomposition reagents, unsafe reagent combinations, and proper usage and handling precautions (2, 4, 7, 8, 38).

Many of these unsafe combinations of dissolution reagents and samples are addressed in Chapter 2. Many hazardous combinations involve the use of strong oxidizing agents or dehydrating agents in the decomposition of organic matrices.

Sparking and Metallic Samples

Many alloys interact with microwave energy and will heat sufficiently to melt plastic vessels; others accumulate large electrical discharge potentials. One case, involving a large alloy sample melting through a container, has been related to us. Several types of ferrous alloys were tested in which spark discharges were quite spectacular. Discrete pieces (>1 mm) of metallic samples should be avoided, because electrical arcs may occur between individual sample pieces or between these alloys and the microwave cavity metallic devices including the walls. The formation and intensity of electrical sparks depend on the composition of the alloy and other conditions such as electric field strength. Electrical arcs from inadequately grounded thermocouples constructed of 316 stainless steel were sufficiently energetic to puncture 1/16 in. Teflon PFA (39, 40). Also, some concentrated solutions also exhibit sparking. Concentrated sodium hydroxide, copper nitrate, and nitric acid have all been observed to discharge electrically within the solution.

Use of Flammable Solvents or Production of Flammables

Another hazard that can be anticipated to occur as chemical applications for microwave equipment diversify is the use of flammable solvents in electromagnetic environments. A documented instance of such a potential problem occurred when students heated diethyl ether in an open container (41). Luckily, no ignition occurred in this instance. Because sparks are common in microwave systems, flammable or explosive substances mixed with oxygen from the air pose especially hazardous situations.

When vessels containing flammable solvents vent from over-pressurization, or vessel failure, the microwave should be equipped with a sensor to terminate power to the magnetron. Flammable vapors must be cleared from the cavity before restarting the microwave system. Most systems capable of microwave extraction, organic synthesis, or other applications based on flammable solvents are equipped with an organic solvent sensor mounted in the cavity or cavity exhaust system.

The mineral–acid decomposition of metal and alloy samples has a unique hazard that must be considered before traditional methods of hot mineral–acid decomposition are applied to the microwave environment. Metals below hydrogen on the electromotive series readily liberate hydrogen when dissolved in acid. If hydrogen is released from the sample into an open beaker or a vessel sealed in air, a potentially flammable or explosive mixture with oxygen may result. In traditional digestion procedures, an ignition source would not be present; however, in microwave equipment, because metallic particles can interact with the strong electromagnetic field, a spark may be generated from the sample, and ignition of the hydrogen may cause a fire or explosion.

To prevent this potentially hazardous situation, closed digestion vessels should be sealed in inert-gas atmospheres to eliminate oxygen. In open-vessel microwave applications, purging the vessel and compartment with inert gases will prevent air from coming in contact with the sample. The limits of flammability of hydrogen in air diluted with an inert gas are presented by Lewis and Von Elbe (42). In open-vessel digestions, rapid removal of hydrogen may also prevent the formation of a hydrogen–oxygen mixture sufficient for ignition. Unfortunately, the amount of hydrogen in air that supports combustion ranges from 4 to 75% (43), and the flammability range of hydrogen in oxygen is 4–94% (42). The energy of activation necessary to ignite this mixture can be achieved by the weakest spark and may also be catalytically ignited; therefore, mixtures of hydrogen and oxygen anywhere in this range are extremely hazardous.

The elimination of oxygen and air by purging the vessel and cavity with an inert gas such as argon or nitrogen is a prudent precaution because hydrogen generation is a potential hazard. The oxidant is the only component that can be eliminated in such an instance.

Scale-Up of Dissolution Procedures

Increasing the sample size can involve significant potential safety problems including solubility problems, the necessity for additional reagents, and the development of too much pressure. The safety and chemistry factors involved in the scaling-up of EPA 3052 digestion method are discussed in detail in Chapter 3 and serve as a guide of good general practice. The most important consideration, based on safety, is the production of gas during dissolution.

Digestion of organic matter generates significant quantities of gaseous decomposition products, primarily CO_2 and NO_x. Always begin acid digestion method development of an organic matrix with a small sample size, less than 0.25 g. Scale-up the sample size after evaluating the final pressure attained during each successive digestion and pressure limitations of the vessels employed. In general, each incremental scale-up step should only be a maximum of 0.2 g.

A simple example of this approach is the digestion of motor oil. A 0.25 g sample was digested with 10 mL nitric acid in 100 mL vessels. The vessels were heated to 180 °C and held for 9.5 min; the final pressure inside the vessels was ~21 atm. The experiment was reproduced with only nitric acid and the final pressure was ~6 atm. After subtracting the pressure produced by the nitric acid, 0.25 g oil produced approximately 15 atm (60 atm/g of motor oil). On the basis of this gas production factor, the approximate maximum sample size can be predicted.

Assume that the digestion vessel is capable of 65 atm. Subtracting a minimum of a 20% safety factor (20% of 65 atm = 13 atm) leaves a safe vessel capacity of 52 atm. Subtracting the pressure that will be produced by heating nitric acid leaves 46 atm of pressure. Dividing 46 atm by 60 atm/g oil results in a maximum safe digestion of approximately 0.75 g of motor oil. Because of numerous assumptions in this calculation, this is meant for a first approximation. Further tests with sample sizes between 0.25 g and 0.75 g are necessary to determine the true safe upper limit of sample size for this particular apparatus.

Labware for Microwave Environment

Microwave sample preparations, including digestion, extractions, organic synthesis, flow reactors, muffle furnaces, and other laboratory operations, require many different types of sample vessels and apparatus. The suitability of a vessel material depends on the specific conditions, use, and reagent for which it is intended. In choosing a container material for open-vessel work, the boiling point of the acid or solvent being used is the maximum temperature to which the material will be exposed. When closed-vessel systems are involved, temperature measurement is necessary to prevent the liquid contents of the vessel from exceeding the upper temperature limit of the vessel material, which is normally the MAWP. At this temperature, a polymeric device will continue to function as intended without change for an indefinite period of time. Above this point, the material may degrade, deform, or melt so that it no longer functions appropriately. For example, a high-boiling solution of concentrated sulfuric acid will melt Teflon but is not a problem when used with quartz or borosilicate glass (*see* Case #13).

After chemical and mechanical stability, the dielectric constant, and microwave energy absorption characteristics of the material are the important physical factors to consider. Many materials are compatible for use in a microwave field because they do not absorb radiation in this frequency range. Among the essentially microwave-transparent materials are all types of fluorocarbons (e.g., Teflon PFA and TFM), quartz, and some glasses, which are all excellent as sample containers because of their exceptional chemical and thermal durability. Other common laboratory polymers, such as polyethylene and polypropylene, are also suitable for use in the microwave environment, but at low temperatures and under mild conditions.

When selecting containers for use in a microwave field, the composition of all parts needs to be considered (e.g., handles or screw-on caps, which tend to be made from different materials than the vessel itself).

Problems can be avoided by selecting materials that do not absorb microwave radiation, or absorb very little. Listed in Table IV are the dielectric constants and melting points of some common laboratory container materials. Fluorinated polymers are excellent materials for microwave containers because their low dielectric constants make them essentially transparent at 2450 MHz. Labware fabricated from or incorporating Bakelite, Lucite, and other thermoplastic resins may be acceptable for some uses, even though they absorb a small amount of energy.

Microwave Safety Illustrative Case Histories

Microwave sample preparation and chemistry procedures impose a unique set of safety considerations in addition to those of good laboratory practice. This is evidenced by the reports of microwave-related incidents described subsequently. Parties involved in some of these illustrative case histories have requested anonymity. In these cases, we have honored their requests and the source of the information is not cited.

The reports below are only brief descriptions of what happened. In some cases, more in-depth descriptions of these events can be found at the SamplePrep Web site (*see* Chapter 15). As other instances of equipment malfunction, runaway chemical reactions, or improper use of microwave equipment are reported, they will be added to the Web site database. Readers are encouraged to contact us with any safety problems that they encounter so that we may quickly report them to the microwave community. The majority of these safety problems and potential hazards are not reported by manufacturers, go unreported by the analyst, or are unavailable. It is our hope that reporting of such problems will lead to their solutions and ultimately to their prevention.

Equipment Failure

Case #1: Chemical Attack of a Safety Interlock

Commercial microwave appliances for cooking food are not designed to withstand chemical attack from corrosive substances. An incident has been documented in which all the safety interlocking devices were rendered inoperative as the result of chemical attack on the metal safety switches. The result was that when the microwave unit door was opened, full power microwave energy emission continued for 1 min directly exposing an operator. This case is the first reported case of human injury from the use of a home microwave unit in the laboratory (case I8-138 reported to Microwave/Acoustic Products Section, FDA, 1986). These units have not been designed for use in a chemical environment and interactions of the instrumentation with chemicals must always be considered before equipment is purchased. The use of appliance-grade microwave equipment in

Table IV. Thermal and Microwave Characteristics of Laboratory Container Materials

Vessel Material	MP (C)	Working Range (C)	Dissipation Factor (tan ∂)	Dielectric Constant
Borosilicate glass	—	~600[a]	0.065–0.11[a]	—
Chlorinated tetrafluoroethylene (CTFE)	>104[b]		0.003–0.032[b]	2.3–2.7[b]
Tetrafluoroethylene (FEP)	252–262[b]		0.0007[b,c]	2.1[b,c]
Nylon 6/10	>215[b]		0.015–0.1[b]	3.1–4.0[b]
Nylon 6/6	>232[b]	160–170[a]	0.010–0.08[b]	3.1–5.2[b]
Perfluoroalkyoxy polymers (PFA)	302[b]	250–260[a]	0.035–0.042[a] <0.0001–0.0003[b] 0.0003[c]	2.1[b,c]
Polycarbonate (PC)	241–266[b]	115–125[a]	0.021–0.023[a] 0.0024–0.01[b]	2.35–3.8[b]
Polystyrene	111[b]	85–95[a]	0.030–0.035[a] 0.0001–0.003[b]	2.52–2.7[b]
Polysulfone		185–190[a]	0.028–0.031[a] <0.002–0.0056[b] 0.030–0.035[a]	3.0–3.8[b]
Polypropylene (PP)	145–182[b]	130–135[a]	0.0001–0.014[b] 0.023–0.024[a]	2.0–3.2[b]
Polytetrafluoroethylene (PTFE)	>327–332[b]	280–380[a]	0.00007–0.0001[b] 0.0002[c]	2.1[b,c]
Quartz		~1000[a]	0.017–0.018[a] 0.005–0.007[a]	

[a] Data are from reference 55.
[b] Data are from reference 56.
[c] Data are from reference 57.

corrosive environments, especially acid vapors, should be avoided. This precaution is a specific recommendation in all U.S. EPA methods using microwave technology.

Case #2: Muffle Furnace Meltdown

Microwave muffle furnaces are constructed of an insulating block containing a high-temperature-resistant and high-microwave-absorbing material (such as silicon carbide) that fits inside the microwave cavity. The absorber block has one side that opens up to allow the introduction or removal of samples. The absorbing block's open side must be reinserted properly onto the block prior to starting microwave heating. Numerous instances have occurred in which the absorbing block door and accompanying insulation were not replaced prior to microwave exposure. The heat from the absorbing block radiated in the direction of the microwave door and initially heated and eventually melted the plastic components of the door. It was not determined whether any microwave exposure to personnel occurred.

Two approaches to this problem are available. One company uses a solid stainless-steel cavity and door with insulation built into them. This unit does not have the potential of a physical meltdown (36). Another manufacturer's approach was the implementation of an IR sensor that audibly signals the operator if the muffle furnace is improperly installed and overheating is occurring (44).

Highly Reactive Samples and Acid Mixtures

Microwave acid digestion accelerates the rate of sample decomposition reactions, but it does not alter the fundamental chemistry involved. Microwave heating and microwave acid digestion of certain chemical compounds, mixtures, and types of samples constitute unreasonable, hazardous misuse of laboratory microwave systems. Explosives, propellants, hypergolic chemical mixtures, and pyrophoric chemicals should never be heated inside a laboratory microwave system. Combinations of reagents that are explosive, or so highly reactive as to be uncontrollable, fall into this category. Several case histories are presented here to underscore this important point.

Case #3: Hypergolic Mixture

A U.S. government laboratory analyst, seeking to analyze the metal content of a candidate drug, attempted to perform acid microwave digestion on the sample dissolved in 3–4 mL of a propylene glycol transdermal patch delivery solution. The sample was dispensed into two vessels and 10 mL of nitric acid was added. A 10 min predigestion step was performed before the vessels were sealed. Both (200 psi) vessels exploded immediately after placement inside the microwave system, without any microwave heating! The reagent and this highly reactive sample are not controllable and are a good example of a classical chemical hazard.

Case #4: Ammonium Perchlorate

A technician, without extensive chemistry training, called the application staff of a microwave-digestion system manufacturer with two questions concerning digestion of samples.

1. How much (what %) ammonium perchlorate can be digested in a wastewater sample? (Some wastewater samples would contain >20% v/v ammonium perchlorate.)
2. A "polymer" sample with the acronym GAP flashes on a hot plate at ~105 °C; is it possible to digest the sample inside a microwave system?

Ammonium perchlorate is an explosive. GAP is an acronym for gelled ammonium perchlorate, a colloidal form used as a solid rocket propellant. Fortunately, the technician was cautious enough to inquire and receive advice that prevented an otherwise inevitable explosion.

Special safety procedures have been developed for handling of explosives and propellants (8, 45–48). Discussions of what constitutes an explosion are also available (49). Certain finely divided metals are pyrophoric (e.g., calcium and zirconium), and spent hydrogenation catalysts are problematic and hazardous because of adsorbed hydrogen that may be present (10).

Case #5: Reactive Sample Component

An explosion was reported during the nitric acid extraction of a Teflon PTFE filter used in air sampling (50). On examination, it was noticed that, in addition to the filter, a plastic ring that held the Teflon filter made of polymethylpentene was decomposed and can react violently in the microwave at elevated temperatures. Subsequently, the laboratory cut the Teflon filter from the polymethylpentene ring and investigated pure Teflon filters to eliminate the problem.

Frequently, other sample components have completely different reaction characteristics from those expected from the major matrix component. All sample components being placed in microwave vessels must be considered during method development.

Case #6: Perchloric Acid and Easily Oxidizable Organic Matter

An analyst called a research laboratory and asked, if he put perchloric acid directly on coal and heated it in a microwave oven, would it be a complete digestion. He was advised an uncontrolled explosion would likely occur due to the combination of easily oxidizable organic matter and hot concentrated perchloric acid. The analyst, frustrated by the ineffectiveness of other methods of digestion, called and reiterated the query and asked if an explosion was a certainty. This analyst was then directed to appropriate literature and assured of this potential and eventuality. Not wanting to abandon the idea, the analyst called another person in the same laboratory, reiterated the question, and received the same answer. Fortunately, the analyst took the advice and did not attempt this reaction (*51*). Perchloric acid is a potentially hazardous reagent that requires a highly experienced analyst for its safe use. Because closed-vessel microwave digestions can achieve relatively high reaction temperatures and the oxidizing power of nitric acid increases with increases in reaction temperature, nitric acid is usually sufficient to oxidize most materials (*see* Chapter 3 for details). Rarely, if ever, is perchloric acid necessary for a microwave digestion. *See* Chapter 2 for general safe use rules.

The potential for runaway or out-of-control reactions in a microwave has been demonstrated. Where exothermic reactions supply enough energy in a very short time frame, they may raise the temperature above that being induced and controlled by the microwave system and the reaction may reach temperatures that are self-sustaining. These reactions are beyond the equipment's ability to control because there is only the cooling of the ambient air on the vessel walls, leaving no way to reduce unwanted thermal accumulation of exothermic reactions (*28*). In addition, perchloric acid has been documented to decompose autocatalytically in a microwave field at 245 °C with no sample present (*52*).

Elevated temperatures increase the oxidizing power of acids such as perchloric and nitric. Figure 9 shows the temperature and pressure curve for the nitric acid–hydrofluoric acid digestion of apple leaves. Even after microwave emission is stopped, the oxidation potential of the digest solution is so high that an exothermic reaction occurs. Similarly, if easily oxidizable organic matter is heated with an acid such as hot concentrated perchloric acid, it will eventually thermally run away and then chemically explode. Many in-depth treatises on perchloric acid have been written and should be consulted (*1, 40*), including Chapter 2 of this text. We cannot recommend the use of perchloric acid in a microwave system with the exception of only the most experienced analyst and under well-known and tightly controlled conditions.

Case #7: Hydrazine Hydrolysis

A technician at a pharmaceutical company was experimenting with a microwave system to develop a method for hydrolyzing a glycoprotein sample for subsequent chromatographic analysis of the constituent sugar units. The technician selected pure hydrazine as the hydrolyzing medium. In less than 30 s after starting the microwave heating program, the cover of the Teflon PFA vessel blew off.

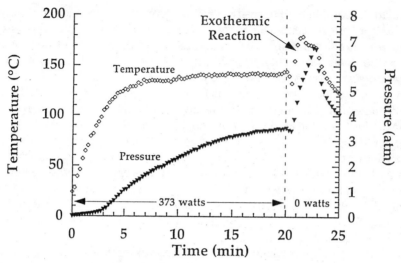

Figure 9. Exothermic run-away reaction during the digestion of 0.25 g SRM 1515 Apple Leaves with 10 g nitric and 0.5 g hydrofluoric acids.

Hydrazinolysis is a relatively common nonmicrowave hydrolysis procedure performed on glycoproteins and carbohydrates. However, this procedure is typically conducted at a temperature of 0 °C. Hydrazine is a liquid rocket propellant with a flash point of 52 °C.

Case #8: Large Organic Sample in a Closed Vessel by EPA Method 3015

A technician from a chemical plant laboratory reported the explosion of a 200 psi microwave-digestion vessel to the applications staff of the manufacturer. The sample involved was stated to be "effluent wastewater" being digested according to U.S. EPA Method 3015 (e.g., 45 mL of water and 5 mL of nitric acid). A duplicate sample of the wastewater was submitted to the microwave system manufacturer for digestion to discern the problem. After analysis, the "effluent wastewater" sample was determined to contain 55% triethanolamine! This concentration equated to ~25 g of pure organic, which was roughly 50 times the vessel's organic sample size limitation.

This over-pressurization of an organic sample would be expected if the liquid or miscible organic portion of the sample was thought of as ~25 g of solid sample. Method 3015 uses the same equipment and temperature profile as EPA Method 3051, which restricts organic sample size to 0.25 g maximum.

Microwave Heating of Exposed Solid Sample Material

Case #9: Microwave Absorbing Viscous Matrix

A U.S. government laboratory technician seeking to analyze peanut butter failed to place a sample in the bottom of the microwave-digestion vessel.

The sample was permitted to adhere to the side wall exposed in air, above the liquid level of nitric acid. During the microwave heating cycle, the peanut butter absorbed microwave energy, charred, and burned into the Teflon PFA vessel wall (the melting point of Teflon PFA is ~300 °C and heat softening occurs above 260 °C). Pressurized nitric acid vapor forcibly vented through a small hole that formed in the region of vessel wall damaged by localized heating and charring of the sample.

Although Teflon is one of the most chemically resistant polymers with excellent thermal properties, it can be damaged by localized overheating. This specific scenario can be avoided by placing the sample under the reagent solution and not permitting it to adhere to the wall of the vessel while open to the air.

Improper Assembly or Usage of Vessels

Case #10: Exposure of Flammable Solvents to a Microwave System

A laboratory technician from a U.S. chemical company reported that a fire had occurred in a new microwave system during its first run. The microwave system had been purchased for this facility to duplicate a solvent extraction procedure previously developed at another U.S. company site. Through discussion, it was learned that the technician had not been trained in operation of the system and had not performed the extraction method before. Vessels containing an hexane–acetone mixture were assembled without rupture membranes installed. During microwave heating, solvent vapor escaped from the open rupture membrane port of the vessel and ignited inside the cavity. The resulting fire went unobserved until extensive damage had occurred to the system.

Solvent sensors are available for the cavities of laboratory microwave systems to detect the presence of flammable organic solvents. They can also be fitted to exhaust ducts of microwave systems. The detection of significant amounts of flammable solvents can be used to automatically stop microwave heating before the lower explosive limit (LEL) concentration in air can be reached. The system reported here was not equipped with a solvent sensor system.

Case #11: Inappropriate Equipment Use

A laboratory technician called the manufacturer to report that a microwave-digestion vessel had failed (melted) during heating. Through discussion it was learned that the technician had heated the microwave-transparent vessel inside a convection oven, not a microwave system for which it was designed. The outer Ultem polyetherimide vessel body was charred and destroyed, without the vessel contents heating. Polyetherimide has a melting point of (~200 °C) and is not meant to be used with convective or conductive heating.

Chemically Dissimilar, Unknown Samples

Case #12: Digestion of Unknown Samples

An environmental laboratory experienced an explosion of a 200 psi microwave-digestion vessel during microwave digestion of two dissimilar samples of unknown composition. The force generated by this vessel failure was sufficient to remove the door of the microwave system, turning it into a secondary missile, which landed approximately 15 feet across the laboratory, where it struck a laboratory bench, and fell to the floor. Figure 10 is a photograph taken of the laboratory immediately following the vessel explosion. Devices are described in Case #15 to modify microwave equipment with inadequate door latches and hinges, with a safety shield. The removal of the microwave doors with vessel ventings has been reported and documented more than four times to the authors with similar equipment.

In this instance, the user had attempted to conduct a microwave digestion run with two samples. The first sample was 0.43 g of an "oily waste" with 5 mL of nitric acid, and the second sample was 45 mL of a "soapy aqueous waste" and 5 mL of nitric acid. The precise chemical composition and content (% organic) of both samples was unknown.

The vessel containing the 0.43 g of oily waste was connected to the pressure control system of the microwave unit. The explosion occurred in the vessel containing the 45 mL of soapy aqueous waste, and resulted in a rupture of the vessel bottom. A program consisting of four, 10 min pressure-controlled microwave heating steps, all at a 60% power setting (600 W unit) was run. The vessel failure occurred in Step 4 with a pressure control setpoint of 180 psi that had not yet been achieved by the monitor vessel.

The pressure generated from the decomposition of the oily waste sample in the control vessel attained the pressure control setpoint and thus caused minimal microwave heating during the first three steps. Microwave heating during the fourth step was apparently sufficient to raise the temperature of the larger (45 mL) volume of the soapy aqueous waste beyond the threshold reaction initiation temperature for decomposition of the organic compounds in the sample. The sample was presumed to contain a mass of organics >0.5 g and rapid evolution of decomposition gases exacerbated the vessel failure.

Organic molecules with different structures are oxidized with varying efficiencies by nitric acid. Thus, samples containing different organic compounds will decompose at different rates, and to different pressures and temperatures (see Chapters 2 and 3). Accordingly, the sample in a pressure–temperature control vessel must be close in character to the sample in all other vessels in a batch or run.

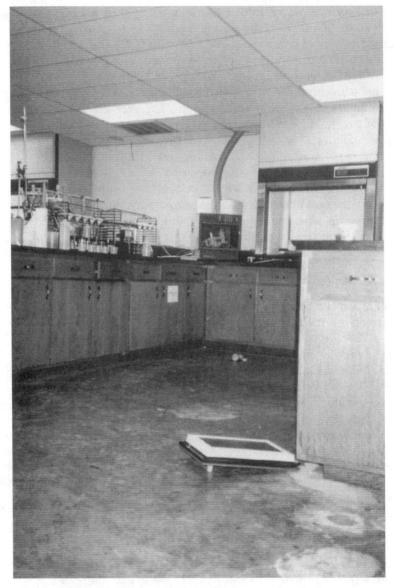

Figure 10. Picture of laboratory after vessel explosion.

High-Boiling-Point Acids and Reagents

Case #13: Melting of Polymer Vessels by Sulfuric Acid

Microwave heating of high-boiling-point acids and reagents (temperatures >300 °C), such as concentrated sulfuric or phosphoric acids is potentially problematic. These acids are strong couplers of microwave energy and their temperatures will rapidly rise to over 300 °C. The boiling points of these acids exceed the melting-point temperatures of polymer materials of microwave vessels and labware. Melting of vessels and release of their contents with continued microwave heating can result in fires and irreparable damage to the microwave system. The problem is most acute in cases where the door to the microwave cavity is also constructed of plastic material. Figure 11 shows the damage incurred to vessels from a long-duration (>120 min) microwave heating test with concentrated H_2SO_4. In this instance, the solid steel door construction safely contained the problem inside the microwave cavity. Although the rotor and vessels were damaged, the unit only suffered surface cosmetic damage. The microwave system used a sensor to detect overheating; when this sensor detected a problem the system was shutdown. Inappropriate or erroneous programmed conditions can lead to microwave system damage or failures.

Case #14: Melted Vessel

A technician at a U.S. chemical company performed an acid digestion procedure on a refractory ceramic by using a 1:1 H_2SO_4:H_3PO_4 mixture in Teflon PFA vessels. After heating for 20 min, the technician removed a sample vessel. The heat softened and melted the bottom of the vessel, which fell apart into the technician's free hand. The technician was not wearing protective gloves and molten plastic caused second-degree burns.

Temperature safeguard mechanisms do exist for heating high-boiling-point acids and reagents inside laboratory microwave systems, and they are currently available on some systems. The external surface temperature of vessels can be monitored by using an IR sensing system (53). An upper temperature limit setpoint is programmed to regulate microwave power delivered to the acids and samples and to prevent heating beyond the service temperature of the polymeric vessel components.

Vessel Failures–Instrument Failures: In-Depth Case Studies

Cases #15 and #16 were reported by the New York State Department of Health's Trace Elements Laboratory, and Case #17 was reported by a Fortune 500 company. They exemplify how improvements in microwave instrument design can come from unfortunate incidents. They also demonstrate why study of the nonintuitive nature of microwave heating interactions is required.

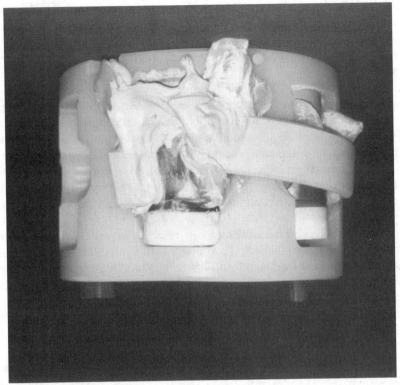

Figure 11. Vessel damage contained within a multimode microwave cavity.

Case #15: Digestion of Lyophilized Human Placental Tissues

The analysis of placental tissue for trace metals presents several problems. Placental tissue can be separated into the three anatomically distinguishable components: membrane, umbilical cord, and placental body. Sample digestion is required prior to instrumental analysis. Previous attempts to achieve sufficiently complete digestion of organic matter by using a standard program for biological materials were unsuccessful. The efficiency of the digestion process was estimated by observing background absorbance signals from placental samples. In this case, determination of the elements lead and cadmium were carried out by using graphite furnace atomic absorption spectrometry with Zeeman background correction.

All tissues were lyophilized to constant weight, and 400–600 mg samples were placed in Teflon PFA-lined microwave-digestion vessels with 10 mL concentrated HNO_3. The Teflon liner was secured inside the Ultem polyetherimide outer body, and the vessel cap was tightened by hand. These vessels have an upper pressure limit of 200 psi. Twelve vessels were placed in the carousel of a 630 W microwave-digestion system. The unit was equipped with a pressure-controller accessory, which was connected to one of the lined vessels with a special double port cap. An eight-step heating program was suggested by the manufacturer's technical support staff (Table V).

After running this heating program with several placenta samples, large background absorption signals were still evident in the furnace, particularly at the cadmium 309 nm line, with cadmium barely detectable. The decision was made to explore additional heating at 100% power and 175 psi cut-off for a further 60 min after allowing the vessels to cool overnight, and manually venting each one before proceeding. This second 60 min heating step, 24 h later, had the effect of considerably reducing the background absorbance and noise, thereby improving the detection limit for cadmium.

During the post-24 h heating stage of a subsequent run, one of the lined digestion vessels suffered a catastrophic failure. The event caused one of the vessels to fragment into 10 pieces, although the cap and rupture membrane remained intact (Figure 12). The force of the event caused the oven door to open, scattering debris and hot acid into the laboratory. Fortunately, no one was near the oven when this event took place. The damage to the oven was limited to a broken mode stirrer and a second cracked vessel.

As follow-up to this event, several freeze-dried placenta samples were provided to the manufacturer's applications laboratory to investigate the conditions that led to vessel failure. Applications staff ran the samples in a 950 W oven with a fiber-optic temperature control system. Placenta samples were digested in 10 mL HNO_3 by using lined digestion vessels (200 psi), and another set was digested in 10 mL HNO_3 in newly developed high-pressure vessels (600 psi) by using the program in Table VI.

The temperature and pressure of one of the lined digestion vessels were monitored and recorded (Figure 13). The data show that a maximum

Table V. Heating Program for 500 mg Placental Tissue and 10 mL Nitric Acid

Step	Time (min)	Pressure Setting (psi)
1	10	20
2	10	40
3	10	60
4	10	80
5	10	100
6	10	120
7	10	140
8	60	175

NOTE: Each step was performed at 100% power of 630 W.

Figure 12. Result of a catastrophic failure of a lined digestion vessel.

Table VI. Heating and Pressure Program for 200 mg Placental Tissue and 10 mL
Nitric Acid

Step	Power (%)	Pressure Setting (psi)	Run Time (min)	Time at Pressure (min)	Temperature (C)	Fan Speed (%)
1	25	20	10	5	120	100
2	35	40	10	5	200	100
3	35	85	10	5	200	100
4	35	150	10	5	200	100
5	35	200	10	5	200	100

NOTE: 100% power was 950 W.

temperature of 194 °C was obtained during Step 4 at a pressure of 200 psi.
One sample was cooled and vented and heated again at 35% power at a
temperature cut-off of 200 °C and a pressure cut-off of 150 psi. Both temperature and pressure were recorded. The temperature inside the vessel
increased quickly and actually exceeded 200 °C at one point at 150 psi.
These data suggest that the temperature of placenta sample digests probably exceeded the recommended vessel operating limit (225 °C). The unit
did not have temperature control. Further experiments digesting

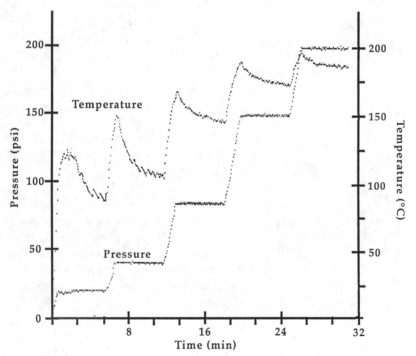

Figure 13. Lined digestion vessel temperature and pressure profiles.

placenta samples by using the high-pressure vessels showed that the maximum pressure obtained was 153 psi, below the stated 200 psi tolerance for lined digestion vessels, and below the final pressure cut-off used in the final step (175 psi). Thus, temperature control is essential for the safe digestion of biological tissue samples such as placenta. Additional aspects of temperature versus pressure control are evaluated in Chapters 2 and 3.

Case #16: Digestion of Fuel Oil

Refined oils are routinely digested for subsequent analysis by spectroscopic and wet chemical techniques. Closed-vessel microwave digestion offers a rapid and convenient technique for achieving digestion. In this case, a standard application method from the manufacturer was followed for digesting samples of #6 Fuel Oil. The application note called for Teflon PFA vessels rated to withstand internal pressures of 120 psi.

A 500 mg #6 Fuel Oil sample was placed in Teflon PFA-lined microwave-digestion vessels with 12 mL concentrated HNO_3. Twelve vessels were placed in the carousel of a 900 W MDS 205 microwave-digestion system. The unit was equipped with a pressure-controller accessory, which was connected to one of the lined vessels with a special double port cap. The twelve-step heating program recommended in the application note was followed with power settings reduced by 30% to compensate for the field intensity of the 900 W microwave system (Table VII).

While performing this routine procedure, one of the fuel samples underwent an uncontrolled exothermic reaction, causing a catastrophic failure of the inner vessel (Figure 14). This failure led to fragmentation of the outer liner, and the force of the event caused the oven door to open. Debris from the fuel oil and vessel were scattered across the laboratory. Again, fortunately no personnel were in the room at the time.

Design and Installation of a Safety Shield. After suffering a catastrophic vessel failure in two different laboratories that resulted in opening the oven door with near disastrous consequences, actions were taken that, at the very least, would ensure the safety of technical personnel working with microwave-digestion procedures. The most important action was to design and install a safety shield that is physically mounted on the oven, and which swings down in front of the oven during use (Figure 15).

The shield design is similar in concept to the safety shield of the Questron 1000, but is adapted for installation on either the CEM MDS 81D and 205 or 2000 microwave oven. The shield was assembled from basic, low-cost workshop materials, and it is easily constructed by any competent,

Table VII. Heating Program for 500 mg Fuel Oil and 12 mL Nitric Acid

Step	Power (%)	Time (min)	Pressure Setting (psi)
1	70	6	170
2	0	2	vent
3	70	2	170
4	0	3	vent
5	70	6	170
6	0	3	vent
7	70	7	170
8	0	4	vent
9	70	8	170
10	0	4	vent
11	70	10	170
12	cool to room temperature		

NOTE: 100% power was 900 W.

well-equipped workshop. The shield is made from 6.4 mm (1/4 inch) polycarbonate supported by 6.4 mm aluminum alloy. A complete schematic of this and other such safety devices can be found on SamplePrep Web (Chapter 15). When the shield is in the lowered position, the distance between the shield and the oven door is 8 cm with 3.5 cm clearance between the oven-door handle and a 6.4 mm aluminum-alloy support bar

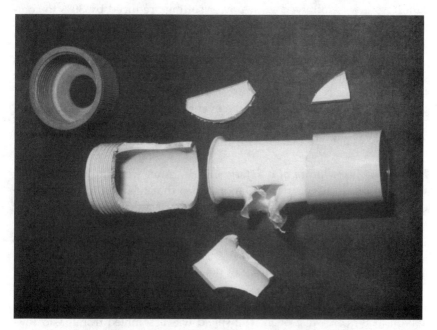

Figure 14. Catastrophic failure of a vessel due to an uncontrolled exothermic reaction.

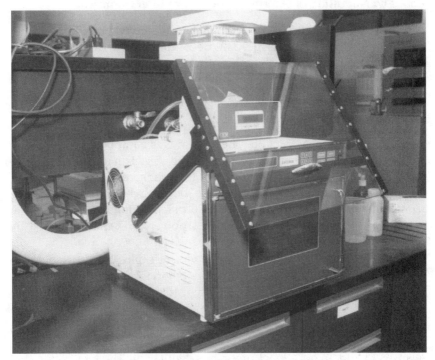

Figure 15. Microwave door safety shield.

that traverses the shield. This support bar is purposely designed to allow the oven door to open slightly during a vessel failure and relieve any pressure buildup in the oven. A plastic stopper on the support bar ensures that, should the oven door burst open during a failure event, the polycarbonate shield will not break.

Outcome. The vessel failures and the decision to install an oven safety shield were reported to the microwave manufacturer in May 1993. They responded by offering a protective shield for retrofitting the unit. We encourage microwave customers to report problems to instrument manufacturers to help improve the quality of future analytical instruments. We also encourage reporting of safety issues, case studies, and preventative methods to the SamplePrep Web site in the hope that this produces effective solutions to prevent additional occurrences. In conclusion, this experience underlined the need to install safety shields on unprotected microwave-digestion systems, to use temperature control when digesting biological tissue samples, and to publicize known safety problems preventing repetition of these events.

Case #17: Nonintuitive Nature of Microwave Interactions

An analytical laboratory at a Fortune 500 company was performing fluoride analysis on plant tissue. The decomposition was a strong base hydrolysis using a relatively concentrated sodium hydroxide solution. Approximately 0.25 g samples of plant tissue were placed in the Teflon liners of seven vessels. Vessels from two different manufacturers were used, one monitoring vessel from the laboratory microwave unit manufacturer, and six microwave vessels from another manufacturer of different design and construction. The system was equipped with pressure feedback control to the monitor vessel. The vessels had drastically different pressure ratings. The unit manufacturer's vessel, used for pressure monitor and control, was rated at 200 psi, whereas the remaining six vessels had a design rating of 1200 psi. The microwave program used in the 600 W microwave was 75% power for 2 min, followed by 40% power for 30 min.

At 10 min into the decomposition, a violent explosion occurred, opening the door, sending a vessel across the laboratory, and destroying the microwave unit. After the incident, the safety officer at the company decided that: "Microwave digestion will no longer be performed in analytical chemistry as a result of this near miss" (54).

The stated purpose of the analyst for using two different types of vessels was to add an additional pressure safety factor. The logic being that because the monitored vessel had a much lower pressure (200 psi) capacity than the majority of his sample vessels (1200 psi) from the second manufacturer, they would have additional pressure capacity. Additional information, obtained from the analyst, indicated that carbonaceous material was found on exploded vessel fragments and inside the microwave unit. Additionally, when asked if sparking was ever heard with this reagent mixture, the analyst reported that indeed he had heard sparking sounds frequently when running this reagent mixture.

Failure analysis in this instance is not simple, nor is it appropriate to use conventional logic without understanding the unique interactions of the microwave and vessel. First, one must refer to the discussion in Chapter 3 dealing with temperature and pressure relationships in closed microwave vessels. Insulated heavy walled vessels, such as the 1200 psi high-pressure vessel used in this example, retain heat and follow conventional temperature–vapor pressure relationships. However, thermal characteristics of the 200 psi pressure-monitoring vessel in this case are more like the example in Chapter 3, where a large amount of heat is transferred and dissipated from the gas phase through the vessel walls. The pressure in this monitor vessel may be only 12–20% that of the insulated vessel at similar temperatures. Thus, monitoring this vessel gives a false value for pressure control in the remaining vessels.

This particular high-pressure vessel has only had two reported uncontrolled reactions where it disintegrated in this manner. The other incident was shown to be a chemical explosion. In this case, the finding of large amounts of carbonaceous material covering the destroyed vessel fragments indicates that a fire may have occurred in the vessel. Because highly concentrated sodium hydroxide solutions can "electrically arc" when irradiated, the solution may have sparked the fire. No digestion vessel is designed to handle a chemical fire or explosion. The kinetics of chemical explosions are outside the design and operations range of safety relief devices for over-pressurization protection. The microwave cavity itself must act as the next barrier with a door or secondary shield appropriately constructed to prevent the escape of such fragment missiles.

Because of the failure to understand the design and performance characteristics and unique interactions of microwave equipment, the analyst created an unsafe situation by attempting to control reaction pressure in an inappropriate manner. This case illustrates the nonintuitive nature of microwave technology that requires study of the subject to enable the analyst to predict important interactions during method development. We strongly encourage analysts to learn the subtleties of this new tool as a prudent measure for efficient and safe operation of microwave instruments.

Epilog

The Fortune 500 company, after analyzing the underlying causes of the explosion, has taken steps to educate its staff and is again using microwave digestion successfully and safely.

Conclusion

Safety of the analyst is the most important consideration in analysis of the sample preparation process. As the result of direct coupling, microwave absorption by chemical reagents, energy is directly transferred to and concentrated on reactants. The advantages of high temperature for chemical procedures are quickly realized; however, control of sample heating and reactions is complicated by this rapid heat-transfer mechanism. As in all good laboratory practices, after becoming informed and removing ignorance as an excuse, there is no substitute for common sense. Common sense dictates careful planning of experiments and cautious experimentation when the results are uncertain and the equipment is unfamiliar. Many of the instrumentation mishaps reported to us involve misuse of equipment. Other unsafe situations arose from failures of equipment that was no longer serviceable. Still others involved equipment that does not perform as specified or as it was expected, because of a lack of thorough testing.

Occasionally, an unanticipated event will occur that results in equipment damage and loss of the sample. These phenomena should be documented and reported to help colleagues in the scientific community avoid these same difficulties. Because science progresses by virtue of the observations and discoveries of our predecessors and peers, we advance science by making our own observations known. This is the purpose of scientific literature and the SamplePrep Web (Chapter 15). As this collection of case studies disappears from common occurrence, newer uses will produce other safety concerns. We hope we have provided some insight into the current state of the art, and we await the new contributions from others to advance the safe use of laboratory microwave systems for chemical analysis and reactions.

Acknowledgments

We thank Timothy Cole (Case #15) and John Orsini (Case #16) of the Wadsworth Center of the New York State Department of Health for technical assistance and for reporting these incidents.

Thanks to Bill Sonnefeld for the photographs of unserviceable vessels, and to the Wadsworth Center's Photographic and Illustrations Unit for the photographs associated with Cases #15 and #16.

We also thank the staff of the Wadsworth Center's Instrumentation and Automation Support Unit for designing and building the microwave safety shield.

References

1. Schilt, A. A. *Perchloric Acid and Perchlorates*; G. F. Smith Chemical Company: Columbus, OH, 1979.
2. Bock, R. *A Handbook of Decomposition Methods in Analytical Chemistry*; Marr, I. L., Ed.; John Wiley and Sons: New York, 1979.
3. Sax, N. I. *Dangerous Properties of Industrial Materials*, 5th ed.; Van Nostrand Reinhold: New York, 1979.
4. Sulcek, Z.; Povondra, P. *Methods of Decomposition in Inorganic Analysis*; CRC: Boca Raton, FL, 1989.
5. Weast, R. C. *CRC Handbook of Chemistry and Physics*, 66th ed.; CRC: Boca Raton, FL, 1986.
6. Bruno, T. J.; Svoronos, P. D. N. *CRC Handbook of Basic Tables for Chemical Analysis*; CRC: Boca Raton, FL, 1989.
7. *Prudent Practices for Handling Hazardous Chemicals in Laboratories*; National Academy Press: Washington, DC, 1981.
8. *Prudent Practices for Disposal of Chemicals from Laboratories*; U.S. National Research Council, National Academy Press: Washington, DC, 1983.

9. Shugar, G. J.; Dean, J. A. *The Chemist's Ready Reference Handbook*; McGraw-Hill: New York, 1989.

10. Young, J. A. *Improving Safety in the Chemical Laboratory: A Practical Guide*, 2nd ed.; John Wiley and Sons: New York, 1991.

11. Steere, N. V. *CRC Handbook of Laboratory Safety*, 3rd ed.; CRC: Boca Raton, FL, 1992.

12. Diberardinis, L. J.; Baum, J. S.; First, M. W.; Gatwood, G. T.; Gordoen, E.; Seth, A. K. *Guidelines for Laboratory Design: Health and Safety Considerations*; John Wiley and Sons: New York, 1987.

13. Thuery, J. *Microwaves: Industrial, Scientific and Medical Applications*; Artech House: Norwood, MA, 1992.

14. Kok, L. P.; Boon, M. E. *Microwave Cookbook for Microscopists: Art and Science of Visualization*, 3rd ed.; Coulomb Press: Leyden, Netherlands, 1992.

15. Foster, K. R.; Guy, A. W. *Sci. Am.* **1986**, *255*, 32–39.

16. Thuery, J. *Microwaves: Industrial, Scientific and Medical Applications*; Artech House: Norwood, MA, 1992; p 562.

17. *International Design Criteria of Boilers and Pressure Vessels*; American Society of Mechanical Engineers: New York, 1984.

18. Criteria for Design of Elevated Temperatue Class 1 Compounds In *ASME Boiler and Pressure Vessel Code*; American Society of Mechanical Engineers: New York, 1989.

19. *Fire Protection for Laboratories Using Chemicals*; NFPA-45; National Fire Protection Association: Quincy, MA, August 16, 1991.

20. Stull, D. R. *Fundamentals of Fire and Explosion*; American Institute of Chemical Engineers: New York, 1977; Vol. 73.

21. Washburn, E. W. *Vapor Pressure of Acids*; McGraw-Hill: New York, 1928; Vol. 3, pp 58–59, 304–305.

22. Floyd, T. *High Temperature and High Pressure Digestion Vessel Assembly*, U.S. Patent 4, 904, 450, 1990.

23. Hukvari, I. S.; Albert, H. J. *Pressure and Temperature Reaction Vessel, Method, and Apparatus*, U.S. Patent 4, 882, 128, 1989.

24. Lautenschläger, W. *Sample Holder for Decomposition or Analysis of Sample Materials*, U.S. Patent 5, 368, 820, 1994.

25. Lautenschläger, W. *Sample Holder for Decomposition or Analysis of Sample Materials*, U.S. Patent 5, 270, 010, 1993.

26. Saville, R. *Microwave Heating Digestion Vessel*, U.S. Patent 4, 613, 738, 1986.

27. Pougnet, M. A. B.; Schnautz, N. G.; Walker, A. M. *S. Afr. Tydskr. Chem.* **1992**, *45*, 86–89.

28. Kingston, H. M.; Jassie, L. B. *Anal. Chem.* **1986**, *58*, 2534–2541.

29. Milestone Patents on Vessel Reclosing Mechanism After Overpressurization **1994**.

30. Singh, J. *Chem. Eng.* **1990**, 104.

31. Lautenschläger, W. *Apparatus for Performing Chemical and Physical Pressure Reactions*, U.S. Patent 5, 382, 414, 1995.

32. Kingston, H. M.; Walter, P. J. In *Inductively Coupled Plasma Mass Spectrometry: From A to Z*; Montaser, A., Ed.; VCH: New York, 1997; Chapter 2.

33. Copson, D. A. *Microwave Heating*, 2nd ed.; Avi Publishing: Westport, CT, 1975.

34. Neas, E. D.; Collins, M. J. In *Introduction to Microwave Sample Preparation: Theory and Practice*; Jassie, L. B.; Kingston, H. M., Eds.; American Chemical Society: Washington, DC, 1988; pp 7–32.

35. Matthes, S. A. In *Introduction to Microwave Sample Preparation: Theory and Practice*; Jassie, L. B.; Kingston, H. M., Eds.; American Chemical Society: Washington, DC, 1988, pp 33–52.

36. "New Product Announcement: MLS-1200 Pyro Sulfate Ashing System"; Milestone Corporation: Sorisol, Italy.

37. Kingston, H. M.; Jassie, L. B. Presentation at the 25th Eastern Analytical Symposium, New York 1986.

38. Gorsuch, T. T. *The Destruction of Organic Matter*, 1st ed.; Pergamon: New York, 1970.

39. Kingston, H. M.; Jassie, L. B.; Fasset, J. D. Presentation at the 190th National Meeting of American Chemical Society, Chicago, IL, 1985; paper ANYL 10.

40. Kingston, H. M.; Jassie, L. B. In *Introduction to Microwave Sample Preparation: Theory and Practice*; Jassie, L. B.; Kingston, H. M., Eds.; American Chemical Society: Washington, DC, 1988; pp 93–154.

41. Bedson, A. *Chem. Br.* **1986**, *22*, 894.

42. Lewis, B.; Von Elbe, G. *Combustion, Flames, and Explosion Gases*, 2nd ed.; Academic: Orlando, FL, 1961, p 695.

43. Weiss, G. *Hazardous Chemicals Data Book*; Noyes Data: Park Ridge, NJ, 1980.

44. "New Product Announcement: MAS-7000 Microwave Muffle Furnace System with IR Detection System"; CEM Corporation: Matthews, NC.

45. *Ammunition and Explosives Safety Standards*; DOD 5184.4S; U.S. Department of Defense: Washington, DC, 1978.

46. Jensen, A. V. *Hazards of Chemical Rockets and Propellants Handbook*; National Technical Information Service: Arlington, VA, 1970; Vol. 1.

47. Benz, F.; Bishop, C. *Ignition and Thermal Hazards of Selected Aerospace Fluids: Overview, Data, Procedures*; White Sands Test Facility: Las Cruces, NM, 1984.

48. *Toxic Chemical and Explosives Facilities: Safety and Engineering Design*; Scott, R. A., Ed.; ACS Symposium Series 96; American Chemical Society: Washington, DC, 1979.

49. Urbanski, T.; Vasudeva, S. K. *J. Sci. Ind. Res.* **1981**, *40*, 512.

50. Weker, R. A.; Sabolefski, M. A. *Chem. Eng. News* **1995**, 4.

51. Kingston, H. M., National Institute of Standards and Technology, personal communication, 1988.

52. Kingston, H. M.; Jassie, L. B. In *Introduction to Microwave Sample Preparation: Theory and Practice*; Jassie, L. B.; Kingston, H. M., Eds.; American Chemical Society: Washington, DC, 1988; p 119.

53. "New Product Announcement: IR-TC-500"; Milestone Corporation: Sorisole, Italy, July 1991.

54. Kingston, H. M., Duquesne University, personal communication, 1992.

55. Lautenschlägen, W.; Schweizer, T. *Labor Praxis* **1990**, 1–5.

56. Howard, M. J. In "The International Plastic Selector, Inc.", 1979.

57. "Berghof/America Teflon Products for Research & Industry", Product Catalog, 1992.

Appendix 1. Hazard Monitors for the Detection of Microwave Leakage

- *Anchor Chemical Australia Pty. Ltd.*, Box 474, P.O. Crow's Nest, N.S.W., 2065 Australia, Tel. (02) 439–2144
- *Applied Microwave Energy Inc.* 31127 Via Colinas, Westlake Village, CA 91362, Tel. (213) 991–4624
- *Bach-Simpson Ltd.*, 1255 Brydges Street, London, Ontario N5W 2C2, Canada, Tel. (519) 452–3200
- *General Microwave Corporation*, 155 Marine Street, Farmingdale, NY 11735, Tel. (516) 694–3600
- *Gerling Laboratories*, 1628 Kansas Ave., Modesto, CA 95351, Tel. (209) 521–6549
- *Holaday Industries Inc.*, 14825 Martin Drive, Eden Prairie, MN 55344, Tel. (612) 934–4920
- *MICOR, INC.*, 3901 Westerly Place, Suite 102, Newport Beach, CA 92660, Tel. (714) 476–0616
- *Microwave Heating Ltd.*, 1 A Heron Trading Estate, Luton Beds, England, LU3 3BB, UK, Tel. (0582) 58474
- *Milestone Inc.*, 160B Shelton Road, Monroe, CT 06468, Tel. (203) 261–6175, Fax (203) 261–6592 (for U.S. and Canada)
- *Milestone s.r.l.*, Via Fatabenefratelli 1/5, 24010 Sorisole (Bg), Italy, Tel. (39) 35–573857, Fax (39) 35–575498 (International)
- *Narda Microwave Corporation*, 435 Moreland Road, Happauge, NY 11788, Tel. (516) 231–1700

Appendix 2. Safety Terminology

1. *Accident.* An unplanned event, sometimes but not necessarily injurious or damaging, that interrupts an activity. A chance occurrence arising from unknown causes, carelessness, ignorance, lack of training, etc.
2. *ACGIH.* The American Conference of Governmental Industrial Hygienists.
3. *Arc.* A high-temperature luminous electric discharge across a gap.
4. *ASME.* American Society of Mechanical Engineers.
5. *ASME Code.* Refers to ASME Section VIII Division 1. Pressure Vessel Code.
6. *BLEVE.* An acronym for a boiling liquid expanding vapor explosion. These explosions involve vessels that contain liquids under pressure at temperatures above their boiling points.
7. *Design Pressure.* Refers to the pressure value used to determine the minimum wall thickness of a pressure-vessel body. In pressure-vessel codes, the design pressure is always higher than the maxi-

mum allowable working pressure (MAWP) or operating pressure. The minimum margin for design pressure is 110% of the MAWP (operating) pressure.

8. *Explosion.* A violent bursting, as of a pressurized vessel. An extremely rapid chemical reaction with the associated production of noise, heat, and violent expansion of gases. Reactive explosions are further categorized as deflagrations, detonations, and thermal explosion.

9. *Failure.* Distortion, breakage, deterioration, or other fault in a structure, component, or system resulting in inoperability or unsatisfactory performance of intended function.

10. *Failure Analysis.* A logical, systematic examination of an item, component, or assembly, and its place and function in a system, to identify and analyze the probability, causes, and consequences of potential and real failures.

11. *HERF.* An acronym for electromagnetic radiation to fuel. The potential for electromagnetic radiation to cause ignition of volatile combustibles such as aircraft fuels.

12. *HERP.* An acronym for hazards of electromagnetic radiation to personnel. The potential for electromagnetic radiation to produce harmful biological effects in humans.

13. *Hypergolic Mixture.* Two chemicals that when mixed create enough heat of reaction to cause ignition, without an external ignition source. (Three examples of chemicals that form hypergolic mixtures with concentrated nitric acid include phenol, acetone, and triethylamine.)

14. *Intrinsically Safe.* The term classifying electrical components or equipment, approved for use in specific hazardous atmospheres, which are incapable of releasing sufficient energy to cause ignition under normal or abnormal conditions. (Microwave systems for laboratories are not designed or rated intrinsically safe.)

15. *IRPA.* International Radiation Protection Association

16. *MAWP.* An acronym for maximum allowable working pressure: the maximum permissible operating pressure inside a vessel at its specified operating temperature.

17. *NEC.* An abbreviation for the *National Electric Code*, issued by the National Fire Protection Association. This code carries the force of law in all 50 states in the United States.

18. *NFPA.* National Fire Protection Association.

19. *Pressure-Relief Valve.* A pressure-relief device that is designed to reclose and prevent the further flow of fluid after normal conditions have been restored.

20. *Pyrophoric Chemicals.* Chemicals that react so rapidly with air and moisture that the ensuing oxidation or hydrolysis leads to ignition.

21. *RADHAZ.* Abbreviation for radiation hazard, term describes hazards of electromagnetic radiation to fuels, electronic hardware, ordinance, and personnel.

22. *Rupture Disk Device.* A nonreclosing pressure-relief device actuated by inlet static pressure and designed to function by the bursting of a pressure-containing disk.

23. *Shall.* Indicates a mandatory requirement in codes and regulations.

24. *Thermal Explosion.* An explosion resulting from an exothermic reaction that generates heat faster than the heat can be dissipated, and thus raising the temperature to a level where the reaction rate is catastrophic.

25. *TÜV.* An abbreviation for Germany's Technische Überwachnung Verein, an authorized inspection agency for industrial safety codes and regulations.

Appendix 3. General Safety References

A.   Steere, N. V. *CRC Handbook of Laboratory Safety*, 2nd ed.; CRC: Boca Raton, FL, 1982.

B.   Stull, D. R. *Fundamentals of Fire and Explosion*; AIChE Monograph Series 10, 1977; Vol. 73.

C.   Brown, S. J. *Impact, Fragmentation and Blast (Vessels, Pipes, Tubes, Equipment)*; PVP Series; American Society of Mechanical Engineers: New York, 1984; Vol. 82.

D.   Chung, H.; Nicholson, D. W. *Advances in Impact, Blast, Ballistics, and Dynamic Analysis of Structures*; PVP Series; American Society of Mechanical Engineers: New York, 1986; Vol. 106.

E.   Kendall, D. P. *High Presseure Technology—Design, Analysis, and Safety of High Pressure Equipment*; PVP Series; American Society of Mechanical Engineers: New York, 1986; Vol. 110.

F.   Mahn, W. J. *Academic Laboratory Chemical Hazards Guidebook*; Van Nostrand Reinhold: New York, 1991.

G.   Mahn, W. J., *Fundamentals of Laboratory Safety—Physical Hazards in the Academic Laboratory*; Van Nostrand Reinhold: New York, 1991.

Index

Index